Advanced Structured Materials

Volume 64

More information about this series at http://www.springer.com/series/8611

Holm Altenbach · Tetsuya Matsuda
Dai Okumura
Editors

From Creep Damage Mechanics to Homogenization Methods

A Liber Amicorum to celebrate the birthday of Nobutada Ohno

Springer

Editors
Holm Altenbach
Lehrstuhl für Technische Mechanik, Fakultät für Maschinenbau, Institut für Mechanik
Otto-von-Guericke-Universität Magdeburg
Magdeburg
Germany

Tetsuya Matsuda
Department of Engineering Mechanics and Energy
University of Tsukuba
Tsukuba
Japan

Dai Okumura
Department of Mechanical Science and Engineering
Nagoya University
Furo-cho, Chikusa-ku, Nagoya
Japan

ISSN 1869-8433 ISSN 1869-8441 (electronic)
Advanced Structured Materials
ISBN 978-3-319-19439-4 ISBN 978-3-319-19440-0 (eBook)
DOI 10.1007/978-3-319-19440-0

Library of Congress Control Number: 2015939823

Springer Cham Heidelberg New York Dordrecht London

Printed on acid-free paper

Springer International Publishing AG Switzerland is part of Springer Science+Business Media (www.springer.com)

Preface

Professor Nobutada Ohno is one of the leading Japanese researchers in solid mechanics and has a worldwide reputation because of his great impact on several research topics. This volume of the Advanced Structured Materials Series has been published to celebrate his 65th birthday, and to express sincere respect and gratitude for his significant achievements and longtime contributions to solid mechanics. Many active researchers in his fields have contributed to this memorial volume, some of them are close to Prof. Ohno, and have also contributed to his published studies. The contents of this book are not limited to one field. Professor Ohno's research fields have been extensive, as seen in the book title "From Creep Damage Mechanics to Homogenization Methods," which will be referred to later in this Preface. We hope that the readers enjoy the variety of the contents of this volume.

Professor Ohno was born in 1950 in Ichinomiya, near Nagoya, and spent his early years there. After graduating from high school, he entered the Department of Mechanical Engineering in Nagoya University, and graduated with top honors in 1973. He then majored in solid mechanics in the graduate course, and received his Ph.D. from the university in 1979. He began his career as an Assistant Professor at Nagoya University before moving to the Toyohashi University of Technology in 1980, where he spent about 8 years. During this time, he spent a year at Harvard University in 1982 as a visiting scholar, and collaborated with Prof. John Hutchinson. In 1988, he returned to Nagoya University as an Associate Professor, and he has been a full Professor there since 1994.

In his academic career, Prof. Ohno has received many awards. He has received international awards such as the K. Washizu Medal (2001) and Khan International Award (2009), and a special issue was published in his honor in the International Journal of Plasticity in 2011. He has also received almost all of the major awards of the Japan Society of Mechanical Engineers (JSME). He has been awarded the

JSME Medal for Outstanding Paper thrice (1977, 1991, 2004), Materials and Mechanics Achievement Award (2004), Computational Mechanics Achievement Award (2006), Computational Mechanics Award (2010), Materials and Mechanics Award (2013), etc. He has been a fellow of JSME since 2002. The Japan Society of Materials Science (JSMS) has also awarded him the JSMS Award for Scientific Papers (1998), JSMS Award for Academic Contribution (2005), etc. He also received the JACM Award from the Japan Association for Computational Mechanics in 2010.

His many awards and honors clearly show his outstanding academic achievements. However, it is difficult to explain what his research field is because his research has been quite wide-ranging, as is often the case with distinguished researchers. We will take this opportunity to summarize his main achievements, although we cannot cover all of his research because of space limitations. The summary below will briefly describe his extensive research, and will help the readers understand the meaning of the title of this book. His papers mentioned in this summary are listed at the end of Preface.

1. Continuum theory of anisotropic creep damage
   Nobutada Ohno investigated continuum modeling of anisotropic creep damage during his doctoral studies under the supervision of Prof. Sumio Murakami. As a result, a continuum theory was developed in which a second order tensor was shown to be an internal variable to represent the anisotropic damage caused by net area reductions in tertiary creep (Murakami and Ohno 1981). This study is highly evaluated as one of the pioneering works on anisotropic damage modeling, and has been cited in many papers and books.
2. Non-hardening region in cyclic plasticity
   Ohno (1982, 1986) postulated that isotropic hardening does not evolve within a plastic strain region after a load reversal, and represented the plastic strain region, referred to as the non-hardening region, by generalizing Chaboche's memory surface. This postulation was used to represent work-hardening stagnation for accurate springback analysis by Yoshida and Uemori in 2002. Professor Ohno thus contributed to the well-known Yoshida-Uemori model, which has been available in LS-DYNA and PAM-STAMP since 2007.
3. Kinematic hardening model for ratcheting
   The nonlinear kinematic hardening model of Armstrong and Frederick is well known, but has the drawback of overpredicting ratcheting and cyclic stress relaxation. Ohno and Wang (1993a) improved the dynamic recovery term in the Armstrong-Frederick model. The developed model, called the Ohno-Wang model, is highly rated and has been used by many researchers to simulate ratcheting and cyclic stress relaxation. The papers of Ohno and Wang (1993a, b) and Abdel-Karim and Ohno (2000) have been cited many times.
4. Homogenization methods for nonlinear time-dependent composites
   Ohno and co-workers extended the computational homogenization method of periodic composites to include nonlinear time-dependent behavior such as creep and viscoplasticity (Wu and Ohno 1999; Ohno et al. 2000). The method was

further extended in the presence of point-symmetric internal structures (Ohno et al. 2001). The extended methods have been verified by simulating the experiments of long fiber-reinforced laminates and plain-woven laminates (Matsuda et al. 2003, 2007, 2014). Recently, the homogenized viscoplastic behavior of pore-pressurized, anisotropic open-porous solids has been studied (Ohno et al. 2012, 2014).

5. Homogenization analysis of cellular material instability
   By developing an updated Lagrangian-type homogenization method based on two scales of periodic structures, the elastic and elastoplastic buckling behavior of hexagonal honeycombs subjected to in-plane biaxial compression were analyzed to elucidate the complex buckling modes observed in experiments (Ohno et al. 2002; Okumura et al. 2002, 2004). These studies have been frequently cited, and have been presented in about twenty invited lectures, including five plenary lectures, at conferences. The method has been applied to other cellular materials (e.g., Ohno et al. 2004; Takahashi et al. 2010).
6. Strain gradient plasticity based on the self-energy of GNDs
   Considering the self-energy of geometrically necessary dislocations (GNDs), Ohno and Okumura (2007) provided an explicit physical basis for strain gradient plasticity theories, and analyzed model crystal grains to derive a closed-form evaluation of initial yield stress. They showed that the self-energy of GNDs explains well the grain-size dependence of initial yield stress in the submicron to several-micron range of grain sizes. This study is highly regarded, and has been presented in about ten invited lectures, including one plenary lecture, at conferences.
7. Implementation of cyclic (visco)plastic models in FEMs
   Implicit stress integration algorithms were developed to implement cyclic (visco)plastic models, including the Ohno-Wang model, using user subroutines in commercial finite element programs (Kobayashi and Ohno 2002; Kobayashi et al. 2003; Akamatsu et al. 2008; Ohno et al. 2013). The subroutine programs developed for cyclic thermomechanical analysis have been used by about ten companies in Japan. Recently, the programs have been integrated into a new version, OLMATS (Ohno Lab. Material Model Software).

As well as performing the above-mentioned research, Prof. Ohno has also been actively involved with scientific journals. He has served as an editorial board member of the International Journal of Plasticity (1991 to present), Computer Modeling in Engineering and Sciences (2004–2010), International Journal of Solids and Structures (2005 to present), Acta Mechanica Solida Sinica (2009 to present), and Bulletin of the JSME (2014 to present). He has also served as the editor-in-chief of the Journal of Solid Mechanics and Materials Engineering (2006–2007), an associate editor of the Materials Science Research International (1997–1999) and JSME International Journal, Series A (2000–2003), a guest co-editor and guest editor of the International Journal of Plasticity (2000, 2008), and a guest editor of the International Journal of Mechanical Sciences (2008).

Moreover, he has played a number of leading roles in academic societies and conference organizations. He has held important positions in academic societies, such as the chair of the JSME Materials and Mechanics Division (2002), chair of the JSME Computational Mechanics Division (2008), executive board member of the JSMS (1996–1997, 2000–2001, 2005–2006), executive board member of the JSME (2013–2014), and vice president of the JSME (2014). He has also made substantial contributions to international and domestic conferences by serving as, for example, the co-chair of the 5th IUTAM Symposium on Creep in Structures (2000), chair of the 8th Asia-Pacific Symposium on Engineering Plasticity and Its Applications (2006), and chair of the 56th JSMS Annual Meeting (2007). All of these conferences took place at Nagoya University, and were very successful owing to his strong leadership.

Professor Nobutada Ohno has had a significant impact on solid mechanics because of his original and unique point of view. His work has spread from academic to industrial fields: your laptop or car may contain proof of his achievements. In addition, he has served in academic societies for decades. His about 20 Ph.D. students have moved into academia and industry. He has hosted several foreign researchers for short- and long-term stays at Nagoya University. All of these contributions have given us great respect for him, and we regard him as an ideal researcher. At the same time, he is a man of heart and humor. He is a husband, father, and grandfather, loves watching movies, and is a fan of the Chunichi Dragons-Nagoya's professional baseball team.

We are grateful to Prof. Ohno for his seminal and longtime contributions to solid mechanics, and believe that he will continue to have an impact in research.

Last but not least, we gratefully acknowledge Dr. Christoph Baumann and Benjamin Feuchter (Springer Publisher) for support of the book project.

Magdeburg
Tsukuba
Nagoya
September 2015

Holm Altenbach
Tetsuya Matsuda
Dai Okumura

Selected Papers of the Main Achievements of Professor Nobutada Ohno

- Abdel-Karim M, Ohno N (2000) Kinematic hardening model suitable for ratchetting with steady-state. Int J Plast 16:225–240
- Akamatsu M, Nakane K, Ohno N (2008) Implicit integration by linearization for high-temperature inelastic constitutive models. J Solid Mech Mater Eng 2:967–980
- Kobayashi M, Mukai M, Takahashi H, Ohno N, Kawakami T, Ishikawa T (2003) Implicit integration and consistent tangent modulus of a time-dependent non-unified constitutive model. Int J Numer Methods Eng 58:1523–1543
- Kobayashi M, Ohno N (2002) Implementation of cyclic plasticity models based on a general form of kinematic hardening. Int J Numer Methods Eng 53:2217–2238
- Matsuda T, Goto K, Kubota N, Ohno N (2014) Negative through-the-thickness Poisson's ratio of elastic-viscoplastic angle-ply carbon fiber-reinforced plastic laminates: Homogenization analysis. Int J Plast 63:152–169
- Matsuda T, Nimiya Y, Ohno N, Tokuda M (2007) Elastic-viscoplastic behavior of plain-woven GFRP laminates: Homogenization using a reduced domain of analysis. Compos Struct 79:493–500
- Matsuda T, Ohno N, Tanaka H, Shimizu T (2003) Effects of fiber distribution on elastic-viscoplastic behavior of long fiber-reinforced laminates. Int J Mech Sci 45:1583–1598
- Murakami S, Ohno N (1981) A continuum theory of creep and creep damage. In: Ponter ARS, Hayhurst DR (Eds.) Creep in Structures, Springer-Verlag:422–443
- Ohno N (1982) A constitutive model of cyclic plasticity with a nonhardening strain region. J Appl Mech-Trans ASME 49:721–727
- Ohno N, Ikenoya K, Okumura D, Matsuda T (2012) Homogenized elastic-viscoplastic behavior of anisotropic open-porous bodies with pore pressure. Int J Solids Struct 49:2799–2806
- Ohno N, Kachi Y (1986) A constitutive model of cyclic plasticity for nonolinear hardening materials. J Appl Mech-Trans ASME 53:395–403

- Ohno N, Matsuda T, Wu X (2001) A homogenization theory for elastic-viscoplastic composites with point symmetry of internal distributions. Int J Solids Struct 38:2867–2878
- Ohno N, Narita K, Okumura D (2014) Homogenized elastic-viscoplastic behavior of plate-fin structures with two pore pressures. Int J Mech Sci 86:18–25
- Ohno N, Okumura D (2007) Higher-order stress and grain size effects due to self-energy of geometrically necessary dislocations. J Mech Phys Solids 55:1879–1898
- Ohno N, Okumura D, Niikawa T (2004) Long-wave buckling of elastic square honeycombs subject to in-plane biaxial compression. Int J Mech Sci 46:1697–1713
- Ohno N, Okumura D, Noguchi H (2002) Microscopic symmetric bifurcation condition of cellular solids based on a homogenization theory of finite deformation. J Mech Phys Solids 50:1125–1153
- Ohno N, Tsuda M, Kamei T (2013) Elastoplastic implicit integration algorithm applicable to both plane stress and three-dimensional stress states. Finite Elem Anal Des 66:1–11
- Ohno N, Wang JD (1993a) Kinematic hardening rules with critical sate of dynamic recovery, Part I: formulation and basic features for ratchetting behavior. Int J Plast 9:375–390
- Ohno N, Wang JD (1993b) Kinematic hardening rules with critical sate of dynamic recovery, Part II: application to experiments of ratchetting behavior. Int J Plast 9:391–403
- Ohno N, Wu X, Matsuda T (2000) Homogenized properties of elastic-viscoplastic composites with periodic internal structures. Int J Mech Sci 42:1519–1536
- Okumura D, Ohno N, Noguchi H (2002) Post-buckling analysis of elastic honeycombs subject to in-plane biaxial compression. Int J Solids Struct 39:3487–3503
- Okumura D, Ohno N, Noguchi H (2004) Elastoplastic microscopic bifurcation and post-bifurcation behavior of periodic cellular solids. J Mech Phys Solids 52:641–666
- Takahashi Y, Okumura D, Ohno N (2010) Yield and buckling behavior of Kelvin open-cell foams subjected to uniaxial compression. Int J Mech Sci 52:377–385
- Wu X, Ohno N (1999) A homogenization theory for time-dependent nonlinear composites with periodic internal structures. Int J Solids Struct 36:4991–5012

Contents

1 **Thermo-Electro-Mechanical Properties of Interpenetrating Phase Composites with Periodic Architectured Reinforcements** . . . 1
Rashid K. Abu Al-Rub, Diab W. Abueidda and Ahmed S. Dalaq

2 **A Continuum Damage Model Based on Experiments and Numerical Simulations—A Review** 19
Michael Brünig

3 **The Multiplicative Decomposition of the Deformation Gradient in Plasticity—Origin and Limitations** 37
Otto T. Bruhns

4 **Effect of Biaxial Work Hardening Modeling for Sheet Metals on the Accuracy of Forming Limit Analyses Using the Marciniak-Kuczyński Approach**. 67
Tomoyuki Hakoyama and Toshihiko Kuwabara

5 **Three-Dimensional FE Analysis Using Homogenization Method for Ductile Polymers Based on Molecular Chain Plasticity Model Considering Craze Evolution** 97
Hideyuki Hara and Kazuyuki Shizawa

6 **Inelastic Deformation and Creep-Fatigue Life of Plate-Fin Structures** 121
Toshihide Igari, Fumiko Kawashima, Yorikata Mizokami and Nobutada Ohno

7 **Review on Spatio-Temporal Multiscale Phenomena in TRIP Steels and Enhancement of Its Energy Absorption** 143
Takeshi Iwamoto and Hang Thi Pham

8 Methods for Creep Rupture Analysis—Previous Attempts and New Challenges 163
Zbigniew L. Kowalewski

9 Strain Gradient Plasticity: A Variety of Treatments and Related Fundamental Issues 199
Mitsutoshi Kuroda

10 Effects of Fiber Arrangement on Negative Poisson's Ratio of Angle-Ply CFRP Laminates: Analysis Based on a Homogenization Theory 219
Tetsuya Matsuda, Keita Goto and Nobutada Ohno

11 Modeling of Internal Damage Evolution of Piezoelectric Ceramics Under Compression-Compression Fatigue Tests 231
Mamoru Mizuno and Ken-ichi Wakui

12 Analysis of Inelastic Behavior for High Temperature Materials and Structures 241
Konstantin Naumenko and Holm Altenbach

13 Onset of Matrix Cracking in Fiber Reinforced Polymer Composites: A Historical Review and a Comparison Between Periodic Unit Cell Analysis and Analytic Failure Criteria 299
Tomonaga Okabe, Yuta Kumagai, Ryo Higuchi and Masaaki Nishikawa

14 Swelling-Induced Buckling Patterns in Gel Films with a Square Lattice of Holes Subjected to In-Plane Uniaxial and Biaxial Pretensions 319
Dai Okumura, Akira Sasaki and Nobutada Ohno

15 A Method to Evaluate Creep Properties of Solder Alloys Using Micro Indentation 335
Katsuhiko Sasaki, Ken-ichi Ohguchi and Atsuko Takita

16 The Behavior of the Graded Cellular Material Under Impact 357
Changjian Shen and Guoxing Lu

17 Fracture Mechanics at Atomic Scales 379
Takahiro Shimada and Takayuki Kitamura

18 Radiation Damage Evolution in Ductile Materials 397
Błażej Skoczeń and Aneta Ustrzycka

19 Capabilities of the Multi-mechanism Model in the Prediction of the Cyclic Behavior of Various Classes of Metals 413
Lakhdar Taleb, Kacem Saï and Georges Cailletaud

20 Phase-Field Modeling for Dynamic Recrystallization 441
Tomohiro Takaki, Akinori Yamanaka and Yoshihiro Tomita

21 Mechanical Properties of Shape Memory Alloy and Polymer 461
Hisaaki Tobushi, Ryosuke Matsui and Kohei Takeda

22 Constitutive Model of Discontinuously-Reinforced Composites Taking Account of Reinforcement Damage and Size Effect and Its Application 489
Keiichiro Tohgo

23 A Study of Metal Fatigue Failure as Inherent Features of Elastoplastic Constitutive Equations 529
Zhao-Ling Wang and Heng Xiao

24 Maximization of Strengthening Effect of Microscopic Morphology in Duplex Steels 541
Ikumu Watanabe, Gaku Nakamura, Kohei Yuge, Daigo Setoyama and Noritoshi Iwata

25 Molecular Dynamics Simulations on Local Lattice Instability at Mode I Crack Tip in BCC Iron 557
Kisaragi Yashiro, Yuta Tsugawa and Hiroshi Katayama

26 Modeling of Large-Strain Cyclic Plasticity Including Description of Anisotropy Evolution for Sheet Metals 571
Fusahito Yoshida, Takeshi Uemori and Hiroshi Hamasaki

27 A New Kinematic Hardening Rule Describing Different Plastic Moduli in Monotonic and Cyclic Deformations 587
Yilin Zhu, Guozheng Kang and Qianhua Kan

Contributors

Rashid K. Abu Al-Rub Mechanical and Materials Engineering Department, Institute Center for Energy, Masdar Institute of Science and Technology, Abu Dhabi, United Arab Emirates

Diab W. Abueidda Mechanical and Materials Engineering Department, Institute Center for Energy, Masdar Institute of Science and Technology, Abu Dhabi, United Arab Emirates

Holm Altenbach Institute of Mechanics, Otto-von-Guericke University Magdeburg, Magdeburg, Germany

Otto T. Bruhns Institute of Mechanics, Ruhr-University Bochum, Bochum, Germany

Michael Brünig Institut Für Mechanik und Statik, Universität der Bundeswehr München, Neubiberg, Germany

Georges Cailletaud MINES ParisTech, Centre des Matériaux, CNRS UMR 7633, Evry Cedex, France

Ahmed S. Dalaq Mechanical and Materials Engineering Department, Institute Center for Energy, Masdar Institute of Science and Technology, Abu Dhabi, United Arab Emirates

Keita Goto Department of Engineering Mechanics and Energy, University of Tsukuba, 1-1-1 Tennodai, Tsukuba 305-8573, Japan

Tomoyuki Hakoyama JSPS Research Fellow (Doctoral Course Students), Department of Mechanical Systems Engineering, Tokyo University of Agriculture and Technology, Koganei-shi, Tokyo, Japan

Hiroshi Hamasaki Department of Mechanical Science and Engineering, Hiroshima University, Higashi-hiroshima, Japan

Hideyuki Hara Graduate School of Science and Technology, Keio University, Yokohama, Japan

Ryo Higuchi Department of Aerospace Engineering, Tohoku University, Aoba-ku, Sendai, Miyagi, Japan

Toshihide Igari Mitsubishi Heavy Industries, Ltd., Nagasaki, Japan

Takeshi Iwamoto Institute of Engineering, Hiroshima University, Higashi-Hiroshima, Hiroshima, Japan

Noritoshi Iwata TOYOTA Central R&D Labs., Inc., Nagakute, Aichi, Japan

Qianhua Kan School of Mechanics and Engineering, Southwest Jiaotong University, Chengdu, People's Republic of China

Guozheng Kang State Key Laboratory of Traction Power and School of Mechanics and Engineering, Southwest Jiaotong University, Chengdu, People's Republic of China

Hiroshi Katayama Graduate School of Engineering, Kobe University, Kobe, Japan

Fumiko Kawashima Department of Advanced Technology, Kumamoto University, Kumamoto, Japan

Takayuki Kitamura Department of Mechanical Engineering and Science, Kyoto University, Kyoto, Japan

Zbigniew L. Kowalewski Institute of Fundamental Technological Research, Warsaw, Poland

Yuta Kumagai Department of Aerospace Engineering, Tohoku University, Aoba-ku, Sendai, Miyagi, Japan

Mitsutoshi Kuroda Graduate School of Science and Engineering, Mechanical Systems Engineering, Yamagata University, Yonezawa, Yamagata, Japan

Toshihiko Kuwabara Division of Advanced Mechanical Systems Engineering, Institute of Engineering, Tokyo University of Agriculture and Technology, Koganei-shi, Tokyo, Japan

Guoxing Lu Faculty of Science, Engineering and Technology, Swinburne University of Technology, Hawthorn Vic, Australia

Tetsuya Matsuda Department of Engineering Mechanics and Energy, University of Tsukuba, Tsukuba, Japan

Ryosuke Matsui Department of Mechanical Engineering, Aichi Institute of Technology, Yakusa-cho, Toyota, Japan

Yorikata Mizokami Mitsubishi Heavy Industries, Ltd., Kobe, Japan

Mamoru Mizuno Akita Prefectural University, Yuri-Honjo, Akita, Japan

Gaku Nakamura Seikei University, Musashino, Tokyo, Japan

Konstantin Naumenko Institute of Mechanics, Otto-von-Guericke University Magdeburg, Magdeburg, Germany

Masaaki Nishikawa Department of Mechanical Engineering and Science, Kyoto University, Nishikyo-ku, Kyoto, Japan

Ken-ichi Ohguchi Akita University, Akita, Japan

Nobutada Ohno Department of Mechanical Science and Engineering, Nagoya University, Chikusa-ku, Nagoya, Japan

Tomonaga Okabe Department of Aerospace Engineering, Tohoku University, Aoba-ku, Sendai, Miyagi, Japan

Dai Okumura Department of Mechanical Science and Engineering, Nagoya University, Furo-cho, Chikusa-ku, Nagoya, Japan

Hang Thi Pham Graduate School of Engineering, Hiroshima University, Higashi-Hiroshima, Hiroshima, Japan; Faculty of Engineering, Vietnam National University of Agriculture, Gialam, Hanoi, Vietnam

Akira Sasaki Department of Computational Science and Engineering, Nagoya University, Furo-cho, Chikusa-ku, Nagoya, Japan

Katsuhiko Sasaki Hokkaido University, Sapporo, Hokkaido

Kacem Saï UGPMM Ecole Nationale d'Ingénieurs de Sfax, Sfax, Tunisia

Daigo Setoyama TOYOTA Central R&D Labs., Inc., Nagakute, Aichi, Japan

Changjian Shen School of Mechanical and Aerospace Engineering, Nanyang Technological University, Nanyang, Singapore

Takahiro Shimada Department of Mechanical Engineering and Science, Kyoto University, Kyoto, Japan

Kazuyuki Shizawa Department of Mechanical Engineering, Keio University, Yokohama, Japan

Błażej Skoczeń Cracow University of Technology, Kraków, Poland

Tomohiro Takaki Faculty of Mechanical Engineering, Kyoto Institute of Technology, Kyoto, Japan

Kohei Takeda Department of Mechanical Engineering, Aichi Institute of Technology, Yakusa-cho, Toyota, Japan

Atsuko Takita Hokkaido University, Sapporo, Hokkaido

Lakhdar Taleb INSA, GPM, CNRS UMR 6634, St. Etienne du Rouvray Cedex, France

Hisaaki Tobushi Department of Mechanical Engineering, Aichi Institute of Technology, Yakusa-cho, Toyota, Japan

Keiichiro Tohgo Department of Mechanical Engineering, Shizuoka University, Nakaku, Hamamatsu, Japan

Yoshihiro Tomita Kobe University, Kobe, Japan

Yuta Tsugawa Graduate School of Engineering, Kobe University, Kobe, Japan

Takeshi Uemori Graduate School of Natural Science and Technology, Okayama University, Okayama, Japan

Aneta Ustrzycka Cracow University of Technology, Kraków, Poland

Ken-ichi Wakui Akita Prefectural University, Yuri-Honjo, Akita, Japan

Zhao-Ling Wang Shanghai Institute of Applied Mathematics and Mechanics, Shanghai University, Shanghai, China; School of Mathematics and Information Sciences, Weifang University, Weifang, Shandong, China

Ikumu Watanabe National Institute for Materials Science, Tsukuba, Ibaraki, Japan

Heng Xiao State Key Laboratory of Advanced Special Steels and Shanghai Institute of Applied Mathematics and Mechanics, Shanghai University, Shanghai, China

Akinori Yamanaka Mechanical Systems Engineering, Tokyo University of Agriculture and Technology, Koganei, Tokyo, Japan

Kisaragi Yashiro Gifu University, Gifu, Japan

Fusahito Yoshida Department of Mechanical Science and Engineering, Hiroshima University, Higashi-hiroshima, Japan

Kohei Yuge Seikei University, Musashino, Tokyo, Japan

Yilin Zhu State Key Laboratory of Traction Power and School of Mechanics and Engineering, Southwest Jiaotong University, Chengdu, People's Republic of China

Chapter 1
Thermo-Electro-Mechanical Properties of Interpenetrating Phase Composites with Periodic Architectured Reinforcements

Rashid K. Abu Al-Rub, Diab W. Abueidda and Ahmed S. Dalaq

Abstract In this study, the multifunctional properties (thermal, electric, and mechanical properties) of a new type of three-dimensional (3D) periodic architectured interpenetrating phase composites (IPCs) are investigated computationally. These new IPCs are created using two interconnected, bicontinuous, and intertwined material phases. The inner reinforcing phase takes the shape of the 3D morphology (architecture) of the mathematically-known triply periodic minimal surfaces (TPMS). The TPMS reinforcements are 3D solid sheet networks with a certain volume fraction and architecture. The interconnectivity of the proposed TPMS-based IPCs provide a novel way of creating multifunctional composites with superior properties. In this study, the effect of six well-known TPMS architectures of various volume fractions on the thermal/electrical conductivity and Young's modulus of the IPCs is investigated using the finite element analysis of a unit cell with periodic boundary conditions. The contrast effect (high and low) between the conductivities and Young's modulus of the two phases is also investigated. The calculated effective properties are compared with some analytical bounds. The proposed TPMS-IPCs possess effective properties close to the upper Hashin-Shtrikman bounds. It is also shown that the effect of TPMS architecture decreases as the contrast decreases. Finally, the manufacturability of these new TPMS-IPCs is demonstrated through using 3D printing technology.

Keywords Interpenetrating phase composites · Architectured materials · Effective properties · Finite element analysis · 3D printing · Multifunctional composites · 3D inclusions

---

R.K. Abu Al-Rub (✉) · D.W. Abueidda · A.S. Dalaq
Mechanical and Materials Engineering Department, Institute Center for Energy, Masdar Institute of Science and Technology, PO Box 54224, Abu Dhabi, United Arab Emirates
e-mail: rabualrub@masdar.ac.ae

D.W. Abueidda
e-mail: dabueidda@masdar.ac.ae

A.S. Dalaq
e-mail: adalaq@masdar.ac.ae

H. Altenbach et al. (eds.), *From Creep Damage Mechanics to Homogenization Methods*, Advanced Structured Materials 64,
DOI 10.1007/978-3-319-19440-0_1

1.1 Introduction

Design of new materials that are durable, lightweight, and environmentally sustainable are commonly inspired by natural composites. For example, bone is strong and tough because its two constituent materials; soft collagen protein and stiff hydroxyapatite mineral, are arranged in complex hierarchical patterns that change at every scale of the composite, from the micro up to the macro (Ashby et al. 2013). However, as engineers we are no longer limited to the natural patterns. We can design our own architectured materials that may perform even better than the ones that already exist in nature. While researchers can come up with hierarchical structures in the design of new materials, going from a computer model to the production of physical artifacts has been a persistent challenge. However, due to the recent advances in three-dimensional (3D) printing (Schaedler et al. 2011; Zheng et al. 2014), researchers can now move from computer-optimized material architectures that are designed using the concepts of structural mechanics directly to fabrication and testing. 3D printing is a process of creating 3D objects from a digital file using a material's printer, in a manner similar to printing images on paper. With the latest 3D printers capable of printing materials with widely contrasting mechanical behavior simultaneously in complex geometries at micrometer resolutions, the potential of this technology is growing. Advanced printing technology now offers the possibility to create complex topologies with fine features composed of a multitude of materials with varying mechanical properties quickly, cheaply and at a large scale. In a matter of few hours, 3D printing can be used to fabricate complex architectures.

The focus of this chapter is on presenting and studying a new type of 3D-printable multifunctional composites that have unique mechanical and physical properties as compared to traditional composites (Abueidda et al. 2015). These composites combine two main ideas; interpenetrating phase composites (IPCs) and the mathematically-known triply periodic minimal surfaces (TPMS). TPMS-based IPCs are virtually created and manufactured using 3D printing technology so that they can be tested and characterized. We will explain next IPCs and TPMS.

IPCs are a new kind of composite microstructures in which both the matrix and reinforcement phases are entirely interconnected and continuous in 3D throughout the microstructure space. Different than traditional composite materials that usually have a continuous phase with one or more discrete (discontinuous) reinforcement phases such as fibers, whiskers, or particles, the continuity of each phase kept in IPCs make them a truly multifunctional and highly durable materials. In IPCs, if any one of the constituent phases were removed, the remaining phase(s) would form a self-supporting foam or porous media. IPCs take the best qualities of the two or several constituents to generate a superior composite material. Periodic (not random or stochastic) 3D IPCs that today do not exist in either engineered or biological forms have not been explored much and this chapter focuses on such composites. It is envisioned that such new materials and composites will transform the way materials are designed for several engineered properties.

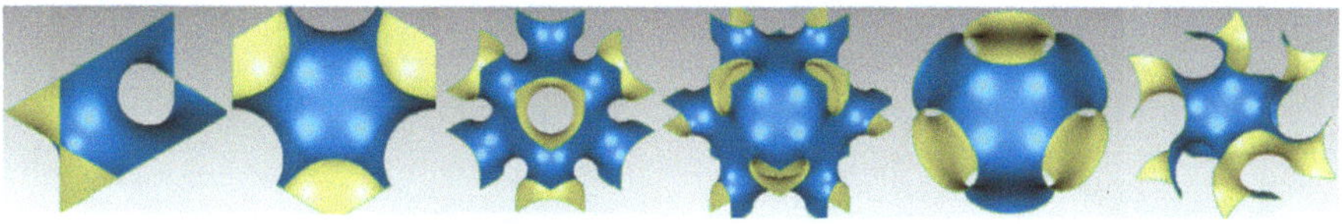

Fig. 1.1 Examples of TPMS, **a** crossed layers of parallels (CLP), **b** diamond in rhombic dodecahedron (diamond), **c** IWP, **d** Neovius, **e** primitive, **f** gyroid

However, the key to engineering the multifunctional properties of IPCs is in the exact way the composite constituents fit together. In this work, this is achieved through creating 3D geometries based on triply periodic minimal surfaces (TPMS) (see Fig. 1.1). Triply periodic means repeating themselves in three dimensions. TPMS are widely used in architectural design due to their remarkable geometric properties. 3D printing provides a novel technique for fabricating such 3D geometric patterns. TPMS are infinite continuous smooth surfaces that separate the space into two intertwined, continuous, and complex regions (Kapfer et al. 2011). Furthermore, TPMS are surfaces that are locally area minimizing, and they are defined as surfaces with zero mean curvature at each point on the surface (Jung and Torquato 2005; Chen et al. 2009). Several TPMS were discovered in the last century, and they were discussed by several investigators (Abueidda et al. 2015; Kapfer et al. 2011; Gandy and Klinowski 2000; Góźdź and Hołyst 1996; Lord 1997; Yoo 2011, 2012). Figure 1.1 illustrates some of the commonly known TPMS. Brakke (2014) developed the software Surface Evolver that generates the TPMS based on minimizing the energy of a surface subjected to constraints. The energy of a surface is proportional to its surface area (Torquato and Donev 2004). The most common natural example of minimal surfaces is the soap films in which the surface tension minimizes the energy of the film; and consequently, minimizes its surface area (Torquato and Donev 2004). TPMS are described in terms of a fundamental patch or asymmetric unit from which the whole minimal surface can be created by its symmetry elements (Gandy and Klinowski 2000).

In the present work, TPMS are utilized to create unique and novel two-phase IPCs and computationally study their effective mechanical, electrical, and thermal properties. These novel IPCs have been discovered recently by Abu Al-Rub and co-workers (Abueidda et al. 2015). The prediction of the macroscopic effective properties of these TPMS-IPCs is based on the finite element analysis of a periodic unit cell. IPCs based on TPMS that are considered in literature are basically IPCs in which the TPMS separate the space into two disjoint but intertwining regions. Each region of those represents a phase of the composite. Furthermore, the two phases of TPMS-IPCs studied in literature such as Torquato and Donev (2004), Challis et al. (2008), Kassner et al. (2005), Torquato et al. (2003) have the same volume fraction (50 % each). In the work of Abu Al-Rub and co-workers (Abueidda et al. 2015), the TPMS are employed differently; TPMS are thickened to create solid shell or sheet networks taking the shape of the TPMS and representing the reinforcing phase with different

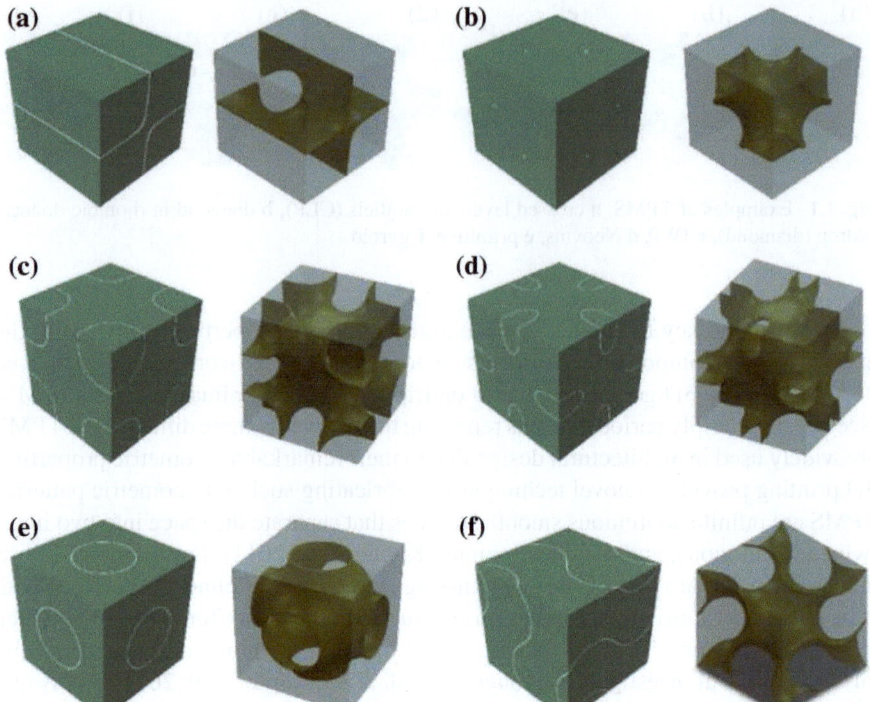

Fig. 1.2 Unit cells of TPMS-IPCs; **a** CLP, **b** diamond, **c** IWP, **d** Neovius, **e** primitive, **f** gyroid. This figure shows a unit cell of the proposed IPC with TPMS as a solid sheet reinforcement

volume fractions (see Fig. 1.2). In fact, since the TPMS are incorporated into the matrix as a 3D solid sheet network, it splits the matrix into two phases such that the TPMS is another phase. However, in this work, two-phase IPCs are considered such that the TPMS solid sheet network is one phase and the matrix material is the other phase.

The most advantageous property of a TPMS-IPC lies in its ability to achieve several desirable properties at once by combining a number of contiguous phases, each of which possesses one of the properties desired. Of course, the goal of research into new composite materials has generally been to achieve combinations of material properties which are not possible with single phase materials. Considering this goal, periodic IPCs may provide strategic solutions for numerous property combinations (i.e., multifunctionality). Moreover, due to the presence of many interfaces in IPCs (i.e., boundaries between dissimilar materials), periodic architectured IPCs are very effective advanced materials for resisting fracture and for distributing damage over a larger area. Hence, IPCs with a large relative fraction of interfaces promise to provide an over an order of magnitude increase in damage-tolerance and, in turn, increased sustainability.

Generally, the effective properties can be quantified using analytical models, empirical equations, and/or numerical modeling. Most of the analytical models developed to predict the behavior of IPCs are based on the volume fractions and the material properties of the individual phases. However, there are other factors influencing the effective properties such as the topology of the phases and the thermal residual stresses (Cheng et al. 2014). In this work, the effect of the architecture or morphology of the microstructure, through using TPMS-type 3D solid sheet reinforcements, the effective properties including the ones related to the transport phenomena (electrical and thermal) and the mechanical elastic properties are investigated computationally using the finite element modeling of a unit cell of the TPMS-IPC.

1.2 Architecture and Numerical Analysis Assumptions

1.2.1 Architectures

The proposed IPCs are composed from solid shells or sheets obtained from thickening the TPMS embodying the reinforcing phase and their complementary cubes representing the matrix. Figure 1.3 schematically shows how the TPMS-IPC is created, for example, using the primitive TPMS-based IPC (P-IPC). The geometries of the TPMS are created using the software *Surface Evolver* to create STL (voxel based) files. After that, these surfaces are transferred to the software *SolidWorks* to be thickened and to create their complementary parts. The thickened TPMS and their complementary parts are then transferred to the commercial finite element software *Abaqus*.

Figure 1.2 shows six unit cells of proposed TPMS-IPCs. From Fig. 1.2, it can be noticed that the primitive-IPC, diamond-IPC, IWP-IPC, Neovius-IPC, and gyroid-IPC are macroscopically isotropic whilst the CLP-IPC is macroscopically transversely-isotropic.

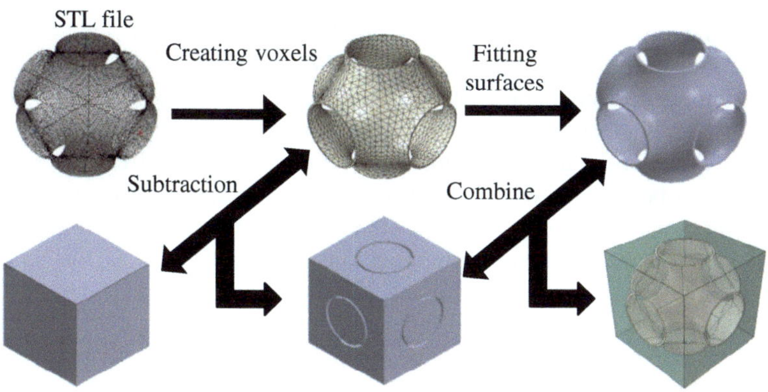

Fig. 1.3 CAD-based TPMS-IPC preparation scheme of the P-IPC composite

1.2.2 Governing Equations and Boundary Conditions

The problem under investigation is a two-phase IPC with geometries based on TPMS as illustrated in Fig. 1.2. Each phase is described by the following differential equation:

$$\mathrm{div}(C\nabla \boldsymbol{u}) = 0 \tag{1.1}$$

with div is the divergence, ∇ is the gradient operator, $\boldsymbol{u}$ is the displacement, temperature, or electric potential, and C is the material property (i.e., elastic, thermal, or electrical property of the material). In the subsequence of this chapter, the mismatch in the property between the two phases in the TPMS-IPC is defined by the ratio $\alpha = C_1/C_2$, where C_2 is the property of TPMS and C_1 is the property of the matrix.

The effective properties of composites in general are evaluated at the macroscopic scale which is large enough so heterogeneities to be smeared out (Michel et al. 1999). The smallest material volume element of the heterogeneous material (composite) that can be used to capture the response and the effective properties is called the representative volume element (RVE) (Kanit et al. 2003). The homogenization theory using RVE is usually carried out to numerically determine the effective properties of composites (Kanit et al. 2003; Muliana 2009; Zohdi and Wriggers 2008). Based on the imposed boundary conditions, either the macroscopic flux or the macroscopic gradient field is calculated by averaging or homogenizing the microscopic counterparts, and then the effective property is calculated by using the macroscopic constitutive relationships (Kushch and Chernobai 2014; Giraud et al. 2007; Lutz and Zimmerman 2005):

$$\bar{\boldsymbol{q}} = \frac{1}{V}\int_V \boldsymbol{q}\mathrm{d}V, \qquad \overline{\nabla u} = \frac{1}{V}\int_V \nabla \boldsymbol{q}\mathrm{d}V, \tag{1.2}$$

$$\bar{\boldsymbol{q}} = C^*\overline{\nabla \boldsymbol{u}} \tag{1.3}$$

where the superimposed bar indicates a spatial average quantity, $\boldsymbol{q}$ is the stress, local heat flux, or electric current, V is the volume, $\overline{\nabla \boldsymbol{u}}$ is the spatial average displacement gradient, temperature gradient, or electrical potential gradient, and C^* is the effective property (i.e., effective thermal or electrical conductivity or the effective Young's modulus). Types of the applied boundary conditions strongly affect the size of the RVE needs to be considered (Jiang et al. 2002). The size of RVE is defined by dimensionless parameter $\delta = D/d$, where d and D are illustrated in Fig. 1.4, and it is assumed to be large enough compared to the atomistic scale, so the continuum mechanics models remain valid (Muliana 2009).

Generally, there are four types of boundary conditions (Jiang et al. 2002; Dalaq et al. 2013; Jiang et al. 2001) that can be applied to the RVE in order to estimate the effective property, which are:

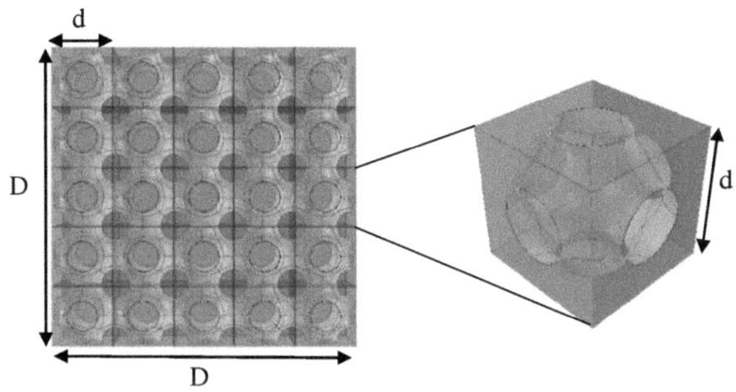

Fig. 1.4 Sample microstructural architecture employed in studying the effect of the boundary conditions

1. Essential Boundary Condition (EBC)

$$\boldsymbol{u} = \overline{\nabla \boldsymbol{u}}\boldsymbol{x}, \qquad \forall \boldsymbol{x} \in \partial B \tag{1.4}$$

2. Natural Boundary Condition (NBC)

$$\boldsymbol{q}\boldsymbol{u} = \overline{\boldsymbol{q}}\boldsymbol{n}, \qquad \forall \boldsymbol{x} \in \partial B \tag{1.5}$$

3. Periodic Boundary Condition (PBC)

$$\boldsymbol{u}(\boldsymbol{x}+\boldsymbol{L}) = \boldsymbol{u}(\boldsymbol{x}) + \overline{\nabla \boldsymbol{u}}\boldsymbol{L}, \qquad \boldsymbol{q}(\boldsymbol{x}+\boldsymbol{L}) = -\boldsymbol{q}(\boldsymbol{x}), \qquad \forall \boldsymbol{x} \in \partial B \tag{1.6}$$

4. Mixed Boundary Condition (MBC)

$$(\boldsymbol{u} - \overline{\nabla \boldsymbol{u}}\boldsymbol{x})(\boldsymbol{q}\boldsymbol{x} - \bar{\boldsymbol{q}}\boldsymbol{n}) = \boldsymbol{0}, \qquad \forall \boldsymbol{x} \in \partial B \tag{1.7}$$

Here ∂B is the boundary of the RVE, $\boldsymbol{n}$ is the outer unit normal vector to ∂B, and $\boldsymbol{L}$ is the length of the periodicity. The effect of boundary conditions is investigated briefly in the present work. In case of PBC and EBC, a given $\overline{\nabla \boldsymbol{u}}$ is applied while a given $\overline{\boldsymbol{q}}$ is used when NBC is applied. When MBC is utilized, EBC is applied on one pair of parallel faces, and NBC is applied on the other pairs.

Figure 1.5 demonstrates the effect of the boundary conditions on the conductivity (thermal or electrical) or the Young's modulus and the required size of the RVE. The results displayed in Fig. 1.5 is for the primitive-IPC, mismatch $\alpha = 1000$, and 2 % volume fraction of the TPMS reinforcing phase. The results obtained from MBC and PBC are independent of while the results obtained from EBC and NBC converge to those of MBC and PBC as δ increases. The architectures of TPMS-IPCs are geometrically periodic. Consequently, periodic boundary conditions can be applied besides mixed ones.

From the argument above, MBC/PBC should be applied to evaluate the effective property and the macroscopic response of TPMS-IPCs due to scale-independence when they are employed. Another advantage of MBC have over the other types of boundary conditions is that they are common in experimental setups for measuring

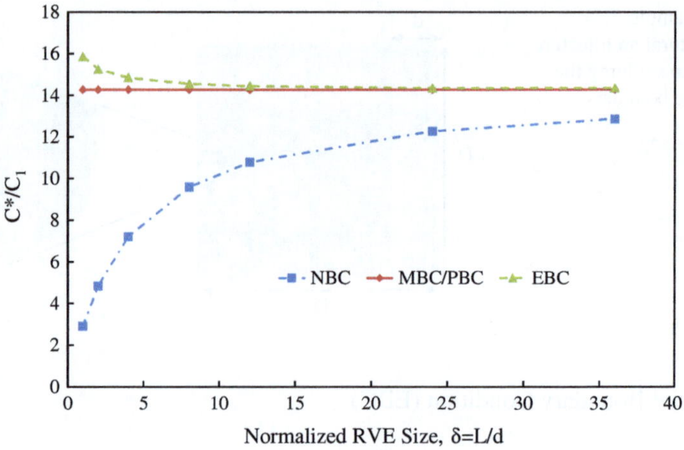

Fig. 1.5 Effect of different types of boundary conditions on effective property and required size of the RVE or unit cell

the effective properties (Jiang et al. 2002). On the other hand, PBCs are essential for the case where the RVE (or the unit cell in this work) is selected to represent the material's microstructure. Therefore, the PBCs will be employed in this work for estimating the effective properties.

1.2.3 Analytical Models for Calculating Effective Properties

Among the first well-known analytical models for estimating the effective properties are those of Reuss and Voigt. The upper (Voigt) and lower (Reuss) bounds for estimating the effective property of any composite (i.e., the so-called rule of mixture), are given by:

$$(\phi_1/C_1 + \phi_2/C_2)^{-1} \leq C^* \leq \phi_1/C_1 + \phi_2/C_2 \tag{1.8}$$

where ϕ_1 and ϕ_2 are the volume fractions of the complimentary part of the IPC (the matrix) and the reinforcing TPMS phase, respectively. The upper and lower bounds are also called the arithmetic and harmonic means, respectively.

The Hashin-Shtrikman bounds (HS bounds) are one of the most common bounds, and they are derived based on the variational principles (Wegner and Gibson 2000). The Hashin-Shtrikman bounds are the tightest bounds (Cadman et al. 2013). The effective property of a two-phase isotropic composite is bounded by these bounds (Torquato et al. 2003), such that:

$$< C > - \frac{\phi_1\phi_2(C_1 - C_2)^2}{< \tilde{C} + 2C_1 >} \leq C^* \leq < C > - \frac{\phi_1\phi_2(C_1 - C_2)^2}{< \tilde{C} + 2C_2 >} \tag{1.9}$$

$$< C >= C_1\phi_1 + C_2\phi_2, \qquad < \tilde{C} >= C_1\phi_2 + C_2\phi_1 \tag{1.10}$$

These bounds are generally applicable to all composites including IPCs in case of isotropic and homogeneous phases when composite with large enough volume is studied (Wegner and Gibson 2000).

It is noteworthy that theory of mixture and Hashin-Shtrikman analytical bounds are prone to deviations from the numerical and experimental results, where at low volume fractions and for high contrasts in the mechanical properties of individual phases the accuracy of these bounds are questionable. There are many other improved analytical models and approximations, but the focus of this study will be on comparing the finite element predictions of effective properties to these well-known bounds.

1.3 Results and Discussions

1.3.1 Effective Thermal/Electrical Conductivity

In this section, we present the predicted effective thermal/electrical conductivities $C^* = k_e$ for the IPCs in Fig. 1.2 for different volume fractions of the TPMS and for various contrast $\alpha = C_2/C_1$, where $k_2 = C_2$ is the conductivity of the TPMS sheet and $k_1 = C_1$ is the conductivity of the surrounding matrix material. Figure 1.6 shows the estimated effective conductivities (thermal or electrical) of the TPMS-IPCs for 2 % TPMS volume fraction and for $\alpha = 1000$. The mismatch $\alpha = 1000$ indicates that the TPMS reinforcing phase is relatively very conductive as compared to the matrix material. This is similar to having the TPMS 3D sheet networks made of carbon nanotubes, graphene, or carbon nanotube/graphene reinforced polymer. Such composites are attractive for various applications such as heat dissipaters in electronics or as self-sensing structural components. Comparison between the predicted effective conductivities and the analytical bounds is also shown in Fig. 1.6.

Figure 1.6 shows that the diamond-IPC has the lowest conductivity among the other isotropic TPMS-IPCs. One main reason for this behavior is that diamond-IPC has poor connectivity at the boundary of the unit cell compared to the others (see Fig. 1.2b). The results obtained for diamond-IPC, IWP-IPC, Neovius-IPC, primitive-IPC, and gyroid-IPC are bounded by the rule of mixtures and the isotropic Hashin-Shtrikman (HS) bounds. In addition, it can be observed that the effective conductivity of CLP-IPC in the longitudinal direction exceeds the upper HS bound. This does not violate the HS bounds since Eq. (1.9) is used for macroscopically isotropic composites, and the CLP-IPC is transversely-isotropic at the macro-scale. In fact, the average normalized conductivity is $k_e/k_1 = 13.6$, which is within the HS bounds. Figure 1.6 shows that IWP-, Neovius-, and primitive-IPCs have comparable effective

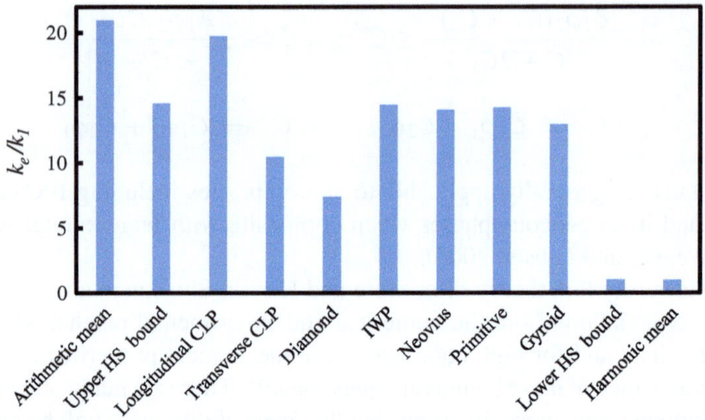

Fig. 1.6 Comparison between the predicted effective conductivity for different TPMS-IPCs with 2 % TPMS volume fraction and mismatch $\alpha = 1000$

conductivity which is close to the HS upper bound. The primitive- and average CLP-IPCs have slightly lower, but sill excellent effective conductivity.

It is also apparent from Fig. 1.6 that the conductivity of CLP-IPC (average value), IWP-IPC, Neovius-IPC, primitive-IPC, and gyroid-IPC are very close to the upper HS bound, and this indicates that the volume of the TPMS is well-distributed when these geometries are used so the effective conductivity is maximized.

Figure 1.7 shows the impact of the mismatch $\alpha = 10$ on the predicted effective conductivity. Figure 1.7 represents a case where the conductivities of both phases are comparable, but the TPMS reinforcement is more conductive. This figure shows that

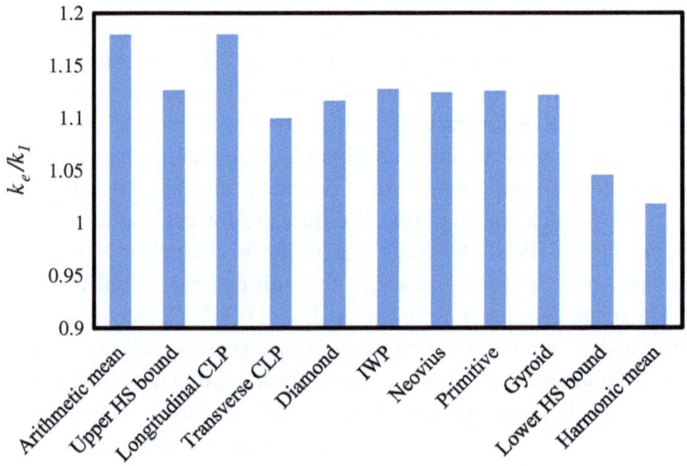

Fig. 1.7 Comparison between the predicted effective conductivity for different TPMS-IPCs with 2 % TPMS volume fraction and mismatch $\alpha = 10$

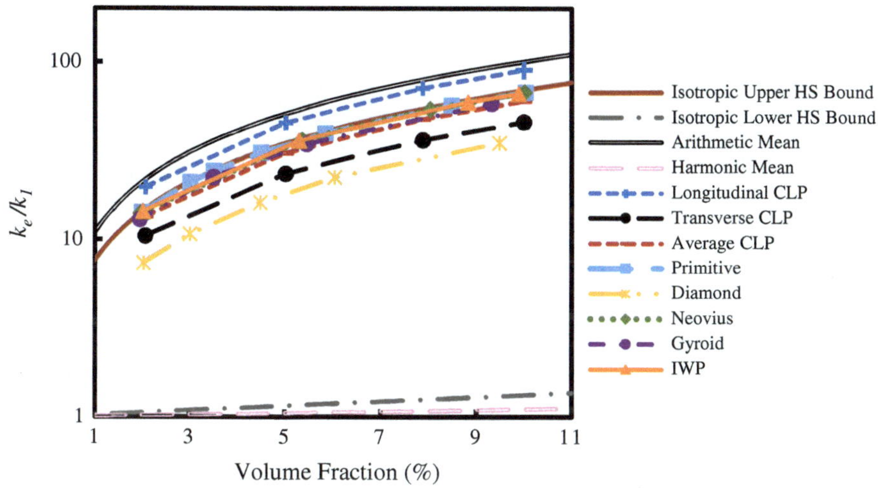

Fig. 1.8 Effect of the TPMS volume fraction on the effective conductivity with mismatches of $\alpha = 1000$

all the TPMS architectures provide almost the same effective conductivity. This indicates that the effect of the architecture is decreasing as the mismatch α is decreasing. The conductivity of the CLP-IPC in the longitudinal direction provides the highest conductivity, whereas it provides the lowest in the transverse direction. However, the average conductivity of CLP-IPC is comparable to the other architectured IPCs.

Figure 1.8 depicts the effect of the reinforcing TPMS phase volume fraction till 10 % for mismatches of 1000. The conclusions from Fig. 1.8 are maintained as the volume fraction of the TPMS reinforcement is increased. This figure shows that the diamond-IPC gives the lowest conductivity whereas the maximum conductivity is achieved though the CLP-IPC in the longitudinal direction.

Figure 1.9 shows the heat flux or the electric current for different IPCs with a 2 % reinforcing phase volume fraction and a mismatch greater than one. The values of the contour in Fig. 1.9 reflect a qualitative distribution. According to Milton (1981), the upper HS bound is attained whenever uniform electric current or heat flux in all the phases except the phase with the highest conductivity (the TPMS in this study) is obtained. This behavior is shown in Fig. 1.9 and obtained with IWP-IPC, Neovius-IPC, primitive-IPC, gyroid-IPC, and CLP-IPC in the longitudinal direction.

1.3.2 Effective Elastic Properties

In this section, we present the enhancement of Young's modulus upon using the TPMS architecture as a reinforcement for $\alpha = C_2/C_1 = E_2/E_1 = 1000$ and 10 where $E_2 = C_2$ is the Young's modulus of the TPMS sheet and $E_1 = C_1$ is the

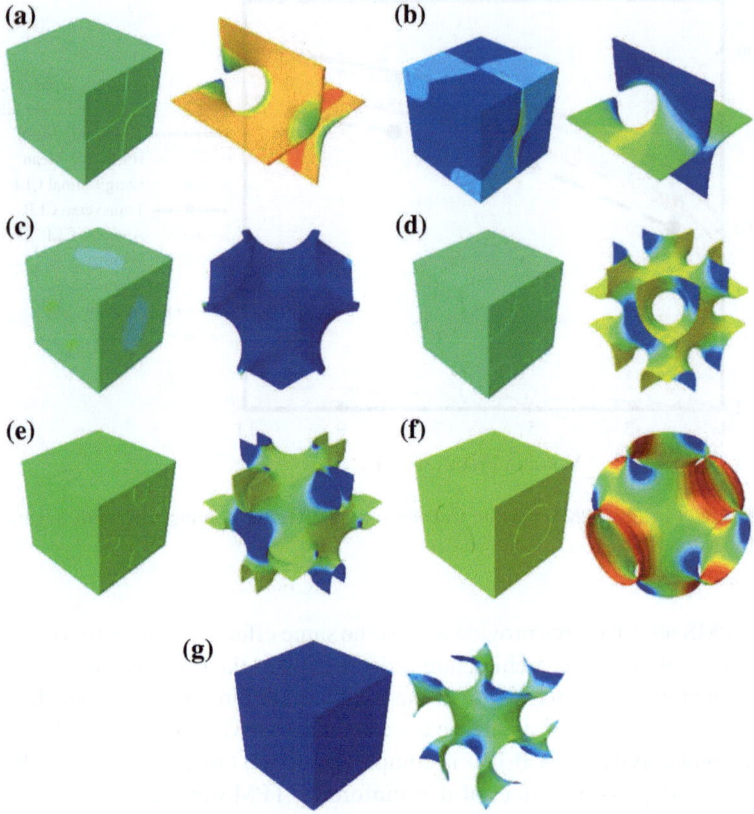

Fig. 1.9 Heat flux/electric current contours in the direction of the imposed temperature gradient in the matrix (*left*) and the TPMS reinforcing-phase (*right*). **a** CLP in the longitudinal direction, **b** CLP in the transverse direction, **c** diamond, **d** IWP, **e** Neovius, **f** primitive, **g** gyroid

Young's modulus of the surrounding matrix material. The volume fraction of the TPMS reinforcement is 5 %. The results are presented in Figs. 1.10 and 1.11.

According to Figs. 1.10 and 1.11, the highest enhancement $C^*/C_1 = E_e/E_l (E_e = C^*$ is the estimated effective Young's modulus) is achieved by the CLP-IPC loaded in the longitudinal orientation, which is similar to that for the effective conductivity. Note that the CLP-IPC in the longitudinal direction coincided with the upper HS bound. The IWP- and Neovius-IPC resulting in comparable enhancements as they both possess similar geometrical morphology (shape). The order continues as primitive-, transverse CLP-, Gyroid and finally the Diamond-IPC being the least performer. At a mismatch of $\alpha = 10$ the enhancement E_e/E_l of each TPMS becomes closer to one another, while the order of Young's modulus is preserved. This is an agreement with the conclusion in the previous section concerning the effective conductivity where the effect of the architecture decreases as the contrast α is decreasing. However, the IWP-IPC gives the highest effective Young's modulus, which is higher than the average Young's modulus of CLP-IPC.

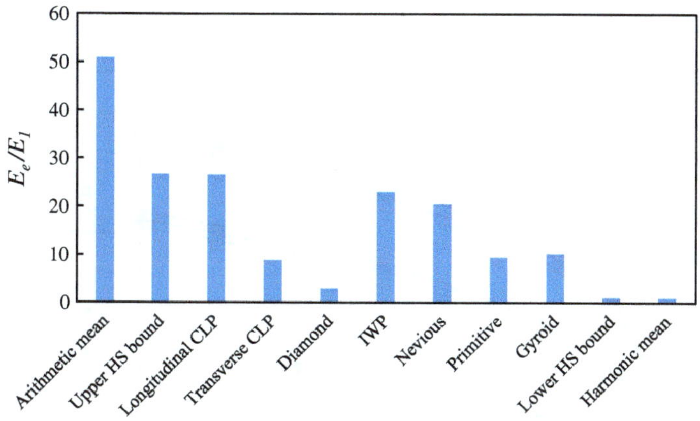

Fig. 1.10 Comparison between the predicted effective Young's modulus for different TPMS-IPCs with 5 % TPMS volume fraction and mismatch $\alpha = 1000$

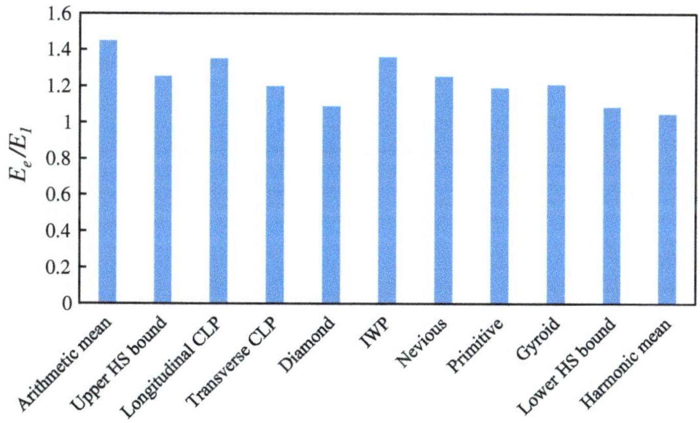

Fig. 1.11 Comparison between the predicted effective Young's modulus for different TPMS-IPCs with 5 % TPMS volume fraction and mismatch $\alpha = 10$

Figure 1.12 shows the variation of the normalized effective Young's modulus as the volume fraction of the TPMS reinforcement is increased. According to Fig. 1.12, the same order of Young's modulus is depicted as a function of TPMS volume fraction. Moreover, the enhancement E_e/E_1 varies linearly with the TPMS volume fraction. It is noticed that as the volume fraction increases, the differences in the effective Young's modulus between different architectures increases. However, still primitive- and gyroid-IPCs show comparable properties, whereas IWP-, Neovius-, and average CLP-IPCs show comparable properties.

Figure 1.13 represents an interesting case for $\alpha = 0$ such that the complementary part is removed creating a cellular material made out of the TPMS only (i.e., TPMS-based foam). The volume fraction of TPMS in Fig. 1.13 is 5 %. Once again the same order of Young's modulus is preserved for different architectured foams indicating

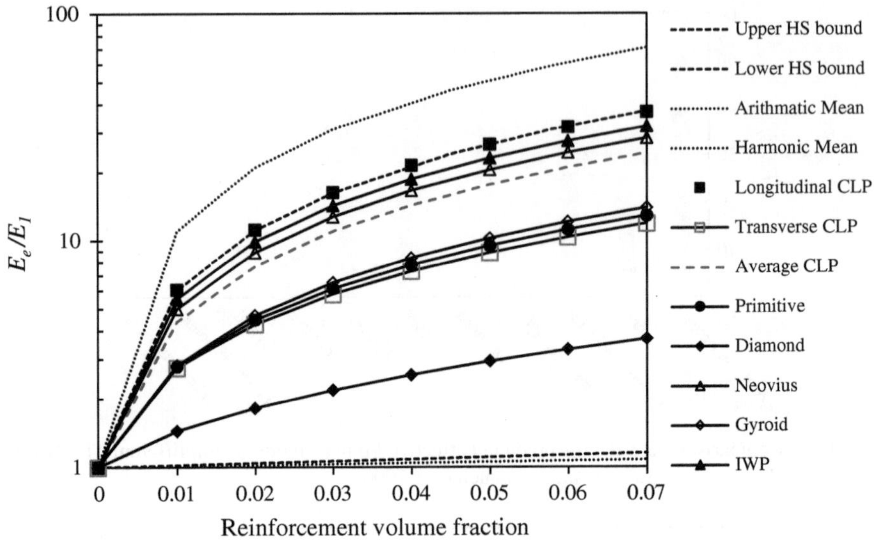

Fig. 1.12 Effect of the TPMS volume fraction on the effective Young's modulus with mismatches of $k_2/k_1 = 1000$

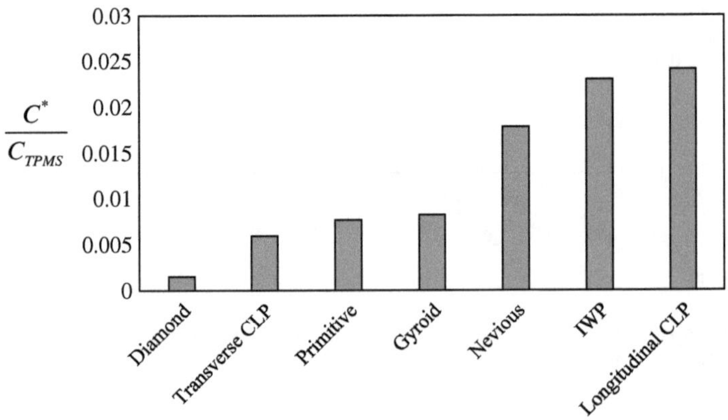

Fig. 1.13 The effective Young's modulus for 5 % TPMS foam (i.e., $\alpha = 0$)

that TPMS reinforcement dictates the overall effective Young's modulus in case of IPCs. This case of foam with TPMS architectures has many applications; mainly as light-weight and tough materials.

Figure 1.14 shows the contours of the maximum principal stress within the IPC, TPMS reinforcement, and the matrix for different architectures. Regions at which debonding and cracks are most likely to initiate are designated by the stress contours. The stress contours within each TPMS show how effective the TPMS as a reinforcement. The distribution of the von Mises is qualitatively similar to that in Fig. 1.14.

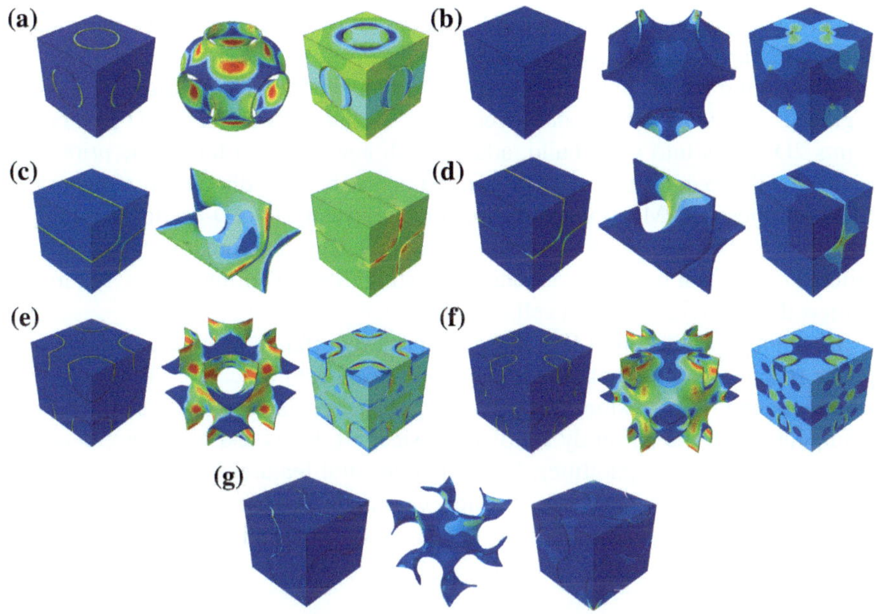

Fig. 1.14 The Maximum principle stresses contours for **a** Primitive-IPC (*left*), Primitive-TPMS (*middle*), and Primitive-complementary (*right*); **b** Diamond-IPC (*left*), Diamond-TPMS (*middle*), and Diamond-complementary (*right*); **c** Longitudinal CLP-IPC (*left*), Longitudinal CLP-TPMS (*middle*), and Longitudinal CLP-complementary (*right*); **d** Transverse CLP-IPC (*left*), Transverse CLP-TPMS (*middle*), and Transverse CLP-complementary (*right*); **e** IWP-IPC (*left*), IWP-TPMS (*middle*), and IWP-complementary (*right*); **f** Neovius-IPC (*left*), Neovius-TPMS (*middle*), and Neovius-complementary (*right*); **g** Gyroid-IPC (*left*), Gyroid-TPMS (*middle*), and Gyroid-complementary (*right*)

1.4 Manufacturability

With recent technologies and advancements in manufacturing techniques, TPMS-IPCs can be manufactured and analyzed utilizing the techniques mentioned in Cooke et al. (2003), Schröder-Turk et al. (2011). However, while researchers can come up with hierarchical structures in the design of new materials, going from a computer model to the production of physical artifacts has been a persistent challenge. Due to the recent advances in 3D printing, researchers can now move from computer-optimized material architectures that are designed using the concepts of structural mechanics directly to fabrication and testing. 3D printing is a process of creating 3D objects from a digital file using a materials' printer, in a manner similar to printing images on a paper. With the latest 3D printers capable of printing materials with widely contrasting mechanical behavior simultaneously in complex geometries at micrometer resolutions, the potential of this technology is growing. Advanced printing technology now offers the possibility to create complex topologies with fine features composed of a multitude of materials with varying mechanical properties quickly, cheaply, and at a large scale. In a matter of few hours, 3D printing can be used to fabricate complex architectures.

3D printing begins with a 3D graphical model of the object. Then the model is digitized and sliced into model layers with special software. The 3D printer then prints 2D layers into a 3D build, adding each new layer on top of the prior layer. Finally, a 3D object is realized that can frequently be used directly from the printer.

The proposed TPMS-IPCs can be manufactured using the 3D-printer such as using the Stratasys Objet260 Connex as shown in Fig. 1.15. Therefore, the current computational results can be easily validated by testing these 3D printed IPCs. Figure 1.15 depicts the manufactured unit cells.

Designing architectured IPC using 3D printing offers a new paradigm for composite functionality. In fact, utilizing periodic architectural features as key elements in defining multi-dimensional material design space promises to enable in-dependent manipulation of the currently coupled physical attributes and to develop materials with unprecedented capabilities. Using architectural features to elicit desired functionality will shift the material creation paradigm from structure→processing→property to property→architecture→fabrication. Feasibility of this "reverse"

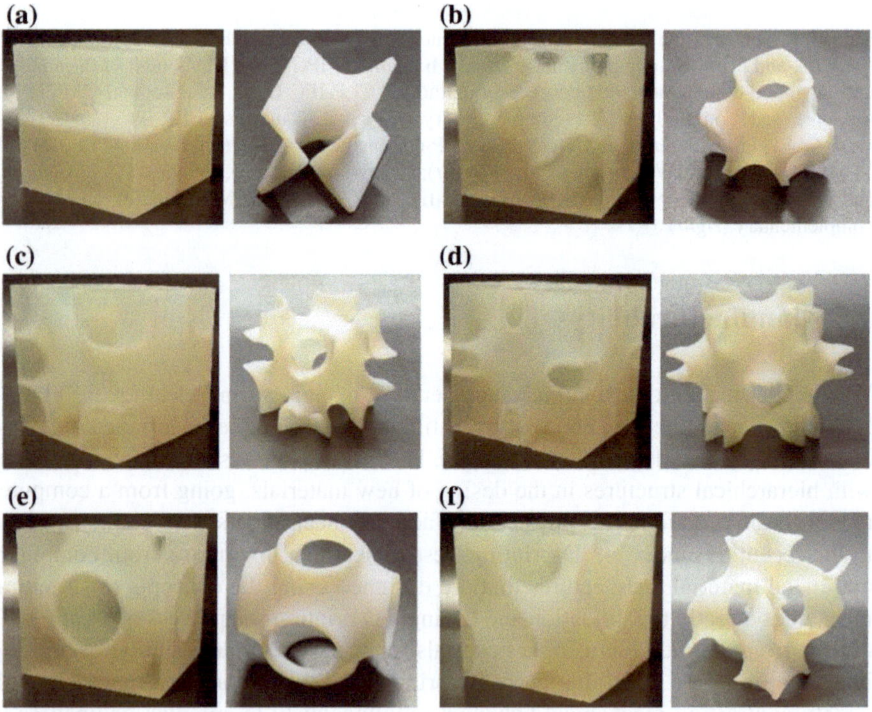

Fig. 1.15 Manufactured architectures of the TPMS-IPCs. **a** CLP-IPC, **b** diamond-IPC, **c** IWP-IPC, **d** Neovius-IPC, **e** primitive-IPC, **f** gyroid-IPC. The *left* pictures show the TPMS-IPCs while the *right* pictures represent the TPMS solids

material construction approach is gated by our ability to understand and predict mechanical and physical responses of these architectured materials (materials whose properties are controlled by their engineered structure rather than by atomic composition alone), where a new design parameter, feature size, plays a critical role.

Potential applications of the proposed TPMS-IPCs are in the energy (energy storage, heat transfer), aerospace (lightweight materials, radiation-tolerant paneling and shielding, high temperature), semiconductor (heat dissipation), and oil and gas (damage resistance, corrosion resistance) sectors. In the energy sector, combining multiple materials in the same IPC would allow us to move from simple to more complex printed objects such as fuel cells, batteries, solar cells, etc. The use of 3D printing technology as a novel technique for fabricating IPCs has the advantages of limited material waste, the ability to build complex geometries, immediate production time, and minimal human involvement.

References

Abueidda DW, Dalaq AS, Al-Rub RKA, Younes HA (2015) Finite element predictions of effective multifunctional properties of interpenetrating phase composites with novel triply periodic solid shell architectured reinforcements. Int J Mech Sci 92:80–89

Ashby MF, Shercliff H, Cebon D (2013) Materials: engineering, science. Processing and design. Butterworth-Heinemann, Oxford

Brakke K (2014) Triply periodic minimal surfaces. Susquehanna University, Selinsgrove, PA. http://www.susqu.edu/brakke/evolver/examples/periodic/periodic.html

Cadman JE, Zhou S, Chen Y, Li Q (2013) On design of multi-functional microstructural materials. J Mater Sci 48(1):51–66

Challis VJ, Roberts AP, Wilkins AH (2008) Design of three dimensional isotropic microstructures for maximized stiffness and conductivity. Int J Solids Struct 45(14):4130–4146

Chen HY, Kwon Y, Thornton K (2009) Multifunctionality of three-dimensional self-assembled composite structure. Scripta Materialia 61(1):52–55

Cheng F, Kim SM, Reddy JN, Abu Al-Rub RK (2014) Modeling of elastoplastic behavior of stainless-steel/bronze interpenetrating phase composites with damage evolution. Int J Plast 61(9):94–111

Cooke MN, Fisher JP, Dean D, Rimnac C, Mikos AG (2003) Use of stereolithography to manufacture critical-sized 3d biodegradable scaffolds for bone ingrowth. J Biomed Mater Res Part B Appl Biomater 64B:65–69

Dalaq AS, Ranganathan SI, Ostoja-Starzewski M (2013) Scaling function in conductivity of planar random checkerboards. Comput Mater Sci 79:252–261

Gandy PJF, Klinowski J (2000) Exact computation of the triply periodic g ('gyroid') minimal surface. Chem Phys Lett 321(5):363–371

Giraud A, Gruescu C, Do DP, Homand F, Kondo D (2007) Effective thermal conductivity of transversely isotropic media with arbitrary oriented ellipsoïdal inhomogeneities. Int J Solids Struct 44(9):2627–2647

Góźdź WT, Hołyst R (1996) Triply periodic surfaces and multiply continuous structures from the Landau model of microemulsions. Phys Rev E 54:5012–5027

Jiang M, Alzebdeh K, Jasiuk I, Ostoja-Starzewski M (2001) Scale and boundary conditions effects in elastic properties of random composites. Acta Mechanica 148(1–4):63–78

Jiang M, Jasiuk I, Ostoja-Starzewski M (2002) Apparent thermal conductivity of periodic two-dimensional composites. Comput Mater Sci 25(3):329–338

Jung Y, Torquato S (2005) Fluid permeabilities of triply periodic minimal surfaces. Phys Rev E Stat Nonlinear Soft Matter Phys 72(5 Pt. 2):056,319

Kanit T, Forest S, Galliet I, Mounoury V, Jeulin D (2003) Determination of the size of the representative volume element for random composites: statistical and numerical approach. Int J Solids Struct 40(13):3647–3679

Kapfer SC, Hyde ST, Mecke K, Arns CH, Schröder-Turk GE (2011) Minimal surface scaffold designs for tissue engineering. Biomaterials 32(29):6875–6882

Kassner ME, Gumbsch P, Kim KS, Knauss W, Kubin L, Langer J, Larson BC, Mahadevan L, Majumdar A, Torquato S, van Swol F, Nemat-Nasser S, Suo Z, Bao G, Barbour JC, Brinson LC, Espinosa H, Gao H, Granick S (2005) New directions in mechanics. Mech Mater 37(2–3):231–259

Kushch VI, Chernobai VS (2014) Transverse conductivity and longitudinal shear of elliptic fiber composite with imperfect interface. Int J Solids Struct 51(13):2529–2538

Lord EA (1997) Triply-periodic balance surfaces. Coll Surf A: Physicochem Eng Aspects 129–130(11):279–295

Lutz MP, Zimmerman RW (2005) Effect of an inhomogeneous interphase zone on the bulk modulus and conductivity of a particulate composite. Int J Solids Struct 42(2):429–437

Michel JC, Moulinec H, Suquet P (1999) Effective properties of composite materials with periodic microstructure: a computational approach. Comput Methods Appl Mech Eng 172(1):109–143

Milton GW (1981) Concerning bounds on the transport and mechanical properties of multicomponent composite materials. Appl Phys A Solids Surf 26:125–130

Muliana AH (2009) A micromechanical model for predicting thermal properties and thermo-viscoelastic responses of functionally graded materials. Int J Solids Struct 46(9):1911–1924

Schaedler T, Jacobsen A, Torrents A, Sorensen A, Lian J, Greer J, Valdevit L, Carter W (2011) Ultralight metallic microlattices. Science 334(6058):962–965

Schröder-Turk GE, Wickham S, Averdunk H, Brink F, Fitz Gerald JD, Poladian L, Large MCJ, Hyde ST (2011) The chiral structure of porous chitin within the wing-scales of Callophrys rubi. J Struct Biol 174(2):290–295

Torquato S, Donev A (2004) Minimal surfaces and multifunctionality. Proc R Soc Lond A 460(2047):1849–1856

Torquato S, Hyun S, Donev A (2003) Optimal design of manufacturable three-dimensional composites with multifunctional characteristics. J Appl Phys 94(9):5748–5755

Wegner L, Gibson L (2000) The mechanical behaviour of interpenetrating phase composites—i: modelling. Int J Mech Sci 42(5):925–942

Yoo D (2011) Porous scaffold design using the distance field and triply periodic minimal surface models. Biomaterials 32(5):7741–7754

Yoo D (2012) Heterogeneous minimal surface porous scaffold design using the distance field and radial basis functions. Med Eng Phys 34(5):625–639

Zheng X, Lee H, Weisgraber TH, Shusteff M, DeOtte J, Duoss EB, Kuntz JD, Biener MM, Ge Q, Jackson JA, Kucheyev SO, Fang NX, Spadaccini CM (2014) Ultralight, ultrastiff mechanical metamaterials. Science 344(6190):1373–1377

Zohdi TI, Wriggers P (2008) Introduction to computational micromechanics. Lecture notes in applied and computational mechanics, vol 20. Springer, Heidelberg

Chapter 2
A Continuum Damage Model Based on Experiments and Numerical Simulations—A Review

Michael Brünig

Abstract The paper summarizes the author's activities in the field of damage mechanics. In this context, a thermodynamically consistent anisotropic continuum damage model is reviewed. The theory is based on consideration of damaged as well as fictitious undamaged configurations. The modular structure of the continuum model is accomplished by kinematic decomposition of strain rates into elastic, plastic and damage parts. A generalized yield condition is used to adequately describe the plastic flow properties of ductile metals and the plastic strain rate tensor is determined by a non-associated flow rule. Furthermore, a stress-state-dependent damage criterion as well as evolution equations of damage strains are proposed. Different branches of the respective criteria are considered corresponding to various damage and failure mechanisms depending on stress state. Since it is not possible to propose and to validate stress-state-dependent criteria only based on tests with uniaxially loaded specimens for a wide range of stress states, numerical calculations on the micro-level have been performed to be able to study stress-state-dependent mechanisms of micro-defects and their effect on macroscopic behavior. In addition, new experiments with two-dimensionally loaded specimens have been developed. Corresponding numerical simulations of these experiments show that they cover a wide range of stress triaxialities and Lode parameters in the tension, shear and compression domains.

Keywords Ductile damage · Stress triaxiality · Lode parameter · Experiments · Numerical simulations

---

M. Brünig (✉)
Institut Für Mechanik und Statik, Universität der Bundeswehr München,
85577 Neubiberg, Germany
e-mail: michael.bruenig@unibw.de

H. Altenbach et al. (eds.), *From Creep Damage Mechanics to Homogenization Methods*, Advanced Structured Materials 64,
DOI 10.1007/978-3-319-19440-0_2

2.1 Introduction

The accurate and realistic modeling of inelastic deformation and failure behavior of engineering materials is essential for the solution of numerous boundary-value problems. For example, microscopic defects cause reduction in strength of materials and shorten the life time of engineering structures. Therefore, a main issue in engineering applications is to provide realistic information on the stress distribution within material elements or assessment of safety factors against failure.

Continuum damage mechanics analyzes systematically the effect of damage on mechanical properties of materials. Critical values of damage variables can be seen as major parameters at the onset of fracture. An important issue in such phenomenological constitutive approaches is the appropriate choice of the physical nature of mechanical variables characterizing the damage state of materials and their tensorial representation. Therefore, to be able to develop a realistic, accurate and efficient phenomenological model it is important to analyze and to understand the complex stress-state-dependent processes of damage and fracture as well as its respective mechanisms acting on different scales. In this context, in the last years various damage models have been published based on experimental observations as well as on multi-scale approaches (Brünig 2001, 2003; Gurson 1977; Murakami and Ohno 1981; Lemaitre 1996; Voyiadjis and Kattan 1999). In this context, Brünig et al. (2011b; 2008); Brünig and Gerke (2011) have proposed a generalized and thermodynamically consistent, phenomenological continuum damage model which has been implemented as user-defined material subroutines in commercial finite element programs allowing analyses of static and dynamic problems in differently loaded metal specimens. To be able to detect stress triaxiality dependence of the constitutive equations tension tests with carefully designed specimens have been developed. For example, differently pre-notched specimens and corresponding numerical simulations have been used by Bai and Wierzbicki (2008); Bao and Wierzbicki (2004); Becker et al. (1988); Bonora et al. (2005); Brünig et al. (2011b; 2008); Dunand and Mohr (2011); Gao et al. (2010). However, these experiments with unnotched and differently notched flat specimens showed stress triaxialities only in a small region of positive values. Larger triaxialities appear in tension tests with cylindrical (axisymmetric) specimens but they cannot be manufactured when the behavior of thin sheets is investigated. Thus, specimens with new geometries have been designed to be able to analyze stress states with small hydrostatic parts. Tension tests with these specimens have been performed (Bao and Wierzbicki 2004; Gao et al. 2010) leading to shear mechanisms in their centers. Similar specimens have been developed and tested by Brünig et al. (2008); Driemeier et al. (2010). Furthermore, to be able to take into account other regions of stress triaxialities butterfly specimens have been manufactured (Bai and Wierzbicki 2008; Dunand and Mohr 2011; Mohr and Henn 2007) which can be tested in different directions using special experimental equipment. Alternatively, series of new tests with biaxially loaded flat specimens taken from thin sheets have been developed by Brünig et al. (2014a; 2015) leading to experimental results on inelastic behavior, damage and fracture of ductile metals for a wide range of stress triaxialities not obtained by the experiments discussed above.

Further information on damage and failure mechanisms can be obtained by performing numerical simulations on the micro-level (Brocks et al. 1995; Brünig et al. 2011a, 2013, 2014b; Chew et al. 2006; Kuna and Sun 1996; Needleman and Kushner 1990; Ohno and Hutchinson 1984; Zhang et al. 2001) considering individual behavior of growth and coalescence of voids and micro-shear-cracks as well as their accumulation to macro-cracks. These numerical calculations have been carried out with different loading conditions covering a wide range of macroscopic stress states. The numerical results elucidated which parameters had remarkable effect on macroscopic stress-strain relations and on evolution equations for the damage variables and which ones only had marginal influence. Thus, it was possible to detect different damage mechanisms which have not been exposed by experiments.

In the present paper main ideas and fundamental governing equations of the phenomenological continuum damage model proposed by the author are reviewed. Some experimental and numerical results are summarized to demonstrate the efficiency and applicability of the approach.

2.2 Continuum Damage Model

Large deformations as well as anisotropic damage and failure of ductile metals are predicted by the continuum model proposed by Brünig (2003) which is based on experimental results and observations as well as on numerical simulations on the micro-level detecting information of microscopic mechanisms due to individual micro-defects and their interactions. Similar to the theories presented by Betten (1982); Grabacki (1991); Murakami and Ohno (1981); Voyiadjis and Park (1999) the phenomenological approach is based on the introduction of damaged and corresponding fictitious undamaged configurations and has been implemented into finite element programs. Extended versions of this model (Brünig et al. 2011b, 2008, 2014a, 2015) propose a stress-state-dependent damage criterion based on different experiments and data from corresponding numerical simulations. Furthermore, it takes into account numerical results of various analyses using unit cell models (Brünig et al. 2011a, 2013, 2014b). Based on these numerical results covering a wide range of stress states it was possible to propose damage equations as functions of the stress triaxiality and the Lode parameter.

2.2.1 *Kinematics*

The kinematic approach is based on the introduction of initial, current and elastically unloaded configurations each defined as damaged and fictitious undamaged configurations, respectively (Brünig 2001, 2003). This leads to the multiplicative decomposition of the metric transformation tensor

$$\mathbf{Q} = \overset{o}{\mathbf{R}}^{-1} \, \bar{\mathbf{Q}}^{pl} \, \overset{*}{\mathbf{R}} \, \mathbf{Q}^{el} \tag{2.1}$$

where $\mathbf{Q}$ describes the total deformation of the material body due to loading, $\overset{o}{\mathbf{R}}$ represents the initial damage, $\bar{\mathbf{Q}}^{pl}$ denotes the plastic deformation of the fictitious undamaged body, $\overset{*}{\mathbf{R}}$ characterizes the deformation induced by evolution of damage and $\mathbf{Q}^{el}$ represents the elastic deformation of the material body. In addition, the elastic strain tensor

$$\mathbf{A}^{el} = \frac{1}{2} \ln \mathbf{Q}^{el} \tag{2.2}$$

and the damage strain tensor

$$\mathbf{A}^{da} = \frac{1}{2} \ln \overset{*}{\mathbf{R}} \tag{2.3}$$

are introduced.

Furthermore, the strain rate tensor

$$\dot{\mathbf{H}} = \frac{1}{2} \mathbf{Q}^{-1} \dot{\mathbf{Q}} \tag{2.4}$$

is defined, which using Eq. (2.1) can be additively decomposed

$$\dot{\mathbf{H}} = \dot{\mathbf{H}}^{el} + \mathbf{R}^{-1} \dot{\bar{\mathbf{H}}}^{pl} \mathbf{R} + \mathbf{Q}^{el\,-1} \dot{\mathbf{H}}^{da} \mathbf{Q}^{el} \tag{2.5}$$

into the elastic

$$\dot{\mathbf{H}}^{el} = \frac{1}{2} \mathbf{Q}^{el-1} \dot{\mathbf{Q}}^{el} , \tag{2.6}$$

the plastic

$$\dot{\bar{\mathbf{H}}}^{pl} = \frac{1}{2} \mathbf{Q}^{el-1} \bar{\mathbf{Q}}^{pl-1} \dot{\bar{\mathbf{Q}}}^{pl} \mathbf{Q}^{el} , \tag{2.7}$$

and the damage strain rate tensor

$$\dot{\mathbf{H}}^{da} = \frac{1}{2} \overset{*}{\mathbf{R}}^{-1} \dot{\overset{*}{\mathbf{R}}} . \tag{2.8}$$

With the identity (see Brünig (2003) for further details)

$$\overset{*}{\mathbf{R}} \mathbf{Q}^{el} = \mathbf{Q}^{el} \mathbf{R} \tag{2.9}$$

the damage metric transformation tensor with respect to the current loaded configurations, $\mathbf{R}$ (see Eq. (2.5)), has been introduced.

2.2.2 Constitutive Equations

Dislocations and growth of micro-defects are the most common modes of irreversible micro-structural deformations at each stage of the loading process. In particular, pure plastic flow is caused by dislocation motion and sliding phenomena along crystallographic planes whereas damage-related irreversible deformations are due to residual opening of micro-defects after unloading. These dissipative processes are distinctly different in their nature as well as in the manner how they affect the compliance of the material and are active on different scales. Consequently, in damage-coupled elastic-plastic theories two sets of internal variables corresponding to formation of dislocations (plastic internal variables) as well as to nucleation and growth of micro-defects (damage internal variables) are separately used to adequately describe the irreversible constitutive behavior and to compute corresponding strain rate tensors.

The effective undamaged configurations are considered to characterize the behavior of the undamaged matrix material. In particular, the elastic behavior of the undamaged matrix material is described by an isotropic hyperelastic law leading to the effective Kirchhoff stress tensor

$$\bar{\mathbf{T}} = 2G\,\mathbf{A}^{el} + \left(K - \frac{2}{3}G\right)\,\mathrm{tr}\mathbf{A}^{el}\,\mathbf{1} \tag{2.10}$$

where G and K represent the constant shear and bulk modulus of the undamaged matrix material, respectively, and $\mathbf{1}$ denotes the second order unit tensor.

In addition, onset and continuation of plastic flow of ductile metals is determined a yield criterion. Experiments on the influence of hydrostatic stress on the behavior of metals carried out by Spitzig et al. (1975; 1976) have shown that the flow stress depends approximately linearly on hydrostatic stress. Additional numerical studies performed by Brünig (1999) elucidated that hydrostatic stress terms in the yield condition affect onset of localization and associated deformation modes leading to notable decrease in ductility. Thus, plastic yielding of the matrix material is governed by the Drucker-Prager-type yield condition

$$f^{pl}\left(\bar{I}_1, \bar{J}_2, c\right) = \sqrt{\bar{J}_2} - c\left(1 - \frac{a}{c}\bar{I}_1\right) = 0\,, \tag{2.11}$$

where $\bar{I}_1 = \mathrm{tr}\bar{\mathbf{T}}$ and $\bar{J}_2 = \frac{1}{2}\,\mathrm{dev}\bar{\mathbf{T}} \cdot \mathrm{dev}\bar{\mathbf{T}}$ denote invariants of the effective stress tensor $\bar{\mathbf{T}}$, c is the strength coefficient of the matrix material and a represents the hydrostatic stress coefficient.

The experiments (Spitzig et al. 1975, 1976) also showed that plastic deformations are nearly isochoric. Thus, the plastic potential function

$$g^{pl}(\bar{\mathbf{T}}) = \sqrt{\bar{J}_2} \tag{2.12}$$

is assumed to depend only on the second invariant of the effective stress deviator leading to the non-associated isochoric effective plastic strain rate

$$\dot{\bar{\mathbf{H}}}^{pl} = \dot{\lambda} \frac{\partial g^{pl}}{\partial \bar{\mathbf{T}}} = \dot{\lambda} \frac{1}{2\sqrt{\bar{J}_2}} \operatorname{dev}\bar{\mathbf{T}} = \dot{\gamma}\, \bar{\mathbf{N}} \,. \tag{2.13}$$

In Eq. (2.13) $\dot{\lambda}$ is a non-negative scalar-valued factor, $\bar{\mathbf{N}} = \frac{1}{\sqrt{2\,\bar{J}_2}} \operatorname{dev}\bar{\mathbf{T}}$ denotes the normalized deviatoric stress tensor and $\dot{\gamma} = \bar{\mathbf{N}} \cdot \dot{\bar{\mathbf{H}}}^{pl} = \frac{1}{\sqrt{2}} \dot{\lambda}$ represents the equivalent plastic strain rate measure used in the present continuum model.

Furthermore, the damaged configurations are considered to characterize the behavior of anisotropically damaged material samples. In particular, since damage remarkably affects the elastic behavior and leads to material softening, the elastic law of the damaged material sample is expressed in terms of both the elastic and the damage strain tensor, (2.2) and (2.3). The Kirchhoff stress tensor of the ductile damaged metal is given by

$$\begin{aligned}\mathbf{T} = {} & 2\left(G + \eta_2 \operatorname{tr}\mathbf{A}^{da}\right)\mathbf{A}^{el} \\ & + \left[\left(K - \frac{2}{3}G + 2\eta_1 \operatorname{tr}\mathbf{A}^{da}\right)\operatorname{tr}\mathbf{A}^{el} + \eta_3\left(\mathbf{A}^{da}\cdot\mathbf{A}^{el}\right)\right]\mathbf{1} \\ & + \eta_3 \operatorname{tr}\mathbf{A}^{el}\mathbf{A}^{da} + \eta_4\left(\mathbf{A}^{el}\mathbf{A}^{da} + \mathbf{A}^{da}\mathbf{A}^{el}\right)\end{aligned} \tag{2.14}$$

where the parameters $\eta_1 ... \eta_4$ describe the deteriorating effect of damage on the elastic material properties. The parameters η_1 and η_2 are related to the isotropic character of damage whereas η_3 and η_4 correspond to anisotropic evolution of damage. With these additional parameters it is possible to simulate the decreases in Young's modulus, Poisson's ratio, shear and bulk moduli measured in experiments (Brünig 2003).

Based on many experiments it is well known that the stress state remarkably affects damage mechanisms occurring in ductile metals which is illustrated in Fig. 2.1 (Brünig et al. 2011a, b). For example, under tension dominated loading conditions (high positive stress triaxialities, $\eta \geq \eta_t$) damage in ductile metals is mainly caused

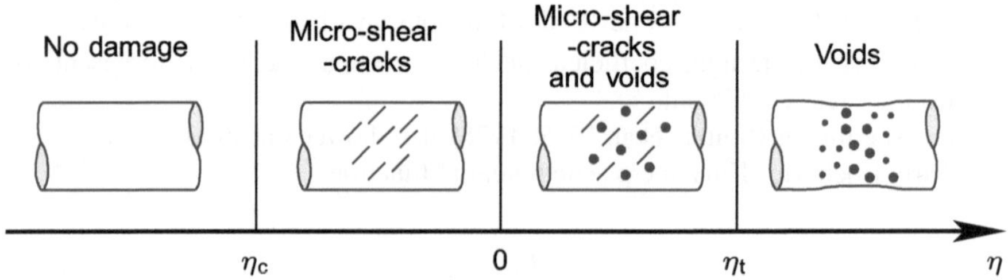

Fig. 2.1 Damage mechanisms depending on stress triaxialities

by nucleation and isotropic growth of voids whereas under shear and compression dominated conditions (negative stress triaxialities, $\eta_c \leq \eta \leq 0$) evolution of micro-shear-cracks is the predominant damage mechanism. Between these regions, $0 < \eta < \eta_t$ damage is a combination of both basic mechanisms with simultaneous growth of voids and formation of micro-shear-cracks. In addition, no damage occurs in ductile metals for high negative stress triaxialities $\eta < \eta_c$. Therefore, to be able to develop a realistic, accurate and efficient phenomenological model it is important to analyze and to understand the complex stress-state-dependent processes of damage as well as its respective mechanisms acting on different scales.

To model onset and continuation of damage the concept of damage surface is adopted. Following the ideas given in Brünig (2003); Chow and Wang (1987) the damage surface is formulated in stress space at the macroscopic damaged continuum level

$$f^{da} = \alpha I_1 + \beta\sqrt{J_2} - \sigma = 0 \tag{2.15}$$

with the stress invariants $I_1 = \mathrm{tr}\mathbf{T}$ and $J_2 = \frac{1}{2}\,\mathrm{dev}\mathbf{T} \cdot \mathrm{dev}\mathbf{T}$ and the damage threshold σ. In Eq. (2.15) the variables α and β represent damage mode parameters depending on the stress intensity (von Mises equivalent stress) $\sigma_{eq} = \sqrt{3J_2}$, the stress triaxiality

$$\eta = \frac{\sigma_m}{\sigma_{eq}} = \frac{I_1}{3\sqrt{3J_2}} \tag{2.16}$$

with the mean stress $\sigma_m = 1/3\, I_1$ as well as on the Lode parameter

$$\omega = \frac{2\tilde{T}_2 - \tilde{T}_1 - \tilde{T}_3}{\tilde{T}_1 - \tilde{T}_3} \quad \text{with } \tilde{T}_1 \geq \tilde{T}_2 \geq \tilde{T}_3 \tag{2.17}$$

expressed in terms of the principal stress components $\tilde{T}_1$, $\tilde{T}_2$ and $\tilde{T}_3$.

Furthermore, evolution of macroscopic irreversible strains caused by the simultaneous nucleation, growth and coalescence of different micro-defects is modeled by a stress-state-dependent damage rule. Thus, the damage potential function

$$g^{da}(\tilde{\mathbf{T}}) = g^{da}(I_1,\ J_2,\ J_3) \tag{2.18}$$

is introduced where $\tilde{\mathbf{T}}$ represents the stress tensor formulated in the damaged configuration which is work-conjugate to the damage strain rate tensor $\dot{\mathbf{H}}^{da}$ (see Brünig 2003, for further details) and I_1, J_2 and J_3 are corresponding invariants. This leads to the damage strain rate tensor

$$\dot{\mathbf{H}}^{da} = \dot{\mu}\,\frac{\partial g^{da}}{\partial \tilde{\mathbf{T}}} = \dot{\mu}\left(\frac{\partial g^{da}}{\partial I_1}\,\mathbf{1} + \frac{\partial g^{da}}{\partial J_2}\,\mathrm{dev}\tilde{\mathbf{T}} + \frac{\partial g^{da}}{\partial J_3}\,\mathrm{dev}\tilde{\mathbf{S}}\right) \tag{2.19}$$

where $\dot{\mu}$ is a non-negative scalar-valued factor and

$$\text{dev}\tilde{\mathbf{S}} = \text{dev}\tilde{\mathbf{T}}\ \text{dev}\tilde{\mathbf{T}} - \frac{2}{3}\ J_2\ \mathbf{1} \qquad (2.20)$$

represents the second order deviatoric stress tensor (Brünig et al. 2013, 2014b). Alternatively, the damage strain rate tensor (2.19) can be written in the form

$$\dot{\mathbf{H}}^{da} = \dot{\mu}\left(\bar{\alpha}\ \frac{1}{\sqrt{3}}\mathbf{1} + \bar{\beta}\ \mathbf{N} + \bar{\delta}\ \mathbf{M}\right) \qquad (2.21)$$

where the normalized deviatoric tensors $\mathbf{N} = \frac{1}{2\sqrt{J_2}}\text{dev}\tilde{\mathbf{T}}$ and $\mathbf{M} = \frac{1}{\|\text{dev}\tilde{\mathbf{S}}\|}\text{dev}\tilde{\mathbf{S}}$ have been used. In Eq. (2.21) the parameters $\bar{\alpha}$, $\bar{\beta}$ and $\bar{\delta}$ are kinematic variables describing the portion of volumetric and isochoric damage-based deformations. The damage rule (2.21) takes into account isotropic and anisotropic parts corresponding to isotropic growth of voids and anisotropic evolution of micro-shear-cracks, respectively. Therefore, both basic damage mechanisms (growth of isotropic voids and evolution of micro-shear-cracks) acting on the micro-level are involved in the damage rule (2.21).

2.3 Uniaxial Tension Tests

Uniaxial tension tests with unnotched specimens taken from thin sheets are used to identify basic elastic-plastic material parameters appearing in the constitutive equations discussed above. From these experiments equivalent stress–equivalent plastic strain curves are easily obtained from load-displacement curves as long as the uniaxial stress field remains homogeneous between the clip gauges fixed on the specimens during the tests (Brünig et al. 2011b, 2008). For example, fitting of numerical curves and experimental data for an aluminum alloy leads to Young's modulus $E = 65000$ MPa and Poisson's ratio is taken to be $\nu = 0.3$. In addition, the power law function for the equivalent stress–equivalent plastic strain function appearing in the yield criterion (2.11)

$$c = c_0 \left(\frac{H\,\gamma}{n\ c_0} + 1\right)^n \qquad (2.22)$$

is used to model the work-hardening behavior. Good agreement of experimental data and numerical results is achieved for the initial yield strength $c_0 = 175$ MPa, the hardening modulus $H = 2100$ MPa and the hardening exponent $n = 0.22$ (Brünig et al. 2015).

Since uniaxial tension tests with differently pre-notched specimens are characterized by different hydrostatic stress states in the small region between the notches (Brünig et al. 2011b, 2008; Driemeier et al. 2010), these experiments can be used to determine the hydrostatic-stress-dependent parameter a/c appearing in the yield criterion (2.11). However, in these tests with pre-notched specimens stress and strain fields are not homogeneous and only quantities in an average sense can be evaluated from these experiments. Therefore, corresponding finite element simulations of these uniaxial tests have been performed. Comparison of the numerically predicted load-displacement curves and the experimental ones leads with an inverse identification procedure to the parameter $a/c = 0.000055\,\text{MPa}^{-1}$.

Moreover, various parameters corresponding to the damage part of the constitutive model have to be identified. For example, onset of damage is determined by comparison of experimental load-displacement curves of the uniaxial tension tests with those predicted by numerical analyses based on the elastic-plastic model only (Brünig et al. 2011b, 2015). First deviation of these curves is an indicator of onset of deformation-induced damage. Using this procedure, the damage threshold appearing in the damage condition (2.15) is $\sigma = 300\,\text{MPa}$ for the aluminum alloy under investigation.

2.4 Numerical Analyses on the Micro-Scale

Identification of further damage parameters is more complicated because most criteria and evolution equations are motivated by observations on the micro-level. Therefore, stress state dependence of the parameters in the damage condition (2.21) and in the damage rule (2.21) has been studied in detail by Brünig et al. (2013; 2014b) performing numerical simulations with micro-void-containing representative volume elements under various loading conditions. In particular, the numerical analyses of differently loaded micro-defect containing unit cells have been performed using the finite element program ANSYS enhanced by an user-defined material subroutine. One eighth of the symmetric cell model as well as the finite element mesh with 3123 3D-elements of type Solid185 are shown in Fig. 2.2. Various loading conditions with

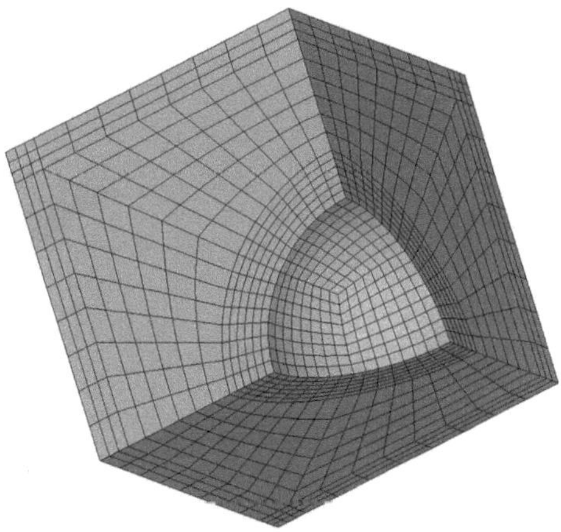

Fig. 2.2 Finite element mesh of one eighth of the unit cell

principal macroscopic stresses acting on the outer bounds of the representative volume elements are taken into account to be able to cover the wide range of stress triaxiality coefficients (16) $\eta = -1, -2/3, -1/3, 0, 1/3, 2/3, 1, 5/3, 2, 7/3$ and 3 as well as of the Lode parameters (2.23) $\omega = -1, -1/2, 0, 1/2$, and 1.

In all micro-mechanical numerical calculations the stress triaxiality coefficient and the Lode parameter are kept constant during the entire loading history of the unit cell to be able to accurately analyze their effect on damage and failure behavior of ductile metals. Based on the numerical results of these unit-cell calculations (Brünig et al. 2013) different damage mode parameters have been identified for the aluminum alloy: In particular, the damage mode parameter α (see Eq. (2.15)) is given by

$$\alpha(\eta) = \begin{cases} 0 \ \text{for} \ \frac{-1}{3} \le \eta \le 0 \\ \frac{1}{3} \ \text{for} \ \eta > 0 \end{cases} \tag{2.23}$$

whereas β is taken to be the non-negative function

$$\beta(\eta, \omega) = \beta_0(\eta, \omega = 0) + \beta_\omega(\omega) \ge 0 \,, \tag{2.24}$$

with

$$\beta_0(\eta) = \begin{cases} -0.45\,\eta + 0.85 \ \text{for} \ \frac{-1}{3} \le \eta \le 0 \\ -1.28\,\eta + 0.85 \ \text{for} \ \eta > 0 \end{cases} \tag{2.25}$$

and

$$\beta_\omega(\omega) = -0.017\,\omega^3 - 0.065\,\omega^2 - 0.078\,\omega \,. \tag{2.26}$$

This function $\beta(\eta, \omega)$ is visualized in Fig. 2.3.

In addition, based on the results of the unit cell analyses the stress-state-dependent parameters $\bar{\alpha}$, $\bar{\beta}$ and $\bar{\delta}$ in the damage rule (2.27) have been identified by Brünig et al. (2013). The non-negative parameter $\bar{\alpha} \ge 0$ characterizing the amount of volumetric damage strain rates caused by isotropic growth of micro-defects is given by the relation

Fig. 2.3 Damage mode parameter β versus stress triaxiality η and Lode parameter ω

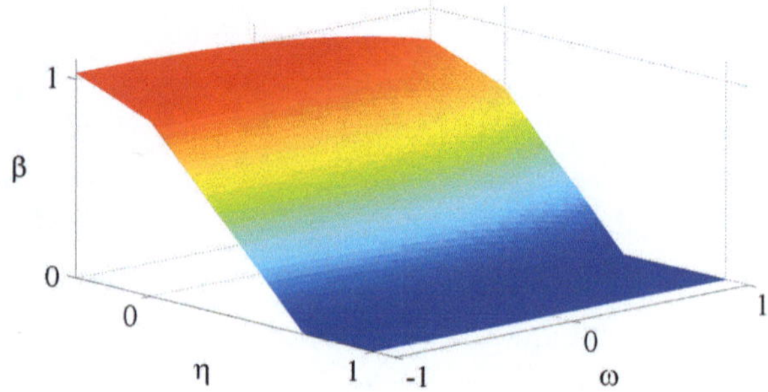

$$\bar{\alpha}(\eta) = \begin{cases} 0 & \text{for } \eta < 0.09864 \\ -0.07903 + 0.80117\,\eta & \text{for } 0.09864 \le \eta \le 1 \\ 0.49428 + 0.22786\,\eta & \text{for } 1 < \eta \le 2 \\ 0.87500 + 0.03750\,\eta & \text{for } 2 < \eta \le 3.33333 \\ 1 & \text{for } \eta > 3.33333 \end{cases} . \tag{2.27}$$

The parameter $\bar{\alpha}$ is high for high stress triaxialities, smaller for moderate ones and zero for negative triaxilities. Dependence on the Lode parameter ω has not been revealed by the numerical simulations on the micro-scale. In addition, the parameter $\bar{\beta}$ characterizing the amount of anisotropic isochoric damage strain rates caused by evolution of micro-shear-cracks is given by the relation

$$\bar{\beta}(\eta,\,\omega) = \bar{\beta}_0(\eta) + f_\beta(\eta)\,\bar{\beta}_\omega(\omega) \tag{2.28}$$

with

$$\bar{\beta}(\eta,\,\omega) = \begin{cases} 0.94840 + 0.11965\,\eta + f_\beta(\eta)\,\bar{\beta}_\omega(\omega) & \text{for } \frac{-1}{3} \le \eta \le \frac{1}{3} \\ 1.14432 - 0.46810\,\eta + f_\beta(\eta)\,\bar{\beta}_\omega(\omega) & \text{for } \frac{1}{3} < \eta \le \frac{2}{3} \\ 1.14432 - 0.46810\,\eta & \text{for } \frac{2}{3} < \eta \le 2 \\ 0.52030 - 0.15609\,\eta & \text{for } 2 < \eta \le \frac{10}{3} \\ 0 & \text{for } \eta > \frac{10}{3} \end{cases} \tag{2.29}$$

with

$$f_\beta(\eta) = -0.0252 + 0.0378\,\eta \tag{2.30}$$

and

$$\bar{\beta}_\omega(\omega) = 1 - \omega^2 \,. \tag{2.31}$$

The parameter $\bar{\delta}$ also corresponding to the anisotropic damage strain rates caused by the formation of micro-shear-cracks is given by the relation

$$\bar{\delta}(\eta,\,\omega) = f_\delta(\eta)\,\bar{\delta}_\omega(\omega) \tag{2.32}$$

with

$$\bar{\delta}(\eta,\,\omega) = \begin{cases} (-0.12936 + 0.19404\,\eta)(1 - \omega^2) & \text{for } \frac{-1}{3} \le \eta \le \frac{2}{3} \\ 0 & \text{for } \eta > \frac{2}{3} \end{cases} \tag{2.33}$$

It is worthy to note that this parameter $\bar{\delta}$ only exists for small stress triaxialities and mainly depends on the Lode parameter ω. Although its magnitude is small in comparison to the parameters $\bar{\alpha}$ and $\bar{\beta}$ its effect on the evolution of macroscopic damage strain rates $\dot{\mathbf{H}}^{da}$ is not marginal (Brünig et al. 2013).

2.5 Experiments and Numerical Simulations with Biaxially Loaded Specimens

The damage related functions discussed above are only based on numerical calculations on the micro-level and their validation has to be realized by experiments also covering a wide range of stress states. This is not possible by the uniaxial tests mentioned above and, therefore, new experiments with biaxially loaded specimens have been developed by Brünig et al. (2014a; 2015). The experiments are performed using the biaxial test machine (Type LFM-BIAX 20 kN from Walter • Bai, Switzerland) shown in Fig. 2.4. It is composed of four electro-mechanically, individually driven cylinders with load maxima and minima of ±20 kN (tension and compression loading is possible). The specimens are fixed in the four heads of the cylinders where clamped or hinged boundary conditions are possible. The geometry of the flat specimens and the loading conditions are shown in Fig. 2.5. The load F_1 will lead to shear mechanisms in the center of the specimen whereas simultaneous loading with F_2 leads to superimposed tension or compression modes leading to shear-tension or shear-compression deformation and failure modes. Therefore, these tests cover the full range of stress states corresponding to the damage and failure mechanisms discussed above with focus on high positive as well as low positive, nearly zero and negative stress triaxialities.

Figure 2.6 clearly shows that different load ratios $F_1 : F_2$ have remarkable influence on the damage and final fracture modes. For example, for the tension test without

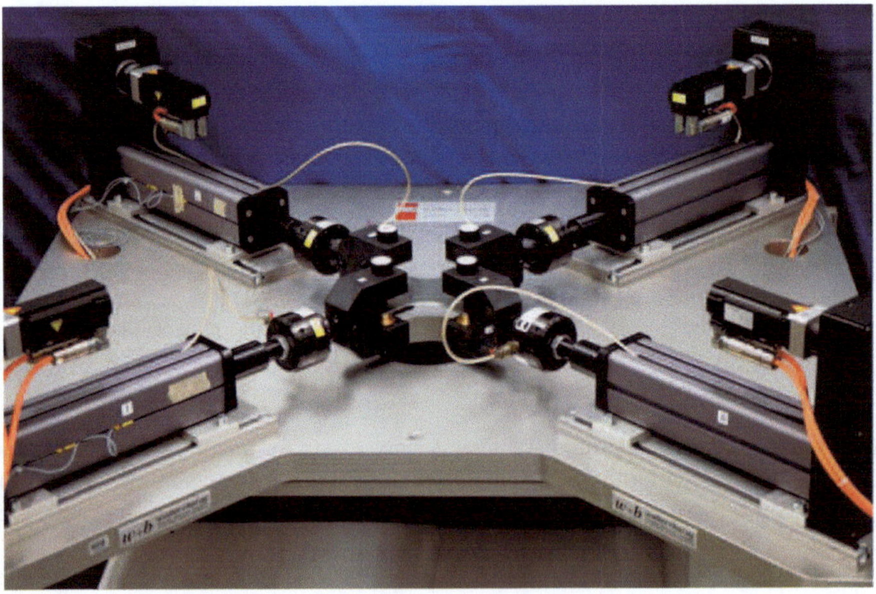

Fig. 2.4 Biaxial test machine

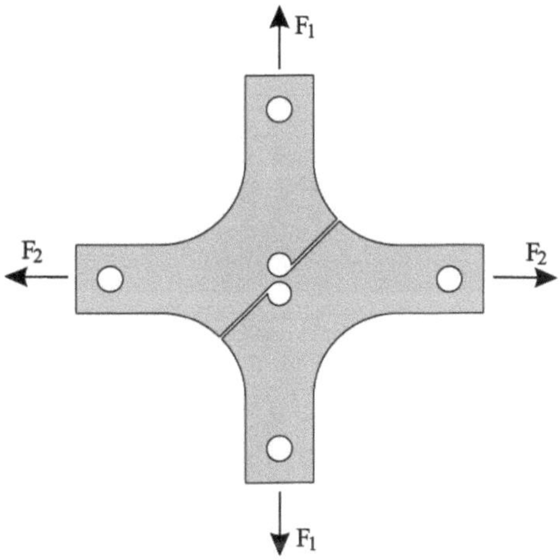

Fig. 2.5 Specimen and loading conditions

shear loading, $F_1 : F_2 = 0 : 1$, a nearly vertical fracture line is obtained. Under this loading condition, damage is mainly caused by growth and coalescence of voids with small influence of micro-shear-cracks leading to the macroscopic tensile fracture mode with small cup-cone fracture effect (Fig. 2.6a). On the other hand, for pure shear loading, $F_1 : F_2 = 1 : 0$, shear fracture is observed where the fracture line has an angle of about 25° with respect to the vertical line. Under this loading condition, damage is mainly caused by formation of micro-shear-cracks leading to the macro-shear-crack shown in Fig. 2.6b. For shear-tension loading, $F_1 : F_2 = 1 : 0.5$, shear-tension fracture is obtained and the fracture line has an angle of only about 10° with respect to the vertical line. Under this loading condition, damage is caused by the simultaneous growth of voids and formation of micro-shear-cracks leading to the macroscopic fracture mode shown in Fig. 2.6c which is between tensile fracture (Fig. 2.6a) and shear fracture (Fig. 2.6b). However, for shear-compression loading, $F_1 : F_2 = 1 : -0.5$, again shear fracture is observed and the fracture line again has an angle of about 25° with respect to the vertical line. Under this loading condition damage seems to be caused only by formation of micro-shear-cracks leading to the macro-shear-crack shown in Fig. 2.6d. It is worthy to note that the damage and failure mechanisms for $F_1 : F_2 = 1 : 0$ (Fig. 2.6b) and $F_1 : F_2 = 1 : -0.5$ (Fig. 2.6d) seem to be very similar and damage modes characteristic for shear loading will not be remarkably affected by superimposed small compression loads.

Furthermore, corresponding numerical simulations have been performed to be able to get detailed information on amounts and distributions of different stress and strain measures as well as further parameters of interest especially in critical regions (Brünig et al. 2015). For example, numerically predicted distributions of the stress triaxiality η at the onset of damage are shown in Fig. 2.7 for different load ratios. In particular, for uniaxial tension loading $F_1 : F_2 = 0 : 1$ remarkable high stress triaxialities up to $\eta = 0.84$ are numerically predicted in the center of the specimen. These high values are caused by the notches in horizontal and thickness direction

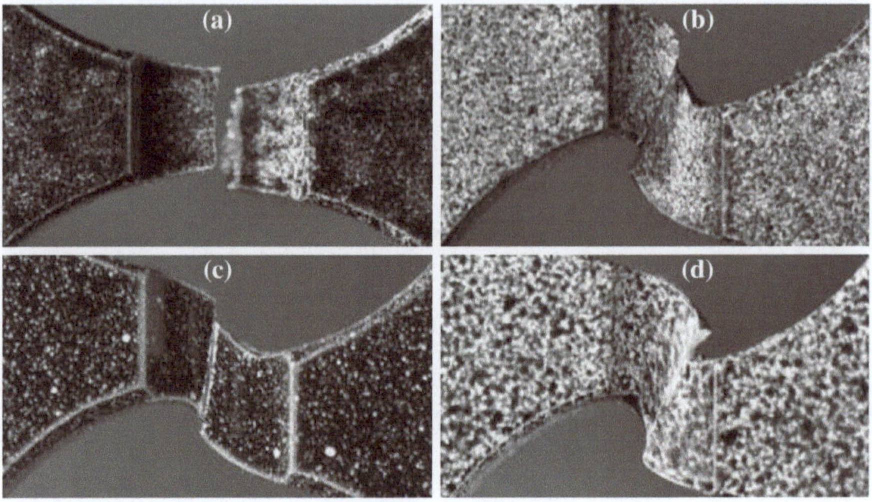

Fig. 2.6 Fracture modes of differently loaded specimens: **a** $F_1 : F_2 = 0 : 1$, **b** $F_1 : F_2 = 1 : 0$, **c** $F_1 : F_2 = 1 : 0.5$, **d** $F_1 : F_2 = 1 : -0.5$

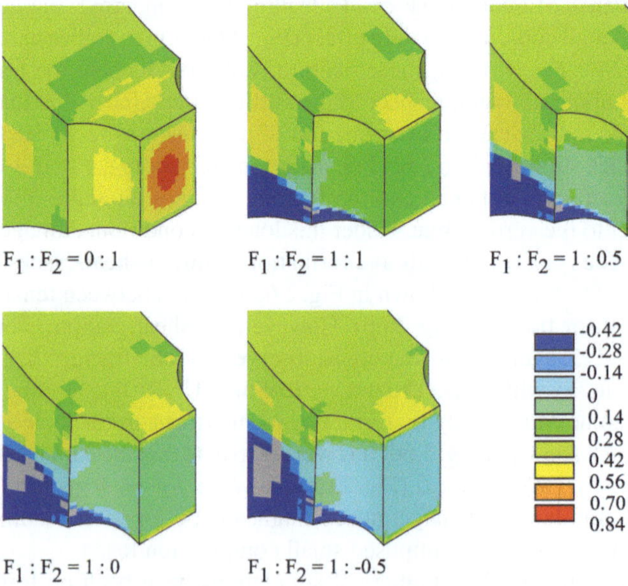

Fig. 2.7 Distribution of the stress triaxiality η for different load ratios

leading to high hydrostatic stress during elongation of the specimen. This will lead to damage and failure mainly due to growth of voids. In addition, when the specimen is only loaded by $F_1 : F_2 = 1 : 0$ (shear loading condition) the stress triaxiality η is numerically predicted to be nearly zero in the whole vertical section shown in Fig. 2.7.

Thus, damage will start in the specimen's center and will be caused by formation and growth of micro-shear-cracks. Furthermore, combined loading in vertical and horizontal direction, F_1 and F_2, will lead to combination of these basic damage modes. For example, for the load ratio $F_1 : F_2 = 1 : 1$ the stress triaxiality is again nearly constant in the vertical section with $\eta = 0.25$, and very similar distribution is numerically predicted for $F_1 : F_2 = 1 : 0.5$ with $\eta = 0.14$. On the other hand, the load ratio $F_1 : F_2 = 1 : -0.5$ represents combined shear-compression loading with the stress triaxiality $\eta = -0.14$ which is nearly constant in the vertical section shown in Fig. 2.7. For this negative stress triaxiality η damage and failure will be caused by formation and growth of micro-shear-cracks only. These numerical results nicely correspond to the fracture modes discussed above and shown in Fig. 2.6.

The stress triaxialities η covered by experiments with different flat specimens manufactured from thin sheets are shown in Fig. 2.8. In particular, for unnotched dog-bone-shape specimens (green) nearly homogeneous stress states occur in the small part during uniaxial tension tests with stress triaxiality $\eta = 1/3$. Higher stress triaxialities can be obtained in uniaxial tension tests when notches with different radii are added in the middle part of the specimens (red). Decrease in notch radius will lead to an increase in stress triaxiality in the specimen's center up to $\eta = 1/\sqrt{3}$. In addition, shear specimens (blue) elongated in uniaxial tension test will lead to stress triaxialities of about $\eta = 0.1$ when notches in thickness direction are added in the central part (Brünig et al. 2011b, 2008; Driemeier et al. 2010) whereas without additional notch they will also lead to onset of damage at nearly $\eta = 1/3$. However, with these flat specimens taken from thin sheets elongated in uniaxial tension tests only the stress triaxialities shown in Fig. 2.8 (green, red and blue points) can be taken into account whereas no information is obtained for high positive ($\eta > 1/\sqrt{3}$), low positive (between $0.1 < \eta < 1/3$) or negative stress triaxialities ($\eta < 0$). However, further experiments with new specimens (grey) tested under biaxial loading conditions will lead to stress triaxialities in the requested regimes. The grey points shown in Fig. 2.8 correspond to the loading conditions discussed above but variation

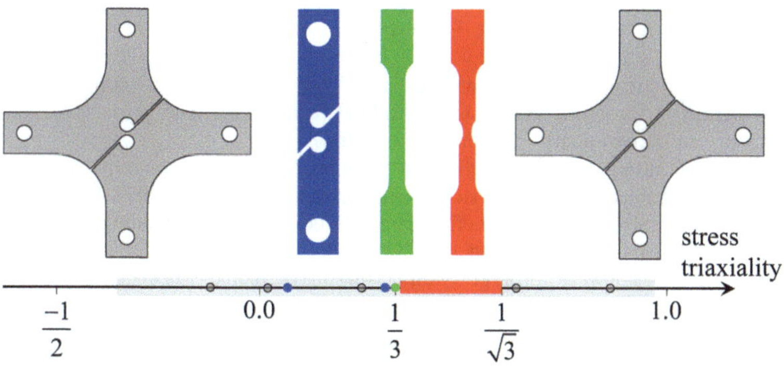

Fig. 2.8 Stress triaxialities covered by different specimens

of the load ratios $F_1 : F_2$ may lead to stress triaxialities marked by the grey zone shown in Fig. 2.8. Therefore, biaxial tests with 2D specimens discussed above cover a wide range of stress triaxialities and Lode parameters.

2.6 Conclusions

The author's activities in the field of damage mechanics have been summarized. The continuum model taking into account stress-state-dependent damage criteria and damage evolution laws has been reviewed. Elastic-plastic material parameters as well as onset of damage have been identified by uniaxial tension tests with smooth and pre-notched flat specimens and corresponding numerical simulations. Further damage parameters depending on stress triaxiality and Lode parameter have been determined using numerical results from unit-cell calculations on the micro-level. Since the proposed functions are only based on numerical analyses validation of the stress-state-dependent model was required. In this context, a series of experiments with biaxially loaded specimens has been presented. Different load ratios led to shear-tension and shear-compression mechanisms with different fracture modes. Corresponding finite element simulations of the experiments revealed a wide range of stress states covered by the tests depending on biaxial loading conditions. They could be used to validate the proposed stress-state-dependent functions of the continuum model.

References

Bai Y, Wierzbicki T (2008) A new model of metal plasticity and fracture with pressure and Lode dependence. Int J Plast 24:1071–1096

Bao Y, Wierzbicki T (2004) On the fracture locus in the equivalent strain and stress triaxiality space. Int J Mech Sci 46:81–98

Becker R, Needleman A, Richmond O, Tvergaard V (1988) Void growth and failure in notched bars. J Mech Phys Solids 36:317–351

Betten J (1982) Net-stress analysis in creep mechanics. Ingenieur-Archiv 52:405–419

Bonora N, Gentile D, Pirondi A, Newaz G (2005) Ductile damage evolution under triaxial state of stress: theory and experiments. Int J Plast 21:981–1007

Brocks W, Sun DZ, Hönig A (1995) Verification of the transferability of micromechanical parameters by cell model calculations with visco-plastic material. Int J Plast 11:971–989

Brünig M (1999) Numerical simulation of the large elastic-plastic deformation behavior of hydrostatic stress-sensitive solids. Int J Plast 15:1237–1264

Brünig M (2001) A framework for large strain elastic-plastic damage mechanics based on metric transformations. Int J Eng Sci 39:1033–1056

Brünig M (2003) An anisotropic ductile damage model based on irreversible thermodynamics. Int J Plast 19:1679–1713

Brünig M, Gerke S (2011) Simulation of damage evolution in ductile metals undergoing dynamic loading conditions. Int J Plast 27:1598–1617

Brünig M, Chyra O, Albrecht D, Driemeier L, Alves M (2008) A ductile damage criterion at various stress triaxialities. Int J Plast 24:1731–1755
Brünig M, Albrecht D, Gerke S (2011a) Modeling of ductile damage and fracture behavior based on different micro-mechanisms. Int J Damage Mech 20:558–577
Brünig M, Albrecht D, Gerke S (2011b) Numerical analyses of stress-triaxiality-dependent inelastic deformation behavior of aluminum alloys. Int J Damage Mech 20:299–317
Brünig M, Gerke S, Hagenbrock V (2013) Micro-mechanical studies on the effect of the stress triaxiality and the Lode parameter on ductile damage. Int J Plast 50:49–65
Brünig M, Gerke S, Brenner D (2014a) New 2D-experiments and numerical simulations on stress-state-dependence of ductile damage and failure. Procedia Mater Sci 3:177–182
Brünig M, Gerke S, Hagenbrock V (2014b) Stress-state-dependence of damage strain rate tensors caused by growth and coalescence of micro-defects. Int J Plast 63:49–63
Brünig M, Gerke S, Brenner D (2015) Experiments and numerical simulations on stress-state-dependence of ductile damage criteria. In: Altenbach H, Brünig M (eds) Inelastic behavior of materials and structures under monotonic and cyclic loading. Advanced Structured Materials Springer, Heidelberg, pp 17–34
Chew H, Guo T, Cheng L (2006) Effects of pressure-sensitivity and plastic dilatancy on void growth and interaction. Int J Solids Struct 43:6380–6397
Chow C, Wang J (1987) An anisotropic theory of continuum damage mechanics for ductile fracture. Eng Fracture Mech 27:547–558
Driemeier L, Brünig M, Micheli G, Alves M (2010) Experiments on stress-triaxiality dependence of mterial behavior of aluminum alloys. Mech Mater 42:207–217
Dunand M, Mohr D (2011) On the predictive capabilities of the shear modified Gurson and the modified Mohr-Coulomb fracture models over a wide range of stress triaxialities and Lode angles. J Mech Phys Solids 59:1374–1394
Gao X, Zhang G, Roe C (2010) A study on the effect of the stress state on ductile fracture. Int J Damage Mech 19:75–94
Grabacki J (1991) On some desciption of damage processes. Eur J Mech A/Solids 10:309–325
Gurson AL (1977) Continuum theory of ductile rupture by void nucleation and growth: part I—yield criteria and flow rules for porous ductile media. J Eng Mater Technol 99:2–15
Kuna M, Sun D (1996) Three-dimensional cell model analyses of void growth in ductile materials. Int J Fracture 81:235–258
Lemaitre J (1996) A course damage mechanics. Springer, Heidelberg
Mohr D, Henn S (2007) Calibration of stress-triaxiality dependent crack formation criteria: a new hybrid experimental-numerical method. Exp Mech 47:805–820
Murakami S, Ohno N (1981) A continuum theory of creep and creep damage. In: Ponter ARS, Hayhorst DR (eds) Creep in structures. Springer, Berlin, pp 422–443
Needleman A, Kushner A (1990) An analysis of void distribution effects on plastic flow in porous solids. Eur J Mech A/Solids 9:193–206
Ohno N, Hutchinson J (1984) Plastic flow localization due to non-uniform void distribution. J Mech Phys Solids 32:63–85
Spitzig WA, Sober RJ, Richmond O (1975) Pressure dependence of yielding and associated volume expansion in tempered martensite. Acta Metallurgica 23:885–893
Spitzig WA, Sober RJ, Richmond O (1976) The effect of hydrostatic pressure on the deformation behaviour of maraging and HY-80 steels and its implications for plasticity theory. Met Trans 7A:1703–1710
Voyiadjis G, Kattan P (1999) Advances in damage mechanics: metals and metal matrix composites. Elsevier, Amsterdam
Voyiadjis G, Park T (1999) The kinematics of damage for finite-strain elasto-plastic solids. Int J Eng Sci 37:803–830
Zhang K, Bai J, Francois D (2001) Numerical analysis of the influence of the Lode parameter on void growth. Int J Solids Struct 38:5847–5856

Chapter 3
The Multiplicative Decomposition of the Deformation Gradient in Plasticity—Origin and Limitations

Otto T. Bruhns

Abstract The history of material equations and hence the development of present material theory as a method to describe the behavior of materials is closely related to the development of continuum theory and associated with the beginning of industrialization towards the end of the 19th century. While on the one hand new concepts such as continuum, stresses and strains, deformable body etc. were introduced by Cauchy, Euler, Leibniz and others and mathematical methods were provided to their description, the pressure of industrialization with the need to ever newer, and likewise reliably secure, developments has led to the fact that more appropriate models for the description of elastic-plastic behavior were introduced. Upon this background, this Chapter wants to introduce into the history of plasticity of the sixties and seventies of last century, and likewise highlight the eminent contributions of A.E. Green and P.M. Naghdi, E.H. Lee and J. Mandel to a modern description of finite plasticity theory.

Keywords Plasticity · Large deformation · Additive decomposition · Multiplicative decomposition · Intermediate configuration · Green-Naghdi theory · Lee-Liu theory · Isoclinic configuration · Director triads

3.1 Introduction

The present article continues a series of contributions intended to highlight the history of plasticity. The first was devoted to the life and works of Heinrich Hencky and the rapid development of plasticity at the beginning of last century (Bruhns 2014a). At that time the main focus of the different considerations was on the mathematical description of the basics of plasticity within the general frame of continuum mechanics, and the combination of the solid-like elastic behavior of most

O.T. Bruhns (✉)
Institute of Mechanics, Ruhr-University Bochum, Universitätsstr. 150, 44780 Bochum, Germany
e-mail: otto.bruhns@rub.de

H. Altenbach et al. (eds.), *From Creep Damage Mechanics to Homogenization Methods*, Advanced Structured Materials 64,
DOI 10.1007/978-3-319-19440-0_3

materials below a certain threshold and the fluid-like plastic behavior beyond this limit. Accordingly in a second contribution (Bruhns 2014b) the significant input of Ludwig Prandtl and Endre Reuss to formulate a unified rate relation for elastic-plastic behavior was underlined. These so-called Prandtl-Reuss equations and several amendments related with the effort to establish more realistic descriptions of several phenomena observed during experiments of plastic flow like the hardening of the material and the Bauschinger effect were rapidly accepted in science and application. In their original forms, however, these relations were formulated within a geometrically linear theory. As for many technological applications, e.g. in metal forming processes such as rolling, extrusion and deep drawing, the deformations and rotations are much larger than a small deformation theory would allow,[1] after World War II this classical plasticity theory was transferred to application within large deformations. Tremendous effort by several German refugees like Hilda Geiringer and William Prager—to mention only two of them—and the excellent foundational works of Drucker (1949, 1950) and Hill (1948, 1950) has led to a at that time well accepted extension of the Prandtl-Reuss theory to finite deformations with the corotational Jaumann rate as objective time derivative replacing the material time derivative of the original formulation.[2]

Nevertheless, in discussion of the development of continuum mechanics, Truesdell (1964) and Truesdell and Noll (1965) have stated that the above classical theory of plasticity fails to meet basic requirements of continuum mechanics, and so they have not included this topic in reviews of the development of non-linear theory: "Various doctrines of plasticity arose in the latter part of the last century and have been cultivated diffusively in this. These theories have always been closely bound in motive, if not in outcome, to engineering needs and have proliferated at once in detailed approximate solutions of boundary-value problems. Their mechanical foundations are insecure to the present day, and they do not furnish representative examples in a program of continuum physics."

This harsh statement may have motivated Green and Naghdi (1965a, b) to establish a rigorous theory of plasticity in the framework of modern continuum mechanics, where the restrictions on the general form of the constitutive relations were derived from thermodynamics principles. The kinematical basis of this work was the assumption that the total strain could be decomposed into the sum of an elastic-like and a plastic strain tensor, respectively.[3]

An alternative approach to a finite deformation theory of plasticity was presented by Lee and Liu (1967) and Lee (1969). This approach was shortly later modified by Mandel (1973a, b) by adding so-called director triads and the notion of an isoclinic

[1]It should be generally accepted that a strain larger than 10 % would violate the requirements of a linear theory.

[2]For details we refer to Bruhns (2014a, b). See also the additional works of Prager (1944) and Edelman and Drucker (1951).

[3]Although in their paper the plastic strain has been introduced as a primitive variable, and the elastic-like strain is merely defined by the difference of the total strain and this plastic variable, it is common practice to refer to this sum as an additive decomposition of the strain tensor into its elastic and plastic parts.

intermediate configuration. In contrast to Green and Naghdi's approach in these papers the multiplicative decomposition of the deformation gradient into elastic and plastic parts is assumed.

The present paper will be restricted to an introduction to these different approaches and a discussion of their pros and cons, thus confining this contribution to the development of plasticity during the sixties and seventies of the last century.

3.2 One-Dimensional Considerations

This introductory Section contains an elementary *one-dimensional* preview on the different ways to describe the complex behavior of a composite elastic-plastic material. It further may be instrumental in understanding the structure of finite deformation plasticity and the various ideas and proposals from different groups that may have contributed to its development.

Let us consider a one-dimensional body, say, a rod. We denote this body by $\mathcal{B}$, and agree on the fact that this body's movement is reduced to axial displacements due to axial forces, one-dimensional temperature treatment, etc. For the sake of simplicity, we here reduce the possible actions to an axial force T. The body consists of a number of elements or particles jointly fixed together. These particles possess specific properties like mass, density, etc. A typical particle may be denoted X.

We refer to X as the *material or Lagrangean coordinate* of the particle. The set of material coordinates is called the *reference configuration.* If t is time, then the *deformation function* is a function $\chi(\bullet, t)$ which, for each t, maps $\mathcal{B}$ into its present configuration

$$x = \chi(X, t), \tag{3.1}$$

where x is the *spatial position or coordinate* of the particle X at time t. These coordinates are also referred to as *Eulerian coordinates.* For a description within continuum mechanics, we will assume here[4] that for each t there exists an inverse χ^{-1} such that

$$X = \chi^{-1}(x, t). \tag{3.2}$$

The velocity of X at time t is

$$\dot{x} = \frac{\partial \chi(X, t)}{\partial t}. \tag{3.3}$$

In the same way, we also may define higher order derivatives, e.g. the acceleration of X at time t.

[4]This insures that X and x are in a one-to-one correspondence, which is reasonable for solid bodies. We note, however, that this assumption also has limitations, e.g. in fracture mechanics, crystallographic slip, lattice distortions, etc., which appear on the microscale level during e.g. plastic deformation.

Because of (3.2) we can regard $\dot{x}$, and the displacement $u = x - X$ as functions of (x, t) or (X, t). The pair (X, t) are called *material variables*, and likewise the pair (x, t) are *spatial variables*. Any function of one set of variables can be converted to a function of the other set.

The deformation gradient is defined by

$$F = \frac{\partial \chi(X, t)}{\partial X} . \tag{3.4}$$

Since χ has an inverse, it is trivially true that $\chi[\chi^{-1}(x, t), t] = x$ and thus,

$$\frac{\partial \chi(X, t)}{\partial X} \frac{\partial \chi^{-1}(x, t)}{\partial x} = 1 . \tag{3.5}$$

Equation (3.5) shows that $F \neq 0$ and thus bound to $F > 0$, and

$$F^{-1} = \frac{\partial \chi^{-1}(x, t)}{\partial x} . \tag{3.6}$$

If $\mathrm{d}X$ denotes the differential element of the body at X, then it follows from Eqs. (3.1) and (3.4) that

$$\mathrm{d}x = F \, \mathrm{d}X . \tag{3.7}$$

Equation (3.7) shows that if $\mathrm{d}X$ is a material element at X then $F \, \mathrm{d}X = \mathrm{d}x$ is the deformed element at x. Thus, F measures the local deformation of the material in the neighborhood of X. It should be easily understood that in our case of homogeneous deformation F is the current stretch ratio. Thus, comparing the lengths of the element in both configurations one reasonable measure to describe the strain E of the body would be

$$E = F - 1 , \tag{3.8}$$

since the strain should vanish for the undeformed body. This measure is sometimes called *nominal strain*.[5]

However, quantity E as defined by Eq. (3.8) is not the only admissible measure to describe the strain of the body. Any function, e. g., of the form

$$E^{(n)} = g(F) = \frac{1}{n}(F^n - 1) \tag{3.9}$$

can replace Eq. (3.8) in the calculations.[6] Here $g(F)$ ought to be a monotonically increasing function, i.e. $g'(F) > 0$ for all $F > 0$, and fulfilling the conditions

[5] Besides this measure is also called Biot strain or Cauchy strain (Ogden 1984; Bertram 2005).

[6] This generalization of the strain measure is mainly due to Doyle and Ericksen (1956); Seth (1964); Hill (1968), and includes also non-integral real values of n.

$$g(1) = 0\,, \quad g'(1) = 1\,.$$

Different results for E as functions of F are given with Fig. 3.1.

We note that with $n = 2$ from Eq. (3.9) we get the well-known Green strain $E^{(2)}$ which frequently is used in descriptions of finite deformation. On the other hand, with $n = -2$ the Almansi strain $E^{(-2)}$ would be given. Moreover, these generalized measures for $n = 0$ also include the so-called natural or *logarithmic strain*, almost 100 years ago introduced by Ludwik (1909); Hencky (1928), and others, viz.

$$H = \lim_{n \to 0} \frac{1}{n}(F^n - 1) = \lim_{n \to 0} \frac{1}{n}\big(\exp(n \ln F) - 1\big) = \ln(F)\,. \tag{3.10}$$

The significant advantage of this logarithmic (Hencky) measure lies in the fact that it tends to infinity as F tends to zero, thus in a very natural way bounding the regime of applicability to the case $F > 0$. This behavior can also be observed for strain measures with negative exponent n. Compared with the latter, however, the logarithmic measure also goes to infinity as F does, whereas it is evident that for negative values of n the strain according to Eq. (3.9) is bound to the limit $-\frac{1}{n}$.

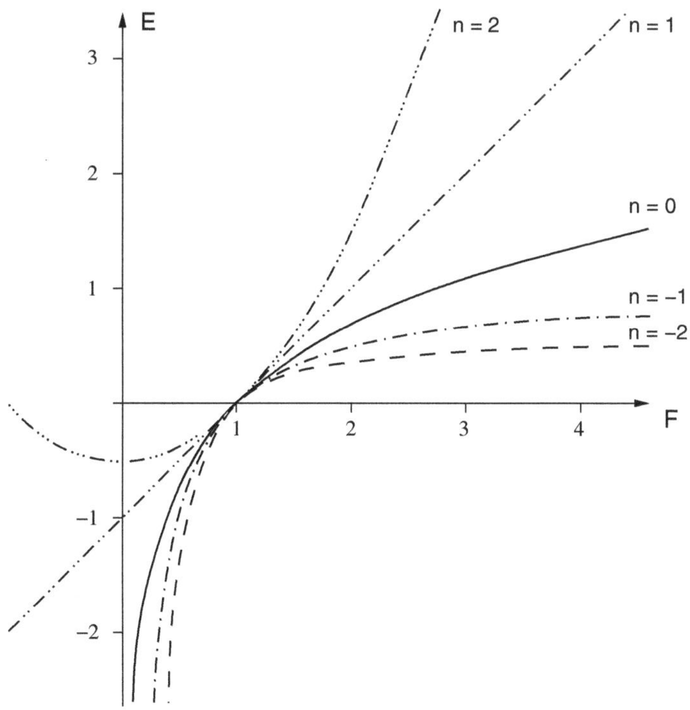

Fig. 3.1 Different strain measures for different values of n and admissible regime of $F > 0$

All measures with positive values of n including the Green strain share the reasonable property of the logarithmic strain for F going to infinity. For F going to zero, however, these measures arrive at finite values for the specific strains, e.g. at $-\frac{1}{2}$ for $n = 2$, which would mean that interpreted from physics a total compression of the rod (to zero length) is related to a finite value of the strain. This awkward result would not agree with our observation—at least what concerns the behavior of solid materials.

This has led Böck and Holzapfel (2004) and somewhat later Darijani and Naghdabadi (2010) to combine the advantages of both groups of measures (with positive and negative values of n) by introducing the following linear combination[7]

$$E^{(m)} = g(F) = \frac{1}{2m}(F^m - F^{-m}) . \tag{3.11}$$

Let us now consider a sequence of two deformations of a material element at X first into a position ξ, and then into the position x, and let us compare the total deformation of this sequence with the deformation of the same element at X in a single step into position x.

According to Eq. (3.7) for the former case, we have to split the total deformation into two parts

$$\mathrm{d}\xi = F_1\,\mathrm{d}X\,, \quad \mathrm{d}x = F_2\,\mathrm{d}\xi\,, \tag{3.12}$$

whereas for the latter Eq. (3.7) holds. Introducing now Eq. (3.12) into Eq. (3.7) results in

$$\mathrm{d}x = F_2 F_1\,\mathrm{d}X\,, \quad F = F_2 F_1\,. \tag{3.13}$$

This is called a *multiplicative decomposition* of the deformation gradient. Now from Eq. (3.9) it follows that

$$E^{(n)} = \frac{1}{n}\left[(F_2 F_1)^n - 1\right] \tag{3.14}$$

for the single step deformation. The corresponding strains for the two sequential steps will be introduced accordingly as

$$E_1^{(n)} = \frac{1}{n}\left(F_1^n - 1\right), \qquad E_2^{(n)} = \frac{1}{n}\left(F_2^n - 1\right) \tag{3.15}$$

for the deformation from $\mathcal{B}_0$ to $\mathcal{B}'$, and from $\mathcal{B}'$ to $\mathcal{B}$, respectively. Introducing these definitions into Eq. (3.14)

$$E^{(n)} = nE_2^{(n)}E_1^{(n)} + E_2^{(n)} + E_1^{(n)} \neq E_2^{(n)} + E_1^{(n)} \tag{3.16}$$

[7] This combination with $m = 1$ was introduced by Böck and Holzapfel (2004) as a two-point deformation tensor. The extension to a more general two-parameter form with independent parameters n and m is due to Darijani and Naghdabadi (2010). We here will confine ourselves to the most simple case when $n = m$.

would prove that the total strain cannot be decomposed additively into the strains of the two sequential steps, unless $n = 0$.

We note, however, that the latter of both strains of Eq. (3.15) is defined relative to the so-called *intermediate configuration* $\mathcal{B}'$. Therefore, to compare both total strains, one might also follow the idea that prior to such a comparison the strain $E_2^{(n)}$ should be transformed into the same reference configuration. With the help of Eq. (3.12)$_1$ this would mean

$$\overline{E}_2^{(n)} = E_2^{(n)} F_1^n = \frac{1}{n} \left(F_2^n - 1 \right) F_1^n \,, \tag{3.17}$$

where $\overline{E}_2^{(n)}$ is now defined relative to the initial configuration. Introducing these results into Eq. (3.14), it turns out that

$$E^{(n)} = \frac{1}{n} \left(n \overline{E}_2^{(n)} + F_1^n - 1 \right) = \overline{E}_2^{(n)} + E_1^{(n)} \,, \tag{3.18}$$

showing that the given strain components are additive irrespective of the chosen number n. Thus, we note further that this result is significantly subject to the definition of the strain of the second step of the deformation. We will see later that this difference in the definition of strains relative to their reference configuration may explain some aspects of the debate on the additivity or non-additivity of the strains during large deformation.

If $\dot{x}$ is expressed as a function of (x, t), then

$$L = \frac{\partial \dot{x}}{\partial x} \tag{3.19}$$

is the velocity gradient. Upon multiplying this expression by F, we find

$$\dot{F} = L F \,. \tag{3.20}$$

Again it turns out that only when introducing here the logarithmic strain (3.10), we can reformulate Eq. (3.20) to get the most simple relation

$$\dot{H} = L \tag{3.21}$$

between the rate of the logarithmic strain H and the velocity gradient. For the more general case of Eq. (3.9), we arrive at a relation

$$\dot{E}^{(n)} = L \left(1 + n E^{(n)} \right) , \tag{3.22}$$

also containing the strain $E^{(n)}$, and the number n.

It should be mentioned that the two-point deformation described with Eq. (3.11) arrives at a much simpler relation, namely

$$\dot{E}^{(m)} = L\,E^{(m)}\,, \tag{3.23}$$

similar to the relation for the deformation gradient F. We note further that this result again appears to be independent of any number m.

We now come back to the above mentioned example which has its own meaning for sequential loading. We will use it here to explain the split of the total strain into its reversible elastic and irreversible plastic parts, respectively, if our rod is stretched beyond a given yield load with $T > T_0$. After reloading this rod we will observe a permanent (stress-free) elongation of the rod which may have led particle X to position ξ_p. According to the above sequential process, we will interpret the first part of the deformation of the material element into position ξ_p as purely plastic, and the second part into position x as purely elastic. Thus we write

$$\mathrm{d}\xi_p = F_p\,\mathrm{d}X\,, \quad \mathrm{d}x = F_e\,\mathrm{d}\xi_p\,, \tag{3.24}$$

$$\mathrm{d}x = F_e F_p\,\mathrm{d}X\,, \quad F = F_e F_p\,. \tag{3.25}$$

This is the well-known and nowadays frequently used multiplicative decomposition of the deformation gradient into elastic and plastic parts of the deformation, respectively, here straightforwardly derived from a combined elastic-plastic loading of a rod under the action of an axial force.

To summarize, we can conclude that for the given one-dimensional problem, we find from Eq. (3.17)

$$\overline{E}_e = E_e^{(n)} F_p^n \tag{3.26}$$

according to the chosen strain measure Eq. (3.9), or in other words

$$E = \overline{E}_e + E_p \neq E_e + E_p\,. \tag{3.27}$$

This means that due to Eq. (3.17) E_e and E_p are far away from additivity to give the total strain E in the presence of finite plastic strain. This, however, would be the case for the logarithmic strain H, where $n = 0$

$$H = H_e + H_p\,. \tag{3.28}$$

We note that due to the underlying transformation into the reference configuration $\overline{E}_e$ comprises a combination of both parts, the purely elastic and the purely plastic part, respectively, of the deformation. This makes it extremely difficult, perhaps even impossible, to attribute in a unique manner the different constitutive relations of elastic and plastic behavior to the allocated strains. As will be seen later this issue may illustrate the crucial point of the Green-Naghdi theory (Green and Naghdi 1965a, b) which has been introduced in 1965.

This problem does not apply for the stretching D, which here due to the symmetry of the underlying one-dimensional problem coincides with the velocity gradient L. As a natural rate quantity characterizing the changing of the deformation state, the stretching is solely related to the current configuration and thus independent of any referential configuration (Haupt 2002). Therefore, the kinematical features of the stretching are essentially different from those of the conventional time rate of any strain tensor (Xiao et al. 2006). As direct extension of the classical separation of the infinitesimal strains into elastic and plastic parts, for finite elastic-plastic deformations we may write

$$D = D_e + D_p \,, \tag{3.29}$$

or equivalently

$$\dot{H} = \dot{H}_e + \dot{H}_p \,, \tag{3.30}$$

provided Hencky strains are used. Moreover, we find

$$D = L = \dot{F}_p F_p^{-1} + \dot{F}_e F_e^{-1} \,, \tag{3.31}$$

for the relation between the stretching and the respective parts of the deformation gradient.

This fast and easy derivation of a concept for separating the kinematics of large elastic-plastic deformations into its reversible elastic and irreversible plastic parts, respectively, may explain why Lee and Liu (1967) and Lee (1969) have used the multiplicative decomposition concept to describe elastic-plastic deformation at finite strain. This nowadays commonly used method to combine different behaviors of components to derive a self-contained constitutive relation for a composite material, became generally accepted over the last almost 50 years. It is the great merit of E.H. Lee having initiated and promoted this development. Indeed, however, he was not the first to use the concept of multiplicative decomposition to combine different disparate material behaviors.[8] We will come back to this point later.

3.3 Basic Facts for a Deforming Continuous Body

We now switch to some relevant facts in finite deformation kinematics of continua. For a deformable body in the pure mechanical sense, as basic field variables the deformation gradient $\boldsymbol{F}$ and the Cauchy or "true" stress $\boldsymbol{\sigma}$ at each particle characterize

[8]The history of the multiplicative decomposition of the total deformation into elastic and plastic parts seems to be somewhat vague or at least unclear. Different authors mention different origins, and the different disciplines continuum mechanics and materials science involved in this discussion seem to have different sources. In their own paper Lee and Liu (1967) mention: "The concept of an unstressed configuration ...has appeared before in the literature, but does not seem to have been put to satisfactory use." All in all, it should be almost clear that Eckart (1948) was the first to use local natural configurations in order to separate the elastic from the plastic part of a total deformation. Backman (1964) introduces three continuous configurations, and represents the elastic and plastic components of the total strains in terms of displacement derivatives, which in general cannot be done. It seems that Kröner (1958, 1960) and Bilby et al. (1957) then were the first to use this expression, with respect to sequences of elastic and plastic distortions.

the local deformation state relative to a reference configuration and the local stressed state, respectively. Let again $\boldsymbol{X}$ and $\boldsymbol{x} = \boldsymbol{\chi}(\boldsymbol{X}, t)$ be the reference and the current position vector of a material particle, respectively. Then, the deformation gradient is given by

$$\boldsymbol{F} = \frac{\partial \boldsymbol{\chi}}{\partial \boldsymbol{X}} . \tag{3.32}$$

We consider the local deformation state occurring at the infinitesimal neighborhood of each particle with $\mathrm{d}\boldsymbol{X}$ and $\mathrm{d}\boldsymbol{x}$ the reference and the current line element, respectively. Then, at the infinitesimal neighborhood of each particle, we have the transformation formula between the line elements

$$\mathrm{d}\boldsymbol{x} = \boldsymbol{F}\,\mathrm{d}\boldsymbol{X}, \quad \det \boldsymbol{F} > 0 . \tag{3.33}$$

The particle velocity and the velocity gradient, respectively, are denoted by

$$\boldsymbol{v} = \dot{\boldsymbol{x}}, \quad \boldsymbol{L} = \frac{\partial \boldsymbol{v}}{\partial \boldsymbol{x}} = \dot{\boldsymbol{F}}\boldsymbol{F}^{-1}, \tag{3.34}$$

where the superposed dot designates the material time derivative.

According to the well-known polar decomposition theorem, the deformation gradient $\boldsymbol{F}$ has the unique left and right multiplicative decompositions

$$\boldsymbol{F} = \boldsymbol{V}\boldsymbol{R} = \boldsymbol{R}\boldsymbol{U}, \quad \boldsymbol{R}^{\mathrm{T}} = \boldsymbol{R}^{-1}, \quad \det \boldsymbol{R} = 1 , \tag{3.35}$$

where $\boldsymbol{V}$ and $\boldsymbol{U}$ are known as the *left and right stretch tensors*, and $\boldsymbol{R}$ as the *rotation tensor.*[9] Both $\boldsymbol{V}$ and $\boldsymbol{U}$ are positive definite symmetric tensors, and determined uniquely by

$$\boldsymbol{B} = \boldsymbol{V}^2 = \boldsymbol{F}\boldsymbol{F}^{\mathrm{T}}, \quad \boldsymbol{C} = \boldsymbol{U}^2 = \boldsymbol{F}^{\mathrm{T}}\boldsymbol{F} , \tag{3.36}$$

where $\boldsymbol{B}$ and $\boldsymbol{C}$ are usually called the *left and right Cauchy-Green tensors*.

On the other hand, the velocity gradient $\boldsymbol{L}$ has the unique additive decomposition

$$\boldsymbol{L} = \boldsymbol{D} + \boldsymbol{W}, \quad \boldsymbol{D} = \frac{1}{2}(\boldsymbol{L} + \boldsymbol{L}^{\mathrm{T}}), \quad \boldsymbol{W} = \frac{1}{2}(\boldsymbol{L} - \boldsymbol{L}^{\mathrm{T}}) , \tag{3.37}$$

where $\boldsymbol{D}$ and $\boldsymbol{W}$ are known as the *stretching* and the *vorticity tensor*, respectively. This relation may suggest the particular role of $\boldsymbol{D}$ as a natural, exact characterization of the rate of change of the local deformation state.

As mentioned before, a general class of strain measures may be defined through a single scale function $g(\xi)$ with $g(1) = g'(1) - 1 = 0$. Their forms can be given by

$$\boldsymbol{E}^{(n)} = \boldsymbol{g}(\boldsymbol{U}) = \frac{1}{n}(\boldsymbol{U}^n - \boldsymbol{I}), \quad \boldsymbol{e}^{(n)} = \boldsymbol{g}(\boldsymbol{V}) = \frac{1}{n}(\boldsymbol{V}^n - \boldsymbol{I}) , \tag{3.38}$$

[9] The symbol $(\bullet)^{\mathrm{T}}$ herein is used to represent the transpose of the second order tensor.

defining Lagrangean and Eulerian strain tensors.[10] Two important examples are the Green strain of Lagrangean type

$$\boldsymbol{E}^{(2)} = \boldsymbol{E} = \frac{1}{2}(\boldsymbol{U}^2 - \boldsymbol{I}) = \frac{1}{2}(\boldsymbol{F}^{\mathrm{T}}\boldsymbol{F} - \boldsymbol{I}) \tag{3.39}$$

and the Hencky strain of Lagrangean and Eulerian type

$$\boldsymbol{E}^{(0)} = \boldsymbol{H} = \ln \boldsymbol{U} = \frac{1}{2}\ln(\boldsymbol{F}^{\mathrm{T}}\boldsymbol{F}), \quad \boldsymbol{e}^{(0)} = \boldsymbol{h} = \ln \boldsymbol{V} = \frac{1}{2}\ln(\boldsymbol{F}\boldsymbol{F}^{\mathrm{T}}). \tag{3.40}$$

With various strain measures, various stress measures may be introduced via the unified concept of work conjugacy.[11] According to this idea, for any given Lagrangean strain $\boldsymbol{A}$ the work conjugacy relation

$$\boldsymbol{T} : \dot{\boldsymbol{A}} = \boldsymbol{\tau} : \boldsymbol{D} \tag{3.41}$$

determines a stress-like symmetric tensor $\boldsymbol{T}$, referred to as the conjugate stress of the strain $\boldsymbol{A}$. Here, $\boldsymbol{\tau}$ is used to designate the *Kirchhoff stress* $\boldsymbol{\tau} = (\det \boldsymbol{F})\boldsymbol{\sigma}$.

One example is the conjugate stress of the Green strain $\boldsymbol{E}$, given by

$$\boldsymbol{S} = \boldsymbol{F}^{-1}\boldsymbol{\tau}(\boldsymbol{F}^{-1})^{\mathrm{T}}, \tag{3.42}$$

which is usually called the *second Piola-Kirchhoff stress.*

3.4 A Short Historical Overview

This article is primarily devoted to the history of plasticity after World War II and before the eighties of last century when with the rapid development of fast and powerful computers in conjunction with efficient numerical methods (e.g. the FEM) new trends in plasticity were initiated. The origins and the development until 1945 with the contributions of the French Barré de Saint-Venant, Maurice Lévy, and Henri Tresca at the end of 19th century, and the great impact of the German-Austrian school around Richard von Mises and Ludwig Prandtl, and Heinrich Hencky have been pointed out in Bruhns (2014a). In addition the effort of the Hungarian group including András (Endre) Reuss and the many contributions of the great European Jewish community has been elaborated in Bruhns (2014a, b). Persecuted by the Nazi regime most of them were forced to emigrate—provided they had this chance to survive.

[10] As usual $\boldsymbol{I}$ is a second order unit tensor.

[11] This idea was exemplified by Macvean (1968), and fully developed by Hill (1968, 1970, 1978). Hill (1978) also introduced the notion of "work conjugacy", although work-rates or stress powers are discussed (see also the Appendix of Hill 1968).

Thus, after World War II the development of plasticity had to be reanimated.[12] However, most of the former German, Austrian and Hungarian scientists now were living and working in Great Britain or in the USA. This may explain why World War II apart from the total destruction of Europe and a large amount of Asia has marked a decisive event in the history of plasticity. One even can realize this from the literature: The German language which primarily has been used before now was replaced by English.

First attempts to describe the behavior of an elastic-plastic material, following the idea of combining the constitutive relations of solid-like elastic and fluid-like plastic materials, were undertaken in the twenties and thirties of last century.

Hencky (1924)—as civil engineer—tried to model this composite behavior as a more solid-like structure. Similar to the proceeding for an elastic material, he started out from an energetic approach of Haar and von Kármán (1909), finally specifying an invertible relationship between stresses and strains:

$$\boldsymbol{\varepsilon} - \frac{1}{3}\mathrm{tr}(\boldsymbol{\varepsilon})\boldsymbol{I} = \frac{1+\varphi}{2\mu}\boldsymbol{\sigma}', \quad \mathrm{tr}(\boldsymbol{\varepsilon}) = \frac{1}{3K}\mathrm{tr}(\boldsymbol{\sigma}), \quad 3K = 3\lambda + 2\mu, \tag{3.43}$$

where here $\boldsymbol{\varepsilon}$ are the infinitesimal strains, λ and μ are the two Lamé's constants of elasticity, and K is the bulk modulus. Moreover, φ is a still undetermined Lagrange parameter, and $\boldsymbol{\sigma}'$ is the stress deviator.[13] It turns out that this later on called "deformation theory" tries to include the plastic part of the behavior via a degradation of the stiffness of the still dominating elastic material behavior. For $\varphi = 0$, this law changes into that of a linear elastic material. The volumetric deformation is purely elastic and at plastic behavior the shear modulus μ is reduced by $(1+\varphi)$, the material thus becomes "softer".

This formulation, though rapidly accepted, soon meets its limitations. A neutral change of the stress, as e.g. occurs during non-proportional loading, cannot be reflected. Moreover, the applicability of the above model was restricted to a monotonic loading of the body. Thus, an improvement was necessary to include also unloading processes. To this end an incremental form of Hooke's law was introduced[14]

$$\dot{\boldsymbol{\varepsilon}} - \frac{1}{3}\mathrm{tr}(\dot{\boldsymbol{\varepsilon}})\boldsymbol{I} = \frac{1}{2\mu}\dot{\boldsymbol{\sigma}}'. \tag{3.44}$$

[12]In a review of the development and the usage of internal variables in inelasticity Horstemeyer and Bammann (2010) stated that—with respect to the period between 1870 and 1945: "Over the next 75 years progress (of plasticity) was slow and spotty,…" It seems to the present author that one should be aware that just this period was marked by two big wars that have reduced large amount of Europe and Asia to wrack and ruin.

[13]Here for simplicity the notation $(\bullet)' = (\bullet) - \frac{1}{3}\mathrm{tr}(\bullet)\boldsymbol{I}$ will be used to mark a deviator of a second order tensor.

[14]This amendment was due to Nádai (1931). Nádai having arrived in the USA in 1927 as well as Hencky himself during his time in the former Soviet Union successfully promoted the distribution of this theory.

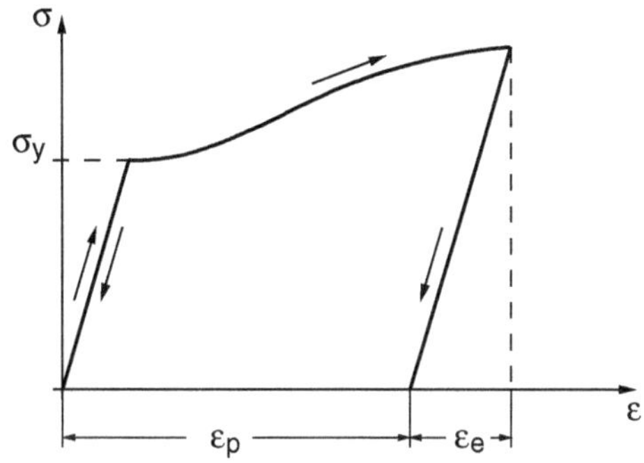

Fig. 3.2 Stress-strain behavior during tension test including unloading

In contrast to this approach, Prandtl (1924) and Reuss (1930, 1932) connected the fluid-like behavior of a plastic material with that of an (incremental) elastic one. Like Hencky, Reuss in his works emanated from the v. Mises yield condition[15] and the observation of a typical tensile test (refer to Fig. 3.2) that the total strain the specimen may achieve can be divided into two parts, a plastic one $\boldsymbol{\varepsilon}_p$ that will remain after a total unloading of the specimen into a stress-free state, and a second reversible elastic part $\boldsymbol{\varepsilon}_e$:

$$\boldsymbol{\varepsilon} = \boldsymbol{\varepsilon}_e + \boldsymbol{\varepsilon}_p \,. \tag{3.45}$$

Herein the fluid-like plastic part will be described by a fluid type law according to the Saint-Venant/Lévy approach

$$\boldsymbol{D}' = c\,\boldsymbol{\sigma}' \,, \tag{3.46}$$

where c is a constant of proportionality. We note that herein the fluid is assumed to be incompressible—what usually applies for the plastic flow of metals.

Thus, Eq. (3.45) may be differentiated to give

$$\dot{\boldsymbol{\varepsilon}} = \dot{\boldsymbol{\varepsilon}}_e + \dot{\boldsymbol{\varepsilon}}_p \,, \tag{3.47}$$

or accordingly

$$\boldsymbol{D} = \boldsymbol{D}_e + \boldsymbol{D}_p \,, \tag{3.48}$$

where $\boldsymbol{D}_e$ and $\boldsymbol{D}_p$ are the respective elastic and plastic parts of the stretching. We note that the latter due to its definition as symmetrical part of the velocity gradient $\boldsymbol{L}$ is also valid for the description of finite deformation, whereas Eqs. (3.45) and (3.47) are bound to small deformations. Combining now Eq. (3.46) with the description of an incremental elastic behavior, Reuss obtained a constitutive law

[15] In a generalized manner this is a surface bounding the domain of elastic deformations. In the classical v. Mises stress space formulation a yield function $\mathrm{f} = \mathrm{f}(\boldsymbol{\sigma}')$ as function of the stress deviator is bound to a so-called yield limit.

$$\boldsymbol{D}' = \boldsymbol{D}_e{}' + \boldsymbol{D}_p = \frac{1}{2\mu}\dot{\boldsymbol{\sigma}}' + \Lambda\boldsymbol{\sigma}', \tag{3.49}$$

that with Λ still contains a yet undetermined function,[16] and where the volumetric part

$$\mathrm{tr}(\boldsymbol{D}) = \mathrm{tr}(\boldsymbol{D}_e) = \frac{1}{3K}\,\mathrm{tr}(\dot{\boldsymbol{\sigma}}) \tag{3.50}$$

remains purely elastic. Here, $\boldsymbol{D}_p$ stands for the (deviatoric) plastic part of the strain rate and $\boldsymbol{D}_e$ for its elastic part. Thus, the basic version of the nowadays commonly used Prandtl-Reuss theory was introduced.

Several steps towards a more general theory to account also for a hardening of the material or to reflect the Bauschinger effect—to mention only a few examples—have been undertaken shortly after.[17] In this respect, we refer to the recently published paper (Bruhns 2014b) which has been devoted to the development of the Prandtl-Reuss theory. This paper also contains some remarks about an extension of these relations towards larger or even finite deformations. In addition, we also should mention the works of Tokuoka (1977, 1978) who following the idea of Prandtl and Reuss, and starting out from the additive decomposition (3.48) has introduced a rate type description of elastic-plastic behavior, where scalar and tensorial internal variables were used to account for the isotropic and kinematic hardening during plastic flow. The elastic part herein is modeled as hypoelastic material with $\boldsymbol{D}_e$ linearly related to an objective corotational rate of the stress $\overset{\circ}{\boldsymbol{\sigma}}$, i.e.,

$$\overset{\circ}{\boldsymbol{\sigma}} = \mathbb{L} : \boldsymbol{D}_e\,. \tag{3.51}$$

Here $\mathbb{L}$ is a fourth order tensor of elasticity, and the colon defines a double contraction according to $(\mathbb{A} : \boldsymbol{B})_{ij} = \mathbb{A}_{ijkl}B_{kl}$.[18]

Whereas the deformation theory of Hencky just from its setting was restricted to applications within small deformations, this was not the case for the Prandtl-Reuss theory. Having in mind its setting as a combination of fluid-like and solid-like materials, its description of the fluid-like plastic part originally was introduced as a relation of the stress $\boldsymbol{\sigma}$ as function of the rate of deformation tensor $\boldsymbol{D}$—without containing any strain measure. Thus, its extension to Eqs. (3.48), and (3.49), respectively, seems to be canonical. Likewise, however, it should be born in mind that neither the material time derivatives incorporated in the solid-like elastic part of Eqs. $(3.49)_1$ and (3.50)

---

[16]This function is in general determined by means of the so-called condition of consistency $\dot{\mathrm{f}}(\boldsymbol{\sigma}') = 0$, which during plastic loading forces the stress to stay on the yield surface.

[17]To this end scalar-valued and second order tensor-valued internal variables have been introduced to model these phenomena by means of the evolution of these variables.

[18]If our notations are used. As objective rate Tokuoka has taken a Jaumann rate—without mentioning this source. Instead he referred to the fundamental work of Truesdell (1955). For the sake of completeness, we add that Truesdell (1952a, b, c, 1955) has introduced the notion of "hypoelasticity" for this specific material, and that in the original forms the tangential stiffness tensor $\mathbb{L}$ has been used in a more general form as function of the stress.

meet the principle of objectivity, nor does a hypoelastic material in general really behave like an elastic material.

It is e.g. observed that in a closed cyclic process even for purely "elastic" deformations a dissipation will occur. This may explain several anomalies and spurious effects like the shear oscillatory phenomenon that were observed when in a first step replacing the material time derivatives in the Prandtl-Reuss equations by objective ones like the Jaumann rate or the Green-Naghdi rate.[19]

3.5 Lagrangean Formulations with Plastic Strain

In contrast to the aforementioned Eulerian formulation, a Lagrangean type formulation of finite deformations was proposed by Green and Naghdi (1965a, b).[20] This theory represents the first effort towards a rigorous treatment of finite elastoplasticity within the framework of continuum thermodynamics.

As extension of the classical small deformation theory to finite deformations, a perhaps more direct idea is to use a finite strain measure and its conjugate stress. The starting point is again the direct extension of the separation (3.45) to finite strain. Let $\boldsymbol{A}$ be any given Lagrangean strain. Then we may define

$$\boldsymbol{A} = \boldsymbol{A}_e + \boldsymbol{A}_p \,, \tag{3.52}$$

where $\boldsymbol{A}_e$ and $\boldsymbol{A}_p$ are labeled elastic and plastic parts of $\boldsymbol{A}$, respectively, with the intention that in conjunction with the conjugate stress of $\boldsymbol{A}$, they will be used as basic variables to formulate an elastic relation and a flow rule.

It is known that the additive separation of a total strain into elastic and plastic parts is restricted to very particular cases. In general, such separation might be of formal sense only.

On account of the complexity and difficulties involved in defining the notion of plastic strain, Green and Naghdi (1965a, b) introduced a strain-like variable of Lagrangean type, denoted $\boldsymbol{E}_p$, and regarded it as a primitive variable, "stating certain of its properties but not defining it explicitly, and thus relegated its explicit identification to special assumptions or situations".

This variable is associated with the total Green strain $\boldsymbol{E}$ as given by Eq. (3.39). Well understanding the limited applicability of the additive separation of $\boldsymbol{E}$, they did

[19]Only recently it was demonstrated that this problem was closely related to a second one, namely to prove the integrability of hypoelastic relation (3.51). It turned out that to prove this a new objective time derivative had to be defined which, moreover applied to a yet undetermined Eulerian strain measure, could give the (Eulerian) stretching. It was shown that this (Eulerian) strain had to be the Hencky strain, and the new (objective) time derivative, in turn, was the logarithmic rate. Only with this logarithmic rate applied to $\boldsymbol{\sigma}$ Eq. (3.51) could be integrated to give a really elastic relation. We therefore refer to Bruhns (2014b) and the numerous references mentioned therein.

[20]Its formulation from a purely mechanical point of view has been developed later by Naghdi and coworkers and others. For details, we refer to Naghdi (1990).

not interpret the difference $\boldsymbol{E} - \boldsymbol{E}_p$ as an elastic strain or part, but as an alternative convenient variable used for well-motivated purposes.

With the plastic strain $\boldsymbol{E}_p$ as additional variable, now the set of basic variables entering the constitutive formulations will be given by $(\boldsymbol{E}, \boldsymbol{E}_p, \boldsymbol{\alpha}, \kappa)$. Note that herein $\boldsymbol{\alpha}$ and κ are internal variables representing the progress of the plastic deformations by suitably defined evolution equations.[21] In accord with the general setting of their theory, the tensor-valued internal variable $\boldsymbol{\alpha}$ is introduced as back stress of Lagrangean type to model the Bauschinger effect, whereas the isotropic hardening is modeled through the evolution of κ.[22] The set $(\boldsymbol{E}, \boldsymbol{E}_p, \boldsymbol{\alpha}, \kappa)$ of basic variables represents a strain space formulation.[23]

In general, it is assumed that the conjugate stress of Green strain $\boldsymbol{E}$, i.e., the 2nd Piola-Kirchhoff stress $\boldsymbol{S}$ given by Eq. (3.42), is determined by the foregoing set of basic variables. This leads to the total stress-strain relation

$$\boldsymbol{S} = \hat{\boldsymbol{S}}(\boldsymbol{E}, \boldsymbol{E}_p, \boldsymbol{\alpha}, \kappa)\,, \tag{3.53}$$

where $\hat{\boldsymbol{S}}$ is a tensor-valued constitutive function relying on all four variables. It is assumed that this tensor function is twice differentiable and establishes a one-one relationship between the stress $\boldsymbol{S}$ and the total strain $\boldsymbol{E}$. Then, the inverted form of Eq. (3.53) yields

$$\boldsymbol{E} = \hat{\boldsymbol{E}}(\boldsymbol{S}, \boldsymbol{E}_p, \boldsymbol{\alpha}, \kappa)\,. \tag{3.54}$$

In addition, $\mathrm{g}(\boldsymbol{E}, \boldsymbol{E}_p, \boldsymbol{\alpha}, \kappa)$ is a yield function in a strain space formulation. It is assumed that the time derivative of each of the three variables $\boldsymbol{E}_p, \boldsymbol{\alpha}, \kappa$ is linear in $\dot{\boldsymbol{E}}$ with coefficients that are functions of the whole set of variables. Thus, the flow rule is given by:

$$\dot{\boldsymbol{E}}_p = \zeta \left(\frac{\partial \mathrm{g}}{\partial \boldsymbol{E}} : \dot{\boldsymbol{E}} \right) \boldsymbol{\varrho}(\boldsymbol{E}, \boldsymbol{E}_p, \boldsymbol{\alpha}, \kappa)\,, \tag{3.55}$$

where $\boldsymbol{\varrho}$ is a tensor-valued constitutive function, and ζ is the so-called plastic multiplier differentiating elastic behavior (including unloading) from elastic-plastic one by taking the values 0 and 1, respectively.

The loading-unloading criterion in strain space is shown to possess a simple, unified form for perfect elastic-plastic as well as hardening and softening behavior (see, e.g., Naghdi 1990). The evolution equations for the hardening variable κ and for the back stress $\boldsymbol{\alpha}$ are given in the forms:

---

[21] Similar internal variables have been first used in line with the amendments of the Prandtl-Reuss relations. We refer e.g. to Bruhns (2014b).

[22] Alternatively, the sets $(\boldsymbol{E} - \boldsymbol{E}_p, \boldsymbol{E}_p, \boldsymbol{\alpha}, \kappa)$ as well as $(\boldsymbol{S}, \boldsymbol{E}_p, \boldsymbol{\alpha}, \kappa)$ may be useful for certain specific purposes. For instance, the last set is used in a stress space formulation.

[23] See, e.g., Naghdi and Trapp (1975a, b, c). For a discussion of the pros and cons of the strain space formulation, we also refer to Naghdi (1990).

$$\dot{\kappa} = \zeta\left(\frac{\partial \mathrm{g}}{\partial \boldsymbol{E}} : \dot{\boldsymbol{E}}\right)\lambda(\boldsymbol{E}, \boldsymbol{E}_p, \boldsymbol{\alpha}, \kappa)\,, \tag{3.56}$$

$$\dot{\boldsymbol{\alpha}} = \zeta\left(\frac{\partial \mathrm{g}}{\partial \boldsymbol{E}} : \dot{\boldsymbol{E}}\right)\boldsymbol{\beta}(\boldsymbol{E}, \boldsymbol{E}_p, \boldsymbol{\alpha}, \kappa)\,. \tag{3.57}$$

Here, λ and $\boldsymbol{\beta}$ are additional scalar- and tensor-valued constitutive functions.

Within the general context of the Lagrangean theory summarized above, Naghdi and coworkers made a rigorous, systematic study of the consequences implied by the work postulate, e.g. Naghdi and Trapp (1975a) broadened the scope of Ilyushin's postulate by introducing

$$\int_{t_0}^{t_f} \boldsymbol{S} : \dot{\boldsymbol{E}}\ \mathrm{d}t \geq 0 \tag{3.58}$$

for every homogeneous finite strain cycle. From this postulate, the essential structure of Green-Naghdi's general theory was derived. For simplicity, these results are presented here in the absence of a back stress $\boldsymbol{\alpha}$.

It is demonstrated that there is a stress potential $\hat{\psi}(\boldsymbol{E}, \boldsymbol{E}_p, \kappa)$ such that the general relation (3.53) for the stress response is reduced to[24]

$$\boldsymbol{S} = \frac{\partial \hat{\psi}}{\partial \boldsymbol{E}}\,. \tag{3.59}$$

Moreover, it is shown that the constitutive function $\boldsymbol{\varrho}(\boldsymbol{E}, \boldsymbol{E}_p, \kappa)$ characterizing the flow rule is related to the yield function $\mathrm{g}(\boldsymbol{E}, \boldsymbol{E}_p, \kappa)$ and the stress potential $\hat{\psi}(\boldsymbol{E}, \boldsymbol{E}_p, \kappa)$ as well as the hardening function $\lambda(\boldsymbol{E}, \boldsymbol{E}_p, \kappa)$ in the following manner:

$$\hat{\boldsymbol{\sigma}} \equiv \frac{\partial^2 \hat{\psi}}{\partial \boldsymbol{E}_p \partial \boldsymbol{E}} : \boldsymbol{\varrho} + \frac{\partial^2 \hat{\psi}}{\partial \kappa \partial \boldsymbol{E}}\lambda = -\gamma\,\frac{\partial \mathrm{g}}{\partial \boldsymbol{E}}\,, \tag{3.60}$$

where γ is a undetermined scalar function relying on $(\boldsymbol{E}, \boldsymbol{E}_p, \kappa)$. From the above, it may be seen that the tensor-valued constitutive function $\boldsymbol{\varrho}(\boldsymbol{E}, \boldsymbol{E}_p, \kappa)$ is obtainable from three scalar constitutive functions, i.e., the yield function $\mathrm{g}(\boldsymbol{E}, \boldsymbol{E}_p, \kappa)$, the stress potential $\hat{\psi}(\boldsymbol{E}, \boldsymbol{E}_p, \kappa)$, and the hardening function $\lambda(\boldsymbol{E}, \boldsymbol{E}_p, \kappa)$, whenever the second gradient $\partial^2\hat{\psi}/\partial \boldsymbol{E}_p \partial \boldsymbol{E}$ as linear transformation over 2nd order symmetric tensor space is invertible.

A direct relation between the stress rate and the total strain rate is derivable

$$\dot{\boldsymbol{S}} = \mathbb{K} : \mathbb{L} : \dot{\boldsymbol{E}} \tag{3.61}$$

[24] In a general context, Hill and Rice (1973) have demonstrated that such a relation holds true with $\hat{\psi}$ being a fully general elastic potential relying on the "prior history of inelastic deformation", and with $(\boldsymbol{S}, \boldsymbol{E})$ any given work-conjugate pair.

and its inverse (see, e.g., Naghdi 1990):

$$\dot{\boldsymbol{E}} = (\mathbb{K} : \mathbb{L})^{-1} : \dot{\boldsymbol{S}}\,, \tag{3.62}$$

where $\mathbb{L}$ is the tensor of elasticity and $\mathbb{K}$ is a second 4th-order material tensor

$$\mathbb{L} = \frac{\partial^2 \hat{\psi}}{\partial \boldsymbol{E}^2}\,, \qquad \mathbb{K} = \mathbb{I} + \zeta \hat{\boldsymbol{\sigma}} \otimes \frac{\partial \mathrm{f}}{\partial \boldsymbol{S}} = \mathbb{I} + \zeta \left(\hat{\boldsymbol{\sigma}} \otimes \frac{\partial \mathrm{g}}{\partial \boldsymbol{E}} \right) : \mathbb{L}^{-1}\,. \tag{3.63}$$

Here, $\mathrm{f} = \mathrm{f}(\boldsymbol{S}, \boldsymbol{E}_p, \kappa)$ is the yield function in stress space, which is obtained by substituting the inverted form (3.54) into the yield function $\mathrm{g} = \mathrm{g}(\boldsymbol{E}, \boldsymbol{E}_p, \kappa)$ in strain space, and the 4th-order tensor $\mathbb{I}$ is the identity transformation over the symmetric 2nd order tensor space.

As has been mentioned in the Introduction, the critical point within this theory is mainly hidden in its setting, namely in the additive splitting of the total strain $\boldsymbol{E}$ to determine the elastic-like quantity $\boldsymbol{E}_e$, if the plastic strain is introduced as a primitive variable $\boldsymbol{E}_p$. This point has been addressed first by Lee and Liu (1967) and has led to a long discussion about the admissibility or even the significance of the additivity of strains in a finite deformation theory, in order to achieve an effective uncoupling of elastic and plastic properties.[25]

A second issue is related with the specific free Helmholtz energy function $\hat{\psi}(\boldsymbol{E}, \boldsymbol{E}_p, \kappa)$ or $\hat{\psi}(\boldsymbol{E} - \boldsymbol{E}_p, \boldsymbol{E}_p, \kappa)$ in an alternative form which is incorporated as an essential quantity of this theory. It might be not clear how this function apart from some very special cases will be determined. While forms of this energy function for elastic and even thermoelastic solids are well established, the introduction of plastic, i.e. irreversible, processes into this function will produce new if not intractable problems.

3.6 Formulations with Unstressed Configurations

In recent years the multiplicative separation of elastic-plastic deformations has become popular and found increasing applications in the phenomenological study of finite elastoplasticity.[26] It is not derived merely from the direct extension of the small deformation case, but motivated by physical considerations. The central idea is the notion of a local intermediate unstressed configuration at each particle defined by an imaginary destressing process. If such configuration could be defined, then

[25] In this respect, we refer, e.g., to Lee (1996) and Lee and Germain (1974), and to the discussion of the different strains for the subsequently loaded rod of Sect. 3.2.

[26] This was initiated by Lee and Liu (1967) and Lee (1969) and use was made by Fox (1968), Willis (1996), Freund (1970), Rice (1971), albeit it may be traced back to earlier works by Eckart (1948), Eglit (1960), Backman (1964), Sedov (1966); see Clayton and McDowell (2003) and the references therein. We also refer to footnote 8.

at each particle elastic and plastic deformations could be separated from the total elastic-plastic deformation in a definite and accurate manner.

Extending our previous considerations, we consider a continuous material body with initial configuration $\mathcal{B}_0$ experiencing finite elastic-plastic deformations in the current configuration $\mathcal{B}$. According to Lee (1996) we may introduce a straining-destressing experiment: "Following elastic-plastic deformation from the undisturbed configuration $\boldsymbol{X}$ to $\boldsymbol{x}$, destressing to zero stress occurs from $\boldsymbol{x}$ to $\boldsymbol{p}$ $\cdots$. Since the configuration $\boldsymbol{p}$ is unstressed, the elastic strain there is zero and the strain in $\boldsymbol{p}$ is therefore totally plastic." Here, $\boldsymbol{X}$ and $\boldsymbol{x}$ are the position vectors of a generic material particle in the initial and current configurations $\mathcal{B}_0$ and $\mathcal{B}$, respectively, and $\boldsymbol{p}$ is the position vector of the same particle in the unstressed configuration (refer to Fig. 3.3).

In the above cited statement, the total elastic-plastic deformation $\boldsymbol{F}$ from $\boldsymbol{X}$ to $\boldsymbol{x}$ is actually undergone by the material body, while the *plastic deformation* from $\boldsymbol{X}$ to $\boldsymbol{p}$ and the *elastic deformation* from $\boldsymbol{p}$ to $\boldsymbol{x}$, denoted $\boldsymbol{F}_e$ and $\boldsymbol{F}_p$, are introduced by an *imaginary destressing procedure* and hence not actually undergone by the material body. $\boldsymbol{F}_e$ and $\boldsymbol{F}_p$ will serve as additional deformation-like variables. The question as to how the destressing procedure is achieved is at the moment left open and will be discussed later.

The deformation gradient $\boldsymbol{F}$ is related to a local affine configuration and based upon the notion of line elements at the infinitesimal neighborhood of a particle. Following Lee (1996), the deformation of a material line element $\mathrm{d}\boldsymbol{X}$ in the aforementioned straining-destressing experiment is given by:

$$\mathrm{d}\boldsymbol{p} = \boldsymbol{F}_p\,\mathrm{d}\boldsymbol{X}\,, \quad \mathrm{d}\boldsymbol{x} = \boldsymbol{F}_e\,\mathrm{d}\boldsymbol{p}\,. \tag{3.64}$$

$$\mathrm{d}\boldsymbol{x} = \boldsymbol{F}\,\mathrm{d}\boldsymbol{X}\,, \tag{3.65}$$

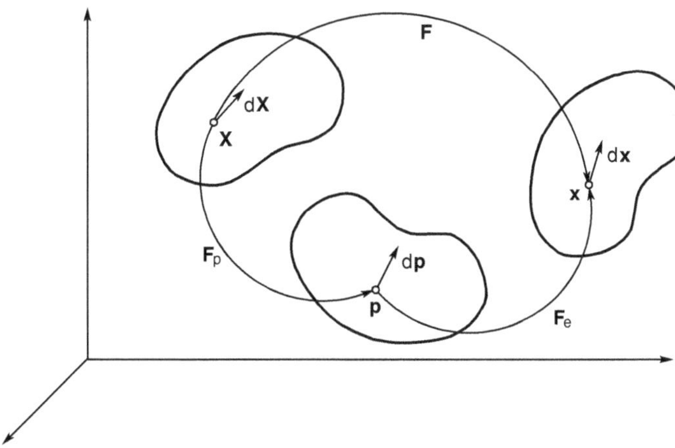

Fig. 3.3 Kinematics of elastic-plastic deformation

In the above, $\mathrm{d}\boldsymbol{x}$ and $\mathrm{d}\boldsymbol{p}$ are the (actual) spatial line element in the current configuration $\mathcal{B}$ and the line element in the fictitious unstressed configuration, which are the counterparts of the material line element $\mathrm{d}\boldsymbol{X}$ in $\mathcal{B}_0$ after experiencing the actual elastic-plastic deformation and the plastic deformation induced by the destressing procedure, respectively. The transformation relations yield the widely used multiplicative separation:

$$\boldsymbol{F} = \boldsymbol{F}_e \boldsymbol{F}_p \,. \tag{3.66}$$

Once the above separation could unambiguously be established, the elastic and plastic deformations $\boldsymbol{F}_e$ and $\boldsymbol{F}_p$ would be separated exactly from the total elastic-plastic deformation $\boldsymbol{F}$ and hence endowed with the desired physical features. Therefore, $\boldsymbol{F}_e$ and $\boldsymbol{F}_p$ could be employed as additional variables to realize physically pertinent formulations of elastic and plastic behavior. However, a central issue with the separation is the non-uniqueness in the following sense: If $\boldsymbol{F}_e$ and $\boldsymbol{F}_p$ obey the separation (3.66), then the same may be true for $\boldsymbol{F}_e \boldsymbol{Q}^{\mathrm{T}}$ and $\boldsymbol{Q}\boldsymbol{F}_p$ with an arbitrary rotation $\boldsymbol{Q}$. This means that the rotational parts of $\boldsymbol{F}_e$ and $\boldsymbol{F}_p$, i.e., $\boldsymbol{R}_e$ and $\boldsymbol{R}_p$, would be rendered indeterminate. In other words, a non-unique separation (3.66) would fail to separate the just-mentioned two rotations.[27]

The decoupling represented by Eq. (3.66) enables us to accomplish a direct formulation of elastic behavior. Now, the Kirchhoff stress $\boldsymbol{\tau}$ may be specified by a single variable, i.e., the elastic deformation $\boldsymbol{F}_e$. Emphasizing the substantial invariance property of the elastic moduli in the process of elastic-plastic deformations, Lee (1969) assumed the following invariable elastic relation

$$\boldsymbol{\tau} = 2\boldsymbol{F}_e \frac{\partial \psi}{\partial \boldsymbol{C}_e} \boldsymbol{F}_e^{\mathrm{T}} \,. \tag{3.67}$$

Note that all elastic domains correspond to the same elastic potential $\psi = \psi(\boldsymbol{C}_e)$ with $\boldsymbol{C}_e = \boldsymbol{F}_e^{\mathrm{T}} \boldsymbol{F}_e$. This means that the elastic behavior will be described by that for the initial elastic domain prior to the occurrence of yielding.

The foregoing non-uniqueness, however, renders relation (3.67) incomplete. To eliminate this, an extra condition was introduced (see, e.g., Lee and Liu 1967; Lee 1969):

$$\boldsymbol{F}_e^{\mathrm{T}} = \boldsymbol{F}_e \,. \tag{3.68}$$

[27] Sometimes the rotational parts incorporated in $\boldsymbol{F}_e$ and $\boldsymbol{F}_p$ and even in $\boldsymbol{F}$ are loosely said to be "superimposed rigid-body rotations". This expression may produce an impression as if these rotations might be not so essential. However, essential difference exists between each such rotation and any truly rigid body rotation. The latter is constant at all points in a body and should have no effect on both basic equations of motion and constitutive formulations of material behavior, whereas the former varies from point to point and exhibits essential effects on both. In fact, the rotational parts of $\boldsymbol{F}$ and $\boldsymbol{F}_e$ and $\boldsymbol{F}_p$ are inseparable parts of deformations and deformation rates, and incorporated in constitutive formulations for both elastic and plastic behavior.

and then the elastic relation (3.67) in the case of isotropy would become[28]

$$\boldsymbol{\tau} = 2\boldsymbol{B}_e \frac{\partial \psi}{\partial \boldsymbol{B}_e}\,, \tag{3.69}$$

The relation (3.67), in particular (3.69), assumes the standard form for the classical hyperelastic relation, which is usually regarded to describe the elastic behavior included in but separated from the elastic-plastic behavior as a single entity.

Towards a physically pertinent formulation of plastic flow, it is desirable to have a proper separation of the total deformation rate $\boldsymbol{D}$ into elastic and plastic parts, as shown by Eq. (3.48). Although the separation (3.66) could realize the decoupling of elastic and plastic deformation except for an arbitrary rotation, a definite deformation rate separation based on it has proved to be not so clear and simple. In fact, we have

$$\boldsymbol{D} = \mathrm{sym}(\dot{\boldsymbol{F}}_e \boldsymbol{F}_e{}^{-1}) + \mathrm{sym}(\boldsymbol{F}_e \dot{\boldsymbol{F}}_p \boldsymbol{F}_p{}^{-1} \boldsymbol{F}_e{}^{-1}) \neq \mathrm{sym}(\dot{\boldsymbol{F}}_e \boldsymbol{F}_e{}^{-1}) + \mathrm{sym}(\dot{\boldsymbol{F}}_p \boldsymbol{F}_p{}^{-1})\,, \tag{3.70}$$

where the last two terms may be called the elastic and plastic deformation rates and will be denoted $\bar{\boldsymbol{D}}_e$ and $\bar{\boldsymbol{D}}_p$.

The above inequality shows that the total deformation rate $\boldsymbol{D}$ can not be split into the sum of the two rates $\bar{\boldsymbol{D}}_e$ and $\bar{\boldsymbol{D}}_p$.[29] Moreover, the non-uniqueness property of Eq. (3.66) leaves each rate term in Eq. (3.70) unspecified. With $\boldsymbol{F}_e = \boldsymbol{V}_e \boldsymbol{R}_e$ and $\boldsymbol{F}_p = \boldsymbol{R}_p \boldsymbol{U}_p$, the following relations make this clear:

$$\dot{\boldsymbol{F}}_e \boldsymbol{F}_e^{-1} = \dot{\boldsymbol{V}}_e \boldsymbol{V}_e^{-1} + \boldsymbol{V}_e \dot{\boldsymbol{R}}_e \boldsymbol{R}_e{}^{\mathrm{T}} \boldsymbol{V}_e^{-1}, \quad \dot{\boldsymbol{F}}_p \boldsymbol{F}_p^{-1} = \dot{\boldsymbol{R}}_p \boldsymbol{R}_p{}^{\mathrm{T}} + \boldsymbol{R}_p \dot{\boldsymbol{U}}_p \boldsymbol{U}_p^{-1} \boldsymbol{R}_p{}^{\mathrm{T}}.$$

They show that each rate term in Eq. (3.70) is essentially dependent on either the elastic rotation $\boldsymbol{R}_e$ or the plastic rotation $\boldsymbol{R}_p$ or on both. To resolve this difficulty, additional assumptions and procedures would be needed. It has been shown (see, Lubarda and Lee 1981; Lee 1996) that the deformation rate separation (3.48) may be re-established by assuming the extra condition (3.68) and defining the elastic and plastic deformation rates by

$$\boldsymbol{D}_e = \hat{\mathbb{K}}^{-1} : (\dot{\boldsymbol{B}}_e + \boldsymbol{B}_e \boldsymbol{W} - \boldsymbol{W} \boldsymbol{B}_e), \qquad \boldsymbol{D}_p = \hat{\mathbb{K}}^{-1} : (2\boldsymbol{V}_e \bar{\boldsymbol{D}}_p \boldsymbol{V}_e)\,, \tag{3.71}$$

with $\hat{\mathbb{K}}$ a 4th-order tensor given by

$$\hat{\mathbb{K}}_{ijkl} = 2(\boldsymbol{B}_e)_{ik}\delta_{jl} = 2(\boldsymbol{B}_e)_{jl}\delta_{ik}\,. \tag{3.72}$$

[28]We note that in this specific case we also have: $\boldsymbol{C}_e = \boldsymbol{V}_e^2 = \boldsymbol{F}_e \boldsymbol{F}_e{}^{\mathrm{T}} = \boldsymbol{B}_e$, i.e. the left and right Cauchy-Green tensors are equal. Moreover, for an isotropic material, $\boldsymbol{V}_e$, $\boldsymbol{B}_e$ and $\partial\psi/\partial\boldsymbol{B}_e$ have the same principal axes so that the products are commutative.

[29]We also refer to the discussions in Nemat-Nasser (1979, 1982); Lee and McMeeking (1980); Lee (1981); Lubarda and Lee (1981); Mandel (1981).

With the deformation rate separation described above, the isotropic elastic relation (3.69) may be reformulated in an equivalent Eulerian rate form as given by

$$\boldsymbol{D}_e = (\hat{\mathbb{L}} : \hat{\mathbb{K}})^{-1} : \overset{\circ}{\boldsymbol{\tau}}{}^J , \tag{3.73}$$

where $\overset{\circ}{\boldsymbol{\tau}}{}^J$ is the Jaumann rate of $\boldsymbol{\tau}$ and

$$\hat{\mathbb{L}}_{ijkl} = 2\delta_{ik}\left(\frac{\partial\psi}{\partial \boldsymbol{B}_e}\right)_{jl} + 2(\boldsymbol{B}_e)_{im}\left(\frac{\partial^2\psi}{\partial \boldsymbol{B}_e^2}\right)_{mjkl} . \tag{3.74}$$

We note that the material tensor

$$\mathbb{N} = \hat{\mathbb{L}} : \hat{\mathbb{K}}$$

herein has the same major and minor symmetry properties as the elasticity tensor $\mathbb{L}$, namely $\mathbb{N}_{ijkl} = \mathbb{N}_{klij} = \mathbb{N}_{ijlk} = \mathbb{N}_{jikl}$.

It appears that within the context of the general constitutive formulation sketched above the consequences implied by the work postulate have not yet been derived.[30] In this case, the normality rule is accepted as a plausible assumption, but the convexity property of the yield surface may or may not be assumed. With the normality rule, the total stress rate-strain rate relationship is established as follows:

$$\boldsymbol{D} = \left(\mathbb{N} + \zeta\,\frac{1}{h}\frac{\partial \mathrm{f}}{\partial \boldsymbol{\tau}} \otimes \frac{\partial \mathrm{f}}{\partial \boldsymbol{\tau}}\right) : \overset{\circ}{\boldsymbol{\tau}}{}^J . \tag{3.75}$$

where $\mathrm{f} = \mathrm{f}(\boldsymbol{\tau}, \boldsymbol{\alpha}, \kappa)$ is the yield surface in stress space, h is a hardening function, which can be derived from the normality rule for the plastic deformation rate $\boldsymbol{D}_p$, the consistency condition of plasticity, and the evolution equations of the internal variables $\boldsymbol{\alpha}$ and κ.[31] As before ζ is a plastic multiplier, differentiating the cases of loading and unloading.

Similar to the situation with the Green-Naghdi theory a long discussion evolved about different issues related with this multiplicative decomposition or the Lee-Liu theory.[32] Three different aspects were mainly under discussion:

(i) The non-uniqueness property of the separation (3.66) would imply that it might be indeterminate not only in mathematical sense, but also in physical sense. It might fail to separate the elastic and plastic deformations from the total elastic-plastic deformation in a full sense. As indicated before, a rotation associated with the unstressed configuration could not be determined. But it is just such a rotation that incorporates the coupling between the elastic and plastic defor-

[30] Results have been given e.g. by Lubliner (1984, 1986), assuming maximal plastic dissipation principles.

[31] For details we refer e.g. to Bruhns (2014b).

[32] This discussion was primarily between the two schools and their followers and lasted several years. We therefore refer to Green and Naghdi (1971); Casey and Naghdi (1980, 1981) and Lee (1982) and the instructive discussion therein.

mations $\boldsymbol{F}_e$ and $\boldsymbol{F}_p$, since both $\boldsymbol{F}_e$ and $\boldsymbol{F}_p$ are related to the same unstressed configuration, as shown in Eq. (3.64). With this understanding in mind, it might be said that, unlike what has been expected, in principle the separation (3.66) could realize only a partial or an incomplete decoupling of the total elastic-plastic deformation, unless the respective rotational parts incorporated in $\boldsymbol{F}_e$ and $\boldsymbol{F}_p$ could be specified in a consistent manner.

The above fact might explain the main reason behind a number of known anomalies accompanying constitutive theories based upon the multiplicative separation. In fact, with a non-unique separation (3.66), the elastic relation (3.67) might be incomplete, except for an isotropic potential ψ, and the two rate quantities $\bar{\boldsymbol{D}}_e$ and $\bar{\boldsymbol{D}}_p$ could not be fully specified. Moreover, the objectivity of the constitutive formulation would be rendered questionable. Under the change of an observing frame, general transformation relations of $\boldsymbol{F}_e$ and $\boldsymbol{F}_p$ were derived (see, e.g., Green and Naghdi (1971); Casey and Naghdi (1980, 1981, 1983)) allowing the unstressed configuration to undergo an arbitrary rotation $\bar{\boldsymbol{Q}}(t)$ independent of the rotation of the observing frame $\boldsymbol{Q}(t)$. Naghdi and coworkers pointed out that a constitutive formulation based upon a *non-unique separation* could not fulfill the objectivity requirement.[33]

To establish a complete constitutive formulation, Lee and coworkers assumed the isotropy of the potential ψ and introduced the particular additional condition (3.68), which neglects the rotational part of $\boldsymbol{F}_e$, i.e., $\boldsymbol{R}_e$. The objectivity may be met *with this ad hoc assumption*,[34] since the separation (3.66) is rendered unique in this particular case. However, this assumption would imply that only particular elastic and plastic rates $\bar{\boldsymbol{D}}_e$ and $\bar{\boldsymbol{D}}_p$ are treated, since the latter two rely essentially on the elastic rotation $\boldsymbol{R}_e$. On the other hand, even from a mathematical standpoint, the assumption (3.68) would be consistent only for an isotropic potential ψ. In this case, the stress $\boldsymbol{\tau}$ does not involve the elastic rotation $\boldsymbol{R}_e$, as shown by Eq. (3.69).

(ii) A further issue is how to separate the total deformation rate $\boldsymbol{D}$ into elastic and plastic parts, namely, how to define proper elastic and plastic deformation rates $\boldsymbol{D}_e$ and $\boldsymbol{D}_p$ in terms of $\boldsymbol{F}_e$ and $\boldsymbol{F}_p$ such that Eq. (3.48) holds. In doing so, again, an additional assumption has to be introduced. The definition given by Eq. (3.71) gives us only one example. It seems that its physical pertinence need to be clarified. For instance, the definition of the purely elastic deformation rate

[33] This aspect was also discussed in the foregoing references and in Sidoroff (1973); Lee (1981, 1996); Lubarda and Lee (1981); Haupt (1985, 2002); Dashner (1986), and many others.

[34] In fact, the assumption (3.68), the separation (3.66) as well as the elastic relation (3.69) become

$$\boldsymbol{F}_e^{*T} = \boldsymbol{F}_e^*, \quad \boldsymbol{F}^* = \boldsymbol{F}_e^* \boldsymbol{F}_p^*, \quad \boldsymbol{\tau}^* = \boldsymbol{\phi}(\boldsymbol{F}_e^*),$$

under the change of frame. Here, Eq. (3.69) is written in the form $\boldsymbol{\tau} = \boldsymbol{\phi}(\boldsymbol{F}_e)$ considering Eq. (3.68). Since $\boldsymbol{\phi}$ is isotropic, the last above and the objectivity of $\boldsymbol{\tau}$ yield $\boldsymbol{\phi}(\boldsymbol{F}_e^*) = \boldsymbol{\phi}(\boldsymbol{Q}^* \boldsymbol{F}_e \boldsymbol{Q}^{*T})$. From this follows that $\boldsymbol{F}_e^* = \boldsymbol{Q}^* \boldsymbol{F}_e \boldsymbol{Q}^{*T}$ for an invertible $\boldsymbol{\phi}$ with a symmetric argument due to Eq. (3.68). Then, $\boldsymbol{F}_p^* = \boldsymbol{Q}^* \boldsymbol{F}_p$ may be derived from $\boldsymbol{F}^* = \boldsymbol{Q}^* \boldsymbol{F}$ and Eq. (3.66), and the second above.

$\boldsymbol{D}_e$ as given by Eq. (3.71)$_1$ involves the total spin $\boldsymbol{W}$ containing both elastic and plastic contributions.

The above mentioned issues of non-uniqueness based on Eq. (3.66) would entail introducing respective additional assumptions. There are many possibilities for either of them. For instance, instead of assumption (3.68), either $\boldsymbol{W}_p = \mathrm{skw}(\dot{\boldsymbol{F}}_p \boldsymbol{F}_p^{-1}) = \boldsymbol{0}$ or $\boldsymbol{F}_p^{\mathrm{T}} = \boldsymbol{F}_p$ may be assumed. On the other hand, as further examples, the elastic deformation rate $\boldsymbol{D}_e$ defined by Eq. (3.71)$_1$ may be replaced by either of the following two definitions:

$$\boldsymbol{D}_e = \frac{1}{2}\boldsymbol{F}^{-\mathrm{T}}\dot{\boldsymbol{C}}_e\boldsymbol{F}^{-1}, \qquad \boldsymbol{D}_e = \dot{\boldsymbol{A}}_e - \boldsymbol{A}_e\boldsymbol{L}^{\mathrm{T}} - \boldsymbol{L}\boldsymbol{A}_e, \quad \boldsymbol{A}_e = \frac{1}{2}(\boldsymbol{B}_e - \boldsymbol{I}).$$

With the above facts in mind, an observation may be made on the conceptual basis of relation (3.67) which is regarded to represent the hyperelastic behavior formulated by the elastic deformation $\boldsymbol{F}_e$. This would become justified, whenever the following two facts would be confirmed true: First, the elastic part of the stress power, i.e., $\dot{w}_e = \boldsymbol{\tau} : \boldsymbol{D}_e$, should be well-defined and, second, the elastic stress work given by the integration of $\dot{w}_e$ should be dissipationless. These two may be evident for an elastic process. However, with the foregoing non-uniqueness properties of the separation (3.66) for an elastic-plastic process, the elastic strain rate $\boldsymbol{D}_e$ could not be well-defined in a unique manner. Furthermore, even if $\boldsymbol{F}_e$ would be rendered unique by a particular assumption, generally the elastic stress work need not be dissipationless for an elastic-plastic process. This observation may be explained by the fact that relation (3.67) which is formally like a hyperelastic equation would be essentially different from the latter, considering that the "elastic deformation" $\boldsymbol{F}_e$ therein is referred to a non-unique fictitious intermediate configuration, whereas a classical hyperelastic equation is formulated by the total deformation $\boldsymbol{F}$ referred to a fixed reference configuration.

(iii) A further fundamental issue is concerned with the limitation of the notion "elastic destressing", as pointed out by Naghdi and coworkers.[35] It is noted that "…the stress at a point in an elastic-plastic material can be reduced to zero without changing plastic strain only if the origin in stress space remains in the region enclosed by the yield surface" Naghdi (1990). Here, one may go a bit further by observing that even if the current stress may be reduced to zero when the origin in stress space stays outside the elastic region, the resultant plastic deformation $\boldsymbol{F}_p$ (and hence $\boldsymbol{F}_e$) might be essentially non-unique in orientation and in magnitude, due to the fact that there exist several elastic-plastic paths or processes from the current to the null stress point. Note that the destressing procedure may be achieved by any possible "thought" experiment, such as by cutting and then destressing infinitesimal material elements out of the material body.

[35] See, e.g., Green and Naghdi (1971), Casey and Naghdi (1980), Naghdi (1990), Naghdi and Casey (1992).

3.7 Director Triads and Isoclinic Configurations

With the particular assumption (3.68), a complete elastic-plastic formulation may be established, but it is confined to the case of an isotropic elastic potential ψ. Towards a more general treatment, one approach is to use *director triads* and *isoclinic configurations*, which originated from Kratochvíl (1971) and Mandel (1972).[36] The main idea is quoted as follows: "...to determine in some way the orientation of the present (stressed or released) configuration, so that an orientation variable must be added to the state variables. We shall use the following mode of orientation. We consider a *material plane* of unit normal **n**, and in this plane a *material direction* **m** (a unit vector). The element is oriented ...by the orthonormal triad formed from **m** and **n**, and which will be called a director triad" Mandel (1974b) and then, "we assume that at time t the material element is very rapidly unloaded .... This unloading process is elastic .... We thus obtain a present released configuration (κ) which is only defined up to an arbitrary rotation." This idea may be realized by selecting a triad formed by three orthonormal vectors, say, $\boldsymbol{\xi} = (\boldsymbol{d}_1, \boldsymbol{d}_2, \boldsymbol{d}_3)$, for each material element. Embedded in the present released configuration, i.e., the unstressed configuration, such a triad $\boldsymbol{\xi}$ is rotating as the former is changing, so that it determines the orientation of the former by specifying the related rotation. Such a triad is said to be a director triad. Further, if a particular director triad $\boldsymbol{\xi}_0 = (\boldsymbol{d}_1^0, \boldsymbol{d}_2^0, \boldsymbol{d}_3^0)$ is chosen in such a manner that it always keeps the same orientation with respect to the fixed axes, then it may be called an isoclinic triad. Accordingly, the unstressed configuration with the orientation specified by an isoclinic triad $\boldsymbol{\xi}_0$ is referred to as an isoclinic configuration.

Evidently, the isoclinic configuration with a well-defined isoclinic triad $\boldsymbol{\xi}_0$ results in a unique separation (3.66) and as such the elastic and plastic deformations $\boldsymbol{F}_e$ and $\boldsymbol{F}_p$ are accordingly specified. Now the constitutive relations are formulated in a somewhat different way. It is assumed that the potential ψ relies on both the elastic Green strain $\boldsymbol{E}_e = \frac{1}{2}(\boldsymbol{F}_e^{\mathrm{T}}\boldsymbol{F}_e - \boldsymbol{I})$ and the internal variables, while the yield function f depends on the stress

$$\boldsymbol{S}_e = \boldsymbol{F}_e^{-1}\boldsymbol{\tau}\boldsymbol{F}_e^{-\mathrm{T}} \tag{3.76}$$

and the internal variables, where $\boldsymbol{S}_e$ is the so-called Mandel stress, acting on the unstressed intermediate configuration. Then, the elastic relation (3.67) is converted to

$$\boldsymbol{S}_e = \frac{\partial\psi}{\partial\boldsymbol{E}_e}\,. \tag{3.77}$$

Besides, the flow rule is non-symmetric and formulated for $\dot{\boldsymbol{F}}_p\boldsymbol{F}_p^{-1}$ in a 9-dimensional space.[37]

---

[36] We also refer to the following works of Mandel (1973a, b, 1974a, b, 1981).

[37] There might be some doubt about the physical pertinence of such a non-symmetric flow rule in 9-dimensional space. It would imply that nine, instead of six, rate equations governing plastic flow

The formulation with an isoclinic triad ξ_0 may, finally, be changed to a more general type with an arbitrary director triad $\xi = Q\xi_0$, and a rotation variable Q. Such general results involve the corotational rate relative to the triad ξ and the rotation Q will enter as an additional variable into each constitutive function.

An observation on the notion of director triad is as follows. Let ξ be an arbitrary director triad embedded in the unstressed configuration. At the initial instant t_0, it is given by the fixed triad ξ_0 in the undeformed configuration $\mathcal{B}_0$. Since a director triad is always orthogonal at any instant t, i.e., $d_i \cdot d_j = 0$ for $i \neq j$, and Eq. (3.64)$_1$ yields $d_i = F_p d_i^0$, we deduce $d_i^0 \cdot (F_p{}^{\mathrm{T}} F_p) d_j^0 = 0$ for $i \neq j$. This implies that the three orthonormal vectors d_i^0 are just the eigenvectors of the plastic stretch $U_p = \sqrt{F_p^{\mathrm{T}} F_p}$. From this and the right polar decomposition of F_p we may infer

$$F_p = R_p U_p, \qquad U_p = \sum_{i=1}^{3} \lambda_i^p d_i^0 \otimes d_i^0 .$$

In the above, the plastic rotation R_p is arbitrary for a general director triad, whereas it is constant for an isoclinic triad.

This result would mean that the principal axes of the plastic stretch U_p always keep unchanged, and this might be a too strong restriction. It could be eliminated by assuming that the rotation of the director triad at each particle to be independent of the deformation of the material element at this particle. But this would not only offer no assistance in clarifying the issues concerning the separation (3.66) but also go beyond the scope of a classical continuum. In fact, no director triad, let alone an isoclinic triad, could be defined in a classical continuum. Generally, any three mutually orthogonal line elements at each particle in the initial configuration could not always maintain the orthogonality property, since any line element in the unstressed configuration, no matter whether real or imaginary, should obey the basic kinematic relation (3.64).

However, the above remark would in no way invalidate the constitutive formulation in the foregoing. Actually, the latter may be independent of the notion of director triad, albeit it was thought to be the starting-point. Here, the essential point might be that three additional conditions or relations would be needed to eliminate the non-uniqueness of the separation (3.66). Now, a flow rule for $\dot{F}_p F_p^{-1}$ in 9-dimensional space already furnish adequate constitutive relations specifying F_p. As such, however, three additional constitutive relations incorporated in the non-symmetric flow rule for $\dot{F}_p F_p^{-1}$ would have to be introduced, as compared to a symmetric flow rule for D_p.

Numerous modifications of the above mentioned concepts have been reported in the literature of the last almost 50 years. Herein also their implementation into robust

(Footnote 37 continued)
should be needed even in the case of infinitesimal deformation, except for some particular cases. We also refer to Mandel (1974a) and the discussion therein.

and efficient numerical codes and the improvement of the latter became increasingly important. This issue, however, would be beyond the scope of the present article, and will be reported elsewhere.

References

Backman M (1964) Form for the relation between stress and finite elastic and plastic strains under impulsive loading. J Appl Phys 35:2524–2533
Bertram A (2005) Elasticity and plasticity of large deformations. Springer, Berlin
Böck N, Holzapfel G (2004) A new two-point deformation tensor and its relation to the classical kinematical framework and the stress concept. Int J Solids Struct 41:7459–7469
Bilby B, Gardner L, Stroh A (1957) Continuous distributions of dislocations and the theory of plasticity. Extrait des Actes du IXe Congrès Intern. de Mécanique Applicqueé, Bruxelles, pp 35–44
Bruhns O (2014a) Some remarks on the history of plasticity—Heinrich Hencky, a pioneer of the early years. In: Stein E (ed) The history of theoretical, material and computational mechanics—mathematics meets mechanics and engineering, vol 1. Springer, Heidelberg, pp 133–152
Bruhns O (2014b) The Prandtl-Reuss equations revisited. Z Angew Math Mech 94:187–202
Casey J, Naghdi P (1980) A remark on the use of the decomposition $\mathbf{F} = \mathbf{F}^e\mathbf{F}^p$ in plasticity. Trans ASME J Appl Mech 47:672–675
Casey J, Naghdi P (1981) Discussion of Lubarda and Lee (1981). Trans ASME J Appl Mech 48:983–984
Casey J, Naghdi P (1983) On the use of invariance requirements for intermediate configurations associated with the polar decomposition of a deformation gradient. Q Appl Math 41:339–342
Clayton J, McDowell D (2003) A multiscale multiplicative decomposition for elastoplasticity of polycrystals. Int J Plast 19:1401–1444
Darijani H, Naghdabadi R (2010) Constitutive modeling of solids at finite deformation using a second-order stress-strain relation. Int J Eng Sci 48:223–236
Dashner P (1986) Invariance considerations in large strain elasto-plasticity. Trans ASME J Appl Mech 53:55–60
Doyle T, Ericksen J (1956) Nonlinear elasticity. Adv Appl Mech 4:53–115
Drucker D (1949) The significance of the criterion for additional plastic deformation of metals. J Colloid Sci 4:299–311
Drucker D (1950) Some implications of work hardening and ideal plasticity. Q Appl Math 7:411–418
Eckart C (1948) The thermodynamics of irreversible processes. IV: The theory of elasticity and anelasticity. Phys Rev 73:373–382
Edelman F, Drucker D (1951) Some extensions of elementary plasticity theory. J Franklin Inst 251:581–605
Eglit M (1960) Tensorial characteristics of finite deformations. Prikl Mat Mekh 24:1432–1438
Fox N (1968) On the continuum theories of dislocations and plasticity. Q J Mech Appl Math 21:67–75
Freund L (1970) Constitutive equations for elastic-plastic materials at finite strain. Int J Solids Struct 6:1193–1209
Green A, Naghdi P (1965a) A general theory of an elastic-plastic continuum. Arch Ration Mech Anal 18:251–281
Green A, Naghdi P (1965b) Corrigenda. Arch Ration Mech Anal 19:408
Green A, Naghdi P (1971) Some remarks on elastic-plastic deformation at finite strain. Int J Eng Sci 9:1219–1229

Haar A, von Kármán T (1909) Zur Theorie der Spannungszustände in plastischen und sandartigen Medien. Nachr Ges Wiss Göttingen, Math-Phys Kl, pp 204–218
Haupt P (1985) On the concept of an intermediate configuration and its application to a representation of viscoelastic-plastic material behavior. Int J Plast 1:303–316
Haupt P (2002) Continuum mechanics and theory of materials, 2nd edn. Springer, Berlin
Hencky H (1924) Zur Theorie plastischer Deformationen und der hierdurch im Material hervorgerufenen Nachspannungen. Z Angew Math Mech 4:323–334
Hencky H (1928) Über die Form des Elastizitätsgesetzes bei ideal elastischen Stoffen. Z Tech Phys 9(215–220):457
Hill R (1948) A variational principle of maximum plastic work in classical plasticity. Q J Mech Appl Math 1:18–28
Hill R (1950) The mathematical theory of plasticity. Clarendon Press, Oxford
Hill R (1968) On constitutive inequalities for simple materials. J Mech Phys Solids 16(229–242):315–322
Hill R (1970) Constitutive inequalities for isotropic elastic solids under finite strain. Proc R Soc London Ser A 314:457–472
Hill R (1978) Aspects of invariance in solid mechanics. Adv Appl Mech 18:1–75
Hill R, Rice J (1973) Elastic potentials and the structure of inelastic constitutive laws. SIAM J Appl Math 25:448–461
Horstemeyer M, Bammann D (2010) Historical review of internal state variable theory for inelasticity. Int J Plast 26:1310–1334
Kratochvíl J (1971) Finite-strain theory of crystalline elastic-inelastic materials. J Appl Phys 42:1104–1108
Kröner E (1958) Kontinuumstheorie der Versetzungen und Eigenspannungen. Springer, Berlin
Kröner E (1960) Allgemeine Kontinuumstheorie der Versetzungen und Eigenspannungen. Arch Ration Mech Anal 4:273–334
Lee E (1969) Elastic-plastic deformation at finite strains. Trans ASME J Appl Mech 36:1–6
Lee E (1981) Some comments on elastic-plastic analysis. Int J Solids Struct 17:859–872
Lee E (1982) Finite deformation theory with nonlinear kinematics. In: Lee E, Mallett R (eds) Plasticity of metals at finite strain: theory. Computation and experiment. Stanford University and RPI, Stanford
Lee E (1996) Some anomalies in the structure of elastic-plastic theory at finite strain. In: Carroll M, Hayes M (eds) Nonlinear effects in fluids and solids. Plenum Press, New York
Lee E, Germain P (1974) Elastic-plastic theory at finite strain. In: Sawczuk A (ed) Problems of plasticity. Noordhoff International Publishing, Leyden
Lee E, Liu D (1967) Finite-strain elastic-plastic theory with application to plane-wave analysis. J Appl Phys 38:19–27
Lee E, McMeeking R (1980) Concerning elastic and plastic components of deformation. Int J Solids Struct 16:715–721
Lubarda V, Lee E (1981) A correct definition of elastic and plastic deformation and its computational significance. Trans ASME J Appl Mech 48:35–40
Lubliner J (1984) A maximal-dissipation principle in generalized plasticity. Acta Mechanica 52:225–237
Lubliner J (1986) Normality rules in large-deformation plasticity. Mech Mater 5:29–34
Ludwik P (1909) Elemente der technologischen Mechanik. Springer, Berlin
Macvean D (1968) Die Elementararbeit in einem Kontinuum und die Zuordnung von Spannungs- und Verzerrungstensoren. Z Angew Math Phys (ZAMP) 19:157–185
Mandel J (1972) Plasticité Classique et Viscoplasticité, CISM Courses and Lectures, vol 97. Springer, Wien
Mandel J (1973a) Equations constitutives et directeurs dans les milieux plastiques et viscoplastiques. Int J Solids Struct 9:725–740

Mandel J (1973b) Relations de comportement des milieux élastiques-viscoplastiques. Notion de répère directeur. In: Sawczuk A (ed) Foundations of plasticity. Noordhoff International Publishing, Leyden, pp 387–399
Mandel J (1974a) Director vectors and constitutive equations for plastic and viscoplastic media. In: Sawczuk A (ed) Problems of plasticity. Noordhoff International Publishing, Leyden, pp 135–143
Mandel J (1974b) Thermodynamics and plasticity. In: Domingos J, Nina M, Whitelaw J (eds) Foundations of continuum thermodynamics. The MacMillan Press, London, pp 283–304
Mandel J (1981) Sur la définition de la vitesse de déformation élastique et sa relation avec la vitesse de contrainte. Int J Solids Struct 17:873–878
Nádai A (1931) Plasticity, a mechanics of the plastic state of matter. McGraw-Hill, New York
Naghdi P (1990) A critical review of the state of finite plasticity. Z Angew Math Phys (ZAMP) 41:315–394
Naghdi P, Casey J (1992) A prescription for the identification of finite plastic strain. Int J Eng Sci 30:1257–1278
Naghdi P, Trapp J (1975a) On the nature of normality of plastic strain rate and convexity of yield surfaces in plasticity. Trans ASME J Appl Mech 42:61–66
Naghdi P, Trapp J (1975b) Restrictions on constitutive equations of finitely deformed elastic-plastic materials. Q J Mech Appl Math 28:25–46
Naghdi P, Trapp J (1975c) The significance of formulating plasticity theory with reference to loading surfaces in strain space. Int J Eng Sci 13:785–797
Nemat-Nasser S (1979) Decomposition of strain measures and their rates in finite deformation elastoplasticity. Int J Solids Struct 15:155–166
Nemat-Nasser S (1982) On finite deformation elasto-plasticity. Int J Solids Struct 18:857–872
Ogden R (1984) Nonlinear elastic deformations. Ellis Harwood, Chichester
Prager W (1944) Exploring stress-strain relations of isotropic plastic solids. J Appl Phys 15:65–71
Prandtl L (1924) Spannungsverteilung in plastischen Körpern. In: Proceedings of 1st international congress on applied mechanics, Delft, pp 43–46
Reuss A (1930) Berücksichtigung der elastischen Formänderung in der Plastizitätstheorie. Z Angew Math Mech 10:266–274
Reuss A (1932) Fließpotential oder Gleitebenen? Z Angew Math Mech 12:15–24
Rice J (1971) Inelastic constitutive relations for solids: an internal-variable theory and its application to metal plasticity. J Mech Phys Solids 19:433–455
Sedov L (1966) Foundations of the non-linear mechanics of continua. Pergamon Press, Oxford
Seth B (1964) Generalized strain measures with applications to physical problems. In: Reiner M, Abir D (eds) Second-order effects in elasticity. Plasticity and fluid dynamics. Pergamon Press, Oxford
Sidoroff F (1973) The geometrical concept of intermediate configuration and elastic-plastic finite strain. Arch Mech 25:299–308
Tokuoka T (1977) Rate type plastic material with kinematic work-hardening. Acta Mechanica 27:145–154
Tokuoka T (1978) Prandtl-Reuss plastic material with scalar and tensor internal variables. Arch Mech 30:801–826
Truesdell C (1952a) The mechanical foundations of elasticity and fluid dynamics. J Ration Mech Anal 1:125–300
Truesdell C (1952b) The mechanical foundations of elasticity and fluid dynamics. J Ration Mech Anal 2:595–616
Truesdell C (1952c) The mechanical foundations of elasticity and fluid dynamics. J Ration Mech Anal 3:801
Truesdell C (1955) Hypo-elasticity. J Ration Mech Anal 4:83–133
Truesdell C (1964) Second-order effects in the mechanics of materials. In: Reiner M, Abir D (eds) Second-order effects in elasticity. Plasticity and fluid dynamics. Pergamon Press, Oxford
Truesdell C, Noll W (1965) The non-linear field theories of mechanics. In: Flügge S (ed) Handbuch der Physik. Springer, Berlin

Willis J (1996) Some constitutive equations applicable to problems of large dynamic plastic deformation. J Mech Phys Solids 17:359–369
Xiao H, Bruhns O, Meyers A (2006) Elastoplasticity beyond small deformations. Acta Mechanica 182:31–111

Chapter 4
Effect of Biaxial Work Hardening Modeling for Sheet Metals on the Accuracy of Forming Limit Analyses Using the Marciniak-Kuczyński Approach

Tomoyuki Hakoyama and Toshihiko Kuwabara

Abstract A servo-controlled tension-internal pressure testing machine with an optical 3D deformation analysis system (ARAMIS®, GOM) was used to measure the multiaxial plastic deformation behavior of a high-strength steel sheet with a tensile strength of 590 MPa for a strain range from initial yield to fracture. Tubular specimens were fabricated by roller bending and laser welding the as-received flat sheet materials. Many linear stress paths in the first quadrant of the stress space were applied to the tubular specimens to measure the forming limit curve (FLC), forming limit stress curve (FLSC), and forming limit plastic work per unit volume (FLPW) of the as-received sheet material in addition to the contours of plastic work and the directions of the plastic strain rates. Differential hardening behavior was observed; the shapes of the work contours constructed in the principal stress space changed with an increase in plastic work. The observed differential hardening behavior was approximated by changing the material parameters and the exponent of the Yld2000-2d yield function as functions of the reference plastic strain. Marciniak-Kuczyński-type forming limit analyses were performed using both the differential hardening model and isotropic hardening models based on the Yld2000-2d yield function. It was found that the material model that is capable of reproducing both the work contours and the directions of the plastic strain rates measured for a strain range close to the fracture limit can give a more effective constitutive model for accurately predicting the FLC, FLSC, and FLPW.

T. Hakoyama
JSPS Research Fellow (Doctoral Course Students), Department of Mechanical Systems Engineering, Tokyo University of Agriculture and Technology,
2-24-16, Nakacho, Koganei-shi, Tokyo 184-8588, Japan
e-mail: 50014833008@st.tuat.ac.jp

T. Kuwabara (✉)
Division of Advanced Mechanical Systems Engineering, Institute of Engineering,
Tokyo University of Agriculture and Technology,
2-24-16, Nakacho, Koganei-shi, Tokyo 184-8588, Japan
e-mail: kuwabara@cc.tuat.ac.jp

H. Altenbach et al. (eds.), *From Creep Damage Mechanics to Homogenization Methods*, Advanced Structured Materials 64,
DOI 10.1007/978-3-319-19440-0_4

Keywords Sheet metal forming · Multiaxial tube expansion test · Differential hardening · Forming limit · Marciniak-Kuczyński model

4.1 Introduction

Lightening the weight of automotive bodies is effective in reducing CO_2 emissions, which is an important step towards the preservation of Earth's environment. High-strength steel sheets are considered a candidate material for reducing the weight of automobiles. However, high-strength steel sheets generally have less ductility than ultralow-carbon steel sheets and are therefore difficult to use for the manufacture of automotive body panels. Finite element analysis (FEA) is a key technology for enhancing the predictive accuracy of forming failures in high-strength steel sheets and realizing trial-and-error-less manufacturing.

A forming limit curve (FLC), which consists of the strains at which localized necking is first observed in sheet metal, is widely used in industry. However, since FLCs are path dependent (Nakazima et al. 1968; Graf and Hosford 1993), they are not effective for predicting the fracture of sheet metals subjected to non-proportional loading. On the other hand, a forming limit stress curve (FLSC), which consists of the stresses at which localized necking is first observed in sheet metal, has been proven to be path independent both experimentally (Yoshida et al. 2005; Yoshida and Kuwabara 2007) and analytically (Stoughton 2000; Yoshida et al. 2007; Yoshida and Suzuki 2008) if the metal sheet obeys the isotropic hardening law. Zimniak (2000a, b); Stoughton and Yoon (2005), and Chen et al. (2007) applied the FLSC to the fracture prediction of sheet metals subjected to proportional and non-proportional loading in FEA and found that the FLSC is useful for improving the predictive accuracy of forming failure in sheet metal parts.

Since the measurement of FLCs and FLSCs takes a significant amount of time in real experiments, it would be helpful if FLCs and FLSCs can be predicted by numerical calculations based on plasticity theories. It is well known that yield functions have a significant effect on the predictive accuracy of FLCs (Kuroda and Tvergaard 2000; Banabic and Dannenmann 2001; Butuc et al. 2002). Therefore, material models (yield functions or polycrystal models) that are capable of accurately reproducing the plastic deformation behavior of real sheet metals are crucial for accurately predicting FLCs and FLSCs.

One of the most popular constitutive models for metals in phenomenological plasticity theories is *isotropic hardening*, which assumes that the yield surface expands with its shape being maintained as plastic deformation progresses. However, the yield surface shape can change even in monotonic loading because of the change in texture with plastic deformation. Hill and Hutchinson (1992) proposed a simple constitutive model to account for progressive changes in the yield locus shape as plastic deformation accumulates. Hill et al. (1994) proposed a new constitutive analysis method by focusing on the contours of equal plastic work (henceforth referred to as *work contour*) in the stress space; the change in shape of successive work contours with increasing deformation was formulated. They further assumed that the successive work contours act instantaneously as plastic potentials; a normality flow rule applies in relation to the work contours. Moreover, they conducted tension-internal pressure tests of 70−30 brass thin-walled tubes with linear stress paths and evaluated the differential hardening behavior as the work contours developed. One of the present authors developed biaxial tensile testing methods for sheet metals using cruciform specimens (Kuwabara et al. 1998) and for tubular specimens (Ishiki et al. 2011; Kuwabara and Sugawara 2013) to observe the differential hardening behavior of cold-rolled ultralow-carbon steel sheets (Kuwabara et al. 1998, 2002; Kuwabara and Sugawara 2013), high-strength steel sheets (Kuwabara et al. 2011; Kuwabara and Nakajima 2011), aluminum alloy sheets (Kuwabara et al. 2006; Yanaga et al. 2012, 2014), pure titanium sheets (Ishiki et al. 2011; Sumita and Kuwabara 2014), and a magnesium alloy sheet (Andar et al. 2012). In these works, it was found that for the most part, a normality flow rule applies in relation to the work contours. Actually, it was experimentally confirmed that the yield functions determined to fit the work contours measured from the biaxial tensile tests gave closer FEA results to the measured data for hole expansion simulations (Hashimoto et al. 2010; Kuwabara et al. 2011) and hydraulic bulge forming simulations (Yanaga et al. 2012, 2014) than other yield functions.

Several authors developed constitutive models that enable us to reproduce the differential hardening behavior in sheet forming simulations. Savoie et al. (1995) investigated the effect of the initial crystallographic textures of three different annealed aluminum alloy sheets on the FLCs using the Marciniak-Kuczyński (M-K) approach (Marciniak and Kuczyński 1967) in conjunction with crystal plasticity models. Xu and Weinmann (2000) calculated the FLCs based on the M-K approach with Hill's 1993 yield criterion (Hill 1993) by changing the anisotropic parameters as functions of the equivalent plastic strain. Aretz (2008) calculated the FLC and FLSC of an aluminum alloy using the M-K approach with the Yld2003 yield function (Aretz 2005) with an exponent of 8; the differential hardening behavior was approximated by changing the anisotropic parameters as functions of the equivalent plastic strain. He found that the differential hardening behavior significantly affects the calculated results for the FLC and FLSC. Stoughton and Yoon (2009) calculated the FLC and FLSC of 5182 aluminum alloy sheet using the M-K model. A non-associated model based on Hill's quadratic yield function (Hill 1948) was used, and the yield stresses ($\sigma_0, \sigma_{45}, \sigma_{90}$, and σ_b) were changed as functions of the equivalent plastic strain. Wang et al. (2012) conducted biaxial tensile tests of aluminum alloy sheets using cruciform specimens and approximated the measured work contours using the Yld2000-2d yield function (Barlat et al. 2003; Yoon et al. 2004) with an exponent of 8 and the material parameters being changed as a function of the equivalent plastic strain. They calculated the FLCs of 5754-O aluminum alloy sheet and compared them with the measured data. Zamiri and Pourboghrat (2007) showed that the r-values of high-purity niobium are very sensitive to plastic strain. They proposed a material model based on Hill's quadratic yield function with the material parameters changing as functions of plastic strain to perform a hydraulic bulge forming simulation. They concluded that the yield function with evolutionary coefficients can correctly predict the strain localization position during the forming of the high-purity niobium sheet. Yoshida et al.

(2014) investigated the differential hardening behavior of an aluminum alloy tube under biaxial and triaxial stress states using a tension-internal pressure-torsion testing machine. They demonstrated that a dislocation-density-based model reproduces the stress-path-dependent work-hardening behavior observed in the experiments.

The objective of the present study is to clarify the effects of the constitutive models (i.e., isotropic hardening models and a differential hardening model) on the accuracy of forming limit predictions based on the M-K analysis. The biaxial plastic deformation behavior of a high-strength steel sheet with a tensile strength of 590 MPa was precisely measured using the multiaxial tube expansion testing (MTET) method proposed by Kuwabara and Sugawara (2013). Many linear stress paths in the first quadrant of the stress space were applied to the test material to measure progressive changes in the work contour shapes in the stress space and the directions of the plastic strain rates as plastic deformation accumulates. Moreover, the FLC, FLSC, and forming limit plastic work per unit volume (FLPW) of the test material were precisely measured. The observed work hardening behavior was approximated using isotropic hardening models and a differential hardening model. The FLC, FLSC, and FLPW of the test material were then calculated using the M-K analyses based on these models and compared with the measured data.

4.2 Constitutive Model

Assuming small elastic and finite plastic deformations, we can write the kinematics in the form

$$\boldsymbol{D} = \boldsymbol{D}^{\mathrm{e}} + \boldsymbol{D}^{\mathrm{p}} = \boldsymbol{D}^{\mathrm{e}} + \dot{\Phi}\boldsymbol{N}^{\mathrm{p}}, \tag{4.1}$$

$$\boldsymbol{W} = \boldsymbol{\omega} + \boldsymbol{W}^{\mathrm{p}} = \boldsymbol{\omega} + \dot{\Phi}\boldsymbol{\Omega}^{\mathrm{p}}, \tag{4.2}$$

where $\boldsymbol{D}$ is the rate-of-deformation tensor (the symmetric part of the velocity gradient tensor $\boldsymbol{L} = \partial v_i/\partial x_j \boldsymbol{e}_i \otimes \boldsymbol{e}_j$, where $\boldsymbol{v}$ is the velocity of a material particle, $\boldsymbol{x}$ is its current position, and $\boldsymbol{e}_i$ is the Cartesian basis), $\boldsymbol{W}$ is the continuum spin tensor (the antisymmetric part of $\boldsymbol{L}$), the superscripts e and p respectively denote the elastic and plastic parts, $\boldsymbol{\omega}$ is the spin of the material substructure, and $\boldsymbol{N}^{\mathrm{p}}$ and $\boldsymbol{\Omega}^{\mathrm{p}}$ define the directions of $\boldsymbol{D}^{\mathrm{p}}$ and $\boldsymbol{W}^{\mathrm{p}}$, respectively. The scalar-valued quantity $\dot{\Phi}$ is a non-negative overstress function for rate-dependent cases.

With a superposed $\circ$ denoting an objective rate with respect to the spin $\boldsymbol{\omega}$ and the superposed dot denoting a material time derivative, the elasticity relation is assumed to be given by Hooke's law as follows:

$$\overset{\circ}{\boldsymbol{\sigma}} = \dot{\boldsymbol{\sigma}} - \boldsymbol{\omega}\cdot\boldsymbol{\sigma} + \boldsymbol{\sigma}\cdot\boldsymbol{\omega} = \boldsymbol{C}^{\mathrm{e}} : \boldsymbol{D}^{\mathrm{e}} = \boldsymbol{C}^{\mathrm{e}} : \boldsymbol{D} - \dot{\Phi}\boldsymbol{C}^{\mathrm{e}} : \boldsymbol{N}^{\mathrm{p}}, \tag{4.3}$$

where $\boldsymbol{\sigma}$ is the Cauchy stress tensor, and $\boldsymbol{C}^{\mathrm{e}}$ is the fourth-order elastic moduli tensor. It is assumed that $\boldsymbol{C}^{\mathrm{e}}$ is determined by Young's modulus E and Poisson's ratio ν.

Orthotropic symmetries in the plasticity are assumed. The structure variables to be considered include two types of equations: the orthonormal unit vectors $\boldsymbol{n}_i$ and the equivalent plastic strain $\bar{\varepsilon}^{\mathrm{p}}$. The orthonormal vectors $\boldsymbol{n}_i$ are defined along the axes of orthotropy $\hat{x}_i$, which evolve according to

$$\overset{\circ}{\boldsymbol{n}}_i = \boldsymbol{\omega} \cdot \boldsymbol{n}_i \tag{4.4}$$

since $\dot{\boldsymbol{n}}_i \equiv \mathbf{0}$. In this paper, the plastic spin is neglected (i.e., $\boldsymbol{\Omega}^{\mathrm{p}} = \mathbf{0}$ for the sake of ease).

The equivalent plastic strain rate $\dot{\bar{\varepsilon}}^{\mathrm{p}}$ is conjugated with an equivalent stress $\bar{\sigma}$ with respect to the plastic stress-power density increment $\dot{w}^{\mathrm{p}}$; i.e.,

$$\dot{w}^{\mathrm{p}} = \boldsymbol{\sigma} : \dot{\Phi} \boldsymbol{N}^{\mathrm{p}} = \bar{\sigma} \dot{\bar{\varepsilon}}^{\mathrm{p}}. \tag{4.5}$$

$\bar{\varepsilon}^{\mathrm{p}}$ is defined as

$$\bar{\varepsilon}^{\mathrm{p}} = \int \dot{\bar{\varepsilon}}^{\mathrm{p}} \mathrm{d}t = \int \dot{\Phi} \frac{\boldsymbol{\sigma} : \boldsymbol{N}^{\mathrm{p}}}{\bar{\sigma}} \mathrm{d}t. \tag{4.6}$$

For rate-dependent viscoplasticity, the dynamic yield surface is assumed to be given by

$$f = \bar{\sigma}(\boldsymbol{\sigma}, \boldsymbol{n}_i, \bar{\varepsilon}^{\mathrm{p}}) - g(\bar{\varepsilon}^{\mathrm{p}}) \left(\frac{\dot{\Phi}}{\dot{\Phi}_0} \right)^m = 0, \tag{4.7}$$

where g is a strain hardening function, $\dot{\Phi}_0$ is a reference value of the overstress function, and m is a rate sensitivity parameter. In this study, in order to express the differential hardening, the dynamic yield function is a function of $\bar{\varepsilon}^{\mathrm{p}}$.

We employ the Yld2000-2d yield function (Barlat et al. 2003; Yoon et al. 2004) to define $\bar{\sigma}$ and Swift's power law to define g; i.e.,

$$\bar{\sigma}(\boldsymbol{\sigma}, \boldsymbol{n}_i, \bar{\varepsilon}^{\mathrm{p}}) = \left\{ \frac{1}{2} \left(|X'_1 - X'_2|^M + |2X''_1 + X''_2|^M + |X''_1 + 2X''_2|^M \right) \right\}^{1/M}, \tag{4.8}$$

$$g = c(\bar{\varepsilon}^{\mathrm{p}} + \alpha)^n \tag{4.9}$$

Here, X'_i and X''_i ($i = 1, 2$) are the principal values of the second-order tensors $\boldsymbol{X}'$ and $\boldsymbol{X}''$, respectively, and are given as

$$X'_i = \frac{1}{2} \left(X'_{xx} + X'_{yy} \pm \sqrt{(X'_{xx} - X'_{yy})^2 + 4{X'_{xy}}^2} \right), \tag{4.10}$$

$$X''_i = \frac{1}{2} \left(X''_{xx} + X''_{yy} \pm \sqrt{(X''_{xx} - X''_{yy})^2 + 4{X''_{xy}}^2} \right). \tag{4.11}$$

$\boldsymbol{X}'$ and $\boldsymbol{X}''$ are obtained from the following linear transformation of $\boldsymbol{\sigma}$:

$$\boldsymbol{X}' = \begin{bmatrix} X'_{xx} \\ X'_{yy} \\ X'_{xy} \end{bmatrix} = \begin{bmatrix} L'_{11} & L'_{12} & 0 \\ L'_{21} & L'_{22} & 0 \\ 0 & 0 & L'_{66} \end{bmatrix} \begin{bmatrix} \sigma_{xx} \\ \sigma_{yy} \\ \sigma_{xy} \end{bmatrix}, \ \boldsymbol{X}'' = \begin{bmatrix} X''_{xx} \\ X''_{yy} \\ X''_{xy} \end{bmatrix} = \begin{bmatrix} L''_{11} & L''_{12} & 0 \\ L''_{21} & L''_{22} & 0 \\ 0 & 0 & L''_{66} \end{bmatrix} \begin{bmatrix} \sigma_{xx} \\ \sigma_{yy} \\ \sigma_{xy} \end{bmatrix}, \tag{4.12}$$

$$\begin{bmatrix} L'_{11} \\ L'_{12} \\ L'_{21} \\ L'_{22} \\ L'_{66} \end{bmatrix} = \begin{bmatrix} 2/3 & 0 & 0 \\ -1/3 & 0 & 0 \\ 0 & -1/3 & 0 \\ 0 & 2/3 & 0 \\ 0 & 0 & 1 \end{bmatrix} \begin{bmatrix} \alpha_1 \\ \alpha_2 \\ \alpha_7 \end{bmatrix}, \begin{bmatrix} L''_{11} \\ L''_{12} \\ L''_{21} \\ L''_{22} \\ L''_{66} \end{bmatrix} = \frac{1}{9} \begin{bmatrix} -2 & 2 & 8 & -2 & 0 \\ 1 & -4 & -4 & 4 & 0 \\ 4 & -4 & -4 & 1 & 0 \\ -2 & 8 & 2 & -2 & 0 \\ 0 & 0 & 0 & 0 & 9 \end{bmatrix} \begin{bmatrix} \alpha_3 \\ \alpha_4 \\ \alpha_5 \\ \alpha_6 \\ \alpha_8 \end{bmatrix}, \tag{4.13}$$

where $\alpha_i\, (i = 1 - 8)$ are the anisotropic parameters.

From Eq. (4.7), the expression for $\dot{\Phi}$ is

$$\dot{\Phi} = \dot{\Phi}_0 \left\{ \frac{\bar{\sigma}(\boldsymbol{\sigma}, \boldsymbol{n}_i, \bar{\varepsilon}^{\mathrm{p}})}{g(\bar{\varepsilon}^{\mathrm{p}})} \right\}^{1/m} . \tag{4.14}$$

For numerical computation, we use a rate tangent modulus method (Peirce et al. 1984). First, we use a linear interpolation within the time increment Δt as follows:

$$\Delta\Phi = \Delta t[(1 - \chi)\dot{\Phi}_t + \chi\dot{\Phi}_{t+\Delta t}], \tag{4.15}$$

where χ ranges from 0 to 1. The subscripts t and $t + \Delta t$ indicate arguments of the functions evaluated at t and $t + \Delta t$, respectively. By substituting the overstress function at $t + \Delta t$ (approximated using the Taylor series) into Eq. (4.15), we obtain

$$\Delta\Phi = \dot{\Phi}\Delta t = \Delta t \left[\frac{\dot{\Phi}_t}{1+\xi} + \frac{\xi}{h(1+\xi)} \boldsymbol{N}^{\mathrm{n}} : \boldsymbol{C}^{\mathrm{e}} : \boldsymbol{D} \right], \tag{4.16}$$

$$\xi = \chi\Delta t \left(\frac{\partial\dot{\Phi}}{\partial\bar{\sigma}} \right)_t h, \tag{4.17}$$

$$h = \boldsymbol{N}^{\mathrm{n}} : \boldsymbol{C}^{\mathrm{e}} : \boldsymbol{N}^{\mathrm{p}} - \left(\frac{\partial\dot{\Phi}}{\partial\bar{\sigma}} \right)_t^{-1} \left(\frac{\partial\dot{\Phi}}{\partial\Phi} \right)_t - \frac{\partial\bar{\sigma}}{\partial\bar{\varepsilon}^{\mathrm{p}}} \frac{\boldsymbol{\sigma} : \boldsymbol{N}^{\mathrm{p}}}{\bar{\sigma}}, \tag{4.18}$$

$$\left(\frac{\partial\dot{\Phi}}{\partial\bar{\sigma}} \right)_t^{-1} \left(\frac{\partial\dot{\Phi}}{\partial\Phi} \right)_t = -\left(\frac{\bar{\sigma}}{g}\frac{\partial g}{\partial\Phi} - \frac{\partial\bar{\sigma}}{\partial\bar{\varepsilon}^{\mathrm{p}}}\frac{\partial\bar{\varepsilon}^{\mathrm{p}}}{\partial\Phi} \right), \tag{4.19}$$

where $\boldsymbol{N}^{\mathrm{n}} \equiv \partial\bar{\sigma}/\partial\boldsymbol{\sigma}$. Here,$\partial\bar{\varepsilon}^{\mathrm{p}}/\partial\Phi$ is given as

$$\frac{\partial\bar{\varepsilon}^{\mathrm{p}}}{\partial\Phi} = \frac{\boldsymbol{\sigma} : \boldsymbol{N}^{\mathrm{p}}}{\bar{\sigma}}. \tag{4.20}$$

In this analysis, we use an associated flow rule; i.e., $\boldsymbol{N}^{\mathrm{p}} = \boldsymbol{N}^{\mathrm{n}} \equiv \partial\bar{\sigma}/\partial\boldsymbol{\sigma}$. Upon substituting Eqs. (4.16)–(4.19) into Eq. (4.3), we finally obtain the rate form constitutive equations as follows:

$$\overset{\circ}{\boldsymbol{\sigma}} = \boldsymbol{C}^{\tan} : \boldsymbol{D} - \frac{\dot{\Phi}_t}{\xi + 1}\boldsymbol{C}^{\mathrm{e}} : \boldsymbol{N}^{\mathrm{p}}, \tag{4.21}$$

$$\boldsymbol{C}^{\tan} = \boldsymbol{C}^{\mathrm{e}} - \frac{\xi}{h(\xi + 1)}(\boldsymbol{C}^{\mathrm{e}} : \boldsymbol{N}^{\mathrm{p}}) \otimes (\boldsymbol{N}^{\mathrm{n}} : \boldsymbol{C}^{\mathrm{e}}). \tag{4.22}$$

4.3 Experimental Methods

4.3.1 Test Material

The test material used in the present study was 1.2 mm thick precipitation hardening steel sheet with a tensile strength of 590 MPa (JSC590R). The work hardening characteristics and r-values at 0°, 45°, and 90° (transverse direction, TD) to the rolling direction (RD) are listed in Table 4.1. Hereafter, the RD, TD, and thickness direction of the material are defined as the x-, y-, and z-axes, respectively.

4.3.2 Biaxial Tensile Testing Methods

Two types of biaxial tensile tests were performed in order to measure the plastic deformation behavior of the test material from initial yield to fracture. Figure 4.1a shows a schematic diagram of the cruciform specimen used for the biaxial tensile tests of the as-received sheet sample. The geometry of the specimen was the same

Table 4.1 Mechanical properties of the test material (JSC590R)

Tensile direction	$\sigma_{0.2}$ (MPa)	c^{a} (MPa)	n^{a}	α^{a}	r-value[b]
0°	458	1054	0.192	−0.0026	0.588
45°	450	1007	0.205	0.0021	1.316
90°	476	1075	0.202	0.0037	0.790

[a]Approximated using $\sigma = c(\alpha + \varepsilon^{\mathrm{p}})^n$ at $\varepsilon^{\mathrm{p}} = 0.02 \sim 0.092$.
[b]Measured at uniaxial nominal strain $\varepsilon_{\mathrm{N}} = 0.1$

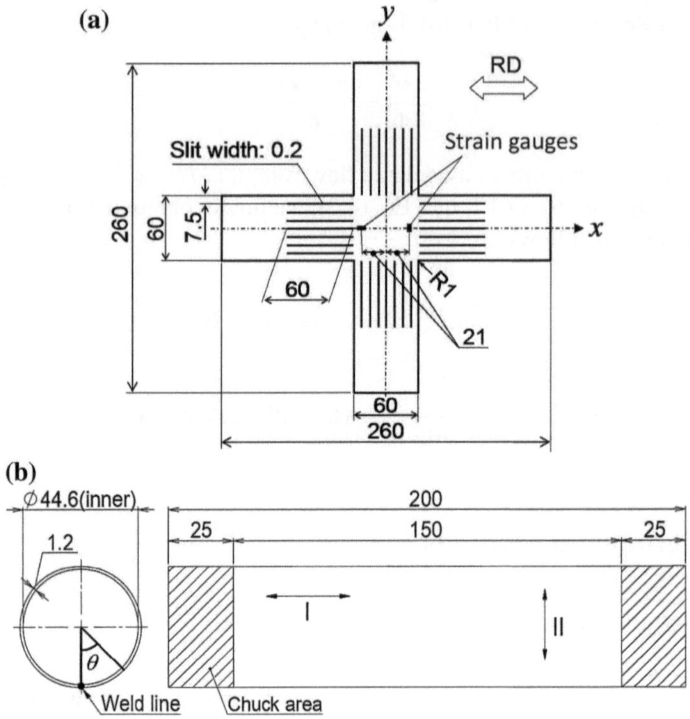

Fig. 4.1 Specimens used for the biaxial tensile tests (units in mm): **a** cruciform specimen and **b** tubular specimen. In **b**, ↔ indicates the rolling direction (RD) of the original sheet sample, and the RD is taken in the axial direction for the type I specimen and in the circumferential direction for the type II specimen; θ indicates the angle from the weld line

as that proposed by Kuwabara et al. (1998). The specimen arms were parallel to the RD and TD of the material. Each arm of the specimen had seven slits 60 mm long and 0.2 mm wide at 7.5 mm intervals to remove the geometric constraint on the deformation of the $60 \times 60\,\text{mm}^2$ square gauge area. The slits were fabricated by laser cutting.

The normal strain components (ε_x, ε_y) were measured using uniaxial strain gauges (YFLA-2, Tokyo Sokki Kenkyujo Co.) mounted at ±21 mm from the center along the maximum loading direction. According to the FEA of the cruciform specimen with the strain measurement positions shown in Fig. 4.1a, the stress measurement error was estimated to be less than 2 % (Hanabusa et al. 2010, 2013). Details of the biaxial tensile testing apparatus and test method are given in Kuwabara et al. (1998) and Kuwabara et al. (2000).

Figure 4.1b shows a schematic of the tubular specimen used for the MTET. The specimens were fabricated by bending a sheet sample into a cylindrical shape and CO_2 laser welding the sheet edges together to fabricate a tubular specimen with an inner diameter of 44.6 mm, a length of 200 mm, and a gauge length (distance between the grips of the testing machine) of 150 mm. The width of the weld line, including that of the heat affected zones, measured approximately 2 mm based on the data for the Vickers hardness distribution across the weld line. Two types of tubular specimens were fabricated; the type I specimen had the RD in the axial direction, and the type II specimen had the RD in the circumferential direction. Type I specimens were used for the tests with $\sigma_x \leq \sigma_y$, and type II specimens were used for the tests with $\sigma_x \geq \sigma_y$; the maximum principal stress direction was always taken to be the circumferential direction.

A servo-controlled tension-internal pressure testing machine was used in the MTET; the testing machine was developed by Kuwabara et al. (2003, 2005). Figure 4.2 shows a schematic diagram of a strain measurement system using a non-contact optical 3D deformation measuring system (ARAMIS®, GOM).

By measuring the coordinates of subsets A, B, C, and D on the surface of the tubular specimen (Fig. 4.2), the axial and circumferential strains $\varepsilon_\phi^{\mathrm{s}}$ and $\varepsilon_\theta^{\mathrm{s}}$ on the outer surface of the specimen were calculated using the following equations:

$$\varepsilon_\phi^{\mathrm{s}} = \ln\left\{\frac{R_\phi}{R_{\phi,0}\sin^{-1}(L_\phi/2R_{\phi,0})}\sin^{-1}\left(\frac{L_\phi}{2R_\phi}\right)\right\}, \tag{4.23}$$

$$\varepsilon_\theta^{\mathrm{s}} = \ln\frac{l_\theta}{L_\theta}, \tag{4.24}$$

where L_ϕ and l_ϕ are the initial and current gauge lengths between A and B, L_θ and l_θ are the initial and current gauge lengths between C and D, and $R_{\phi,0}$ and R_ϕ are the initial and current radii of axial curvature, respectively. In this study, the subset size of the measuring point was 25 pixels (approximately 2 mm). The initial gauge lengths L_ϕ and L_θ were chosen to be 15 mm.

By measuring the coordinates of subsets P, O, and Q aligned parallel to the axial direction of the tubular specimen (with O being taken at the center of the bulging specimen), the radius of axial curvature was calculated as

$$R_\phi = \pm\frac{o}{2\sqrt{1-((p^2+q^2-o^2)/2pq)^2}}, \tag{4.25}$$

where o, p, and q are the current distances between P and Q, O and Q, and P and O, respectively. In order to reduce the measurement error for R_ϕ as much as possible, the gauge length between P and Q was automatically changed during the tests to be

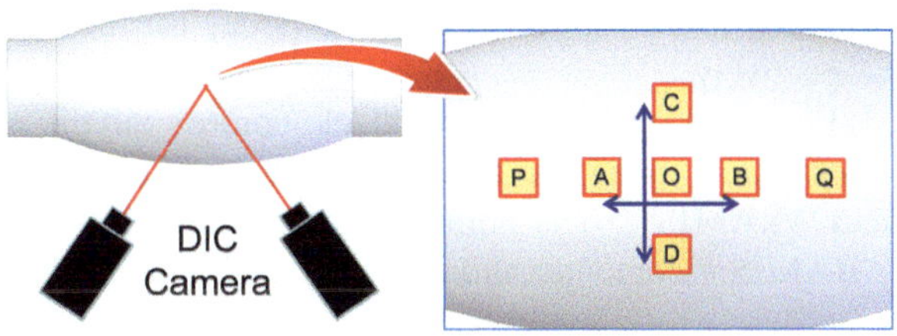

Fig. 4.2 Conceptual diagram of the measurement method for the axial and circumferential strains and the radius of axial curvature of a bulging specimen

46 mm when $|R_\phi| > 1000$ mm, 26 mm when $|R_\phi| \le 1000$ mm, and 13 mm when $|R_\phi| \le 150$ mm.

From Eqs. (4.23) and (4.24), the axial and circumferential strains ε_ϕ and ε_θ at the mid-wall of the specimen were evaluated using the following equations:

$$\varepsilon_\phi = \varepsilon_\phi^{\mathrm{s}} - \ln \frac{R_\phi}{R_\phi - (t/2)}, \tag{4.26}$$

$$\varepsilon_\theta = \ln \frac{D - t}{D_0 - t_0}, \tag{4.27}$$

where D_0 and t_0 are the initial outer diameter and wall thickness of the specimen, respectively. The current outer diameter D at the mid-section of the bulging specimen is determined as $D = D_0 \exp(\varepsilon_\theta^{\mathrm{s}})$. From the assumption of a constant volume and neglecting elastic strains, the current wall thickness t at the mid-section was calculated as

$$t = t_0 \exp(-\varepsilon_\phi - \varepsilon_\theta). \tag{4.28}$$

Equation (4.28) can only be calculated implicitly. Therefore, during the testing, the current wall thickness t was calculated using the following equation:

$$\begin{aligned} t &= \frac{-3a^2 + 2DR_\phi \pm \sqrt{(3a^2 - 2DR_\phi)^2 - 4(-3a + D + 2R_\phi)\Gamma}}{2(-3a + D + 2R_\phi)}, \\ \Gamma &= (-a^3 + 2D_0 R_\phi t_0 \exp(-\varepsilon_\phi^{\mathrm{s}}) - 2R_\phi t_0^2 \exp(-\varepsilon_\phi^{\mathrm{s}})). \end{aligned} \tag{4.29}$$

Equation (4.29) was derived from Eq. (4.28) using a Taylor series at an initial thickness a.

The axial and circumferential stresses σ_ϕ and σ_θ at the mid-section of the bulging specimen were calculated as those at the mid-wall using the following equations based on the equilibrium requirements for a material element at the mid-section of the specimen:

$$\sigma_\phi = \frac{P\pi(D/2 - t)^2 + T}{\pi(D - t)t}, \tag{4.30}$$

$$\sigma_\theta = \frac{(R_\phi - t)(D - 2t)}{(2R_\phi - t)t} P - \frac{D - t}{2R_\phi - t} \sigma_\phi, \tag{4.31}$$

where T and P are the measured values for the axial load and internal pressure, respectively.

Figure 4.3 shows the feedback control circuit used for controlling the true stress paths using the servo-controlled tension-internal pressure testing machine and ARAMIS®. The nominal strains and axial curvature were calculated on the ARAMIS® computer and output as analog data from 0 to 10 V at 15 Hz. A Bessel

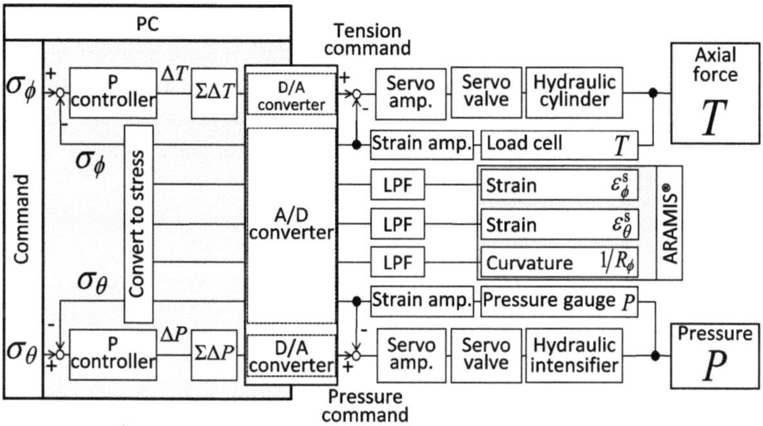

Fig. 4.3 Electrical feedback circuit for servo-controlled tension-internal pressure testing machine. LPF indicates a low-pass filter

low-pass filter whose filter frequency was 30 Hz was used between the ARAMIS® computer and a personal computer in order to reduce the line noise.

The measured values of T, P, ε_ϕ^s, ε_θ^s, and $1/R_\phi$ were input into the personal computer, and σ_ϕ and σ_θ were calculated using Eqs. (4.30) and (4.31). The calculated values of σ_ϕ and σ_θ were then compared with the command values of σ_ϕ and σ_θ. According to the discrepancy between the measured and commanded stress values, two P-controllers output command signals for controlling T and P every 0.01 s. The command signals for T and P were then compared with the measured values, and the servo valves were controlled according to the difference between the measured and command values of T and P so as to activate the hydraulic cylinders and intensifier such that the desired values of σ_ϕ and σ_θ are applied to the specimen. The resolution in the axial load measurements was 4 N, and that in the internal pressure measurements was 1 kPa. The outputs of T, P, ε_ϕ, ε_θ, σ_ϕ, σ_θ, and R_ϕ were recorded every 0.01 s with A/D data conversion and a personal computer and were saved on a disk.

Linear stress paths were applied to the cruciform and tubular specimens; the true stress ratios $\sigma_x : \sigma_y$ were chosen to be 4:1, 2:1, 4:3, 1:1, 3:4, 1:2, and 1:4. Standard uniaxial tensile specimens (JIS 13 B-type) were used for the uniaxial tensile tests with $\sigma_x : \sigma_y = 1 : 0$ and 0:1. True stress increments were controlled and applied to the specimens so that the von Mises equivalent plastic strain rate became approximately constant at (3–7) $\times 10^{-4}$ s^{-1} for all stress paths. Two specimens were used for each stress path.

The concept of the work contour in the stress space (Hill and Hutchinson 1992; Hill et al. 1994) was introduced to evaluate the work hardening behavior of the test material under biaxial tension. The true stress-logarithmic plastic strain curve obtained from a uniaxial tensile test along the RD was selected as a reference datum for work hardening; the uniaxial true stresses σ_0 and the plastic work per unit volume W_0 performed during the test up to the instant when a particular value of uniaxial logarithmic plastic strain $\varepsilon_0^{\mathrm{p}}$ (henceforth referred to as *reference plastic strain*) was reached were determined first. The uniaxial true stress σ_{90} measured from a tensile test in the TD and the biaxial true stress components (σ_x, σ_y) measured from biaxial tensile tests were then determined for the same plastic work as W_0. The stress points $(\sigma_0, 0)$, $(0, \sigma_{90})$, and (σ_x, σ_y) plotted in the principal stress space form a work contour associated with $\varepsilon_0^{\mathrm{p}}$.

4.3.3 Measurement of Forming Limit Strains, Stresses, and Plastic Work per Unit Volume

Forming limit strains, forming limit stresses, and forming limit plastic work per unit volume were measured using MTET. The forming limit strains were defined as the maximum logarithmic plastic strain components $(\varepsilon_x^{\mathrm{p}}, \varepsilon_y^{\mathrm{p}})$ achieved at the instant of fracture of the tubular specimens for the respective linear stress paths. The forming limit stresses were defined as the measured stress components (σ_x, σ_y) at the instant when the specimen reached the forming limit strain. The forming limit plastic work per unit volume was defined as the total plastic work performed during the test up to the instant of fracture of the tubular specimen.

Standard uniaxial tensile specimens were used for $\sigma_x : \sigma_y = 1 : 0$ and 0:1. The forming limit strains were determined by measuring the deformation of a 2 mm square grid pattern printed on the top surface of the specimen adjacent to the localized neck. The forming limit stresses were determined by extrapolating Swift's power law.

As described later in Sect. 4.4.1, the maximum value of $\varepsilon_0^{\mathrm{p}}$ attained using a tubular specimen for equibiaxial tension ($\sigma_x : \sigma_y = 1{:}1$) was 0.16 because of fracture occurring in the weld line. Therefore, hydraulic bulge tests were additionally performed to measure the forming limit strain and stress for equibiaxial tension. See Kuwabara and Sugawara (2013) for details on the hydraulic bulge testing method. The diameter and shoulder radius of the die opening were 150 and 8 mm, respectively. The forming limit strains were determined by measuring the deformation of 4 mm grid squares printed on the top surface of the specimen. The forming limit stress for the equibiaxial tension was determined by extrapolating Swift's power law determined for the measured equibiaxial stress-thickness strain curve.

In order to clarify the effect of the specimen geometry on the measurement of the forming limit strains and stresses, biaxial stretching tests using a flat-head punch proposed by Marciniak and Kuczyński (1967) were conducted for $\varepsilon_x^{\mathrm{p}} : \varepsilon_y^{\mathrm{p}} \approx 1{:}0$, $1 : 1$, and 0:1. The diameter and shoulder radius of the punch were 100 and 15 mm, respectively. For the strain measurement, a 4 mm square grid was printed on the top surface of the specimen. A Teflon sheet lubricated with Vaseline and a driving plate with a 30 mm diameter hole in the center were inserted between the specimen and the punch. The time average of the von Mises equivalent plastic strain rate was approximately constant at $(3\text{–}7) \times 10^{-4}\,\mathrm{s}^{-1}$ for each specimen. Two specimens were used for each strain path.

4.4 Material Modeling

4.4.1 Results of Biaxial Tensile Tests

The true stress-logarithmic plastic strain (s-s) curves measured from the MTET were offset and smoothly connected to those measured using the cruciform specimens to compensate for the effect of prestrain (bending strain) applied to the tubular specimens during their fabrication. For details on the offset method, see Kuwabara and Sugawara (2013).

Figure 4.4a shows the measured stress points forming the work contours. Each stress point represents an average of two specimen data; the difference between the two was less than 2 % of the flow stress for all data points. The maximum value of $\varepsilon_0^{\mathrm{p}}$ for which the work contour has a full set of nine stress points was $\varepsilon_0^{\mathrm{p}} = 0.16$, which is approximately three times larger than that obtained using a cruciform specimen. Moreover, it is noteworthy that a maximum strain of $\varepsilon_0^{\mathrm{p}} = 0.285$ was attained for $\sigma_x : \sigma_y = 4 : 3$. For only $\sigma_x : \sigma_y = 1 : 1$, fracture occurred at the weld line of the tubular specimens at $\varepsilon_0^{\mathrm{p}} = 0.16$.

Figure 4.4b shows the measured stress points forming the work contours for $\varepsilon_0^{\mathrm{p}} \leq 0.16$. All of the stress values forming a work contour were normalized by σ_0 associated with specific values of $\varepsilon_0^{\mathrm{p}}$. The shapes of the work contours slightly changed with work hardening or equivalently with $\varepsilon_0^{\mathrm{p}}$; thus, the test material exhibited differential hardening.

The shape ratios of the work contours normalized by σ_0 were determined and are shown in Fig. 4.5 for a quantitative evaluation of the amount of differential hardening. The shape ratio is defined as $r/r_{0.2}$, where $r_{0.2}$ is the distance between the origin in the principal stress space and a stress point that forms the work contour for $\varepsilon_0^{\mathrm{p}} = 0.002$, and r is the distance between the origin in the principal stress space and a

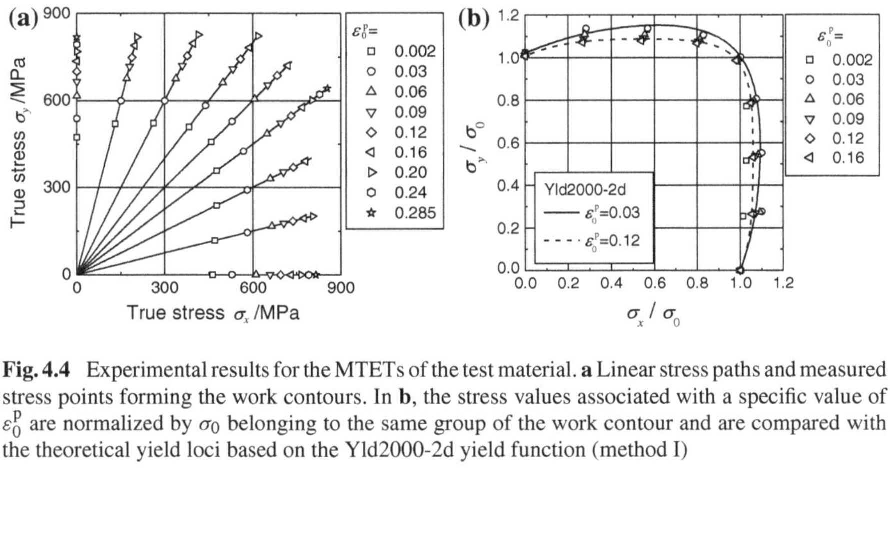

Fig. 4.4 Experimental results for the MTETs of the test material. **a** Linear stress paths and measured stress points forming the work contours. In **b**, the stress values associated with a specific value of $\varepsilon_0^{\mathrm{p}}$ are normalized by σ_0 belonging to the same group of the work contour and are compared with the theoretical yield loci based on the Yld2000-2d yield function (method I)

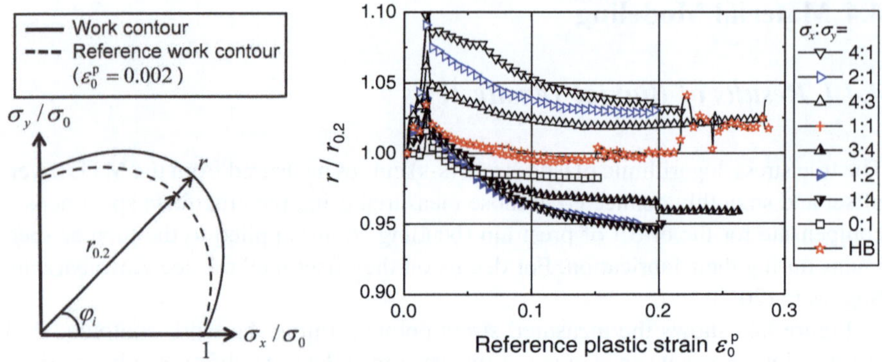

Fig. 4.5 Variation in the shape ratios $r/r_{0.2}$ of the work contours with ε_0^p for linear stress paths

stress point that forms a work contour for a particular value of $\varepsilon_0^p(\geq 0.002)$. Here, $(r/r_{0.2}) > 1$ and $(r/r_{0.2}) < 1$ indicate the expansion and shrinkage of the work contour, respectively. The values of $r/r_{0.2}$ for all stress ratios increase steeply for a strain range of $0 \leq \varepsilon_0^p \leq 0.02$. The value of $r/r_{0.2}$ for $\sigma_x : \sigma_y = 0 : 1$ decreases steeply for $0.02 \leq \varepsilon_0^p \leq 0.06$, while the values of $r/r_{0.2}$ for other stress ratios gradually decrease for $0.02 \leq \varepsilon_0^p \leq 0.12$. For $0.12 \leq \varepsilon_0^p$, the values of $r/r_{0.2}$ are almost constant for all stress ratios; thus, the material can be viewed as exhibiting isotropic hardening.

4.4.2 Isotropic Hardening Model

The Yld2000-2d yield function was used to approximate the work contours for $\varepsilon_0^p = 0.03$ and 0.12. $\varepsilon_0^p = 0.03$ is the strain at which yield elongation almost terminates, and $\varepsilon_0^p = 0.12$ is a representative strain value at which the test material can be viewed to initiate isotropic hardening. The material parameters $\alpha_i\,(i = 1-8)$ and exponent M of the Yld2000-2d yield function were determined following methods I, II, III, and IV:

Method I: $r_0, r_{45}, r_{90}, r_b, \sigma_0/\sigma_0, \sigma_{45}/\sigma_0, \sigma_{90}/\sigma_0$, and σ_b/σ_0 were used, where $r_\#$ and $\sigma_\#$ are the r-value and tensile flow stress measured at an angle of # from the RD, respectively, and r_b and σ_b are the plastic strain rate ratio $d\varepsilon_y^p/d\varepsilon_x^p$ and the flow stress at equibiaxial tension $\sigma_x : \sigma_y = 1 : 1$, respectively. The values of r_0, r_{45}, and r_{90} used were the same as those in Table 4.1. An exponent M was selected to minimize the root mean square error between the work contour and the calculated yield locus (see Appendix A1).

Methods II, III, and IV: The anisotropic parameters and exponent M were determined to minimize the cost function as given by Eq. (4.32)

$$F = \sum_{j=1}^{N} w_{j,\sigma}(r' - r)^2 + \sum_{j=1}^{N} w_{j,\beta}(\beta' - \beta)^2, \tag{4.32}$$

with the weighted parameters $w_{j,\sigma} = 1$ and $w_{j,\beta} = 0$ for method II, $w_{j,\sigma} = 0$ and $w_{j,\beta} = 1$ for method III, and $w_{j,\sigma} = 1$ and $w_{j,\beta} = 0.01$ for method IV. Here, $N(= 9)$ is the number of stress points forming a work contour, r is the distance between the origin of the principal stress space and the jth stress point, r' is the distance between the origin of the principal stress space and the calculated yield locus along the stress path to the jth stress point, β is the direction of the plastic strain rate measured for the jth stress path, and β' is the predicted direction of the plastic strain rate calculated using the yield function and the associated flow rule for the jth stress path. Real-coded genetic algorithm was used to minimize the cost function and to avoid a local optimum solution. The value of $w_{1,\sigma}$ (for $\sigma_x : \sigma_y = 1 : 0$) was set to 100 because the equivalent stress calculated using the yield function should coincide with the flow stress calculated using the strain hardening function determined for the RD.

The calculated values of $\alpha_i (i = 1 - 8)$ and M determined using methods I to IV are shown in Table 4.2. The yield loci determined using method I are shown in Fig. 4.4b.

The yield loci and the directions of the plastic strain rates determined using methods II, III, and IV are shown in Fig. 4.6. The yield loci calculated using methods II and IV are in good agreement with the measured work contours, while the yield loci calculated using method III slightly deviate from the measured work contours (Fig. 4.6a). The directions of the plastic strain rates calculated using method II show significant deviation from the measured values for $\varepsilon_0^{\mathrm{p}} = 0.03$, while the directions of the plastic strain rates calculated using methods III and IV are in good agreement with the measured values for both $\varepsilon_0^{\mathrm{p}} = 0.03$ and 0.12 (Fig. 4.6b and c). Consequently, it is concluded that the Yld2000-2d yield function determined by method IV gave the closest agreement with the experimental data for both the work contours and the directions of the plastic strain rates.

4.4.3 Differential Hardening Model

In order to reproduce the differential hardening behavior of the test material, the work contours and the directions of the plastic strain rates were measured for every $\varepsilon_0^{\mathrm{p}}$ at an increment of 0.01. Then, $\alpha_i (i = 1 - 8)$ and M of the Yld2000-2d yield function were determined for respective work contours for $0.002 \le \varepsilon_0^{\mathrm{p}} \le 0.16$ using method

Table 4.2 Material parameters of the Yld2000-2d yield function

	$\varepsilon_0^{\mathrm{p}}$	α_1	α_2	α_3	α_4	α_5	α_6	α_7	α_8	M
I	0.03	0.93554	0.93755	0.89222	0.98440	1.04281	1.01981	1.06043	1.02391	5.10
	0.12	0.94098	0.99332	0.99380	1.00324	1.03008	1.02355	1.05868	1.03563	7.89
II	0.03	1.03392	1.01019	1.04628	0.95493	0.98977	1.05448	1.05030	1.04397	7.964
	0.12	0.84686	1.04505	1.03359	1.01111	1.03504	0.97177	1.06087	1.02778	7.179
III	0.03	0.97283	0.92462	0.96162	0.98951	1.04438	1.06963	1.06145	0.99805	5.658
	0.12	0.93486	0.97949	1.00250	1.00061	1.03776	1.03242	1.06382	1.01864	6.355
IV	0.03	1.00220	0.90211	0.95049	0.98728	1.02704	1.02661	1.05854	1.02116	5.679
	0.12	0.90909	1.02101	1.02507	1.01407	1.03836	1.02953	1.06434	1.00938	6.508

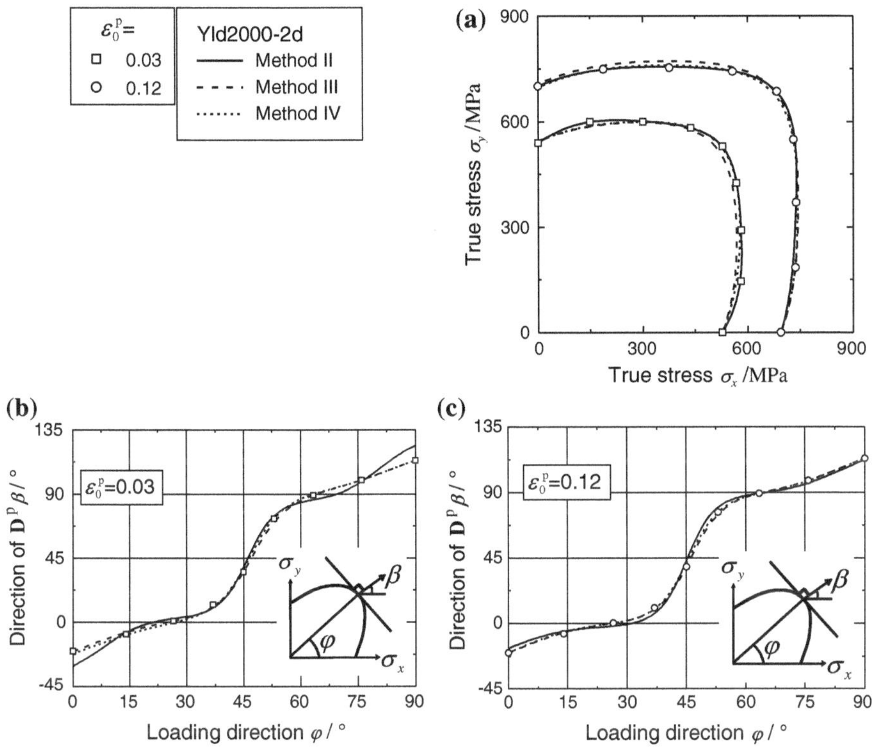

Fig. 4.6 Calculated results based on the Yld2000-2d yield function determined using methods II, III, and IV compared with the experimental data. **a** Work contours. **b** and **c** Directions of $\boldsymbol{D}^{\mathrm{p}}$

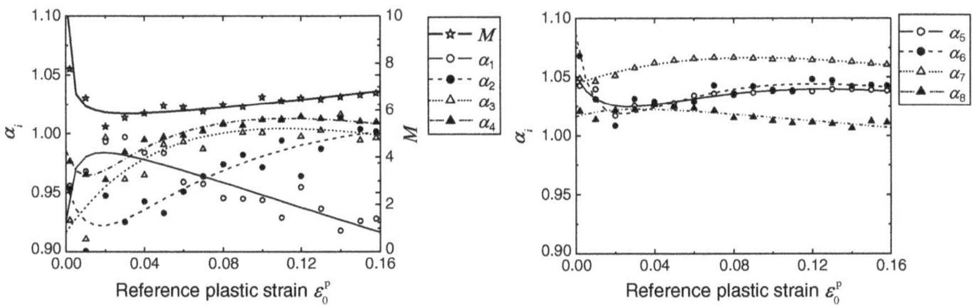

Fig. 4.7 Variations in M and $\alpha_i\,(i = 1-8)$ of the Yld2000-2d yield function with $\varepsilon_0^{\mathrm{p}}$, approximated using Eqs. (4.33) and (4.34)

IV. Figure 4.7 shows the calculated values of $\alpha_i\,(i = 1 - 8)$ and M for respective values of $\varepsilon_0^{\mathrm{p}}$. Moreover, the variations in α_i and M with $\varepsilon_0^{\mathrm{p}}$ were approximated by the following equations:

Table 4.3 The values of the parameters A, B, C, and D in Eqs. (4.33) and (4.34)

	M^{a}	α_1^{a}	α_2^{b}	α_3^{b}	α_4^{b}	α_5^{b}	α_6^{b}	α_7^{a}	α_8^{a}
A	5.34646	1.00780	−0.51967	−0.82622	−0.50889	−0.65138	−0.54399	−0.08112	−0.01441
B	−1.46978	0.58170	0.01440	0.07765	0.01159	0.02437	0.01004	11.73891	48.14033
C	0.00859	−0.00029	1.00253	0.84759	0.86743	0.95039	0.91931	3.62983	7.01315
D	0.00157	0.00354	0.10397	0.12084	0.06438	0.11174	0.07624	3.22601	6.80547

[a]Approximated using $M, \alpha_i = A\exp(-B\varepsilon_0^{p}) + C/(\varepsilon_0^{p} + D), (i = 1, 7, 8)$

[b]Approximated using $\alpha_i = A\log(B + \varepsilon_0^{p}) + C\varepsilon_0^{p}/(\varepsilon_0^{p} + D), (i = 2 - 6)$

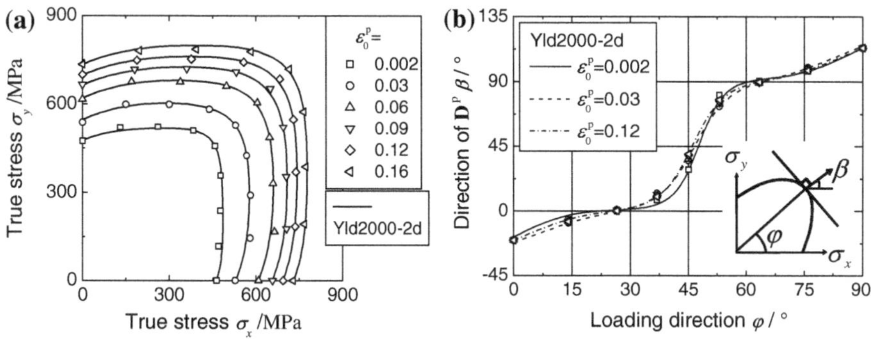

Fig. 4.8 Calculated results obtained from the differential hardening model for the test material compared with the experimental data. **a** Work contours. **b** Directions of $\boldsymbol{D}^{\mathrm{p}}$

$$M, \alpha_i = A \exp(-B\varepsilon_0^{\mathrm{p}}) + \frac{C}{\varepsilon_0^{\mathrm{p}} + D} \qquad (i = 1, 7, 8), \tag{4.33}$$

$$\alpha_i = A \log(B + \varepsilon_0^{\mathrm{p}}) + \frac{C\varepsilon_0^{\mathrm{p}}}{\varepsilon_0^{\mathrm{p}} + D} \qquad (i = 2 - 6). \tag{4.34}$$

The parameters A, B, C, and D for $\alpha_i\,(i = 1 - 8)$ and M are listed in Table 4.3. The approximation curves for $\alpha_i\,(i = 1 - 8)$ and M are also shown in Fig. 4.7. For $\varepsilon_0^{\mathrm{p}} \geq 0.16$, isotropic hardening was assumed ($\alpha_i\,(i = 1 - 8)$ and M were assumed to be constant). See Appendix A2 for details on the formulation method for the differential hardening.

Figure 4.8 shows the yield loci and the directions of the plastic strain rates calculated using the differential hardening model. The calculated results are in good agreement with the measured results for a strain range of $0.002 \leq \varepsilon_0^{\mathrm{p}} \leq 0.16$. Thus, it is concluded that the successive work contours determined using method IV act instantaneously as plastic potentials. See also Fig. 4.15.

4.5 Forming Limit Analysis

4.5.1 Conditions of Analysis

The forming limit analysis based on the M-K model was performed using the elasto-viscoplasticity model, as described in Sect. 4.2. The forming limit strain for one strain rate ratio was determined as follows: (i) calculate the strain components $\varepsilon_{11}^{\mathrm{L}}$ and $\varepsilon_{22}^{\mathrm{L}}$ outside the band at the onset of localized necking for an initial band angle of ψ_0; (ii) find the minimum value of $\varepsilon_{11}^{\mathrm{L}}$ by varying ψ_0 from 0° to 90° at an interval of 1°; and (iii) determine the corresponding strain components $\varepsilon_{11}^{\mathrm{L}}$ and $\varepsilon_{22}^{\mathrm{L}}$ and the initial band

Table 4.4 Parameters used in Swift's power law for strain hardening

Strain range	c^a (MPa)	n^a	α^a
$\varepsilon_0^p \leq 0.0185$	29.1	2.767	2.7115
$0.0185 \leq \varepsilon_0^p \leq 0.1$	1034	0.183	−0.0047
$0.1 \leq \varepsilon_0^p$	968	0.140	−0.0254

[a]Approximated using $\sigma = c(\alpha + \varepsilon_0^p)^n$

angle ψ_0 as the forming limit strains ε_{11}^* and ε_{22}^* and the critical initial band angle ψ^*, respectively. Please refer to Yoshida et al. (2007) for details on the calculation procedures of the M-K analysis.

The M-K analyses were performed using the Yld2000-2d yield function whose parameters were determined from methods I-IV to evaluate the effect of the material modeling methods on the accuracy of the forming limit predictions. The M-K analysis based on the Yld2000-2d yield function whose parameters were determined for the work contour with $\varepsilon_0^p = 0.03$ using methods I-IV are denoted as MK1-I-IV, respectively. The M-K analysis based on the Yld2000-2d yield function whose parameters were determined for the work contour with $\varepsilon_0^p = 0.12$ using methods I-IV are denoted as MK2-I-IV, respectively. The M-K analysis based on the differential hardening model as determined in Sect. 4.4.3 is denoted as MK3. The other material parameters assumed in the M-K analyses are as follows: the elastic modulus is 210 GPa, the Poisson's ratio is 0.3, and the strain rate sensitivity exponent (m-value) is 0.01. The magnitude of the initial imperfection was assumed to be 0.996, which was estimated from the variation in measured thickness. Swift's power law was applied to the strain hardening function, the parameters of which are listed in Table 4.4. The parameters were changed depending on the strain ranges in order to achieve an accurate reproduction of the strain hardening behavior. The strain rate of the maximum principal strain was assumed to be $5.0 \times 10^{-4}\ \text{s}^{-1}$.

4.5.2 Results and Discussion

Figure 4.9a shows the fractured specimens for respective proportional stress paths. Localized necks appeared all around the central bulged area of the respective specimens for the stress ratios $\sigma_x : \sigma_y = 1:4,\ 1:2,\ 3:4$, as shown in Fig. 4.9b.

Figures 4.10 and 4.11 show the FLC, FLSC, and FLPW calculated from MK1-I-IV and MK2-I-IV, respectively, compared with the measured data. MK2 had closer agreement with the measurement than MK1. This means that the forming limit strains and stresses are mainly dominated by the plastic deformation behavior for a strain range close to the fracture limit. Moreover, from the fact that MK2-IV had the closest

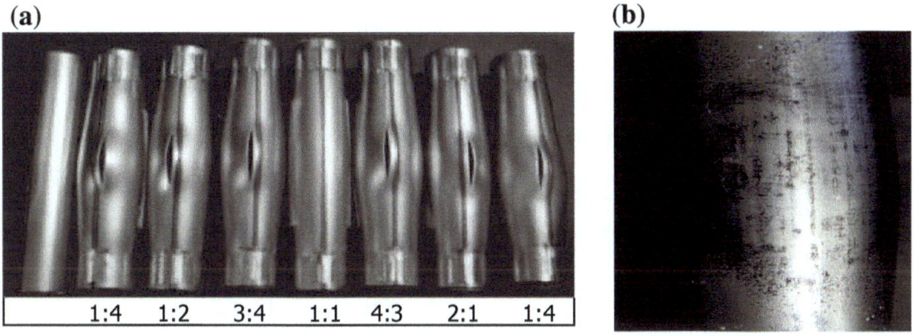

Fig. 4.9 Specimens after fracture. **a** Specimens obtained for each linear stress path. **b** Localized necks observed for $\sigma_x : \sigma_y = 3 : 4$; *black* paint was applied to the localized necks for good visibility

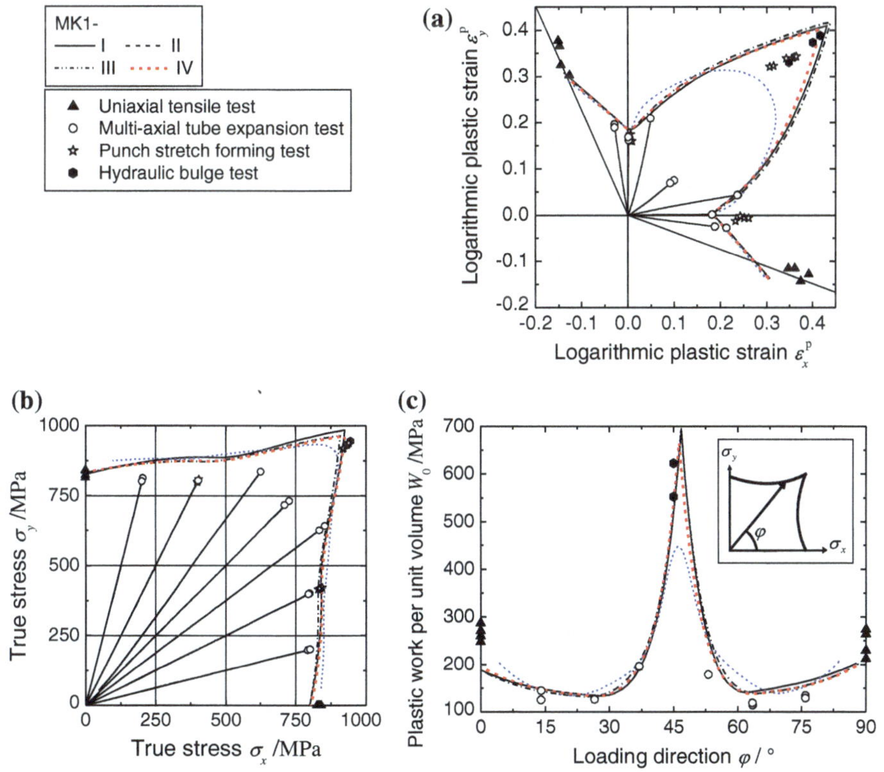

Fig. 4.10 Calculated results obtained from MK1-I-IV models: **a** FLC, **b** FLSC, and **c** FLPW

agreement with the measurement for all three cases (i.e., FLC, FLSC, and FLPW), it is concluded that the yield function that is capable of reproducing both the work contours and the directions of the plastic strain rates measured for a strain range

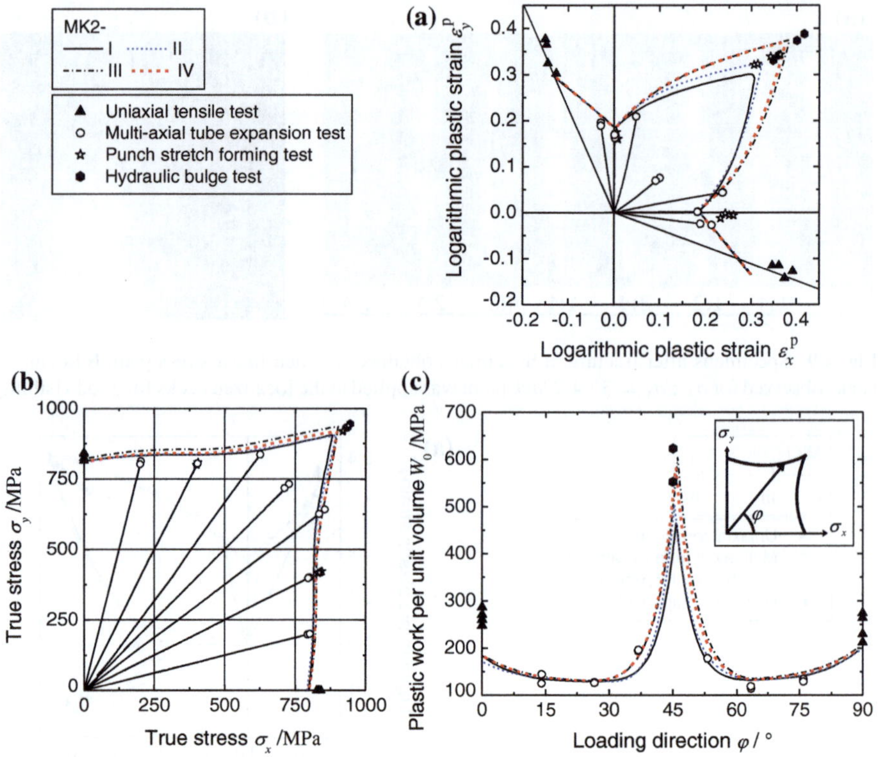

Fig. 4.11 Calculated results obtained from MK2-I-IV models: **a** FLC, **b** FLSC, and **c** FLPW

close to the fracture limit can give a more effective constitutive model for accurately predicting the FLC, FLSC, and FLPW than those determined for a smaller strain range.

Figure 4.12 compares the measured FLC, FLSC, and FLPW with those calculated using MK3. The calculated FLC, FLSC, and FLPW based on MK2-VI are also depicted in the figure. The FLC, FLSC, and FLPW based on MK2-VI are almost identical to those based on MK3. Thus, it is concluded that the differential hardening behavior for $\varepsilon_0^{\mathrm{p}} < 0.12$ has little effect on the forming limits of the test material.

For $\sigma_x : \sigma_y = 1 : 0$ and $0 : 1$, the measured forming limit strains, stresses, and plastic work per unit volume are significantly larger than the calculated results. This could be attributed to the fact that the material was assumed to follow isotropic hardening for $\varepsilon_0^{\mathrm{p}} > 0.16$ and that the extrapolation of the strain hardening function for that strain range including the diffuse necking phenomenon was erroneous.

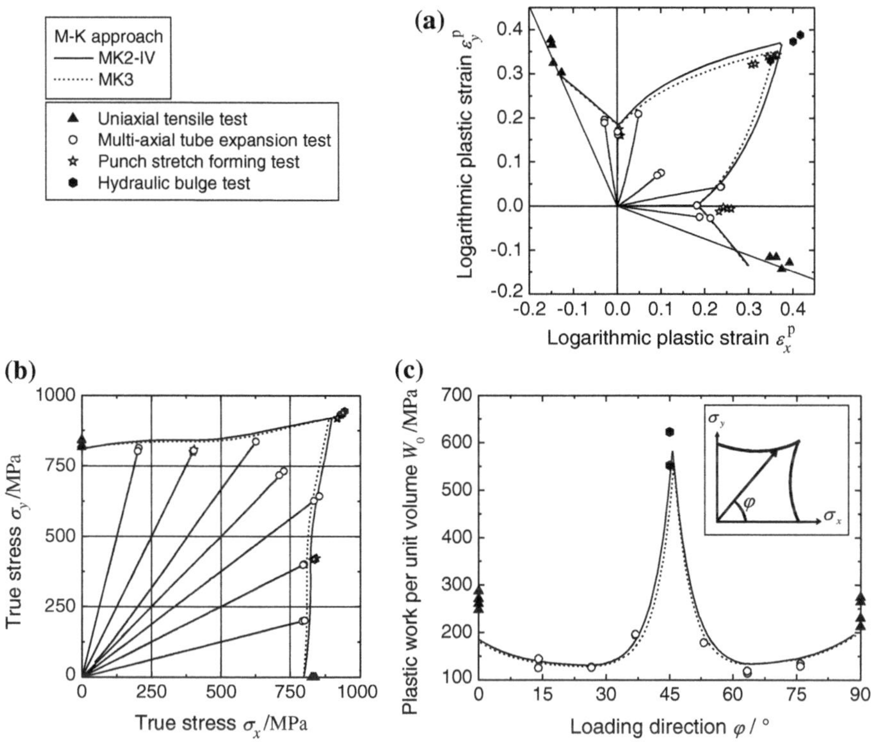

Fig. 4.12 Calculated results obtained from MK2-IV and MK3 models: **a** FLC, **b** FLSC, and **c** FLPW

4.6 Conclusions

A MTET system combined with a digital image correlation (DIC) system as a strain measurement method was developed for the first time. MTET was performed to measure the deformation behavior and forming limits of a high-strength steel sheet subjected to linear stress paths. The isotropic and differential hardening models determined from the MTET results were applied to the forming limit analyses based on the M-K approach, and the effect of the material model on the accuracy of the forming limit predictions was investigated. The conclusions of this study are summarized as follows:

1. The work contours for the test material were successfully measured for a reference plastic strain range of $0.002 \leq \varepsilon_0^p \leq 0.16$. A maximum strain of $\varepsilon_0^p = 0.285$ was attained for $\sigma_x : \sigma_y = 4 : 3$. The test material exhibited significant differential hardening for $\varepsilon_0^p \leq 0.12$, while the hardening behavior was almost isotropic for $\varepsilon_0^p \geq 0.12$; the shapes of the work contours were almost similar.

2. A material modeling method for reproducing the differential hardening using the Yld2000-2d yield function was developed, in which the material parameters $\alpha_i (i = 1 - 8)$ and the exponent M change as functions of $\varepsilon_0^{\mathrm{p}}$.
3. The forming limit strains, stresses, and plastic work per unit volume of the test material were calculated using the M-K approach based on the isotropic and differential hardening models. It is concluded that a yield function that is capable of reproducing both the work contours and the directions of the plastic strain rates measured for a strain range close to the fracture limit can give a more effective constitutive model for accurately predicting the FLC, FLSC, and FLPW than those determined for a smaller strain range.

Acknowledgments The authors are grateful to JFE Steel Co. and Nippon Steel &Sumitomo Metal Co. for providing the test material. This work was supported by the Committee for Standardization of the Testing Methods for the Evaluation of Steel Tube Formability organized by The Iron and Steel Institute of Japan. This work was also supported by a Grant-in-Aid for JSPS Fellows (266118).

Appendix A1

To quantitatively evaluate the difference between the shapes of the theoretical yield loci and the measured work contours, the root mean square error δ_r was calculated using the following equation:

$$\delta_r = \sqrt{\frac{\sum_i \{r'(\varphi_i) - r(\varphi_i)\}^2}{N}}, \tag{4.35}$$

where $\varphi_i (i = 1 \text{ to } N)$ is the loading angle of the ith stress point from the x-axis in the principal stress space, $r(\varphi_i)$ is the distance between the origin of the principal stress space and the ith stress point, and $r'(\varphi_i)$ is the distance between the origin of the principal stress space and the theoretical yield locus along the loading direction φ_i, as shown by the schematic in Fig. 4.13a.

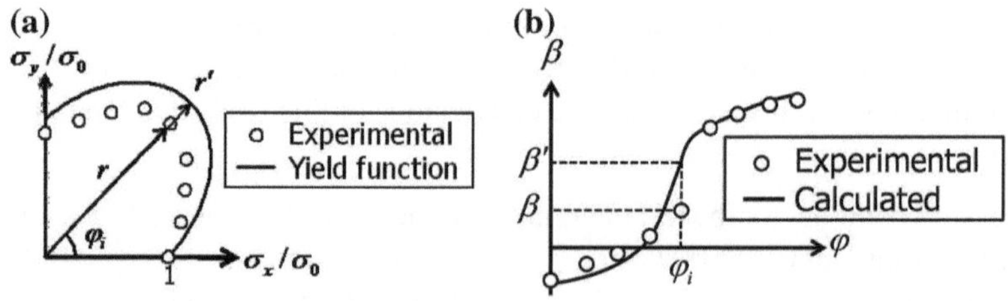

Fig. 4.13 Schematics for the calculation of **a** the root mean square error δ_r of a calculated yield locus from the measured work contour and **b** the root mean square error δ_β of the calculated directions of the plastic strain rates based on the normality flow rule for a calculated yield locus from those measured

To quantitatively evaluate the difference between the measured directions of the plastic strain rates and those predicted using the selected yield functions, the root mean square error δ_β was calculated using the following equation:

$$\delta_\beta = \sqrt{\frac{\sum_i \{\beta'(\varphi_i) - \beta(\varphi_i)\}^2}{N}}, \quad (4.36)$$

where $\beta(\varphi_i)$ is the direction of the plastic strain rate measured for the ith stress path, and $\beta'(\varphi_i)$ is that predicted using a selected yield function for the ith stress path, as shown by the schematic in Fig. 4.13b. In method I, only δ_r was taken into account, and δ_β was not considered.

Appendix A2

In order to formulate the differential hardening model for the test material, the material parameters $\alpha_i (i = 1 - 8)$ and exponent M of the Yld2000-2d yield function were determined as functions of ε_0^p by the following calculation procedures:

1. Determine $\alpha_i (i = 1 - 8)$ and M for the measured work contours corresponding to particular values of ε_0^p by method IV.
2. Approximate the variation in M with ε_0^p using Eq. (4.33).
3. Recalculate $\alpha_i (i = 1 - 8)$ using M determined in 2. by method IV for the measured work contours corresponding to particular values of ε_0^p.
4. Approximate the variations in $\alpha_i (i = 1 - 8)$ with ε_0^p obtained in 3. using Eqs. (4.33) and (4.34).

Appendix A3

In order to verify the computer program of the M-K analysis based on the differential hardening model (MK3), the biaxial stress-strain (s-s) curves for proportional loading were calculated, and the calculated s-s curves were compared with those measured. Figure 4.14 shows the biaxial s-s curves for $\sigma_x : \sigma_y = 2 : 1$ and $3 : 4$. Figure 4.15 shows the variations in the work contours and the directions of the plastic strain rates. The calculated results are in good agreement with the measured data; thus, the M-K analysis program has been validated.

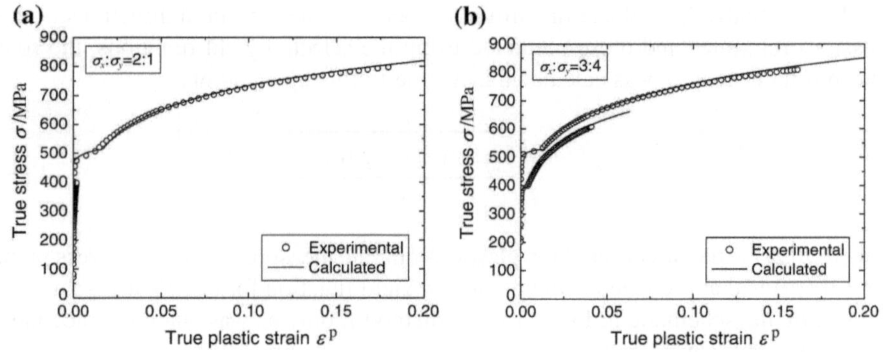

Fig. 4.14 Measured true stress-logarithmic plastic strain curves for $\sigma_x : \sigma_y =$. **a** 2 : 1 and **b** 3 : 4 compared with those calculated using MK3 model

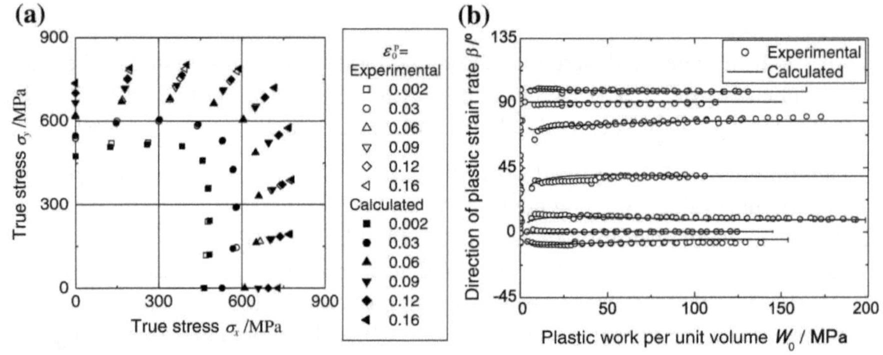

Fig. 4.15 The calculated results obtained from MK3 model compared with the experimental data. **a** Stress points forming the work contours. **b** Directions of $\boldsymbol{D}^p$

Appendix A4

In order to check the accuracy of the strain measurement method using the DIC, the logarithmic thickness plastic strain ε_z^p and the axial curvature R_ϕ^{-1} measured using the DIC were compared with those measured using a micrometer (PMUD150-25MJ, Mitsutoyo Co.) and a 2D contour tester (Contourecord 1700SD3, Tokyo Seimitsu Co.), respectively. The values of R_ϕ^{-1} were approximated using circular arcs for a range of $\pm$13 mm at the mid-section of the tubular specimens. The test material used was a cold-rolled ultralow-carbon steel sheet (SPCG), and the initial thickness was 0.6 mm. The sheet sample was bent and welded using a YAG laser to fabricate tubular specimens with an inner diameter of 44.6 mm, a length of 200 mm, and a gauge length (distance between the grips of the testing machine) of 150 mm.

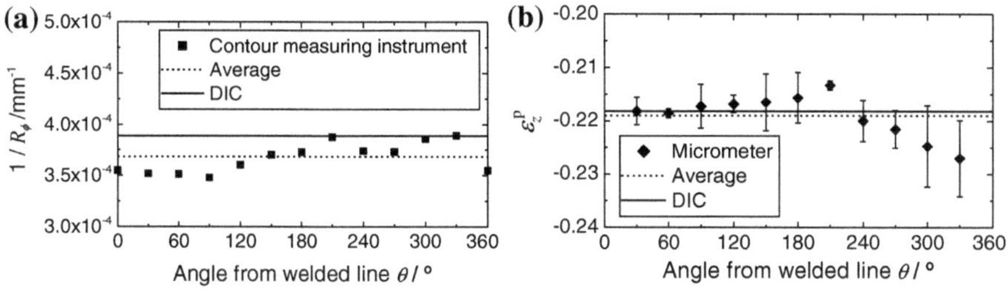

Fig. 4.16 Validation of the accuracy of the DIC measurement: **a** R_{ϕ}^{-1} and **b** $\varepsilon_z^{\mathrm{p}}$

Figure 4.16 compares the measured data for $\sigma_{\phi} : \sigma_{\theta} = 1 : 1$ and $\varepsilon_{\theta} = 0.10$. The relative measurement error was 5.6 % for R_{ϕ}^{-1} and 0.39 % for $\varepsilon_z^{\mathrm{p}}$.

References

Andar MO, Kuwabara T, Steglich D (2012) Material modeling of AZ31 Mg sheet considering variation of r-values and asymmetry of the yield locus. Mater Sci Eng A 549:82–89

Aretz H (2005) A non-quadratic plane stress yield function for orthotropic sheet metals. J Mater Process Technol 168(1):1–9

Aretz H (2008) A simple isotropic-distortional hardening model and its application in elastic-plastic analysis of localized necking in orthotropic sheet metals. Int J Plast 24(9):1457–1480

Banabic D, Dannenmann E (2001) Prediction of the influence of yield locus on the limit strains in sheet metals. J Mater Process Technol 109:9–12

Barlat F, Brem JC, Yoon JW, Chung K, Dick RE, Choi SH, Pourboghrat F, Chu E, Lege DJ (2003) Plane stress yield function for aluminum alloy sheets—part 1: theory. Int J Plast 19(9):1297–1319

Butuc MC, Banabic D, Barata da Rocha A, Gracio JJ, Ferreira Duarte J, Jurco P, Comsa DS (2002) The performance of Yld96 and BBC2000 yield functions in forming limit prediction. J Mater Process Technol 125–126:281–286

Chen MH, Gao L, Zuo DW, Wang M (2007) Application of the forming limit stress diagram to forming limit prediction for the multi-step forming of auto panels. J Mater Process Technol 178–188:173–177

Graf A, Hosford WF (1993) Effect of changing strain paths on forming limit diagrams of Al 2008–T4. Met Trans 24A:2503–2512

Hanabusa Y, Takizawa H, Kuwabara T (2010) Evaluation of accuracy of stress measurements determined in biaxial stress tests with cruciform specimen using numerical method. Steel Res Int 81(9):1376–1379

Hanabusa Y, Takizawa H, Kuwabara T (2013) Numerical verification of a biaxial tensile test method using a cruciform specimen. J Mater Process Technol 213(6):961–970

Hashimoto K, Kuwabara T, Iizuka E, Yoon JW (2010) Effects of anisotropic yield functions on the accuracy of forming simulations of bore expansion (in Japanese). Tetsu-to-Hagané 96:557–563

Hill R (1948) A theory of the yielding and plastic flow of anisotropic metals. Proc Roy Soc London A 193:281–297

Hill R (1993) A user-friendly theory of orthotropic plasticity in sheet metals. Int J Mech Sci 35:19–25

Hill R, Hutchinson JW (1992) Differential hardening in sheet metal under biaxial loading: a theoretical framework. Trans ASME J Appl Mech 59:S1–S9

Hill R, Hecker SS, Stout MG (1994) An investigation of plastic flow and differential work hardening in orthotropic brass tubes under fluid pressure and axial load. Int J Solids Struct 31:2999–3021
Ishiki M, Kuwabara T, Hayashida Y (2011) Measurement and analysis of differential work hardening behavior of pure titanium sheet using spline function. Int J Mater Form 4(2):193–204
Kuroda M, Tvergaard V (2000) Forming limit diagrams for anisotropic metal sheets with different yield criteria. Int J Solids Struct 37(37):5037–5059
Kuwabara T, Ikeda S, Kuroda K (1998) Measurement and analysis of differential work-hardening in cold-rolled steel sheet under biaxial tension. J Mater Process Technol 80–81:517–523
Kuwabara T, Kuroda K, Tvergaard V, Nomura K (2000) Use of abrupt strain path change for determining subsequent yield surface: experimental study with metal sheets. Acta Mater 48:2071–2079
Kuwabara T, van Bael A, Iizuka E (2002) Measurement and analysis of yield locus and work-hardening characteristics of steel sheets with different R-values. Acta Mater 50(14):3717–3729
Kuwabara T, Ishiki M, Kuroda M, Takahashi S (2003) Yield locus and work-hardening behavior of a thin-walled steel tube subjected to combined tension-internal pressure. J de Phys IV 105:347–354
Kuwabara T, Yoshida K, Narihara K, Takahashi S (2005) Anisotropic plastic deformation of extruded aluminum alloy tube under axial forces and internal pressure. Int J Plast 21:101–117
Kuwabara T, Umemura M, Yoshida K, Kuroda M, Hirano S, Kikuta Y (2006) Forming limit strains of 5000 series aluminum alloys with different magnesium contents (in Japanese). J Japan Inst Light Met 56:323–328
Kuwabara T, Hashimoto K, Iizuka E, Yoon JW (2011) Effect of anisotropic yield functions on the accuracy of hole expansion simulations. J Mater Process Technol 211(3):475–481
Kuwabara T, Nakajima T (2011) Material modeling of 980 MPa dual phase steel sheet based on biaxial tensile test and in-plane stress reversal test. J Solid Mech Mater Eng 5(12):709–720
Kuwabara T, Sugawara F (2013) Multiaxial tube expansion test method for measurement of sheet metal deformation behavior under biaxial tension for a large strain range. Int J Plast 45:103–118
Marciniak Z, Kuczyński K (1967) Limit strains in the processes of stretch-forming sheet metal. Int J Mech Sci 9:609–620
Nakazima K, Kikuma T, Hasuka K (1968) Study on the formability of steel sheets. Yawata Technical Rep 264:8517–8530
Peirce D, Shih CF, Needleman A (1984) A tangent modulus method for rate dependent solids. Comp Struct 18:875–887
Savoie J, Jonas JJ, Macewen SR, Perrin R (1995) Evolution of r-value during the tensile deformation of aluminium. Textures Microstruct 23:149–171
Stoughton TB (2000) A general forming limit criterion for sheet metal forming. Int J Plast 42(1): 1–27
Stoughton TB, Yoon JW (2005) Sheet metal formability analysis for anisotropic materials under non-proportional loading. Int J Plast 47(12):1972–2002
Stoughton TB, Yoon JW (2009) Anisotropic hardening and non-associated flow in proportional loading of sheet metals. Int J Plast 25(9):1777–1817
Sumita T, Kuwabara T (2014) Measurement and material modeling of biaxial work-hardening behavior for pure titanium sheet. In: NUMISHEET 2014, Melbourne, January 2014. AIP Conference Proceedings, AIP Publishing, vol 1567, pp 516–519
Wang H, Wan M, Yan Y (2012) Effect of flow stress-strain relation on forming limit of 57540 aluminum alloy. Trans Nonferr Met Soc China 22(10):2370–2378
Xu S, Weinmann KJ (2000) Effect of deformation-dependent material parameters on forming limits of thin sheets. Int J Mech Sci 42(4):677–692
Yanaga D, Kuwabara T, Uema N, Asano M (2012) Material modeling of 6000 series aluminum alloy sheets with different density cube textures and effect on the accuracy of finite element simulation. Int J Solids Struct 49(25):3488–3495
Yanaga D, Takizawa H, Kuwabara T (2014) Formulation of differential work hardening of 6000 series aluminum alloy sheet and application to finite element analysis (in Japanese). J Japan Soc Technol Plast 55(636):55–61

Yoon JW, Barlat F, Dick RE, Chung K, Kang TJ (2004) Plane stress yield function for alumi-num alloy sheets-Part II: FE formulation and its implementation. Int J Plast 20:495–522
Yoshida K, Kuwabara T, Narihara K, Takahashi S (2005) Experimental verification of the path-independence of forming limit stresses. Int J Form Process 8:283–298
Yoshida K, Kuwabara T (2007) Effect of strain hardening behavior on forming limit stresses of steel tube subjected to nonproportional loading paths. Int J Plast 23:1260–1284
Yoshida K, Kuwabara T, Kuroda M (2007) Path-dependence of the forming limit stresses in a sheet metal. Int J Plast 23:361–384
Yoshida K, Suzuki N (2008) Forming limit stresses predicted by phenomenological plasticity theories with anisotropic work-hardening behavior. Int J Plast 24:118–139
Yoshida K, Ishii A, Tadano Y (2014) Work-hardening behavior of polycrystalline aluminum alloy under multiaxial stress paths. Int J Plast 53:17–39
Zamiri A, Pourboghrat F (2007) Characterization and development of an evolutionary yield function for the superconducting niobium sheet. Int J Solids Struct 44:8627–8647
Zimniak Z (2000a) Application of a system for sheet metal forming design. J Mater Process Technol 106:159–162
Zimniak Z (2000b) Implementation of the forming limit stress diagram in FEM simulations. J Mater Process 106:261–266

Chapter 5
Three-Dimensional FE Analysis Using Homogenization Method for Ductile Polymers Based on Molecular Chain Plasticity Model Considering Craze Evolution

Hideyuki Hara and Kazuyuki Shizawa

Abstract Thermoplastic polymers can be classified into glassy polymers and crystalline polymers depending on their internal structures. Glassy polymers have a random coil structure in which molecular chains are irregularly entangled. Crystalline polymers can be regarded as a mixture consisting of glassy and crystalline phases where molecular chains are regularly folded. Moreover, the fracture of ductile polymers occurs at the boundary between regions with oriented and non-oriented molecular chains after neck propagation. This behavior stems from the concentration of craze, which is a type of microscopic damage typically observed in polymers. In this study, three-dimensional FE simulations coupled with a craze evolution equation are carried out for glassy and crystalline polymers using a homogenization method and models of ductile polymers based on crystal plasticity theory. We attempt to numerically represent the propagation of a high-strain-rate shear band and a high-craze-density region in the macroscopic structure and to directly visualize the orientation of molecular chains in glassy and crystalline phases. In addition, differences between the deformation behavior of glassy and crystalline polymers at both the macroscopic and microscopic scales are investigated.

Keywords Ductile polymer · Craze · Homogenization · FEM

H. Hara
Graduate School of Science and Technology, Keio University,
3-14-1 Hiyoshi Kohoku-ku, Yokohama 223-8522, Japan
e-mail: hideyuki@shizawa.mech.keio.ac.jp

K. Shizawa (✉)
Department of Mechanical Engineering, Keio University,
3-14-1 Hiyoshi Kohoku-ku, Yokohama 223-8522, Japan
e-mail: shizawa@mech.keio.ac.jp

H. Altenbach et al. (eds.), *From Creep Damage Mechanics to Homogenization Methods*, Advanced Structured Materials 64,
DOI 10.1007/978-3-319-19440-0_5

5.1 Introduction

Polymers form different internal structures depending on the cooling rate in their production process. In general, polymers tend to form a glassy state at a high cooling rate and a crystalline state at a low cooling rate. Glassy polymers such as polymethyl methacrylate (PMMA) and polycarbonate (P) have a random coil structure in which molecular chains are irregularly entangled as shown in Fig. 5.1a. Crystalline polymers such as polypropylene (PP) and polyethylene (PE) can be regarded as a mixture of glassy and lamellar crystalline phases, where molecular chains are regularly folded in the latter as shown in Fig. 5.1b. A spherulite structure is formed at a lower cooling rate by the radial growth of a twisted lamellar crystalline phase. However, because it is difficult to model such a complicated structure, we employ a mixed structure consisting of glassy and crystalline phases as shown in Fig. 5.1b formed at a middle cooling rate as the internal structure of crystalline polymers. Thermoplastic polymers have many useful properties such as high ductility and a light weight. In particular, composite materials with a polymer matrix are industrially applicable and have been widely used as structural materials in severe mechanical environments involving a large strain. Therefore, it is highly desirable to develop a material model that can precisely express the mechanical deformation responses and fracture behavior peculiar to polymers. On the other hand, it is known that the ductile fracture of polymers occurs along the high-strain-rate shear band at the front edge of a propagating neck, i.e., at the boundary between the region with oriented and non-oriented molecular chains after neck propagation as shown in Fig. 5.2. This distinct behavior in ductile fracture stems from the accumulation of craze, which is a type of microscopic damage typically observed in polymers. A craze, which is different from conventional damage, consists of voids and fibrils, which are aggregates of oriented molecular chains. When stress exceeding the fibril strength is applied to a region where sufficient craze has accumulated, ductile fracture is caused by the breaking of the fibrils.

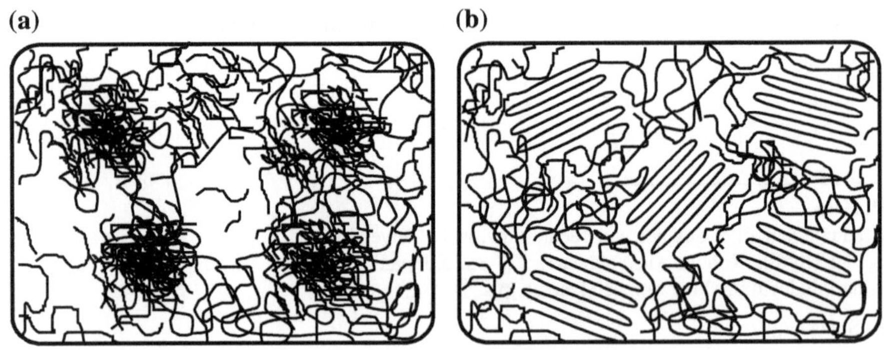

Fig. 5.1 Internal structures of polymers, **a** General glassy polymer, **b** General crystalline polymer

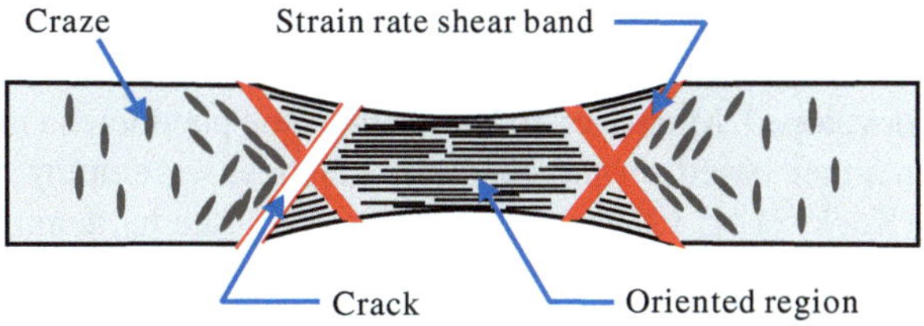

Fig. 5.2 Ductile fracture mechanism of polymers

The mechanical properties of polymers, such as strain softening after yielding, and the formation and propagation of the neck, have conventionally been expressed in terms of the phenomenological theory of plasticity (Tuğcu and Neale 1990; Tomita and Hayashi 1993; Murakami et al. 2002). However, in this theory, it is undesirable that the constitutive equations become complicated when developing a rigorous model. Recently, multiscale models considering the knowledge obtained from the materials science to solid mechanics have been actively studied as a means of describing the mechanical response of polymers more precisely. As a typical example of such attempts, the molecular chain network model (Boyce et al. 1988; Arruda and Boyce 1993; Wu and Giessen 1993; Tomita and Tanaka 1995), which employs Argon's hardening law (Argon 1973) and back stress based on the eight-chain model (Arruda and Boyce 1993), was proposed. However, these models cannot directly express the deformation-induced orientation of molecular chains from information of the microscopic structure because the plastic constitutive equation used is based on J_2-flow theory or J_2-deformation theory (Murakami et al. 2002). On the other hand, one of the authors numerically reproduced crazing behavior with the formation of the high-strain-rate shear band in terms of continuum mechanics and predicted ductile fracture at the front edge of a propagating neck by evaluating the stress applied to fibrils (Takahashi et al. 2010). However, the craze evolution equations proposed in that paper were developed macroscopically. Thus, they should not be applied to a multiscale model such as the molecular chain plasticity model (Nada et al. 2015) used in this study. In our previous paper (Hara and Shizawa 2013), we proposed a multiscale material model for crystalline polymers, referred to as the homogenized molecular chain plasticity model, that homogenizes the mixed structure of the glassy phases expressed by the molecular chain plasticity model and the crystalline phases represented by the conventional crystal plasticity model (Peirce et al. 1983) for metals in unit cell. Moreover, a craze evolution equation for the glassy phase was newly developed in terms of chemical kinetics (Glasstone et al. 1941). The propagation of the craze-concentration region and high-strain-rate shear band in the macroscopic specimen were numerically reproduced using these models. However, these analyses were conducted using simplified two-dimensional slip systems and the plane strain condition. Thus, the material parameters were not suitably determined since the experimental data used to identify the parameters were obtained using a thin plate. In addition, the three-dimensional orientation state of the molecular chains was not visualized. In this paper, the slip systems in the crystal-plasticity-based models and computational models used in finite element (FE) analyses are extended to three-dimensional structures. Orientation parameters, which can express the three-dimensional orientation of molecular chains in the glassy phase, are newly developed. Moreover, to handle inelastic phenomena in ductile polymers in a unified manner based on the chemical kinetics, we employ the thermal activation model (Kocks et al. 1975) instead of the strain rate hardening law of Pan and Rice (1983) for calculation of the inelastic shear strain rate in the crystalline phase. FE simulations using the present model are carried out for ductile polymers. We attempt to computationally represent the propagation of the high-strain-rate shear band and high-craze-density region in a three-dimensional structure and to directly visualize the orientation of molecular chains in the glassy and crystalline phases. Furthermore, by adopting the homogenization method (Ohno et al. 2002; Nakamachi et al. 2007) for unit cells consisting of only the glassy phase or both glassy and crystalline phases, differences between the deformation behaviors of glassy and crystalline polymers at both the macroscopic and microscopic scales are discussed.

5.2 Material Models for Ductile Polymers

In this section, we summarize the material models for glassy and crystalline polymers proposed in our previous paper (Hara and Shizawa 2013). Note that the visualization methods of the orientation of molecular chains in the glassy phase and the slip systems in the crystalline phase are extended to enable three-dimensional analysis, in contrast to in our previous paper.

5.2.1 Configurations with Damage

A fictitious configuration neglecting the change in the cross-sectional area storing the stress is called the pseudo-undamaged configuration, although damage such as craze exists inside the structure. A real configuration that takes account of the change in the cross-sectional area is called the damaged configuration. By using the fraction of the damaged area ω, corresponding to the craze density in this paper, the stress σ^* in the damaged configuration, called the effective stress, and Young's modulus E in the pseudo-undamaged configuration can be expressed as

$$\sigma^* = \sigma/(1-\omega), \tag{5.1}$$

$$E = (1-\omega)E_0, \tag{5.2}$$

where σ denotes the stress in the pseudo-undamaged configuration and E_0 denotes Young's modulus in the damaged configuration, which is a constant. In the following section, constitutive equations are derived by considering the apparent reduction of Young's modulus in the pseudo-undamaged configuration as given by Eq. (5.2).

5.2.2 Model for Glassy Phase

For the glassy phase, the molecular chain plasticity model proposed in our previous paper (Nada et al. 2015) is employed. We replace the chain structure neighboring an entangled point in the glassy phase with an eight-chain model as shown in Fig. 5.3a. Assuming that four molecular chains sharing an entangled point exist on the same plane, the shear deformation of a molecular chain due to the kink rotation of a segment occurs in the diagonal direction of the plane as shown in Fig. 5.3b. Thus, there are four slip systems per unit block. In this case, the molecular chain basis tensor is defined by $\boldsymbol{P}^{(\alpha)} = \boldsymbol{s}^{(\alpha)} \otimes \boldsymbol{m}^{(\alpha)}$, where $\boldsymbol{s}^{(\alpha)}$ and $\boldsymbol{m}^{(\alpha)}$ are the molecular chain basis vectors in the shear direction and the normal direction to the slip plane of slip system α, respectively. Using these vectors, we can express the inelastic deformation rate $\boldsymbol{D}^i$ and inelastic spin $\boldsymbol{W}^i$ as

$$\boldsymbol{D}^i = \sum_\alpha \boldsymbol{D}^{i(\alpha)} = \sum_\alpha \dot{\gamma}^{(\alpha)} \boldsymbol{P}_{\mathrm{S}}^{(\alpha)}, \tag{5.3}$$

$$\boldsymbol{W}^i = \sum_\alpha \boldsymbol{W}^{i(\alpha)} = \sum_\alpha \dot{\gamma}^{(\alpha)} \boldsymbol{P}_{\mathrm{A}}^{(\alpha)}, \tag{5.4}$$

where $(\ldots)_{\mathrm{S}}$ and $(\ldots)_{\mathrm{A}}$ denote the symmetric and anti-symmetric parts of the second-order tensor, respectively, and $\dot{\gamma}^{(\alpha)}$ is the inelastic shear strain rate. On the other hand, the temporal evolutions of the molecular chain basis vectors are given by

$$\dot{\boldsymbol{s}}^{(\alpha)} = \left(\boldsymbol{W} - \boldsymbol{W}^{i(\alpha)}\right) \boldsymbol{s}^{(\alpha)}, \tag{5.5}$$

$$\dot{\boldsymbol{m}}^{(\alpha)} = \left(\boldsymbol{W} - \boldsymbol{W}^{i(\alpha)}\right) \boldsymbol{m}^{(\alpha)}, \tag{5.6}$$

where $\boldsymbol{W}$ denotes the continuum spin. From Eqs. (5.5) and (5.6), each slip system rotates independently in accordance with its own inelastic spin $\boldsymbol{W}^{i(\alpha)}$ not with the continuum spin $\boldsymbol{W}$ common to all slip systems. To express the random coil structure in the glassy phase, a number of slip system sets are assigned at each material point

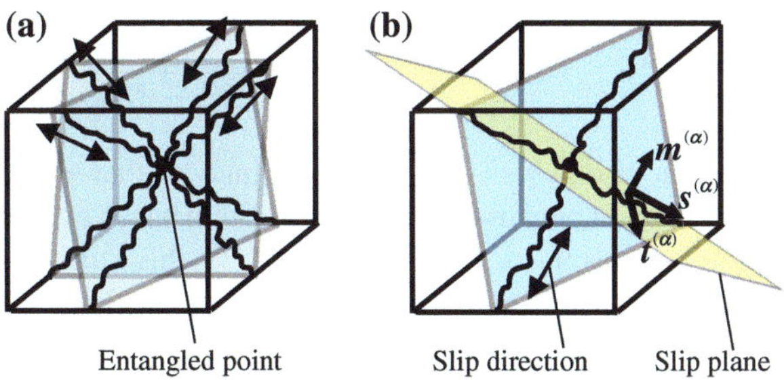

Fig. 5.3 Model of internal structure in glassy phase, **a** Modified eight-chain model, **b** Slip systems

by using the extended Taylor model (Asaro and Needleman 1985). Therefore, the molecular chain plasticity model has the following elasto-viscoplastic constitutive equation for the glassy phase:

$$\frac{1}{N_g}\sum_k \overset{\circ}{\boldsymbol{T}}{}^{[k]} = \frac{1}{N_g}\sum_k \left((1-\omega^{[k]})\boldsymbol{C}^{\mathrm{e}}_0 : \boldsymbol{D} - \sum_\alpha \dot{\gamma}^{(\alpha)[k]}\boldsymbol{\Omega}^{(\alpha)[k]} - \frac{\dot{\omega}^{[k]}}{1-\omega^{[k]}}\boldsymbol{T}^{[k]} \right), \tag{5.7}$$

$$\boldsymbol{\Omega}^{(\alpha)[k]} \equiv (1-\omega^{[k]})\boldsymbol{C}^{\mathrm{e}}_0 : \boldsymbol{P}^{(\alpha)[k]}_{\mathrm{S}} + \boldsymbol{\beta}^{(\alpha)[k]}, \qquad \boldsymbol{\beta}^{(\alpha)[k]} \equiv \boldsymbol{P}^{(\alpha)[k]}_{\mathrm{A}}\boldsymbol{T}^{[k]} - \boldsymbol{T}^{[k]}\boldsymbol{P}^{(\alpha)[k]}_{\mathrm{A}}, \tag{5.8}$$

where $\overset{\circ}{\boldsymbol{T}} \equiv \dot{\boldsymbol{T}} - \boldsymbol{W}\boldsymbol{T} + \boldsymbol{T}\boldsymbol{W}$ is the Jaumann rate of the Cauchy stress $\boldsymbol{T}$, $\boldsymbol{D}$ the total deformation rate, k the number of entangled points, N_g the total number of entangled points and $\boldsymbol{C}^{\mathrm{e}}_0$ is the fourth-order tensor of isotropic elastic moduli.

Subsequently, orientation parameters that can be used to visualize the three-dimensional orientation of molecular chains are newly proposed. The orientation direction is expressed by averaging the directions of all molecular chains as follows:

$$\bar{\boldsymbol{s}} = \frac{\sum_k \sum_\alpha \boldsymbol{s}^{(\alpha)[k]}}{\left\| \sum_k \sum_\alpha \boldsymbol{s}^{(\alpha)[k]} \right\|}. \tag{5.9}$$

To calculate the intensity of the orientation, we consider the average projection of each molecular chain base vector $\boldsymbol{s}^{(\alpha)[k]}$ in the average direction $\bar{\boldsymbol{s}}$ as

$$\hat{s} = \frac{1}{N_g}\frac{1}{N_s}\sum_k \sum_\alpha \bar{\boldsymbol{s}} \cdot \boldsymbol{s}^{(\alpha)[k]}. \tag{5.10}$$

Here, $\hat{s}$ takes a value of 1 when all molecular chains have the same direction and a value of $\hat{s}_{\min}$ in the perfectly isotropic case; thus, the orientation intensity can be expressed as a value from 0 to 1 if we define the orientation intensity parameter Θ as

$$\Theta \equiv \frac{\hat{s} - \hat{s}_{\min}}{1 - \hat{s}_{\min}}. \tag{5.11}$$

Note that Θ and $\bar{\boldsymbol{s}}$ should be determined so that Θ takes a maximum value at each material point because the molecular chain base vector $\boldsymbol{s}^{(\alpha)[k]}$ is identical to $(-\boldsymbol{s}^{(\alpha)[k]})$ in the crystal-plasticity-based model. In this study, $\hat{s}_{\min}$ is set to a constant value regardless of the total number of molecular chain slip systems per material point for convenience. By assuming the existence of consecutive molecular chains in a hemispherical region to be the most isotropic state, $\hat{s}_{\min}$ can be calculated as

$$\hat{s}_{\min} = \frac{1}{\pi^2}\int_0^{\pi}\int_0^{\pi} \bar{\boldsymbol{s}} \cdot \boldsymbol{s}(\theta, \phi)\mathrm{d}\phi\mathrm{d}\theta = \frac{1}{\pi^2}\int_0^{\pi}\int_0^{\pi} (\sin\theta \sin\phi)\mathrm{d}\phi\mathrm{d}\theta = \frac{4}{\pi^2}. \tag{5.12}$$

5.2.3 Model for Crystalline Phase

In this paper, we assume the orthorhombic structure as the internal structure of the crystalline phase for simplicity, although the crystalline phases of crystalline polymers have several types of the crystal structure (Ahzi et al. 1994). Thus, the model of the crystalline phase has eight slip systems consisting of chain slips $\widehat{\alpha} =$ 1–4 and transverse slips $\widehat{\alpha} =$ 5–8 as shown in Fig. 5.4 (Uchida and Tada 2013). The temporal evolution of the crystal basis is represented by the following equations so as to retain the angle between the slip systems in the same way as in conventional crystal plasticity theory:

$$\dot{\widehat{\boldsymbol{s}}}^{(\widehat{\alpha})} = (\widehat{\boldsymbol{W}} - \widehat{\boldsymbol{W}}^{i})\, \widehat{\boldsymbol{s}}^{(\widehat{\alpha})}, \tag{5.13}$$

$$\dot{\widehat{\boldsymbol{m}}}^{(\widehat{\alpha})} = (\widehat{\boldsymbol{W}} - \widehat{\boldsymbol{W}}^{i})\, \widehat{\boldsymbol{m}}^{(\widehat{\alpha})}, \tag{5.14}$$

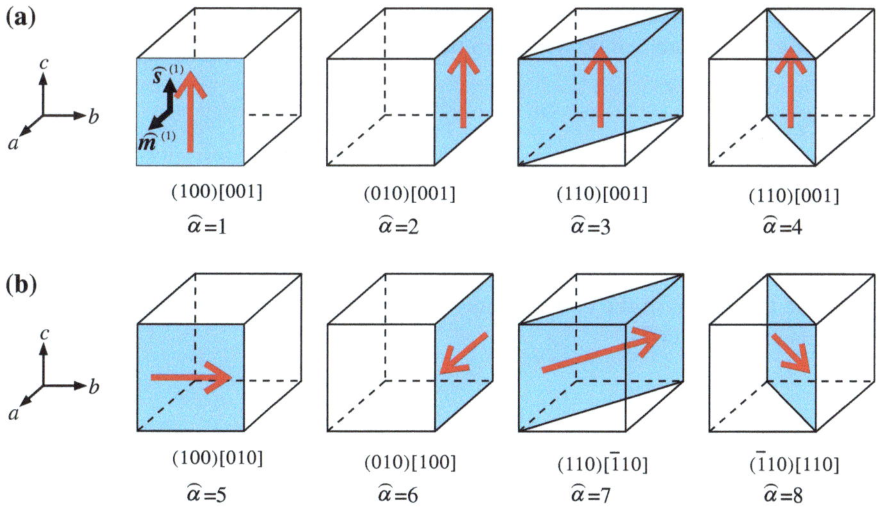

Fig. 5.4 Slip systems in crystalline phase (c-axis denotes molecular chain direction), **a** Chain slip, **b** Transverse slip

where $(\widehat{\ldots})$ denotes the quantity of the crystalline phase. The crystalline phase is inextensible in the chain direction. This property, expressed as $\widehat{\boldsymbol{S}}^{(1)} \cdot \widehat{\boldsymbol{D}} = 0$, is approximately represented by using a penalty method (Uchida and Tada 2013). Then, the penalty constant ψ_0 is introduced into the elastic modulus tensor in the constitutive equation of crystal plasticity theory. Furthermore, by considering $\omega = 0$ in the crystalline phase and rewriting the corotational rate of the Cauchy stress as the Jaumann rate, we obtain the constitutive equation for the crystalline phase as

$$\overset{\circ}{\widehat{\boldsymbol{T}}} = (\widehat{\boldsymbol{C}}{}^{\mathrm{e}}_{0} + \psi_0 \widehat{\boldsymbol{S}}^{(1)} \otimes \widehat{\boldsymbol{S}}^{(1)}) : \widehat{\boldsymbol{D}} - \sum_{\alpha} \dot{\widehat{\gamma}}^{(\alpha)} \widehat{\boldsymbol{\Omega}}_0^{(\alpha)}, \tag{5.15}$$

$$\widehat{\boldsymbol{\Omega}}_0^{(\alpha)} \equiv \widehat{\boldsymbol{C}}{}^{\mathrm{e}}_{0} : \widehat{\boldsymbol{P}}_S^{(\alpha)} + \widehat{\boldsymbol{\beta}}^{(\alpha)}, \tag{5.16}$$

where $\widehat{\boldsymbol{S}}^{(1)}$ is defined by $\widehat{\boldsymbol{S}}^{(1)} = \widehat{\boldsymbol{s}}^{(1)} \otimes \widehat{\boldsymbol{s}}^{(1)}$.

5.2.4 Homogenization Method

We suppose that the material velocity $\boldsymbol{v}$ is decomposed into $\boldsymbol{v}^0$ and $\boldsymbol{v}^\dagger$ such that $\boldsymbol{v} = \boldsymbol{v}^0 + \boldsymbol{v}^\dagger$, where $(\ldots)^0$ and $(\ldots)^\dagger$ denote the macroscopic component and microscopic disturbance component, respectively. In addition, $\boldsymbol{v}^\dagger$ for a heterogeneous microstructure can be written in the form

$$v_i^\dagger = -\chi_i^{kl} D_{kl}^0 + \phi_i, \tag{5.17}$$

where χ_i^{kl} and ϕ_i are the characteristic velocities and D_{kl}^0 is the macroscopic deformation rate. Noting that there is a characteristic velocity ϕ_i originating from the strain rate dependence, i.e., the viscoplasticity in Eq. (5.17), the homogenized-crystal-plasticity constitutive equation in the macroscopic structure can be derived in the same manner as in the conventional homogenization method (Ohno et al. 2002; Nakamachi et al. 2007).

5.3 Material Response Law

5.3.1 Inelastic Response Law for Glassy Phase

To calculate the inelastic shear strain rate $\dot{\gamma}^{(\alpha)}$ in Eq. (5.7), we employ the inelastic response law based on the change in the local free volume proposed by Hasan and Boyce (1995). Introducing a probability density function $\varphi^{(\alpha)}$ for the activation energy ΔF_0, Hasan and Boyce expressed the local distribution of the high-free-volume-fraction region that can be shear deformed and formulated the inelastic shear strain rate as

$$\begin{aligned}\dot{\gamma}^{(\alpha)} = \dot{\gamma}_{0H}\,\mathrm{sgn}(\tau^{(\alpha)}) \int_0^{\infty} \phi^{(\alpha)} \Bigg[& \exp\left(\frac{-\Delta F_0\{1-|\tau^{(\alpha)}/\tau_{0H}|\}}{k_{\mathrm{B}}\theta}\right) \\ & - \exp\left(\frac{-\Delta F_0\{1+|\tau^{(\alpha)}/\tau_{0H}|\}+S^{(\alpha)}}{k_{\mathrm{B}}\theta}\right)\Bigg] \mathrm{d}\Delta F_0,\end{aligned} \tag{5.18}$$

where $\dot{\gamma}_{0H}$ denotes the reference strain rate, k_{B} the Boltzmann constant, θ the ambient temperature, $\tau^{(\alpha)}$ the resolved shear stress, τ_{0H} the critical strength and $S^{(\alpha)}$ is the molecular-chain friction energy, which affects the nonlinear strain recovery in the unloading process. Moreover, $\phi^{(\alpha)}$ in Eq. (5.18) is expressed as

$$\phi^{(\alpha)} = \begin{Bmatrix} \dfrac{1}{2A_1^{(\alpha)}} \exp\left(\dfrac{\Delta F_0 - a^{(\alpha)}}{\alpha^{(\alpha)}}\right) \sin\left(\dfrac{\Delta F_0 - a^{(\alpha)}}{\alpha^{(\alpha)}}\right) & \\ \qquad (a^{(\alpha)} \le \Delta F_0 \le a^{(\alpha)} + 3\pi\alpha^{(\alpha)}/4) & \\ \dfrac{1}{2A_2^{(\alpha)}} \exp\left(-\dfrac{\Delta F_0 - a'^{(\alpha)}}{\alpha^{(\alpha)}}\right) \sin\left(\dfrac{\Delta F_0 - a'^{(\alpha)}}{\alpha^{(\alpha)}}\right) & \\ \qquad (a^{(\alpha)} + 3\pi\alpha^{(\alpha)}/4 \le \Delta F_0 \le a^{(\alpha)} + 3\pi\alpha^{(\alpha)}/2) & \\ 0 \qquad (\text{otherwise}) & \end{Bmatrix}, \tag{5.19}$$

where $a^{(\alpha)}$ and $\alpha^{(\alpha)}$ are the minimum value and standard deviation of the probability distribution of the activation energy, respectively. Their temporal equations are expressed as

$$\dot{a}^{(\alpha)} = -(a^{(\alpha)} - a_{\mathrm{eq}}) \exp[-\zeta \exp(-\zeta \gamma_a)] \sum_{\beta} \varpi^{(\beta)}, \tag{5.20}$$

$$\dot{\alpha}^{(\alpha)} = -(\alpha^{(\alpha)} - \alpha_{\mathrm{eq}}) \sum_{\beta} \varpi^{(\beta)}, \tag{5.21}$$

where a_{eq} and α_{eq} are the equilibrium values of $a^{(\alpha)}$ and $\alpha^{(\alpha)}$, respectively, ζ is a material parameter and $\gamma_a \equiv \sum_{\alpha} |\gamma^{(\alpha)}|$. Using the frequency factor ϖ_0, $\varpi^{(\alpha)}$ is expressed as $\varpi^{(\alpha)} = (\varpi_0/\dot{\gamma}_{0H})|\dot{\gamma}^{(\alpha)}|$. In addition, $A_1^{(\alpha)}$, $A_2^{(\alpha)}$ and $a'^{(\alpha)}$ in Eq. (5.19) are calculated as

$$A_1^{(\alpha)} = \frac{\alpha^{(\alpha)}}{2}\left[1 + \sqrt{2}\exp\left(\frac{3\pi}{4}\right)\right] = \mathrm{e}^{\pi} A_2^{(\alpha)}, \tag{5.22}$$

$$a'^{(\alpha)} = a^{(\alpha)} + \frac{\pi\alpha^{(\alpha)}}{2}. \tag{5.23}$$

5.3.2 Hardening Law for Crystalline Phase

To handle the inelastic phenomena of polymers in a unified manner in terms of the chemical kinetics, we adopt the following thermal activation model (Kocks et al. 1975) for the calculation of $\dot{\widehat{\gamma}}^{(\widehat{\alpha})}$ in Eq. (5.15), which can specifically express the dependence of the yield stress on the ambient temperature differently from the strain rate hardening law of Pan and Rice (1983) employed in Hara and Shizawa (2013). The inelastic shear strain rate $\dot{\widehat{\gamma}}^{(\widehat{\alpha})}$ is expressed as

$$\dot{\widehat{\gamma}}^{(\widehat{\alpha})} = \dot{\gamma}_{0E}\,\mathrm{sgn}(\widehat{\tau}^{(\widehat{\alpha})})\exp\left(\frac{-\Delta\widehat{F}_0\{1-|\widehat{\tau}^{(\widehat{\alpha})}/\tau_{0E}^{(\widehat{\alpha})}|\}}{k_{\mathrm{B}}\theta}\right), \tag{5.24}$$

where $\dot{\gamma}_{0E}$ denotes the reference strain rate and $\tau_{0E}^{(\widehat{\alpha})}$ is the shear strength for each slip system. Conventionally, material constants p and q are introduced in the term $\{1-|\widehat{\tau}^{(\widehat{\alpha})}/\tau_{0E}^{(\widehat{\alpha})}|^p\}^q$. In this paper, p and q are set to 1 and are not specified as variables for simplicity. Moreover, we assume the relation $\tau_{0E}^{(\widehat{\alpha})} = \kappa g^{(\widehat{\alpha})}$ between $\tau_{0E}^{(\widehat{\alpha})}$ and the flow stress $g^{(\widehat{\alpha})}$, where κ is a material parameter (see Sect. 5.5.2.1). The flow stress $g^{(\widehat{\alpha})}$ is assumed to be constant for each slip system.

5.4 Craze Evolution Equation

We adopt the craze evolution equation proposed in Hara and Shizawa (2013) for the glassy phase. The craze density rate in each slip system $\dot{\omega}^{(\alpha)}$ is decomposed into a craze nucleation rate $\dot{\omega}_{\mathrm{n}}^{(\alpha)}$ and craze growth rate $\dot{\omega}_{\mathrm{g}}^{(\alpha)}$, such that $\dot{\omega}^{(\alpha)} = \dot{\omega}_{\mathrm{n}}^{(\alpha)} + \dot{\omega}_{\mathrm{g}}^{(\alpha)}$. Also, $\dot{\omega}_{\mathrm{n}}^{(\alpha)}$ and $\dot{\omega}_{\mathrm{g}}^{(\alpha)}$ are expressed in terms of the chemical kinetics as follows:

$$\dot{\omega}_{\mathrm{n}}^{(\alpha)} = \dot{\omega}_{0\mathrm{n}}^{(\alpha)}\int_0^\infty \phi^{(\alpha)}(\Delta F_0)\exp\left(\frac{-\Delta G_1(\Delta F_0)}{k_{\mathrm{B}}\theta}\right)\mathrm{d}\Delta F_0, \tag{5.25}$$

$$\dot{\omega}_{\mathrm{g}}^{(\alpha)} = \dot{\omega}_{0\mathrm{g}}^{(\alpha)}\int_0^\infty \phi^{(\alpha)}(\Delta F_0)\exp\left(\frac{-\Delta G_2(\Delta F_0)}{k_{\mathrm{B}}\theta}\right)\mathrm{d}\Delta F_0, \tag{5.26}$$

where ΔG_1 and ΔG_2 denote the activation energies required for craze nucleation and growth, respectively, which are given as

$$\Delta G_i = C_i \Delta F_0 (1 - |\tau^{(\alpha)}/\tau_{0H}|) + D_i (\gamma_a/\gamma_c)^{p_i}, \tag{5.27}$$

where C_i, D_i and p_i $(i = 1, 2)$ are material parameters and γ_c is the oriented-hardening strain. It is assumed that the craze nucleation rate is proportional to the area of the remaining region where the craze can nucleate, given as the difference between the saturated craze nucleation density $\omega_{\rm ns}^{(\alpha)}$ and the present craze nucleation density $\omega_{\rm n}^{(\alpha)}$. It is also assumed that the craze growth rate is proportional to the craze nucleation density $\omega_{\rm n}^{(\alpha)}$. Therefore, the reference craze nucleation rate $\dot{\omega}_{\rm 0n}^{(\alpha)}$ and growth rate $\dot{\omega}_{\rm 0g}^{(\alpha)}$ in Eqs. (5.25) and (5.26), respectively, can be represented as

$$\dot{\omega}_{\rm 0n}^{(\alpha)} = n_r < \omega_{\rm ns}^{(\alpha)} - \omega_{\rm n}^{(\alpha)} >, \tag{5.28}$$

$$\dot{\omega}_{\rm 0g}^{(\alpha)} = g_r \omega_{\rm n}^{(\alpha)}, \tag{5.29}$$

where n_r and g_r are material constants and $< \ldots >$ are the Macaulay brackets, which give a value of zero when $(\omega_{\rm ns}^{(\alpha)} - \omega_{\rm n}^{(\alpha)})$ takes a negative value. By modifying the equation for the saturated craze number density proposed by Kitagawa and Kawagoe (1978) to make it applicable to our craze evolution model, the saturated craze nucleation density $\omega_{\rm ns}^{(\alpha)}$ is expressed as

$$\omega_{\rm ns}^{(\alpha)} = \omega^* \int_0^\infty \phi^{(\alpha)}(\Delta F_0) \exp\left(\frac{-\Delta G_1(\Delta F_0)}{k_{\rm B}\theta}\right) {\rm d}\Delta F_0, \tag{5.30}$$

where ω^* denotes the total area density where the craze can nucleate.

5.5 FE Simulation and Discussion

5.5.1 Analysis with Specimen Model for Glassy Polymer

We conduct an FE simulation for a glassy polymer using the molecular chain plasticity model and the craze evolution equation described in Sects. 5.2.2 and 5.4, respectively. Note that the homogenization method is not used in this section.

5.5.1.1 Computational Model for Glassy Polymer

In our computational model, we consider a PMMA plate whose length L is 60.0 mm, initial mean width is W and thickness is H as shown in Fig. 5.5a. The aspect ratio is

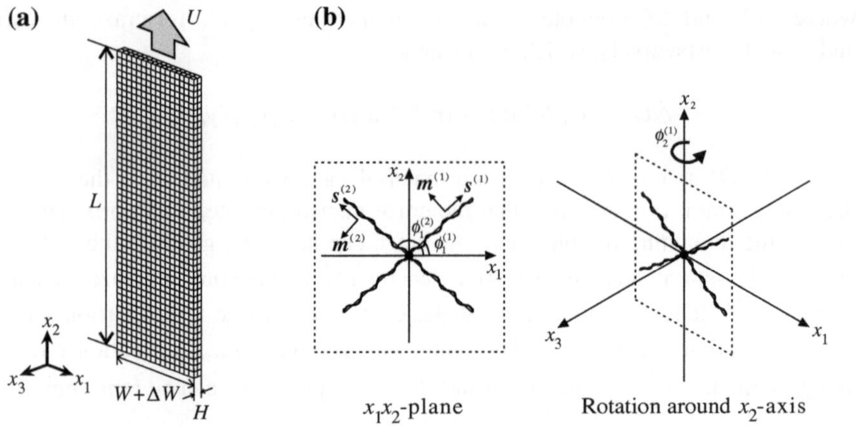

Fig. 5.5 Analysis condition for glassy polymer, **a** Mesh of glassy polymer specimen, **b** Determination process of initial direction of molecular chain base vectors

$L/W = 3$ and $H/W = 0.1$. The variable ΔW is the initial geometrical imperfection, which expresses the geometrical heterogeneity of the material and is written in the form (Tvergaard et al. 1981)

$$\Delta W = W[-W_1 \cos(\pi x_2/L) + W_2 \cos(m_w \pi x_2/L)], \tag{5.31}$$

where $W_1 = 0.00375$ and $W_2 = 0.00150$ are the amplitudes of the imperfection and $m_w = 4.00$ is the number of waves. The boundary conditions of the displacement u_i are given as follows:

$$\begin{cases} u_1 = u_2 = u_3 = 0 \text{ at } x_1 = x_2 = x_3 = 0 \\ u_2 = 0 \quad \text{at} \quad x_2 = 0 \\ u_2 = U \quad \text{at} \quad x_2 = L \end{cases} \tag{5.32}$$

The strain rate and ambient temperature are fixed to $\dot{U}/L = 0.550\,\text{ks}^{-1}$ and $\theta = 338\,\text{K}$, respectively, which are the same values as the experiment (Inoue and Yamamoto 1982). Eight-node hexagonal elements are used and the number of elements is 1536. Forty sets of slip systems are assigned to each integration point using the extended Taylor model (Asaro and Needleman 1985).

The initial directions of molecular chain base vectors are determined by the following procedure. First, $\phi_1^{(1)}$ and $\phi_1^{(2)}$ are given so that slip system 1 is symmetric with respect to slip system 2 about the x_2-axis in the x_1x_2-plane as shown in Fig. 5.5b. Next, slip systems 1 and 2 are rotated equally around the x_2-axis by the angle $\phi_2^{(1)} = \phi_2^{(2)}$. The directions of slip systems 3 and 4 are determined by the same process as for slip systems 1 and 2. The angles $\phi_1^{(1)}$ and $\phi_1^{(3)}$ are given by uniformly distributed random numbers in the range of 10°–70°. The angle $\phi_2^{(1)}$ is a uniformly

Table 5.1 Material constants and numerical parameters used in glassy phase

Elastic parameters	$E_0 = 700\,\text{MPa}$, $\nu_0 = 0.350$
Inelastic response law of Hasan and Boyce	$\dot{\gamma}_{0\text{H}} = 15.0\,\text{ns}^{-1}$, $\tau_{0\text{H}} = 42.5\,\text{MPa}$, $\varpi = 600\,\text{ns}^{-1}$, $\zeta = 5.00$ $a_0 = 1.06\,\text{eV}$, $a_{\text{eq}} = 1.00\,\text{eV}$, $\alpha_0 = 0.0435\,\text{eV}$, $\alpha_{\text{eq}} = 1.00\,\text{eV}$
Craze evolution equation	$n_{\text{r}} = 10.0\,\text{ns}^{-1}$, $g_{\text{r}} = 1.00\,\text{ps}^{-1}$, $\gamma_{\text{c}} = 0.800$, $\omega^* = 1.00$ $C_1 = C_2 = 0.500$, $D_1 = D_2 = 0.0252\,\text{eV}$, $p_1 = p_2 = 50.0$

distributed random number in the range of 0°–90°, and $\phi_2^{(3)} = \phi_2^{(1)} + 90°$. Note that the same set of initial directions is assigned for each integration point to make the initial state homogeneous. The material constants used here are listed in Table 5.1, where ν_0 denotes Poisson's ratio and a_0 and α_0 are the initial values of $a^{(\alpha)}$ and $\alpha^{(\alpha)}$, respectively. Moreover, $S^{(\alpha)}$ in Eq. (5.18) is set to 0.00 eV for all analyses because reverse-loading analysis is not conducted in this study.

5.5.1.2 Computational Results for Glassy Polymer

Figure 5.6 shows experimental data of the uniaxial tension for PMMA obtained by Inoue and Yamamoto (1982) and the numerically obtained nominal stress-strain curve. We can confirm the validity of the present models since the numerical stress-strain curve closely fits the experimental plots. Moreover, the distributions of the equivalent inelastic strain rate, those of the craze density and the orientation state of molecular chains at 30, 50 and 70 % elongations are shown in Fig. 5.7. Note that the material constants ϖ_0 and a_{eq} used in Fig. 5.7 have been modified to $\varpi_0 = 1.20$ ps^{-1} and $a_{\text{eq}} = 0.860\,\text{eV}$, respectively, so that the neck formation after yielding is relatively strengthened. In Fig. 5.7c, the length and color of the line segment denote the orientation intensity Θ, and the direction of the line segment expresses the orientation direction $\bar{\boldsymbol{s}}$. First, we can see from Fig. 5.7a that the neck forms in the center

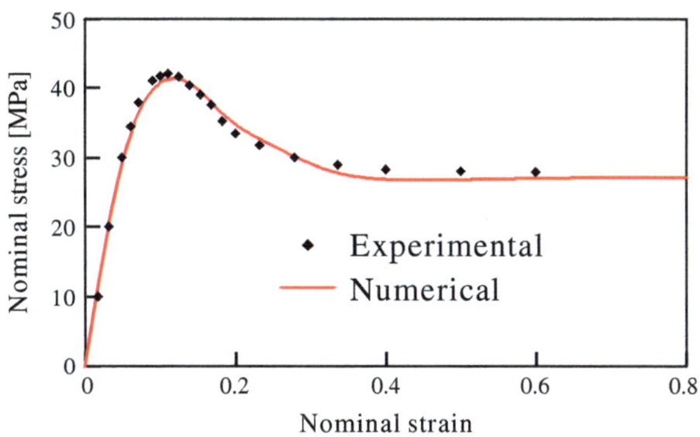

Fig. 5.6 Experimental plots and numerically obtained nominal stress-strain curve obtained by FE simulation for PMMA using present model for glassy phase and material constants shown in Table 5.1

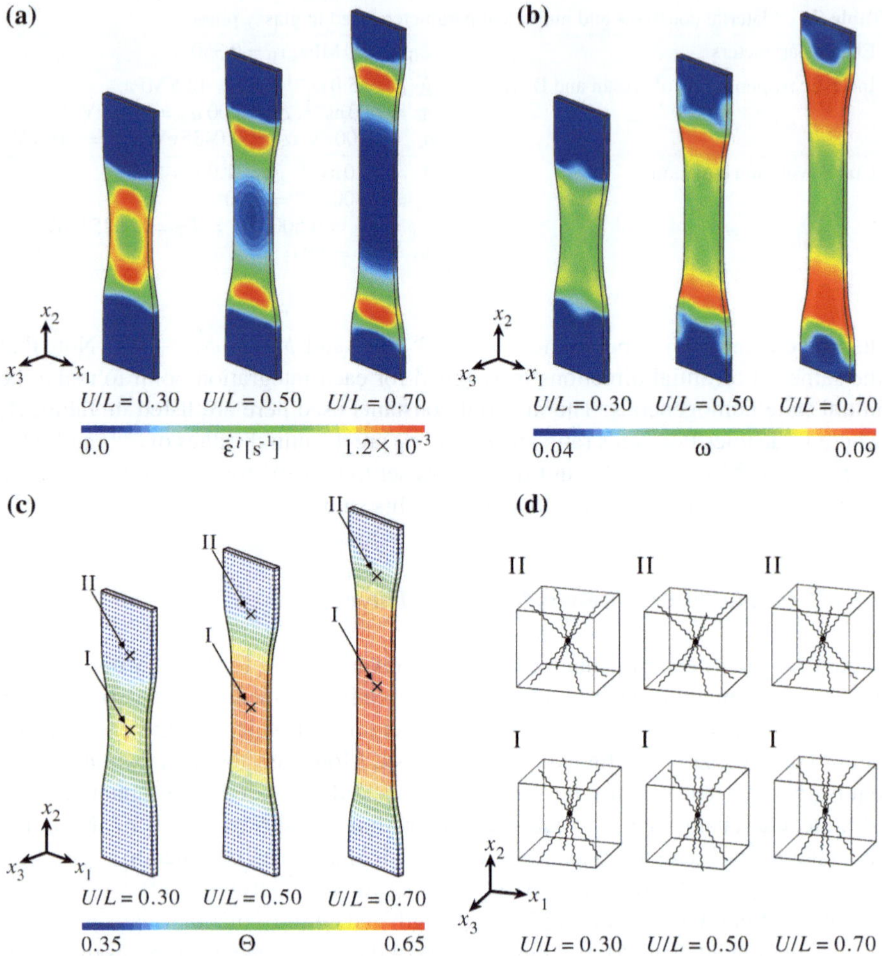

Fig. 5.7 Computational results for glassy polymer, **a** Distributions of equivalent inelastic strain rate, **b** Distributions of craze density, **c** Orientation of molecular chains, **d** Direction of molecular chains in eight-chain model at sampling points I and II

region of the specimen at 30 % elongation and propagates in the tensile direction with the high-strain-rate shear band. The region with high craze density also propagates with increasing the craze density shown in Fig. 5.7b. Furthermore, the propagation of the high-strain-rate shear band coincides with the propagation of the region with high craze density. Comparing Fig. 5.7c, d, we find that the directions of the molecular chains approach each other in the eight-chain model at sampling point I in the specimen, where the orientation intensity is high. On the other hand, the directions of the molecular chains are almost unchanged from the initial state in the eight-chain model at point II, located in a region of the specimen with a low orientation intensity.

Therefore, the anisotropy of the molecular chains can be expressed by Eq. (5.11), which is newly proposed in this paper. In addition, the orientation of molecular chains in the necking region can be observed in Fig. 5.7c. Consequently, the characteristic deformation and crazing behaviors of ductile polymers can be clarified by three-dimensional analysis similar to the two-dimensional analysis conducted in Hara and Shizawa (2013).

5.5.2 Single-Phase Analysis for Crystalline Phase

In this section, by conducting FE simulations for the crystalline phase of a crystalline polymer using the crystal plasticity model described in Sect. 5.2.3, the orientations of the molecular chains and the stress-strain responses are investigated.

5.5.2.1 Computational Model for Crystalline Phase

We assume a microscopic region in the crystalline phase of PP undergoing uniaxial tension and apply a periodic boundary condition to the analytical region as shown in Fig. 5.8. The analytical region is cubic and divided into eight elements by eight-node hexagonal elements. A constant deformation rate of $D^0_{22} = 1.00\,\mathrm{s}^{-1}$ is applied to the analytical region. The ambient temperature θ is set to 293 K to represent room temperature. Furthermore, the initial directions of the crystal base vectors in the crystalline phase are determined by the Euler angle as follows:

$$\overset{\frown}{\boldsymbol{s}}^{(\overset{\frown}{\alpha})} = \boldsymbol{R}_{y_1}\boldsymbol{R}_{y_2}\boldsymbol{R}_{y_3}\,\overset{\frown}{\boldsymbol{s}}_0^{(\overset{\frown}{\alpha})}, \tag{5.33}$$

$$\overset{\frown}{\boldsymbol{m}}^{(\overset{\frown}{\alpha})} = \boldsymbol{R}_{y_1}\boldsymbol{R}_{y_2}\boldsymbol{R}_{y_3}\,\overset{\frown}{\boldsymbol{m}}_0^{(\overset{\frown}{\alpha})}, \tag{5.34}$$

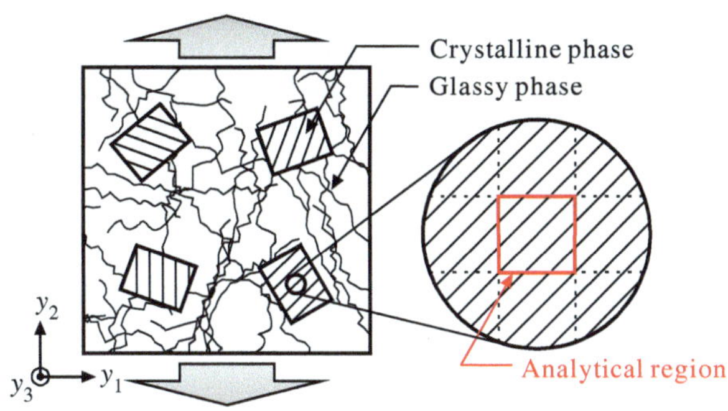

Fig. 5.8 Schematic diagram of computational model for crystalline phase (mixed lamellar structure and a microscopic region in the crystalline phase)

where R_{y_i} denotes the rotation matrix around the y_i-axis, and $\widehat{\boldsymbol{s}_0}^{(\alpha)}$ and $\widehat{\boldsymbol{m}_0}^{(\alpha)}$ are the crystal base vectors when the a, b and c axes of the crystal lattice coincide with the y_1, y_2 and y_3 axes of the microstructure, respectively.

Here, it is desirable that the material parameters $\Delta \widehat{F}_0$ and κ in Eq. (5.24) are determined from the material parameters in the strain rate hardening law by Pan and Rice (1983), which is described as

$$\dot{\widehat{\gamma}}^{(\alpha)} = \dot{\gamma}_{0P} \operatorname{sgn}(\widehat{\tau}^{(\alpha)}) \left| \frac{\widehat{\tau}^{(\alpha)}}{g^{(\alpha)}} \right|^{\frac{1}{m}}, \tag{5.35}$$

where $\dot{\gamma}_{0P}$ denotes the reference strain rate and m is the strain rate sensitivity. We consider the case in which the resolved shear stress takes a positive value in the following formulation for convenience. Assuming that the resolved shear stress $\widehat{\tau}^{(\alpha)}$ in Eqs. (5.24) and (5.35) takes the same value and that the strain rate sensitivities $(\ln \widehat{\tau}^{(\alpha)} / \ln \dot{\widehat{\gamma}}^{(\alpha)})$ of Eqs. (5.24) and (5.35) are identical, we can derive the following relations:

$$\kappa = \left(\frac{\dot{\widehat{\gamma}}^{(\alpha)}}{\dot{\gamma}_{0P}} \right)^m \left[1 - m \ln \left(\frac{\dot{\widehat{\gamma}}^{(\alpha)}}{\dot{\gamma}_{0E}} \right) \right], \tag{5.36}$$

$$\Delta \widehat{F}_0 = \kappa \frac{k_B \theta}{m} \left(\frac{\dot{\widehat{\gamma}}^{(\alpha)}}{\dot{\gamma}_{0P}} \right)^{-m}. \tag{5.37}$$

Substituting the material parameters $\dot{\gamma}_{0P}$ and m into Eqs. (5.36) and (5.37), we can obtain κ and $\Delta \widehat{F}_0$ corresponding to the substituted parameters. Note that $\dot{\widehat{\gamma}}^{(\alpha)}$ is set to $\dot{\gamma}_{0P}$ for all slip systems and that θ denotes the temperature observed during the experimental calibration of m. Therefore, $\dot{\gamma}_{0E}$ is the only arbitrary value in the above formulation. In this paper, $\dot{\gamma}_{0P} = 1.00\,\mathrm{s}^{-1}$, $m = 0.09$ and $\theta = 343\,\mathrm{K}$ are substituted into Eqs. (5.36) and (5.37) on the basis of the paper by G'sell et al. (1997), and $\dot{\gamma}_{0E}$ is set to $15.0\,\mathrm{ns}^{-1}$. The material parameters used for the crystalline phase are summarized in Table 5.2.

5.5.2.2 Computational Results for Crystalline Phase

Figure 5.9 illustrates numerically obtained true stress-strain curves whose initial crystal basis vectors are rotated around only one axis, where ϕ_{y_i} denotes the rotation angle around the y_i-axis. In the numerical results for rotation around the y_1-axis, i.e., Fig. 5.9a, the stress increases sharply after yielding in all analyses, whereas the stress has a constant value after strain hardening in Fig. 5.9c. In Fig. 5.9b, the stress takes a constant value immediately after yielding regardless of the initial direction of the crystal basis.

To investigate the relation between these stress-strain responses and the direction of molecular chains, we conducted further analyses with the initial directions of crystal base vectors that allow the easy visualization of molecular chains. The numerical stress-strain curves for the initial directions $(\phi_{y_1}, \phi_{y_2}, \phi_{y_3}) = (90°, 135°, 0°)$ and $(0°, 90°, 60°)$ are shown in Fig. 5.10a, b, respectively. From Fig. 5.10a, we can see that the molecular chains are oriented in the tensile direction, i.e., the y_2-axis direction during plastic deformation and that the stress rapidly increases owing to the orientation of the molecular chains. This is caused by a decrease in the resolved shear stress for the chain slip systems, and the penalty constant facilitates a sharp increase in the stress. On the other hand, In Fig. 5.10b, the molecular chains are rotated around the direction of the molecular chains, i.e., the c-axis direction of the crystal lattice, because the slip system that can be most easily activated among the transverse slip systems, i.e., slip system 6 as shown in Table 5.2, is oriented in the tensile direction. After the orientation, the slip continuously occurs by activation of slip systems 7

Table 5.2 Material constants and numerical parameters used in crystalline phase

Elastic parameters	$\widehat{E}_0 = 4.00\,\text{GPa}$, $\widehat{\nu}_0 = 0.330$, $\psi_0 = 100\,\text{GPa}$
Thermal activation model	$\dot{\gamma}_{0E} = 15.0\,\text{ns}^{-1}$, $\Delta \widehat{F}_0 = 1.02\,\text{eV}$, $\kappa = 3.10$
Flow stress [MPa] (G'sell et al. 1997)	$g^{(1)} = 12.8$, $g^{(2)} = 4.00$, $g^{(3)} = 13.4$, $g^{(4)} = 13.4$ $g^{(5)} = 12.8$, $g^{(6)} = 4.00$, $g^{(7)} = 13.4$, $g^{(8)} = 13.4$

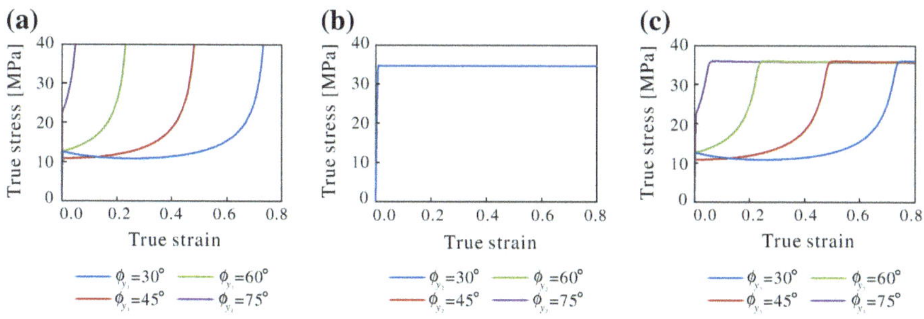

Fig. 5.9 Numerical stress-strain responses in crystalline phase with various initial directions of crystal base vectors rotated around only one axis, **a** Rotated around y_1-axis ($\phi_{y_2} = \phi_{y_3} = 0°$), **b** Rotated around y_2-axis ($\phi_{y_3} = \phi_{y_1} = 0°$), **c** Rotated around y_3-axis ($\phi_{y_1} = \phi_{y_2} = 0°$)

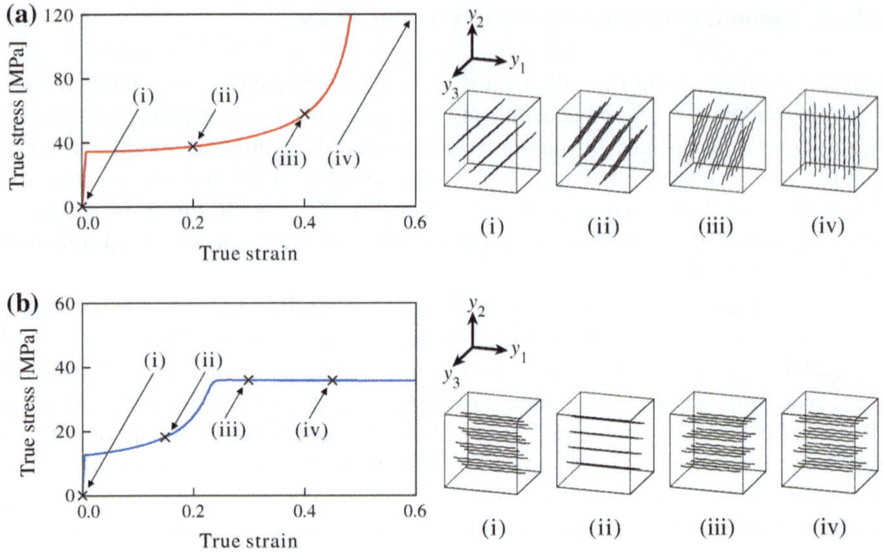

Fig. 5.10 Typical orientation modes of crystalline phase under uniaxial tension (*left* true stress-strain curve, *right* orientation state of molecular chains), **a** Mode 1: primary slip system of chain slip is oriented in tensile direction, **b** Mode 2: primary slip system of transverse slip is oriented in tensile direction

and 8, which are still subject to the resolved shear stress. It results in a plastic flow state with a constant stress after the hardening. Therefore, we can conclude that the orientation behavior of the crystalline phase, which has an orthorhombic crystal structure, can be classified into two modes depending on whether its primary slip system becomes a chain slip or transverse slip. This is the main difference from the analysis using the two-dimensional slip system in our previous paper. The stress-strain responses in Fig. 5.9a results from orientation mode 1, as shown in Fig. 5.10a, and those in Fig. 5.9b, c are caused by orientation mode 2 as shown in Fig. 5.10b. Although the crystal base vectors are only rotated around the y_2-axis, the chain direction is perpendicular to the tensile direction regardless of the angle of rotation, similarly to in Fig. 5.10b-(iii), (iv). Thus, a hardening stage does not appear, and the stress always takes a constant value during plastic deformation in Fig. 5.9b.

5.5.3 Multiscale Analysis for Ductile Polymers

In this section, we carry out multiscale FE simulations for ductile polymers using all the models mentioned in Sects. 5.2–5.4 including the homogenization method. Unit cells consisting of only the glassy phase or both the glassy and crystalline phases are employed as microscopic structures to discuss an effect of the crystalline phase on deformation behaviors in macroscopic and microscopic structures.

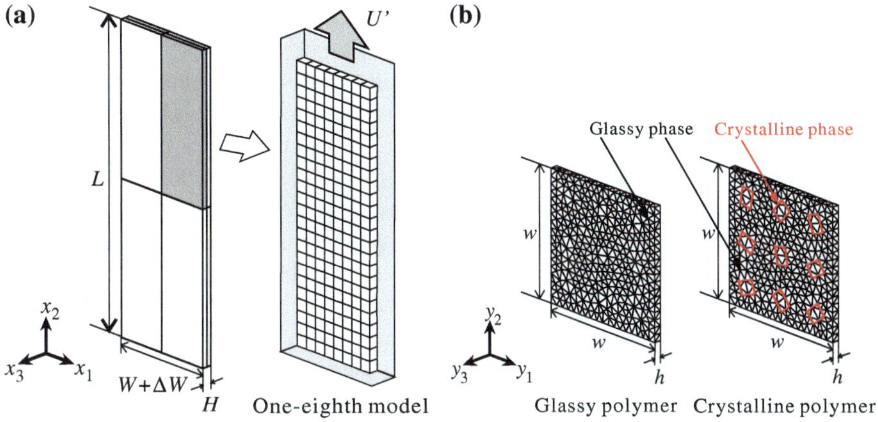

Fig. 5.11 Computational models of macro-micro coupling analysis, **a** Computational model and mesh of macroscopic specimen, **b** Mesh of unit cell for glassy polymer and crystalline polymer

5.5.3.1 Computational Model for Ductile Polymers

We assume a PP plate, which has the same size and aspect ratio as the computational model in Sect. 5.5.1.1 as shown in Fig. 5.11a. Here, only one-eighth of the specimen model is calculated assuming symmetric deformation to reduce the computational cost. A constant strain rate of $2\dot{U}'/L = 1.00\,\mathrm{s}^{-1}$ is applied to the edge of the specimen. Eight-node hexagonal elements are used and the number of elements is 192. As the microscopic structure, we assume a square plate of width $w = 0.900\,\mu\mathrm{m}$ and thickness $h = 0.0500\,\mu\mathrm{m}$ as shown in Fig. 5.11b. A periodic boundary condition and the macroscopic deformation rate $\boldsymbol{D}^0$ obtained by macroscopic analysis are applied to the unit cells. The unit cell comprising only the glassy phase is used to analyze a glassy polymer, and the unit cell including the nine crystalline phases in the glassy matrix is employed in the analysis of a crystalline polymer. Each unit cell is divided into 2096 four-node tetrahedron elements. The ambient temperature θ is set to 293 K. One entangled point is assigned to each integration point in the glassy phase of both polymers since the microscopic heterogeneity is satisfactorily expressed by introducing the homogenization method. The initial directions of the molecular chain base vectors are determined in the same way as in Sect. 5.5.1.1. Note that the angles $\phi_1^{(1)}$ and $\phi_1^{(3)}$ are given by uniformly distributed random numbers in the range of 10°–75°, differently from in Sect. 5.5.1.1. The Euler angles $(\phi_{y1}, \phi_{y2}, \phi_{y3})$ for the crystalline phases are given by random numbers in the range of 0°–180° that are distributed uniformly at intervals of 10°. The material constants used for the glassy phases are given in Table 5.1 but with the modified values $E_0 = 1.80\,\mathrm{GPa}$, $\nu_0 = 0.330$, $\tau_0 = 19.0\,\mathrm{MPa}$, $\varpi = 1.00\,\mathrm{ps}^{-1}$ and $a_{eq} = 0.900\,\mathrm{eV}$, and for the crystalline phases, the values given in Table 5.2 are used.

5.5.3.2 Computational Results for Ductile Polymers

The numerically obtained nominal stress-strain curves for the glassy polymer and crystalline polymer are shown with experimental plots for PP (Murakami et al. 2002) in Fig. 5.12. It is found that the present model accurately predicts the experimental results for the crystalline polymer. The level stress after strain softening of the crystalline polymer is higher than that for the glassy polymer because the flow stress of the crystalline phase is greater than that of the glassy phase. The distributions of the equivalent inelastic strain rate and the craze density in the glassy polymer and crystalline polymer at 25 % elongation are exhibited in Figs. 5.13 and 5.14, respectively. Note that the computational results for the macroscopic specimen are expressed as those for the full-specimen model by copying the one-eighth model. In the glassy polymer, the high-strain-rate shear band and craze-concentration region propagate in the tensile direction, although those in the crystalline polymer become difficult to

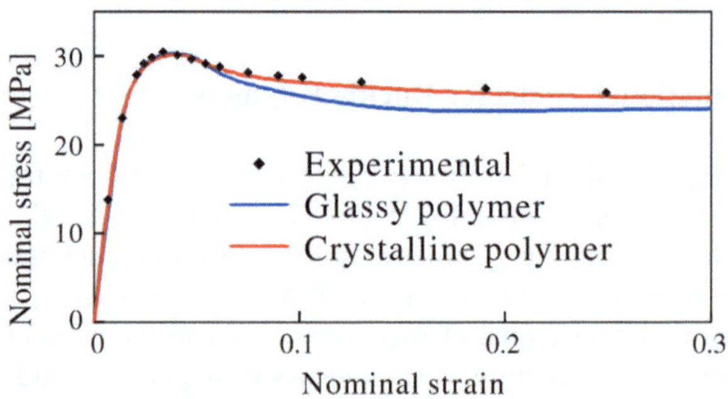

Fig. 5.12 Experimental plots and numerical nominal stress-strain curves obtained by FE simulation for PP using present model for glassy polymer and crystalline polymer

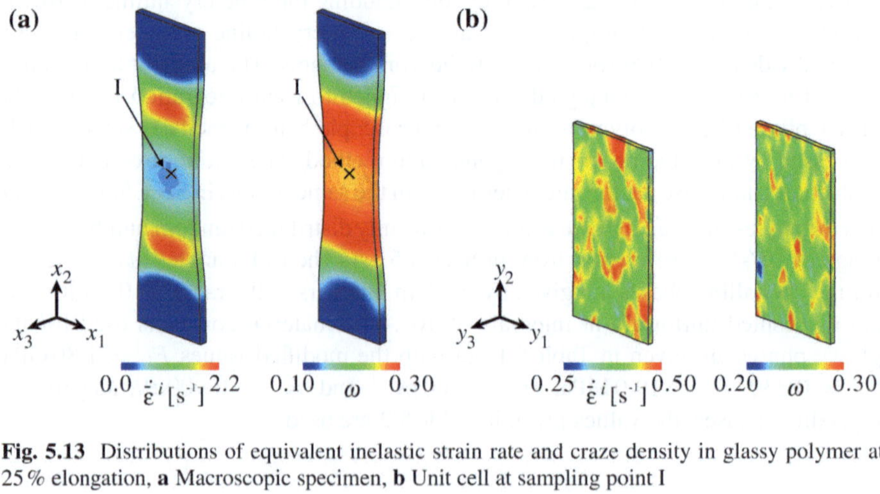

Fig. 5.13 Distributions of equivalent inelastic strain rate and craze density in glassy polymer at 25 % elongation, **a** Macroscopic specimen, **b** Unit cell at sampling point I

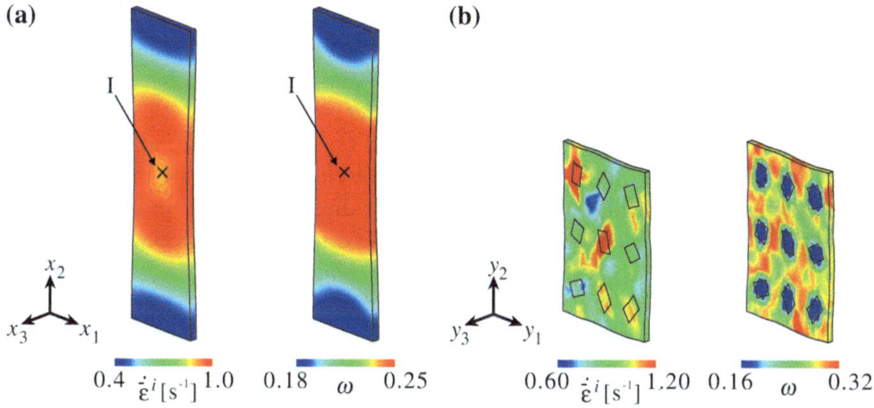

Fig. 5.14 Distributions of equivalent inelastic strain rate and craze density in crystalline polymer at 25 % elongation, **a** Macroscopic specimen, **b** Unit cell at sampling point I

localize, corresponding to the increase in the level stress shown in Fig. 5.12. Moreover, from the distribution of the equivalent inelastic strain rate in the unit cell of the glassy polymer, we can confirm that the inhomogeneous inelastic deformation, i.e., the micro-shear band, appears differently from the computational results in Sect. 5.5.1.2. In the crystalline polymer, microscopic deformation is concentrated at some of the crystalline phases since the crystalline phases become softer in the tensile direction than the glassy phase depending on the initial crystal direction. On the other hand, the craze density in the unit cell is not clearly distributed in the glassy phase, while the craze density becomes zero in the crystalline phases. Therefore, the craze density develops so that the craze threads between the crystalline phases in the crystalline polymer.

5.6 Conclusions

In this paper, we employed the molecular chain plasticity and crystal plasticity models for the glassy and crystalline phases in ductile polymers, respectively. The scalar quantities for inelastic phenomena including craze evolution are expressed on the basis of the chemical kinetics. We developed a material model that can handle the deformation and damage behaviors of ductile polymers in the unified manner of the crystal-plasticity-based model and chemical kinetics, and the model was extended to have three-dimensional slip systems. Three-dimensional FE simulations using the present model were carried out for ductile polymers under a uniaxial tensile condition. The conclusions obtained in this study are summarized as follows.

1. Characteristic behaviors in ductile polymers under uniaxial tension, i.e., the formation and propagation of a neck with a high-strain-rate shear band and a high-craze-density region with increasing a craze density, can be reproduced by the three-dimensional analysis of a glassy polymer.
2. By performing the single-phase analysis of a crystalline phase, we find that the orientation behavior of molecular chains can be classified into two modes, i.e., the orientation of molecular chains in the tensile direction, resulting in rapid strain hardening, and the orientation of perpendicular direction to molecular chains, i.e., a-axis direction of a crystal lattice, in the tensile direction, yielding a constant stress.
3. As a result of the multiscale simulations using the homogenization method that couples a macroscopic specimen with unit cells assuming glassy and crystalline polymers, it is suggested that the deformation rate and craze density of the crystalline polymer are difficult to localize in the macroscopic specimen compared with those of the glassy polymer owing to inclusion of crystalline phases.

References

Ahzi S, Lee B, Asaro R (1994) Plasticity and anisotropy evolution in crystalline polymers. Mater Sci Eng Ser A 189:35–44

Argon A (1973) A theory for the low-temperature plastic deformation of glassy polymers. Philos Mag 28(4):839–865

Arruda E, Boyce M (1993) A three-dimensional constitutive model for the large stretch behavior of rubber elastic materials. J Mech Phys Solids 41(2):389–412

Asaro R, Needleman A (1985) Texture development and strain hardening in rate dependent polycrystals. Acta Metallurgica 33(6):923–953

Boyce M, Parks D, Argon A (1988) Large inelastic deformation of glassy polymers: Part I—Rate dependent constitutive model. Mech Mater 7(1):15–33

Glasstone S, Laider K, Eyring H (1941) The theory of rate processes—the kinetics of chemical reactions, viscosity. Diffusion and electrochemical phenomena. McGraw-hill, New York

G'sell C, Dahoun A, Favier V, Hiver J, Philippe M, Canova G (1997) Microstructure transformation and stress-strain behavior of isotactic polypropylene under large plastic deformation. Polymer Eng Sci 37(10):1702–1711

Hara H, Shizawa K (2013) Modeling of craze evolution equation based on chemical kinetics and craze evolution simulation for crystalline polymer using homogenization method of molecular chain plasticity (in Japanese). Trans Jpn Soc Mech Eng Ser A 79(807):1604–1619

Hasan O, Boyce M (1995) A constitutive model for the nonlinear viscoelastic viscoplastic behavior of glassy polymers. Polymer Eng Sci 35(4):331–344

Inoue T, Yamamoto I (1982) Tensor expression for crazing of polymer with inelastic deformation and its characteristics. Trans Jpn Soc Mech Eng Ser A 48(426):221–229 (in Japanese)

Kitagawa M, Kawagoe M (1978) Statistical aspect of crazes in PC and PVC. J Soc Mater Sci Jpn 27(301):995–1000 (in Japanese)

Kocks U, Argon A, Ashby M (1975) Thermodynamics and kinetics of slip. Progress in materials science. Pergamon Press, Oxford

Murakami D, Kobayashi S, Torigaki T, Shizawa K (2002) A thermomechanical modeling and simulation of viscoplastic large deformation behavior for polymeric materials [2nd report, Vertex model based on flow rule and its finite element analysis, in Japanese]. Trans Jpn Soc Mech Eng Ser A 68(668):682–689

Nada H, Hara H, Tadano Y, Shizawa K (2015) Molecular chain plasticity model similar to crystal plasticity theory based on change in local free volume and FE simulation of glassy polymer. Int J Mech Sci 93:120–135
Nakamachi E, Tam N, Morimoto H (2007) Multi-scale finite element analyses of sheet metals by using SEM-EBSD mesured crystallographic RVE models. Int J Plast 23:450–489
Ohno N, Okumura D, Noguchi H (2002) Microscopic symmetric bifurcation condition of cellular solids based on a homogenization theory of finite deformation. J Mech Phys Solids 50:1125–1153
Pan J, Rice J (1983) Rate sensitivity of plastic flow and implications for yield-surface vertices. Int J Solids Struct 19(11):973–987
Peirce D, Asaro R, Needleman A (1983) Material rate dependence and localized defor-mation in crystalline solids. Acta Metallurgica 31(12):1951–1976
Takahashi J, Yamamoto T, Shizawa K (2010) Modeling and simulation for ductile fracture prediction of crystalline polymer based on craze behavior. Int J Mech Sci 52(2):266–276
Tomita Y, Hayashi K (1993) Thermo-elasto-viscoplastic deformation of polymeric bars under tension. Int J Solids Struct 30(2):225–235
Tomita Y, Tanaka S (1995) Prediction of deformation behavior of glassy polymers based on molecular chain network model. Int J Solids Struct 32(23):3423–3434
Tuğcu P, Neale K (1990) Cold drawing of polymers with rate and temperature dependent properties. Int J Mech Sci 32(5):405–416
Tvergaard V, Needleman A, Lo K (1981) Flow localization in the plane strain tensile test. J Mech Phys Solids 29(2):115–142
Uchida M, Tada N (2013) Micro-, meso- to macroscopic modeling of deformation behavior of semi-crystalline polymer. Int J Plast 49:164–184
Wu P, van der Giessen E (1993) On improved network models for rubber elasticity and their applications to orientation hardening in glassy polymers. J Mech Phys Solids 41(3):427–456

Chapter 6
Inelastic Deformation and Creep-Fatigue Life of Plate-Fin Structures

Toshihide Igari, Fumiko Kawashima, Yorikata Mizokami and Nobutada Ohno

Abstract The equivalent-homogeneous-solid concept, i.e., the homogenization approach, is often applied to inelastic analysis and creep-fatigue life prediction for plate-fin heat exchangers such as those used in high-temperature gas cooled reactors. The applicability of this approach to actual plate-fin structures manufactured by means of brazing still remains to be clarified. Firstly in this paper, basic features of the inelastic behavior of ultra-fine plate-fin structures with offset were numerically clarified by homogenization FEM analysis for a unit cell model. Secondly, considering the weakest loading direction, uniaxial tensile, creep and creep-fatigue tests on small-size plate-fin structures made of SUS316 were carried out at 600 °C. Lastly, thermal fatigue tests using the partial model of heat exchangers were performed under cyclic loading of temperature distribution. The results obtained can be summarized as follows. The macroscopic stress-strain behavior and macroscopic strength of plate-fin structures were successfully predicted by the homogenization approach. Fatigue failure, on the other hand, is sensitive to local structures such as the brazing filler metal shape in plate-fins, and experiments are necessary to obtain the fatigue strength reduction of plate-fin structures. Thermal fatigue life prediction based on anisotropic elastic-plastic FEM analysis yielded good results showing predicted life of 2,100 cycles, or 60 % of the observed life, thus within a factor of 2.

T. Igari (✉)
Mitsubishi Heavy Industries, Ltd., Nagasaki 851-0392, Japan
e-mail: toshihide_igari@mhi.co.jp

F. Kawashima
Department of Advanced Technology, Kumamoto University,
Kumamoto 860-8555, Japan
e-mail: kawashima@mech.kumamoto-u.ac.jp

Y. Mizokami
Mitsubishi Heavy Industries, Ltd., Kobe 652-8585, Japan
e-mail: yorikata_mizokami@mhi.co.jp

N. Ohno
Department of Mechanical Science and Engineering, Nagoya University,
Chikusa-ku, Nagoya 464-8603, Japan
e-mail: ohno@mech.nagoya-u.ac.jp

H. Altenbach et al. (eds.), *From Creep Damage Mechanics to Homogenization Methods*, Advanced Structured Materials 64,
DOI 10.1007/978-3-319-19440-0_6

Keywords Plate-fin · Homogenization · Inelasticity · Anisotropy · Thermal fatigue · HTGR · Recuperative heat exchanger

6.1 Introduction

When undertaking the structural design of a plate-fin-type heat exchanger such as for recuperative heat exchangers (RHX) used in high-temperature gas cooled reactors (HTGR) (Ishiyama et al. 2001; Kawashima et al. 2007), the equivalent-homogeneous-solid concept, i.e., the homogenization approach, is necessary for modeling large-scale heat exchangers with numerous periodic local structures. Figure 6.1 shows the outline of the equivalent-homogeneous-solid concept applied to the structural design of a plate-fin structure. Both the macroscopic cross section area and the macroscopic length of the equivalent-homogeneous-solid are the same as those of a unit cell of the periodical structure. The strength of the base material of the plate and fin is the base line for strength evaluation. In the structural design taking into account primary loads such as internal pressure, strength reduction in tensile and creep loading in the plate-fin structure is considered by multiplying the "stress magnification factor" with the macroscopic stress. With respect to thermal fatigue life prediction for cyclic loading of temperature distribution, fatigue strength reduction in plate-fin structures is considered by multiplying the "strain range magnification factor" with the macroscopic strain range from structural analysis based

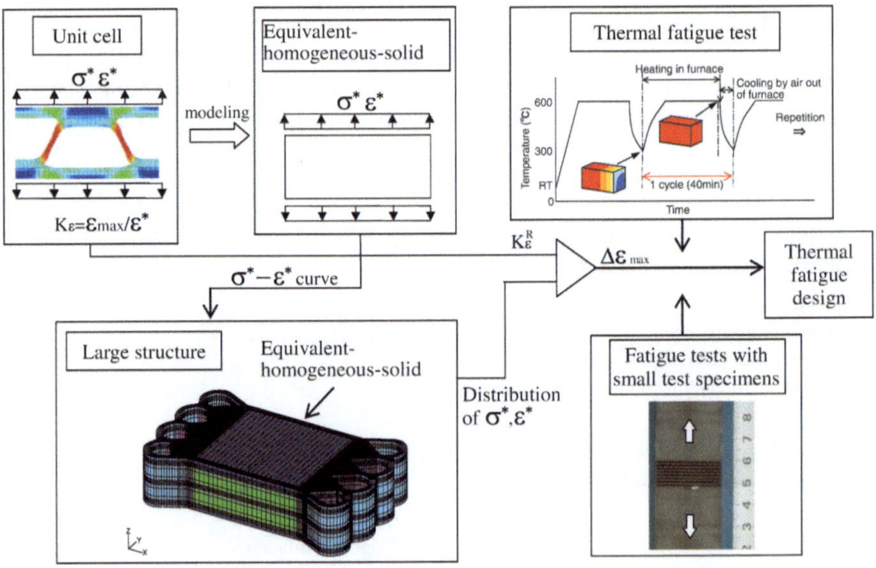

Fig. 6.1 Outline of the equivalent-homogeneous-solid concept

on "homogenization modeling". The essential features of homogenization modeling, the stress magnification and strain concentration factors, can be studied by FEM analysis for the unit cell model. Experimental verification is required because the local configuration of brazing filler material can have an influence on deformation and failure of plate-fin structures.

The equivalent-homogeneous solid concept, i.e., the homogenization approach, was itself originally adopted for the structural design of perforated structures with triangular patterns of circular holes in heat exchangers, such as elastic-plastic design procedure (O'Donnell and Porowski 1982), creep deformation and creep rupture life prediction (Igari et al. 1986, 2001) and application to thermal fatigue life prediction (Tokiyoshi et al. 2001; Igari et al. 2003).

Regarding the application of the homogenization approach to plate-fin structures, on the other hand, several works taking both experimental and the numerical approaches have been published in the last decade, such as pressure burst tests of small-size core models (Ishiyama et al. 2001), high-temperature strength behavior (Kawashima et al. 2007; Jiang et al. 2008), fabrication process by means of brazing (Mizokami et al. 2010) and high-temperature structural design procedure (Mizokami et al. 2013). In the homogenization analysis in previous works (Kawashima et al. 2007), both isotropy and the conventional assumption of volume-incompressibility in macroscopic inelastic deformation were incorporated. Recently, a material model was proposed considering both anisotropy in macroscopic power-law creep and volume-compressibility in macroscopic creep strain (Tsuda et al. 2010; Ohno et al. 2012), based on the numerical simulation of an idealized configuration of a unit cell of ultra-fine plate-fin structures without offset (the meaning of offset is discussed in Sect. 6.3).

Firstly in this paper, basic features of elastic-plastic behavior of an idealized ultra-fine plate-fin structure with offset were clarified by homogenization FEM analysis, such as multi-axial anisotropy of macroscopic deformation and stress-strain concentration. Secondly, uniaxial tensile, creep and creep-fatigue tests were carried out at 600 °C, using small-size plate-fin specimens of SUS316 (JIS SUS316 equivalent to ASME Type 316 stainless steel) to clarify the actual strength characteristics of the brazed structure. Lastly, a thermal fatigue test using a structural model with a 100 mm cubic core corresponding to the plate-fin was undertaken to obtain the crack initiation life under a cyclically loaded temperature gradient. Inelastic analysis for the thermal fatigue test was performed to predict fatigue life, based on homogenization modeling for plate-fin structures.

6.2 Outline of a Plate-Fin Heat Exchanger

Figure 6.2 shows the outline of the planning design (Mizokami et al. 2010) of RHX for HTGR whose core is composed of a periodic structure of SUS316 plate and fin. During steady state operation, the maximum temperature of helium gas is 600 °C, and the pressure difference between the two counter flows is 4.7 MPa. Fins are manufactured

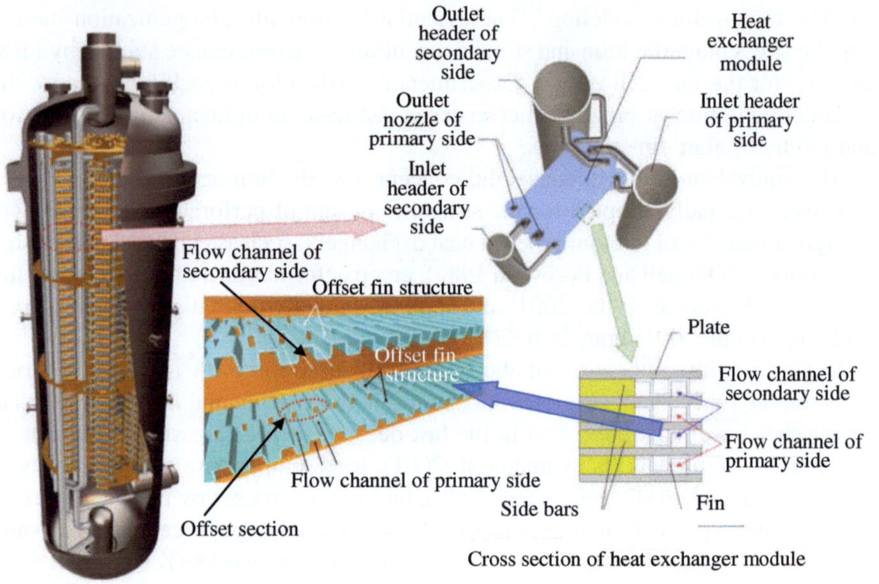

Fig. 6.2 Outline of the plate-fin heat exchanger

by means of a forming press from a flat plate with a thickness of 0.2 mm; the thickness of the pressed fins varies from 0.15 to 0.2 mm; the height is 1.0 mm; the pitch, i.e., half the distance between the tops of the neighboring fins, is 1.2 mm; the length of fins in the flow direction is 5.0 mm; and there is an offset of 5.0 mm for neighboring fins in the flow direction to increase the heat transfer performance.

Both the flat plates with a thickness of 0.5 mm and the fins are vertically stacked, and are brazed into periodic plate-fin structures in vacuum conditions in an electric furnace. The melting point of the brazing filler metal is 1080 °C. During the brazing, a pressure of 0.2 MPa is loaded towards the stacking direction on the work to avoid incomplete bonding.

6.3 Elastic-Plastic Homogenization Analysis of Unit Cell Model

An example of a unit cell model, considering the periodic repetition of configuration, is shown in Fig. 6.3a. In this figure, two layers of plate and fin are modeled, where the thickness of plate is 0.5 mm, and the height and thickness of fin are 1.0 and 0.2 mm, respectively. The three dimensional size of the model is set as follows: the width x_1 is 2.4 mm, i.e., double the pitch of fin; the height y_1 is 3 mm, i.e. two fins and two plates, where the upper and bottom plates are of half thickness in consideration of symmetry; and the length z_1 is 10 mm, i.e., two pitches of offset length. Figure 6.3b

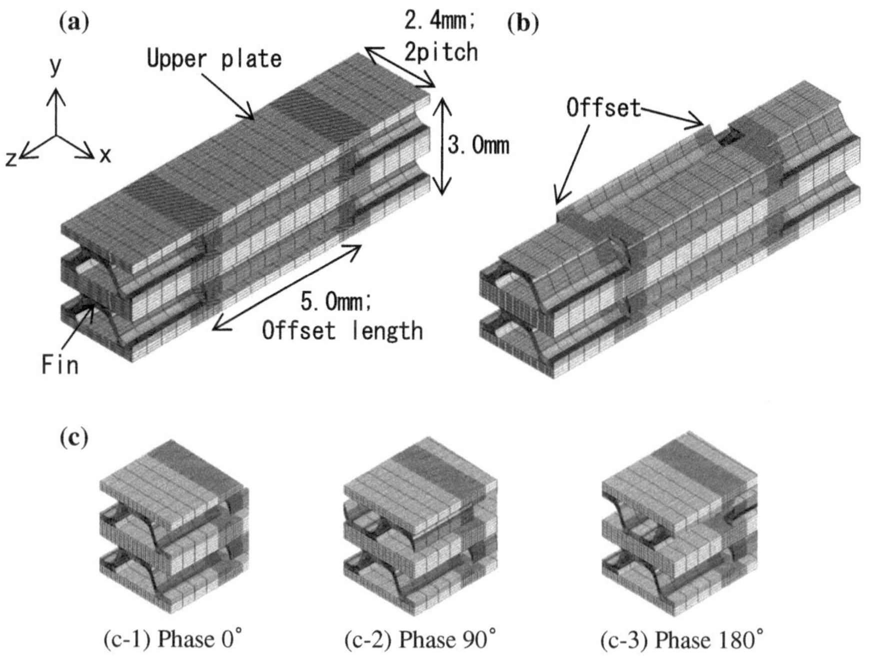

Fig. 6.3 FEM model for unit cell. **a** A unit cell model (Phase 0°) with upper plate. **b** A unit cell model (Phase 0°) without upper plate. **c** 3 types of phase in the unit cell model (a part of each model is shown)

shows the configuration of an offset fin by removing the upper plate from Fig. 6.3a. At the both ends of the offset length of 5 mm, offset of 0.6 mm in the x-direction is seen. To simulate the periodic repetition of the plate-fin structure, periodic boundary conditions (Tsuda et al. 2010) were employed in the unit cell model. Figure 6.3c shows a part of the unit cell model with three types of phase angle. The phase angle expresses the difference in the stacking pattern of the upper and lower layers. Here, three types of phase angle (0°, 90° and 180°) were considered, as shown in the figure. 0° means that the location of the two peaks in the upper and lower fins are the same; 180° means that the location of the peaks in the upper fins and lower fins are opposite. Since the shape of the brazing filler metal determined by the surface tension of the melted filler metal is not known before brazing, the shape of the brazing filler metal is not considered in this modeling. As shown in Fig. 6.12 for the actual configuration of brazing filler metal shape, the influence of the filler metal is large at both ends of the offset length, and is small at the center of the offset length.

The elastic-plastic stress-strain curve for the base material itself was that for SUS316 at 600 °C, as expressed in the following equation (Kawashima et al. 2007)

$$\varepsilon = \frac{\sigma}{E} + \left(\frac{\sigma - \sigma_p}{K}\right)^{\frac{1}{m}}, \tag{6.1}$$

where Young's modulus $E = 1.488 \times 10^5$ MPa, Poisson's ratio $\nu = 0.31$, proportional limit $\sigma_p = 106.2$ MPa, coefficient $K = 335.9$ MPa and exponent $m = 0.363$. The uniaxial stress-strain relation indicated in Eq. (6.1) was multi-linearly approximated to use the cyclic plasticity model of Ohno and Wang (1993). Three types of loading in the x, y and z-directions were applied to the unit cell model to obtain the relation between macroscopic stress σ^* and macroscopic strain ε^* in each loading direction. The definition of the macroscopic stress σ^* and strain ε^* for the case of loading in the y-direction, for example, is considered for the area of "$x_1 \times z_1$" and the gage length of y_1, respectively.

Figure 6.4 shows the macroscopic stress-strain behavior of a unit cell model with a phase angle of 0 °. When compared with the flow stress of the base material itself, the macroscopic flow stresses in the x and z-directions were approximately 50 %, and that in the y-direction was approximately 8 %. This anisotropy in macroscopic stress-strain behavior was nearly the same as in the case without offset (Tsuda et al. 2010; Ohno et al. 2012).

Figure 6.5 shows the macroscopic strain response in the x- and z-directions in the case of loading in the y-direction. Along with the increase of ε_y^*, both ε_x^* and ε_z^* showed gradual reduction. This reduction, however, was much smaller than the conventional assumption of volume-incompressibility in plastic strain (expressed as the broken line). At the point of $\varepsilon_y^* = 0.5$ %, both ε_x^* and ε_z^* in the isotropic and incompressible material should be approximately −0.25 %, but the actual values were below −0.02 %. The reason for this tendency is that the plate itself cannot easily shrink in the x- and z-directions with loading in the y-direction.

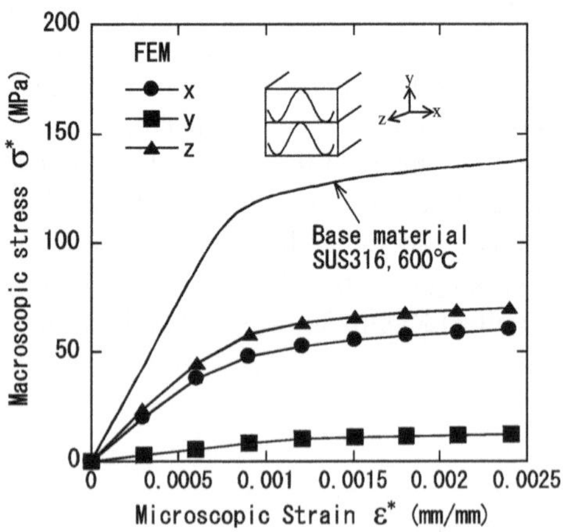

Fig. 6.4 Macroscopic behavior of unit cell (pitch 1.2 mm, phase 0 °)

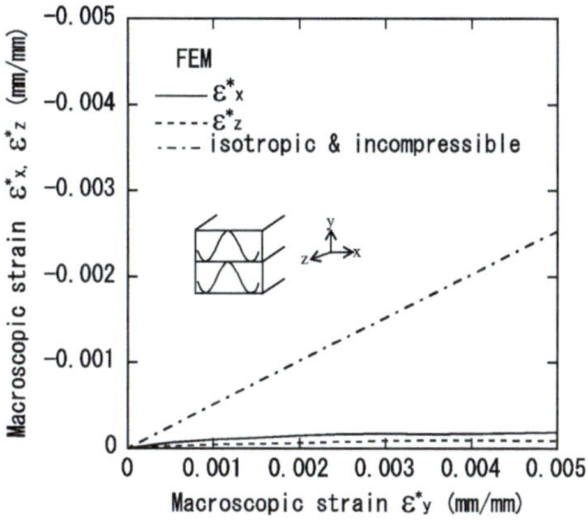

Fig. 6.5 Macroscopic behavior of unit cell (pitch 1.2 mm, phase 0°)

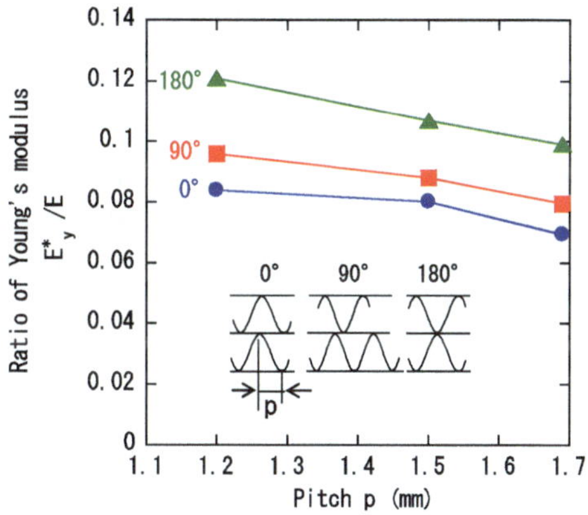

Fig. 6.6 Effects of fin-pitch and phase on macroscopic Young's modulus

Figure 6.6 shows the influence of both fin-pitch and phase angle on the macroscopic Young's modulus with loading in y-direction. Smaller pitch and larger phase angle are found to give the larger macroscopic Young's modulus. In the case of pitch of 1.2 mm, for example, the ratio E^*/E is in the range of 0.08–0.12, depending on the phase angle. The macroscopic Young's modulus for a phase angle of 180° is approximately 40 % larger than that of 0°, and that for a phase angle of 90° is between the other two phase angles. The influence of phase angle was found to be negligible in the range of plastic deformation.

Figure 6.7 shows the distribution of the local strain concentration factor K_ε in Eq. (6.2), along the fin length AB in the figure, at a macroscopic strain of 0.15 % in the y-direction loading for the model with a phase angle of 0°, as shown in Fig. 6.3.

$$K_\varepsilon = \frac{\varepsilon_{\max}}{\varepsilon^*} \qquad (6.2)$$

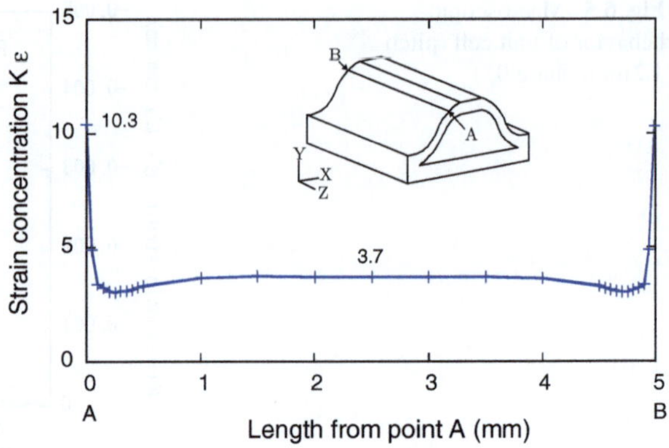

Fig. 6.7 Strain concentration in unit cell

The local strain concentration factor K_ε at the center position of the line AB was 2.1 in the elastic range, and 3.7 in the elastic-plastic range. The local strain concentration factor in the elastic-plastic range at ends A and B showed a peak of 10.3, almost three times higher than that at the center of the fin length. As shown in Fig. 6.12 in the next section, the strain concentration in the actual plate-fin shows a different tendency from the results in this section, because the existence of brazing filler metal at both ends of offset length (at ends A and B) drastically reduces the strain concentration. At the center of the offset length of 5 mm, i.e., the center of A and B, the influence of brazing filler metal is small.

Five kinds of stress and strain concentration factors appear in this paper; Table 6.1 summarizes these factors, including the strain concentration factor in Eq. (6.2).

Table 6.1 Summary of stress and strain concentration factors

Notation	Eq. no.	Definition	Note
K_ε	(6.2)	$\frac{\varepsilon_{max}}{\varepsilon^*}$	Monotonic loading in Fig. 6.7. ε_{max} is the maximum local strain on a fin surface along AB line in Fig. 6.7
K_σ	(6.3)	$\frac{\sigma_{failure}}{\sigma^*_{failure}}$	Monotonic loading. $\sigma^*_{failure}$ is an experimental failure strength of plate-fin. $\sigma_{failure}$ is the failure strength of base metal
K_σ^E	(6.4)	$\frac{E}{E^*}$	Monotonic loading. Simplified prediction of K_σ based on the Young's modulus of base metal and plate-fin
K_ε^R	(6.5)	$\frac{\Delta\varepsilon_{max}}{\Delta\varepsilon^*}$	Cyclic loading, i.e. fatigue loading. Strain ranges of base metal and plate-fin at a given fatigue life
K_t	(6.10)	$\frac{\sigma_{max\,e}}{\sigma_{nom}}$	Monotonic loading. $\sigma_{nom} = \frac{E}{E^*}\sigma^*$ in Eq. (6.8). $\sigma_{max\,e}$ is an elastic maximum local stress on the fin surface

6.4 Uniaxial Strength Tests with Small Plate-Fin Specimens

6.4.1 Tensile and Creep Tests

Figure 6.8 shows a small plate-fin specimen with a rectangular cross-section of 30×10 mm, with eight layers of fins and nine layers of plates, and two solid pieces on the top and bottom. All materials are made of SUS316. These small specimens were subjected to tensile load through pins and holes located in the two solid pieces. The gage length of the tensile and fatigue tests was 15 mm, and that of the creep test was 25 mm. The macroscopic stress σ^* was determined by dividing the tensile load by the cross section area (30×10 mm). The macroscopic strain ε^* was determined by dividing the displacement of gage length by the height of the plate-fin part (12 mm). All tests were performed in an air environment. Test specimens were heated to a uniform temperature using an electric furnace in both the tensile and creep tests. In the fatigue test shown in the next section, however, the specimens were heated by high frequency.

Figure 6.9 summarizes the results of the tensile test in the temperature range from room temperature to 700 °C. Four types of brazing conditions, from No. 1 to 4 were adopted (Kawashima 2008), although condition No. 3 (4 days at 1080 °C with the filler metal thickness of 100 μm) is the main focus. In the figure, the open circle and open square symbols depict brazing condition No. 3. The failure location was a fin, and the failure mode was ductile failure in all conditions for No. 3. In the case of No. 3 at 600 °C, strength reduction from tensile strength and 0.2 % yield stress of the base material were 1/(12.4) and 1/(7.64), respectively (Kawashima 2008). The tensile strength σ_B and 0.2 % yield stress σ_y of the base material multiplied by the reduction are respectively shown as solid and broken lines in Fig. 6.9. Although there is scattering depending on the brazing condition, experimental data for No. 3 corresponded fairly well to the data trend of the solid and broken lines.

Two types of the stress magnification factor K_σ, i.e., inverse values of the strength reduction, were determined as shown in Eq. (6.3) for tensile strength and 0.2 % yield stress

$$K_\sigma = \frac{\sigma_{\text{failure}}}{\sigma^*_{\text{failure}}} \tag{6.3}$$

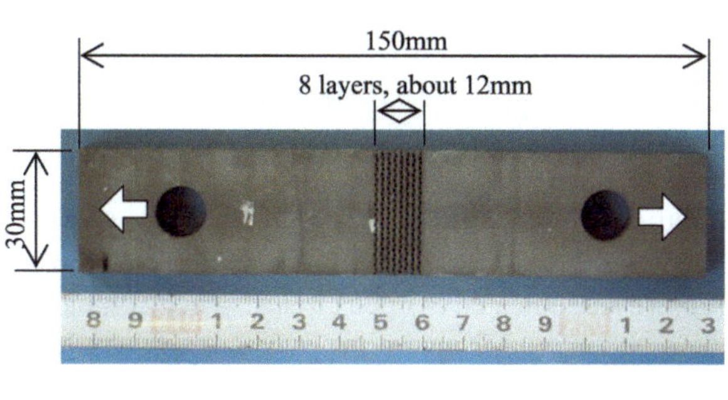

Fig. 6.8 Plate-fin small specimen

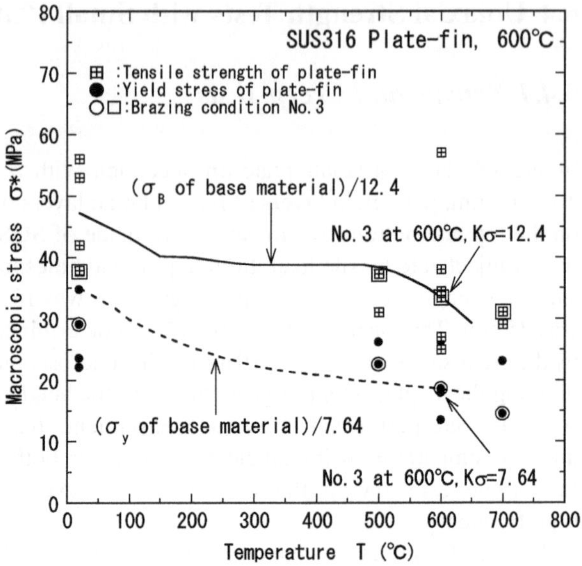

Fig. 6.9 Results of the tensile tests and the stress magnification factor

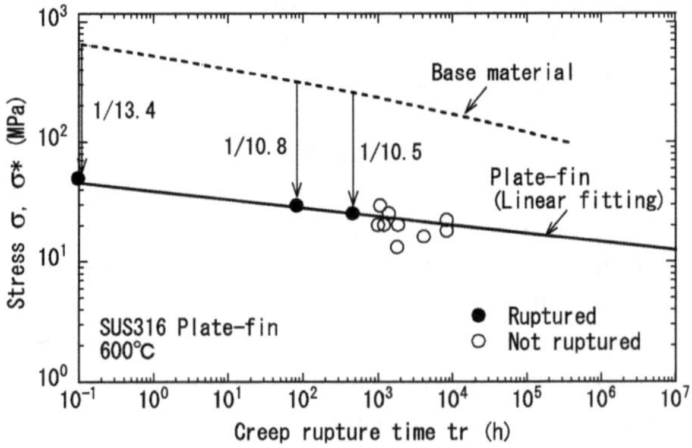

Fig. 6.10 Results of the creep tests and the stress magnification factor

The inverse value of “12.4” for the case of tensile strength is 8 %, nearly the same reduction as the plastic flow stress from homogenization FEM in the previous section.

Figure 6.10 shows the results of the creep test at 600 °C. The test was carried out up until final rupture in the three loading cases. Among these three cases, the test with macroscopic stress of 49 MPa ruptured almost instantaneously. The other two cases with macroscopic stresses of 29 and 25 MPa ruptured respectively at 83.5 and 467.8 h. The failure location was at the fin, and failure mode was ductile creep failure in all three cases. In Fig. 6.10, rupture times for these three cases are presented as solid circles, and those for the other nine conditions being interrupted before final rupture are indicated as open circles. When compared with the creep rupture data for the base material (Kawashima 2008), strength reduction for the three ruptured cases was from 1/(13.4) to 1/(10.5), as shown in Fig. 6.10. The stress magnification factor K_σ in Eq. (6.3) for the case of the creep test is the inverse value of these strength reductions, in the range of 10.5–13.4 depending on macroscopic stress and rupture time.

When the creep test results are extrapolated to the long-term side, the macroscopic stress corresponding to the RHX design rupture life of 2.5×10^5 h is predicted to be 13.1 MPa. Long-term creep rupture data for the plate-fin specimen are necessary to confirm the allowable stress in the structural design for RHX.

In order to discuss the stress dependence of strength reduction in the tensile and creep tests, the stress magnification factors from these tests are plotted against the applied stress as shown in Fig. 6.11. The stress magnification factor as shown in Eq. (6.4) is also included here, where the ratio of Young's modulus for the base material and plate-fin structure is obtained from the homogenization FEM of the unit cell.

$$K_\sigma^E = \frac{E}{E^*} \tag{6.4}$$

Equation (6.4) is a simplified prediction of K_σ in Eq. (6.3). The idea of using the ratio of Young's modulus for undamaged and damaged material is shown in Lemaitre (1985) as describing the effective stress in a damage mechanics approach for ductile fracture. Equation (6.4) gives values of 8.3 to 12.2 for phase angles 180° and 0°, respectively. This range is in basic accordance with those from the tensile and creep tests of the plate-fin specimens. Although the phase angle of the fin is not controlled in the manufacturing process, Eq. (6.4), based on the ratio of Young's modulus from FEM, can serve as a preliminary method for predicting the strength of the plate-fin structure.

6.4.2 Fatigue Tests

Fatigue tests of the plate-fin specimens were carried out at 600 °C under strain-controlled conditions with a macroscopic strain range from 0.121 to 0.577 %. The

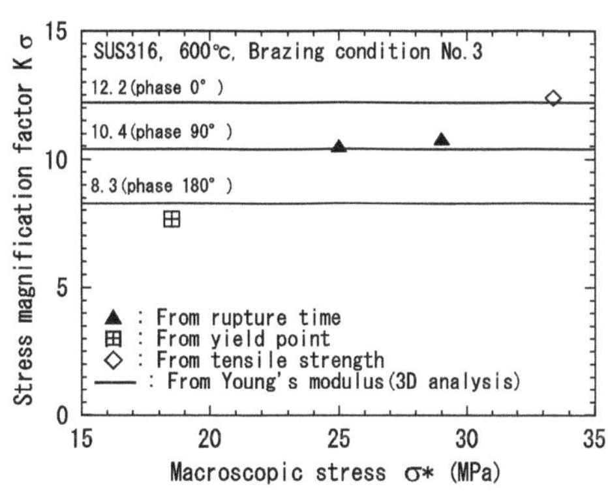

Fig. 6.11 Stress level and stress magnification factor

brazing condition for the fatigue test specimens was No. 3 (4 days at 1080 °C with filler metal thickness of 100 μm). Three cases in the tests include a strain hold at the maximum tensile strain for 10 or 60 min to examine the creep-fatigue interaction. The strain waveform was triangular and the macroscopic strain rate was 6 cpm. Fatigue life N_f was determined as a cycle showing a 25 % decrease in macroscopic peak stress from the steady cycles. An observed image of crack initiation in the cross section of the plate-fin specimen is shown in Fig. 6.12. The failure location was in the fillet part of the fin, i.e., line AB as shown in Fig. 6.7 with respect to homogenization FEM. Although the strain concentration at the two ends of AB was higher than that at the center of line AB (as shown in Fig. 6.7 in the case without brazing filler metal), cracks initiated at the center of the fin length in the actual configuration that had brazing filler metal. The decrease of 25 % in peak stress for defining the fatigue life of the plate-fin specimen is considered to represent a 25 % decrease of the total cross-sectional area in all fins due to cracks.

Figure 6.13 shows the results of the fatigue test of the plate-fin specimen. Since the results of the creep-fatigue test for the plate-fin specimen shown by the solid circles nearly agreed with those for the fatigue test shown by the open circles, the life reduction by strain hold was found to be negligible in the tested conditions, having a rather short hold time duration. When comparing the fatigue data of the plate-fin specimen with that of the base material (Kawashima 2008) shown as a broken line, a reduction of fatigue strength, i.e., a reduction in strain range at a prescribed cycle is found. The fatigue curve of the plate-fin specimen in the figure was found to have an almost linear shape, differing from that of base material. As can be seen from Fig. 6.13, the strain range magnification factor in Eq. (6.5) shows a dependence on fatigue life, showing 2.9–5.7 for fatigue life of 300–70,000 cycles

$$K_\varepsilon^R = \frac{\Delta\varepsilon}{\Delta\varepsilon^*} \tag{6.5}$$

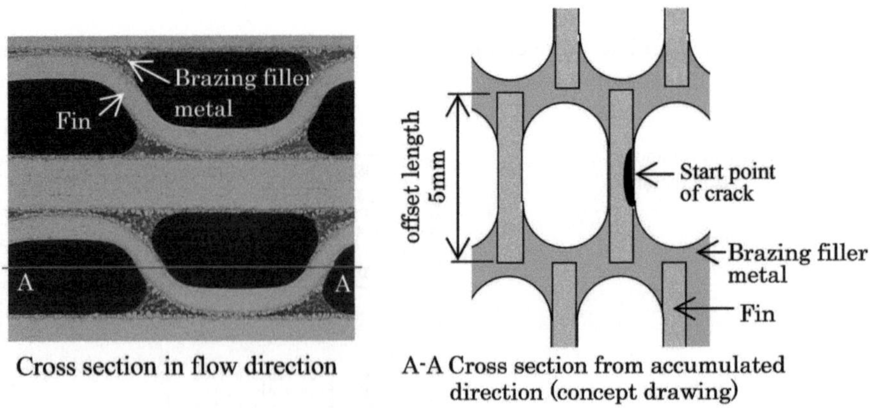

Fig. 6.12 Conceptual drawing of the start point of fatigue crack

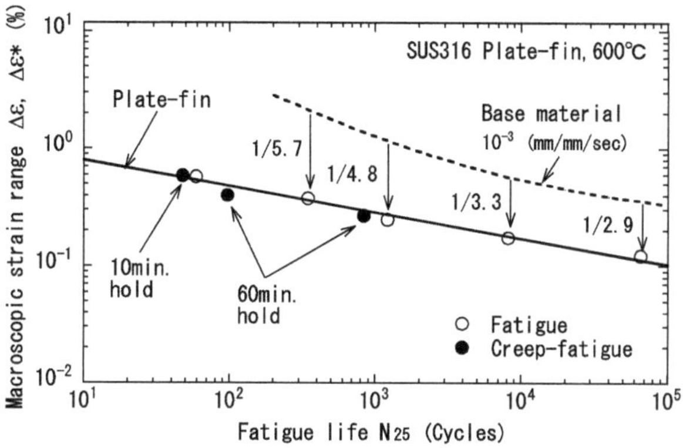

Fig. 6.13 Results of the fatigue tests and the strain concentration factor

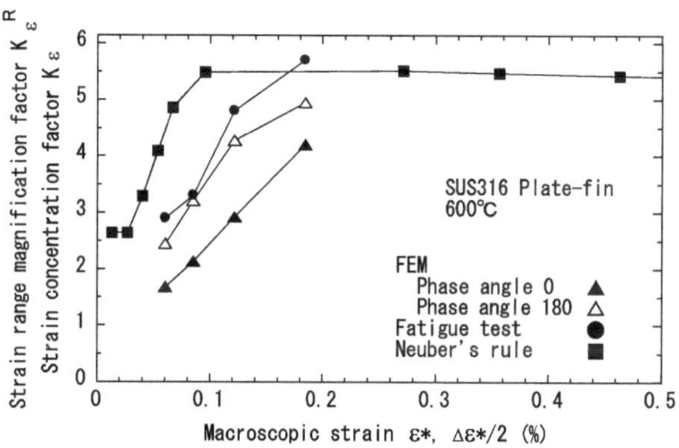

Fig. 6.14 Macroscopic strain and strain concentration factor

When considering the observed image of crack initiation in Fig. 6.12, an initial crack area or a crack growth rate near the center of the fin length could be different depending on the fatigue life or the strain level. Figure 6.14 summarizes the relationship between the experimentally obtained strain range magnification factor in Eq. (6.5), and the strain amplitude. In this figure, the results for strain concentration at the center of fin length by elastic-plastic analysis of the unit cell model in Fig. 6.3 are also depicted for the phase angles of 0° and 180°. As can be seen from the figure, an increasing tendency of the strain concentration factor with increased strain amplitude is similar for both the fatigue test and FEM results. Regarding the influence of phase angle, results for the phase angle of 180° nearly coincides with those for the fatigue test. Experimental results with the strain amplitude above 0.1 % show slightly higher values than those by FEM for the phase angle of 180°. In this figure, a simplified prediction of the strain concentration factor by Neuber's rule is included. A discussion on this prediction appears in the next section.

6.4.3 Comparison with FEM Results

Figure 6.15 shows an example of the macroscopic stress-strain curve in the first cycle of the fatigue test, together with simulation results by FEM for the unit cell with phases of 0° and 180° in Fig. 6.3. Young's modulus from the experiment nearly coincided with the FEM result of 180° in the elastic range. Plastic flow stress, on the other hand, coincided with the averaged value for the two phase angles. Although the actual shape of the brazing filler metal was neglected in FEM modeling of the unit cell in Fig. 6.3, the influence of neglecting the local structure is found to be small in describing the macroscopic stress-strain behavior of the plate-fin structure.

As reported in Kawashima et al. (2007), a simplified prediction of the macroscopic stress-strain behavior of the plate-fin structure was tried by way of the effective stress concept. The following process is used in this approach: the stress-strain curve of the base material in Eq. (6.6) is assumed; the macroscopic stress-strain curve of the plate-fin structure is predicted by Eq. (6.7) using the same function as Eq. (6.6); and the effective stress σ_{nom} is defined in Eq. (6.8) by the ratio of Young's modulus of the base material and the equivalent-homogeneous solid. This is known as the reference stress approach (Boyle and Spence 1983), in which the meaning of effective stress is the stress of the base material giving the same macroscopic strain as the plate-fin structure. Determination of the effective stress using the ratio of Young's modulus is based on the damage mechanics approach (Lemaitre 1985) for ductile fracture, which defines the damage using the ratio of Young's modulus of undamaged material and damaged material with an array of defects

$$\varepsilon = \varepsilon_e + \varepsilon_p = \frac{\sigma}{E} + f(\sigma), \tag{6.6}$$

$$\varepsilon^* = \frac{\sigma_{nom}}{E} + f(\sigma_{nom}), \tag{6.7}$$

$$\sigma_{nom} = \frac{E}{E^*}\sigma^* \tag{6.8}$$

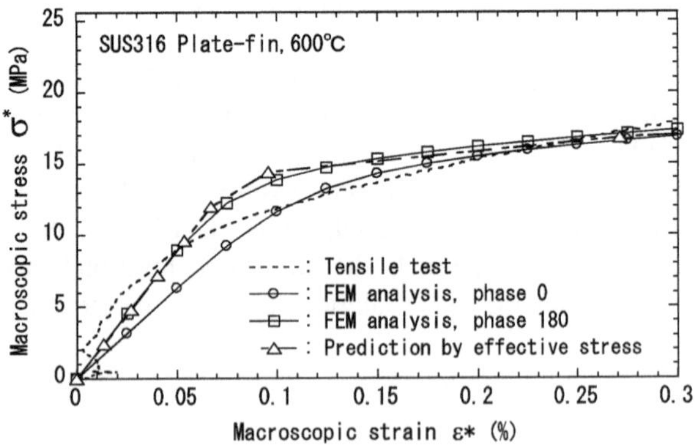

Fig. 6.15 Macroscopic stress-strain curve

The ratio of Young's modulus E/E^* in Eq. (6.8) is in the range of 8.3–12.2 depending on the phase angle, as shown in Fig. 6.11. Predicted results of the macroscopic stress-strain curve using Young's modulus for the phase angle of 180°, $E/E^* = 8.3$, are shown in Fig. 6.15. A simplified prediction of the effective stress by means of Eq. (6.8) in the reference stress approach makes sense in describing macroscopic stress-strain behavior.

The other viewpoint is the relation between the macroscopic stress-strain behavior from the homogenization FEM and experimentally-obtained tensile and creep strength. In the homogenization FEM for loading in the stacking-direction of the plate-fin structure, plastic flow stress in the stacking direction was almost 8 % of that for the base material. The inverse value of 8 % is 12.5 and almost the same value as the strength magnification factor 12.4 from the tensile strength of the plate-fin specimen. The reason is that the failure mode of the plate-fin specimen is ductile failure in the fins; in other words, the effective stress based on the ratio of Young's modulus in Eq. (6.8), or that based on the plastic flow stress, governed the ductile failure in the fins.

A simplified evaluation method of K_ε in Eq. (6.2) on the basis of elastic analysis is useful in the initial stage of structural design. Here, a simplified evaluation of K_ε based on Neuber's rule is examined. Neuber's rule is defined in Eq. (6.9) for the case of plate-fin structures (Kawashima et al. 2007)

$$\sigma_{\max}\varepsilon_{\max} = K_t^2\sigma_{\mathrm{nom}}\varepsilon_{\mathrm{nom}}, \tag{6.9}$$

$$K_t = \frac{\sigma_{\max\,\mathrm{e}}}{\sigma_{\mathrm{nom}}}, \tag{6.10}$$

where $\sigma_{\max}$, $\varepsilon_{\max}$, K_t, σ_{nom} and $\varepsilon_{\mathrm{nom}}$ are, respectively, elastic-plastic maximum local stress and strain at the cross section of the fillet in the fin, the elastic stress concentration factor in Eq. (6.10) and the above-mentioned effective stress and strain. Considering the relation of Eqs. (6.8) and (6.11), Eq. (6.9) can be rewritten as shown by Eq. (6.12)

$$\varepsilon_{\mathrm{nom}} = \varepsilon^*, \tag{6.11}$$

$$\sigma_{\max}\varepsilon_{\max} = K_t^2\frac{E}{E^*}\sigma^*\varepsilon^* \tag{6.12}$$

The strain concentration factor K_ε in Eq. (6.2) is obtained using Eqs. (6.9) and (6.11), where both pairs of "$\sigma_{\max}$ and $\varepsilon_{\max}$" and "σ_{nom} and $\varepsilon_{\mathrm{nom}}$" obey the stress-strain relationship of the base material in Eq. (6.1). The elastic stress concentration factor K_t of 2.64 at the center of the offset length by FEM for the unit cell model with a phase angle of 180° in Fig. 6.3 is also used. The difference between K_t of 2.64 and K_ε of 2.1 in Sect. 6.3 above is due to the difference of definition. Predicted results for the strain concentration factor K_ε for a given macroscopic strain are shown in Fig. 6.14. Predicted results by Neuber's rule were found to be conservative when compared with the FEM results for a phase angle of 180°. From the standpoint of a comparison with the experimental results, the increasing tendency in the experimental data differs from the predicted results based on the elastic stress concentration at the center of the fin length. The reason is considered to be that the influence of the actual distribution of strain concentration along the fin length is reflected in the experimental data. In the range of strain amplitude below 0.2 %, however, the results predicted by Neuber's rule serve as a conservative estimate for the strain range magnification factor in the plate-fin specimen fatigue experiment.

6.5 Thermal Fatigue Test of Plate-Fin Structure

6.5.1 Method and Results of Thermal Fatigue Test

Figure 6.16 shows the heat exchanger model for the thermal fatigue test. The model having a cubic shape was made of SUS316, with the width, length and height of the model being the same at 100 mm. The bottom and upper plate thickness was 6 mm, and the thickness of side walls was 10 mm. Additional cooling fins were set on the cooled surface in order to increase the cooling efficiency. The loading condition is schematically depicted in the figure; first the heat exchanger model was heated to a uniform temperature of 600 °C in an electric furnace, after which the model was taken out of the furnace for cooling by blowing air into the core. In this process, the temperature of the cooled surface fell rapidly, while the other side of the cooled surface was kept at a high temperature. This resulted in the temperature distribution in the plate-fin portion as schematically indicated in Fig. 6.16. One cycle of heating and cooling was 40 min, and was repeated until cracks were observed on the cooled surface.

Visual inspection using a digital microscope was carried out periodically. While no cracks were observed at 3000 cycles, indication was seen after 3200 cycles. After 3500 cycles several cracks were observed at the center of the cooled surface. After the thermal fatigue test, the heat exchanger model was dissected for detailed investigation. Figure 6.17 shows the cross-section area A-A near the center of the

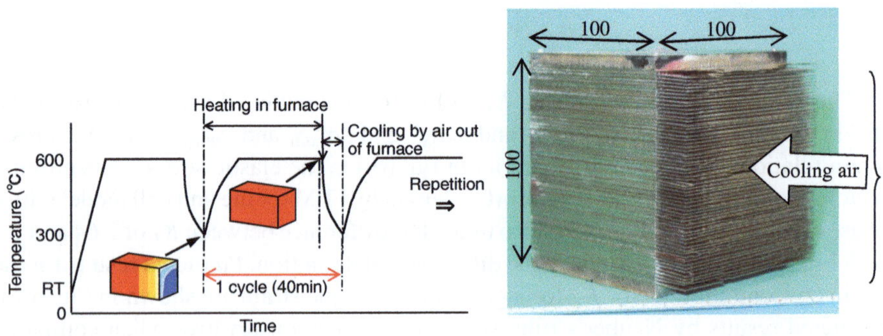

Fig. 6.16 Small heat exchanger model for thermal fatigue test

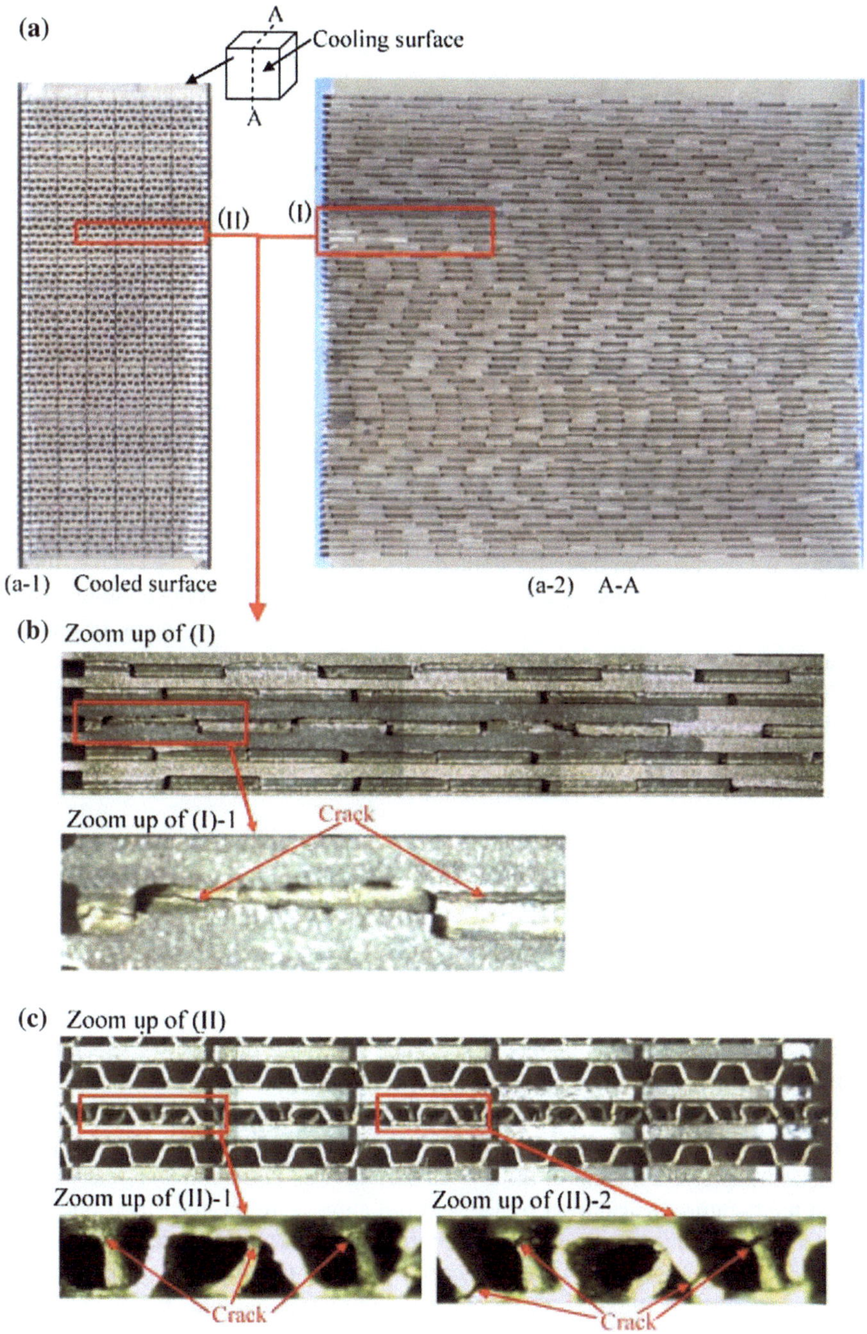

Fig. 6.17 Cross section of small heat exchanger model after thermal fatigue test. **a** Overview. **b** Cross section perpendicular to cooled surface. **c** Cooled surface

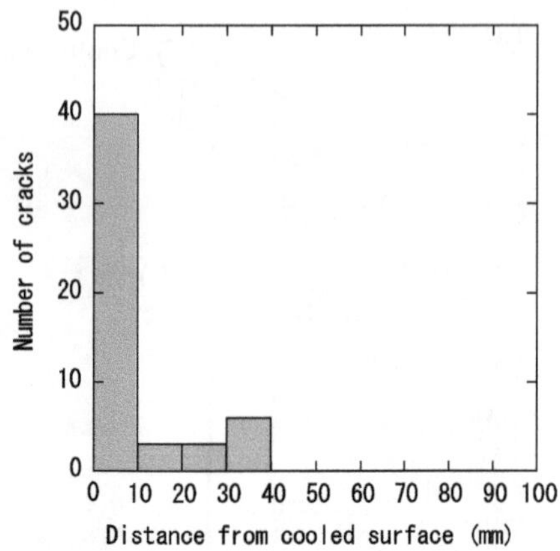

Fig. 6.18 Distribution of number of cracks

cooled surface; Fig. 6.17(a-1) is a view of the cooled surface; Fig. 6.17(a-2) is cut section A-A; Fig. 6.17b is a zoom up of the cut section (I) showing cracks in the fins; Fig. 6.17c is a zoom up of the cooled surface (II) showing cracks in the fillet of the fins. Almost 50 cracks were observed and the distance from the cooling surface was measured for each crack. Figure 6.18 summarizes the distribution of the distance from the cooled surface for all observe cracks. The distance of cracks from the cooled surface was in the range of 0–40 mm, and 40 cracks, or 80 % of the total number of cracks, were found to be in the range of 0–10 mm. These 40 cracks were considered to have initiated at around 3200 cycles, and the other short cracks at distances of over 10 mm were considered to have initiated at around 3500 cycles.

6.5.2 Fatigue Life Prediction Based on Homogenization FEM Analysis

Transient temperature distribution during the thermal fatigue test was obtained by heat conduction analysis using the FEM model. The plate-fin structure was modeled as the equivalent-homogeneous-solid material with both anisotropic heat conduction coefficient and density, considering the ratio of actual and macroscopic cross-sectional area in the three directions of width, stacking and flow (Kawashima 2008). The heat transfer coefficient at the cooled surface was determined so that the simulated temperature distribution by FEM could agree with the measured one in the test. Considering the symmetry of the configuration of the heat exchanger model, a quarter-size FEM model (with respect to the actual test model) was prepared. A time history of the simulated temperature at representative points by FEM analysis showed good agreement with the measured results.

Based on the calculated time history of the temperature distribution in the model, the following three cases of FEM analysis were carried out using ABAQUS ver.6.2.

- Case 1: Isotropic elastic analysis
- Case 2: Anisotropic elastic analysis
- Case 3: Anisotropic elastic-plastic analysis considering volume-compressibility in plasticity

Anisotropic material constants at 600 °C were determined from the FEM results by the unit cell model with a phase angle of 0 ° and a fin pitch of 1.2 mm (Kawashima 2008). The reason for choosing 0 ° is to examine the case with a lower Young's modulus yielding a larger strain range. Temperature dependence of the material constants were considered, except Poisson's ratio. In Case 2, the macroscopic Young's modulus values for the stacking, width and flow directions in the plate-fin structure are set respectively as 8, 42 and 51 % of that of the base material. In Case 1, Young's modulus in all directions is 8 % of that for the base material. Regarding Case 3, the generalized quadratic equivalent stress of von Mises (von Mises 1928) was used to represent the plastic anisotropy and volume-compressibility at 0.2 % macroscopic plastic strain in the homogenization FEM analysis in Sect. 6.3.

An example of the distribution of stacking-direction stress in Case 2 at 20 s after the start of cooling is shown in Fig. 6.19. The maximum stress was located at the center of the cooled surface, where cracks were found in the thermal fatigue test. Distribution of the stacking-direction strain at this time is shown in Fig. 6.20 for the

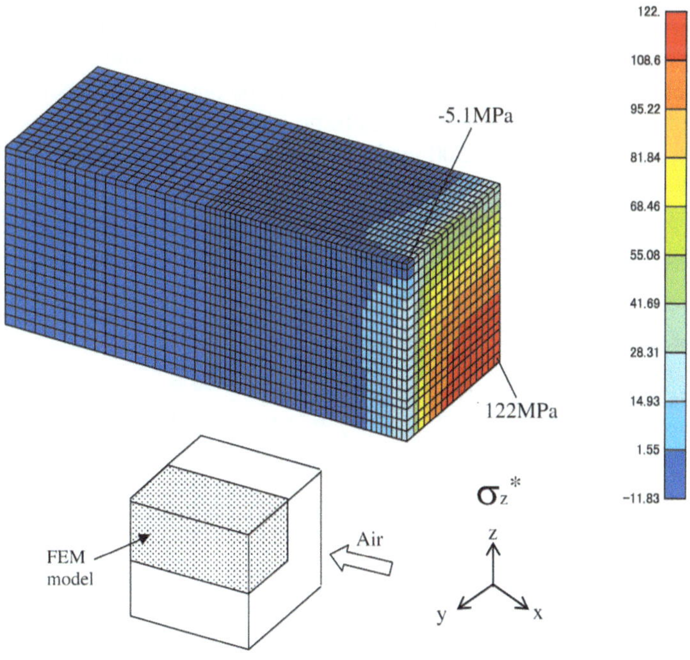

Fig. 6.19 Distribution of macroscopic stress in stacking direction (20 s after cooling, anisotropic elastic analysis)

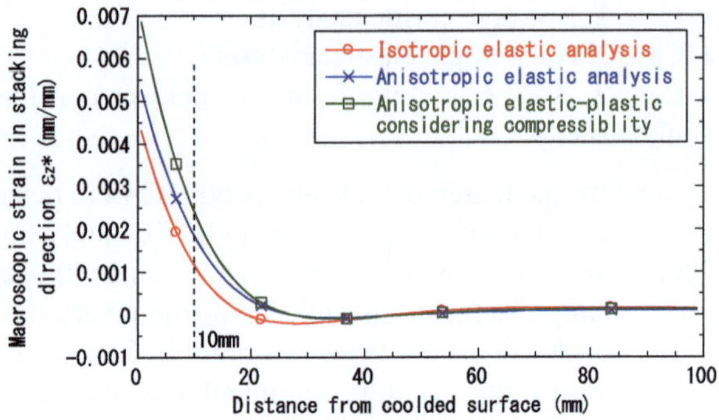

Fig. 6.20 Distribution of macroscopic strain in thermal fatigue test (20 s after cooling)

above three cases, along the flow direction from the center of the cooled surface. Results of the comparison between Cases 1, 2 and 3 are summarized as follows: Case 3 gives the maximum, and Case 1 shows the minimum strain; Case 3 gives 1.3 times larger strain than that of Case 2.

The results of fatigue life prediction for the thermal fatigue test are shown in Fig. 6.21, where the fatigue curve of the base material and that of the small plate-fin specimen are the same as shown in Fig. 6.13. The three kinds of macroscopic strain range at 10 mm from the cooled surface are shown.

As a result of the combination of the macroscopic strain range from FEM and the fatigue curve of the small plate-fin specimen, predicted fatigue life values at 10 mm from the cooled surface in the thermal fatigue test are as follows: 60,000 cycles by isotropic elastic analysis for Case 1, 6,900 cycles by anisotropic elastic analysis for Case 2 and 2,100 cycles by anisotropic elastic-plastic analysis for Case 3. Since the cracks were found at 3200–3500 cycles, Case 3 was found to give a conservative fatigue life prediction. Regarding the elastic analysis, there is a possibility for Case 2 to be applied to the preliminary design analysis of the plate-fin heat exchanger by considering additional factors such as elastic follow up.

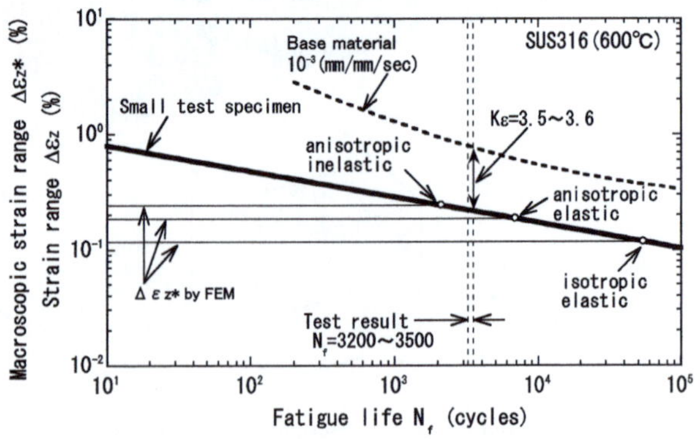

Fig. 6.21 Evaluation of fatigue life of thermal fatigue test by strain range in stacking direction (10 mm from cooled surface)

The other point of view of fatigue life prediction is to combine the fatigue curve of the base material and the predicted local strain range considering the strain range magnification factor. As can be seen from Fig. 6.13, the strain range magnification factor, i.e., the ratio of the strain range of the base material and the plate-fin at a prescribed number of cycles, takes different values depending on the number of cycles. If the strain range magnification factor is 3.6 as a representative value, the predicted fatigue life is 21,000, 5,000 and 3,000 cycles respectively for Cases 1 to 3. This tendency is similar to that for the case combining the macroscopic strain range and the fatigue curve for the small plate-fin specimen.

6.6 Conclusions

(1) Anisotropy and volume-compressibility in macroscopic plastic deformation were found by the homogenization FEM analysis of the unit cell with offset. The deformation in the stacking direction exhibited the weakest behavior, showing nearly 8 % of macroscopic flow stress when compared with those of the base material. The influence of local structure such as the brazing filler metal shape was small in describing the macroscopic stress-stress behavior obtained in the uniaxial tensile test of small plate-fin specimens.

(2) Tensile and creep strength in the stacking direction of the small plate-fin specimen showed a reduction from the strength of the base material, and this reduction showed similar values to the results of the homogenization FEM of the unit cell, including reduction of plastic flow stress. The failure mode of the fins in the tensile and creep tests was ductile, and this is considered to be the reason that the strength in the experiment and the plastic flow stress in FEM showed similar reductions to each other.

(3) Regarding the fatigue test in the stacking direction of the small plate-fin specimen, a reduction of strain range from that of the base material at prescribed cycles was found. Since the fatigue strength of the plate-fin structure is very sensitive to local structures such as the brazing filler metal shape, an experimental approach is required so as to quantify the fatigue curve of plate-fin structures.

(4) The homogenization approach, combining homogenization FEM and the experimentally obtained fatigue curve of the plate-fin structure, was successfully applied to life prediction for the thermal fatigue test of the heat exchanger partial model. Thermal fatigue life prediction based on anisotropic elastic-plastic FEM analysis gave a predicted life of 2,100 cycles, or 60 % of the observed life, thus within a factor of 2.

Acknowledgments This study shows the R&D results obtained by Mitsubishi Heavy Industries during 2005–2007 in the program known as "The Technology Development of High Efficiency and High Capacity Gas/Gas Heat Exchanger Necessary to the Practical Application of Gas-cooled Reactor", sponsored by the Ministry of Education, Culture, Sports, Science and Technology of Japan.

References

Boyle JT, Spence J (1983) Stress Analysis for Creep. Butterworth, London

Igari T, Setoguchi K, Nomura S (1986) Simplified creep analysis of perforated plates under steady creep condition. Nucl Eng Des 97:161–166

Igari T, Tokiyoshi T, Mizokami Y (2001) Prediction of macroscopic and local stress-strain behavior of perforated plates under primary and secondary creep conditions. JSME Int J 44:237–245

Igari T, Tokiyoshi T, Mizokami Y (2003) Inelastic analysis and thermal fatigue life prediction of a perforated cylinder based on the effective stress concept. Mater Sci Res Int 9:9–15

Ishiyama S, Mutoh Y, Tanihira M, Ogata K, Kamito Y, Igari T (2001) Development of the compact heat exchanger for the HTGR, (I) Fabrication of the ultra-fine offset fin. J Japan Nucl Power Soc 43:603–611

Jiang W, Gong J, Chen H, Tu ST (2008) The effect of filler metal thickness on residual stress and creep for stainless-steel plate-fin structure. Int J Press Vessel and Pip 85:569–574

Kawashima F (2008) High-temperature strength evaluation considering microscopic and macroscopic structures. PhD thesis, Kumamoto University

Kawashima F, Igari T, Miyoshi Y, Kamito Y, Tanihira M (2007) High temperature strength and inelastic behavior of plate-fin structures for HTGR. Nucl Eng Des 237:591–599

Lemaitre J (1985) A continuum damage mechanics model for ductile fracture. Trans ASME J Eng Mater Technol 107:83–89

Mizokami Y, Igari T, Nakashima K, Kawashima F, Sakakibara N, Kishikawa R, Tanihira M (2010) Development of plate-fin heat exchanger for intermediate heat exchanger of high temperature gas cooled reactor - fabrication process, high-temperature strength and creep fatigue life prediction of plate-fin structure made of Hastelloy X. Trans At Energ Soc Japan 9:219–232

Mizokami Y, Igari T, Kawashima F, Sakakibara N, Tanihira M, Yuhara T, Hiroe T (2013) Development of structural design procedure of plate-fin heat exchanger for HTGR. Nucl Eng Des 255:248–262

O'Donnell WJ, Porowski JS (1982) Design analysis of perforated plates and shells. In: Zamrik SY, Dietrich D (eds) Pressure Vessels and Piping: Design Technology. A Decade of Progress, ASME, New York

Ohno N, Wang JD (1993) Kinematic hardening rules with critical state of dynamic recovery, part I: formulation and basic features for ratchetting behavior. Int J Plast 9:375–390

Ohno N, Ikenoya K, Okumura D, Matsuda T (2012) Homogenized elastic-viscoplastic behavior of anisotropic open-porous bodies with pore pressure. Int J Solids Struct 49:2799–2806

Tokiyoshi T, Kawashima F, Igari T, Kino H (2001) Crack propagation life prediction of perforated plate under thermal fatigue. Int J Press Vessel Pip 78:837–845

Tsuda M, Takemura E, Asada T, Ohno N, Igari T (2010) Homogenized elastic-viscoplastic behavior of plate-fin structures at high temperatures: numerical analysis and macroscopic constitutive modeling. Int J Mech Sci 52:648–656

von Mises R (1928) Mechanik der plastischen Formänderung von Kristallen. Zeitschrift für Angewandte Mathematik und Mechanik 8:161–185

Chapter 7
Review on Spatio-Temporal Multiscale Phenomena in TRIP Steels and Enhancement of Its Energy Absorption

Takeshi Iwamoto and Hang Thi Pham

Abstract Among various materials used for structures, TRIP steel has attracted the interest of the scientific community because TRIP steel indicates better performances such as formability, toughness and energy absorption as well as strength and ductility than other kinds of advanced high strength steel because of strain-induced martensitic transformation. From a microstructural point of view, two mechanisms are usually considered to explain the high performances: the Magee and the Greenwood-Johnson effects: however, it has not been proven yet. On the other hand, even though TRIP steel shows the excellent energy absorption under the high-speed deformation, the amount of martensite is very small by an influence of the heat generation with plastic transformation. In order to solve this paradox and explain macroscopic performances from the microscopic deformation mechanism, a concept of a spatio-temporal multiscale should be introduced. In this paper, research works done by the one of the authors on the mechanism of high functionalization in TRIP steel from the viewpoint of a spatio-temporal hierarchy are summarized with reviewing some related papers.

Keywords TRIP steels · Strain-induced martensitic transformation · Multiscale · Multiphysics · Spatio-temporal Hierarchy

T. Iwamoto (✉)
Institute of Engineering, Hiroshima University, 1-4-1 Kagamiyama,
Higashi-Hiroshima, Hiroshima 739-8527, Japan
e-mail: iwamoto@mec.hiroshima-u.ac.jp

H.T. Pham
Graduate School of Engineering, Hiroshima University, 1-4-1 Kagamiyama,
Higashi-Hiroshima, Hiroshima 739-8527, Japan
e-mail: d133655@hiroshima-u.ac.jp

H.T. Pham
Faculty of Engineering, Vietnam National University of Agriculture,
Ngoxuanquang Street, Trauquy, Gialam, Hanoi 131004, Vietnam
e-mail: pthang@vnua.edu.vn

H. Altenbach et al. (eds.), *From Creep Damage Mechanics to Homogenization Methods*, Advanced Structured Materials 64,
DOI 10.1007/978-3-319-19440-0_7

7.1 Introduction

In temperature range from martensitic start temperature without applied stress M_s to temperature M_d determined by its chemical composition, martensitic transformation occurs in steel with a metastable austenitic microstructure when it undergoes plastic deformation. As a result, volume fraction of martensite in the steel increases with promotion of deformation. This phase transformation is called strain-induced martensitic transformation (SIMT) (Tamura 1982). In the steel with SIMT, it can be easily imagined that its strength increases because of formation of hard martensite. However, not only the strength but also its ductility increase in a certain range of temperature near room temperature. This phenomenon with high ductility is called transformation-induced plasticity (TRIP) (Tamura 1982; Zackay et al. 1967). The steel with a fully austenitic microstructure by arrangement of chemical compositions with an appropriate thermo-mechanical treatment is called TRIP steel (Zackay et al. 1967).

Among various materials used for structures, TRIP steel has attracted the interest of the scientific community. There are two types of such steels: TRIP steels and TRIP-assisted steels. TRIP steel tends to be rich in nickel and other expensive austenite stabilizing elements. By contrast, austenite is only a minor phase in the overall microstructures of TRIP-assisted steels (Matsumura et al. 1987). Allotriomorphic ferrite comprises about 50–60 vol.% of the microstructures of these materials and the residue becomes a mixture of bainite and retained austenite with higher carbon content (Chatterjee 2006). Comparing with TRIP-assisted steels, TRIP steels have particularly attracted the more considerable attention of researchers because of its simplicity on the initial microstructure (Fischer et al. 2000). The well-known relationship between strength and ductility parameters for conventional low and advanced high strength steels for automobile components is described in Fig. 7.1. It is clearly seen that the elongation decreases with the increase in the ultimate tensile strength. However, TRIP steel including TRIP-assisted steel indicates better performance than

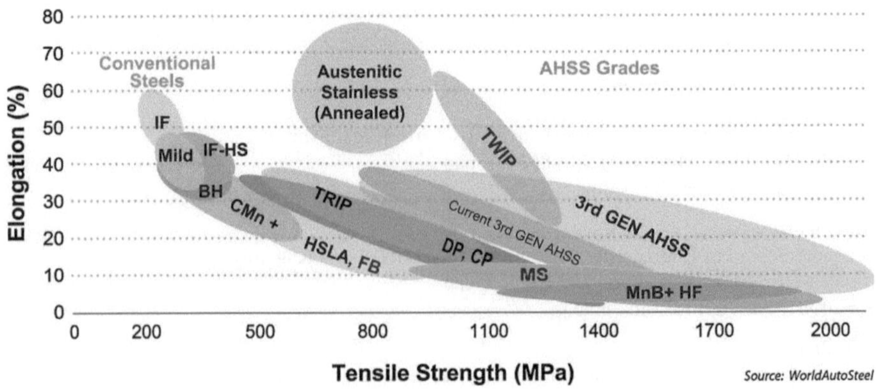

Fig. 7.1 Total elongation versus tensile strength for advanced high strength steels (Keeler and Kimchi 2014)

other kinds of advanced high strength steel. Therefore, it can be considered that TRIP steels possess favorable mechanical properties such as excellent formability and toughness as well as high strength and ductility. These excellent properties may be appeared because of SIMT (Iwamoto et al. 1998; Fischer et al. 2000). At the same time, the process of microscopic shape and volume change due to SIMT can dissipate the significant amount of energy. As a result, TRIP steel might indicate the excellent characteristic of energy absorption. It can be considered that in TRIP steel, huge amount of kinetic energy may be absorbed during the collision (Kazanci and Bathe 2012) when TRIP steel is applied to the components of the vehicles. Therefore, TRIP steel might be suited for the automotive structures and safety components, such as crash-boxes, front side members, bumpers and side panels. They are nowadays in widespread use in the automotive industry.

From a microstructural point of view, two mechanisms are usually considered to explain TRIP: the Magee mechanism (Magee 1966) and the Greenwood-Johnson effect (Greenwood and Johnson 1965). The first effect accounts for the preferred orientation of martensitic plates caused by the applied force which is equivalent to an internal resultant force of microscopic shear stress and an average macroscopic shape change. The Greenwood-Johnson effect accounts for the instantaneous volume expansion and shape change produced by the martensitic transformation. This stress-free strain of the formed martensite is accommodated in the surrounding austenitic region, leading to redundant plastic deformation and, consequently, to extra strain hardening. Moreover, according to the Greenwood and Johnson effect (Greenwood and Johnson 1965) for an explanation of TRIP phenomenon, martensitic transformation induces the instantaneous volume expansion and shape change. It is considered that the volume expansion due to SIMT relaxes the negative hydrostatic pressure which may be a cause of damage or micro-cracking (Delannay et al. 2005). Therefore, these microscopic effects are caused by delay the onset of necking. As a result, macroscopic performances such as the work hardenability of the steel related to the formability can be enhanced. This discussion can be accepted logically; however, it has not been proven yet. In order to prove the result of this discussion, a computational method with considering the spatial multiscale should be established.

On the other hand, He and Sun (2010) recently studied about an existence of the temporal hierarchy as well as the spatial one and a relationship between different temporal and spatial scales by conducting experiments under the various conditions for a Ni–Ti shape memory alloy in order to consider the strain rate sensitivity and effects of the thermal conductivity and the heat transfer on plateau and hysteresis which appear in the stress-strain curve. In this study, it is attempted to make clear the multi-time-scale problem on the mechanical behavior of metallic materials by an experimental observation. At the same time, the direction of the study is quite important to focus on the multi-time-scale. Focusing upon the energy absorption in TRIP steel, TRIP steel may be excellent on the energy absorption due to a formation of more martensitic phase because SIMT occurs during a high-speed deformation with consumption of the energy by itself. This discussion can be also accepted logically; however, it has not been proven yet as well as the above-mentioned multi-spatial-scale problem. Generally, martensitic transformation is a strongly-temperature-dependent phenomenon and much heat generates by an irreversible work with plastic transformation. As a result, the probability of the formation of martensitic phase decreases. In other words, it can be possible to appear a paradoxical phenomenon which the amount of martensite is very small by an influence of the heat generation with plastic transformation even though TRIP steel shows the excellent energy absorption in the high-speed deformation. It is quite important to elucidate the transformation-thermomechanical of TRIP steel behavior under the high strain rate for this paradox. To understand a mechanism of this paradox essentially, let consider time scale of transformation-thermomechanical fields in TRIP steel as shown schematically in Fig. 7.2. As shown in this figure, a change in the characteristic variables in each field is measured with respect to time and an initial slope of the respective diagram is expressed. The slope, i.e. rate in generation of deformation, martensitic phase and heat, represents the characteristic time scale however correlations between slopes have not been clarified as an experimental evidence because their measurements at higher strain rate are not established and quite difficult.

The aim of this paper is to summarize research works done by the one of the authors on the mechanism of high functionalization in TRIP steel from the viewpoint of spatio-temporal hierarchy with reviewing some related papers. First, a spatio-temporal macroscopic model for TRIP steel proposed by one of authors (Tomita and Iwamoto 1995; Iwamoto et al. 1998; Iwamoto and Tsuta 2000, 2002b) are described and two spatial multiscale models (Iwamoto 2004; Iwamoto and Tsuta 2004) also proposed by the one of authors are reviewed. Next, a validity by comparing with the experimental results is confirmed and the applicability of the model is discussed.

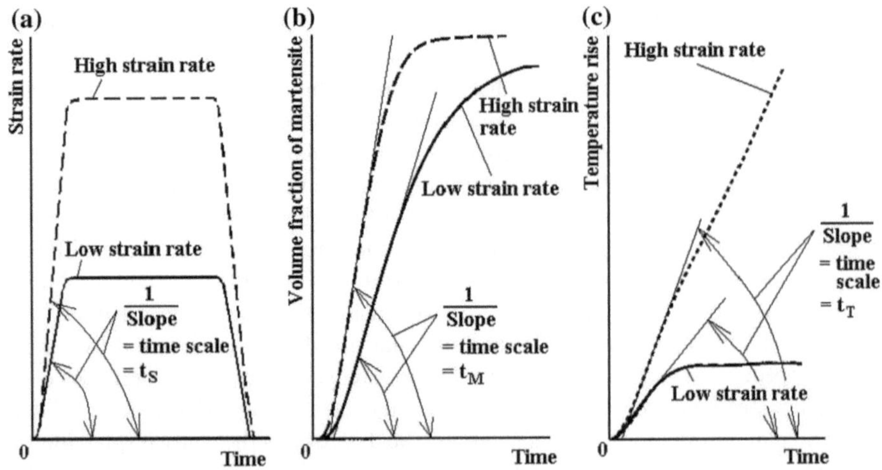

Fig. 7.2 Schematic illustration on time scale in different fields to observe a special phenomenon in steels; **a** strain rate versus time, **b** volume fraction of martensite versus time, and **c** temperature rise versus time

Related to the experiment, results on geometrically orientation distribution function of martensite in micrographs are shown by an image analysis based on the Fourier transformation (Iwamoto and Tsuta 2002a) and geometrical orientation of martensite is discussed. Then, some computational results with the different spatial length scales are introduced and make some discussions. Finally, an importance of the temporal multiscale modeling is addressed with introducing a quite interested experimental result on the energy absorption of TRIP steel.

7.2 Constitutive Model for TRIP Steel

The influence of strain and temperature on the deformation behavior and the microscopic transformation mechanism in TRIP steels are quite complicated, and it may be very difficult to determine a method for the evaluation of the mechanical properties merely through experiments (Tomita and Shibutani 2000). From this point of view, several macro- and micromechanical models have been proposed to account for the TRIP effect. Most of the models are based on an idea of the Greenwood and Johnson effect (Greenwood and Johnson 1965) from the microscopic point of view even though the model is established macroscopically (Dan et al. 2007). On the other hand, physical and phenomenological models of SIMT have been proposed and examined by some researchers.

7.2.1 Macroscopic Model

Among the models of SIMT, the model by Olson and Cohen (1975) has been the most widely-applied. They found the fact that a mesoscopic shear-band intersection is the dominant mechanism of nucleation in SIMT, and proposed a semi-physical model to predict the temperature-dependent SIMT well. Then, phenomenological phase transformation model was developed by Cortes et al. (1992) and the energy balance equation was formulated to calculate flow stress of each phase. Furthermore, Stringfellow et al. (1992) incorporated the effect of mechanical driving force on the martensitic transformation into the Olson and Cohen model (Olson and Cohen 1975) and predicted the stress-strain behavior as the two-phase composite. The contribution of the martensitic phase for strengthening is also examined to establish the constitutive equation. Based on the proposed constitutive model, they performed finite element analysis of the necking behavior of the circular cylinder and clarified the effect of the stress state on the strain-induced martensitic transformation (Stringfellow et al. 1992). The thermo-coupled effects and strain rate sensitivity was taken into transformation kinetics model proposed by Tomita and Iwamoto (1995). After that, Iwamoto et al. (1998) generalized the Tomita and Iwamoto model to incorporate stress state dependence and formulated the constitutive equation for TRIP steels. The validity of the modeling approach was further investigated by Tomita and Shibutani (2000), by comparing the empirical results with those of simulations of laboratory test on samples with varying geometries under various loading conditions. Moreover, further models of the computational simulation were developed by various authors (Cortes et al. 1992; Sachdev 1983) incorporating the various condition in order to predict the evolution of the SIMT. In addition, Lichtenfeld et al. (2006) applied the Olson-Cohen model to high strain rate deformation by taking into account the variation of the parameters α and β with temperature. A new kinetic model, which considers the SIMT as a relaxation process of internal strain energy, aiding the chemical driving force was proposed by Shin et al. (2001). Then, Zaera et al. (2012) has suggested a constitutive model for analyzing martensite formation in austenitic steels deforming at high strain rate. This model includes temperature effects in the phase transformation kinetics, and in the softening of each solid phase through the use of a homogenization technique.

The macroscopic constitutive equation (Tomita and Iwamoto 1995; Iwamoto and Tsuta 2000, 2002b) expressed thermo-elasto-visco-plastic behavior as two-phase aggregate material of austenite and martensite can be derived to account for finite deformation using the standard notation of the updated Lagrangian formulation as follows

$$
\begin{aligned}
&\overset{\nabla}{S}_{ij} = D^e_{ijkl}\dot{\varepsilon}_{kl} - B^e_{ij}\dot{T} - P_{ij}\dot{\bar{\varepsilon}}^p - Q_{ij}\Delta v \dot{f}^{\alpha'}, \\
&B^e_{ij} = \frac{E}{1-2\nu}\alpha_T\delta_{ij}, \quad P_{ij} = \frac{3E}{2\bar{\sigma}(1+\nu)}\frac{\partial f}{\partial \sigma_{ij}}, \quad Q_{ij} = -P_{ij}\Sigma + \frac{1}{3}\delta_{ij}\frac{E}{1-2\nu}, \\
&\dot{\bar{\varepsilon}}^p = \dot{\bar{\varepsilon}}^{pslip} + R\dot{f}^{\alpha'} + \Sigma\Delta v\dot{f}^{\alpha'}, \quad \frac{\partial f}{\partial \sigma_{ij}} = \left(1 + \kappa\frac{J_3}{2J_2^{3/2}}\right)\sigma'_{ij} - \frac{\kappa}{J_2^{1/2}}t_{ij}, \\
&t_{ij} = \sigma'_{im}\sigma'_{mj} - \frac{2}{3}J_2\delta_{ij}, \quad \dot{\kappa} = C_k(1-\kappa)\dot{\bar{\varepsilon}}^{pslip}, \bar{\sigma} = \sqrt{3\left(J_2 - \kappa\frac{J_3}{J_2{}^{1/2}}\right)},
\end{aligned}
\tag{7.1}
$$

where $\overset{\nabla}{S}_{ij}$ is the Jaumann rate of Kirchhoff stress, D^e_{ijkl} is the elastic stiffness tensor, α_T is the thermal expansion coefficient, E is Young's modulus and ν is Poisson's ratio, J_2 and J_3 are the 2nd and the 3rd invariants of the deviatoric stress, Δv and R are the parameters accounting for the magnitudes of volume change and shape changes, κ is the parameter which reflects the development of microstructures inside the material such as the texture, C_k is a constant, and $\bar{\sigma}$ is the equivalent stress which express the tension-compression asymmetry in the uniaxial stress-strain curve (Iwamoto and Tsuta 2002b; Altenbach and Kolupaev 2015). The forward gradient method (Peirce et al. 1984) is introduced to improve the computational efficiency.

The volume fraction of martensite is regarded as an internal state variable. The macroscopic model proposed by Olson and Cohen (1975), and Stringfellow et al. (1992) as the evolution equation is generalized by considering the strain rate sensitivity and pressure dependency of shear band formation, the dependence of grain size as follows (Iwamoto et al. 1998; Iwamoto and Tsuta 2000)

$$
\begin{aligned}
&\dot{f}^{\alpha'} = (1 - f^{\alpha'})\left[A\dot{\bar{\varepsilon}}^{pslip}_{(\gamma)} + B\dot{g}\right],\\
&A = \alpha\beta n(f^{sb})^{n-1}(1 - f^{sb}),\quad B = \eta n_v(g)(f^{sb})^n H(\dot{g}),\\
&\alpha = (\alpha_1 T^2 + \alpha_2 T + \alpha_3 - \alpha_4 \Sigma)\left[\frac{\dot{\bar{\varepsilon}}^{pslip}_{(\gamma)}}{\dot{\varepsilon}_y}\right]^M,\quad \beta = \eta\left(\frac{d}{d_0}\right)^2 p,\\
&p = \int_{-\infty}^{g} n_v(g')\,dg',\quad n_v(g) = \frac{1}{\sqrt{2\pi}\sigma_g}\exp\left\{-\frac{(g-g_0)^2}{2\sigma_g^2}\right\},\\
&g = -T + g_1\Sigma,\quad \Sigma = \frac{\sigma_{ii}}{3\bar{\sigma}},
\end{aligned}
\tag{7.2}
$$

where $\dot{\bar{\varepsilon}}^{pslip}_{(\gamma)}$ is the equivalent strain rate of the slip deformation in the austenite, f^{sb} is the volume fraction of the shear band, p is the probability that an intersection forms a martensitic embryo (Stringfellow et al. 1992; Tomita and Iwamoto 1995), g denotes the driving force for martensitic transformation, σ_g, g_0 and g_1 are the standard deviation, the mean value of g and a constant, respectively. $H(\dot{g})$ is the Heaviside step function with respect to $\dot{g}$ which describes the irreversible process of martensitic transformation, n represents a geometrical constant and has a value of 4.5 (Olson and Cohen 1975) determined by stereology, η is a geometric coefficient, σ_{ij} is Cauchy stress, T is the absolute temperature, d is the initial grain size of the austenite (Iwamoto and Tsuta 2000), d_0 is the reference grain size, M is the strain rate sensitivity exponent for nucleation sites, $\alpha_1, \alpha_2, \alpha_3, \alpha_4$ are material parameters, and $\dot{\varepsilon}_y$ is the reference strain rate.

In addition, the heat conduction equation considering latent heat by means of martensitic transformation can be expressed as

$$
\rho C\dot{T} = \kappa_T \nabla^2 T + \zeta\sigma\dot{\varepsilon}^p_{ij} - \rho\ell^{\alpha'}\dot{f}^{\alpha'} \tag{7.3}
$$

where ρ is density, C is specific heat, κ_T is heat conductivity coefficient, ζ is a Taylor-Quinney coefficient, $\ell^{\alpha'}$ is latent heat of martensitic transformation. The robust computational protocol is established by implementing Eqs. (7.1)–(7.3) into the finite element method.

7.2.2 A Bridging Method Between the Spatial Scales and Microscopic Model

The most noticeable research works about a microscopic model are done by Leblond et al. (1986), Taleb and Sidoroff (2003), and Fischer et al. (2000). Leblond's model

(Leblond et al. 1986) is proposed to take account of the interaction between classical plasticity and TRIP, however, its experimental validation has not yet been performed (Taleb and Petit 2006). Then, Taleb and Sidoroff (2003) investigated some discrepancies from a re-evaluation of the micromechanical model as originally developed by Leblond et al. (1986). In their research work, a more complete formulation taking into account the elasticity in both phases is developed.

Because phase transformation occurs in the crystal lattice scale basically, the continuum crystal plasticity theory established by Asaro (1983) due to a decomposition of the deformation gradient is expanded to treat phase transformation including Magee and Greenwood-Johnson effects within the crystal plasticity theory. The following constitutive equation (Iwamoto and Tsuta 2004) is formulated as

$$\begin{aligned}
\overset{\nabla}{S}_{ij} &= D^{e}_{ijkl}\dot{\varepsilon}_{kl} - \sum_{a} R^{(a)}_{ij}\dot{\gamma}^{(a)} - \sum_{I} R^{t(I)}_{ij}\dot{\gamma}^{t(I)} - B^{e}_{ij}\dot{T} \\
R^{(a)}_{ij} &= D^{e}_{ijkl}P^{(a)}_{kl} + \beta^{(a)}_{ij}, \quad R^{t(I)}_{ij} = D^{e}_{ijkl}Q^{(I)}_{kl} + \beta^{t(I)}_{ij}, \\
\beta^{(a)}_{ij} &= W^{(a)}_{ik}\sigma_{kj} - \sigma_{ik}W^{(a)}_{kj}, \quad \beta^{t(I)}_{ij} = \Omega^{(I)}_{ik}\sigma_{kj} - \sigma_{ik}\Omega^{(I)}_{kj} \\
P^{(a)}_{ij} &= \frac{1}{2}\left(s^{et(a)}_{i}m^{et(a)}_{j} + m^{et(a)}_{i}s^{et(a)}_{j}\right), \quad W^{(a)}_{ij} = \frac{1}{2}\left(s^{et(a)}_{i}m^{et(a)}_{j} - m^{et(a)}_{i}s^{et(a)}_{j}\right), \\
Q^{(I)}_{ij} &= \frac{1}{2}\left(\ell^{e(I)}_{i}n^{e(I)}_{j} + n^{e(I)}_{i}\ell^{e(I)}_{j}\right), \quad \Omega^{(I)}_{ij} = \frac{1}{2}\left(\ell^{e(I)}_{i}n^{e(I)}_{j} - n^{e(I)}_{i}\ell^{e(I)}_{j}\right),
\end{aligned} \tag{7.4}$$

where $s^{et(a)}_{i}$ is a unit vector after elastic deformation and martensitic transformation in the direction of slip deformation lying on the slip plane, $m^{et(a)}_{i}$ is the normal vector on the slip plane after elastic deformation and martensitic transformation, $\dot{\gamma}^{(a)}$ is shear strain rate on the slip plane and index (a) means the ath slip system. $\ell^{e(I)}_{i}$ is a unit vector in the direction of deformation due to martensitic transformation after elastic deformation, $n^{e(I)}_{i}$ is the normal vector on a habit plane after elastic deformation and $\dot{\gamma}^{t(I)}$ is transformation strain rate. To elucidate a mechanism of the occurrence of martensitic transformation in the microstructure, the model is combined with the cellular automata approach using a local rule on the basis of thermodynamics-based transformation drive force and new computational method by a use of the finite element method is established.

To link the above-mentioned macroscopic model to phenomena in a lower spatial hierarchy, the asymptotic homogenization method is introduced (Iwamoto 2004). Obeying the conventional homogenization procedure on velocity and temperature fields, rearranging obtained governing equations with respect to the same order in η, which is the non-dimensional scale parameter, and taking the limit of $\eta \rightarrow +0$, the following homogenized expressions of a rate form of the virtual work principle and weighted residual method for velocity and temperature fields can be obtained as

$$\int_{\Omega} \left(L^H_{ijkl} + \sigma^H_{ijkl}\right) \dot{E}^0_{kl} \frac{\partial \delta v_i}{\partial x_j} d\Omega = \int_{\Omega} \left(A^H_{ij} + B^H_{ij} \dot{T}^0\right) \frac{\partial \delta v_i}{\partial x_j} d\Omega + \int_{S_t} \dot{F}_i \delta v_i dS + \int_{\Omega} \rho^H \dot{G}^H_i \delta v_i d\Omega, \quad (7.5)$$

$$\int_{\Omega} J^H \dot{T}^0 \delta T d\Omega + \int_{\Omega} K^H_{ij} \frac{\partial T^0}{\partial x_j} \frac{\partial \delta T}{\partial x_i} d\Omega = \int_{\Omega} F^H \delta T d\Omega + \int_{S_q} \bar{Q} \delta T dS \quad (7.6)$$

where the notation $\dot{E}^0_{ij}$ indicates the macroscopic strain rate. L^H_{ijkl}, σ^H_{ijkl}, A^H_{ij}, B^H_{ij}, K^H_{ij}, J^H, ρ^H and F^H are the homogenized moduli given by the following equations as

$$L^H_{ijkl} = \frac{1}{|Y|} \int_Y \left(L_{ijkl} - L_{ijpq} \frac{\partial \chi^{kl}_p}{\partial y_q}\right) dY, \quad L_{ijkl} = D^v_{ijkl} - F_{ijkl},$$

$$F_{ijkl} = \frac{1}{2} \left(\sigma_{lj}\delta_{ki} + \sigma_{kj}\delta_{li} + \sigma_{li}\delta_{kj} + \sigma_{ki}\delta_{lj}\right)$$

$$\sigma^H_{ijkl} = \frac{1}{|Y|} \int_Y \left(\sigma_{lj}\delta_{ki} - \sigma_{qj}\delta_{pi} \frac{\partial \chi^{kl}_p}{\partial y_q}\right) dY,$$

$$A^H_{ij} = \frac{1}{|Y|} \int_Y \left\{ P_{ij}\Theta_1 + \left(D^v_{ijkl} - \sigma_{ij}\delta_{kl}\right) \dot{\varepsilon}^{tr}_{kl} - \left(L_{ijkl} + \sigma_{lj}\delta_{ki}\right) \frac{\partial \varphi_k}{\partial y_l} \right\} dY,$$

$$B^H_{ij} = \frac{1}{|Y|} \int_Y \left\{ \left(B^e_{ij} + P_{ij}\Theta_2 - 3\alpha_T \sigma_{ij}\right) - \left(L_{ijkl} + \sigma_{lj}\delta_{ki}\right) \frac{\partial \Phi_k}{\partial y_l} \right\} dY,$$

$$K^H_{ij} = \frac{1}{|Y|} \int_Y \kappa_T \left\{ \delta_{ij} - \frac{1}{2} \left(\frac{\partial \phi_i}{\partial y_j} + \frac{\partial \phi_j}{\partial y_i} \right) \right\} dY$$

$$\rho^H = \frac{1}{|Y|} \int_Y \rho dY, \quad J^H = \frac{1}{|Y|} \int_Y \rho C dY, \quad F^H = \frac{1}{|Y|} \int_Y \left(\zeta \sigma_{ij} \dot{\varepsilon}^p_{ij} + \dot{\gamma}^i \right) dY,$$

Equations (7.5) and (7.6) are the governing equations for macroscopic velocity and temperature fields, respectively. χ^{kl}_i, φ_i, Φ_i and ϕ_i are the characteristic functions defined in the unit cell, which satisfy the Y—periodic condition. As a result, it becomes possible to estimate deformation and transformation behavior in both macro and microscopic regions.

7.3 Validation by Experiments

7.3.1 Macroscopic Model

The one of authors conducted uniaxial tensile and compressive tests in the temperature range from liquid nitrogen temperature to the room temperature. As a result, macroscopic uniaxial stress and the martensitic volume fraction-strain relationships can be obtained. From the obtained experimental results, material constants and parameters in the above-mentioned models including uniaxial stress-strain relationship of martensitic phase, which is more difficult to determine, based on the optimization technique by a combination between the nonlinear least-square method and the finite element method (Iwamoto et al. 2001). Figure 7.3 indicates (a) the true stress—plastic strain and the (b) martensitic volume fraction—plastic strain simulated using the parameters in the constitutive models obtained by the least-square technique at each test temperature. As shown in this figure, the developed model is possible to reproduce experimental results successfully.

To measure strain-induced martensitic transformation behavior under high-speed compression in real time, a method to capture both the temperature change and the microstructural change macroscopically has been developed during an impact compression test using the split Hopkinson pressure bar (SHPB) method (Iwamoto et al. 2008). Figure 7.4 shows a (a) the nominal compressive stress—nominal strain obtained by the FE simulation and test based on SHPB method and (b) the change in volume resistivity measured during compressive test—nominal strain relationship.

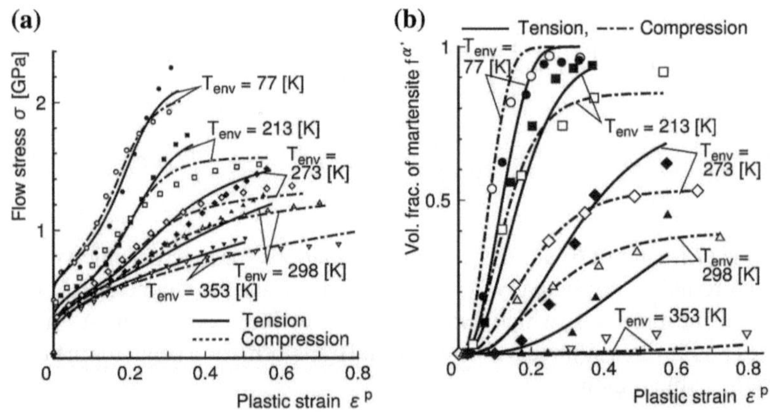

Fig. 7.3 Computational results of uniaxial tensile and compressive tests for TRIP steel using the parameters in the constitutive models obtained by the least-square technique at each test temperature with experimental results; **a** the true stress—plastic strain, and **b** volume fraction of martensite—plastic strain.

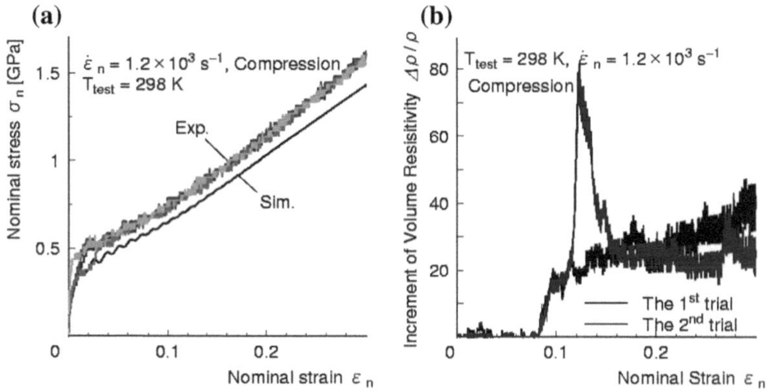

Fig. 7.4 Computational and experimental result of dynamic compressive behavior of TRIP steel; **a** the nominal compressive stress—nominal strain obtained by the FE simulation and test based on SHPB method, and **b** the change in volume resistivity measured during compressive test—nominal strain relationship (Iwamoto et al. 2008)

From this figure (a), the developed model (7.1) can reproduce experimental results for not only quasi static but also impact tests successfully as similar to Fig. 7.3. Additionally, from this figure (b), it can be observed that the change in measured volume resistivity is increased with increase in nominal strain as similar to Fig. 7.4b. Furthermore, the computational results obtained by FE simulations of fracture behavior of CT specimen under mode I loading (Iwamoto and Tsuta 2002b) and cyclic deformation behavior at two different temperature Tomita and Iwamoto (2001) indicate qualitative agreement with experimental results. From these simulations, it is clarified that higher functionalization can be realized by a similar mechanism as above-mentioned for the uniaxial deformation.

7.3.2 Analyzes of Micrographs to Discuss the Microstructural Change

On the other hand, to clarify the relationship between macroscopic mechanical properties of TRIP steel and geometrical features of the transformed phase in micrographs, micrographs of the specimen after the experiment are taken and geometrical features in microregions are analyzed by an image processing analysis method based on the Fourier transformation (Iwamoto and Tsuta 2003). As a result, it becomes possible to set up a guideline of how to control the microstructure of the transformed phase by an arrangement of the huge amount of data by processing the images. Figure 7.5 indicates orientation distribution functions obtained by the analysis of micrographs of the specimens after tensile deformation at 77 K when the plastic strain are (a) 0.03 and (b) 0.28. In this figure, the angle where a peak can be seen in the relative

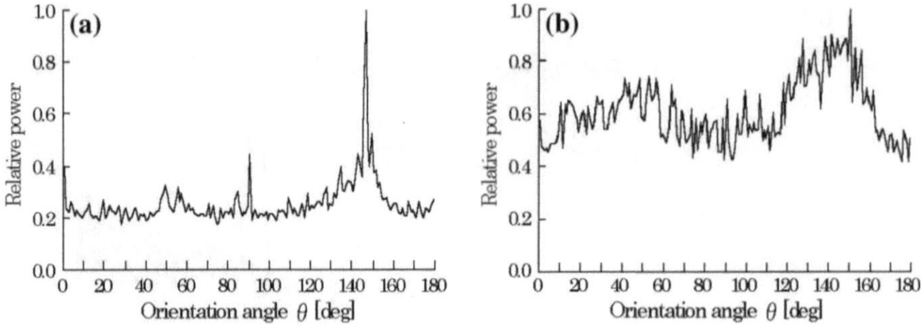

Fig. 7.5 Dependence of the plastic strain ε^p on orientation distribution functions; **a** for 0.03 of the plastic strain, and **b** for 0.28 of the plastic strain

power corresponds the orientation angle of martensitic phase. It can be understood that martensitic phase orients in the direction of 145° because a peak around 145° can be observed in the early stage of deformation as shown in this figure (a). Then, as shown in the figure (b), an another peak appears around 45° with progressive deformation and the magnitude of the peak becomes the similar level with the peak in the orientation of 145°. Therefore, it can be summarized that at first martensitic phase orients in the direction of 145° and then martensitic phase is also transformed with the orientation around 45° with promotion of deformation.

7.4 Obtained Results and Discussions

7.4.1 Spatial Multiscale Phenomena

Figure 7.6 shows (a) true stress and work hardening rate— true strain and (b) volume fraction of martensite—nominal strain with each test temperature for tensile deformation of a rounded bar simulated using the above-mentioned macroscopic model (7.1) $\sim$ (7.3). It can be found that the amount of the martensitic phase is increased with tensile deformation of the bar as shown in this figure (b). Additionally, it can be observed that the relationship between the volume fraction and strain becomes linear around room temperature. This means continuous transformation during deformation around the room temperature can be found. On the other hand, if a satisfaction of Considere's condition which the work-hardening rate becomes equal to true stress is obeyed, a white circle expresses an onset point of necking in the point in this figure (a). Conventional metallic materials indicate a decrease in ductility with decreasing temperature, however, TRIP steel shows that ductility is increased suddenly in the temperature range around the room temperature as shown in this figure (a). It is also able to reproduce the phenomenon observed already in the experiment by the numerical simulation. Actually, it cannot be denied about the suppression of progressed

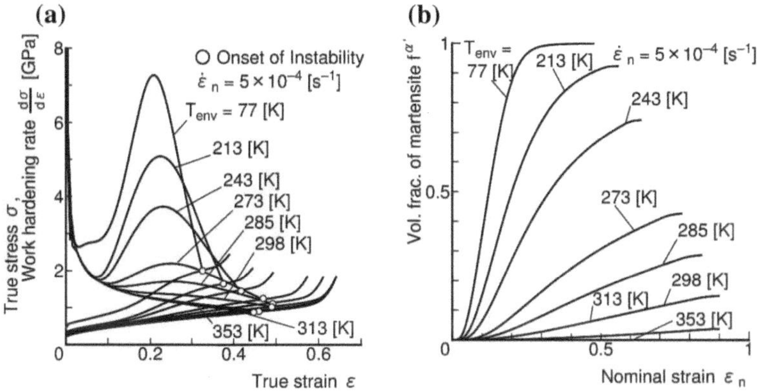

Fig. 7.6 Computational results of tensile deformation of a rounded bar simulated using the above-mentioned macroscopic model with each test temperature; **a** true stress and work hardening rate—true strain, and **b** volume fraction of martensite—nominal strain

necking by an interspersing effect of hard martensitic phase on the neck as the former experimental results of microscopic observations. However, it can be considered that transformed martensitic phase contributes to a retarding effect of an onset of necking by generating continuous transformation with promotion of deformation from this macroscopic computational result (Tomita and Iwamoto 1995).

Figure 7.7 presents distribution of equivalent plastic strain $\bar{\varepsilon}^p$ at environmental temperature T_{env} = 150 K with nominal strain rate $\dot{\varepsilon}_n = 5 \times 10^{-4}$ s^{-1} for nominal strain $\varepsilon_n = 0.06$ in the cases of tensile directions $\Theta = 0$ and 45° during a growth process of a ellipsoidal martensitic phase in a unit cell simulated using the asymptotic homogenization technique expressed in Eqs. (7.5) and (7.6) (Iwamoto 2004). In this figure, an interior side in the ellipsoid indicates martensitic phase. In the case of Θ= 45°, $\bar{\varepsilon}^p$ becomes larger around the martensitic particles and $\bar{\varepsilon}^p$ in both phases is higher than that in the case of $\Theta = 0$. A region of high $\bar{\varepsilon}^p$ can be observed particularly near the boundary in the direction perpendicular to Θ. In the case of $\Theta = 45°$, the localization of $\bar{\varepsilon}^p$ in the austenite region can be observed across the two corners. Therefore, $f^{\alpha'}$

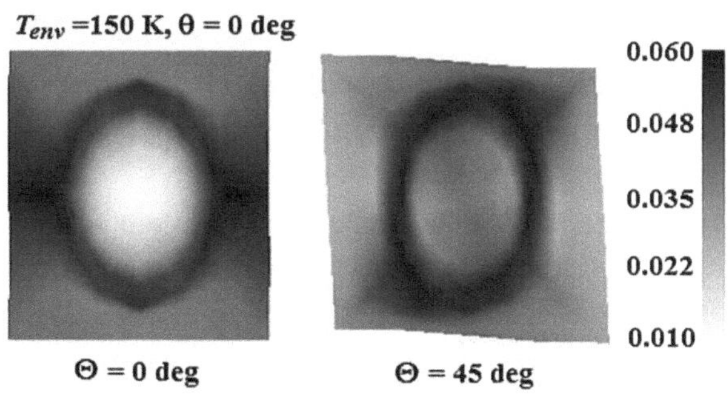

Fig. 7.7 The distribution of equivalent plastic strain $\bar{\varepsilon}^p$ in the unit cell at T_{env} = 150 K with $\dot{\varepsilon}_n = 5 \times 10^{-4}$ s^{-1} for $\varepsilon_n = 0.06$ in the cases of tensile directions $\Theta = 0$ and 45° (Iwamoto 2004)

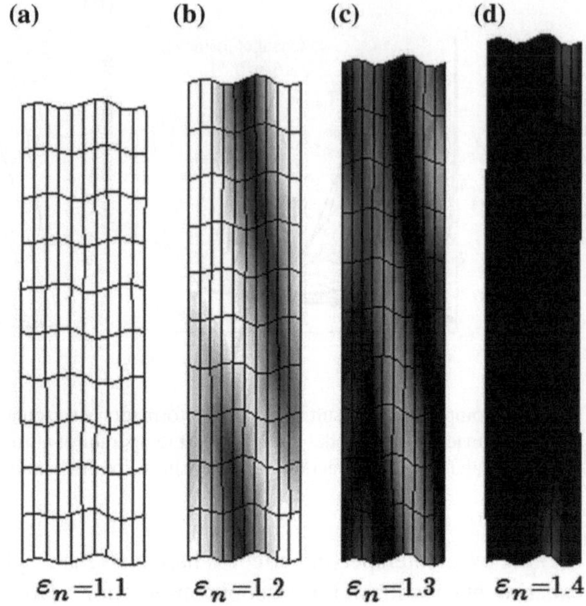

Fig. 7.8 Distributions of martensitic phase with promotion of deformation in the case of initial crystal orientation $\phi = 60°$ for the nominal strain of **a** 1.1, **b** 1.2, **c** 1.3, and **d** 1.4

and the size of elliptic martensite with $\Theta = 45°$ increases. Figure 7.8 shows a process of the microstructural change at each ε_n in the single crystal TRIP steel simulated using the crystal plasticity theory including SIMT expressed in Eq. (7.4) (Iwamoto and Tsuta 2004) when an initial crystal orientation of parent phase is 60° against the horizontal direction. In the figure, martensitic and austenitic phases indicate black and white, respectively. As shown in this figure, nothing happens until the certain level of macroscopic strain and then the periodic band-like structure is suddenly appeared along the perpendicular direction to the initial orientation. After that, the bands of the product phase are extending to the direction. Finally, the region almost transforms to the fully martensitic state. It can be considered that the simulated band-like microstructure reproduces geometric characteristic of the micrstructure usually observed by a microscope.

7.4.2 *Temporal Multiscale Phenomenon on the Energy Absorption Characteristic*

In order to discuss the temporal multiscale phenomenon in TRIP steel, the energy absorption is focused upon here. Generally, energy absorption of materials themselves can be evaluated from the stress-strain curve obtained by a tensile test. Nevertheless, at the high deformation rate, an interaction between a plastic wave and unstable deformation during tension induces the difficulties to avoid an onset of a necking in a specimen. As a result, the specimen is always fractured earlier at a high velocity of tensile loading than a required period (Nemes and Eftis 1993). Thus, the mechanism of the capability of absorbing energy for the tensile deformation is quite complicated, especially at high strain rate.

As the above-mentioned from Fig. 7.2, the strain rate sensitivity represents the time scale and becomes one of the deformation characteristic to discuss the energy absorption because it can be calculated by an increase in the external force since the area surrounding the normalized force-normalized deflection curve until a certain level of normalized deflection increases because of a rate-sensitive hardening effect. Normally, temperature rise in the material due to adiabatic heating by an inelastic irreversible work is occurred. At the same time, this heating phenomenon introduces thermal softening of the material. The rate sensitivity in TRIP steel includes the effects of the rate-sensitive hardening, SIMT and temperature change as shown in Fig. 7.2.

7.4.2.1 Strain Rate Sensitivity

The rate sensitivity in TRIP steel can be divided by two factors. One is the effect of temperature on the kinetics of SIMT. Temperature rise with an increase in strain rate induces stabilization of the austenitic phase. Thus, the condition prevents further martensitic evolution (Hecker et al. 1982; Murr et al. 1982; Nanga et al. 2009; Talonen et al. 2014).

The other is the rate sensitivity of a microscopic shear band. This part is already understood and reported for fully austenitic stainless steels. The increase in strain rate results in an acceleration of the martensitic transformation at low strain as reported by Hecker et al. (1982) and Murr et al. (1982). They studied the strain rate sensitivity on the stress-strain response of type 304 stainless steel over a range of strain rate of 10^{-3} to 10^{3} s^{-1}. It is observed that the number of transformation site as well as the α'-martensite content at low strains increases with increasing strain rate. However, the adiabatic heating at high strain rate results in lower martensite content at strain above 0.25 as above-mentioned. Talonen et al. (2014) and Nanga et al. (2009) studied the effect of strain rate on martensitic transformation in type 301LN stainless steel over a range of strain rate from 3×10^{-4} to 200 s^{-1}. Both authors reported that increasing the strain rate halts the martensite transformation because of stabilizing austenite. The strain rates during the crushing of automotive components may be up to approximately 1000 s^{-1}, while forming operations involve lower intermediate strain rates (Talonen et al. 2014).

7.4.2.2 Three-Point Bending Test for the Pre-cracked Specimen of TRIP Steel

The bending deformation mode is considered to be relatively simple than tensile test. In fact, the smooth specimen made of ductile materials can be deformed at high deflection rate without fracture and any plastic instabilities. Thus, energy can be dissipated just only into inelastic bending deformation with phase transformation process during bending deformation of TRIP steel and energy absorption capability can be evaluated through an investigation of bending deformation behavior of smooth specimen. At the same time, energy absorption capability during fracture process of materials is considered to be associate d with fracture toughness (Nakayama 1964). Thus, the evaluation of the energy absorption is equivalent to measure J-integral from force-displacement curve at an onset point of a crack extension. The fracture toughness can be determined as the critical value of fracture mechanical parameters such as stress-intensity factor, J-integral, etc.

On the other hand, three-point bending test under impact loading has become an important tool in the determination fracture toughness. The pre-cracked three-point bending specimen is widely used to measure the fracture toughness of materials under static (ASTM Standard E399-90 2002) and dynamic loading (Kobayashi et al. 1986; Kalthoff 1985; Server et al. 1977). Figure 7.9 shows a relationship between the J-integral and the normalized deflection rate obtained from the three-point bending test of TRIP steel (Shi et al. 2013). It can be observed that there is an approximately linear relationship between J-integral and the normalized deflection rate in the semi-logarithmic plot. The positive rate-sensitivity can be observed that the J-integral in TRIP steel increases with increasing the deflection rate.

In three-point bending test, the inertia influence the stiffness and the natural frequency of pre-cracked specimen. In addition, the inertia effect can be an influence on the behavior of the pre-cracked bending specimens under impact loading, in particular at the very beginning of the impact process. Therefore, it should be considered in calculating the dynamic fracture toughness (Jiang et al. 2004; Atluri et al. 1984; Iwamoto and Tsuta 2002b). The observed finding in the paper by Bohme and Kalthoff (1982) illustrated the quite complicated influences of dynamic effects on the behavior of impacted specimens.

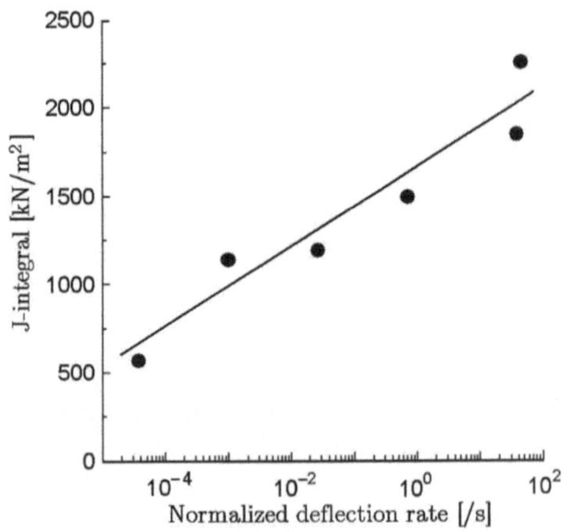

Fig. 7.9 Relationship between the J-integral and the normalized deflection rate of the three-point bending test (Shi et al. 2013)

7.5 Summary

In order to summarize research works done by the one of the authors on the mechanism of high functionalization in TRIP steel from the viewpoint of spatio-temporal hierarchy, first, a spatio-temporal macroscopic model for TRIP steel proposed by one of authors were described and two spatial multiscale models also proposed by the one of author were reviewed. After a validation by comparing with the experimental results was confirmed and the applicability of the model was discussed, results on geometrically orientation distribution function of martensite in micrographs were shown by an image analysis based on the Fourier transformation to discuss the geometrical orientation of martensite. Then, some computational results with the different spatial length scales were introduced. Finally, an importance of the temporal multiscale modeling was addressed with introducing a quite interested experimental result on the energy absorption of TRIP steel.

The authors believe that new findings related to understand quite complicated phenomena with the plastic deformation and enhancement of mechanical properties by means of the martensitic transformation have already been provided on the basis of considering a hierarchical structure of space and time. In addition, advanced techniques are proposed to control the mechanical properties and the validity and applicability of the techniques are shown by the computational simulation and experiment. Obtained fruitful outcomes can be sufficiently exploited for requests from the industry to predict the performance of a material and a process design on various kind of work.

Acknowledgments The authors gratefully acknowledge Professor Nobutada Ohno for his encouragement on our project all the time.

References

Altenbach H, Kolupaev V (2015) Classical and non-classical failure criteria. In: Altenbach H, Sadowski T (eds) Failure and damage analysis of advanced materials, CISM courses and lectures no. 560. Springer, Wien, New York, pp 1–66

Asaro R (1983) Crystal plasticity. Trans ASME J Appl Mech 50:921–934

ASTM Standard E399-90 (2002) Standard test method for plane strain fracture toughness of metallic materials. Annual book of ASTM standards. ASTM International, Baltimore, pp 443–473

Atluri S, Nishioka T, Nakagaki M (1984) Incremental path-independent integrals in inelastic and dynamic fracture mechanics. Eng Frac Mech 20:209–244

Bohme W, Kalthoff J (1982) The behavior of notched bend specimens in impact testing. Int J Fract 20:139–143

Chatterjee S (2006) Transformations in TRIP-assisted steels: microstructure and properties. PhD thesis, Darwin College, University of Cambridge

Cortes J, Tsuta T, Mitani Y, Osakada K (1992) Flow stress and phase transformation analyses in the austenitic stainless steel under cold working (part I, phase transformation characteristics and constitutive formulation by energetic criterion). JSME Int J 35:201–209

Dan W, Zhang W, Li S, Lin Z (2007) A model for strain-induced martensitic transformation of TRIP steel with strain rate. Comp Mater Sci 40:101–107

Delannay L, Lani F, Jacques P (2005) Simulation of deep-drawing of TRIP-assisted multiphase steel based on a micro-macro modelling. In: Boudeau N (ed) 24th international deep-drawing research group Congress, Kluwer, pp 1–10

Fischer F, Reisner G, Werner E, Tanaka K, Cailletaud G, Antretter T (2000) A new view on transformation induced plasticity (TRIP). Int J Plast 16:723–748

Greenwood G, Johnson R (1965) The deformation of metals under small stresses during phase transformation. Proc Roy Soc A 283:403–422

He Y, Sun Q (2010) Rate-dependent domain spacing in a stretched NiTi strip. Int J Solids Struct 47:2775–2783

Hecker S, Stout M, Staudhammer K, Smith L (1982) Effects of strain state and strain rate on deformation-induced transformation in 304 stainless steel: part 1. Magnetic measurements and mechanical behavior. Metall Trans A 13A:619–626

Iwamoto T (2004) Multiscale computational simulation of deformation behavior of TRIP steel with growth of martensitic particles in unit cell by asymptotic homogenization method. Int J Plast 20:841–869

Iwamoto T, Tsuta T (2000) Computational simulation of the dependence of the austenitic grain size on the deformation behavior of TRIP steels. Int J Plast 16:791–804

Iwamoto T, Tsuta T (2002a) Assessment of geometrical orientation of martensitic particles in TRIP steel by Fourier and wavelet transformation image analyses. Key Eng Mater 233–236:627–636

Iwamoto T, Tsuta T (2002b) Computational simulation on deformation behavior of CT specimens of TRIP steels under mode I loading for evaluation of fracture toughness. Int J Plast 18:1583–1606

Iwamoto T, Tsuta T (2003) Assessment geometrical orientation of martensitic particles in TRIP steel by Fourier and wavelet transformation image analysis. Key Eng Mater 233–236:627–636

Iwamoto T, Tsuta T (2004) Finite element simulation of martensitic transformation in single-crystal TRIP steel based on crystal plasticity with cellular automata approach. Key Eng Mater 274–276:679–684

Iwamoto T, Tsuta T, Tomita Y (1998) Investigation on deformation mode dependence of strain-induced martensitic transformation in TRIP steels and modeling of transformation kinetics. Int J Mech Sci 40:173–182

Iwamoto T, Kawagishi Y, Tsuta T, Morita S (2001) Identification of constitutive equation for TRIP steel and its application to improve mechanical properties. JSME Int J Ser A 44:443–452

Iwamoto T, Cherkaoui M, Sawa T (2008) A study on impact deformation and transformation behavior of TRIP steel by finite element simulation and experiment. Int J Mod Phys B 22:5985–5990

Jiang F, Rohatgi A, Vecchio K, Cheney J (2004) Analysis of the dynamic responses for a pre-cracked three-point bend specimen. Int J Fract 127:147–165

Kalthoff J (1985) On the measurement of dynamic fracture toughness—a review of recent work. Int J Fract 27:277–298

Kazanci Z, Bathe K (2012) Crushing and crashing of tubes with implicit time integration. Int J Impact Eng 42:80–88

Keeler S, Kimchi S (2014) Advanced high-strength steels application guidelines version 5.0. Technical report 1–3, World Auto Steel

Kobayashi T, Yamamoto I, Ninomi M (1986) Evaluation of dynamic fracture toughness parameters by instrumented Charpy impact test. Eng Frac Mech 24:773–782

Leblond J, Mottet G, Devaux J (1986) A theoretical and numerical approach to the plastic behavior of steels during phase transformations II: study of classical plasticity for ideal-plastic phases. J Mech Phys Solids 34:411–432

Lichtenfeld J, Tyne C, Mataya M (2006) Effect of strain rate on stress-strain behavior of alloy 309 and 304L austenitic stainless steel. Metall Mater Trans A 37:147–161

Magee C (1966) Transformation kinetics, micro-plasticity and aging of martensite in Fe—31Ni. PhD thesis, Carnegie Institute of Technology, Pittsburgh, PA

Matsumura O, Sakuma Y, Takechi H (1987) Enhancement of elongation by retained austenite in inter critical annealed 0.4 C-1.5 Si-0.8 Mn steel. Trans Iron Steel Inst Jpn 27:570–579
Murr L, Staudhammer K, Hecker S (1982) Effects of strain state and strain rate on deformation-induced transformation in 304 stainless steel: part 2. Microstructural study. Metall Trans A 13A:626–635
Nakayama J (1964) A bending method for direct measurement of fracture energy of brittle material. Jpn J Appl Phys 3:422–423
Nanga S, Pineau A, Tanguy B, Nazé L, Santacreu P (2009) Plasticity and strain induced martensitic transformation in two austenitic stainless steels. In: Olson G, Lieberman D, Saxena A (eds) Proceedings of international conference on martensitic transformations (ICOMAT). Wiley, Hoboken, NJ, pp 341–348
Nemes J, Eftis J (1993) Constitutive modelling on the dynamic fracture of smooth tensile bars. Int J Plast 9:243–270
Olson G, Cohen M (1975) Kinetics of strain-induced martensitic nucleation. Metall Trans A 6A:791–795
Peirce D, Shin C, Needleman A (1984) A tangent modulus method for rate dependent solids. Comp Struct 18:875–887
Sachdev A (1983) Effect of retained austenite on the yielding and deformation behavior of a dual phase steel. Acta Metall 31:2037–2042
Server W, Wullaert R, Sheckherd J (1977) Evaluation of current procedures for dynamic fracture toughness testing. In: Barsom J (ed) Flaw growth and fracture, vol STP 631. ASTM, Philadelphia, pp 446–461
Shi L, Hashimoto S, Iwamoto T (2013) An experimental study on rate sensitivity of J-integral and its evaluation by small punch test for TRIP steel. Eng Trans 61:119–136
Shin H, Ha T, Chang Y (2001) Kinetics of deformation induced martensitic transformation in a 304 stainless steel. Scr Mater 45:823–829
Stringfellow R, Parks D, Olson G (1992) A constitutive model for transformation plasticity accompanying strain-induced martensitic transformations in metastable austenitic steels. Acta Metall 40:1703–1716
Taleb L, Petit S (2006) New investigations on transformation induced plasticity and its interaction with classical plasticity. Int J Plast 22:110–130
Taleb L, Sidoroff F (2003) A micromechanical modeling of the Greenwood-Johnson mechanism in transformation induced plasticity. Int J Plast 19:1821–1842
Talonen J, Aspegren P, Hanninen H (2014) Comparison of different methods for measuring strain induced α'—martensite content in austenitic steels. J Mater Sci Tech 20:1506–1512
Tamura I (1982) Deformation-induced martensitic transformation and transformation-induced plasticity in steels. Met Sci 16:245–253
Tomita Y, Iwamoto T (1995) Constitutive modeling of TRIP steel and its application to the improvement of mechanical properties. Int J Mech Sci 37:1295–1305
Tomita Y, Iwamoto T (2001) Computational prediction of deformation behavior of TRIP steels under cyclic loading. Int J Mech Sci 43:2017–2034
Tomita Y, Shibutani Y (2000) Estimation of deformation behavior of TRIP steels-smooth/ringed-notched specimens under monotonic and cyclic loading. Int J Plast 16:769–789
Zackay V, Parker E, Fahr D, Busch R (1967) Enhancement of ductility in high strength steels. ASTM Trans 60:252–259
Zaera R, Rodriguez-Martinez J, Casado A, Fernandez-Saez J, Rusinek A, Pesci R (2012) A constitutive model for analyzing martensite formation in austenitic steels deforming at high strain rates. Int J Plast 29:77–101

Chapter 8
Methods for Creep Rupture Analysis—Previous Attempts and New Challenges

Zbigniew L. Kowalewski

Abstract The chapter presents selected methods of creep analysis with special emphasis on damage development. It is divided into three main sections. In the first one some previous methods of creep rupture analysis are described. The attention is focused on certain kind of uniaxial creep characterisation of materials, namely, an influence of prior deformation on creep behaviour. Subsequently, the results from creep tests under complex stress states are presented together with theoretical approaches commonly used to their description. In the second section a comprehensive historical survey concerning advances in modelling of creep constitutive equations is discussed. The third section illustrates selected new concepts of damage development due to creep on the basis of data captured from the own experimental programme.

Keywords Creep · Creep damage · Prior deformation · Constitutive equations · Complex stress states · Non-destructive testing

8.1 Introduction

Typical creep phenomenon occurs as a result of long term exposure to high levels of stress that are below the yield point of the material. It is more severe in materials that are subjected to elevated temperature for long periods, and near melting point. It always becomes faster with temperature increase. The rate of this deformation is a function of the material properties, exposure time, exposure temperature and the applied structural load. Depending on the magnitude of the applied stress and its duration, the deformation may become so large that a component can no longer

Z.L. Kowalewski (✉)
Institute of Fundamental Technological Research,
Ul. Pawinskiego 5B, 02-106 Warsaw, Poland
e-mail: zkowalew@ippt.pan.pl

H. Altenbach et al. (eds.), *From Creep Damage Mechanics to Homogenization Methods*, Advanced Structured Materials 64,
DOI 10.1007/978-3-319-19440-0_8

perform its function. Creep is usually of concern to engineers and metallurgists when evaluating components that operate under high stresses or high temperatures. The temperature range in which creep deformation may occur differs in various materials. For example, tungsten requires a temperature in the thousands of degrees before creep deformation can occur while lead will creep near the room temperature 20 °C. The effects of creep deformation generally become noticeable at approximately 30 % of the melting point (as measured on a thermodynamic temperature scale such as Kelvin) for metals and 40 % of melting point for alloys. For typical creep curve one can distinguish three stages. In the initial stage, or primary creep, the strain rate is relatively high, but slows with increasing time. This is due to work hardening. The strain rate eventually reaches a minimum and becomes near constant. This is due to the balance between work hardening and thermal softening. This stage is known as secondary or steady-state creep. In tertiary creep, the strain rate exponentially increases with stress because of necking phenomena.

Creep leads to the development of material damage process. There are two essential periods of such process:

- damage developing without microscopically visible cracks due to the nucleation process and growth of the microvoids, and
- propagation of the dominant fissure up to failure.

At the end of the first stage, the macroscopically observed crack takes place in form of one or several fissures. In the second stage of the rupture process the dominant fissure propagates decreasing, as a consequence, loading admissible capacity of a construction element and leading finally to its failure. In most cases the duration of the second stage of damage process is negligible short in comparison to the exploitation time of an element. Experiments concerning the processes of microcrack nucleation and growth, which are responsible for the failure of materials during creep, exhibit that failure mechanisms can be divided into the three following types (Hayhurst 1972, 1983; Dyson and Gibbons 1987; Abo El Ata and Finnie 1972; Browne et al. 1981; Ashby et al. 1979): brittle, ductile and mechanism being their combination. For brittle failure mechanism the microdefects are created and developed on the grain boundaries perpendicular to the maximum principal tension stress. During ductile failure mechanism the microdefects are created on the grain boundaries and they are developing due to grain boundary slides. Brittle failure mechanism is usually dominant in the case of polycrystalline materials tested at low levels of the uniaxial stress states. Material degradation during this mechanism has the intergranular character. At high stress levels the rupture takes place mainly due to the ductile failure mechanism, for which the damages have a transgranular character and develop due to the slides passing through the grains. It is well known that there are no exact values, which can be treated as the limits for particular failure mechanism domination. For majority of real exploitation loading conditions the failure mechanism seems to be a combination of the simultaneously developing brittle and ductile failure mechanisms.

8.1.1 Uniaxial Creep Tests—Tool for Initial Material Characterization

During manufacturing and exploitation processes most engineering structures or some their elements are subjected to deformation. Therefore, it is important from engineering point of view to know the influence of this deformation on such different material properties at high temperatures as minimum creep rate, ductility, lifetime, rupture and crack propagation. It has been found that plastic deformation at both room and elevated temperatures prior to creep testing has either beneficial or detrimental effect on the material properties (Dyson and Rodgers 1974; Dyson et al. 1976; Kowalewski 1991a, 1992; Marlin et al. 1980; Murakami et al. 1990; Ohashi et al. 1986; Pandey et al. 1984; Rees 1981; Trąmpczyński 1982; Wilson 1973; Xia and Ellyin 1993). Although the problem has been previously studied experimentally for several materials, only limited amount of available data reflects the influence of plastic predeformation on creep process up to rupture (Dyson and Rodgers 1974; Dyson et al. 1976; Marlin et al. 1980; Pandey et al. 1984; Trąmpczyński 1982). It is well known that the problem is particularly important during fabrication or assembly processes, where a number of materials used in critical elements of engineering structures may receive such cold work, and as a consequence, it may change significantly their lifetime. Up to now the amount of experimental data is still insufficient to estimate exactly whether the increase or decrease of creep strengthening occurs up to a certain amount of prior deformation only, or whether this creep property is in some way proportional to the amount of predeformation. Thus, in order to achieve better understanding of this problem further systematic investigations are required.

In this paper in order to identify an influence of prior plastic deformation on the basic creep parameters the results of uniaxial tensile tests obtained for aluminium alloy will be presented.

8.1.2 Multiaxial Creep Tests—Advanced Characterization of Materials

The results from uniaxial creep tests are not able to reflect complex material behaviour. Therefore, many efforts are focused on tests carrying out under multiaxial loading conditions. Such experiments are very difficult not only in execution but also in elaboration of the results.

A description of creep process requires the essential interrelations among stress, strain, and time. The well known method depicting these interrelation-ships under complex stress states, first proposed by McVetty (1934), is through isochronous stress-strain curves obtainable from the standard creep curves. Since that time, many graphical methods of the creep data presentation have been elaborated. It has been found that multi-axial creep rupture results are conveniently plotted in terms of isochronous surfaces (Piechnik and Chrzanowski 1970; Leckie and Hayhurst 1977; Chrzanowski and Madej 1980; Hayhurst et al. 1980; Litewka and

Hult 1989; Kowalewski et al. 1994; Kowalewski 2004) being loci of constant rupture time in a stress space. Such approach especially simplifies theoretical and experimental creep results analysis giving comprehensive graphical representation of material lifetime, and therefore, it will be used as a tool for the creep data presentation. The paper demonstrates comparison of experimental creep data achieved for pure copper (Kowalewski 1995, 1996; Lin et al. 2005) with the results for 2017 aluminium alloy obtained. Theoretical approach for determination of the isochronous surfaces will be discussed, and the data from creep tests will be used to illustrate how to construct an experimental form of such surfaces.

8.1.3 New Concepts of Creep Analysis

Nowadays many new approaches and testing techniques are used for damage assessments. Among them one can generally distinguish destructive (Hayhurst 1972, 1983; Krauss 1996; Lin 2003; Trąmpczyński and Kowalewski 1986; Dietrich and Kowalewski 1997; Kowalewski 2002), and non-destructive methods (Sablik and Augustyniak 1999; Narayan GR 1975; Fel et al. 2001; Martínez-Ona and Pérez 2000; Ogi et al. 2000). Having the parameters of destructive and non-destructive methods for damage development evaluation it is instructive to analyze their variation in order to find possible correlations. This is because of the fact that typical destructive investigations, like creep or standard tension tests, give the macroscopic parameters characterizing the lifetime, strain rate, yield point, ultimate tensile stress, ductility, etc. without sufficient knowledge concerning microstructural damage development and material microstructure variation. On the other hand, non-destructive methods provide information about damage at a particular time of the entire working period of an element, however, without sufficient information about the microstructure and how it varies with time. Therefore, it seems reasonable to plan future damage development investigations in the form of interdisciplinary tests connecting results achieved using destructive and non-destructive methods with microscopic observations in order to find mutual correlations between their parameters. This issue will be demonstrated on the basis of last own results.

8.2 Previous Attempts of Creep Analysis—Selected Examples of Uniaxial and Biaxial Tests

8.2.1 Analysis of Prior Deformation Effect on Creep Under Uniaxial Loading Conditions

It is commonly known that standard tensile creep tests are most often used to characterize a majority of engineering materials. In this section such kind of material testing was applied to identify an influence of prior cold work on creep of 2017 aluminium alloy. Thin-walled tubular specimens of 40 [mm] gauge length, 22 [mm] internal diameter, 0.75 [mm] wall thickness and 140 [mm] total length were used in all tests. The experimental programme comprised creep tests under uniaxial tension carried out at two different temperatures 423 and 473 K. Creep tests were performed for the material in the as-received state and for the same material plastically prestrained at the room temperature. The aluminium alloy specimens were prestrained up to 1.0, 2.0, 6.0 and 8.0 % for both creep test temperatures taken into account.

Investigations of the effect of prior plastic deformation on subsequent creep process were carried out according to the following procedure. First of all, each thin-walled tubular specimen was proportionally deformed up to the selected value of plastic prestrain by uniaxial tension at the room temperature using an Instron testing machine, and then unloaded. Subsequently, each specimen was mounted at the standard creep testing machine, heated uniformly at the chosen test temperature for 24 [h] prior to creep testing, and then subjected to the constant stress level depending on the creep testing temperature. Both creep stress levels selected for tested material were smaller than the value of yield point of the material at the considered temperatures. Diagram of the experimental procedure is schematically presented in Fig. 8.1.

The experimental results for aluminium alloy are presented in Fig. 8.2. As it is clearly seen from this figure, creep process under constant stress is generally affected by prior plastic strain at the room temperature. Cold work preceding the creep induced hardening effect expressed by significant decrease of the minimum creep rate, Fig. 8.3. Similar effect was earlier observed by Trąmpczyński (1982) and Kowalewski (1991a) who tested copper. Taking into account the recovery creep theory based on the Orowan's equation in the following form

$$d\sigma = \left(\frac{\partial \sigma}{\partial \varepsilon}\right) d\varepsilon + \left(\frac{\partial \sigma}{\partial t}\right) dt \tag{8.1}$$

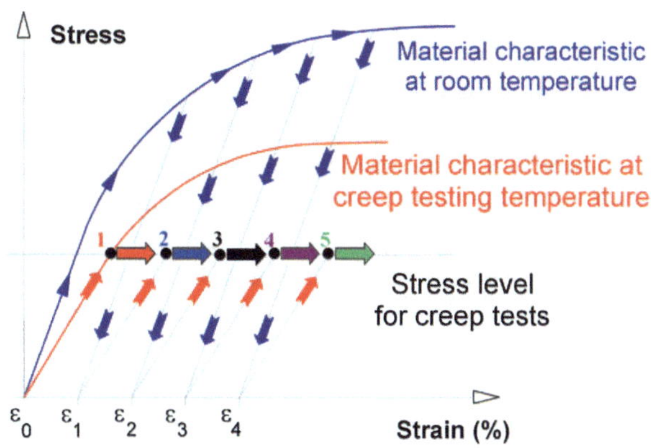

Fig. 8.1 Scheme of the experimental programme (2017 aluminium alloy tested under 300 MPa at 423 K, and under 200 MPa at 473 K—$\varepsilon_0 = 0\,\%$, $\varepsilon_1 = 1.0\,\%$, $\varepsilon_2 = 2.0\,\%$, $\varepsilon_3 = 6.0\,\%$, $\varepsilon_4 = 8.0\,\%$)

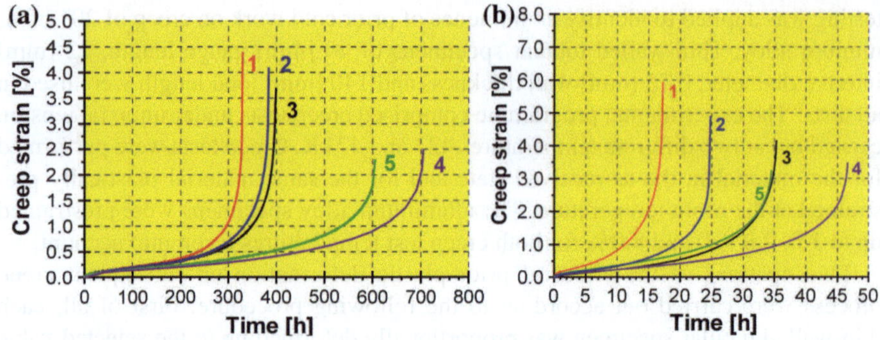

Fig. 8.2 Creep curves of 2017 aluminium alloy at: **a** $\sigma = 300\,\text{MPa}$, $T = 423\,\text{K}$; **b** $\sigma = 200\,\text{MPa}$, $T = 473\,\text{K}$ (*1* material in the as-received state; *2–5* material prestrained up to 1.0, 2.0, 6.0, 8.0 %, respectively (Kowalewski 2005)

Fig. 8.3 Variation of the dimensionless minimum creep rate due to prior plastic deformation for aluminium alloy (Minimum creep rates of nonprestrained material are used as the reference values (Kowalewski 2005))

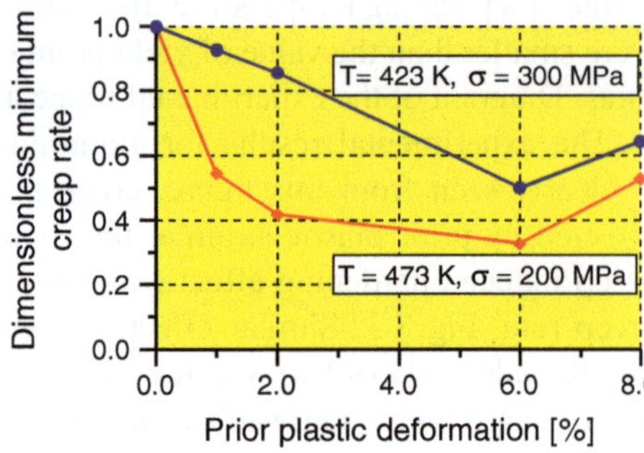

this is an expectable effect. According to this theory a balance between the recovery rate $\partial\sigma/\partial t$ and the rate of strain hardening $\partial\sigma/\partial\varepsilon$ is responsible for constant value of strain rate observed in the second period of the creep process.

Plastic predeformation of a material generates dislocations, the density of which depends on the prestrain amount. Therefore, the plastically prestrained material should creep at lower rate during second period of the process than the nonprestrained one. Taking into account the results of tests carried out at 423 K, Fig. 8.3, it is easy to note that the strain hardening effect observed exhibits gradual increase with the plastic predeformation increase only up to the prior plastic deformation close to 6 %. Over this value the hardening effect expressed by decrease of the secondary creep rate was also remarkable, but its amount was not proportional to the magnitude of prestraining. In the tests carried out for copper at 473 K similar tendency can be observed.

On the basis of the results achieved for 2017 aluminium alloy it may be concluded that the tensile plastic prestrains decrease the secondary creep rate, but the magnitude of this decrease is not proportional to the amount of tensile plastic prestrain. Such behaviour cannot be predicted by the recovery creep theory.

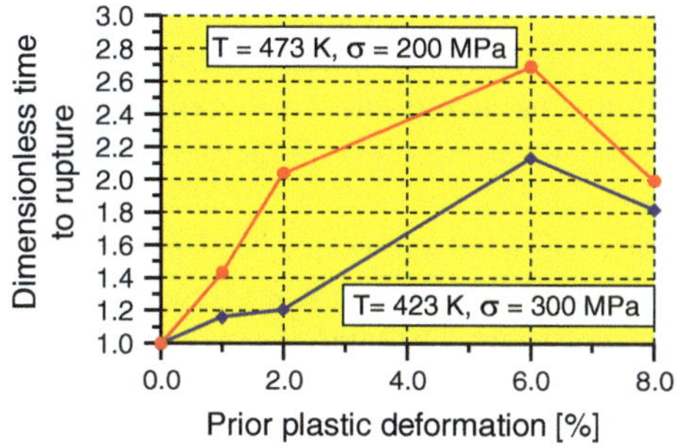

Fig. 8.4 Variation of the dimensionless time to rupture due to prior plastic deformation for aluminium alloy (Times to creep rupture of nonprestrained material are used as the reference values (Kowalewski 2005))

Lifetime extension was obtained for prestrained aluminium alloy at both temperatures considered, Fig. 8.4. For relatively small values of prior plastic deformation (up to 6 %) the mutual relation between the lifetime and the amount of prior plastic deformation was almost proportional. For higher values of plastic prestraining the lifetime extension can be also observed in comparison to the lifetime achieved for the nonprestrained material, however, in these cases mutual relation between the lifetime and the amount of prior plastic deformation was not proportional. It means that for higher magnitudes of plastic deformation (>6 %) the creep lifetime becomes to be smaller, and for sufficiently high magnitude it can reach the lower value than that obtained for the material tested in the as-received state.

Prior plastic deformation also can change the duration of typical creep stages (Table 8.1). The duration of the primary creep period was reduced, in practice, independently on the amount of prior plastic deformation. The duration of secondary creep stage was increased with the increase of the plastic prestrain magnitude.

The ductility during creep was also strongly affected by the prior plastic deformation at room temperature. For both temperatures considered an essential reduction of the total creep strain at rupture was observed.

8.2.2 Creep Tests Under Complex Stress States

The vast majority of the creep-to-rupture investigations have been carried out under uniaxial stress states (Norton 1929; Malinin and Rżysko 1981; Rabotnov 1969; Gittus 1975). Results of such tests have been subsequently used to deter-

Table 8.1 Creep parameters determined from tensile creep tests of 2017 aluminium alloy

	$\sigma = 300$ (MPa), $T = 432$ (K)					$\sigma = 200$ (MPa), $T = 432$ (K)				
ε (%)	0	1.0	2.0	6.0	8.0	0	1.0	2.0	6.0	8.0
$\dot{\varepsilon} \times 10^{-5}$ (1/h)	1.4	1.3	1.2	0.7	09	5.5	3.0	2.3	1.8	2.9
t_I (h)	70	60	50	50	40	1	2	2	4	3
t_{II} (h)	160	180	200	260	250	6	9	12.5	17	12
t_R (h)	330	384	399	705	601	17.3	24.8	35.3	46.6	34.6

mine material constants existing in different theoretical models with the objective to precisely reflect creep behaviour of the material considered. These models are often generalized into multi-axial stress states under assumption of the isotropy of a body examined (Hayhurst 1983; Garofalo 1965; Kachanov 1958; Odqvist 1966). However, the isotropic materials exist in practice rather seldom since manufacturing processes used to produce semi-manufactures, such as rods, tubes, sheets etc., induce anisotropy which cannot be often remove by any heat treatment subsequently applied. In some cases the material can be isotropic in sense of plastic parameters such as yield limit and ultimate tensile strength, but during creep can exhibit anisotropic properties (Kowalewski 1991a, b). In these situations carrying out only uni-axial creep tests to obtain material constants for constitutive model describing material behaviour may lead to significant errors.

In this section the results of biaxial creep tests will be presented to identify phenomena that should be reflected during elaboration of reasonable constitutive equations.

The materials investigated were electrolytic copper of 99.9 % purity and 2017 aluminium alloy (notation according to ASTM). Creep investigations were carried out on thin-walled tubular specimens (40 mm gauge length, 140 mm total length, 22 mm internal diameter, 1.5 mm (copper specimens) or 0.75 mm (aluminium alloy specimens) wall thickness in the gauge length region) with the use of the biaxial creep testing machine enabling realisation of plane stress conditions by simultaneous loading of the specimens by an axial force and twisting moment at elevated temperature.

The experimental programme comprised creep tests up to rupture for copper and aluminium alloy specimens subjected to biaxial stress state obtained by various combinations of tensile and torsional stresses: ($\sigma_{12}/\sigma_{11} = 0, \sigma_{12}/\sigma_{11} = \sqrt{3}/3, \sigma_{12}/\sigma_{11} = \infty$. For each material tests were carried out at three effective stress levels (σ_e): 70.0; 72.5; and 75.0 [MPa] in the case of copper, and 280.0; 300.0; and 320.0 [MPa] in the case of aluminium alloy. The effective stress was defined in the following form:

$$\sigma_e = \left(\frac{3}{2} S_{ij} S_{ij}\right)^{\frac{1}{2}} = \left(\sigma_{11}^2 + 3\sigma_{12}^2\right)^{\frac{1}{2}}, \tag{8.2}$$

where S_{ij}—stress deviator, σ_{11}—axial stress, σ_{12}—shear stress.

Before creep test each specimen was heated uniformly at the test temperature (523 K in the case of copper, and 423 K in the case of aluminium alloy) for 24 h. Creep investigations were carried out until rupture of the specimens was achieved giving as a consequence whole creep curves.

8.2.2.1 Creep Results of Pure Copper

The creep curves up to rupture for copper are presented in Fig. 8.5. The effective creep strain was defined by the relation in the following form:

$$\varepsilon_{\mathrm{e}} = \left(\frac{2}{3}\varepsilon_{ij}\varepsilon_{ij}\right)^{\frac{1}{2}} = \sqrt{\varepsilon_{11}^2 + \frac{4}{3}\varepsilon_{12}^2}, \tag{8.3}$$

where ε_{11} and ε_{12} denote axial and shear strain, respectively.

The creep characteristics obtained at the same effective stress but under different stress states exhibit drastic differences for all stress levels considered. In all cases the shortest lifetimes, and moreover, the lowest ductility have been achieved for

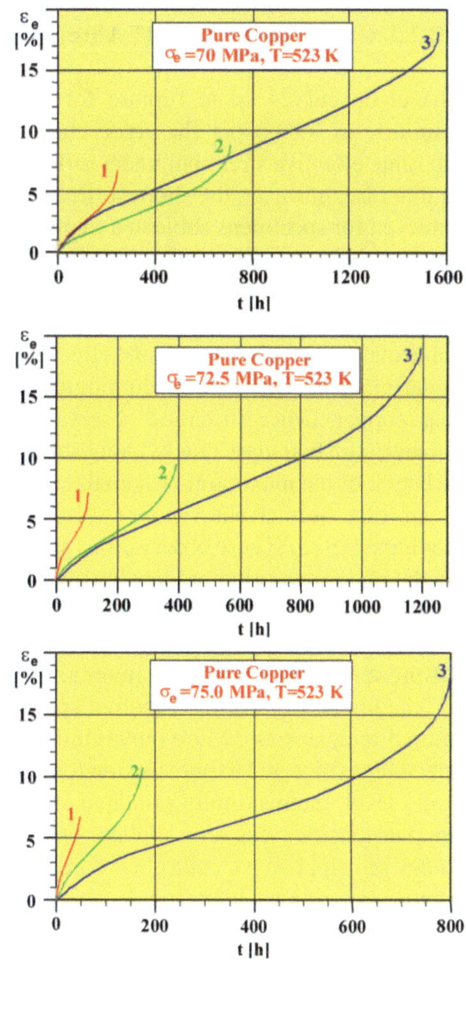

Fig. 8.5 Creep curves for copper: *1* $(\sigma_{12}/\sigma_{11}) = 0$, *2* $(\sigma_{12}/\sigma_{11}) = \sqrt{3}/3$, 3 $(\sigma_{12}/\sigma_{11}) = \infty$

tensioned specimens. The opposite effect was observed for specimens subjected to pure torsion. It has to be emphasized that differences in creep curves due to different loading types applied are great, and they are reflected by variations of the basic creep parameters. Microscopic examination showed that the majority of microcracks were observed at those grain boundaries which were perpendicular to the maximum principal stress. It was confirmed by the shapes of the specimen cross-section in places where rupture occurred. The failure line in each creep rupture test was perpendicular to the maximum principal stress. Since the maximum principal stress was not the same for the same effective stress creep tests, it can be concluded that the resulting variations in lifetimes for the same effective stress tests follow from the differences in magnitude of the maximum principal stress. The longest lifetimes were achieved for pure torsion creep tests for which the maximum principal stresses had the lowest values.

8.2.2.2 Creep Results of 2017 Aluminium Alloy

The creep curves up to rupture for aluminium alloy are presented in Fig. 8.6. Similarly as for copper the creep characteristics of aluminium alloy, obtained at the same effective stress but under different stress states, exhibit drastic differences. In this case, however, the shortest lifetimes, and moreover, the lowest ductility were achieved for specimens subjected to pure torsion. All of the creep parameters which characterise macroscopically creep behaviour prove that the process is a stress state sensitive. More importantly, it has to be noticed that creep behaviour depends on the material type. Analysis of the results for both materials allows to conclude that for some materials tested at the same effective stress the longest lifetime can be achieved under uniaxial tension (e.g. aluminium alloy) whereas for the others under torsion (e.g. copper). Microstructural observations of damage in aluminium alloy showed narrow grain boundary cracks along some grain boundary facets perpendicular to the direction of the maximum principal tension stress in both uniaxial and biaxial stress creep tests. Similar observations were made by Johnson et al. (1962) and later by Hayhurst (1972). These observations suggest that the growth of damage is dependent on the maximum principal tension stress. However, the biaxial tests carried out by both Hayhurst (1972); Johnson et al. (1962) unambiguously showed that the aluminium alloy studied did obey an effective stress criterion. Also, certain aspects of the presented results for aluminium alloy support the latter thesis.

The failure lines of the ruptured specimens were not perpendicular to the maximum principal stress. More importantly, the shortest lifetimes were achieved for the specimens subjected to pure torsion for which, taking into account the same effective stress level, the maximum principal stresses were significantly lower than those at uniaxial tension creep tests applied. The dichotomy between the observation of maximum principal stress controlled damage growth and the observed effective stress rupture criterion is still being discussed, although a suggestion has been made that it is a consequence of tertiary creep being controlled by more than a single damage state variable only (Dyson and Gibbons 1987).

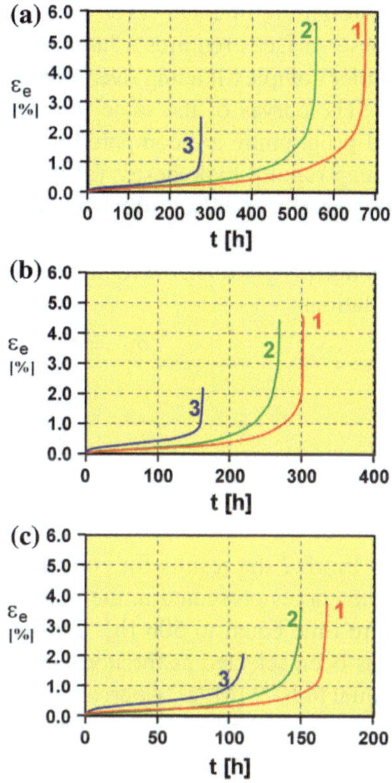

Fig. 8.6 Creep curves for aluminium alloy at 423 K under stress equal to: **a** 280.0 MPa; **b** 300 MPa; **c** 320 MPa; *1* $(\sigma_{12}/\sigma_{11}) = 0$; *2* $(\sigma_{12}/\sigma_{11}) = \sqrt{3}/3$; *3* $(\sigma_{12}/\sigma_{11}) = \infty$

8.2.2.3 Creep Rupture Data Analysis Using Isochronous Surface Concept

The comprehensive presentation and comparison of the experimental data from tests performed at complex stress states procure many difficulties, particularly for stress states being a combination of tension and torsion. In these cases, the data comparison is usually carried out for the effective strains defined in the form of a function of the second invariant of strain tensor, since effects of the first as well as the third invariants are relatively small and they can be often neglected.

Although creep curves in diagrams representing effective strains versus time can be compared, it is difficult to evaluate precisely all differences in material response due to the action of different stress state types. To overcome this deficiency, creep rupture results are commonly presented in the form of isochronous surfaces (Hayhurst 1972, 1983; McVetty 1934; Piechnik and Chrzanowski 1970; Leckie and Hayhurst 1977; Chrzanowski and Madej 1980; Hayhurst et al. 1980; Litewka and Hult 1989; Kowalewski et al. 1994; Kowalewski 1996, 2004; Lin et al. 2005), being loci of constant rupture time in a stress space. This approach especially simplifies theoretical creep results analysis giving the comprehensive graphical representation of the material lifetime. However, the accurate experimental determination of the shape of these surfaces requires a large number of creep rupture data from tests carried out under complex loading over a wide range of stress levels.

The curves of the same time to rupture determined on the basis of experimental programme are compared with theoretical predictions of the three well known creep rupture hypotheses: (a) the maximum principal stress rupture criterion (8.4), (b) the Huber-Mises effective stress rupture criterion (8.5), (c) the Sdobyrev creep rupture criterion (8.6). For the biaxial stress state conditions, realised in the experimental programme, the rupture criteria mentioned above are defined by the following relations:

$$\sigma_{\mathrm{R}} = \sigma_{\max} = \frac{1}{2}(\sigma_{11} + \sqrt{\sigma_{11}^2 + 4\sigma_{12}^2}), \tag{8.4}$$

$$\sigma_{\mathrm{R}} = \sigma_{\mathrm{e}} = \sqrt{\sigma_{11}^2 + 3\sigma_{12}^2}, \tag{8.5}$$

$$\sigma_{\mathrm{R}} = \beta\sigma_{\max} + (1 - \beta)\sigma_{\mathrm{e}} \tag{8.6}$$

In Fig. 8.7 the results for copper are shown, while in Fig. 8.8 for aluminium alloy. The curves presented in the normalised co-ordinate system are referred to the rupture time equal to 500 [h]. Tensile stress corresponding to the lifetime of 500 [h] has been selected as the normalisation factor ($\sigma_{\mathrm{R}\ 500}$). In the case of copper it was equal to 67.9 [MPa], whereas for aluminium alloy—288 [MPa]. As it is clearly seen for copper, the best description of the experimental data has been achieved for the Sdobyrev creep rupture criterion taken with the coefficient $\beta = 0.4$, calculated on the basis of creep data from tests carried out. The value of β indicates that the damage mechanism governed by the effective stress as well as the maximum principal stress played a considerable role in the creep rupture of the copper tested. Contrary to the results achieved for copper, the best fit of the aluminium alloy data is obtained using the effective stress rupture criterion. It has to be noting however, that the lifetimes predicted by this criterion are still quite far from experimental data.

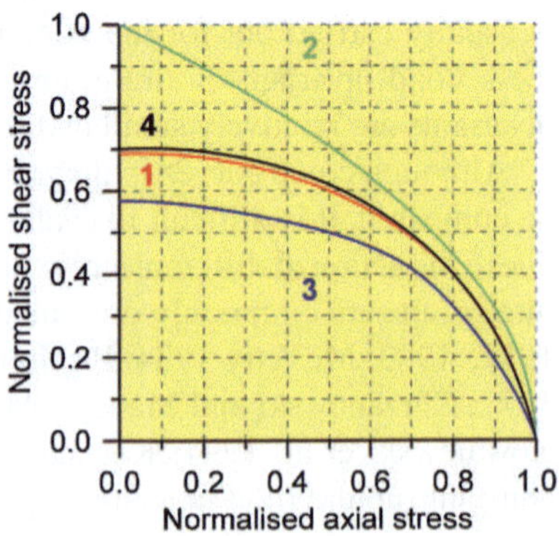

Fig. 8.7 Comparison of the isochronous creep rupture surfaces ($t_{\mathrm{R}} = 500$ [h]) determined for copper (*1* experimental results; *2–4* theoretical predictions using the maximum principal stress criterion; the effective stress criterion; and the Sdobyrev criterion, respectively)

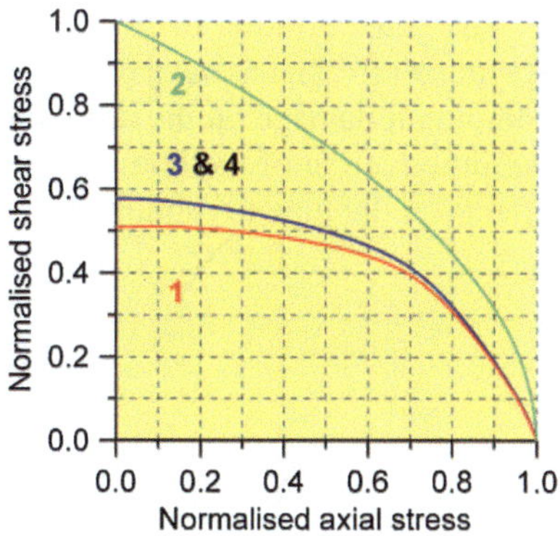

Fig. 8.8 Comparison of the isochronous creep rupture surfaces ($t_R = 500$ [h]) determined for aluminium alloy (*1* experimental results; *2*–*4* theoretical predictions using the maximum principal stress criterion; the effective stress criterion; and the Sdobyrev criterion, respectively)

8.3 A Short Survey on Advances in Modelling of Creep Damage Development

A definition of damage measure is treated as the essential problem taking place in creep rupture analysis. In 1958 Kachanov has introduced a scalar measure of damage in the form of parameter of cross-section continuity, which becomes to be 1 at the beginning of the deformation process and 0 at a localized failure of the material (Kachanov 1958). It corresponds to the assumption that the load is only carrying by the effective part of the specimen cross-section being a difference between the initial cross-section and the damage area, i.e. the area resulted from the sum of all voids or fissures areas. Rabotnov (1969) modified the Kachanov's damage measure giving more convenient measure being the complementary parameter to that proposed by Kachanov. It is defined in the following form

$$\omega = 1 - \psi = \frac{A_0 - A}{A_0}, \qquad 0 \le \omega \le 1 \tag{8.7}$$

and physically can be interpreted as the area of all defects referred to the undamaged initial cross-sectional area. Using this damage parameter the creep constitutive equation set for uniaxial stress state can be written in the following normalised form:

$$\frac{\dot{\varepsilon}}{\dot{\varepsilon}_0} = \frac{1}{(1-\omega)^m}\left(\frac{\sigma}{\sigma_0}\right)^n, \qquad \frac{\dot{\omega}}{\dot{\omega}_0} = \frac{1}{(1-\omega)^\eta}\left(\frac{\sigma}{\sigma_0}\right)^\nu, \tag{8.8}$$

where $n, m, \nu, \eta, \dot{\varepsilon}_0, \dot{\omega}_0, \sigma_0$ are material constants.

For constant stress level it is easy to integrate the equations in set (8.8) to give the time variations of strain and damage. By applying the rupture condition $\omega = 1$, it is possible to determine time to rupture t_R (Leckie and Hayhurst 1977).

The generalisation of Eq. (8.8) to multiaxial stresses, proposed by Leckie and Hayhurst (1977), has been achieved by making the assumption that the influence of continuum damage on the deformation rate process is of a scalar character, and by the introduction of the homogeneous stress function which reflects the stress state effects on the time to rupture. Equation (8.8) can then be written as:

$$\frac{\dot{\varepsilon}_{ij}}{\dot{\varepsilon}_0} = \frac{3}{2}\left(\frac{\sigma_{\mathbf{e}}}{\sigma_0}\right)^{n-1}\left(\frac{S_{ij}}{\sigma_0}\right)\frac{1}{(1-\omega)^n}, \tag{8.9}$$

$$\frac{\dot{\omega}}{\dot{\omega}_0} = \Delta^{\nu}\frac{1}{(1+\eta)(1-\omega)^{\eta}}, \tag{8.10}$$

where $\Delta = \Delta(\sigma_{ij}/\sigma_0) = \sigma_{\max}/\sigma_0$ for copper, and $\Delta = \Delta(\sigma_{ij}/\sigma_0) = \sigma_e/\sigma_0$ for aluminium alloys. Integration of the damage evolution equation (8.10) for the following boundary conditions: $\omega = 0, t = 0$ and $\omega = 1, t = t_{\mathbf{R}}$, yields after normalisation to the relation describing time to rupture in the form:

$$\frac{t_R}{t_0} = \frac{1}{\Delta^{\nu}} \tag{8.11}$$

Substitution of $t_R = t_0$ in Eq. (8.11) gives the equation of the isochronous surface.

It has been found convenient to present the rupture results in terms of the isochronous surface representing stress states with the same rupture times. According to (Johnson et al. 1956, 1962), the rupture criteria for aluminium alloy and pure copper appear to represent the extremes of material behaviour, since the isochronous surface for many metals lies somewhere between these criteria. They have shown that the dependence of the rupture time upon the nature of the applied stress system for an aluminium alloy can be described by the octahedral shear stress criterion, whereas for pure copper—by the maximum principal stress criterion. In spite of the fact that these observations have been made on the basis of a relatively limited amount of the experimental data, and in certain cases did not give precise description of rupture, they are still influencing the process of developing new creep damage models (Hayhurst et al. 1980; Litewka and Hult 1989; Kowalewski et al. 1994; Kowalewski 2004; Lin et al. 2005; Dyson and McLean 1977; Sdobyrev 1959).

Multiaxial creep constitutive equations (8.9) and (8.10) describe phenomenological aspect of the process. The material constants in this set do not have clear physical meaning. Therefore, the physically-based constitutive equations have been developed in the last decades. Typical example of such equation set has been proposed by Kowalewski et al. (1994) in the following form:

$$
\begin{aligned}
\frac{\mathrm{d}\varepsilon_{ij}}{\mathrm{d}t} &= \frac{3}{2}\frac{A}{(1-\omega_2)^n}\frac{\bar{S}_{ij}}{\sigma_\mathrm{e}}\sinh\left(\frac{B\sigma_\mathrm{e}(1-H)}{1-\Phi}\right),\\
\frac{\mathrm{d}H}{\mathrm{d}t} &= \frac{h}{\sigma_\mathrm{e}}\frac{A}{1-\omega_2}\sinh\left(\frac{B\sigma_\mathrm{e}(1-H)}{1-\Phi}\right)\left(1-\frac{H}{H^*}\right),\\
\frac{\mathrm{d}\Phi}{\mathrm{d}t} &= \frac{K_\mathrm{c}}{3}(1-\Phi)^4,\\
\frac{\mathrm{d}\omega_2}{\mathrm{d}t} &= \frac{DA}{(1-\omega_2)^n}\left(\frac{\sigma_1}{\sigma_\mathrm{e}}\right)^n N\sinh\left(\frac{B\sigma_\mathrm{e}(1-H)}{1-\Phi}\right),
\end{aligned}
\qquad (8.12)
$$

where A, B, H^*, h, K_c, D—material constants and n is given by

$$
n = \frac{B\sigma_\mathrm{e}(1-H)}{1-\Phi}\coth\left(\frac{B\sigma_\mathrm{e}(1-H)}{1-\Phi}\right)
$$

The stress level dependence of creep rate is described by a sinh function. Material parameters which appear in this model may be divided into three groups, i.e.

- the constants h, H^* which describe primary creep;
- the constants A and B which characterise secondary creep;
- the constants K_c and D responsible for damage evolution and failure.

The second equation in set (8.12) describes primary creep using variable H, which varies from 0 at the beginning of the creep process to H^*, where H^* is the saturation value of H at the end of primary period and subsequently maintains this value until failure.

The equation set contains two damage state variables used to model tertiary softening mechanisms:

- Φ, which is described by the third equation in set (8.12), is defined from physics of ageing to lie within the range 0–1 for mathematical convenience,
- ω_2, which is defined by the fourth equation in set (8.12), describes grain boundary creep constrained cavitation, the magnitude of which is strongly sensitive to alloy composition and the processing route.

The parameter N is used to indicate the state of loading; e.g. for σ_1 tensile $N = 1$; and for σ_1 compressive, $N = 0$. In the equation set (8.12) a damage evolution depends on the maximum principal stress as well as the effective stress. After appropriate integration of the normalised form of equation set (8.12) the isochronous surfaces can be achieved (Kowalewski et al. 1994). It has been shown that the shape of the isochronous rupture loci is independent of the damage level (ω_2) for which they are determined, but is dependent on the stress level. At lower stress levels the curves become more dependent on the maximum principal stress.

It has been found from the experimental investigations that the minimum creep rate, which is directly related to the primary creep controlled by $\dot{H}$ in equation set (8.12), varies with stress-states for both materials. In the second equation of set (8.12) $\dot{H}$ is only a function of σ_e and could not model the feature. In addition, the rupture lifetime is not a constant of σ_1/σ_e for a given effective stress for both tested materials. This indicates that both the primary hardening and damage rate equations in (8.12) need to be modified. The modified equations are formulated based on set of equations (8.12) by taking into account the influence of stress-states on primary, secondary and tertiary creep. For the simplicity, only one damage state variable is used here to model grain boundary creep constrained cavitation. The evolutionary equations are given in the following form (Lin et al. 2005)

$$\begin{aligned}
\dot{\varepsilon}_e &= \frac{A}{(1-\omega)^n}\sinh\left(B\sigma_e(1-H)\right),\\
\dot{\varepsilon}_{ij} &= \frac{3}{2}\dot{\varepsilon}_e\left(\frac{\bar{S}_{ij}}{\sigma_e}\right),\\
\frac{dH}{dt} &= h(Q-H)\dot{\varepsilon}_e,\\
\frac{d\omega}{dt} &= D\left(\frac{\sigma_1}{\sigma_e}\right)^{\gamma} N\dot{\varepsilon}_e,
\end{aligned} \tag{8.13}$$

where $Q = Q_0(\sigma_1/\sigma_e)^{\phi}$ and $\gamma = \beta\sigma_1$. Parameter γ varies linearly with the maximum principal stress. The constant β is used to express the stress-state effects on the damage evolution of materials, and moreover, to model lifetimes and tertiary creep deformation behaviour of materials. Relation $\beta < 0$ indicates that the damage evolution of the material exceeds the effective stress control (a case typical for aluminium alloys), and, the presence of a low value of σ_1 would reduce the lifetime. If $\beta > 0$ then the damage evolution is under controlled by the effective stress. In the case of $\beta = 0$ the lifetime and tertiary creep of the material is controlled by effective stress only. The parameter N is introduced in (8.13) to ensure $\dot{\omega} = 0$, when σ_1 is compressive.

The evolution of the variable H in equation set (8.13) represents the primary hardening of the materials, which is mainly due to the accumulation of dislocation density during the primary creep process. As creep deformation proceeds, the increment of dislocation density and its recovery under elevated temperature reaches a dynamic balance condition. This is the steady-state, or, secondary creep, which is one of the most important properties in creep deformation. In the equation, the parameter Q, which indicates the end of primary creep and controls the secondary creep rate, is stress-state dependent and defined as $Q = Q_0(\sigma_1/\sigma_e)^{\phi}$. For the stress-state independent material, the constant $\phi = 0$ and $Q = Q_0$. Thus, Q is the saturation value of the primary hardening variable H and also determines the secondary creep rates, i.e. the minimum creep rate, $\dot{\varepsilon}_{min}$. However, if a material is stress-state dependent, $\phi \neq 0$, the value of Q varies with the ratio of the maximum principal stress and effective stress σ_1/σ_e. In consequence, the saturation value of the variable H changes with a variation of the stress-state. Thus, the minimum effective creep rates can be controlled according to the first equation of set (8.13). In this way, both primary and secondary creep periods are modelled by the introduction of the internal variable H.

The other material constants have the similar meanings as discussed for equation set (8.12). Optimisation techniques for the determination of the material constants arising in the constitutive equations are based on minimising the sum of the errors between the computed and experimental data using Evolutionary Algorithms (EA) (Lin and Yang 1999; Li et al. 2002). The fitness function used here for the optimisation based on the concept developed by Li et al. (2002). In this method, errors are defined by the shortest distance between computational and experimental data. An EA-based optimisation software package was developed using C++ based (Li et al. 2002). The multiaxial creep damage constitutive equations (equation set (8.13)) are implemented into the optimisation software package through a user-defined subroutine.

Figures 8.9 and 8.10 show the comparison of the experimental (symbols), and computed (solid curves) effective creep curves for the three stress-states for copper and aluminium alloy, respectively. The curves are computed using the determined material constants. It can be seen that there are some differences between the computed and experimental data, although the overall fitting quality is good. The difference might be due to the errors of the experimental results coming from always possible specimen-to-specimen variations of the material.

Presented here attempts for creep damage analysis reflect only advances in constitutive equations development in which the scalar damage measures are used. There are many papers devoted to creep damage where vector or tensor measures of damage were applied. Due to limits required for this chapter such issue is not discussed here.

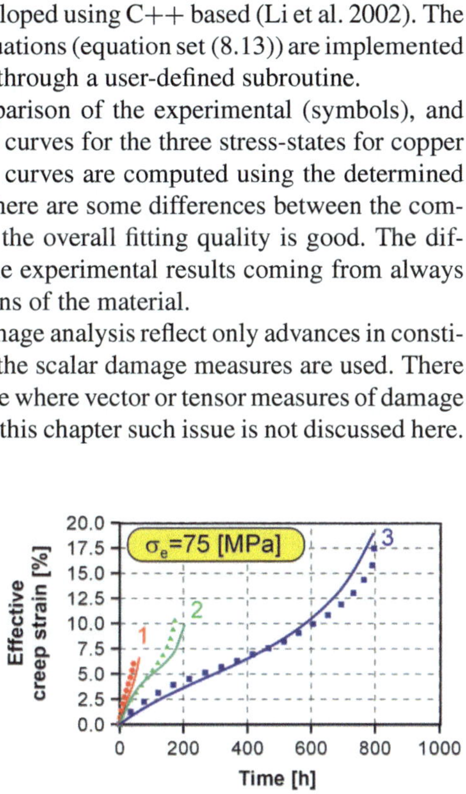

Fig. 8.9 Comparison of experimental and theoretical (equation set (8.13)) creep curves for pure copper at 423 K under stress equal to 75 MPa; *1* $(\sigma_{12}/\sigma_{11}) = 0$; *2* $(\sigma_{12}/\sigma_{11}) = \sqrt{3}/3$; *3* $(\sigma_{12}/\sigma_{11}) = \infty$

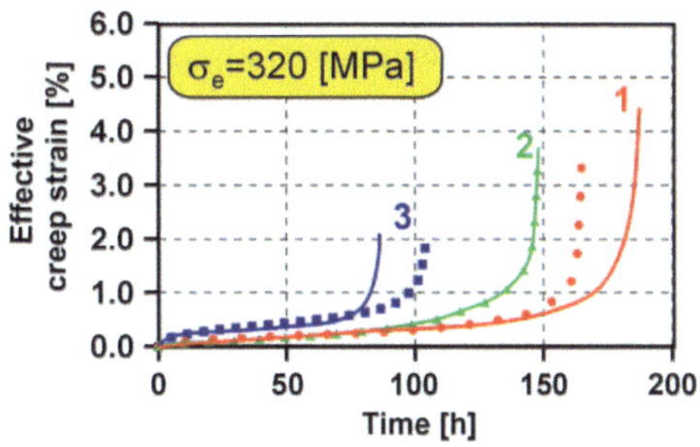

Fig. 8.10 Comparison of experimental and theoretical (equation set (8.13)) creep curves for pure copper at 423 K under stress equal to 320 MPa; *1* $(\sigma_{12}/\sigma_{11}) = 0$; *2* $(\sigma_{12}/\sigma_{11}) = \sqrt{3}/3$; *3* $(\sigma_{12}/\sigma_{11}) = \infty$

8.4 New Attempts for Damage Development During Creep

There are many testing techniques for creep damage analysis. They can be generally divided into destructive and non-destructive methods. To assess damage using destructive method the specimens after different amounts of prestraining were stretched to failure (Kowalewski et al. 2008, 2009; Makowska et al. 2014). Afterwards, the selected tension parameters were determined and their variations were used for identification of damage development. Ultrasonic and magnetic investigations were selected as the non-destructive methods for damage development evaluation. For the ultrasonic method, the acoustic birefringence coefficient was used to identify damage development in the tested steel. Two magnetic techniques for non-destructive testing were applied, i.e. measurement of the Barkhausen effect (HBE) and the magneto-acoustic emission (MAE). Both effects are due to an abrupt movement of the magnetic domain walls depicted from microstructural defects when the specimen is magnetised. The laboratory test specimens were magnetised by a solenoid and the magnetic flux generated in the specimen was closed by a C-core shaped yoke. The magnetizing current (delivered by a current source) had a triangular like waveform and frequency of order 0.1 Hz. Its intensity was proportional to the voltage Ug. Two sensors were used: (a) a pickup coil (PC), and (b) an acoustic emission transducer (AET). A voltage signal induced in the PC was used for the magnetic hysteresis loop $B(H)$ evaluation (low frequency component) as well as for the HBE analysis (high frequency component). The intensity of the HBE was given by the rms (root mean square) voltage Ub envelopes. The maximal values ($Ubpp$) of Ub for one period of magnetisation were compared. An analogous analysis was performed for the MAE voltage signal from the AET. In this case the maximal values ($Uapp$) of the Ua voltage envelopes were compared. The magnetic coercivity Hc, evaluated from the $B(H)$ hysteresis loop plots, was also compared.

8.4.1 Experimental Details

The X10CrMoVNb9-1 steel commonly used in selected elements of Polish power plants was investigated. Its chemical composition is presented in Table 8.2.

The experimental programme comprised tests for the material in the as-received state and for the same material subjected to a range of selected magnitudes of prior deformation due to creep at elevated temperatures, Fig. 8.11, and due to plastic flow at room temperature, Fig. 8.12. Uniaxial tension creep tests were carried out for the

Table 8.2 Chemical composition of the X10CrMoVNb9-1 steel

C	Mn	Nb	P	S	Cr	Ni	Mo	V	Cu
0.10	0.70	0.07	0.01	0.01	8.50	0.30	0.94	0.22	0.20

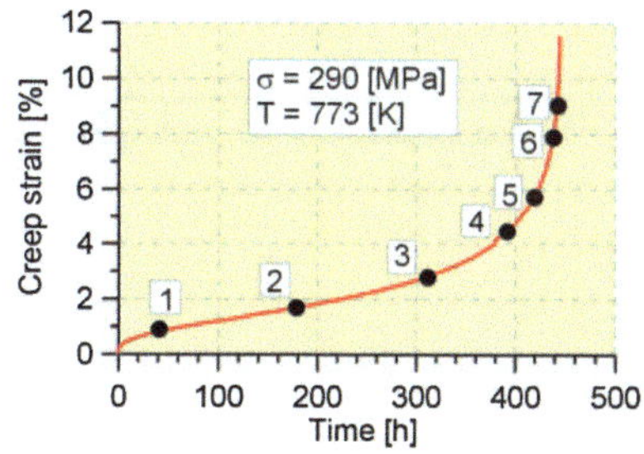

Fig. 8.11 Creep curve of X10CrMoVNb9-1 steel with points representing interrupted creep tests

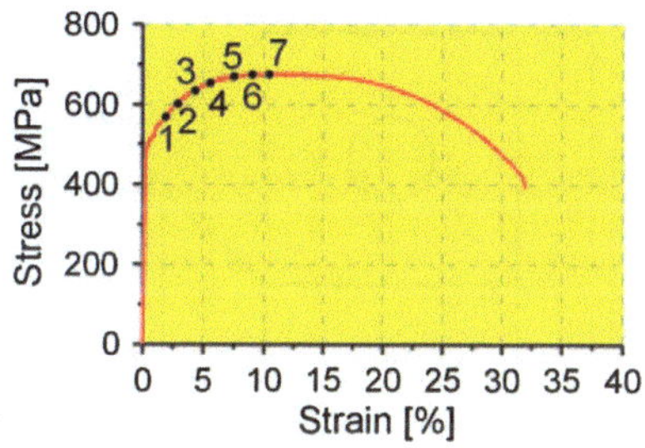

Fig. 8.12 Tension curve of X10CrMoVNb9-1 steel with points representing interrupted tensile tests

X10CrMoVNb9-1 steel using plane specimens, Fig. 8.13. All tests were conducted in the same conditions: i.e. the stress level was 290 MPa, and the temperature was 773 K. Details of the destructive tests programme as well as its main results are presented by Kowalewski et al. (2008).

In order to assess damage development during creep the tests for the X10CrMoV-Nb9-1 steel were interrupted after 40 h (0.85 %), 180 h (1.85 %), 310 h (3.15 %), 390 h (4.6 %), 425 h (5.9 %), 440 h (7.9 %) and 445 h (9.3 %), which correspond to increasing amounts of creep strain (values are presented in brackets). To check how deformation type changes damage development, almost the same prestraining levels as those under creep were induced by means of plastic flow: 2, 3, 4.5, 5.5, 7.5, 9, and 10.5 %, Fig. 8.12. After each prestraining test the specimen damage was assessed using the non-destructive methods. Two non-destructive methods were applied: magnetic (Augustyniak 2003) and ultrasonic (Szelążek 2001). In the next step of the experimental procedure, the same specimens were mounted on a hydraulic servo-controlled MTS testing machine and then stretched until failure was achieved.

8.4.1.1 Non-destructive Techniques

Magnetic properties were measured using the standard laboratory method of magnetisation, where hysteresis loops with the HBE and also the MAE can be tested (Augustyniak 2003; Augustyniak et al. 2000). A block diagram of the magnetising circuit is shown in Fig. 8.14. A specimen (1) was magnetised with the driving coil (2). A current amplifier provided a triangular wave-form with a frequency of about

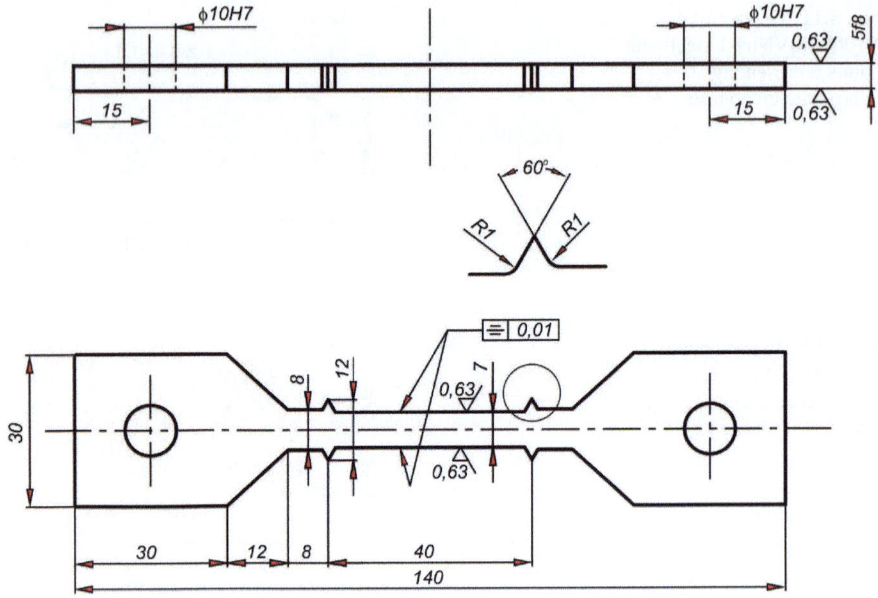

Fig. 8.13 Specimen for creep and plastic f low tests

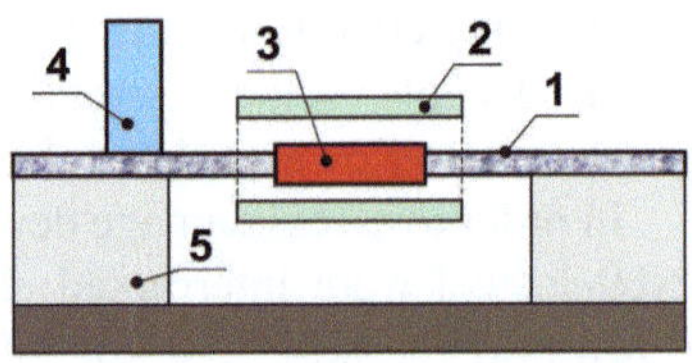

Fig. 8.14 Measuring set for magnetic properties: *1* specimen, *2* driving coil, *3* pick-up coil, *4* MAE detector, *5* magnetic core (Augustyniak 2003)

1 Hz. Magnetic measurements were made with a pickup coil of 200 turns wound directly on the central part of the specimen. The voltage U induced in this coil was used to determine the hysteresis loop and the HBE signal. The HBE signal was separated from the U signal using an ac amplifier. The output voltage was transformed to the rms like voltage Ub (intensity envelope). This envelope of the HBE intensity is presented when the HBE envelopes are compared. The MAE signal was detected with a resonant PZT transducer. The output of the MAE voltage signal was amplified and then transformed to the Ua voltage using the analogous rms integral circuit.

The HBE properties along the specimen were also investigated as well as the relationships between the HBE and static load by means of bending. Figure 8.15 illustrates the experimental setup used for measurement of the HBE stress dependence. Here, the HBE intensity was measured not at the specimen central part, but near the specimen end. Each specimen was loaded by means of bending. A probe during the HBE tests contains ferrite with a pick-up coil. The probe was used for the HBE intensity measurement along the specimen. It was connected to the MEB-1

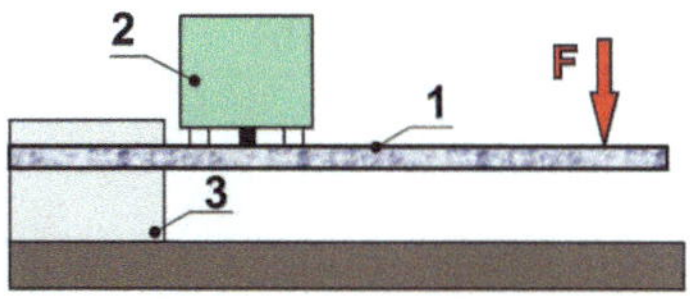

Fig. 8.15 Measuring set for stress dependence of HBE: *1* specimen, *2* HBE probe, *3* metal support, *F* applied force (Augustyniak 2003)

meter. This meter provides not only an analog rms voltage, but also a signal proportional to the pulse counting rate and total number of detected pulses (Nc) with amplitudes higher than a certain threshold level.

Ultrasonic wave velocity and attenuation are acoustic parameters most often used to assess material damage due to creep or fatigue. The results of investigations (Fel et al. 2001; Martínez-Ona and Pérez 2000) show that the attenuation of ultrasonic waves is in practice stable until the last creep or fatigue stages. It was also observed that velocity changes due to creep or fatigue are small, and therefore, an application of velocity measurement for damage evaluation, in industrial conditions, is very difficult. Difficulties in the attenuation and velocity measurements, or their combinations, are caused by the heterogeneous acoustic properties of technical materials, such as steel. The second reason is a dependence of both the attenuation and velocity of ultrasonic waves on numerous factors other than material damage. This observation is confirmed by the results of tests (Martínez-Ona and Pérez 2000) where the steel specimens were subjected to 10 % plastic deformation and subjected to loading for a period of 140,000 h at elevated temperature. The results showed that the ultrasonic wave attenuation was not influenced by the plastic deformation or long term, high temperature load exposure.

In order to evaluate damage progress in specimens made of X10CrMoVNb9-1 steel, instead of the velocity and attenuation measurement, the acoustic birefringence B was measured (Kowalewski et al. 2008, 2009). Specimens were subjected to creep according to the programme presented earlier in this paper.

The acoustic birefringence B is a measure of material acoustic anisotropy. It is based on the velocity difference of two shear waves polarized in the perpendicular directions. In specimens subjected to creep the shear waves were propagated in the specimen thickness direction and were polarized along its axis and in the perpendicular direction. The birefringence was measured in the fixtures, where a texture of material was assumed to be unchanged during a creep test, and in the working part of the specimen, Fig. 8.16. The birefringence B was calculated using the following expression (Szelążek 2001):

$$B = \frac{2(t_l - t_p)}{t_l + t_p} = B_0 + B_P \tag{8.14}$$

where: t_l—time of flight of ultrasonic shear wave pulse for the wave polarization direction parallel to the sample axis, t_p—time of flight of ultrasonic shear wave

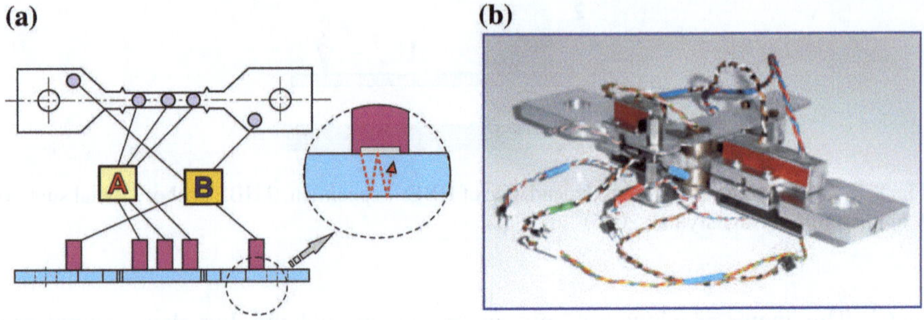

Fig. 8.16 Set-up for measurements of the birefringence coefficient: **a** scheme showing positions of probes (*B* probes at the griping part of specimen, *A* probes distributed along gauge length of specimen); **b** general view of specimen before ultrasonic testing

pulse for the wave polarization perpendicular to the sample axis, B_0—acoustic birefringence for the material in the virgin state (before creep test), B_P—acoustic birefringence for the material after deformation.

8.4.2 Experimental Results and Discussion

8.4.2.1 Evaluation of Damage Development Using Destructive Tests

The tensile characteristics for the material after prestraining are presented in Fig. 8.17. In diagrams the characteristics for the prestrained steel are compared to the tensile curve of steel in the as-received state.

On the basis of these tensile characteristics, Fig. 8.17, variations of the basic mechanical properties of steel, due to deformation achieved by prior creep or plas-

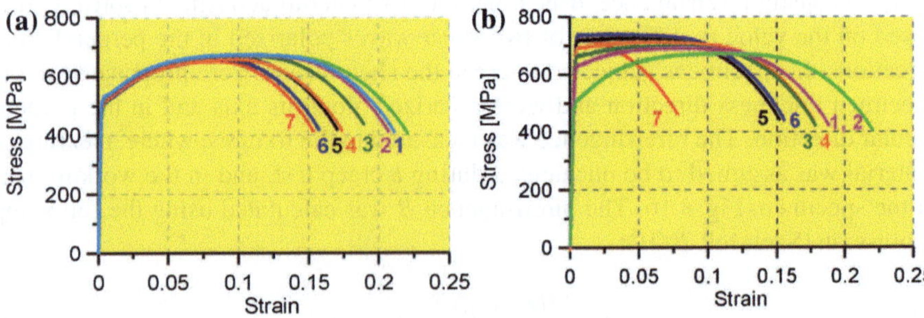

Fig. 8.17 Tensile characteristics of the X10CrMoVNb9-1 steel: **a** material after prior deformation due to creep interrupted in different phases of damage development, **b** material after prior deformation due to plastic flow interrupted in different phases of the process, (numbers correspond to those in Figs. 8.11 and 8.12 presented)

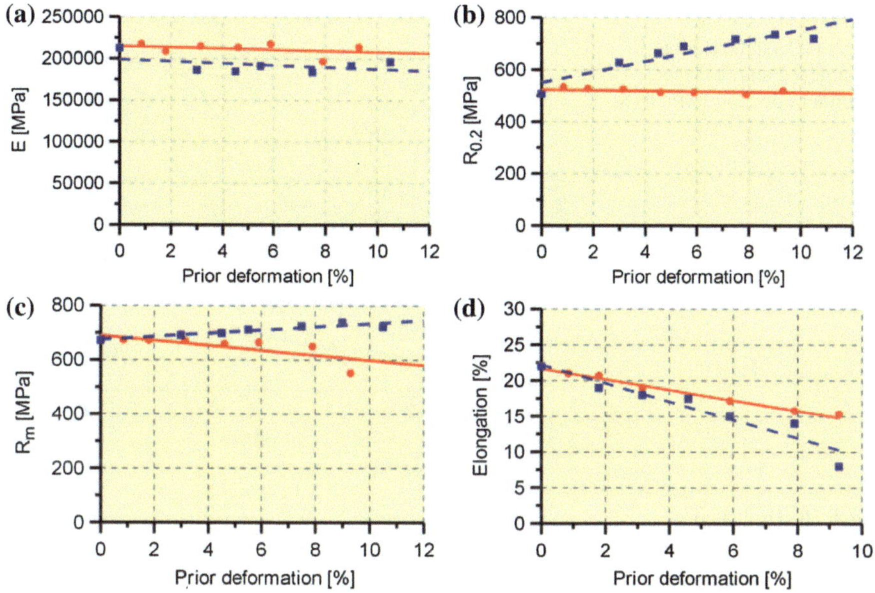

Fig. 8.18 Variation of tensile parameters of the X10CrMoVNb9-1 steel due to creep (*solid lines*) and plastic (*broken lines*) deformations: **a** Young's modulus; **b** yield point; **c** ultimate tensile strength; **d** elongation

tic flow were determined, Fig. 8.18. It was observed that the Young's modulus, Fig. 8.18a, is almost insensitive to the magnitude of creep and plastic deformations.

Contrary to the Young's modulus the other considered tension test parameters, especially the yield point, Fig. 8.18b, and the ultimate tensile strength, Fig. 8.18c, exhibit clear dependence on the level of prestraining. Taking into account the results presented for the steel a difference between magnitudes of such parameters as the yield point or ultimate tensile strength observed for the same value of prior deformation induced by creep and plastic flow is quite significant. Prior plastic deformation caused the hardening of the steel, while creep prestraining led to its softening. It is important to note that the observed softening effect is only expressed on the basis of the ultimate tensile strength variations since for the testing conditions applied in these investigations the magnitude of the yield point is not sensitive to the amount of prior creep deformation.

8.4.2.2 Evaluation of Damage Development Using Magnetic Techniques

An influence of plastic flow and creep damage on the basic magnetic properties can be analysed using $B(H)$ hysteresis loops. Representative results are presented in Figs. 8.19 and 8.20 for the X10CrMoVNb9-1 steel. The curves obtained for an undamaged specimen ($\varepsilon = 0\,\%$) and for the specimens after prior deformation are

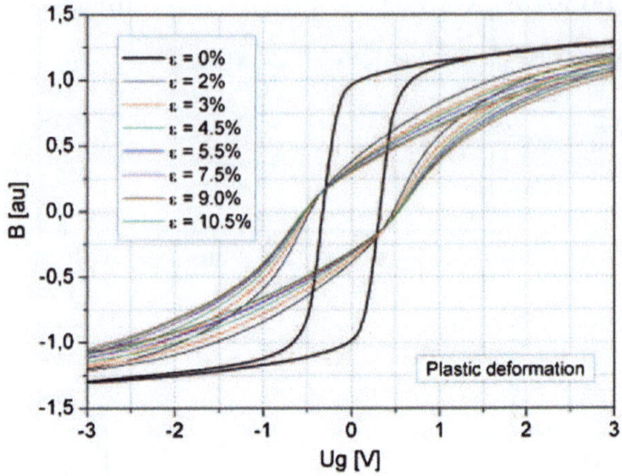

Fig. 8.19 Evaluated magnetic hysteresis loops of undamaged and damaged specimens due to plastic flow

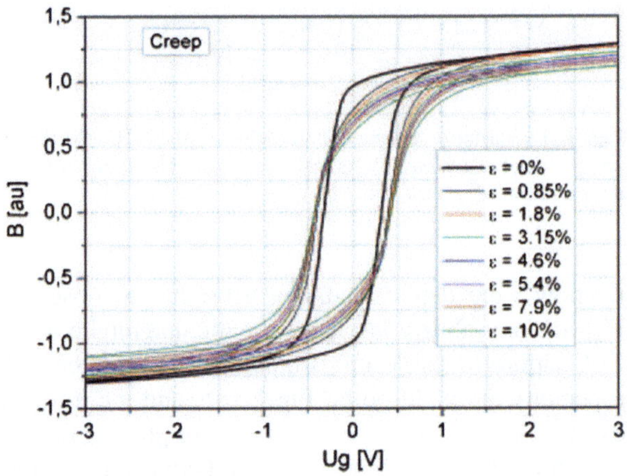

Fig. 8.20 Evaluated magnetic hysteresis loops of undamaged and damaged specimens due to creep

compared. The quantity Ug denotes the voltage proportional to the driving current intensity, and hence—magnetic field strength H.

The broadening as well as the decrease of the slope of the $B(H)$ loops for both cases is evident. Taking into account the same prestrain levels the effects are much more intensive in the case of plastic flow than those after creep. The coercivity Hc was evaluated from the width of $B(Ug)$ plot at $B = 0$, Fig. 8.21. The observed systematic increase of coercivity is due to an increase of pining force of the 180° magnetic domain walls by damage induced modifications of microstructure. It should be emphasised that there is a two times higher increase of the coercivity (about +60 %) for the specimens after plastic flow than for the specimens after creep (about +30 %).

Modification of the hysteresis properties, as shown by the HBE intensity envelopes, can be deduced from the series of plots presented in Figs. 8.22 and 8.23 for specimens

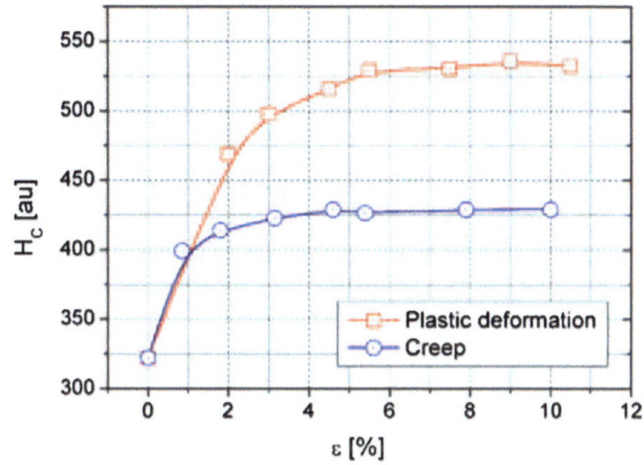

Fig. 8.21 Dependence between the coercivity and deformation for specimens after plastic flow (*squares*) and after creep (*circles*)

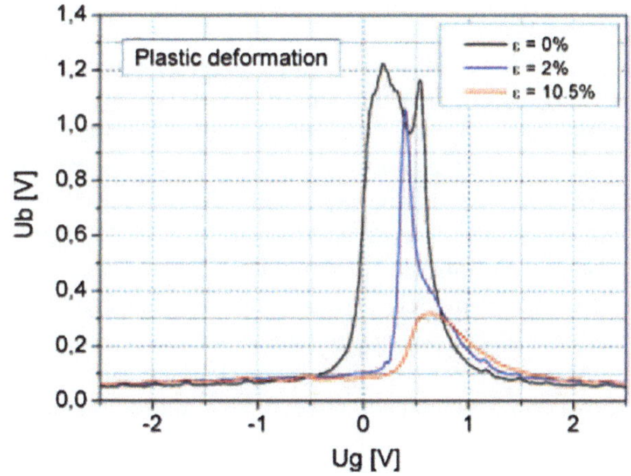

Fig. 8.22 Envelopes of the HBE intensity in function of increasing field strength for the undamaged specimen ($\varepsilon = 0$) and for two specimens after plastic flow

after plastic flow and after creep, respectively. It has to be noticed that there is an increase of the scale of the Ub values in Fig. 8.23. Such a presentation of the results was made in order to show the very high increase of the HBE intensity after the first step of the creep damage experiment in comparison to the initial stage signal. The HBE intensity envelope for an undamaged sample is characterised by a shape consisting of two peaks. The X10CrMoVNb9-1 steel has a martensitic microstructure, and thus, the first peak (positioned at lower field strength) can be attributed formally to the 'soft' component of the alloy while the second one to the 'hard' component without detailed discussion about the microstructure reference, which can be done after microscopic inspection of the magnetic domain structure.

The increase of the plastic strain after plastic flow leads to some general decrease of the HBE intensity: for the first tested stage of deformation ($\varepsilon = 2\,\%$) one narrow peak appears and further plastic flow leads to a monotonic decrease of its amplitude as well as to the decrease of the area under the signal envelope. A shift of the

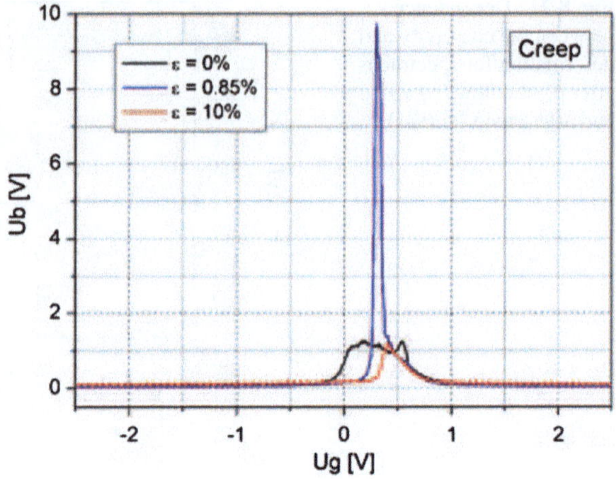

Fig. 8.23 Envelopes of the HBE intensity in function of increasing field strength for the undamaged specimen (1) ($\varepsilon = 0$) and for two specimens after creep damage

maximum towards a higher level of the magnetic field strength can also be clearly seen. However, the behaviour of the Ub properties when the samples are subjected to creep damage is very different. Figure 8.23 shows that for a low level of creep damage ($\varepsilon = 0.85\,\%$) the plot of the Ub envelope appears as one very high, narrow peak. Such behaviour can be explained by the increase of the 180° domain walls mobility due to the decrease of internal stress level resulting from an annealing or—more probably, by anisotropy of magnetic domain structure due to the tensile load applied, enhanced by the high temperature. Further creep damage leads to the systematic decrease of this peak amplitude as well as to its shift toward higher field strength.

The as described features of the HBE intensity are well presented by means of plots showing a dependence between the amplitudes of Ub envelopes and magnitudes of prior deformation—peak to peak values Ub_{pp} in Fig. 8.24, and a dependence between the integrals of the Ub envelopes and prior deformation (Fig. 8.25) for specimens after plastic flow (squares) and after creep (circles). Thus, one can say that the HBE intensity as a function of the resulting prestrain either decreases monotonically (for integrals) or peaks when amplitudes of the Ub envelopes are compared. These two sets of plots reveal also that creep damage leads (at its final stage) to a 'decrease' of the HBE intensity which is much lower than that observed for specimens after plastic flow. Comparing two plots in each figure it can be seen that the Ub signal properties such as the amplitude or integral for the highest strain after creep damage are roughly the same as for the analogous signals for the first stage of plastic flow.

The main features of the magnetoacoustic effect are shown in Fig. 8.26 (after plastic flow) and in Fig. 8.27 (after creep).

The MAE intensity envelope for undamaged specimens is also characterised by the existence of two peaks. Comparing plots in Figs. 8.22 and 8.26 one can easily check that the plot of the MAE intensity envelope (Ua) is much broader than the plot of the Ub intensity. It is due to the fact that the MAE is caused mainly by an abrupt movement of 'not' 180° domain walls. The first peak is usually attributed mainly

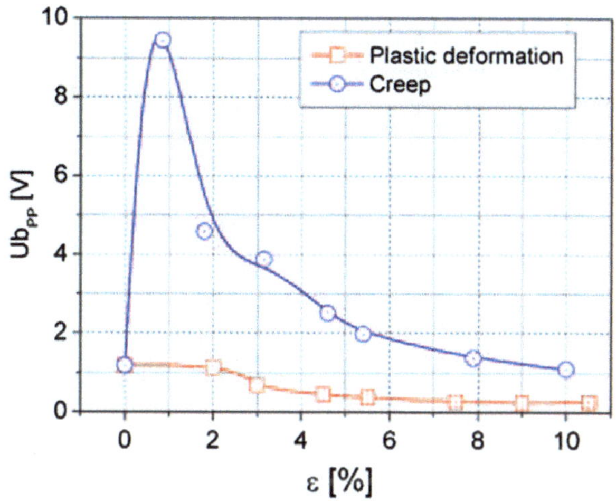

Fig. 8.24 Dependence between the amplitudes of Ub envelopes and deformation for specimens after plastic flow (*squares*) and after creep (*circles*)

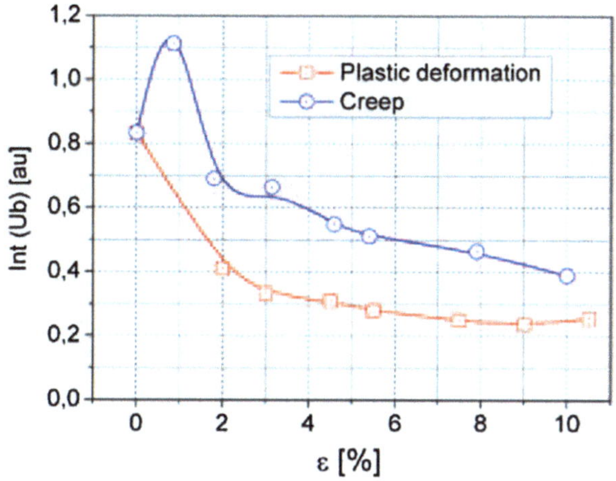

Fig. 8.25 Dependence between the integrals of Ub envelopes and deformation for specimens after plastic flow (*squares*) and after creep (*circles*)

to a creation of the magnetic domains, and the second one—to their annihilation. Both processes are characterised by a high contribution of this type of domain walls (Augustyniak et al. 2000). These pictures also show that plastic flow as well as creep damage modifies the MAE intensity significantly.

The plastic flow modifies the MAE intensity in two ways: the two peaks observed after the first step of flow ($\varepsilon = 2\,\%$) are broader and their amplitudes are much smaller. This means that the produced dislocations tangles have strongly blocked the mobility of 'not' 180° domain walls. Further plastic flow leads to a monotonic decrease of the MAE intensity.

Again, the results of creep damage show an influence of prior deformation on the MAE properties, as shown by the plots in Fig. 8.27. The stage with small creep deformation level ($\varepsilon = 0,85\,\%$) is characterised by a single, narrow peak. This

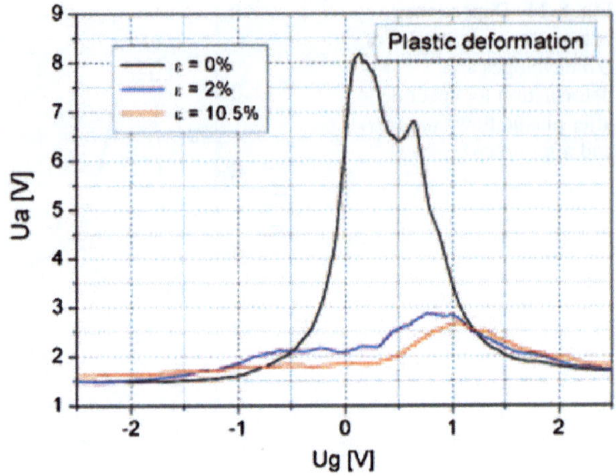

Fig. 8.26 Envelopes of the MAE intensity as a function of increasing field strength for the undamaged specimen (plot 1, $\varepsilon = 0$) and for two specimens after plastic flow

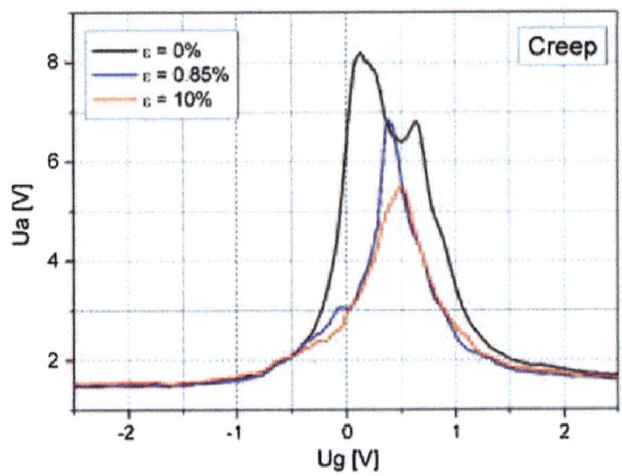

Fig. 8.27 Envelopes of the MAE intensity as a function of increasing field strength for the undamaged specimen (plot 1, $\varepsilon = 0$) and for two specimens after creep damage

means that now the displacement of 'not' 180° domain walls may become the main contributor to the MAE instead of the creation and annihilation processes. A synthetic description of the MAE properties as a function of prior deformation is given by the two sets of plots shown in Fig. 8.28 (amplitudes of the MAE envelopes) and in Fig. 8.29 (integrals of the MAE envelopes). Amplitudes of the MAE intensity decrease for both cases, but the dynamics of their change is different, as is evident from Fig. 8.28. Moreover, amplitudes of the MAE intensity envelopes do not decrease so abruptly for the creep prestrained specimens, and do not reach the level obtained for the first step of plastic deformation due to plastic flow at room temperature.

Figure 8.29 shows how integrals of the MAE intensity vary with the increasing prior deformation. The dynamic of the integrals decrease is not as high as that observed in the case of amplitudes. However, a difference between both types of damage is still visible. It is easy to observe that a level of the MAE intensity (esti-

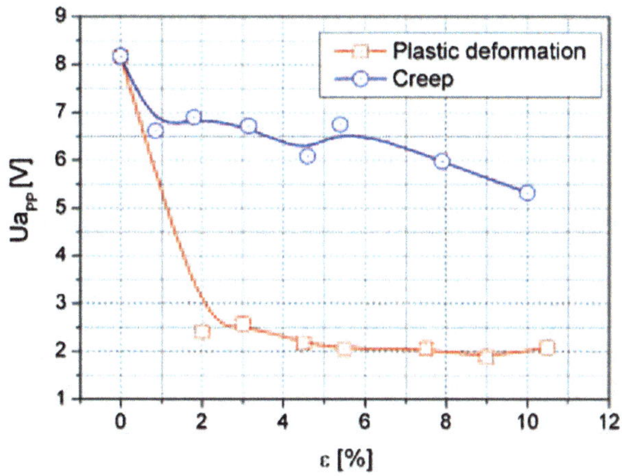

Fig. 8.28 Dependence between amplitudes of *Ua* envelopes and prior deformation due to plastic flow (*squares*) and creep (*circles*)

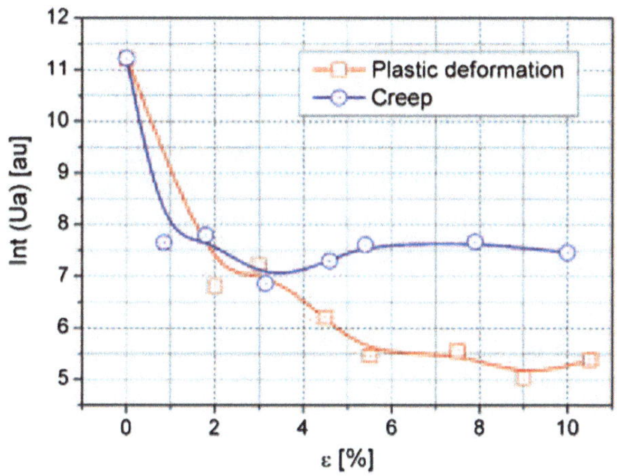

Fig. 8.29 Dependence between integrals of *Ua* envelopes and prior deformation due to plastic flow (*squares*) and creep (*circles*)

mated by means of the integral) for the specimen after creep with a strain of order $\varepsilon = 10\,\%$ is nearly the same as that detected for specimens after plastic deformation with a strain level of order $\varepsilon = 2\,\%$. However, it does not mean that these two specimens have the same microstructure. A difference in the microstructure for these two stages is demonstrated well by different shapes of the Ua envelopes for $\varepsilon = 2\,\%$ (Fig. 8.26) and for $\varepsilon = 10\,\%$ (Fig. 8.27). The stresses which influence the MAE activity are located inside the cells made by dislocation tangles. They are 'created' by these tangles and can be highly compressive in the case of plastic deformation. The MAE intensity is also influenced by the precipitates developing mainly at grain boundaries.

8.4.2.3 Evaluation of Damage Development Using Ultrasonic Technique and Correlations of Damage Sensitive Parameters

Figure 8.30 presents mean values of the acoustic birefringence measured in specimens after creep or plastic deformation.

The birefringence was measured in the fixtures, where the texture of the material was assumed to be unchanged during creep testing, and in the working part of the specimen. The plots presented in Fig. 8.30 indicate that the acoustic birefringence is sensitive to the amount of prior deformation. Another advantage of this parameter is also well represented in Fig. 8.30. Namely, it is sensitive to the form of prior deformation. For specimens prestrained due to plastic flow a decrease of this parameter is observed with the increase of prior deformation. In the case of prior creep also decrease of acoustic birefringence is observed, however, it is not as large as that after plastic deformation obtained. The results show that the acoustic birefringence can be a quite sensitive indicator of material degradation and can help to locate the regions where material properties are changed due to creep. Measurements of the ultrasonic wave attenuation and velocities carried out on the same steel did not exhibit such a good sensitivity in the material damage assessments.

In the next step of analysis possible relations between the mechanical and magnetic parameters were evaluated Figs. 8.31, 8.32, 8.33, 8.34 and 8.35.

Figures 8.31, 8.32, 8.33 and 8.34 show relationships between two magnetic parameters of MBE: i.e. Ub_{pp} and Int(U_b) and two mechanical parameters: i.e. yield point and ultimate tensile strength. Figures 8.31 and 8.32 do not include results of

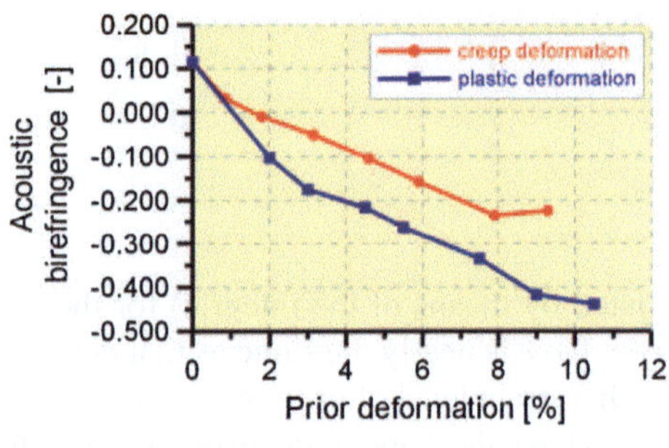

Fig. 8.30 Acoustic birefringence *B* variations due to prior deformation of the X10CrMoVNb9-1 steel

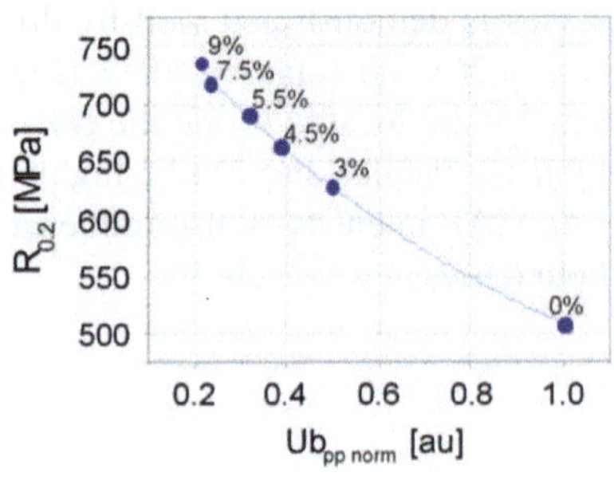

Fig. 8.31 Variation of yield point of the X10CrMoVNb9-1 steel versus amplitude of the magnetic Barkhausen emission (results for steel prestrained due to plastic flow)

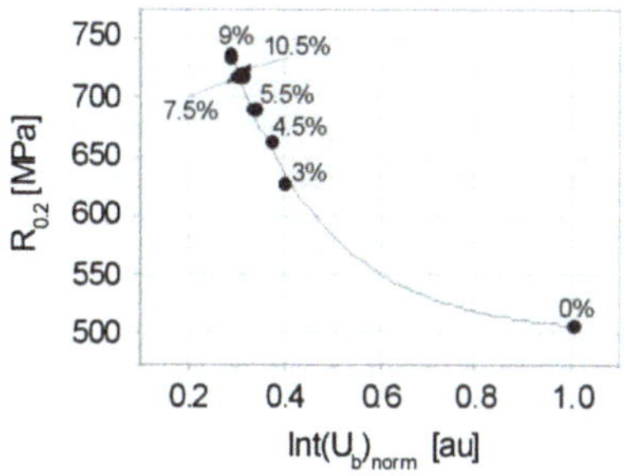

Fig. 8.32 Variation of yield point of the X10CrMoVNb9-1 steel versus integral over half-period voltage signal of the magnetic Barkhausen emission (results for steel prestrained due to plastic flow)

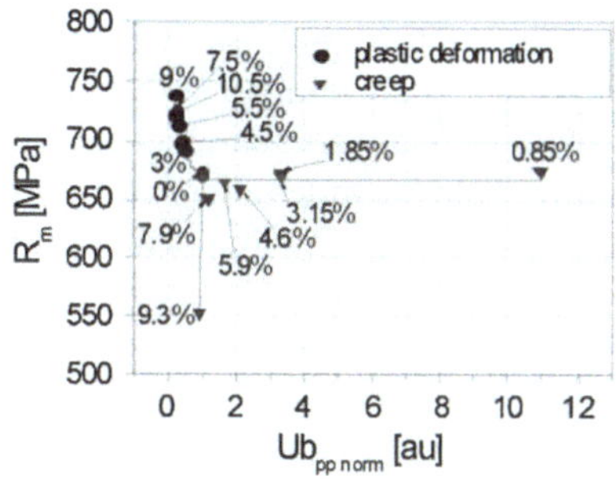

Fig. 8.33 Variation of ultimate tensile stress of the X10CrMoVNb9-1 steel versus amplitude of the magnetic Barkhausen emission

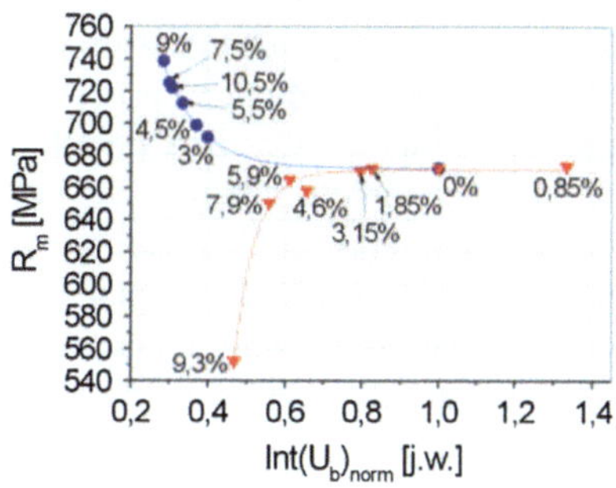

Fig. 8.34 Variation of ultimate tensile stress of the X10CrMoVNb9-1 steel versus integral over half-period voltage signal of the magnetic Barkhausen emission (*triangles* steel after creep; *circles* steel after plastic flow)

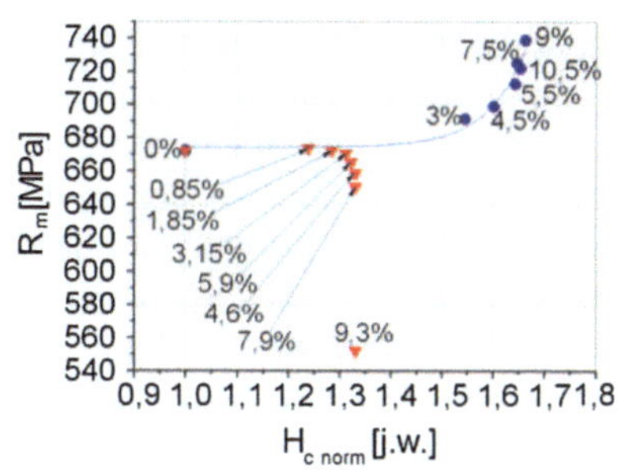

Fig. 8.35 Variation of ultimate tensile stress of the X10CrMoVNb9-1 steel versus coercivity (*triangles* steel after creep; *circles* steel after plastic flow)

the material prestrained due to creep, since the yield point of the X10CrMoVNb9-1 steel subjected to creep was insensitive to deformation level induced by this process. The magnetic parameters are normalized to values captured for the non-deformed specimen. Numbers in figures denote the level of prior deformation. Figures 8.31 and 8.32 allow concluding that both parameters (Ub_{pp} and Int(U_b)) of the Barkhausen noise may be used to estimate a level of the yield point of plastically deformed specimens. Also, the ultimate tensile strength of the X10CrMoVNb9-1 steel subjected to prior plastic flow may be assessed using relationships between R_m and $Ub_{pp\,norm}$ or $Int(Ub)_{norm}$ (Figs. 8.33 and 8.34), however, only for the material prestrained due to plastic flow, since non-unique relationships between R_m and $Ub_{pp\,norm}$ or $Int(Ub)_{norm}$ were found for the steel pre-strained by creep.

The relations in Figs. 8.31, 8.32, 8.33 and 8.34 indicate that the steel after plastic deformation, leading to higher values of $R_{0.2}$ and R_m, can be characterised by lower values of magnetic parameters. This is because the prestrained material contains more dislocation tangles that impede domain walls movement. On the other hand the higher values of magnetic parameters can be attributed to the lower magnitudes of R_m for the steel after creep.

The results make evident that the MBE intensity varies significantly due to microstructure modification, however, in different ways depending on prior deformation type. This intensity decreases after plastic flow (for deformation higher than 2 %) and increases after creep. Strongly non-linear character of plots in Figs. 8.33 and 8.34 makes impossible direct estimation of mechanical parameters when only single magnetic parameter is used. Addressing the issue for practical application of the MBE measurement in assessment of mechanical properties for damaged steel one can conclude that it is possible only then if at least two magnetic parameters will be taken into account. It can be seen in Figs. 8.33 and 8.34 that relative decrease of the Ub_{pp} and Int(U_b) with prestraining denotes plastic deformation while rapid increase of the Int(U_b) associated with prestrain increase is observed for early stage of creep damage development. The most difficult case for interpretation takes place when advanced creep is in question. It should be emphasized that such analysis can be done by simultaneous analysis of plots in Figs. 8.33 and 8.34 and the MBE peak shape variations. The results obtained for such case are not consistent, i.e. points representing subsequent magnitudes of prior deformation are not placed in order, and therefore, cannot be described be an adequate function.

Better correlation was achieved between R_m and coercivity H_c, Fig. 8.35. As it is seen, except specimen prestrained up to 10.5 % due to plastic flow, all results are ordered, and as a consequence, they can be well described by adequate functions depending on the type of prior deformation. The main disadvantage of the relationships between R_m and H_c, is related to the fact that it cannot distinguish a type of prior deformation for small prestrain magnitudes.

Similar remarks can be formulated for the relationships between R_m and acoustic birefringence coefficient B, Fig. 8.36.

The relationships between selected destructive and non-destructive parameters sensitive for damage development show a new feature that may improve damage identification. In order to provide more thorough analysis reflecting physical

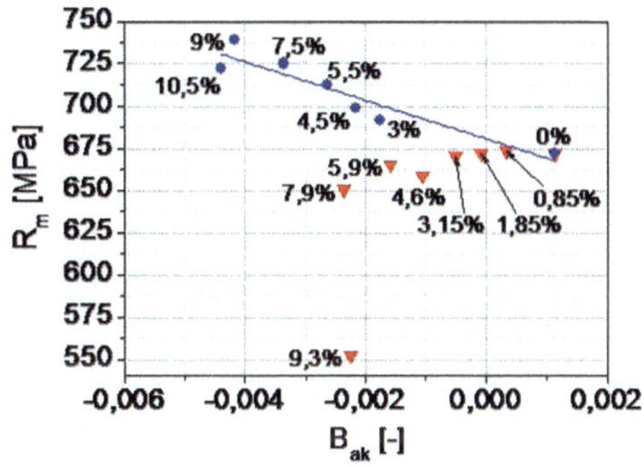

Fig. 8.36 Variation of ultimate tensile strength of the X10CrMoVNb9-1 steel versus acoustic birefringence (*triangles* steel after creep; *circles* steel after plastic flow)

interpretation of the relationships obtained further investigations are necessary. Programmes of such tests should contain advanced microscopic observations using not only optical techniques, but also SEM and TEM.

8.5 Concluding Remarks

This chapter is devoted to creep analysis using selected experimental methods. A short survey dealing with theoretical aspects of creep investigations is also presented.

It is shown that prior plastic deformation changes significantly values of the typical creep parameters. Depending of the magnitude of prestraining some of these parameters can be improved, the others however, become to be weaker than those for the nonprestrained material achieved. The tensile creep resistance measured as the value of steady creep rate was generally enhanced by plastic prestrain, which was expressed by significant decrease of the steady creep rate. The effect has proportional character up to certain limit value of plastic deformation, only. The creep data for aluminium alloy exhibit essential lifetime variation due to prestraining in both temperatures in question (423 and 473 K), namely, an extension of lifetime proportional to the magnitude of plastic prestrain. It has to be noted however, that plastic prestrain magnitudes greater than 6 % led to the opposite effect, i.e. lifetime reduction. The amount of creep deformation for both temperatures considered was markedly reduced by prior tensile plastic strain, yielding very low levels. Elongation of the testpieces was proportionally decreased when the magnitude of plastic prestrain was increased.

The chapter emphasises significance of the multiaxial creep testing, and identifies procedures for elaboration of data captured from such investigations.

This study also presents the results of interdisciplinary tests for damage assessments as a new promising tool for damage identification.

It is shown that the same level of deformation induced due to different processes does not guarantee the same mechanical properties of a material.

The results clearly indicate that selected ultrasonic and magnetic parameters can be good indicators of material degradation and can help to locate the regions where material properties are changed due to prestraining. In order to evaluate damage progress in specimens made of the X10CrMoVNb9-1 steel, instead of velocity and attenuation measurements frequently applied, the acoustic birefringence B measurements were successfully applied. In the case of magnetic investigations for damage identification the measurements of the Barkhausen effect (HBE) and the magnetoacoustic emission (MAE) were applied. Both effects show that the magnetic properties are highly influenced by prior deformation, and moreover, they are sensitive not only to the magnitude of prior deformation, but also to the way it is introduced.

The results suggest that experimental investigations concerning creep problems should be based on the interdisciplinary tests giving a chance to find mutual correlations between parameters assessed by classical macroscopic destructive investigations and parameters coming from the non-destructive experiments. Such relationships should be supported by thorough microscopic tests, thus giving more complete understanding of the phenomena observed during creep damage development.

References

Abo El Ata MM, Finnie I (1972) On the prediction of creep-rupture life of components under multiaxial stress. In: Hult J (ed) Proceedings of IUTAM symposium on creep in structures 1970, Gothenburg. Springer, Berlin, pp 80–95

Ashby MF, Gandhi C, Taplin DMR (1979) Fracture-mechanism maps and their construction for f.c.c. metals and alloys. Acta Metallurgica 27:699–729

Augustyniak B (2003) Magnetoelastic effects and their application in the non-destructive testing of materials (in Polish). Publishing House of Gdańsk University of Technology, Gdańsk

Augustyniak B, Chmielewski M, Sablik MJ (2000) Multiparameter magnetomechanical NDE. IEEE Trans Magn 36(5):3624–3626

Browne RJ, Lonsdale D, Flewitt PEJ (1981) The role of stress state on the creep rupture of 1%Cr1/2%Mo and 12%Cr1%MoVW tube steels. In: Wilshire B, Owen D (eds) Creep and fracture of engineering materials and structures. Pineridge Press, Swansea, pp 545–557

Chrzanowski M, Madej J (1980) The construction of failure limit curves by means of a damage. J Theor Appl Mech 18:587–601 (in Polish)

Dietrich L, Kowalewski ZL (1997) Experimental investigation of an anisotropy in copper subjected to predeformation due to constant and monotonic loadings. Int J Plast 13:87–109

Dyson BF, Gibbons TB (1987) Tertiary creep in nickel-base superalloys: analysis of experimental data and theoretical synthesis. Acta Metallurgica 35:2355–2369

Dyson BF, McLean D (1977) Creep of nimonic 80A in torsion and tension. Metal Sci 11:37–45

Dyson BF, Rodgers MJ (1974) Prestrain, cavitation and creep ductility. Metal Sci 8:261–266

Dyson BF, Loveday MS, Rodgers MJ (1976) Grain boundary cavitation under various states of applied stress. Proc R Soc Lond A349:245–259

Fel D, Hsu DK, Warchol M (2001) Simultaneous velocity, thickness and profile imaging by ultrasonic scan. J Nondest Eval 8:95–112

Garofalo F (1965) Fundamentals of creep and creep rupture in metals. Macmillan, New York

Gittus J (1975) Creep. Viscoelasticity and creep fracture in solids. Wiley, New York

Hayhurst D, Trąmpczyński WA, Leckie FA (1980) Creep rupture under non-proportional loading. Acta Metallurgica 28:1171–1183

Hayhurst DR (1972) Creep rupture under multi-axial states of stress. J Mech Phys Solids 20:381–390
Hayhurst DR (1983) On the role of creep continuum damage in structural mechanics. In: Wilshire B, Owen D (eds) Engineering approaches to high temperature design. Pineridge Press, Swansea, pp 85–176
Johnson AE, Henderson J, Mathur VD (1956) Combined stress fracture of commercial copper at 250 C. The Eng 202:261
Johnson AE, Henderson J, Khan B (1962) Complex-stress creep. Relaxation and fracture of metallic alloys. H.M.S.O, Edinburgh
Kachanov LM (1958) The theory of creep (English Translation Edited by Kennedy AJ). National Lending Library, Boston Spa
Kowalewski Z (1991a) Creep behavior of copper under plane stress state. Int J Plast 7:387–404
Kowalewski Z (1991b) The influence of deformation history on creep of pure copper. In: Życzkowski M (ed) Creep in structures. Springer, Berlin, pp 115–122
Kowalewski ZL (1992) The role of grain size on creep of copper under uniaxial tension. Arch Metall 37:65–76
Kowalewski ZL (1995) Experimental evaluation of the influence of stress state type on creep characteristics of copper at 523K. Arch Mech 47:13–26
Kowalewski ZL (1996) Biaxial creep study of copper on the basis of isochronous creep surfaces. Arch Mech 48:89–109
Kowalewski ZL (2002) Creep rupture analysis of metals under complex stress state. In: Skrzypek J, Ganczarski A (eds) Anisotropic behaviour of damaged materials. Springer, Kraków-Przegorzały, pp 79–92
Kowalewski ZL (2004) Isochronous creep rupture loci for metals under biaxial stress. J Strain Anal Eng Des 39:581–593
Kowalewski ZL (2005) Creep analysis of M1E copper and PA6 aluminium alloy subjected to prior plastic deformation. J Theor Appl Mech 43:241–256
Kowalewski ZL, Hayhurst DR, Dyson BF (1994) Mechanisms-based creep constitutive equations for an aluminium alloy. J Strain Anal 29:309–316
Kowalewski ZL, Szelążek J, Mackiewicz S, Pietrzak K, Augustyniak B (2008) Evaluation of damage in steels subjected to exploitation loading—destructive and non-destructive methods. Int J Modern Phys Lett B 22:5533–5538
Kowalewski ZL, Szelążek J, Mackiewicz S, Pietrzak K, Augustyniak B (2009) Evaluation of damage development in steels subjected to exploitation loading—destructive and nondestructive techniques. J Multiscale Model 1:479–499
Krauss G (1996) Fatigue and fracture. In: ASM Handbook. ASM International, OH, pp 680–690
Leckie FA, Hayhurst DR (1977) Constitutive equations for creep rupture. Acta Metallurgica 25:1059–1070
Li B, Lin J, Yao X (2002) A novel evolutionary algorithm for determining unified creep damage constitutive equations. Int J Mech Sci 44:987–1002
Lin J (2003) Damage mechanisms, models and calibration techniques. In: Dietrich L (ed) Training course on mechanics—Proceedings of mechanical investigations of material properties and structures. IPPT PAN, Warszawa, pp 123–143
Lin J, Yang J (1999) GA based multiple objective optimization for determining viscoplastic constitutive equations for superplastic alloys. Int J Plast 15:1181–1196
Lin J, Kowalewski ZL, Cao J (2005) Creep rupture of copper and aluminium alloy under combined loadings—experiments and their various descriptions. Int J Mech Sci 47:1038–1058
Litewka A, Hult J (1989) One parameter CDM model for creep rupture prediction. Eur J Mech A/Solids 8:185–200
Makowska K, Kowalewski ZL, Augustyniak B, Piotrowski L (2014) Determination of mechanical properties of P91 steel by means of magnetic Barkhausen emission. J Theor Appl Mech 52:181–188
Malinin N, Rżysko J (1981) Mechanics of materials. PWN, Warszawa (in Polish)

Marlin RT, Cosandey F, Tien JK (1980) The effect of predeformation on the creep and stress rupture of an oxide dispersion strengthened mechanical alloy. Metall Trans A 11A:1771–1775

Martínez-Ona R, Pérez MC (2000) Research on creep damage detection in reformer tubes by ultrasonic testing. In: Proceedings of 15th World congress on nondestructive testing. AIPnD, Roma

McVetty PG (1934) Working stresses for high temperature service. Mech Eng 56:149

Murakami S, Kawai M, Yamada Y (1990) Creep after cyclic-plasticity under multiaxial conditions for type 316 stainless steel at elevated temperature. J Eng Mater Tech 112:346–352

Narayan R, Green RE Jr (1975) Ultrasonic attenuation monitoring of fatigue damage in nuclear pressure vessel steel at high temperature. Materials evaluation, pp 25–36

Norton FH (1929) Creep of steel at high temperatures. McGraw-Hill, New York

Odqvist FKG (1966) Mathematical theory of creep and creep rupture. Clarendon Press, Oxford

Ogi H, Minami Y, Aoki S, Hirao M (2000) Contactless monitoring of surface-wave attenuation and nonlinearity for evaluating remaining life of fatigued steels. In: Proceedings of 15th World congress on nondestructive testing. AIPnD, Roma

Ohashi Y, Kawai M, Momose T (1986) Effects of prior plasticity on subsequent creep of type 316 stainless steel at elevated temperature. J Eng Mater Tech 108:68–74

Pandey MC, Mukherjee AK, Taplin DMR (1984) Prior deformation effects on creep and fracture in Inconel alloy X-750. Metall Trans A 15A:1437–1441

Piechnik S, Chrzanowski M (1970) Time of total creep rupture of a beam under combined tension bending. Inst J Solids Struct 6:453–477

Rabotnov YN (1969) Creep problems in structural members. North Holland Publishing Company, Amsterdam

Rees DWA (1981) Effects of plastic prestrain on the creep of aluminium under biaxial stress. In: Wilshire B, Owen DR (eds) Creep and fracture of engineering materials and structures. Pineridge Press, Swansea, pp 559–572

Sablik MJ, Augustyniak B (1999) Encyclopedia of electrical and electronics engineering. Magnetics methods on nondestructive evaluation. Wiley, Chichester

Sdobyrev VP (1959) Creep criterion for some high-temperature alloys in complex stress state. Izv AN SSSR Mekh and Mashinostr 6:12–19 (in Russian)

Szelążek J (2001) Advances in ultrasonic investigations of stresses. Report 4, IFTR, Warszawa

Trąmpczyński W, Kowalewski Z, (1986) A tension-torsion testing technique. In: Dyson B, Loveday M (eds) Techniques for multiaxial creep testing. Elsevier Applied Science, London and New York, pp 79–92

Trąmpczyński WA (1982) The influence of cold work on the creep of copper under biaxial states of stress. Acta Metall 30:1035–1041

Wilson RN (1973) The influence of 3% prestrain on the creep strength of Al-2.5% Cu-1.2% Mg alloys at 150 °C. J Inst Metals 101:188–196

Xia Z, Ellyin F (1993) An experimental study on the effect of prior plastic straining on creep behavior of 304 stainless steel. J Eng Mater Tech 115:200–203

Chapter 9
Strain Gradient Plasticity: A Variety of Treatments and Related Fundamental Issues

Mitsutoshi Kuroda

Abstract Different theoretical interpretations and possible mathematical expressions for the higher-order strain gradient plasticity theory initiated by Aifantis are investigated. These different interpretations of the theory result in different computational procedures. The effects of the orders of finite-element shape functions and the number of Gaussian quadrature points on the qualities of numerical solutions are examined for different formulations.

Keywords Size effect · Length scale · Phenomenological plasticity · Finite element method · Constitutive relations

9.1 Introduction

Conventional plasticity theories can determine the onset of strain localization, but they cannot deal with issues associated with the size of localization regions such as the width of shear band (Aifantis 1984, 1987). Furthermore, the conventional theories cannot characterize experimental results on micron-size specimens that exhibit a significant size-dependent mechanical response in the presence of plastic strain gradients (e.g., Fleck et al. 1994; Stölken and Evans 1998). These are serious theoretical contradictions in the conventional plasticity theories, which originate from the lack of intrinsic length-scale effects. Aifantis (1984, 1987) first modified the conventional theories by introducing plastic strain gradient terms into the yield function, which naturally account for the length-scale effects in bodies undergoing nonuniform deformation. Since the studies of Aifantis (1984, 1987), a considerable number of investigations on strain gradient plasticity theories have been conducted (e.g., Fleck et al. 1994; Mühlhaus and Aifantis 1991; Fleck and Hutchinson 2001; Gudmundson 2004; Gurtin and Anand 2009; Kuroda and Tvergaard 2010; Hutchinson 2012).

M. Kuroda (✉)
Graduate School of Science and Engineering, Mechanical Systems Engineering, Yamagata University, Jonan 4-3-16, Yonezawa, Yamagata 992-8510, Japan
e-mail: kuroda@yz.yamagata-u.ac.jp

H. Altenbach et al. (eds.), *From Creep Damage Mechanics to Homogenization Methods*, Advanced Structured Materials 64,
DOI 10.1007/978-3-319-19440-0_9

It is now widely recognized that strain gradient plasticity theories must be higher-order in the sense that it should be possible to impose extra boundary conditions with respect to plastic strains or their gradients. The theories in Fleck et al. (1994); Mühlhaus and Aifantis (1991); Fleck and Hutchinson (2001); Gudmundson (2004); Gurtin and Anand (2009); Kuroda and Tvergaard (2010); Hutchinson (2012) are all explicitly higher-order, but their apparent mathematical forms appear to be different. While some of them involve a set of higher-order stress quantities as a central part of the theory (Fleck et al. 1994; Fleck and Hutchinson 2001; Gudmundson 2004; Gurtin and Anand 2009; Hutchinson 2012), the others do not refer to such extra stress quantities (Mühlhaus and Aifantis 1991; Kuroda and Tvergaard 2010). Although the connections between such different treatments of strain gradient plasticity have been discussed to some extent in Gudmundson (2004); Gurtin and Anand (2009); Kuroda and Tvergaard (2010); Kuroda (2015) from their respective points of view, a more systematic and unifying investigation seems to be needed for the further understanding and development of scale-dependent strain gradient theories of plasticity.

In the present paper, different theoretical interpretations and possible mathematical expressions for the higher-order strain gradient plasticity theory initiated by Aifantis (1984, 1987) are reviewed and systematically reexamined in detail. These different interpretations of the theory result in different numerical procedures. In the present study, the effects of the order of finite-element shape functions and the number of Gaussian quadrature points on the qualities of numerical solutions are investigated for different formulations of strain gradient plasticity.

9.2 Basic Relations Unchanged from Classical J_2 Theory

In a small strain context, the total strain rate is decomposed into elastic and plastic parts as in classical theories of elastoplasticity:

$$\dot{\boldsymbol{E}} = (\dot{\boldsymbol{u}} \otimes \nabla)_{\text{sym}} = \dot{\boldsymbol{E}}^{\text{e}} + \dot{\boldsymbol{E}}^{\text{p}}, \tag{9.1}$$

where superscripts e and p denote elastic and plastic parts, respectively, $\dot{\boldsymbol{u}}$ is the displacement rate vector, $\nabla(= \partial/\partial x_i \boldsymbol{e}_i$; $\boldsymbol{e}_i$ are Cartesian bases and x_i are Cartesian coordinates) is the gradient operator, $\otimes$ is the tensor product, $(\ldots)_{\text{sym}}$ denotes the symmetric part of the tensor, and a superposed dot denotes the material-time derivative (assumed equal to the time derivative in the case of a small strain). Hooke's law is applied to model elasticity, and a simple coaxial flow rule is used for plasticity:

$$\dot{\boldsymbol{E}}^{\text{e}} = \boldsymbol{C}^{-1} : \dot{\boldsymbol{\sigma}}; \qquad \dot{\boldsymbol{E}}^{\text{p}} = \dot{\phi}^{\text{p}} \boldsymbol{N}^{\text{p}}; \qquad \boldsymbol{N}^{\text{p}} = \frac{\boldsymbol{\sigma}'}{|\boldsymbol{\sigma}'|}, \tag{9.2}$$

where $\boldsymbol{\sigma}$ is a symmetric (Cauchy) stress tensor, $\boldsymbol{C}$ is a fourth-order elasticity tensor, $\boldsymbol{\sigma}'$ is the deviatoric stress tensor, $\dot{\phi}^{\text{p}}$ is a non-negative plastic multiplier, and $|(\ldots)|$

is the norm of a tensor, i.e., $|(\ldots)| = \sqrt{(\ldots):(\ldots)}$. The equivalent plastic strain is defined as follows:

$$\varepsilon^{\mathrm{p}} = \int_0^t \dot{\varepsilon}^{\mathrm{p}} \mathrm{d}t; \qquad \dot{\varepsilon}^{\mathrm{p}} = \sqrt{\frac{2}{3}\dot{\boldsymbol{E}}^{\mathrm{p}} : \dot{\boldsymbol{E}}^{\mathrm{p}}} = \sqrt{\frac{2}{3}}\dot{\phi}^{\mathrm{p}}, \tag{9.3}$$

where t is the time.

The virtual work principle (VWP) is introduced simply as a weak form of the equilibrium equation ($\nabla \cdot \boldsymbol{\sigma} = \mathbf{0}$) with mechanical (Cauchy's formula) and displacement boundary conditions (neglect of the body force effect). An incremental version of the VWP is written as

$$\int_V \dot{\boldsymbol{\sigma}} : \delta\dot{\boldsymbol{E}} \mathrm{d}V = \int_S \underline{\dot{\boldsymbol{t}}} \cdot \delta\dot{\boldsymbol{u}} \mathrm{d}S, \tag{9.4}$$

where V is the volume of the solid considered, S is the surface of the solid, and $\underline{\dot{\boldsymbol{t}}}$ is the surface traction. An underbar $\underline{(\ldots)}$ emphasizes that the quantity is prescribed on the boundary. This notation will be used throughout the paper.

9.3 Introduction of Plastic Strain Gradient into Yield Condition

Aifantis (1984, 1987) proposed a modified yield condition with a plastic strain gradient effect as

$$\sigma_{\mathrm{e}} + \beta \nabla^2 \varepsilon^{\mathrm{p}} - R(\varepsilon^{\mathrm{p}}) = 0 \tag{9.5}$$

for a plastic loading state, where $\sigma_{\mathrm{e}} = \sqrt{3/2}|\boldsymbol{\sigma}'|$, β is a positive coefficient, $R(\varepsilon^{\mathrm{p}})$ is a positive strain hardening function, and ∇^2 is the Laplacian operator ($\nabla^2 = \mathrm{div}\nabla = \nabla \cdot \nabla$). The introduction of the gradient term into conventional plasticity theory was mainly motivated by its ability to predict the postlocalization feature of material behavior; i.e., the inclusion of the gradient term is necessary to determine the shear band width in a softening-type material.

The physical basis for the introduction of $\beta \nabla^2 \varepsilon^{\mathrm{p}}$ is strengthened by an argument based on dislocation theory as discussed in Kuroda and Tvergaard (2010); Kuroda (2015); Kuroda and Tvergaard (2006). That is, a dislocation-induced long-range internal stress arises in response to spatial gradients of the geometrically necessary dislocation (GND) density (Groma et al. 2003; Evers et al. 2004), and the GND density is equated with the spatial gradient of the crystallographic slip (Ashby 1970). Thus, the internal stresses correspond to the second gradients of crystallographic slips. The introduction of $\beta \nabla^2 \varepsilon^{\mathrm{p}}$ is consistent with this argument. The terms $\sigma_{\mathrm{e}} + \beta \nabla^2 \varepsilon^{\mathrm{p}}$, which can be interpreted as the superposition of stress due to external forces and the effect of internal stress due to the distributions of GNDs, could represent a net stress that activates plastic straining. Based on this view, the plastic dissipation

D should be accounted for using $D = (\sigma_e + \beta\nabla^2\varepsilon^p)\dot{\varepsilon}^p = R\dot{\varepsilon}^p \geq 0$. Thus, the gradient term $\beta\nabla^2\varepsilon^p$ is naturally interpreted as a recoverable or energetic quantity, as also discussed in Kuroda and Tvergaard (2010).

The above simplest model uses the effective stress σ_e and equivalent plastic strain ε^p to describe the primary effects of stresses due to external forces and internal stresses originating from the non-uniform distribution of GNDs, respectively. Consequently, the positive or negative direction of the resolved shear stress and slip, which would be accounted for in the context of crystal plasticity, is not considered. Thus, a Bauschinger-like effect cannot be represented using Eq. (9.5). More generalized strain gradient formulations in which the plastic strain gradients are accounted for by a third-order tensor $\boldsymbol{E}^p \otimes \nabla$, where

$$\boldsymbol{E}^p = \int_0^t \dot{\boldsymbol{E}}^p \mathrm{d}t,$$

instead of $\nabla\varepsilon^p$, can represent such unsymmetric mechanical behavior (e.g., Gurtin and Anand 2005a; Niordson and Legarth 2010). The aim of the present study is not to generalize strain gradient plasticity theories. Instead, the study mainly focuses on issues related to a variety of theoretical interpretations of strain gradient plasticity that give the same solution but appear to have different mathematical expressions. For this purpose, we consider the simplest model given by Eq. (9.5).

Before proceeding, Eq. (9.5) is modified to the form

$$\sigma_e + \nabla \cdot \boldsymbol{g}^p - R(\varepsilon^p) = 0 \tag{9.6}$$

with the definition[1]

$$\boldsymbol{g}^p = \beta\nabla\varepsilon^p. \tag{9.7}$$

This modification is necessary to transform one treatment to the other treatments. When β is chosen to be constant, Eqs. (9.5) and (9.6) are identical.

9.4 Different Treatments of Strain Gradient Plasticity

In this section, different interpretations and treatments for the simplest higher-order strain gradient plasticity theory are examined. It is emphasized that these different treatments give the same solution to the same problem provided that extra (unconventional) boundary conditions with respect to $\dot{\varepsilon}^p$ or $\nabla\dot{\varepsilon}^p$ are equivalently imposed.

[1] In Kuroda and Tvergaard (2010), somewhat different definitions for $\boldsymbol{g}^p$ were used; i.e., $\boldsymbol{g}^p = -\nabla\varepsilon^p$ without the inclusion of β for the simplest theory, and $\boldsymbol{g}^p = -l_*^2 h\nabla\varepsilon^p$ to discuss an alternative formulation (corresponding to Treatment 4 in the present study) of Fleck-Hutchinson theory (Fleck and Hutchinson 2001), where l_* is a length-scale parameter. In the present paper, to enable a broader and extended discussion, the more general definition given by Eq. (9.7) is employed.

9.4.1 Treatment 1: Eq. (9.6) is Simply an Extra Balance Law

We consider that Eq. (9.6) cannot be simply constitutive and should be viewed as a balance law. The physical origin of the gradient term in this balance law is sought in configurational stresses produced by GNDs as discussed in Sect. 9.3. Equation (9.6) should be associated with concomitant boundary conditions since Eq. (9.6) is viewed as a PDE. In order to deduce such boundary conditions and obtain numerical solutions to Eq. (9.6), its weak form is derived, starting from the following self-evident integral relation:

$$\int_V (\sigma_\mathrm{e} + \nabla \cdot \boldsymbol{g}^\mathrm{p} - R)\delta\dot{\varepsilon}^\mathrm{p} \mathrm{d}V = 0, \tag{9.8}$$

where $\delta\dot{\varepsilon}^\mathrm{p}$ is a scalar weighting function, which is arbitrary and thus may be regarded as a *virtual plastic strain rate*, and V is the volume of the solid considered. Applying integration by parts and the divergence theorem followed by time differentiation, Eq. (9.8) becomes

$$\int_V \left\{(\dot{R} - \dot{\sigma}_\mathrm{e})\delta\dot{\varepsilon}^\mathrm{p} + \boldsymbol{g}^\mathrm{p} \cdot \nabla\delta\dot{\varepsilon}^\mathrm{p}\right\} \mathrm{d}V = \int_S \underline{\boldsymbol{n} \cdot \dot{\boldsymbol{g}}^\mathrm{p}} \delta\dot{\varepsilon}^\mathrm{p} \mathrm{d}S, \tag{9.9}$$

where S is the surface of the solid and $\boldsymbol{n}$ is a unit normal to the surface. In the following, β is taken to be constant for simplicity, although it could be a variable. Substituting the relation

$$\dot{\sigma}_\mathrm{e} = \sqrt{\frac{3}{2}} \boldsymbol{N}^\mathrm{p} : \dot{\boldsymbol{\sigma}}$$

and using the constitutive relations (Eqs. (9.1) and (9.2)), Eq. (9.9) is rewritten as

$$\begin{aligned} &- \int_V \sqrt{\frac{3}{2}} \delta\dot{\varepsilon}^\mathrm{p} \boldsymbol{N}^\mathrm{p} : \boldsymbol{C} : \dot{\boldsymbol{E}} \mathrm{d}V \\ &+ \int_V \left\{\delta\dot{\varepsilon}^\mathrm{p} \left(\frac{3}{2}\boldsymbol{N}^\mathrm{p} : \boldsymbol{C} : \boldsymbol{N}^\mathrm{p} + h\right)\dot{\varepsilon}^\mathrm{p} + \beta\nabla\delta\dot{\varepsilon}^\mathrm{p} \cdot \nabla\dot{\varepsilon}^\mathrm{p}\right\} \mathrm{d}V = \int_S \delta\dot{\varepsilon}^\mathrm{p} \underline{\boldsymbol{n} \cdot \dot{\boldsymbol{g}}^\mathrm{p}} \mathrm{d}S, \end{aligned} \tag{9.10}$$

where $h = \mathrm{d}R/\mathrm{d}\varepsilon^\mathrm{p}$.

Substituting the constitutive relations (9.2) and (9.1) into the incremental VWP (Eq. (9.4)) gives

$$\int_V \delta\dot{\boldsymbol{E}} : \boldsymbol{C} : \dot{\boldsymbol{E}} \mathrm{d}V - \int_V \sqrt{\frac{3}{2}} \delta\dot{\boldsymbol{E}} : \boldsymbol{C} : \boldsymbol{N}^\mathrm{p} \dot{\varepsilon}^\mathrm{p} \mathrm{d}V = \int_S \delta\dot{\boldsymbol{u}} \cdot \underline{\dot{\boldsymbol{t}}} \mathrm{d}S. \tag{9.11}$$

In the present treatment, Eqs. (9.11) and (9.10) are viewed as a set of simultaneous equations to be solved to obtain the two independent variables, $\dot{\boldsymbol{u}}$ and $\dot{\varepsilon}^\mathrm{p}$. This type of treatment has recently been employed in Kuroda (2015) in the context of finite-deformation strain localization analysis incorporating a corner-like plasticity effect. In this treatment, neither the introduction nor the recognition of higher-order stresses is necessary.

9.4.2 Treatment 2: A Virtual Work Principle with Higher-Order Quantities Is Introduced as the Major Premise

In this treatment, unconventional internal virtual work and external virtual work (Fleck and Hutchinson 2001; Gurtin and Anand 2009) are considered *a priori* without the introduction of any balance law in PDE form. The argument of internal stresses based on dislocation theory (Sect. 9.3) is not quoted at the starting point of the formulation.

The internal virtual work is assumed to be

$$\delta W_{\text{int}} = \int_V \left\{ \boldsymbol{\sigma} : \delta \dot{\boldsymbol{E}}^{\text{e}} + Q \delta \dot{\varepsilon}^{\text{p}} + \boldsymbol{\tau} \cdot \nabla \delta \dot{\varepsilon}^{\text{p}} \right\} \mathrm{d}V, \tag{9.12}$$

where $\boldsymbol{\sigma} : \delta \dot{\boldsymbol{E}}^{\text{e}}$ is the conventional elastic virtual work, Q is defined as the work conjugate to $\dot{\varepsilon}^{\text{p}}$, and $\boldsymbol{\tau}$ is introduced as a higher-order stress vector quantity that is work-conjugate to $\nabla \dot{\varepsilon}^{\text{p}}$. The external virtual work for incremental problems is assumed to be

$$\delta W_{\text{ext}} = \int_S \left\{ \underline{\boldsymbol{t}} \cdot \delta \dot{\boldsymbol{u}} + \underline{\chi} \delta \dot{\varepsilon}^{\text{p}} \right\} \mathrm{d}S, \tag{9.13}$$

where, as in conventional theories, the surface traction $\underline{\boldsymbol{t}}$ is work-conjugate to $\dot{\boldsymbol{u}}$, and $\underline{\chi}$ is unconventionally introduced as a higher-order traction scalar quantity that is work-conjugate to $\dot{\varepsilon}^{\text{p}}$ on the surface. Then, the internal virtual work is equated to the external virtual work, i.e., $\delta W_{\text{int}} = \delta W_{\text{ext}}$, in order to construct an extended VWP as

$$\int_V \left\{ \boldsymbol{\sigma} : \delta \dot{\boldsymbol{E}}^{\text{e}} + Q \delta \dot{\varepsilon}^{\text{p}} + \boldsymbol{\tau} \cdot \nabla \delta \dot{\varepsilon}^{\text{p}} \right\} \mathrm{d}V = \int_S \left\{ \underline{\boldsymbol{t}} \cdot \delta \dot{\boldsymbol{u}} + \underline{\chi} \delta \dot{\varepsilon}^{\text{p}} \right\} \mathrm{d}S. \tag{9.14}$$

Noting the relation $\boldsymbol{\sigma} : \delta \dot{\boldsymbol{E}}^{\text{p}} = \sigma_{\text{e}} \delta \dot{\varepsilon}^{\text{p}}$, Eq. (9.12) is integrated by parts and rewritten as

$$\begin{aligned} \delta W_{\text{int}} = & \int_V \left\{ -\nabla \cdot \boldsymbol{\sigma} \cdot \delta \dot{\boldsymbol{u}} + (Q - \sigma_{\text{e}} - \nabla \cdot \boldsymbol{\tau}) \delta \dot{\varepsilon}^{\text{p}} \right\} \mathrm{d}V \\ & + \int_S \left\{ \boldsymbol{n} \cdot \boldsymbol{\sigma} \cdot \delta \dot{\boldsymbol{u}} + \boldsymbol{n} \cdot \boldsymbol{\tau} \delta \dot{\varepsilon}^{\text{p}} \right\} \mathrm{d}S. \end{aligned} \tag{9.15}$$

Subtracting Eq. (9.13) from Eq. (9.15) (cf. $\delta W_{\text{int}} - \delta W_{\text{ext}} = 0$) gives

$$\nabla \cdot \boldsymbol{\sigma} = \boldsymbol{0}, \tag{9.16}$$

$$\sigma_{\text{e}} + \nabla \cdot \boldsymbol{\tau} - Q = 0 \tag{9.17}$$

on V and

$$\underline{t} = \boldsymbol{n} \cdot \boldsymbol{\sigma}, \tag{9.18}$$

$$\underline{\chi} = \boldsymbol{n} \cdot \boldsymbol{\tau} \tag{9.19}$$

on S.

To complete the theory, the constitutive relation for the higher-order stress $\boldsymbol{\tau}$ needs to be specified. Gudmundson (2004) and Gurtin and Anand (2009) proposed the employment of the thermodynamic consistency.[2] Here, it is assumed that the free energy ψ is composed of conventional elastic strain energy and unconventional defect energy. The latter is postulated to be a quadratic function of the plastic strain gradient as one of the simplest models,

$$\psi = \frac{1}{2}\boldsymbol{E}^{\mathrm{e}} : \boldsymbol{C} : \boldsymbol{E}^{\mathrm{e}} + \frac{1}{2}\alpha|\nabla\varepsilon^{\mathrm{p}}|^2; \qquad \alpha \geq 0, \tag{9.20}$$

where α is a constant for dimensional consistency. The defect energy associated with the plastic strain gradients is modeled as a recoverable quantity. From a physical point of view, $\nabla\varepsilon^{\mathrm{p}}$ is used as a macroscopic measure of the stored GNDs whose energy can be released by elimination of the plastic strain gradients (Gurtin 2002; Hutchinson 2012). This thought is consistent with the argument given in Sect. 9.3.

By the second law of thermodynamics, the temporal increase in the free energy is less than or equal to the power expended. This leads to the inequality

$$\dot{\psi} - (\boldsymbol{\sigma} : \dot{\boldsymbol{E}}^{\mathrm{e}} + Q\dot{\varepsilon}^{\mathrm{p}} + \boldsymbol{\tau} \cdot \nabla\dot{\varepsilon}^{\mathrm{p}}) \leq 0. \tag{9.21}$$

Guided by this inequality, constitutive relations for the stress quantities and a condition of plastic dissipation are suggested to be

$$\boldsymbol{\sigma} = \boldsymbol{C} : \boldsymbol{E}^{\mathrm{e}}; \qquad \boldsymbol{\tau} = \alpha\nabla\varepsilon^{\mathrm{p}}; \qquad Q\dot{\varepsilon}^{\mathrm{p}} \geq 0. \tag{9.22}$$

Comparing Eqs. (9.22) and (9.7), the higher order stress $\boldsymbol{\tau}$ is identified with $\boldsymbol{g}^{\mathrm{p}}$ in Eqs. (9.6) and (9.7) provided that the defect energy is assumed to be given by $\frac{1}{2}\alpha|\nabla\varepsilon^{\mathrm{p}}|^2$, as in Eq. (9.20), and $\alpha = \beta$. It is also obvious that Q is identified with R. It is clear from the derivation that the quantity $\boldsymbol{\tau}$ and its divergence $\nabla \cdot \boldsymbol{\tau}$ are taken as recoverable or energetic quantities (Gudmundson 2004; Gurtin and Anand 2009).

An incremental version of Eq. (9.14) is

$$\int_V \{\dot{\boldsymbol{\sigma}} : \delta\dot{\boldsymbol{E}}^{\mathrm{e}} + \dot{Q}\delta\dot{\varepsilon}^{\mathrm{p}} + \dot{\boldsymbol{\tau}} \cdot \nabla\delta\dot{\varepsilon}^{\mathrm{p}}\}\mathrm{d}V = \int_S \{\dot{\underline{t}} \cdot \delta\dot{\boldsymbol{u}} + \dot{\underline{\chi}}\delta\dot{\varepsilon}^{\mathrm{p}}\}\mathrm{d}S \tag{9.23}$$

---

[2]This type of derivation of constitutive relations for the higher-order stresses was presented in the context of gradient crystal plasticity theory (Gurtin 2002). The crystal plasticity version of Eq. (9.14) also appeared in Gurtin (2002). Thermodynamical formulations of crystal plasticity were also discussed in Forest et al. (2002).

In Eq. (9.23), if we consider that $\delta\dot{\boldsymbol{u}} = \boldsymbol{0}$ but $\delta\dot{\varepsilon}^{\mathrm{p}} \neq 0$, and consequently

$$\dot{\boldsymbol{\sigma}} : \delta\dot{\boldsymbol{E}}^{\mathrm{e}} = -\dot{\boldsymbol{\sigma}} : \delta\dot{\boldsymbol{E}}^{\mathrm{p}} = -\dot{\sigma}_{\mathrm{e}}\delta\dot{\varepsilon}^{\mathrm{p}},$$

then Eq. (9.23) reduces to

$$\int_V \{(\dot{Q} - \dot{\sigma}_{\mathrm{e}})\delta\dot{\varepsilon}^{\mathrm{p}} + \dot{\boldsymbol{\tau}} \cdot \nabla\delta\dot{\varepsilon}^{\mathrm{p}}\}\mathrm{d}V = \int_S \underline{\dot{\chi}}\delta\dot{\varepsilon}^{\mathrm{p}}\mathrm{d}S. \tag{9.24}$$

This is identical to Eq. (9.9) upon noting that $\dot{Q} = \dot{R}$, $\boldsymbol{\tau} = \dot{\boldsymbol{g}}^{\mathrm{p}}$ and $\underline{\dot{\chi}} = \boldsymbol{n} \cdot \dot{\boldsymbol{\tau}} = \boldsymbol{n} \cdot \dot{\boldsymbol{g}}^{\mathrm{p}}$. Instead, if we consider that $\delta\dot{\varepsilon}^{\mathrm{p}} = 0$ (consequently, $\delta\dot{\boldsymbol{E}}^{\mathrm{e}} = \delta\dot{\boldsymbol{E}}$), Eq. (9.23) reduces to the standard incremental VWP, Eq. (9.4). Therefore, Eqs. (9.10) and (9.11) for Treatment 1 are also valid for Treatment 2. Thus, the numerical strategy for Treatment 2 can be fundamentally the same as that for Treatment 1.

In summary, the extended VWP of (9.14) is introduced as the major premise in this treatment. This statement of the VWP *yields* both the conventional equilibrium equation (9.16) and the additional balance law (9.17) that is equivalent to the yield function (9.6). The constitutive assumptions with respect to the gradient effect are given in the representation of the free energy. The thermodynamic inequality with the specific free energy representation can lead to constitutive relations for the standard stress and nonstandard higher-order stress. The solution strategy can be the same as that for Treatment 1.

9.4.3 Treatment 3: A Variational Principle Is Utilized

Following Mühlhaus and Aifantis (1991), a potential functional F is considered for incremental boundary value problems as

$$F = \int_V \Phi(\dot{\boldsymbol{E}}^{\mathrm{e}}, \dot{\varepsilon}^{\mathrm{p}}, \nabla\dot{\varepsilon}^{\mathrm{p}})\mathrm{d}V - \int_S (\underline{\dot{\boldsymbol{t}}} \cdot \dot{\boldsymbol{u}} + \underline{\dot{\chi}}\dot{\varepsilon}^{\mathrm{p}})\mathrm{d}S \tag{9.25}$$

with

$$\Phi(\dot{\boldsymbol{E}}^{\mathrm{e}}, \dot{\varepsilon}^{\mathrm{p}}, \nabla\dot{\varepsilon}^{\mathrm{p}}) = \frac{1}{2}\dot{\boldsymbol{E}}^{\mathrm{e}} : \boldsymbol{C} : \dot{\boldsymbol{E}}^{\mathrm{e}} + \frac{1}{2}h(\dot{\varepsilon}^{\mathrm{p}})^2 + \frac{1}{2}\eta|\nabla\dot{\varepsilon}^{\mathrm{p}}|^2, \tag{9.26}$$

where η is a coefficient for dimensional consistency. The first term of Eq. (9.26) is the incremental elastic work rate density. The second term is the conventional plastic work rate density. The third term introduces the effect of the plastic strain gradient, which takes the simplest quadratic form following the first two terms. The condition $\delta F = 0$ gives

$$\int_V \dot{\boldsymbol{\sigma}} : \delta\dot{\boldsymbol{E}} \mathrm{d}V - \int_S \underline{\dot{\boldsymbol{t}}} \cdot \delta\dot{\boldsymbol{u}} \mathrm{d}S + \int_V \{(h\dot{\varepsilon}^{\mathrm{p}} - \dot{\sigma}_{\mathrm{e}})\delta\dot{\varepsilon}^{\mathrm{p}} + (\eta\nabla\dot{\varepsilon}^{\mathrm{p}}) \cdot \nabla\delta\dot{\varepsilon}^{\mathrm{p}}\}\mathrm{d}V - \int_S \underline{\dot{\chi}}\delta\dot{\varepsilon}^{\mathrm{p}} \mathrm{d}S = 0, \tag{9.27}$$

and integration by parts and the divergence theorem lead to

$$\begin{aligned} &-\int_V (\nabla \cdot \dot{\boldsymbol{\sigma}}) \cdot \delta\dot{\boldsymbol{u}} \mathrm{d}V + \int_S (\boldsymbol{n} \cdot \dot{\boldsymbol{\sigma}} - \underline{\dot{\boldsymbol{t}}}) \cdot \delta\dot{\boldsymbol{u}} \mathrm{d}S \\ &- \int_V \{\dot{\sigma}_{\mathrm{e}} + \nabla \cdot (\eta\nabla\dot{\varepsilon}^{\mathrm{p}}) - (h\dot{\varepsilon}^{\mathrm{p}})\}\delta\dot{\varepsilon}^{\mathrm{p}} \mathrm{d}V + \int_S \{\boldsymbol{n} \cdot (\eta\nabla\dot{\varepsilon}^{\mathrm{p}}) - \underline{\dot{\chi}}\}\delta\dot{\varepsilon}^{\mathrm{p}} \mathrm{d}S = 0. \end{aligned} \tag{9.28}$$

The first and second terms correspond to the conventional equilibrium equation and the mechanical boundary condition, respectively. Thus, the additional balance law and the associated boundary condition are deduced to be

$$\dot{\sigma}_{\mathrm{e}} + \nabla \cdot (\eta\nabla\dot{\varepsilon}^{\mathrm{p}}) - h\dot{\varepsilon}^{\mathrm{p}} = 0, \tag{9.29}$$

$$\underline{\dot{\chi}} = \boldsymbol{n} \cdot (\eta\nabla\dot{\varepsilon}^{\mathrm{p}}). \tag{9.30}$$

Aside from the inclusion of the higher-order traction term, $\underline{\dot{\chi}}\dot{\varepsilon}^{\mathrm{p}}$, on the surface, Eq. (9.25) with (9.26) is fundamentally the same as the equation introduced by Mühlhaus and Aifantis (1991) with η taken as a constant. They did not explicitly define higher-order stress quantities and consequently only considered situations of $\boldsymbol{n} \cdot \nabla\dot{\varepsilon}^{\mathrm{p}} = 0$ or $\dot{\varepsilon}^{\mathrm{p}} = 0$ on the boundary.

Obviously, the function Φ in Eq. (9.26) involves all the factors of constitutive modeling, which, in fact, leads to Eq. (9.29) that is the incremental form of Eq. (9.6) with (9.7) (under the assumption that $\eta(= \beta$ is constant). In general, it may not be so simple to obtain sufficient insight into the physics (e.g., dislocation theory or thermodynamics) when formulating a functional such as F. Mühlhaus and Aifantis (1991) remarked on their variational principle that:"…we do not wish to provide any specific physical meaning or interpretation to the functional. We simply treat it as an intermediate quantity which can motivate the extra boundary conditions and facilitate the finite element formulation of the problem."

The principle of Eq. (9.25) with (9.26) yields Eq. (9.27) that is equivalent to Eqs. (9.4) and (9.9). Thus, the numerical strategy can be fundamentally the same as that for Treatment 1.

For the specific choice of $\eta = hl_*^2$, the potential functional F in Eq. (9.26) coincides with that introduced by Fleck and Hutchinson (2001). In their formulation, the terms with respect to plastic strain were collected to give

$$h(\dot{\varepsilon}^{\mathrm{p}})^2 + \eta|\nabla\dot{\varepsilon}^{\mathrm{p}}|^2 = h[(\dot{\varepsilon}^{\mathrm{p}})^2 + l_*^2|\nabla\dot{\varepsilon}^{\mathrm{p}}|^2] = h\dot{E}_{\mathrm{p}}^2,$$

where

$$\dot{E}_{\mathrm{p}} = \sqrt{(\dot{\varepsilon}^{\mathrm{p}})^2 + l_*^2|\nabla\dot{\varepsilon}^{\mathrm{p}}|^2}$$

was defined *a priori* as a gradient dependent effective plastic strain with a length scale l_* for dimensional consistency. In this case, $\delta F = 0$ leads to

$$\dot{\sigma}_e + \nabla \cdot (l_*^2 h \nabla \dot{\varepsilon}^p) - h\dot{\varepsilon}^p = 0 \tag{9.31}$$

$$\underline{\dot{\chi}} = \boldsymbol{n} \cdot (l_*^2 h \nabla \dot{\varepsilon}^p). \tag{9.32}$$

In Fleck and Hutchinson (2001), the constitutive relations for $\dot{\boldsymbol{\tau}}$ and $\dot{Q}$ were determined as $\dot{\boldsymbol{\tau}} = l_*^2 h \nabla \dot{\varepsilon}^p$ and $\dot{Q} = h\dot{\varepsilon}^p$, respectively, through a comparison between Eq. (9.31) deduced from the variational principle with the specific gradient-dependent quantity $\dot{E}_p$ and the incremental form of Eq. (9.17) that was derived from the extended virtual work assumption. In Fleck and Hutchinson (2001), the higher-order stress $\boldsymbol{\tau}$ was implicitly considered as a dissipative quantity. It has been noted by Gudmundson (2004); Gurtin and Anand (2009); Hutchinson (2012) that $\boldsymbol{\tau} \cdot \nabla \dot{\varepsilon}^p$ can be negative for certain non-proportional strain histories and this violates thermodynamic restrictions.

The introduction of a specific choice of $\dot{E}_p$ might have some quantitative effect on predictions, but it is not critical in the fundamental structure of the formulation as stated in Fleck and Hutchinson (2001) and also in Gurtin and Anand (2009). This choice yields a consequence that the gradient effect is directly related to the work-hardening rate, and for a non-hardening material the gradient effect completely vanishes. This may not be acceptable from a physical point of view. It is doubtful that the internal stresses due to the GND distribution have such a strong dependence on the degree of work hardening that should be accounted for by statistically stored dislocations (SSDs). We will not pursue the use of $\dot{E}_p$ in the present paper for this reason and also for brevity.

To summarize this section, the functional given by Eq. (9.25) with (9.26) in the quadratic form gives a relation equivalent to Eq. (9.6). Fleck-Hutchinson theory (Fleck and Hutchinson 2001) is fundamentally based on the major premise of the virtual work statement, Eq. (9.14) or (9.23), in Treatment 2. However, in the derivation of the constitutive relations for higher-order stresses, the variational principle for incremental problems is invoked in parallel.

9.4.4 Treatment 4: Eq. (9.6) is Merely a Constitutive Relation

In the present treatment, Eq. (9.6) is considered simply as a constitutive relation, i.e., a yield function (Kuroda and Tvergaard 2010). We seek the physical origin of the gradient term in the argument that internal stresses are caused by the GND density distribution as discussed in Sect. 9.3. As in the standard procedure used in the conventional plasticity theories, the consistency condition is first applied to Eq. (9.6), and then using Eq. (9.2), the plastic multiplier $\dot{\phi}^p$ is calculated as

$$\dot{\phi}^p = \frac{(\partial\sigma_e/\partial\boldsymbol{\sigma}) : \boldsymbol{C} : \dot{\boldsymbol{E}} + \nabla \cdot \dot{\boldsymbol{g}}^p}{A} > 0 \quad \text{for plastic loading,} \tag{9.33}$$

where

$$A = (\partial\sigma_e/\partial\boldsymbol{\sigma}) : \boldsymbol{C} : \boldsymbol{N}^p + \sqrt{\frac{2}{3}}h; \qquad h = dR/d\varepsilon^p. \tag{9.34}$$

Considering Eqs. (9.1), (9.2), and (9.33), the VWP is rewritten as

$$\int_V \delta\dot{\boldsymbol{E}} : \boldsymbol{C}^{ep} : \dot{\boldsymbol{E}} dV - \int_V \delta\dot{\boldsymbol{E}} : \boldsymbol{C} : \boldsymbol{N}^p \frac{1}{A} \nabla \cdot \dot{\boldsymbol{g}}^p dV = \int_S \delta\dot{\boldsymbol{u}} \cdot \underline{\dot{\boldsymbol{t}}} dS \tag{9.35}$$

with

$$\boldsymbol{C}^{ep} = \boldsymbol{C} - \frac{1}{A}(\boldsymbol{C} : \boldsymbol{N}^p) \otimes (\partial\sigma_e/\partial\boldsymbol{\sigma}) : \boldsymbol{C}. \tag{9.36}$$

The elastic-plastic moduli tensor, $\boldsymbol{C}^{ep}$, in Eq. (9.36) is the same as that appearing in conventional J_2 theory. In Eq. (9.35), $\dot{\boldsymbol{g}}^p$ is an extra independent variable in addition to $\dot{\boldsymbol{u}}$. To determine both $\dot{\boldsymbol{u}}$ and $\dot{\boldsymbol{g}}^p$, an additional governing equation to be solved simultaneously with Eq. (9.35) is needed.

The $\boldsymbol{g}^p$ has already been defined in Eq. (9.7). Now, this equation cannot be simply viewed as a definition. It should be viewed as a balance law; i.e., the quantity $\boldsymbol{g}^p$ always exists in balance with $\beta\nabla\varepsilon^p$,

$$\boldsymbol{g}^p - \beta\nabla\varepsilon^p = \boldsymbol{0}. \tag{9.37}$$

One can imagine an analogy between this equation and the relation for the GND density $\rho_G - \frac{1}{b}\partial\gamma/\partial x = 0$ (Fleck et al. 1994; Ashby 1970), where ρ_G is the density of GNDs with the edge character, b is the magnitude of the Burgers vector, γ is a crystallographic slip on a slip system, and x is taken along the slip direction.

The weak form of the incremental version of Eq. (9.37) is derived starting from

$$\int_V (\dot{\boldsymbol{g}}^p - \beta\nabla\dot{\varepsilon}^p) \cdot \delta\boldsymbol{w} dV = \boldsymbol{0}, \tag{9.38}$$

where $\delta\boldsymbol{w}$ is an arbitrary vector-valued weighting function and β is again taken as a constant. Applying integration by parts and the divergence theorem, and using Eq. (9.3) with (9.33), Eq. (9.38) becomes

$$\int_V \sqrt{\frac{2}{3}}\frac{\beta}{A} \nabla \cdot \delta\boldsymbol{w} (\partial\sigma_e/\partial\boldsymbol{\sigma}) : \boldsymbol{C} : \dot{\boldsymbol{E}} dV + \int_V (\delta\boldsymbol{w} \cdot \dot{\boldsymbol{g}}^p + \sqrt{\frac{2}{3}}\frac{\beta}{A} \nabla \cdot \delta\boldsymbol{w} \nabla \cdot \dot{\boldsymbol{g}}^p) dV$$
$$= \int_S \beta\delta\boldsymbol{w} \cdot \boldsymbol{n}\underline{\dot{\varepsilon}}^p dS. \tag{9.39}$$

Equations (9.35) and (9.39) are solved simultaneously to determine $\dot{\boldsymbol{u}}$ and $\dot{\boldsymbol{g}}^{\mathrm{p}}$. No higher-order stress quantity appears in this formulation. A treatment similar to the above has been used in the context of crystal plasticity by Evers et al. (2004) and Kuroda and Tvergaard (2006).

9.5 Computational Aspects

9.5.1 Finite Element Procedure

As discussed in the previous section, in Treatments 1–3, $\dot{\varepsilon}^{\mathrm{p}}$ is the additional independent variable, while in Treatment 4 $\dot{\boldsymbol{g}}^{\mathrm{p}}$ is the additional independent variable. In this section, two benchmark problems, a constrained simple shear and a plane strain tension of a passivated strip, are solved using finite element method. The effects of the orders of shape functions used in the interpolation of the displacement rate ($\dot{\boldsymbol{u}}$) and the additional nodal variable ($\dot{\varepsilon}^{\mathrm{p}}$ or $\dot{\boldsymbol{g}}^{\mathrm{p}}$) on the qualities of the numerical solutions are examined. We henceforth refer to the numerical method in which the additional nodal variable is chosen to be $\dot{\varepsilon}^{\mathrm{p}}$ as the "*e*-formulation" (Treatments 1–3), while the method in which the additional nodal variable is chosen to be $\dot{\boldsymbol{g}}^{\mathrm{p}}$ is referred to as the "*g*-formulation" (Treatment 4). Derivation of finite element equations that are similar to those for the *e*-formulation can be found in Mühlhaus and Aifantis (1991); Kuroda (2015); Niordson and Hutchinson (2003); Fredriksson et al. (2009); Niordson and Redanz (2004), and the derivation of finite element equations for the *g*-formulation has been given in Kuroda and Tvergaard (2010).

All numerical computations are done with quadrilateral isoparametric elements under plane strain conditions. The types of finite element employed are symbolized as, for example, "4FI-4FI", where the former "4FI" means that the four-node linear element with full integration (2×2) is used for displacement rate analysis and the latter "4FI" denotes that the four-node linear element with full integration is also used for the additional variable analysis (i.e., $\dot{\varepsilon}^{\mathrm{p}}$ in the *e*-formulation or $\dot{\boldsymbol{g}}^{\mathrm{p}}$ in the *g*-formulation). In addition to "4FI", the element types "8FI" and "8RI" are also employed. The former denotes the eight-node quadratic (serendipity) element with full integration (3×3), and the latter denotes the same element with reduced integration (2×2). The actual incremental numerical computations are performed using a forward Euler time integration scheme.

9.5.2 Numerical Issues

The first problem is a constrained simple shear. A strip with thickness H in the x_2-direction and infinite length in the x_1-direction is subjected to simple shear in the lateral (x_1-) direction. The top and bottom surfaces of the strip are assumed to be bounded by hard materials impenetrable to plastic flow (i.e. $\dot{\varepsilon}^{\rm p} = 0$). This problem has been frequently used as a benchmark test problem in the previous gradient plasticity studies. In Kuroda and Tvergaard (2010), an analytical solution for a linear strain hardening material has been given, and finite element computations were carried out using the g-formulation with 4FI-4FI elements for a constant length-scale coefficient of $\beta = l^2\sigma_0$ with length scales $l/H = 0.3$ and 1 (σ_0 the initial yield stress). Here, both the e- and g-formulations with various types of finite element are applied to the same problem using the same material parameter values. For this problem, finite element discretization is only needed in the thickness direction, i.e., only one column of finite elements is used. In the lateral (x_1-) direction, periodicity is assumed. All the formulations, except for the g-formulation with 8RI-8RI, produce almost identical solutions for the equivalent plastic strain distribution in the thickness direction, which fundamentally coincide with the analytical solution (Kuroda and Tvergaard 2010), when discretizations of at least thirty elements for 8FI- and 8RI- cases and at least forty elements for 4FI- cases are employed in the thickness direction. These results are not depicted here because they are indistinguishable from those shown in Kuroda and Tvergaard (2010). In the g-formulation, the 8RI-8RI elements yield a singular matrix, which leads to fundamental numerical failure. In all the numerical solutions obtained, the distributions of the equivalent plastic strain in the thickness direction are smooth, while local stress oscillation occurs in computations using the e-formulation with 4FI-4FI or 8FI-8FI elements. Theoretically, the shear stress, which is only the nonzero stress component, must be uniform throughout the thickness due to the equilibrium requirement.

Figure 9.1a shows the stress oscillation behavior at an overall shear strain of 3 % for the e-formulation with forty 4FI-4FI elements for the length scale of $l/H = 1$ (all other material constants are the same as those used in Kuroda and Tvergaard (2010)).

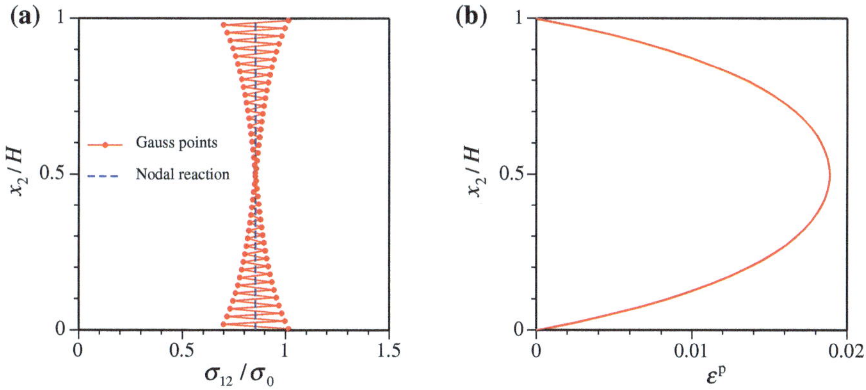

Fig. 9.1 **a** Oscillatory stress behavior in constrained simple shear problem ($l/H = 1$) computed using the e-formulation with forty 4FI-4FI elements. The shear stress value indicated by a *dashed line* is evaluated with nodal reaction forces at the *top* or *bottom* surface. **b** corresponding equivalent plastic strain distribution in the thickness direction

Table 9.1 Summary of qualities of numerical solutions for various finite element formulations for constrained simple shear problem

	4FI-4FI	8RI-8RI	8FI-8FI	8RI-4FI
e-formulation[a]	Large stress oscillation[c]	Satisfactory	Small stress oscillation	Satisfactory
g-formulation[b]	Satisfactory	Did not work	Satisfactory	Did not work

[a]The additional independent variable is $\dot{\varepsilon}^{\mathrm{p}}$
[b]The additional independent variable is $\dot{\boldsymbol{g}}^{\mathrm{p}}$
[c]Figure 9.1a

The shear stress value computed from the nodal reaction forces on the bottom or top surface coincides with the average value of oscillating stresses at the Gauss points, as illustrated in Fig. 9.1a, and thus the overall mechanical response seems not to be affected by these oscillations of stress. Figure 9.1b shows a plot of nodal values of the equivalent plastic strain computed with 4FI-4FI elements, which forms a smooth curve. Similar spurious stress oscillation behavior for particular choices of shape functions and quadrature orders has also been reported in Fredriksson et al. (2009). Table 9.1 summarizes the qualities of the numerical solutions for the different formulations. As shown in the table, the 8RI-4FI elements produce satisfactory numerical solutions for both the e- and g-formulations.

The second problem is the plane strain tension of a passivated strip. The finite element model, which is basically the same as that used in the constrained simple shear problem, is subjected to uniform tension in the x_1-direction. The top and bottom surfaces are assumed to be impenetrable to plastic flow ($\dot{\varepsilon}^{\mathrm{p}} = 0$), mimicking surface passivation. Again all the material constants except for the length scale are the same as those used in Kuroda and Tvergaard (2010) and in the above constrained simple shear computations. In this problem, the nonzero stress component is only σ_{11}, which is non-uniform for finite-length-scale settings. All the formulations, except for the g-formulation with 8RI-8RI elements, which leads to numerical failure, produce fundamentally equivalent numerical solutions whose differences are not resolved on the scale plotted when discretizations of at least thirty elements for 8FI- and 8RI-cases and at least forty elements for 4FI-cases in the thickness direction are employed. As an example, Fig. 9.2 depicts numerical results obtained using the e-formulation with thirty 8RI-4FI elements for length scales of $l/H = 0$ (size-independent), 1 and 2. The qualities of numerical solutions are summarized in Table 9.2. Spurious nonzero transverse stresses, σ_{22}, on the order of 10^{-2} to 10^{-3} relative to σ_{11} (for the case of $l/H = 2$) are generated in the results obtained using the e-formulations with 4FI-4FI and 8FI-8FI elements and the g-formulation with 8FI-8FI elements at a tensile strain of 3 %. However, the average value of these spurious σ_{22} within each element is zero in all the cases. Thus, they do not appear to affect the overall mechanical response. Again, the 8RI-4FI elements produce satisfactory solutions both for the e- and g-formulations without any numerical contradiction.

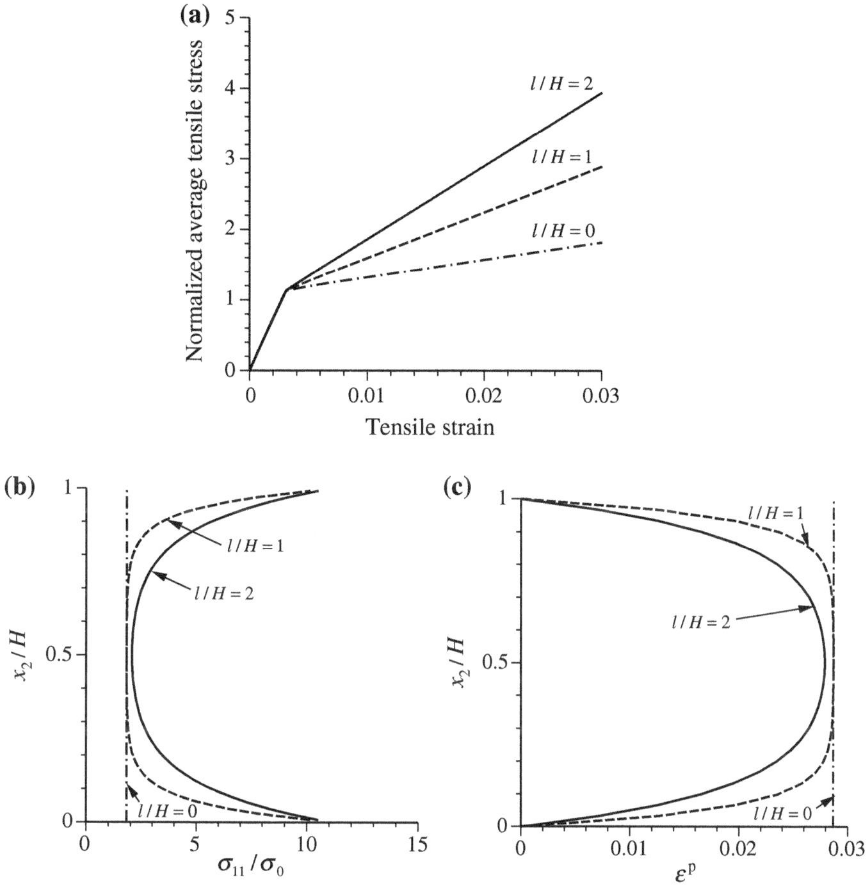

Fig. 9.2 Example of numerical solutions for plane strain tension of a passivated strip obtained with the e-formulation using thirty 8RI-4FI elements. **a** Curves of normalized average tensile stress (average of σ_{11} over the thickness divided by σ_0) versus tensile strain E_{11}. **b** Distributions of the tensile stress σ_{11}. **c** Distributions of the equivalent plastic strain ε^p

Table 9.2 Summary of qualities of numerical solutions for various finite element formulations for plane strain tension of a passivated strip

	4FI-4FI	8RI-8RI	8FI-8FI	8RI-4FI
e-formulation	$\mathrm{Max}[\lvert\sigma_{22}\rvert/\sigma_{11}] \approx 10^{-2}$ [a]	Satisfactory	$\mathrm{Max}[\lvert\sigma_{22}\rvert/\sigma_{11}] \approx 10^{-3}$ [a]	Satisfactory
g-formulation	Satisfactory	Did not work	$\mathrm{Max}[\lvert\sigma_{22}\rvert/\sigma_{11}] \approx 10^{-3}$ [a]	Satisfactory

[a]For case of $l/H = 2$. The average σ_{22} within each element is zero

9.6 Discussion

9.6.1 Plastic Dissipation

Based on the argument regarding the physical origin of the plastic strain gradient term in Sects. 9.3 and 9.4.2, $\boldsymbol{g}^{\mathrm{p}}$ is considered as a recoverable (or energetic) quantity. Here, the constitutive relation for $\boldsymbol{g}^{\mathrm{p}}$ considered so far in the present study is redefined as

$$\boldsymbol{g}^{\mathrm{p}}_{\mathrm{en}} = \beta \nabla \varepsilon^{\mathrm{p}}. \tag{9.40}$$

Some authors (e.g. Gurtin and Anand 2005a; Fleck and Willis 2009) consider that $\boldsymbol{g}^{\mathrm{p}}$ in Eq. (9.6) should include a "dissipative" part. Thus, it has been proposed that $\boldsymbol{g}^{\mathrm{p}}$ is additively decomposed into a dissipative part $\boldsymbol{g}^{\mathrm{p}}_{\mathrm{dis}}$ and an energetic part $\boldsymbol{g}^{\mathrm{p}}_{\mathrm{en}}$ as

$$\boldsymbol{g}^{\mathrm{p}} = \boldsymbol{g}^{\mathrm{p}}_{\mathrm{dis}} + \boldsymbol{g}^{\mathrm{p}}_{\mathrm{en}}. \tag{9.41}$$

For the dissipative part, the following type of formulation[3] has been proposed (Gurtin and Anand 2005a):

$$\boldsymbol{g}^{\mathrm{p}}_{\mathrm{dis}} = \lambda \frac{l^2_{\mathrm{dis}} \nabla \dot{\varepsilon}^{\mathrm{p}}}{\dot{\varepsilon}^{\mathrm{p}}}, \tag{9.42}$$

where l_{dis} is a dissipative length scale and λ is a positive coefficient. Note that assumed additional dissipation, $\boldsymbol{g}^{\mathrm{p}}_{\mathrm{dis}} \cdot \nabla \dot{\varepsilon}^{\mathrm{p}}$, is nonnegative.

These assumptions give the following interpretation of Eq. (9.6):

$$\sigma_{\mathrm{e}} + \underbrace{\nabla \cdot (\beta \nabla \varepsilon^{\mathrm{p}})}_{\text{energetic}} = \underbrace{R(\varepsilon^{\mathrm{p}}) - \nabla \cdot \left[\lambda \frac{l^2_{\mathrm{dis}} \nabla \dot{\varepsilon}^{\mathrm{p}}}{\dot{\varepsilon}^{\mathrm{p}}} \right]}_{\text{dissipative}}. \tag{9.43}$$

The last term on the right-hand side, i.e., $\nabla \cdot \boldsymbol{g}^{\mathrm{p}}_{\mathrm{dis}}$, is not an accumulated quantity and is an unknown in the current state. At the initial yielding of a body undergoing non-uniform deformation, this term immediately and discontinuously takes a finite value, and leads to an apparent "strengthening effect" at the initial stage of plastic deformation, which is not predicted from the accumulated quantity $\boldsymbol{g}^{\mathrm{p}}_{\mathrm{en}}$ at least with a constant β, as illustrated in Fig. 9.2a. In general, $\boldsymbol{g}^{\mathrm{p}}_{\mathrm{dis}}$ changes discontinuously when the boundary conditions and/or loading conditions change in the incremental problem. At that moment, one inevitably observes a "stress gap" if the material behavior follows Eq. (9.43) (Hutchinson 2012; Fleck et al. 2014, 2015). Whether a formulation such as Eq. (9.43) associated with the stress gap is physically acceptable cannot be judged at present because experiments on micron-size specimens with

---

[3]Gurtin and Anand (2005a) used the quantity $\dot{E}_{\mathrm{p}} = \sqrt{(\dot{\varepsilon}^{\mathrm{p}})^2 + l^2_{\mathrm{dis}} |\nabla \dot{\varepsilon}^{\mathrm{p}}|^2}$ for the denominator of Eq. (9.42).

a nonproportional loading history and/or abrupt changes of the loading directions have not yet been reported. The mass of GNDs associated with the spatial gradient of their density inevitably produces an internal stress at the material point around which they exist. The term $\nabla \cdot \boldsymbol{g}^{\mathrm{p}}_{\mathrm{en}} (= \nabla \cdot (\beta \nabla \varepsilon^{\mathrm{p}}))$ in Eq. (9.43) represents this effect. Then, a question arises: what does the quantity $\nabla \cdot \boldsymbol{g}^{\mathrm{p}}_{\mathrm{dis}}$ physically mean? To answer this fundamental question, further micromechanical dislocation studies are needed.

Aside from the physical meaning of $\nabla \cdot \boldsymbol{g}^{\mathrm{p}}_{\mathrm{dis}}$ or $\boldsymbol{g}^{\mathrm{p}}_{\mathrm{dis}}$, it seems that one motivation to introduce a model such as Eq. (9.42) involving the plastic strain rate gradient would be an inefficiency of the recoverable model like Eq. (9.40) for representing the strengthening phenomenon at an early stage of plastic deformation, which is reminiscent of the Hall-Petch effect. Recently, Fleck et al. (2015) have proposed a model that is free from the stress gap, in which the contribution from the plastic strain gradient is entirely recoverable, but the early flow strength elevation can be reproduced. In this model, a non-quadratic defect energy of the form $\sigma_0 k (N+1)^{-1} (l|\nabla \varepsilon^{\mathrm{p}}|)^{N+1}$ is phenomenologically assumed in the free energy function instead of the quadratic defect energy of $\frac{1}{2}\alpha |\nabla \varepsilon^{\mathrm{p}}|^2$ in Eq. (9.20), where N and k are the hardening exponent (being 0.1–0.5 for usual metals) and a non-dimensional coefficient used to represent the uniaxial stress-strain curve, respectively, i.e., $R(\varepsilon^{\mathrm{p}}) = \sigma_0 (1 + k \varepsilon^{\mathrm{p}N})$. According to this choice, $\boldsymbol{g}^{\mathrm{p}} (= \boldsymbol{g}^{\mathrm{p}}_{\mathrm{en}} = \boldsymbol{\tau})$ is derived via Eq. (9.21) as $\boldsymbol{g}^{\mathrm{p}} = \beta_{\mathrm{FHW}} \nabla \varepsilon^{\mathrm{p}}$ with $\beta_{\mathrm{FHW}} = \sigma_0 k l (l |\nabla \varepsilon^{\mathrm{p}}|)^N |\nabla \varepsilon^{\mathrm{p}}|^{-1}$. A physical interpretation of this variable β has not yet been proposed. When we assume that $N = 1$ and $k = 1$, this model is exactly the same as that used in Sect. 9.5.2 (i.e., $\beta_{\mathrm{FHW}} = \beta = \sigma_0 l^2$). Within the context of crystal plasticity, Ohno and Okumura (2007) proposed a defect energy accounting the self-energy of GNDs. This model gives a strictly linear dependence of the free energy on the GND densities, which corresponds to $N = 0$ in the above non-quadratic model and again produces a stress gap at initial yield.

9.6.2 Large Strains

In the present study, small strain conditions are considered throughout in order to focus our attention on the fundamental nature of strain gradient plasticity. The extension from small strain to large strain formulations is reasonably straightforward. Niordson and Redanz (2004) have extended Fleck and Hutchinson (2001) theory to a large-strain version based on an updated Lagrangian strategy. Gurtin and Anand (2005b) proposed a finite-deformation version of their small-deformation strain gradient plasticity theory for isotropic materials. Recently, Kuroda (2015) proposed a large-strain theory, which is categorized into Treatment 1, with a corner-like plasticity model, and showed its connection to a VWP-based formulation corresponding to Treatment 2.

9.6.3 Crystal Plasticity

Within the context of crystal plasticity, Evers et al. (2004) first presented a treatment equivalent to Treatment 4. In the case of crystal plasticity, the GND densities are naturally chosen as additional independent variables. The connection between this treatment and the virtual work-based formulation proposed by Gurtin (2002), which is equivalent to Treatment 2, was explored in Kuroda and Tvergaard (2006, 2008b). For the treatment in which the GND densities are nodal unknowns, the effects of finite-element shape functions and the number of Gaussian quadrature points on qualities of numerical solutions have been systematically investigated in Kuroda (2011) using a large-strain version of the theory (Kuroda and Tvergaard 2008a). It was shown that for plane strain problems the 8RI-4FI combination also gives satisfactory numerical solutions without any contradictions (e.g., no emergence of stress oscillations and/or fictitious stress components). An application of the large strain version of higher-order gradient crystal plasticity (Kuroda and Tvergaard 2008a) to a three-dimensional problem, i.e. a single-crystal micropillar compression test, has been conducted with 3D 20RI-8FI combination in which the serendipity shape functions were used for the displacement rate analysis with reduced integration and the bilinear shape functions were used for the GND density rate analysis with full integration (Kuroda 2013). This led to satisfactory solutions for the three-dimensional problem.

9.7 Conclusions

Among the different treatments for strain gradient plasticity covered in the present paper, the main difference is the places where the constitutive assumptions are made. In Treatments 1 and 4, the physical basis of the plastic strain gradient term in the yield function is sought in dislocation theory, and thus the constitutive assumption based on it is directly applied to the yield function. In these treatments, the existence of higher-order stress quantities is not emphasized. Meanwhile, in Treatment 2, the higher-order stress is introduced as a work conjugate to the plastic strain gradient. The free energy is assumed to be augmented by the defect energy involving the plastic strain gradients. The constitutive arbitrariness is placed in the formulation of the free energy function, which determines a specific relation for the higher-order stress. Both the considerations claim that the quantities contributed from the plastic strain gradients should be entirely recoverable. How the plastic dissipation can be quantified without any omission or double accounting is still an important open question. In Treatment 3, the constitutive assumptions are put into the potential functional F. In general, it may not be so simple to obtain sufficient insight into the physics (e.g. dislocation theory or thermodynamics) when formulating F.

In addition to the theoretical aspect, the present numerical investigation provides guidelines for the orders of finite-element shape functions that lead to satisfactory numerical solutions for strain gradient plasticity.

References

Aifantis EC (1984) On the microstructural origin of certain inelastic models. Trans ASME J Eng Mater Technol 106:326–330

Aifantis EC (1987) The physics of plastic deformation. Int J Plast 3:211–247

Ashby MF (1970) The deformation of plastically non-homogeneous alloys. Phil Mag 21:399–424

Evers LP, Brekelmans WAM, Geers MGD (2004) Non-local crystal plasticity model with intrinsic SSD and GND effects. J Mech Phys Solids 52:2379–2401

Fleck NA, Hutchinson JW (2001) A reformulation of strain gradient plasticity. J Mech Phys Solids 49:2245–2271

Fleck NA, Willis JR (2009) A mathematical basis for strain gradient plasticity. Part 1: scalar plastic multiplier. J Mech Phys Solids 57:161–177

Fleck NA, Muller GM, Ashby MF, Hutchinson JW (1994) Strain gradient plasticity: theory and experiment. Acta Metall Mater 42:475–487

Fleck NA, Hutchinson JW, Willis JR (2014) Strain gradient plasticity under non-proportional loading. Proc R Soc London A 470:20140267

Fleck NA, Hutchinson JW, Willis JR (2015) Guidelines for constructing strain gradient plasticity theories. J Appl Mech (in press), doi:10.1115/1.4030323

Forest S, Sievert R, Aifantis EC (2002) Strain gradient crystal plasticity: thermomechanical formulations and applications. J Mech Behav Mater 13:219–232

Fredriksson P, Gudmundson P, Mikkelsen LP (2009) Finite element implementation and numerical issues of strain gradient plasticity with application to metal matrix composites. Int J Solids Struct 46:3977–3987

Groma I, Csikor FF, Zaiser M (2003) Spatial correlations and higher-order gradient terms in a continuum description of dislocation dynamics. Acta Mater 51:1271–1281

Gudmundson P (2004) Unified treatment of strain gradient plasticity. J Mech Phys Solids 52:1379–1406

Gurtin ME (2002) A gradient theory of single-crystal viscoplasticity that accounts for geometrically necessary dislocations. J Mech Phys Solids 50:5–32

Gurtin ME, Anand L (2005a) A theory of strain-gradient plasticity for isotropic, plastically irrotational materials. Part I: small deformations. J Mech Phys Solids 53:1624–1649

Gurtin ME, Anand L (2005b) A theory of strain-gradient plasticity for isotropic, plastically irrotational materials. Part II: finite deformations. Int J Plast 21:2297–2318

Gurtin ME, Anand L (2009) Thermodynamics applied to gradient theories involving the accumulated plastic strain: the theories of aifantis and fleck and hutchinson and their generalization. J Mech Phys Solids 57:405–421

Hutchinson JW (2012) Generalizing J_2 flow theory: fundamental issues in strain gradient plasticity. Acta Mech Sin 28:1078–1086

Kuroda M (2011) On large-strain finite element solutions of higher-order gradient crystal plasticity. Int J Solids Struct 48:3382–3394

Kuroda M (2013) Higher-order gradient effects in micropillar compression. Acta Mater 61:2283–2297

Kuroda M (2015) A higher-order strain gradient plasticity theory with a corner-like effect. Int J Solids Struct 58:62–72

Kuroda M, Tvergaard V (2006) Studies of scale dependent crystal viscoplasticity models. J Mech Phys Solids 54:1789–1810

Kuroda M, Tvergaard V (2008a) A finite deformation theory of higher-order gradient crystal plasticity. J Mech Phys Solids 56:2573–2584

Kuroda M, Tvergaard V (2008b) On the formulations of higher-order gradient crystal plasticity models. J Mech Phys Solids 56:1591–1608

Kuroda M, Tvergaard V (2010) An alternative treatment of phenomenological higher-order strain-gradient plasticity theory. Int J Plast 26:507–515

Mühlhaus HB, Aifantis EC (1991) A variational principle for gradient plasticity. Int J Solids Struct 28:845–857
Niordson CF, Hutchinson JW (2003) Non-uniform plastic deformation of micron scale object. Int J Numer Methods Eng 56:961–975
Niordson CF, Legarth BN (2010) Strain gradient effects on cyclic plasticity. J Mech Phys Solids 58:542–557
Niordson CF, Redanz P (2004) Size-effects in plane strain sheet-necking. J Mech Phys Solids 52:2431–2454
Ohno N, Okumura D (2007) Higher-order stress and grain size effect due to self-energy of geometrically necessary dislocations. J Mech Phys Solids 55:1879–1898
Stölken JS, Evans AG (1998) A microbend test method for measuring the plasticity length scale. Acta Mater 46:5109–5115

Chapter 10
Effects of Fiber Arrangement on Negative Poisson's Ratio of Angle-Ply CFRP Laminates: Analysis Based on a Homogenization Theory

Tetsuya Matsuda, Keita Goto and Nobutada Ohno

Abstract The effects of fiber arrangement on the negative through-the-thickness Poisson's ratios of angle-ply carbon fiber-reinforced plastic (CFRP) laminates are investigated based on a homogenization theory. First, angle-ply CFRP laminates are modeled three-dimensionally with $+\theta$- and $-\theta$-laminae in which carbon fibers are distributed in a hexagonal arrangement. Then, a homogenization theory for nonlinear time-dependent composites is applied to the laminates. Using the present method, the elastic-viscoplastic Poisson's ratios of angle-ply carbon fiber/epoxy laminates are analyzed. The analysis results are then compared with those of the previous study, in which a square fiber arrangement in each lamina was assumed. It is shown that the fiber arrangement affects the negative through-the-thickness Poisson's ratios of the laminates in both elastic and viscoplastic regions, though they have little effect on the macroscopic stress-strain relationships of the laminates.

Keywords Negative Poisson's ratio · CFRP · Composite · Viscoplasticity · Homogenization

T. Matsuda (✉) · K. Goto
Department of Engineering Mechanics and Energy, University of Tsukuba,
1-1-1 Tennodai, Tsukuba 305-8573, Japan
e-mail: matsuda@kz.tsukuba.ac.jp

K. Goto
e-mail: s1230204@u.tsukuba.ac.jp

N. Ohno
Department of Mechanical Science and Engineering, Nagoya University,
Chikusa-ku, Nagoya 464-8603, Japan
e-mail: ohno@mech.nagoya-u.ac.jp

H. Altenbach et al. (eds.), *From Creep Damage Mechanics to Homogenization Methods*, Advanced Structured Materials 64,
DOI 10.1007/978-3-319-19440-0_10

10.1 Introduction

Carbon fiber-reinforced plastic (CFRP) laminates can exhibit negative Poisson's ratios depending on their laminate configurations. This phenomenon was originally noted by Herakovich (1984), who analytically showed that the through-the-thickness Poisson's ratios of angle-ply T300/5208 carbon fiber/epoxy laminates strongly depended on the laminate configuration; the maximum value was over 0.4, whereas the minimum value reached about −0.2. This drew attention as an important issue in material design as well as from a mechanical perspective, and was followed by several analytical and experimental studies (Clarke et al. 1994; Harkati et al. 2007).

The studies mentioned above, however, were limited to elastic analysis. Moreover, they were based on the classical lamination theory in which each lamina in a laminate was regarded as a homogeneous elastic material, even though laminae are intrinsically heterogeneous materials composed of fibers and a matrix. Thus, Matsuda et al. (2014) recently investigated the negative through-the-thickness Poisson's ratios of angle-ply CFRP laminates in the viscoplastic region based on a homogenization theory for nonlinear time-dependent composites (Wu and Ohno 1999; Ohno et al. 2000, 2001). In this study, the microscopic structures of laminae with fibers and a matrix, and the viscoplasticity of the matrix were explicitly taken into account. The analysis results revealed that the through-the-thickness Poisson's ratios of the laminates exhibited negative values not only in the elastic region but also in the viscoplastic region. Moreover, their negativity significantly increased with the progress of viscoplastic deformation in the laminates.

In the above-mentioned study, the authors assumed that each lamina in angle-ply CFRP laminates had a square fiber arrangement. Such an assumption of square fiber arrangements has been often adopted in unit cell-based analyses of fiber-reinforced composites because of its simplicity. It is known, however, that square fiber arrangements in laminae make them transversely anisotropic in both elastic and inelastic regions (Ohno et al. 2000), although actual laminae which have random fiber arrangements are transversely isotropic. In contrast, laminae having hexagonal fiber arrangements exhibit almost isotropic elastic/inelastic transverse behavior (Ohno et al. 2000), suggesting that hexagonal fiber arrangements are better than square ones for simulating the mechanical behavior of CFRP laminates. For more quantitative investigation, it is therefore worthwhile considering not only square but also hexagonal fiber arrangements in laminae when analyzing the negative Poisson's ratios of angle-ply CFRP laminates.

In this study, the effects of fiber arrangement on the elastic-viscoplastic negative Poisson's ratios of angle-ply CFRP laminates are investigated based on a homogenization theory. First, a modeling of angle-ply CFRP laminates is conducted in which each lamina in a laminate has a hexagonal fiber arrangement. Then, the homogenization theory for nonlinear time-dependent composites previously developed by the authors (Wu and Ohno 1999; Ohno et al. 2000, 2001) is applied to analyze the elastic-viscoplastic Poisson's ratios of angle-ply carbon fiber/epoxy laminates with various laminate configurations. The results obtained are compared with those obtained for a square fiber arrangement shown in the previous study (Matsuda et al. 2014) to discuss the effects of fiber arrangement.

10.2 Analysis Method

10.2.1 Modeling of Angle-Ply CFRP Laminates

Let us consider an angle-ply CFRP laminate consisting of $+\theta$- and $-\theta$-laminae, each of which has a thickness of l and is composed of carbon fibers and a matrix (Fig. 10.1). The fibers are assumed to have a hexagonal arrangement in the cross-sectional plane (transverse plane) of each lamina. Each lamina contains 16 (or 15) fibers in the through-the-thickness (stacking) direction in accordance with typical prepreg sheets (Matsuda et al. 2007, 2014). The laminate is subjected to macroscopically uniform loading, and exhibits infinitesimal strain both macroscopically and microscopically.

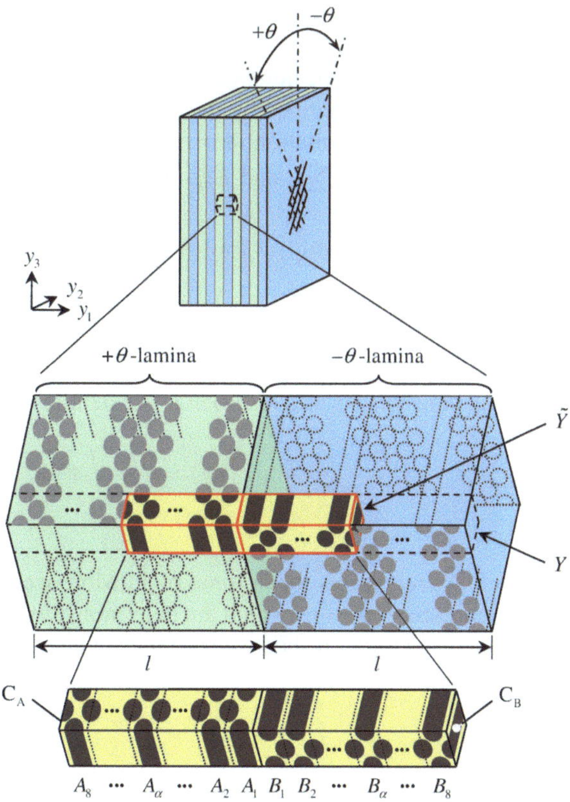

Fig. 10.1 $[\pm\theta]$ angle-ply CFRP laminate, unit cell Y, semiunit cell $\tilde{Y}$ and divided substructures A_α and B_α ($\alpha = 1, 2, \ldots, 8$)

10.2.2 Homogenization Theory for Nonlinear Time-Dependent Composites

For the angle-ply laminate shown above, a unit cell Y can be defined as indicated by the dashed lines in Fig. 10.1. However, half of Y (i.e., $\tilde{Y}$ in the figure) can also be adopted as an analysis domain using the point-symmetry with respect to the centers of the lateral facets of $\tilde{Y}$, C_A and C_B in the figure (Ohno et al. 2001; Matsuda et al. 2007, 2014). Therefore, in this study, $\tilde{Y}$ is employed as an analysis domain, and is hereafter called a semiunit cell. Stress and strain in $\tilde{Y}$ are expressed as σ_{ij} and ε_{kl}, respectively. For $\tilde{Y}$, the Cartesian coordinates $y_i (i = 1, 2, 3)$ are defined (Fig. 10.1).

Constituents in $\tilde{Y}$ are assumed to obey the following elastic-viscoplastic constitutive equation:

$$\dot{\sigma}_{ij} = c_{ijkl}(\dot{\varepsilon}_{kl} - \beta_{kl}), \tag{10.1}$$

where c_{ijkl} and β_{kl} represent the elastic stiffness and viscoplastic strain rate of the constituents, respectively, satisfying $c_{ijkl} = c_{jikl} = c_{ijlk} = c_{klij}$ and $\beta_{kl} = \beta_{lk}$. In this study, carbon fibers are regarded as elastic materials, meaning that the term β_{kl} becomes zero for the fibers. The matrix material, on the other hand, exhibits viscoplasticity, which plays a dominant role in the nonlinear behavior of CFRP laminates.

Then, according to the homogenization theory for nonlinear time-dependent composites previously developed by the authors (Wu and Ohno 1999; Ohno et al. 2000, 2001), the evolution equation of σ_{ij} and the relationship between the macroscopic stress and strain rates, $\dot{\Sigma}_{ij}$ and $\dot{E}_{kl}$, are derived as follows:

$$\dot{\sigma}_{ij} = c_{ijpq}(\delta_{pk}\delta_{ql} + \chi^{kl}_{p,q})\dot{E}_{kl} - c_{ijkl}(\beta_{kl} - \varphi_{k,l}), \tag{10.2}$$

$$\dot{\Sigma}_{ij} = \left\langle c_{ijpq}(\delta_{pk}\delta_{ql} + \chi^{kl}_{p,q})\right\rangle \dot{E}_{kl} - \left\langle c_{ijkl}(\beta_{kl} - \varphi_{k,l})\right\rangle, \tag{10.3}$$

where δ_{ij} indicates Kronecker's delta, $\langle\ \ \rangle$ denotes the volume average in $\tilde{Y}$ defined as

$$\langle \# \rangle = \frac{1}{|\tilde{Y}|}\int_{\tilde{Y}} \# \mathrm{d}Y, \tag{10.4}$$

in which $|\tilde{Y}|$ represents the volume of $\tilde{Y}$. Moreover, χ^{kl}_i and φ_i in Eqs. (10.2) and (10.3) represent functions which are determined by solving the following boundary value problems for $\tilde{Y}$, respectively (Wu and Ohno 1999; Ohno et al. 2000, 2001):

$$\int_{\tilde{Y}} c_{ijpq}\chi^{kl}_{p,q}v_{i,j}\mathrm{d}Y = -\int_{\tilde{Y}} c_{ijkl}v_{i,j}\mathrm{d}Y, \tag{10.5}$$

$$\int_{\tilde{Y}} c_{ijpq}\varphi_{p,q}v_{i,j}\mathrm{d}Y = \int_{\tilde{Y}} c_{ijkl}\beta_{kl}v_{i,j}\mathrm{d}Y \tag{10.6}$$

The above boundary value problems can be generally solved using the finite element method. However, solving these problems involves numerical difficulty because $\tilde{Y}$ must be considerably large, even though the analysis domain is already reduced by half. Therefore, $\tilde{Y}$ is divided into the substructures A_α and B_α ($\alpha = 1, 2, \ldots, 8$) as shown in Fig. 10.1, and the substructure method (Zienkiewicz and Taylor 2000) is applied as in the previous study (Matsuda et al. 2014). This method reduces the degrees of freedom of Eqs. (10.5) and (10.6), leading to a marked reduction in the computational costs.

10.3 Analysis

In this section, the elastic-viscoplastic Poisson's ratios of angle-ply carbon fiber/epoxy laminates are investigated using the analysis method described in Sect. 10.2. The results are compared with those obtained by assuming a square arrangement of carbon fibers in the laminae (Matsuda et al. 2014) to discuss the effects of fiber arrangements on the elastic-viscoplastic Poisson's ratios of angle-ply laminates.

10.3.1 Analysis Conditions

In the present analysis, $[\pm\theta]$ angle-ply CFRP laminates with 17 laminate configurations, from $[\pm 5]$ to $[\pm 85]$ in 5° increments, were considered. The semiunit cell $\tilde{Y}$ for each laminate was divided into the substructures A_α and B_α ($\alpha = 1, 2, \ldots, 8$) as illustrated in Fig. 10.1, which were then discretized into eight-node isoparametric elements. Representative examples of finite element meshes are depicted in Figs. 10.2, 10.3 and 10.4 for $[\pm 30]$, $[\pm 45]$ and $[\pm 60]$, respectively, each of which has 21,504 elements and 23,693 nodes. The fiber volume fraction was prescribed to be 56 % in each lamina (Matsuda et al. 2002, 2003). The carbon fibers were considered as transversely isotropic elastic materials, while the epoxy was assumed to be an isotropic elastic-viscoplastic material characterized by

$$\dot{\varepsilon}_{ij} = \frac{1+\nu_m}{E_m}\dot{\sigma}_{ij} - \frac{\nu_m}{E_m}\dot{\sigma}_{kk}\delta_{ij} + \frac{3}{2}\dot{\varepsilon}_0^{\mathrm{p}}\left[\frac{\sigma_{\mathrm{eq}}}{g(\bar{\varepsilon}^{\mathrm{p}})}\right]^n \frac{s_{ij}}{\sigma_{\mathrm{eq}}}, \tag{10.7}$$

where E_m, ν_m, and n are material parameters, $g(\bar{\varepsilon}^{\mathrm{p}})$ is a hardening function dependent on the equivalent viscoplastic strain $\bar{\varepsilon}^{\mathrm{p}}$, $\dot{\varepsilon}_0^{\mathrm{p}}$ indicates a reference strain rate, s_{ij} denotes the deviatoric part of σ_{ij}, and

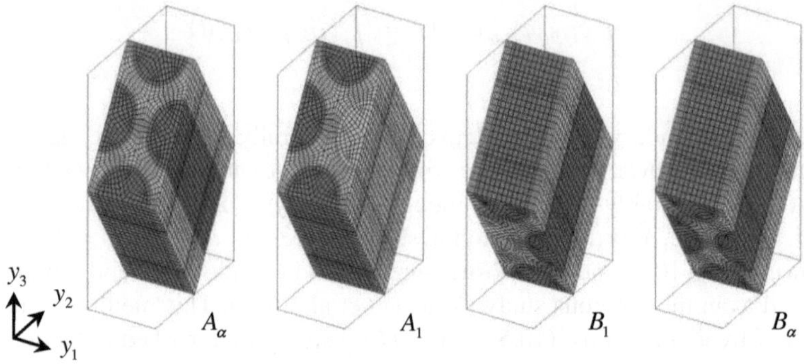

Fig. 10.2 Finite element meshes of substructures for [±30] angle-ply CFRP laminate

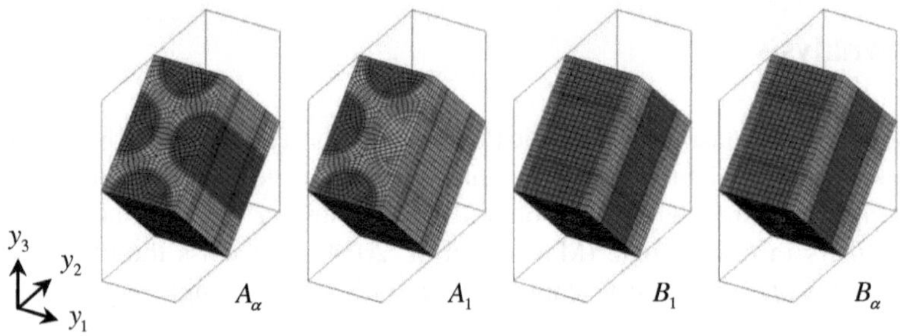

Fig. 10.3 Finite element meshes of substructures for [±45] angle-ply CFRP laminate

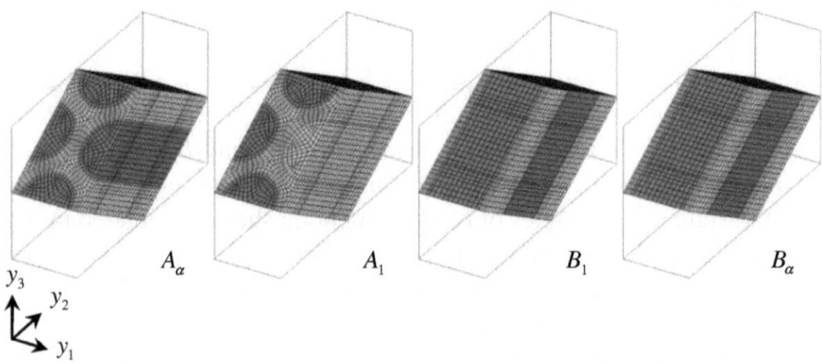

Fig. 10.4 Finite element meshes of substructures for [±60] angle-ply CFRP laminate

$$\sigma_{eq} = \sqrt{\frac{3}{2} s_{ij} s_{ij}}. \tag{10.8}$$

The material parameters of the carbon fibers and the epoxy were set as listed in Table 10.1 (Matsuda et al. 2002, 2003). A constant macroscopic tensile strain rate, $\dot{E}_{33} = 10^{-5}\ \text{s}^{-1}$, was applied to the laminates.

Table 10.1 Material parameters of carbon fibers and epoxy (Matsuda et al. 2002, 2003)

Carbon fiber	$E_{LL} = 240$ [GPa], $E_{TT} = 15.5$ [GPa], $G_{LT} = 24.7$ [GPa], $\nu_{TT} = 0.49$, $\nu_{LT} = 0.28$
Epoxy	$E = 3.5$ [GPa], $\nu = 0.35$, $\dot{\varepsilon}_0^{\mathrm{p}} = 10^{-5}$ [s^{-1}], $n = 35$, $g(\bar{\varepsilon}^{\mathrm{p}}) = 141.8(\bar{\varepsilon}^{\mathrm{p}})^{0.165} + 10$ [MPa]

10.3.2 Results of Analysis: Macroscopic Stress-Strain Relationships

Figure 10.5 shows the macroscopic stress-strain relationships for the representative laminate configurations of [±30], [±45] and [±60] angle-ply CFRP laminates. In Fig. 10.5, the solid lines indicate the results of the present analysis (hexagonal fiber arrangement), while the dashed lines denote those obtained in the previous study for a square fiber arrangement (Matsuda et al. 2014). Clear nonlinearity and remarkable anisotropy of angle-ply CFRP laminates are observed. However, little difference between the two fiber arrangements is observed for all laminate configurations. This suggests that the fiber arrangements in the laminae have a weak influence on the macroscopic stress-strain relationships of angle-ply CFRP laminates. Experimental data for the [±45] configuration (Matsuda et al. 2002) are also plotted in Fig. 10.5 (open circles) and are in good agreement with the analysis results.

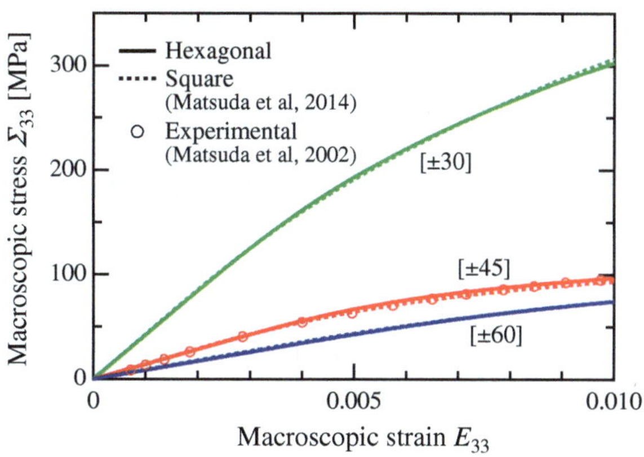

Fig. 10.5 Macroscopic stress-strain relationships of [±30], [±45] and [±60] angle-ply CFRP laminates

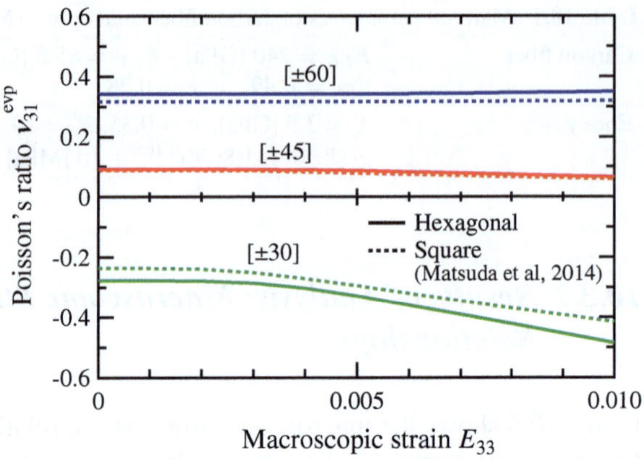

Fig. 10.6 Relationships between the elastic-viscoplastic Poisson's ratio ν_{31}^{evp} and the macroscopic strain E_{33} for [±30], [±45] and [±60] angle-ply CFRP laminates

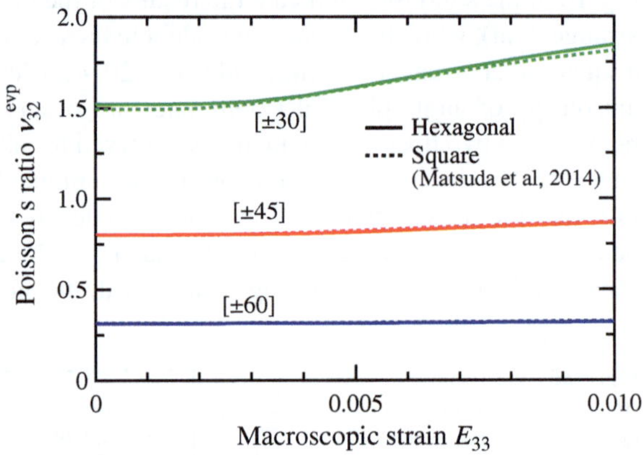

Fig. 10.7 Relationships between the elastic-viscoplastic Poisson's ratio ν_{32}^{evp} and the macroscopic strain E_{33} for [±30], [±45] and [±60] angle-ply CFRP laminates

10.3.3 Results of Analysis: Elastic-Viscoplastic Poisson's Ratios

The relationships between the elastic-viscoplastic Poisson's ratios, ν_{31}^{evp} or ν_{32}^{evp}, and the macroscopic tensile strain E_{33} are depicted in Figs. 10.6 and 10.7, respectively, for the [±30], [±45] and [±60] configurations, where

$$\nu_{31}^{evp} = -\frac{E_{11}}{E_{33}}, \qquad \nu_{32}^{evp} = -\frac{E_{22}}{E_{33}}. \tag{10.9}$$

The elastic Poisson's ratios ν_{31}^{e} and ν_{32}^{e} are plotted at $E_{33}=0$. The solid and dashed lines in the figures indicate the results for the hexagonal and square fiber arrangements, respectively. From Fig. 10.6, the through-the-thickness Poisson's ratio ν_{31}^{evp} for the [±30] configuration is negative not only in the elastic region but also in the viscoplastic region, and the negativity of ν_{31}^{evp} progressively becomes greater with increasing E_{33}, irrespective of the fiber arrangement. However, the negativity of ν_{31}^{evp}

for the hexagonal arrangement is clearly stronger than that for the square one through all regions. Especially in the viscoplastic region, the gradient of ν_{31}^{evp} for the hexagonal arrangement is larger than that for the square one. The difference between ν_{31}^{evp} for the hexagonal and square arrangements thus enlarges from about 0.04 to 0.08. In the case of ν_{31}^{evp} for the [±60] configuration, which is always positive and varies little as E_{33} increases, the effect of the fiber arrangement is relatively small; ν_{31}^{evp} for the hexagonal arrangement is about 0.03 higher than that for square one in all regions. In contrast, ν_{31}^{evp} for the [±45] configuration exhibits almost no dependence on the fiber arrangement.

In contrast, as shown in Fig. 10.7, the effects of the fiber arrangements on the in-plane Poisson's ratio ν_{32}^{evp} are quite small compared with ν_{31}^{evp}, though it should be noted that ν_{32}^{evp} of the [±30] configuration is considerably high and varies noticeably from about 1.5 to 1.8.

In the same manner as described above, ν_{31}^{evp} and ν_{32}^{evp} were examined for the different laminate configurations from [±5] to [±85] in 5° increments. In Figs. 10.8 and 10.9, respectively, ν_{31}^{evp} and ν_{32}^{evp} at $E_{33} = 0.01$ are plotted for all laminate

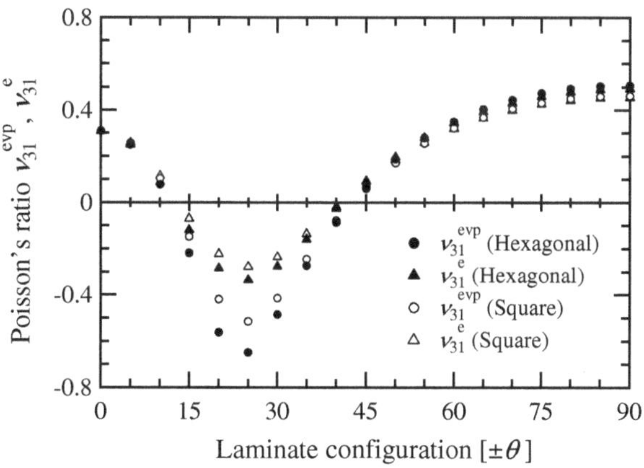

Fig. 10.8 Variations in the elastic-viscoplastic Poisson's ratio ν_{31}^{evp} depending on the laminate configuration [$\pm\theta$] ($E_{33} = 0.01$)

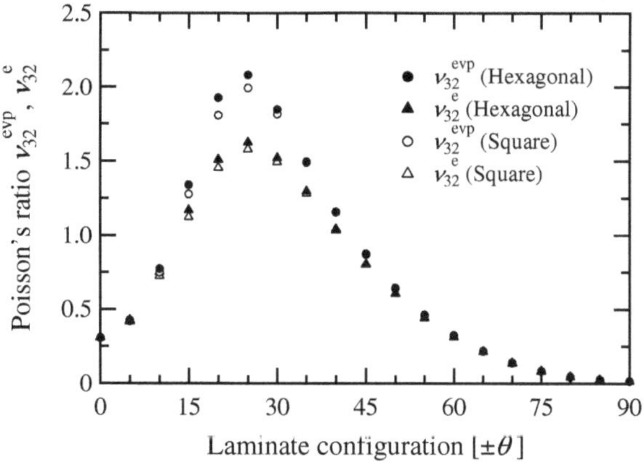

Fig. 10.9 Variations in the elastic-viscoplastic Poisson's ratio ν_{32}^{evp} depending on the laminate configuration [$\pm\theta$] ($E_{33} = 0.01$)

configurations. The solid circles indicate the analysis results for the hexagonal arrangement, while the open circles represent those for the square one. For reference, ν_{31}^{e} and ν_{32}^{e} are also plotted in the figures, where the solid and open triangles correspond to the analysis results for the hexagonal and square arrangements, respectively. Here, $\theta = 0°$ and $90°$ represent unidirectional laminates in the fiber longitudinal and transverse directions, respectively. Figure 10.8 shows that ν_{31}^{evp} becomes negative in the range between about 15° and 40°, and that it becomes significantly more negative than ν_{31}^{e}, as already pointed out in the previous study (Matsuda et al. 2014). It should be emphasized here that the negativity of ν_{31}^{evp} for the hexagonal arrangement becomes significantly greater than that for the square one in the range between 15° and 30°. In contrast, as seen from Fig. 10.9, ν_{32}^{evp} is always positive irrespective of the laminate configuration, and the difference between ν_{32}^{evp} for the hexagonal and square arrangements is not so significant compared with that seen for ν_{31}^{evp}.

10.3.4 Results of Analysis: Microscopic Mechanisms

The $T - L$ rectangular coordinates are introduced by rotating the $y_2 - y_3$ coordinates with θ (Fig. 10.10). The T- and L-directions coincide with the fiber transverse direction and the fiber longitudinal direction, respectively. The distributions of the microscopic stress σ_T at E_{33}=0.01 in A_8, which is the farthest substructure from the interlaminar plane, are depicted in Fig. 10.10 for the [±30], [±45] and [±60] configurations. It can be seen that very high compressive stress in the T-direction is distributed in A_8 for the [±30] configuration, as in the case of a square fiber arrangement (Matsuda et al. 2014). Such compressive stress induces positive strain

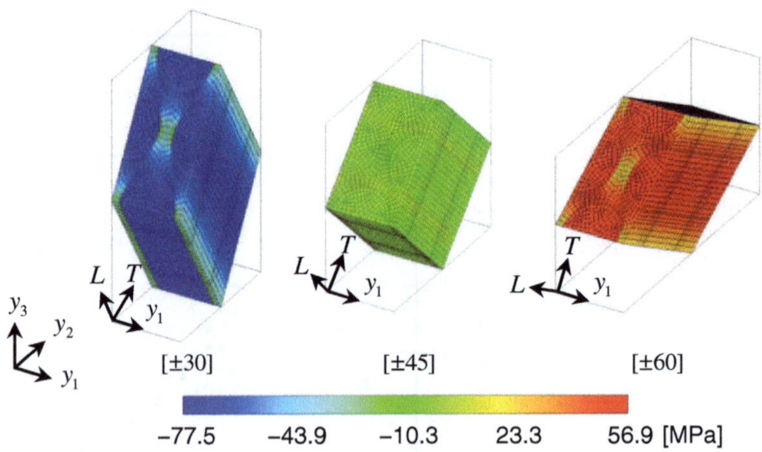

Fig. 10.10 Distributions of microscopic normal stress perpendicular to the fiber, σ_T, in A_8 (E_{33}=0.01)

Table 10.2 Elastic constants of carbon fiber/epoxy unidirectional laminae

Hexagonal	$E_{LL} = 136$ [GPa], $E_{TT} = 7.60$ [GPa], $G_{LT} = 3.93$ [GPa], $\nu_{TT} = 0.492$, $\nu_{LT} = 0.308$
Square	$E_{LL} = 136$ [GPa], $E_{TT} = 8.19$ [GPa], $G_{LT} = 4.06$ [GPa], $\nu_{TT} = 0.455$, $\nu_{LT} = 0.308$

in the through-the-thickness direction. In particular, the epoxy exhibits more positive through-the-thickness strain in the viscoplastic region than in the elastic region because of its viscoplastic deformation. Thus, the negativity of the through-the-thickness Poisson's ratio for the [±30] configuration becomes stronger with progressing viscoplastic deformation of the laminate (Fig. 10.6).

According to the mechanism mentioned above, we can say that the transverse behavior of laminae plays an important role in determining the through-the-thickness Poisson's ratios of angle-ply CFRP laminates. Table 10.2 shows the elastic constants of carbon fiber/epoxy unidirectional laminae having hexagonal and square fiber arrangements with 56 % fiber volume fraction. The data were calculated using the homogenization theory, where the same elastic constants as in Table 10.1 were used for the carbon fibers and the epoxy. It is noted that the Poisson's ratio ν_{TT} of the hexagonal arrangement is larger than that of the square one, implying that laminae with the hexagonal arrangement relatively easily exhibit positive through-the-thickness strain when compressed as shown in Fig. 10.10. This leads to the greater negativity of the through-the-thickness Poisson's ratio in the elastic region of the [±30] configuration with the hexagonal arrangement (Fig. 10.6). Also, it is known that laminae with the hexagonal fiber arrangement subjected to transverse loading exhibit more significant viscoplasticity than those with the square fiber arrangement subjected to transverse loading in the nearest-fiber direction (Ohno et al. 2000). This causes the larger gradient of the through-the-thickness Poisson's ratio of the [±30] configuration in the viscoplastic region (Fig. 10.6).

In contrast, for the macroscopic stress-strain relationships shown in Fig. 10.5, the dominant deformation mechanism is not the transverse behavior of laminae but the shear deformation of the matrix along the fiber longitudinal direction. Thus, the difference in fiber arrangement had almost no influence on the stress-strain relationships of the angle-ply CFRP laminates.

10.4 Conclusions

In this study, the effects of fiber arrangement in laminae on the negative through-the-thickness Poisson's ratios of angle-ply CFRP laminates were investigated based on a homogenization theory. For the analysis, angle-ply CFRP laminates consisting of laminae with a hexagonal fiber arrangement were considered, and a homogenization theory for nonlinear time-dependent composites was applied to the laminates.

Using the present method, the elastic-viscoplastic Poisson's ratios of angle-ply carbon fiber/epoxy laminates with various laminate configurations were analyzed. The results were compared with those obtained in a previous study for laminae containing a square fiber arrangement. The through-the-thickness Poisson's ratio of angle-ply CFRP laminates with the hexagonal fiber arrangement became more negative than that of the square one, although there was little difference between the macroscopic stress-strain relationships of the two fiber arrangements. This suggests that it is important to pay attention to the effects of fiber arrangements in laminae when analyzing the through-the-thickness Poisson's ratios of CFRP laminates. Quantitative validation of the present analysis will be experimentally performed in future work.

Acknowledgments The authors are grateful for partial support from a Japan Society for the Promotion of Science (JSPS) Grant-in-Aid for Young Scientists (B) (No. 25870101) and Grant-in-Aid for JSPS Fellows (No. 26·2028).

References

Clarke JF, Duckett RA, Hine PJ, Hutchinson IJ, Ward IM (1994) Negative Poisson's ratios in angle-ply laminates: theory and experiment. Compos 25:863–868

Harkati EH, Bezazi A, Scarpa F, Alderson K, Alderson A (2007) Modelling the influence of the orientation and fibre reinforcement on the negative Poisson's ratio in composite laminates. Phys Stat Sol (b) 244:883–892

Herakovich CT (1984) Composite laminates with negative through-the-thickness Poisson's ratios. J Compos Mater 18:447–455

Matsuda T, Ohno N, Tanaka H, Shimizu T (2002) Homogenized in-plane elastic-viscoplastic behavior of long fiber-reinforced laminates. JSME Int J Ser A 45:538–544

Matsuda T, Ohno N, Tanaka H, Shimizu T (2003) Effects of fiber distribution on elastic-viscoplastic behavior of long fiber-reinforced laminates. Int J Mech Sci 45:1583–1598

Matsuda T, Okumura D, Ohno N, Kawai M (2007) Three-dimensional microscopic interlaminar analysis of cross-ply laminates based on a homogenization theory. Int J Solids Struct 44:8274–8284

Matsuda T, Goto K, Kubota N, Ohno N (2014) Negative through-the-thickness Poisson's ratio of elastic-viscoplastic angle-ply carbon fiber-reinforced plastic laminates: homogenization analysis. Int J Plast 63:152–169

Ohno N, Wu X, Matsuda T (2000) Homogenized properties of elastic-viscoplastic composites with periodic internal structures. Int J Mech Sci 42:1519–1536

Ohno N, Matsuda T, Wu X (2001) A homogenization theory for elastic-viscoplastic composites with point-symmetry of internal distributions. Int J Solids Struct 38:2867–2878

Wu X, Ohno N (1999) A homogenization theory for time-dependent nonlinear composites with periodic internal structures. Int J Solids Struct 36:4991–5012

Zienkiewicz OC, Taylor RL (2000) The finite element method, 5th edn. Butterworth-Heinemann, Oxford

Chapter 11
Modeling of Internal Damage Evolution of Piezoelectric Ceramics Under Compression-Compression Fatigue Tests

Mamoru Mizuno and Ken-ichi Wakui

Abstract A damage evolution equation for piezoelectric ceramics under compression-compression fatigue tests were formulated on the basis of previous experimental results within the framework of the continuum damage mechanics. In the previous experiments, columnar specimens of piezoelectric ceramics were subjected to compression-compression fatigue loading. The fatigue tests were suspended at specified intervals, and the resonance and anti-resonance frequencies and the electrostatic capacity of specimens were measured by an impedance analyzer. The loading and measurement were repeated up to the fatigue fracture. Material properties of piezoelectric ceramic specimens were calculated from the resonance and anti-resonance frequencies and the electrostatic capacity, and the variation of material properties during fatigue tests was elucidated. Development of internal damage within piezoelectric ceramics was evaluated as a damage variable by the variation of elastic coefficient on the basis of the continuum damage mechanics. An evolution equation of the damage variable was formulated by using fatigue life which was formulated as functions of the static fracture strength, the fatigue limit and the mean stress. Prediction of damage development was compared with experimental results and the validity of the formulation was verified.

Keywords Piezoelectric ceramics · Fatigue · Damage evolution equation

11.1 Introduction

Piezoelectric ceramics possess the piezoelectric effects and the converse piezoelectric effects, which are applied to sensors, actuators, ultrasonic vibrators, energy harvesters, etc. The advantages of piezoelectric ceramics are high energy conversion

M. Mizuno (✉) · K. Wakui
Akita Prefectural University, Yuri-Honjo, Akita 015-0055, Japan
e-mail: mizuno@akita-pu.ac.jp

K. Wakui
e-mail: m13a023@akita-pu.ac.jp

H. Altenbach et al. (eds.), *From Creep Damage Mechanics to Homogenization Methods*, Advanced Structured Materials 64,
DOI 10.1007/978-3-319-19440-0_11

efficiency, high speed response, possibility of miniaturization and generation of large force, while the disadvantages are small displacement and brittleness. In order to take measures to improve the displacement of actuators, bimorph actuators and laminated actuators are developed (Uchino 1986). Anyhow, piezoelectric ceramics are utilized in many devises with the aim of precise control and accurate measurement.

The piezoelectric effect (or the converse piezoelectric effect) in piezoelectric ceramics is revealed after poling process, by which polarization is almost aligned in one direction. Accordingly, piezoelectric ceramics reveal transverse isotropy after the poling process. Piezoelectric ceramics are brittle materials basically, and their deformation in their practical use is described by linear equations, in which an additional term to represent the piezoelectric effect (or the converse piezoelectric effect) is included. Since the nonlinearity of the deformation in piezoelectric ceramics under excessive loading is caused by domain switching (change in poling direction of domain), the behavior of polarization is modeled theoretically (Joshi 1992; Hwang et al. 1995, 1998; Shindo et al. 2003). Since the piezoelectric properties depend on internal structures, molecular dynamic simulations are applied to piezoelectric ceramics in order to predict material properties (Nakamachi et al. 2013).

On the other hand, fracture of piezoelectric ceramics has been investigated experimentally and theoretically, and many papers have been published (Mehta and Virkar 1990; Suo et al. 1992; Sosa 1992; Schneider and Heyer 1999; Shindo et al. 2001, 1997, 2005; Cao and Evans 1994; Makino and Kamiya 1994, 1998; Shang and Tan 2001; Koh et al. 2004; Freiman and White 1995; Busche and Hsia 2001; Ueda 2002; Chen and Hasebe 2005). For instance, behaviors of a macroscopic crack taking into account effects of applied electric field were elucidated and the fracture mechanics was applied to the crack (Mehta and Virkar 1990; Suo et al. 1992; Sosa 1992; Schneider and Heyer 1999; Shindo et al. 2001). Then numerical simulation such as finite element method was also performed to deal with macroscopic cracks (Shindo et al. 1997, 2005). The effects of applied electric field on a fatigue crack and fatigue life was also discussed (Cao and Evans 1994; Makino and Kamiya 1994, 1998; Shang and Tan 2001). In the case of laminated actuators, it is reported that the stress singularity at the tip of an inner electrode due to mismatch between active and inactive regions causes crack initiation (Koh et al. 2004).

Piezoelectric ceramics are brittle materials, so that it is important to investigate the development of internal damage prior to macroscopic crack initiation rather than behaviors of a macroscopic crack. The internal damage affects material properties and strength. Since piezoelectric ceramics are utilized in devices with the aim of precise control and accurate measurement, it is necessary to elucidate the variation of material properties due to the internal damage development and the effects of internal damage on material properties.

One of the present authors is dealing with the internal damage of piezoelectric ceramics within the framework of the continuum damage mechanics (Lemaitre and Chaboche 1990; Lemaitre 1992; Skrzypek and Ganczarski 1999; Altenbach and Skrzypek 1999; Naumenko and Altenbach 2007; Murakami 2012; Murakami and Ohno 1981; Murakami 1987; Murakami and Mizuno 1992) in a series of works (Mizuno et al. 2008a, b, 2010a, b, 2013). In the continuum damage mechanics

(Lemaitre and Chaboche 1990; Lemaitre 1992; Skrzypek and Ganczarski 1999; Altenbach and Skrzypek 1999; Naumenko and Altenbach 2007; Murakami 2012; Murakami and Ohno 1981; Murakami 1987; Murakami and Mizuno 1992), internal damage is represented by a damage variable, and the internal damage development is modeled by an evolution equation of the damage variable mathematically. On the other hand, if the variation of material properties is described by material functions of the damage variable, the effects of internal damage on mechanical behaviors can be incorporated into the piezoelectric constitutive equations through the material functions. By using the constitutive equations as well as the damage evolution equation, numerical simulation taking into account internal damage development can be performed, and accuracy of the predictions is improved.

In a series of experiments made by one of the present authors, the variation of material properties and the development of internal damage have been elucidated for piezoelectric ceramics (Mizuno et al. 2008a, b, 2010a, b). Static compression tests and compression-compression fatigue tests have been made for columnar specimens. Those tests were suspended at specified intervals. Then the resonance and anti-resonance frequencies and the electrostatic capacity of specimens were measured by an impedance analyzer. The loading and measurement were repeated up to the fatigue fracture. Material properties of piezoelectric ceramics can be calculated from those values (Mizuno et al. 2010b). Then the variation of material properties during tests up to fracture has been clarified. Development of internal damage was evaluated by the variation of elastic coefficient from its initial value as a scalar damage variable on the basis of the continuum damage mechanics (Lemaitre 1992).

In Mizuno et al. (2013), on the basis of the static experimental results, the variation of material properties was rearranged with respect to the damage variable, and material functions of the damage variable were proposed in order to describe the variation of material properties due to internal damage. Moreover, the development of internal damage was formulated as an evolution equation of the damage variable. Then a set of piezoelectric constitutive equations taking into account internal damage development due to static compressive loading were proposed. Stress-strain curves were predicted by the piezoelectric constitutive equations, and the validity of the equations was verified by comparing the predictions with experimental results.

In the present paper, on the basis of the fatigue experimental results (Mizuno et al. 2008a, b), an evolution equation of the damage variable under compression-compression fatigue loading is proposed. For the sake of the formulation of the evolution equation, first of all, previous experimental results are introduced briefly. Then fatigue life is expressed by functions of the static fracture strength, the fatigue limit and the mean stress (Lemaitre and Chaboche 1990), and the damage evolution equation is formulated by using the fatigue life. The development of internal damage depending on the applied stress is predicted, and the validity of the formulation is verified by comparing the predictions with experimental results.

11.2 Experimental Results

Columnar specimens for compression-compression fatigue tests were made of lead zirconate titanate (PZT) piezoelectric ceramics (Fuji ceramics Co.). Dimensions of the specimen were 3.0 mm in diameter and 7.5 mm in length. Specimens which were poled in the longitudinal direction were purchased.

Columnar specimens were subjected to compression-compression fatigue loading of which stress ratio was fixed to $R = 0.05$, and the maximum compressive stress σ_{max} was changed in the fatigue tests. The fatigue loading was suspended at specified intervals, and the resonance and anti-resonance frequencies, f_{r} and f_{a}, and the electrostatic capacity C^{T} of specimens were measured by an impedance analyzer. The loading and measurement were repeated up to the fatigue fracture of specimens.

Material properties of piezoelectric ceramics in the poling direction are calculated by following equations (Mizuno et al. 2010b):

$$k_{33} = \sqrt{\frac{1}{0.405\dfrac{f_{\mathrm{r}}}{f_{\mathrm{a}} - f_{\mathrm{r}}} + 0.810}}, \tag{11.1}$$

$$\varepsilon_{33} = \frac{C^{\mathrm{T}} t}{A}, \tag{11.2}$$

$$c_{33}^{E} = \rho(2\ell f_{\mathrm{r}})^2, \qquad c_{33}^{D} = \frac{c_{33}^{E}}{1 - k_{33}^2}, \tag{11.3}$$

$$d_{33} = k_{33}\sqrt{\frac{\varepsilon_{33}}{c_{33}^{E}}}, \tag{11.4}$$

where k_{33} and ε_{33} represent an electromechanical coupling coefficient and a dielectric constant, while c_{33}^{E} and c_{33}^{D} are an elastic coefficient with constant electric field E and an elastic coefficient with constant electric displacement D, respectively. Moreover, d_{33} indicates a piezoelectric constant. On the other hand, A and t represent the area of an electrode and the distance between electrodes to measure the electrostatic capacity C^{T}, while ℓ and ρ are the length and density of specimens, respectively.

Figure 11.1 shows fatigue life for columnar specimens of piezoelectric ceramics in compression-compression fatigue tests (Mizuno et al. 2008a, b, 2010a). Fatigue life N_{f} increases as the maximum compressive stress σ_{max} decreases with constant stress ratio $R = 0.05$, and it is found that the correlation between the maximum compressive stress σ_{max} and fatigue life N_{f} is similar to metals.

Figures 11.2 show the variation of material properties under the compression-compression fatigue tests (Mizuno et al. 2008a, b). Values in the right side of symbols in the legend within the figures are the maximum compressive stress σ_{max} by which the compression-compression fatigue test was carried out. In the Fig. 11.2b, close-up

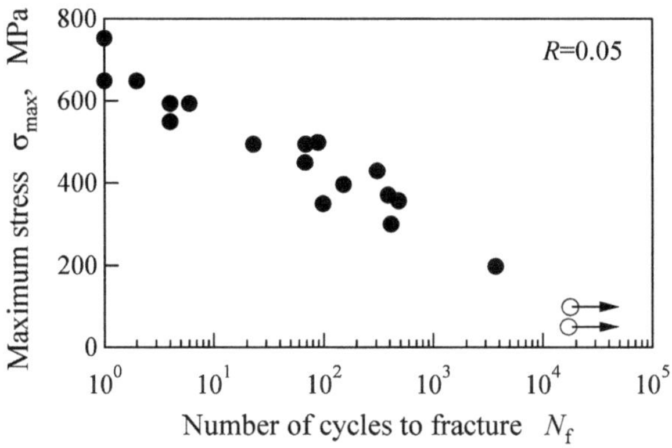

Fig. 11.1 Compression-compression fatigue life of piezoelectric ceramics when the stress ratio is fixed to $R = 0.05$ and the maximum stress σ_{max} is changed (Mizuno et al. 2008a, b, 2010a)

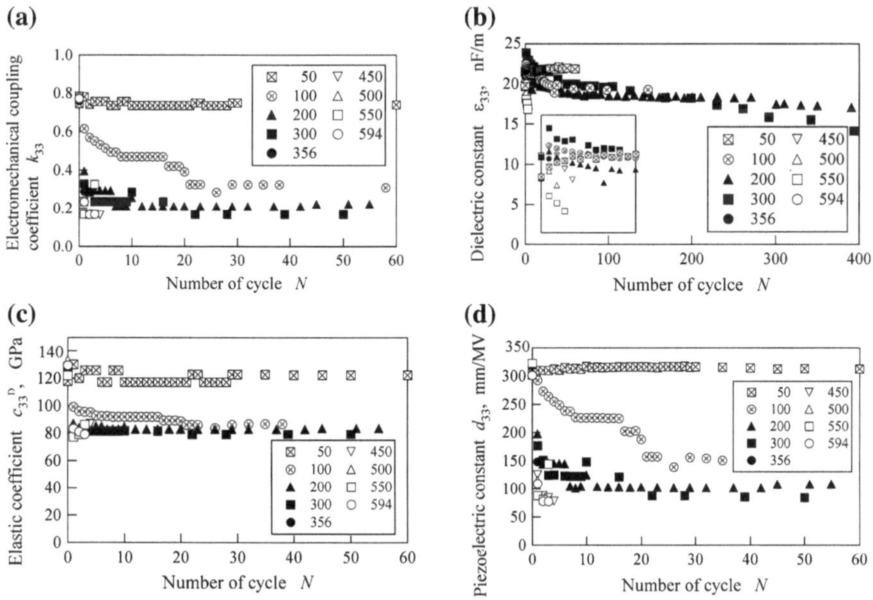

Fig. 11.2 The variation of material properties of piezoelectric ceramics under compression-compression fatigue tests (Mizuno et al. 2008a, b): **a** an electromechanical coupling coefficient k_{33}, **b** a dielectric constant ε_{33}, **c** an elastic coefficient c_{33}^D with constant electric displacement D, **d** a piezoelectric constant d_{33}

of first 12 cycles is depicted within the figure. The dielectric constant ε_{33} increases at first, then it decreases during fatigue tests. The electromechanical coupling coefficient k_{33}, the elastic coefficient c_{33}^D with constant electric displacement D and the piezoelectric constant d_{33} decrease from the beginning according to the increase of number of cycle. The decrease is more rapid as the maximum compressive stress

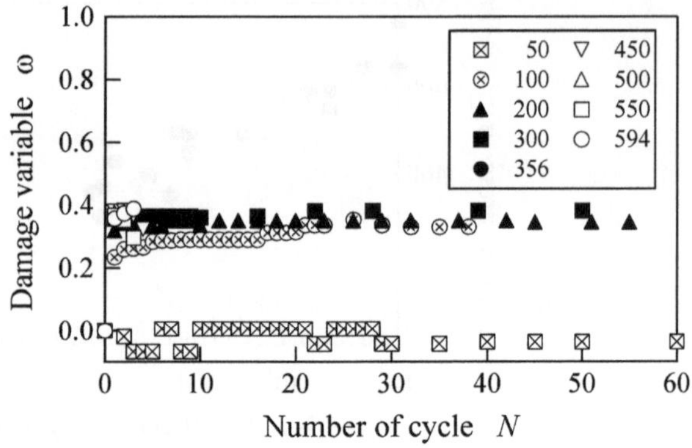

Fig. 11.3 The development of internal damage evaluated as a damage variable ω by the variation of elastic coefficient c_{33}^D from its initial value (Mizuno et al. 2008a, b, 2010a) on the basis of the continuum damage mechanics (Lemaitre 1992)

σ_{max} is larger. However, when the maximum compressive stress is $\sigma_{max} = 50$ MPa, there is no change in any material property.

The development of internal damage was evaluated indirectly as a scalar damage variable ω by the variation of the elastic coefficient c_{33}^D on the basis of the continuum damage mechanics (Lemaitre 1992) by following equation:

$$\omega = 1 - \frac{c_{33}^D}{(c_{33}^D)_0}, \tag{11.5}$$

where $(c_{33}^D)_0$ is an initial value of c_{33}^D.

Figure 11.3 shows the development of internal damage evaluated as a scalar damage variable ω by the variation of the elastic coefficient c_{33}^D (Mizuno et al. 2008a, b, 2010a). When the maximum compressive stress is larger than $\sigma_{max} = 50$ MPa, the damage variable ω increases as the number of cycle increases, and the increase is saturated to the constant value. On the other hand, when the maximum compressive stress is $\sigma_{max} = 50$ MPa, there is no change in the damage variable ω.

11.3 Modeling of Fatigue Damage Evolution

The continuum damage mechanics was developed for creep originally (Murakami and Ohno 1981; Murakami 1987; Murakami and Mizuno 1992), and it has been applied to many fields including numerical simulation for crack behaviors (Hayhurst et al. 1984; Murakami et al. 2000). Fatigue damage is also treated within the framework of the continuum damage mechanics (Lemaitre and Chaboche 1990; Lemaitre 1992), and the damage evolution equations are formulated with respect to the number of cycle. However, the fatigue is discussed for metals mainly. In their theoretical approach, the fatigue damage evolution equation is formulated by using fatigue life and the nonlinearity of the evolution is discussed phenomenologically.

Referring experimental results for piezoelectric ceramics shown in the previous section, the S-N curve for piezoelectric ceramics is similar to that for metals. However, the damage development is brittle one and it is quite different from that for ductile materials such as metals. Therefore, in the present paper, equations to prescribe the fatigue life for metals is used for piezoelectric ceramics, and the fatigue damage evolution equation for metals is modified in order to describe the damage development for piezoelectric ceramics.

Fatigue life for piezoelectric ceramics is described by the Woehler-Miner law (Lemaitre and Chaboche 1990) as follows:

$$N_{\mathrm{f}} = \frac{\sigma_{\mathrm{u}} - \sigma_{\max}}{\sigma_{\max} - \sigma_{\ell}(\bar{\sigma})} \left(\frac{\sigma_{\max} - \bar{\sigma}}{B(\bar{\sigma})} \right)^{-\beta}, \tag{11.6}$$

$$\sigma_{\ell}(\bar{\sigma}) = \bar{\sigma} + \sigma_{\ell_0}(1 - b\bar{\sigma}), \tag{11.7}$$

$$B(\bar{\sigma}) = B_0(1 - b\bar{\sigma}), \tag{11.8}$$

where σ_{u} and $\sigma_{\ell}(\bar{\sigma})$ represent the static fracture strength and the fatigue limit, respectively, and the fatigue limit $\sigma_{\ell}(\bar{\sigma})$ is formulated by taking into account the effect of mean stress $\bar{\sigma}(= (1 - R)\sigma_{\max}/2)$ on the fatigue limit by Eq. (11.7) according to Goodman's linear relation (Lemaitre and Chaboche 1990). In Eq. (11.7), σ_{ℓ_0} is the fatigue limit at $\bar{\sigma} = 0$. b, B_0 and β are material constants. In the present paper, since only compression-compression fatigue tests are dealt with, absolute value of compressive stress is applied to equations.

For the fatigue damage for metals, a nonlinear evolution with respect to the number of cycle N is expressed by following equation (Lemaitre and Chaboche 1990):

$$\frac{\delta\omega}{\delta N} = \frac{(1 - \omega)^{-k}}{k + 1} \frac{1}{N_{\mathrm{f}}}, \tag{11.9}$$

where k is a material constant. However, it describes the damage development which is downward convex for ductile materials, and it can not describe the fatigue damage development for piezoelectric ceramics which is upward convex as shown in Fig. 11.3. Accordingly, the damage evolution equation is modified to describe the fatigue damage development of piezoelectric ceramics as follows:

$$\frac{\delta\omega}{\delta N} = \frac{A}{\exp([\omega/k]^m)} \frac{1}{(N_{\mathrm{f}})^n}, \tag{11.10}$$

where A, k, m and n are material constants.

Material constants are determined on the basis of experimental results as follows:

$$\begin{aligned} &\sigma_{\mathrm{u}} = 640\,\mathrm{MPa}, \beta = 1.95, \sigma_{\ell_0} = 58\,\mathrm{MPa}, b = 6.5 \times 10^{-4}, \\ &B_0 = 2.3 \times 10^3, A = 0.76, k = 0.145, m = 3, n = 0.11 \end{aligned} \tag{11.11}$$

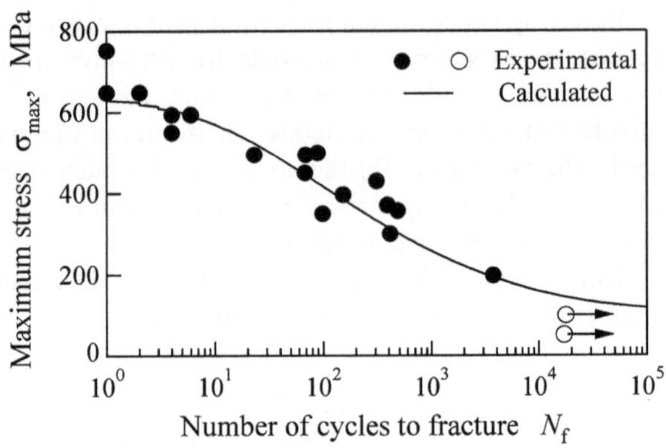

Fig. 11.4 Prediction of fatigue life N_f by Eqs. (11.6)–(11.8) compared with experimental results shown in Fig. 11.1

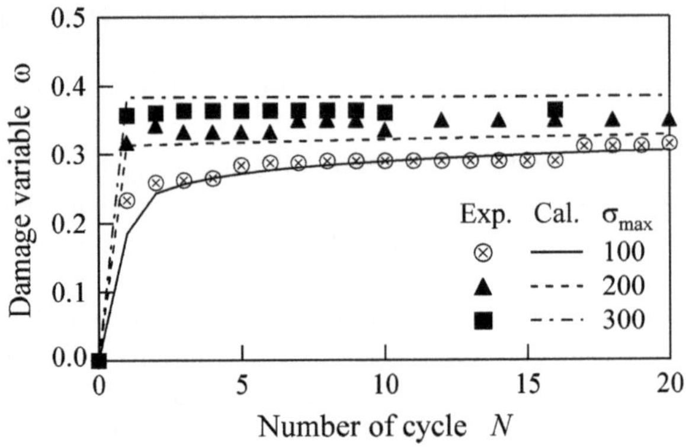

Fig. 11.5 Predictions of internal damage development by Eq. (11.10) compared with experimental results shown in Fig. 11.3

Figure 11.4 shows prediction of fatigue life N_f by Eqs. (11.6)–(11.8) compared with experimental results. Prediction agrees with experimental results from the static fracture strength σ_u through the fatigue limit $\sigma_\ell(\bar{\sigma})$.

Figure 11.5 shows predictions of internal damage development depending on the maximum compressive stress σ_{max} by Eq. (11.10) compared with experimental results. Experimental results indicate that internal damage development is rapid and asymptotic value is larger when the maximum compressive stress σ_{max} is larger. Those internal damage development is described by Eq. (11.10).

11.4 Concluding Remarks

In the present paper, in order to formulate internal damage development of piezoelectric ceramics under compression-compression fatigue tests, experimental results in the previous research were introduced briefly. In the experiments, the variation of material properties during fatigue tests was measured and internal damage was evaluated by the variation of elastic coefficient as a scalar damage variable on the basis of the continuum damage mechanics. Then an evolution equation of the damage variable was formulated with respect to the number of cycle by using fatigue life. The fatigue life was expressed by functions of the static fracture strength, the fatigue limit and the mean stress. The predictions of fatigue life and internal damage development were compared with experimental results, and the validity of the formulation was verified.

As the next step, in the near future, the variation of material properties are rearranged with respect to the damage variable, and the variation is formulated as a material function of the damage variable. When internal damage is calculated by the evolution equation of the damage variable under fatigue conditions, the variation of material properties due to the internal damage is described by material functions and the variation of material properties is reflected on numerical simulation through the constitutive equation of piezoelectric ceramics using the functions.

References

Altenbach H, Skrzypek JJ (eds) (1999) Creep and damage in materials and structures. Springer, Wien

Busche MJ, Hsia KJ (2001) Fracture and domain switching by indentation in barium titanate single crystals. Scripta Mater 44:207–212

Cao H, Evans AG (1994) Electric-field-induced fatigue crack growth in piezoelectrics. J Am Ceram Soc 77:1783–1786

Chen YH, Hasebe N (2005) Current understanding on fracture behaviors of ferroelectric/piezoelectric materials. J Intel Mater Sys Struct 16:673–687

Freiman SW, White GS (1995) Intelligent ceramic materials: issues of brittle fracture. J Intel Mater Sys Struct 6:49–54

Hayhurst D, Dimmer P, Morrison C (1984) Development of continuum damage in the creep rupture of notched bars. Phil Trans R Soc Lond A 311:103–129

Hwang SC, Lynch CS, McMeeking RM (1995) Ferroelectric/ferroelastic inter-actions and a polarization switching model. Acta Metall Mater 43:2073–2084

Hwang SC, Huber JE, McMeeking RM, Fleck NA (1998) The simulation of switching in polycrystalline ferroelectric ceramics. J Appl Phys 84:1530–1540

Joshi SP (1992) Non-linear constitutive relations for piezoelectric materials. Smart Mater Struct 1:80–83

Koh JH, Jeong SJ, Ha MS, Song JS (2004) Electric field induced fracture mechanism and aging of piezoelectric behavior in $Pb(MgNb)O_3$-$Pb(ZrTi)O_3$ multilayer ceramic actuators. Ceram Int 30:1863–1867

Lemaitre J (1992) A course on damage mechanics. Springer, Berlin

Lemaitre J, Chaboche JL (1990) Mechanics of solid materials. Cambridge University Press, Cambridge

Makino H, Kamiya N (1994) Effects of dc electric field on mechanical properties of piezoelectric ceramics. Jpn J Appl Phys 33:5323–5327

Makino H, Kamiya N (1998) Electromechanical fatigue of lead zirconate titanate ceramics. Jpn J Appl Phys 37:5301–5305

Mehta K, Virkar AV (1990) Fracture mechanics in ferroelectric-ferroelastic lead zirconate titanate (Zr:Ti=0.54:0.46) ceramics. J Am Ceram Soc 73:567–574

Mizuno M, Odagiri N, Okayasu M (2008a) Evaluation of damage development in piezoelectric ceramics under repeated mechanical loading. In: Proceedings of the XXII international congress of theoretical and applied mechanics, Adelaide Convention Centre, Adelaide, 25–29 Aug 2008
Mizuno M, Odagiri N, Okayasu M (2008b) Variation of material properties of piezoelectric ceramics due to mechanical loading and evaluation of internal damage (in Japanes). Trans Japan Soc Mech Eng A-74:467–472
Mizuno M, Enomoto Y, Okayasu M (2010a) Fatigue life of piezoelectric ceramics and evaluation of internal damage. Procedia Eng 2:291–297
Mizuno M, Okayasu M, Odagiri N (2010b) Damage evaluation of piezoelectric ceramics from the variation of the elastic coefficient under static compressive stress. Int J Damage Mech 19:375–390
Mizuno M, Nishikata T, Okayasu M (2013) Modeling of material properties of piezoelectric ceramics taking into account damage development under static compression. Smart Mater Struct 22:105002
Murakami S (1987) Progress of continuum damage mechanics. JSME Int J 30:701–710
Murakami S (2012) Continuum damage mechanics. Springer, Dordrecht
Murakami S, Mizuno M (1992) A constitutive equation of creep, swelling and damage under neutron irradiation applicable to multiaxial and variable states of stress. Int J Solids Struct 29:2319–2328
Murakami S, Ohno N (1981) A continuum theory of creep and creep damage. In: Ponter ARS, Hayhurst D (eds) Creep in structures. Springer, Berlin, pp 422–444
Murakami S, Liu Y, Mizuno M (2000) Computational methods for creep fracture analysis by damage mechanics. Comput Methods Appl Mech Eng 183:15–33
Nakamachi E, Uetsuji Y, Kuramae H, Tsuchiya K, Hwang H (2013) Process crystallographic simulation for biocompatible piezoelectric material design and generation. Arch Comput Methods Eng 20:155–183
Naumenko K, Altenbach H (2007) Modeling of creep for structural analysis. Springer, Berlin
Schneider GA, Heyer V (1999) Influence of the electric field on Vickers indentation crack growth in $BaTiO_3$. J Europ Ceram Soc 19(6):1299–1306
Shang JK, Tan X (2001) A maximum strain criterion for electric-field-induced fatigue crack propagation in ferroelectric ceramics. Mater Sci Eng A301:131–139
Shindo Y, Tanaka K, Narita F (1997) Singular stress and electric fields of a piezoelectric ceramic strip with a finite crack under longitudinal shear. Acta Mech 120:31–45
Shindo Y, Oka M, Horiguchi K (2001) Analysis and testing of indentation fracture behavior of piezoelectric ceramics under an electric field. J Eng Mater Tech 123:293–300
Shindo Y, Narita F, Horiguchi K, Magara Y, Yoshida M (2003) Electric fracture and polarization switching properties of piezoelectric ceramic PZT studied by the modified small punch test. Acta Mater 51:4773–4782
Shindo Y, Narita F, Mikami M (2005) Electroelastic fracture mechanics of piezoelectric layered composites. J Intel Mater Sys Struct 16:573–582
Skrzypek J, Ganczarski A (eds) (1999) Modeling of material damage and failure of structures. Springer, Berlin
Sosa H (1992) On the fracture mechanics of piezoelectric solids. Int J Solids Struct 29:2613–2622
Suo Z, Kuo CM, Barnett DM, Willis JR (1992) Fracture mechanics for piezoelectric ceramics. J Mech Phys Solids 40:739–765
Uchino K (1986) Electrostrictive actuators: materials and applications. Ceram Bull 65:647–652
Ueda S (2002) The mode I crack problem for layered piezoelectric plates. Int J Fract 114:63–86

Chapter 12
Analysis of Inelastic Behavior for High Temperature Materials and Structures

Konstantin Naumenko and Holm Altenbach

Abstract This review provides a current status in modeling and analysis of structures for high-temperature applications. Basic features of inelastic behavior of heat resistant alloys are discussed. Typical responses for stationary and varying loading and temperature are presented and classified. Microstructural features and microstructural changes in the course of inelastic deformation at high temperature are discussed. The state of the art on material modeling and structural analysis in the inelastic range at high temperature is resented.

Keywords Creep · Low cycle fatigue · Damage mechanics · Length scales · Temporal scales · Structural analysis

12.1 Introduction

The aim of this contribution is to give an overview of experimental and theoretical approaches to analyze the behavior of materials and structures subjected to mechanical loading and "high-temperature" environment. The definition of "high-temperature" materials and "high-temperature" structures can be related to the value of the homologous temperature, that is $T/T_{\rm m}$, where T is the absolute temperature and $T_{\rm m}$ is the melting point of the considered material. Materials that can be efficiently used within the temperature range $0.3 < T/T_{\rm m} < 0.7$ are called high-temperature materials. Examples include heat resistant steels, nickel-bases alloys, age-hardened aluminum alloys, cast iron materials and metal matrix composites. Structures that operate in the temperature range $0.3 < T/T_{\rm m} < 0.7$ over a long period of time are called high-temperature structures. Examples include turbine

K. Naumenko (✉) · H. Altenbach
Institute of Mechanics, Otto-von-Guericke University Magdeburg,
06099 Magdeburg, Germany
e-mail: konstantin.naumenko@ovgu.de

H. Altenbach
e-mail: holm.altenbach@ovgu.de

H. Altenbach et al. (eds.), *From Creep Damage Mechanics to Homogenization Methods*, Advanced Structured Materials 64,
DOI 10.1007/978-3-319-19440-0_12

blades, turbine housings, rotors, turbochargers, steam pipework, microelectronics components etc. This book deals with high-temperature material behavior with the emphasis on modeling, analysis and design of structures.

Despite the structural analysis one has often to simulate hot deformation processes. Examples include friction stir welding and hot forming. Here materials are subjected to higher temperatures but still below the melting point, usually in the range $0.7 < T/T_{\mathrm{m}} < 0.9$. The material behavior under the condition of long-term high-temperature service and hot working operations appear to be closely related, since they share similar temperature properties and thermally activated deformation mechanisms. However, materials under hot working conditions are subjected to much higher strain rates and strains.

In what follows we discuss basic features of the inelastic behavior of materials and structures at high temperatures. Typical responses for various loading paths are presented and classified. Microstructural features and microstructural changes in the course of inelastic deformation at high temperature are discussed. Furthermore the state of the art on material modeling and structural analysis in the inelastic range at high temperature is presented.

12.2 High-Temperature Inelasticity in Structural Materials

Material behavior can be examined with different experimental observations, for example, macroscopic and microscopic. The engineering approach is related to the analysis of stress and strain states in structures and mostly based on the standard mechanical tests. In this section we discuss basic features of inelastic material behavior according to published results of material testing under uni-axial and multi-axial stress states.

12.2.1 Uni-axial Stress State

12.2.1.1 Tensile Behavior

Uni-axial tensile tests are basic experiments of the material behavior evaluation. A standard tension specimen is subjected to the uniform elongation with a constant rate. From the measured force and the given elongation the stress and the strain are computed. Many materials exhibits a linear stress-strain relationship up to a yield point. The linear portion of the curve is the elastic region and the slope is the modulus of elasticity or Young's modulus. After the loading and subsequent unloading within the elastic range the specimen takes the original length. The loading and unloading paths coincide, and the work done to deform the specimen is stored as the elastic energy. Beyond the yield point the strain increases for almost constant stress. Here the material is undergoing a rearrangement of microstructure, such that atoms are being moved to new equilibrium positions. In crystalline materials the plastic flow is explained by motion of dislocations—line defects of the crystal structure. Within the hardening regime the stress-strain curve goes up such that the material supports the additional increasing load. The hardening is usually explained as an increasing resistance against the plastic flow in the course of deformation. For example, the plastic straining generates dislocations. With the increase in dislocation density, the dislocation movement becomes more difficult. Another example is the micro-stress fields generated during the plastic flow, as a result of heterogeneous deformation on the micro-scale. Several microstructural zones, for example slip planes, grains with certain crystallographic orientations exhibit higher levels of inelastic strain rate. The remaining part of microstructure behaves more or less elastically. This leads to changes of micro-stress states and to formation of residual stresses upon unloading. Residual micro-stress fields affect the overall deformation rate and provide an additional hardening.

In the course of deformation the cross section of the specimen shrinks, starting from the Poisson effect in the elastic regime. For many materials the inelastic deformation does not produce an essential change in volume, such that the uniform change in the cross section can be easily related to the elongation. As deformation proceeds the geometric instability causes strain to localize in a small region—the phenomenon called necking until the final stage of rupture. Within the inelastic regime an essential part of the work done to deform the material dissipates (usually as heat). After the removing the force the specimen does not return to its original shape—after the elastic springback the permanent plastic strain remains. From a stress-strain diagram several material characteristics, important for design of structures, can be identified. They include the Young's modulus E, the yield limit σ_{y} and the ultimate tensile strength σ_{u}. The yield limit is often not well defined from the shape of the stress-strain curve. Instead of yield point the upper and the lower yield points as well as an offset yield point $R_{\mathrm{p}0.2}$ are usually introduced. The latter is the stress value, for which the permanent plastic strain is 0.2 %.

Tension tests are frequently performed at elevated temperature. The specimen is uniformly heated up to a certain temperature, usually in the range $T = (0.3-0.7)\,T_{\mathrm{m}}$ and then subjected to elongation, to examine material properties for high-temperature applications. Let us discuss basic features of hot deformation observed from tensile tests. Figure 12.1 shows schematically stress-strain diagrams obtained at high temperature. Here the yield point cannot be defined, the $R_{\mathrm{p}0.2}$ stress is used instead in most cases. The hardening is usually accompanied by recovery phenomena. For example the increase of dislocation density in a course of inelastic straining—a hardening phenomenon—is accompanied by annihilation of dislocations—a recovery phenomenon observed at high temperature. Internal stresses, generated due to heterogeneous inelastic deformation, relax as a result of microstructure rearrangements at high temperature—for example diffusion of vacancies. Therefore, the stress-strain diagram of several materials shows a well-defined horizontal shape, the so-called saturation or steady-state flow regime, for which the hardening and recovery processes are in equilibrium. An important material characteristic is the steady-state (saturated) stress level, Fig. 12.1. Examples for hardening/recovery materials include pure

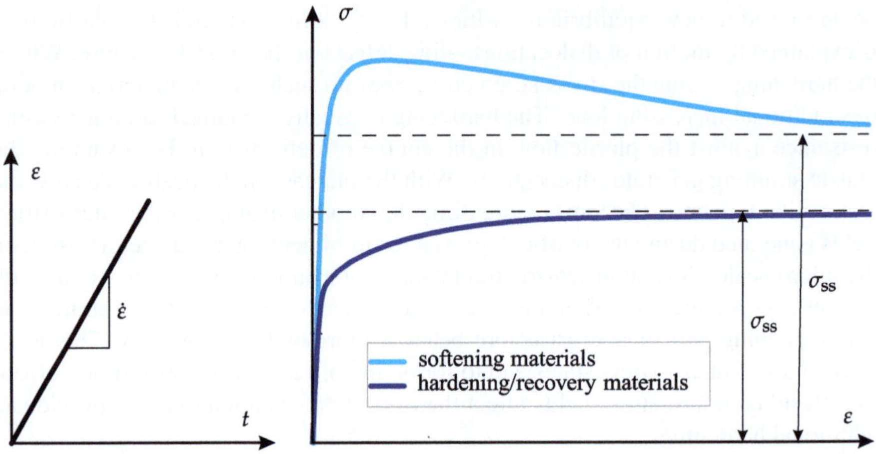

Fig. 12.1 Stress-strain diagrams for materials with hardening and softening at high temperature

metals, solid solution alloys with a relatively low initial density of dislocations and 300-series austenitic steels. Experimental data for 304 and 316 steels are presented in Gorash et al. (2012); Kawai (1989), among others.

Many materials contain relatively high dislocation density at the initial (virgin) state after the processing. Examples include 9–12 % Cr ferritic steels, where a high density of dislocations is induced after martensitic transformation. For these materials the inelastic deformation is accompanied by the recovery of dislocation substructures such as coarsening of subgrains (Blum 2008). The stress-strain curve shows a descending (softening) branch, Fig. 12.1. Experimental data for 9–12 % Cr steels are presented in Naumenko et al. (2011a); Röttger (1997); Yaguchi and Takahashi (2005), among others.

Material properties like Young's modulus and yield limit as well as, hardening, recovery and softening processes strongly depend on the temperature level. As an example, Fig. 12.2 shows stress-strain diagrams for 12 % Cr steel X20CrMoV12-1 steel for different values of the absolute temperature (Röttger 1997). At room temperature the material shows a typical tensile behavior with the elastic range followed by hardening and necking regimes up to final fracture. In contrast, at 600 °C after a relatively short hardening range the stress strain curve shows a clear descending branch. This softening regime is observed for small strains (lower than 2 %) and is related to coarsening of dislocation substructures.

Inelastic response depends on the rate of the applied loading. For many materials the rate dependence becomes essential for high temperatures, i.e. for $T/T_{\mathrm{m}} > 0.3$. As an example, Fig. 12.3 shows stress-strain curves for modified 9Cr-1Mo steel at 550 °C and different strain rates after Yaguchi and Takahashi (2005). Increase of the strain rate leads to an increase in the saturation stress value. The corresponding dependency of σ_{ss} on $\dot{\varepsilon}$ is called strain rate sensitivity of the tensile response and is an important characteristic for design and analysis of structures and processes.

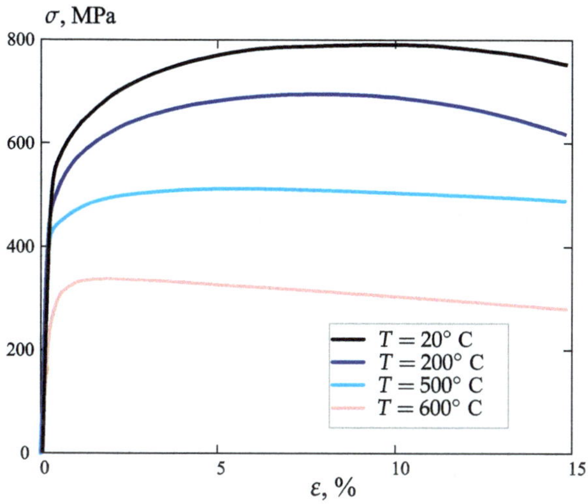

Fig. 12.2 Stress-strain diagrams for X20CrMoV12-1 steel under strain rate $\dot{\varepsilon} = 2.63 \times 10^{-4}\ \mathrm{s}^{-1}$ and different temperature levels, after (Röttger 1997)

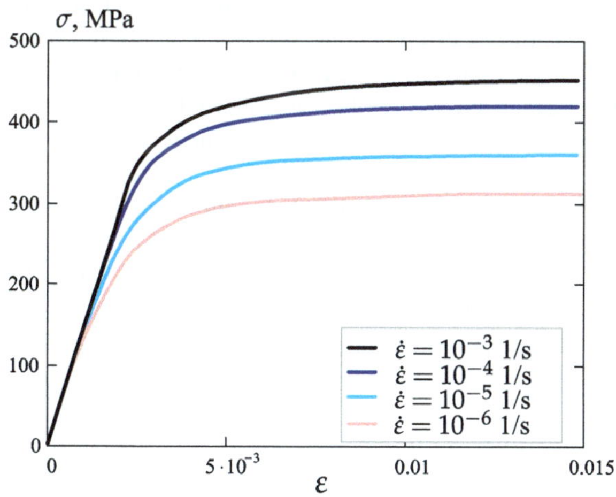

Fig. 12.3 Stress-strain diagrams for modified 9Cr-1Mo steel at $T = 550\,°\mathrm{C}$ and different strain rates, after Yaguchi and Takahashi (2005)

12.2.1.2 Creep and Relaxation

Uni-axial creep test is another experiment to examine material behavior at high temperature. A standard cylindrical tension specimen is heated up to the temperature $T = (0.3 - 0.5)\ T_{\mathrm{m}}$ and loaded by a tensile force F. The value of the normal stress in the specimen σ is usually less than the yield point σ_{y} or offset yield point $R_{\mathrm{p}0.2}$ of the material at the given temperature. The instantaneous material response is therefore elastic. The load and the temperature are kept constant during the test and the axial engineering strain ε is plotted versus time t. The instantaneous response can be characterized by the strain value ε^{el}. The time-dependent response is the slow increase of the strain ε with a variable rate. Following Andrade (1910), three stages can be considered in a typical creep curve: the first stage (primary or reduced creep), the second stage (secondary or stationary creep) and the third stage (tertiary or accelerated creep). During the primary creep stage the creep rate decreases to a certain value (minimum creep rate). The secondary stage is characterized by the approximately constant creep rate. During the tertiary stage the strain rate increases. At the end of the tertiary stage creep rupture of the specimen occurs. A number of properties, important for design of structures, can be deduced from the uni-axial creep curve. These are the duration of the stages, the value of minimum creep rate, the time to fracture and the strain value before fracture. The shape of the creep curve, the duration of the creep stages the creep rate and the time to fracture depend strongly on the stress and temperature values.

For the analysis of creep behavior it is convenient to introduce the inelastic (creep) strain ε^{cr} as the difference between the measured strain ε and the calculated elastic strain $\varepsilon^{el} = \sigma/E$, where E is the Young's modulus. From the original strain versus time curve the creep rates can be computed and plotted as a functions of creep strain and/or time. Two examples of creep rate versus creep strain curves are presented schematically in Fig. 12.4. For several materials, for example pure metals, the classical creep with three creep stages are observed. Here the creep rate is nearly constant over a certain range of creep strain values. The secondary creep stage is characterized by the steady state creep rate $\dot{\varepsilon}^{cr}_{ss}$, Fig. 12.4. For a range of temperatures and stress levels an important deformation regime is the dislocation creep. Here the deformation is controlled by the movement of mobile dislocations (Illschner 1973; Blum 2001; Frost and Ashby 1982). The creep rate is related to the velocity of mobile dislocations. The dislocation velocity decreases with increasing overall dislocation density Blum (2001). The creep deformation produces dislocations and the creep rate is expected to decrease with an increase in dislocation density. On the other hand, annihilation

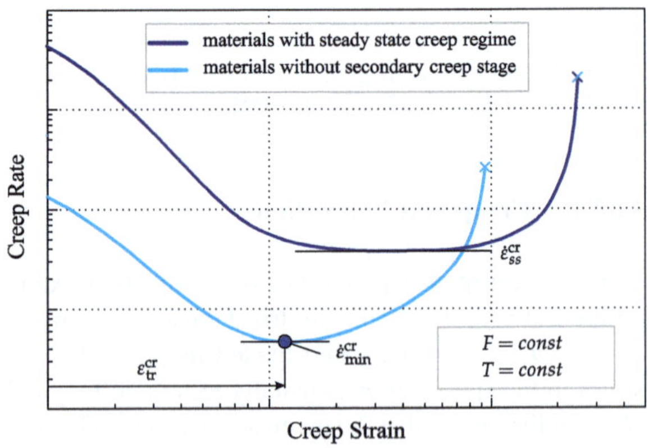

Fig. 12.4 Creep rate versus creep strain curves with and without steady state creep regime

of dislocations at high temperature, for example, due to diffusion of vacancies take place leading to a decrease (recovery) of dislocation density. In a steady state creep regime the dislocation density does not change, that is, the rate of production and the rate of annihilation are the same (Blum 2001). For many structural materials, for example advanced heat resistant steel, a pronounced secondary creep stage is not observable. The creep rate decreases at the beginning of the creep process, attains the minimum value at a certain value of the strain $\varepsilon_{\rm tr}^{\rm cr}$ and immediately increases after that. The corresponding characteristic is the minimum creep rate $\dot{\varepsilon}_{\rm min}^{\rm cr}$, Fig. 12.4.

The dependencies of creep rate on stress and temperature are of primary interest to an engineer designing some structure or machine. In order to obtain mechanical properties of the material, series of creep tests are usually performed for different stress and temperature values. From the resulting families of creep curves one can obtain the minimum creep rate versus stress curve, the minimum creep rate versus temperature curve, the creep rate versus time curve and the stress versus time to fracture curve (long term strength curve). The ranges of stress and temperature should be specified according to the ranges expected in the structure during the service.

Figure 12.5a illustrates experimental data for minimum creep rate as a function of stress for steel 316 for the temperature 600 °C (Rieth et al. 2004). To generate reliable data for the minimum (secondary) creep rate for the stress values less than 100 MPa creep tests with the long duration (approx. 10 years) are required (Rieth et al. 2004). The data for this range of moderate and low stress levels are of interest for the analysis of power plant components operating in high-temperature range over long period of time. Special uni-axial specimen with an increased gauge length were designed in Rieth et al. (2004) to improve the resolution of creep strain measurements for low stress levels. In Kloc and Sklenička (1997, 2004); Kloc et al. (2001) the testing facilities and experimental data for a 9 % Cr steel are presented. Creep tests for the stress levels below 20 MPa were performed by the use of helical springs while for the stress levels over approx. 100 MPa standard uni-axial specimen were applied. Figure 12.5b shows the experimental data for the temperature 600 °C. Examples for minimum creep rate versus stress curves for various materials can be found in Boyle and Spence (1983); Illschner (1973); Kassner and Pérez-Prado (2004); Kraus (1980); Malinin (1981); Naumenko et al. (2009); Odqvist and Hult (1962); Odqvist (1974); Penny and Mariott (1995) and many papers related to the experimental analysis of creep, e.g. Altenbach et al. (2013); Gariboldi and Casaro (2007); El-Magd et al. (1996); Hyde et al. (1997, 1999); Kimura et al. (2009); Längler et al. (2014). To discuss basic features of tertiary creep consider a sketch of the creep rate versus creep strain curves usually observed for 9–12 % Cr steels, Fig. 12.6. Experimental creep curves are published in Kimura et al. (2009); Naumenko et al. (2011a); Straub (1995). The secondary or the steady state creep stage is usually not observable. The creep strain at which the transition from primary to the tertiary creep takes place is dependent on the stress level. The tertiary creep rate is primarily controlled by the softening processes. Two important examples are the coarsening of the subgrain microstructure and coarsening of carbide precipitates Abe (2009); Dyson and McLean (1998); Blum (2008); Straub (1995). An additional factor is the cross-section shrinkage of the specimen as a result of essential creep deformation. As the force is kept constant

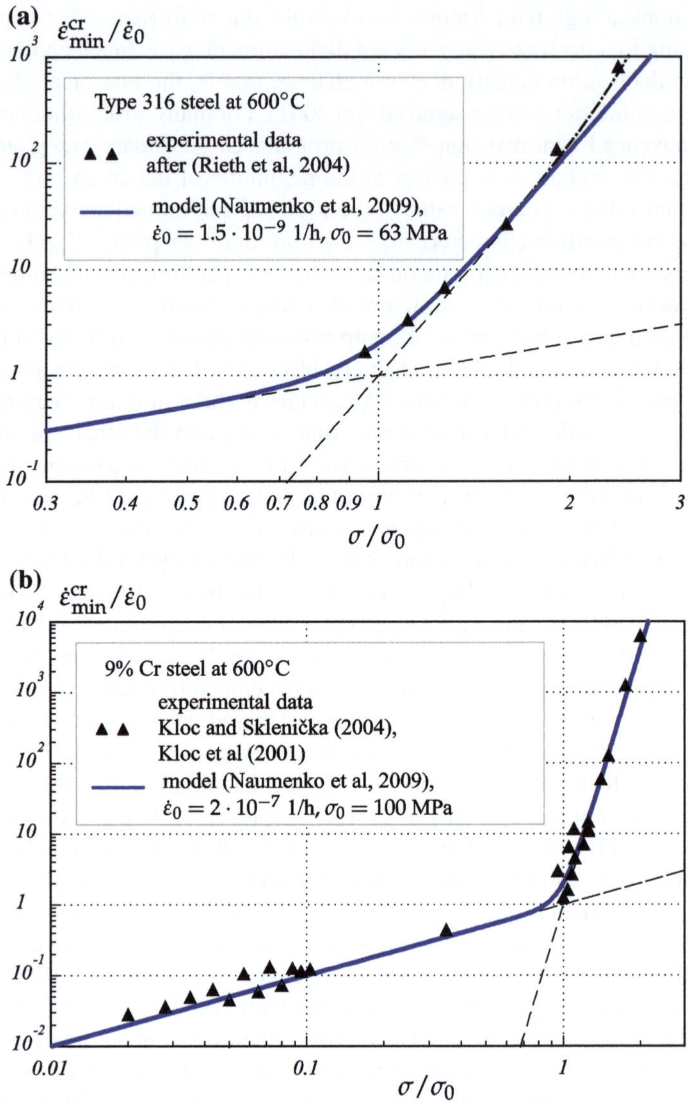

Fig. 12.5 Normalized minimum creep rate versus normalized stress. **a** Type 316 steel at 600 °C, **b** 9 % Cr steel at 600 °C

during the test the cross section reduction leads to an increase in the true stress value. The essential non-linearity of the creep rate with respect to the true stress leads to an additional acceleration of creep. The final stage of the creep curve is affected by damage processes. The principal damage mechanism is the formation and growth of cavities on grain boundaries, subgrain boundaries or carbides. This mechanism of creep cavitation is common for many polycrystalline materials. Processes including

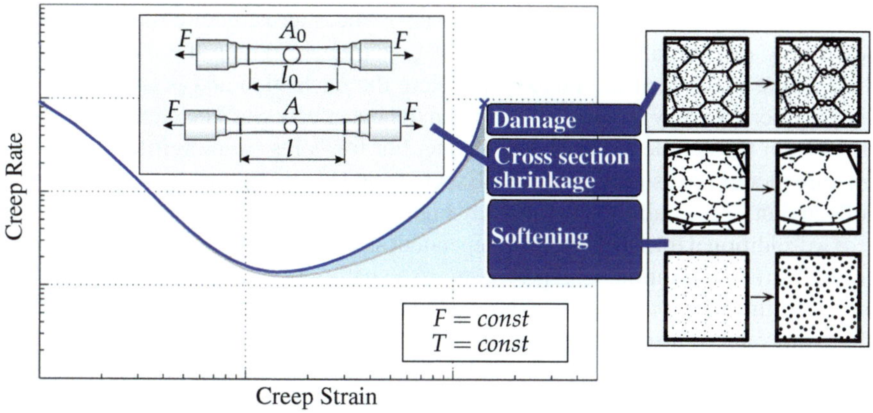

Fig. 12.6 Mechanisms of tertiary creep in advanced steel

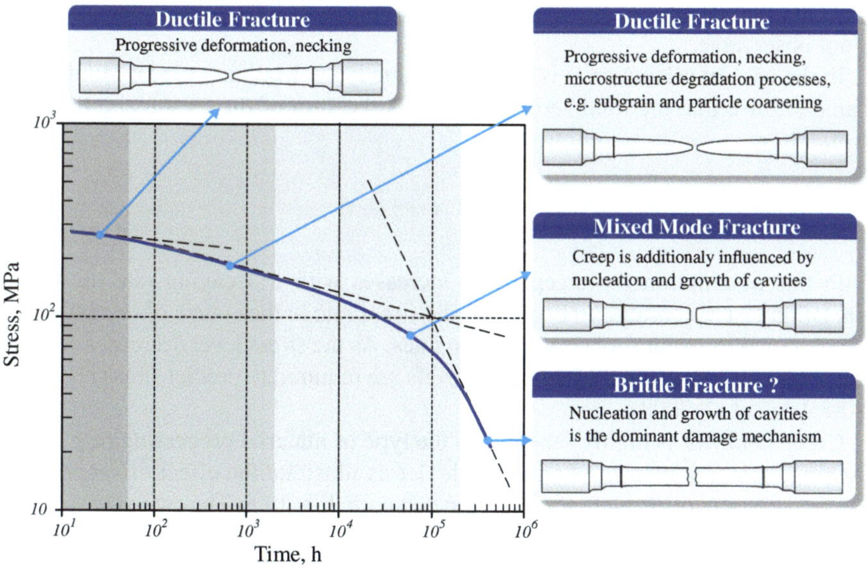

Fig. 12.7 Sketch for a long term strength *curve* for advanced heat resistant steel (Naumenko and Kostenko 2009)

softening, damage and cross-section shrinkage are more or less dominant depending on the stress level and temperature. To explain the influence of these mechanisms let us consider a typical long term strength (creep rupture strength) curve, Fig. 12.7. Here the value of the applied stress is plotted as a function of creep life deduced from uni-axial creep tests. For high stress values the fracture mode is ductile and the uni-axial specimen necks down after a certain time as a result of excessive deformation. For lower stress values the necking is still observed, but the slope of the curve decreases. The origin of this decrease is the microstructure degradation like coarsening of subgrains or coarsening of precipitates. Within the transition range the fracture mode is of the mixed type. Here the nucleation and growth of cavities and microcracks may have an influence on the creep process. The curve changes the slope from the ductile to the brittle regime. For low stresses the brittle damage and fracture modes are usually observed. Experimental creep rupture strength curves are collected in Yagi et al. (2004) for many high temperature materials.

Two additional forms of the time-dependent stress-strain behavior are creep recovery and stress relaxation. Creep recovery is usually observed, when after a certain period of time the load is spontaneously removed. After unloading the strain drops about the value ε^{el} (recovery of the elastic strain). Then the strain slowly decreases down the permanent (irrecoverable) value ε^{pm}. Stress relaxation is observed when the strain is held constant in time ($\varepsilon = const$). A uni-axial specimen is instantaneously deformed to the strain value $\varepsilon^{\text{el}} = \sigma/E$, where E is the Young's modulus. During the test the load is continuously decreased in such a way that the initial strain remains constant. A threshold of the initial stress (strain) exists below which the relaxation is not observable.

In the case of relaxation it is usually assumed, e.g. Malinin (1981); Stouffer and Dame (1996), that the total zero strain rate is the sum of the elastic and the creep strain rates

$$\dot{\varepsilon} = \frac{\dot{\sigma}}{E} + \dot{\varepsilon}^{\text{cr}} = 0 \tag{12.1}$$

With this assumption the creep strain increases with a decaying rate during the relaxation test. According to Eq. (12.1) the stress rate in the course of relaxation can be computed from data concerning creep rates. As the stress level decreases accurate creep data for low and moderate stress levels are required to predict stress relaxation (Altenbach et al. 2008).

Creep behavior is highly sensitive to the type of material processing (e.g. plastic forming, heat treatment). As an example, let us illustrate the effect of spontaneous plastic pre-strain on the subsequent creep behavior, Fig. 12.8. The creep curve shown by solid line is obtained under the constant stress σ_0. The dotted lines present the second and the third creep curves after spontaneous loading to the stresses σ_1 and $\sigma_2 > \sigma_1$ leading to small plastic strains $\varepsilon_1^{\text{pl}}$ and $\varepsilon_2^{\text{pl}} > \varepsilon_1^{\text{pl}}$, respectively, and subsequent unloading to the stress σ_0. The creep rate after the loading in the inelastic range is significantly lower compared to the creep rate of the "virgin" material. The effect of reduction in creep rate becomes stronger with the increase of the prior plastic strain. Effects of this type are sometimes called "plasticity-creep" or "creep-plasticity" interactions, e.g. Inoue (1988); Krausz and Krausz (1996); Krempl (1999); Miller (1987).

Many materials exhibit anisotropic creep behavior. Examples are: directionally solidified nickel-based superalloys, e.g. Winstone (1998), fiber reinforced materials, e.g. Robinson et al. (2003a, b), deep drawing sheets, e.g. Betten (1976, 2001), forgings Naumenko and Gariboldi (2014) and multi-pass weld metals Hyde et al. (2003). In

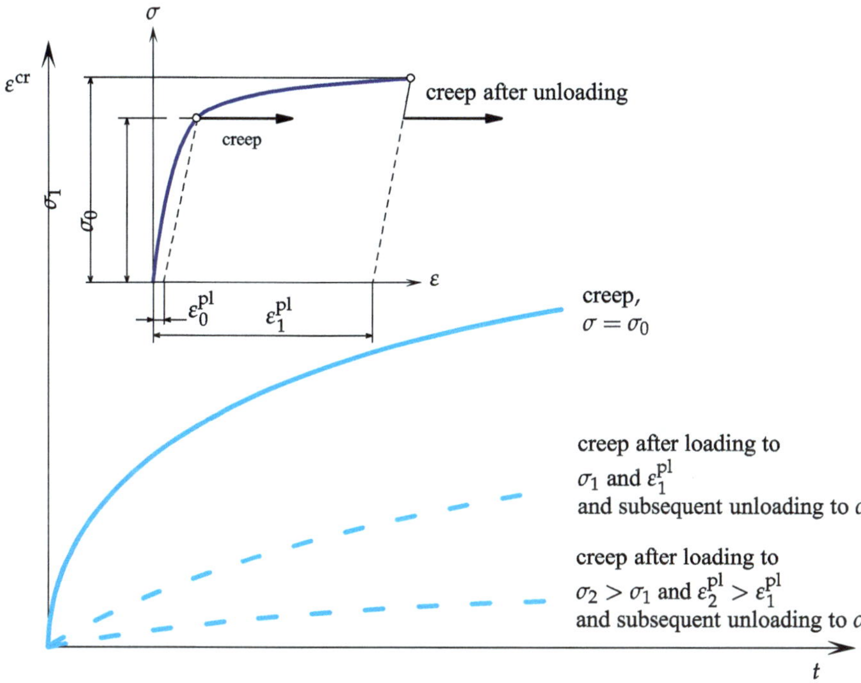

Fig. 12.8 Effect of initial plastic strain on creep behavior, after Kawai (1989)

these cases series of uni-axial creep tests for specific loading directions are performed in order to establish the material behavior. The number of the required tests and the corresponding loading directions are dictated according to the assumed symmetries of the material microstructure.

12.2.1.3 Creep Under Varying Load and Cyclic Creep

To investigate transient creep effects tests under non-stationary loading under constant high temperature are performed. Creep curves under stepwise loading are presented in Faruque et al. (1996); Malinin (1981), for example. The creep test starts under a certain value of the load. After reaching steady-state creep the load is rapidly increased (decreased) and kept constant over a period of time (hold time). Such tests are useful to analyze hardening, recovery and softening processes after the rapid changes of loading. Figure 12.9 shows experimental data from creep test under compressive stress changes published in Straub (1995). During the test the specimen was initially subjected to the constant compressive true stress with the value of 196 MPa. After a certain hold time the stress was rapidly reduced to the value of 150 MPa. Several hold and loading/ unloading phases were performed. As Fig. 12.9 shows the first unloading was performed after the reaching the tertiary creep stage. For the loading with the constant compressive true stress the main mechanism for acceler-

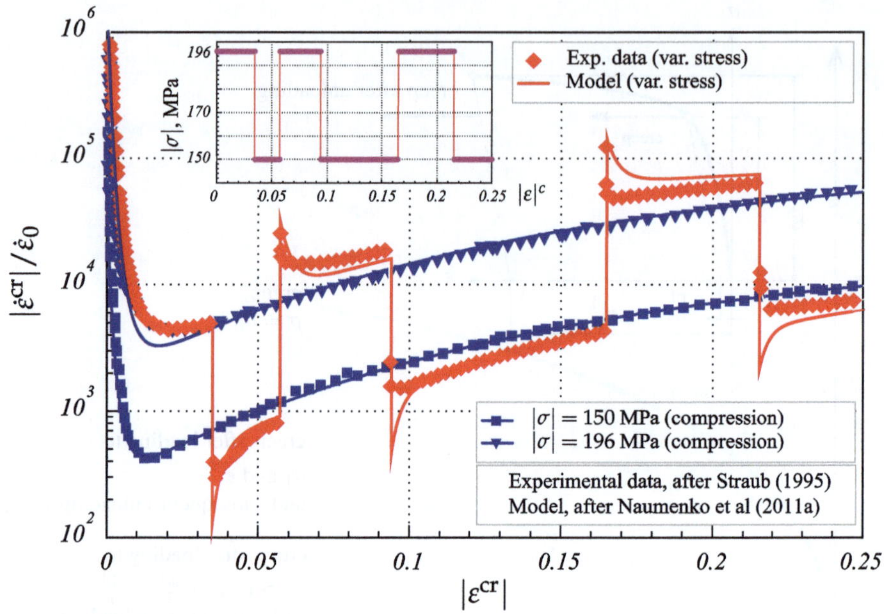

Fig. 12.9 Normalized creep rate versus creep strain for X20CrMoV 12-1 Steel at T = 600 °C and variable compressive stress

ated creep is the softening process, associated with coarsening of subgrain structure (Straub 1995). One additional feature influenced by the softening is observed for variable loading conditions. After the stress decrease the creep rate remains lower, while after the stress increase it becomes higher than the corresponding creep rate under the constant stress level, Fig. 12.9.

Components are often subjected to cyclic loading at high temperature. To analyze the structural behavior material tests for cyclic force or displacement controls are required. Periodically varied force causes cyclic creep response. The applied periodic stress can be characterized by the amplitude σ_a, the period τ_c and the mean stress σ_m. The following stress ratios are used to indicate the kind of cyclic loading

$$\hat{A} = \frac{\sigma_\mathrm{a}}{\sigma_\mathrm{m}}, \quad R = \frac{\sigma_\mathrm{max}}{\sigma_\mathrm{min}}, \tag{12.2}$$

where $\sigma_\mathrm{max} = \sigma_\mathrm{m} + \sigma_\mathrm{a}$ and $\sigma_\mathrm{min} = \sigma_\mathrm{m} - \sigma_\mathrm{a}$. Two cases of the periodic loading are presented in Fig. 12.10a, b. Let us assume that the value of the maximum stress σ_max is lower than the yield point (yield offset) of the material at the testing temperature. Creep behavior for the case of periodic loading with hold time is schematically illustrated in Fig. 12.10c. Here the mean stress σ_m, the amplitude σ_a, the rate of loading/unloading and the hold time influence the creep response. The case of harmonic loading is shown in Fig. 12.10b. Such loading is important in those engineering applications, where technological or operational conditions (non-stationary

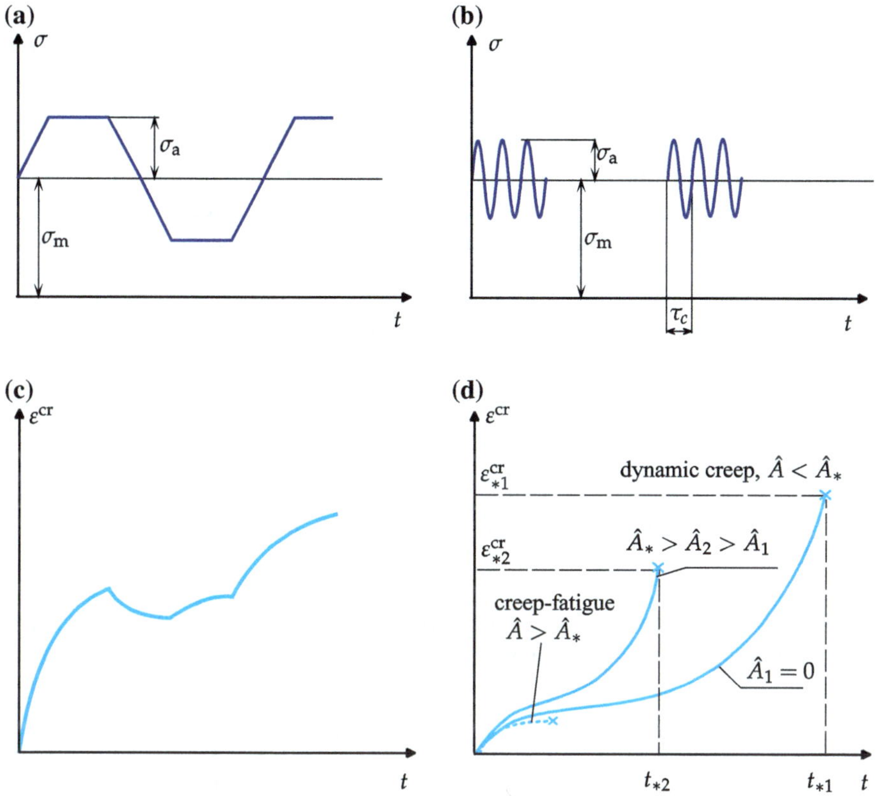

Fig. 12.10 Types of loading and corresponding cyclic creep *curves*. **a** Cyclic loading with hold time, **b** harmonic loading with high frequency, **c** creep response for cyclic loading with hold time, **d** different responses for loading with high frequency

flow, combustion, acoustic action, etc.) cause the development of forced vibrations. The harmonic stress variation can be described as follows

$$\sigma = \sigma_{\mathrm{m}}(1 + \hat{A}\sin\Omega t), \quad \Omega = \frac{2\pi}{\tau_{\mathrm{c}}} = 2\pi, \tag{12.3}$$

Creep behavior under harmonic loading (12.3) with frequencies $f > 1 \ldots 2$ Hz has been studied in Bernhardt and Hanemann (1938); Lazan (1949); Taira (1962); Taira and Koterazawa (1962). For this cyclic loading condition primary, secondary and tertiary stages can be observed similarly to the static case, Fig. 12.10d. Furthermore, the shape of the cyclic creep curve is geometrically similar to the static one caused by the stress $\sigma = \sigma_{\mathrm{m}}$, but the creep rate is rather higher and the time to fracture is essentially smaller. It was found that creep under fast cyclic loading is not sensitive to the frequency of stress variation, e.g. Taira and Ohtani (1986). In contrast, the stress cycle asymmetry parameter $\hat{A}$ has significant influence on the creep rate. For a number of investigated materials a material property $\hat{A}_*$ has been found which is termed as the critical value of the stress cycle asymmetry parameter. For $\hat{A} < \hat{A}_*$ the high cyclic creep process is similar to the static one with increased creep rate and decreased time to fracture. For $\hat{A} > \hat{A}_*$ such a behavior is not observable, and fracture takes place as a consequence of creep-fatigue interaction. Following Lazan (1949); Rabotnov (1969); Taira and Ohtani (1986), the processes of high-frequency cyclic creep are classified as: dynamic creep for $\hat{A} < \hat{A}_*$ and high cyclic creep for $\hat{A} > \hat{A}_*$. Creep curves for both cases are schematically presented in Fig. 12.10d.

The cycle by cycle accumulation of inelastic strain under the applied cyclic force is called ratchetting. Experimental data show that the ratchetting deformation depends on the mean stress, the stress ratio as well as on the loading rate and hold time (Altenbach et al. 2013; Ohno et al. 1998; Ohno 1998). For advanced 9–12 % Cr steels anomalous ratchetting behavior is documented—in cyclic force controlled tests with zero mean force progressive deformation in the tensile direction is observed (Yaguchi and Takahashi 2005; Röttger 1997; Bunch and McEvily 1987).

12.2.1.4 Low Cycle Fatigue, Creep Fatigue and Thermo-Mechanical Fatigue

Figure 12.11a shows basic parameters of the strain controlled low cycle fatigue (LCF) test. The loading is characterized by the strain amplitude ε_a, strain range $\Delta\varepsilon = 2\varepsilon_a$,

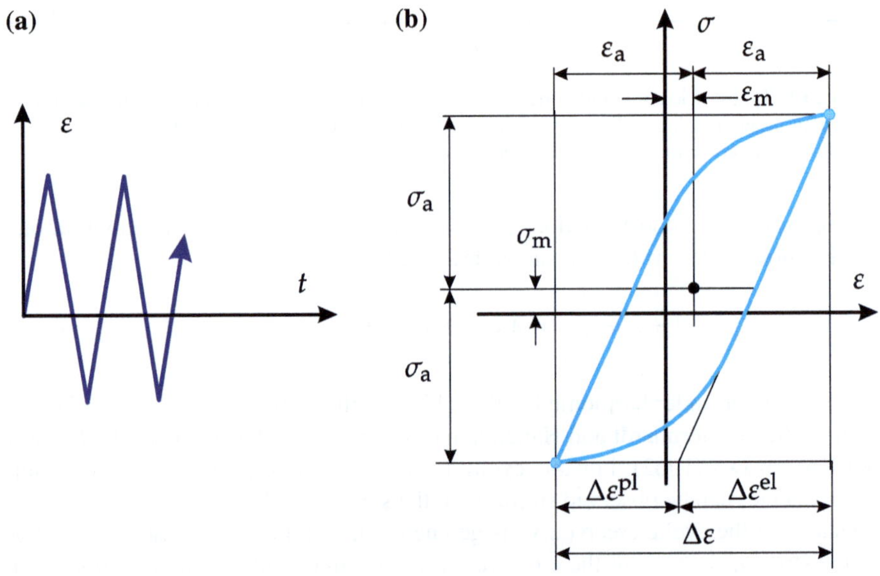

Fig. 12.11 Loading and stress response parameters in LCF regime. **a** Triangular strain waveform, **b** hysteresis loop

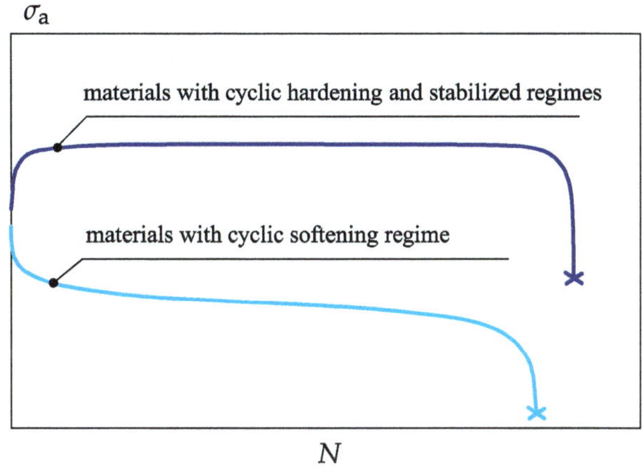

Fig. 12.12 Stress amplitude versus number of cycles curves with different cyclic regimes for $R_\varepsilon = -1$

and the mean strain ε_m. Similarly to the stress controlled tests the following ratios can be introduced

$$\hat{A}_\varepsilon = \frac{\varepsilon_a}{\varepsilon_m}, \quad R_\varepsilon = \frac{\varepsilon_{max}}{\varepsilon_{min}}, \tag{12.4}$$

where $\varepsilon_{max} = \varepsilon_m + \varepsilon_a$ and $\varepsilon_{min} = \varepsilon_m - \varepsilon_a$. During the test the stress as a function of time is recorded. The basic stress response parameters are the stress amplitude σ_a and the mean stress σ_m. From a hysteresis loop for a certain loading cycle, Fig. 12.11b one may also compute the elastic strain range $\Delta\varepsilon^{el}$ and the plastic strain rage $\Delta\varepsilon^{pl}$. Typical stress responses obtained from LCF tests with $R_\varepsilon = -1$ are schematically shown in Fig. 12.12. Here the stress amplitude is presented as a function of the cycle number. For a class of materials, for example 300-series austenitic steels the cyclic hardening followed by the stabilized response is observed. The stress amplitude increases over a number of cycles (cyclic hardening stage), attains a steady-state value (stabilized response stage) and decreases rapidly at the final stage of fatigue failure. On contrary, many materials, for example 9–12 % Cr steels exhibit cyclic softening, that is the decrease of the stress amplitude starting from the first loading cycles over the whole fatigue life. In strain controlled fatigue tests with $R_\varepsilon = -1$ at low homologous temperature the strain amplitude is the important loading input parameter that determines the fatigue life. Strain amplitude versus cycles to failure diagrams are widely used to characterize fatigue strength of materials.

Fatigue damage evolution is a sequence of several microstructural evens. The total fatigue life can be separated into the stage of macrocrack initiation and the stage of macrocrack growth until reaching the critical crack length and final failure. The first initiation stage can include processes of increase in dislocation density, formation of dislocation substructures, localization of inelastic strain along persistent slip bands, surface relief evolution, formation and early growth stage of surface microcracks (Mughrabi 2009; Polák 2003). This early stage ist mostly controlled by the non-homogeneous cycle by cycle inelastic deformation. The accumulation of irreversible slip steps on the surface leads to the surface roughening, with sites of local stress concentrations at which microcracks can form (Mughrabi 2009). Short cracks usually grow by the sliding mode (mode II in the sense of fracture mechanics). Progressively, after crossing several grains the main crack propagates, usually in the plane orthogonal to the loading axis.

The inelastic response at high temperature depends essentially on the rate of loading, cp. Sect. 12.2.1.1. For static and cyclic tests under low strain rates creep regime becomes essential. Decrease in the strain rate by keeping the total strain range fixed leads to an increase in the inelastic strain range and the dissipated work within the cycle. Fatigue life usually decreases with a decrease of the strain rate (loading frequency) for constant strain range, e.g. (Skelton 2003).

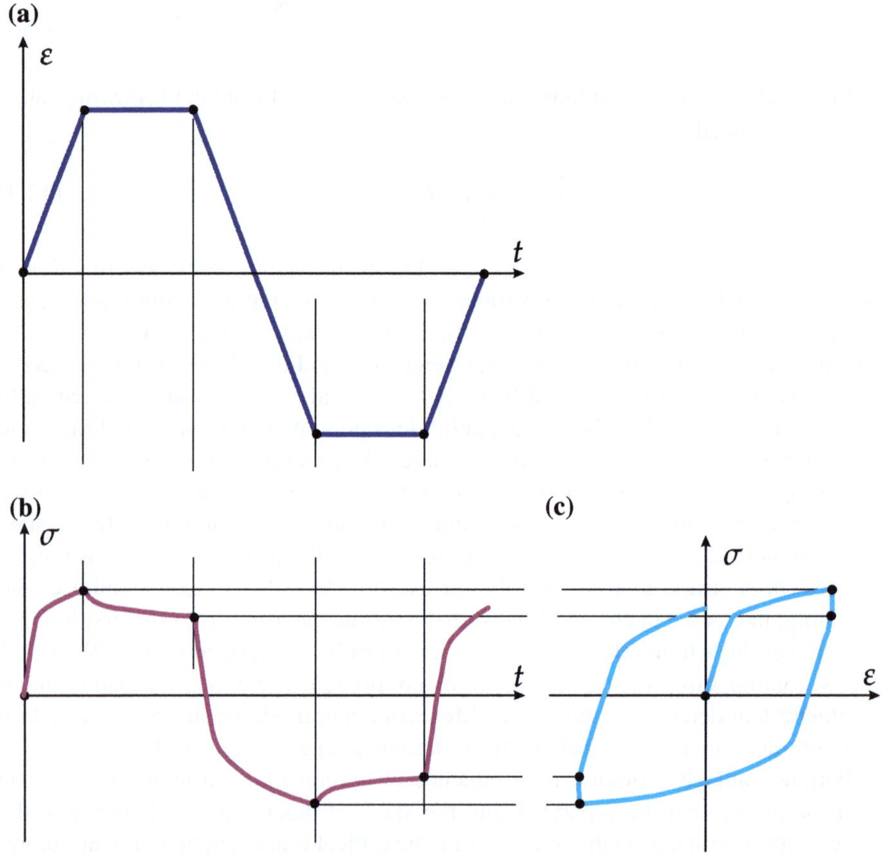

Fig. 12.13 Low cycle fatigue with hold time. **a** Loading, **b** stress response, **c** hysteresis loop

Test performed under cyclic strain with hold times at high temperature are usually performed to examine creep-fatigue (relaxation-fatigue) behavior. Figure 12.13a provides a sketch of a cyclic strain control with hold phases under tension and compression. The stress response over time, Fig. 12.13b and the hysteresis loop, Fig. 12.13c illustrate the stress relaxation regimes during hold phases. Various controls of creep-fatigue tests can be applied to analyze material behavior. Examples include cycles with tension-compression holds, Fig. 12.13, cycles with tensile holds, cycles with compressive holds, etc. Tensile and compressive holds with the same duration have usually different influence on the creep-fatigue life for many materials. For example 9–12 % Cr steels compressive hold phases more detrimental than tensile ones, in the sense that the fatigue life is more severely reduced under compressive holds than under tensile holds (Aktaa and Petersen 2009; Fournier et al. 2008). This effect might be surprising as one expects creep damage evolution under tension rather than under compression hold phases. However, analysis of the deformation process and the changes in the hysteresis loop from cycle to cycle as a result of softening and relaxation. Tensile holds lead to a reduction in the tensile peak stress in the cycle as observed for a half of the fatigue life (Aktaa and Petersen 2009). As a result, cycling with compressive holds leads to much higher peaks in tensile stress. Assuming that tensile peaks promote fatigue damage from cycle to cycle, the shortening of fatigue life can be explained (Aktaa and Petersen 2009).

Many components operate under changing temperature environment and mechanical loading over a long period. Critical positions may be subjected to fatigue damage due to thermal transients and/or creep damage during exposure at high temperature. Generally, the structural integrity of a component is ensured through reliable design, precise manufacturing, definition of allowable operational modes, and timely inspection. However, permanently changing economic situations and environmental conditions require more flexible operation modes in service, e.g. daily start-up and shut-down and/or increase of steam pressure and temperature. An important step in the life-time assessment is to analyze material behavior for service-type loading and temperature profiles. To this end an inelastic structural analysis that takes into account both the slow changes in the stress and strain states during hold (running) phases and transient behavior during start-ups and shut-downs is required. Hysteresis loops obtained from structural analysis of results of real components provide an input to generate service-type thermo-mechanical fatigue (TMF) loading profiles for laboratory testing of materials. For example, strain and temperature variations on the surface of real components were used for TMF testing of uni-axial specimens in (Cui et al. 2009; Holdsworth et al. 2007). A two-bar system subjected to thermal cycling will be presented in Sect. 12.3.2.

Figure 12.14 illustrates schematically strain and temperature profiles and the stress response over one cycle of TMF loading. Experimental data for such strain and temperature controls are presented in Cui et al. (2009); Kostenko et al. (2013); Cui and Wang (2014) for uni-axial specimens from 10 % Cr steel and in Samir et al. (2005) for cruciform specimens from 1 % Cr steel under biaxial loading. Following the classification by Berger et al. (2008) the loading profile presented in Fig. 12.14 corresponds to the hot start situation. Further examples include cold and warm start profiles with higher differences in the maximum and minimum absolute temperatures within the cycles (Holdsworth et al. 2007).

Lifetime of specimens or components under TMF loading conditions is limited by fatigue damage, creep damage, environmental damage or in many cases by complex interaction of different damage mechanisms. In Holdsworth et al. (2007) results of uni-axial TMF tests for 1CrMoV forged steel under service-type loading conditions are discussed. Specimens were subjected to three heat profiles with the same maximum temperature but different temperatures at the beginning and the end of the cycles. All three types of tests were conducted under the same mechanical strain control that corresponds to the start-up situation. Post test inspection has been employed to characterize the associated damage mechanisms. It was observed that the increase of the temperature difference within the cycle leads to reduction in cycles to crack initiation endurance. Two dominant types of damage are usually observed in TMF tests of the considered low-alloy steel. Fatigue damage develops as a relatively uniform distribution of short transgranular cracks at the surface along and around the gauge section. Creep damage evolution is the nucleation and coalescence of cavities at grain boundaries and propagation of intergranular cracks across the specimen section. In tests with low difference between the minimum and the maximum temperatures in a cycle fatigue-dominated damage evolution was observed. Surface fatigue cracks were initiated and a main crack propagated towards the cross section of the specimen, where creep cavitation was also observed. Damage development in tests with moderate difference in temperature levels was creep dominated. At the end of test there was a high intensity of relatively fine intergranular microcracks in the gauge section. In tests with high temperature differences, the extent of surface oxidation and

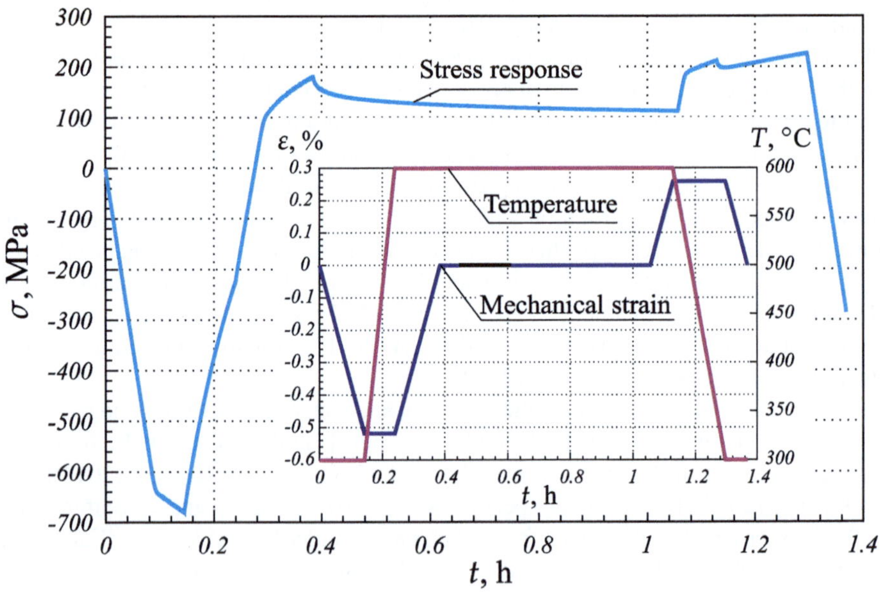

Fig. 12.14 Loading profiles and stress response for one cycle of TMF loading, after Naumenko et al. (2011b)

spallation was high due to the high thermal transient and the rate of metal removal at the surface appeared to be greater than the rate of crack development.

Several elementary rules are available to estimate the LCF life based on the loading characteristics within the cycle. An example is the Coffin-Manson equation Coffin (1954); Manson (1953) that relates the plastic strain amplitude within a cycle with the number of cycles to fatigue failure. More advanced rules were developed to account for strain rate and hold time dependencies under isothermal loading conditions. They use data from stress-strain loops within a characteristic cycle, for example, the steady state stress and/or strain rate values at the end of the hold phase, strain ranges the dissipated mechanical work together with experimental data, for example creep strength and fatigue endurance. For reviews we refer to Viswanathan (1989); Penny and Mariott (1995); Berger et al. (2008). These approaches are mostly applicable to situations, where stabilized stress-strain behavior after a certain number of loading cycles is expected. For materials with cyclic softening the shape of the stress-strain loop changes continuously from cycle to cycle such that the whole deformation process should be analyzed.

Alternatively, advanced constitutive equations for inelastic deformation and kinetic laws for hardening/recovery, softening and damage can be developed to capture the whole cycle by cycle material behavior (Aktaa and Petersen 2009; Chaboche 2008; Kostenko et al. 2013). Development of kinetic laws for interaction of different damage mechanisms and efficient numerical methods to solve kinetic equations over many loading cycles are still challenging problems.

12.2.2 *Multi-axial and Stress State Effects*

Experimental data obtained from uni-axial tests allow us to establish basic features of inelastic behavior and to find relations between strain rate, stress, temperature and time. However, most structural members are subjected to multi-axial stress conditions. In order to analyze the influence of the stress state on the material behavior, multi-axial tests are required.

Various techniques have been developed to test materials under multi-axial loading conditions. Examples are: thin-walled tubes subjected to axial force and torque, e.g. Kowalewski (1995); Kawai (1989), two- and three-dimensional cruciform specimens subjected to axial forces, e.g. Sakane and Hosokawa (2001); Sakane and Tokura (2002); Samir et al. (2005), circumferentially notched specimens subjected to axial force, e.g. Hyde et al. (1996); Perrin and Hayhurst (1994); Simon (2007); Cui et al. (2009) and three-dimensional component-like test pieces Colombo et al. (2008).

Figure 12.15 shows a thin-walled tube under the axial force and torque with the magnitudes F and M, respectively. Let r_{m} be the mean radius of the cross section, h the wall thickness and L the gauge length. With the local cylindrical basis $\boldsymbol{e}_r$, $\boldsymbol{e}_\varphi$ and $\boldsymbol{k}$, as shown in Fig. 12.15, the stress state can be characterized by the following tensor

$$\boldsymbol{\sigma} = \sigma \boldsymbol{k} \otimes \boldsymbol{k} + \tau(\boldsymbol{e}_\varphi \otimes \boldsymbol{k} + \boldsymbol{k} \otimes \boldsymbol{e}_\varphi), \quad \sigma = \frac{F}{2\pi r_\mathrm{m} h}, \quad \tau = \frac{M}{2\pi r_\mathrm{m}^2 h} \tag{12.5}$$

The deviatoric part of the stress tensor is

$$\boldsymbol{s} = \sigma(\boldsymbol{k} \otimes \boldsymbol{k} - \frac{1}{3}\boldsymbol{I}) + \tau(\boldsymbol{e}_\varphi \otimes \boldsymbol{k} + \boldsymbol{k} \otimes \boldsymbol{e}_\varphi), \tag{12.6}$$

where $\boldsymbol{I}$ is the second rank unit tensor. As a stress measure which can be used to compare different multi-axial tests is the von Mises equivalent stress σ_vM which is defined as follows

$$\sigma_\mathrm{vM} = \sqrt{\frac{3}{2}\boldsymbol{s} \cdot\cdot \boldsymbol{s}} = \sqrt{\sigma^2 + 3\tau^2}$$

From the measured elongation ΔL and the angle of twist ϕ_T the axial strain ε_L and the shear strain γ can be computed

$$\varepsilon_\mathrm{L} = \frac{\Delta L}{L}, \quad \gamma = \frac{r_\mathrm{m}\phi_\mathrm{T}}{L}$$

Assuming that the material behavior is isotropic, the strain state in a tube can be characterized by the following tensor

$$\boldsymbol{\varepsilon} = \varepsilon_\mathrm{L}\boldsymbol{k} \otimes \boldsymbol{k} + \varepsilon_Q(\boldsymbol{I} - \boldsymbol{k} \otimes \boldsymbol{k}) + \frac{1}{2}\gamma(\boldsymbol{e}_\varphi \otimes \boldsymbol{k} + \boldsymbol{k} \otimes \boldsymbol{e}_\varphi),$$

where $\varepsilon_\mathrm{Q} = \Delta r_\mathrm{m}/r_\mathrm{m}$ is the transverse normal strain.

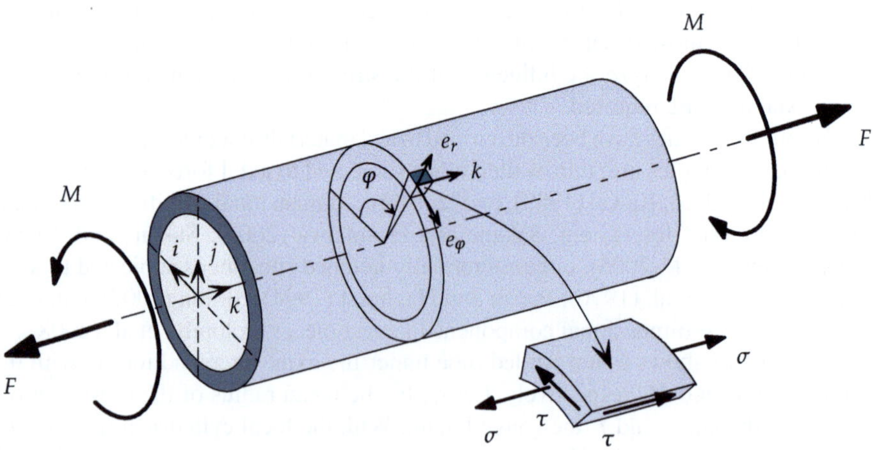

Fig. 12.15 Thin-walled tube for multi-axial creep tests

In what follows let us limit to the stress controlled tests and analyze creep behavior under multi-axial stress states. The inelastic (creep) strain tensor is defined as the difference between the strain tensor $\boldsymbol{\varepsilon}$ which includes the measurable quantities and the tensor of initial elastic strains which can be calculated from Hooke's law. As a result we obtain

$$\begin{aligned}\boldsymbol{\varepsilon}^{\mathrm{cr}} &= \left(\varepsilon_{\mathrm{L}} + 2\varepsilon_{\mathrm{Q}} - \frac{1-2\nu}{E}\sigma\right)\frac{1}{3}\boldsymbol{I} + \left(\varepsilon_{\mathrm{L}} - \varepsilon_{\mathrm{Q}} - \frac{(1+\nu)}{E}\sigma\right)(\boldsymbol{k}\otimes\boldsymbol{k} - \frac{1}{3}\boldsymbol{I})\\ &+ \frac{1}{2}\left(\gamma - \frac{2(1+\nu)}{E}\tau\right)(\boldsymbol{k}\otimes\boldsymbol{e}_{\varphi} + \boldsymbol{e}_{\varphi}\otimes\boldsymbol{k}),\end{aligned} \tag{12.7}$$

where ν is the Poisson's ratio. The basic assumption related to the multi-axial creep behavior is the volume constancy during the creep deformation, e.g. Odqvist (1974); Odqvist and Hult (1962). In this case the following relations should be satisfied

$$\mathrm{tr}\boldsymbol{\varepsilon} = \mathrm{tr}\boldsymbol{\varepsilon}^{\mathrm{el}} \quad\Rightarrow\quad \varepsilon_{\mathrm{L}} + 2\varepsilon_{\mathrm{Q}} = \frac{1-2\nu}{E}\sigma$$

From (12.7) it follows

$$\boldsymbol{\varepsilon}^{\mathrm{cr}} = \frac{3}{2}\left(\varepsilon_{\mathrm{L}} - \frac{1}{E}\sigma\right)(\boldsymbol{k}\otimes\boldsymbol{k} - \frac{1}{3}\boldsymbol{I}) + \frac{1}{2}\left(\gamma - \frac{2(1+\nu)}{E}\tau\right)(\boldsymbol{k}\otimes\boldsymbol{e}_{\varphi} + \boldsymbol{e}_{\varphi}\otimes\boldsymbol{k})$$

Under the condition of stationary loading the creep rate tensor is

$$\dot{\boldsymbol{\varepsilon}} = \dot{\boldsymbol{\varepsilon}}^{\mathrm{cr}} = \frac{3}{2}\dot{\varepsilon}_{\mathrm{L}}(\boldsymbol{k}\otimes\boldsymbol{k} - \frac{1}{3}\boldsymbol{I}) + \frac{1}{2}\dot{\gamma}(\boldsymbol{k}\otimes\boldsymbol{e}_{\varphi} + \boldsymbol{e}_{\varphi}\otimes\boldsymbol{k}) \tag{12.8}$$

The von Mises equivalent creep rate is defined by

$$\dot{\varepsilon}_{\mathrm{vM}} = \sqrt{\frac{2}{3}\dot{\boldsymbol{\varepsilon}} \cdot\cdot\, \dot{\boldsymbol{\varepsilon}}} = \sqrt{\dot{\varepsilon}_{\mathrm{L}}^2 + \frac{1}{3}\dot{\gamma}^2}$$

The results of creep tests on tubes are usually presented as: strains ε_{L} and γ versus time curves, e.g. Hayhurst and Leckie (1990); Inoue (1988); Kawai (1989), creep strains

$$\varepsilon_{\mathrm{L}}^{\mathrm{cr}} = \varepsilon_{\mathrm{L}} - \frac{\sigma}{E}, \quad \gamma^{\mathrm{cr}} = \gamma - \frac{2(1+\nu)}{E}\tau$$

versus time curves, e.g. Murakami and Sanomura (1985); Penkalla et al. (1988); Ohno et al. (1990), von Mises equivalent creep strain

$$\varepsilon_{\mathrm{vM}}^{\mathrm{cr}} = \sqrt{\frac{2}{3}\boldsymbol{\varepsilon}^{\mathrm{cr}} \cdot\cdot\, \boldsymbol{\varepsilon}^{\mathrm{cr}}} = \sqrt{(\varepsilon_{\mathrm{L}}^{\mathrm{cr}})^2 + \frac{1}{3}(\gamma^{\mathrm{cr}})^2}$$

versus time curves, e.g. Kowalewski (1995, 2001), and the so-called specific dissipation work

$$q(\bar{t}) = \int_0^{\bar{t}} \dot{\boldsymbol{\varepsilon}} \cdot\cdot\, \boldsymbol{s} dt = \int_0^{\bar{t}} (\dot{\varepsilon}_{\mathrm{L}}\sigma + \dot{\gamma}\tau) dt$$

versus time curves (Sosnin 1974; Sosnin et al. 1986).

Figure 12.16 illustrates typical results of creep testing under constant von Mises stress σ_{vM}. Sketches of creep curves are presented for the case of tension under the normal stress $\sigma = \sigma_{\mathrm{vM}}$ and torsion under the shear stress $\tau = \sigma_{\mathrm{vM}}/\sqrt{3}$. For many structural materials the kind of the stress state (e.g. tension or torsion) has negligible influence on the primary and secondary creep behavior. However, this is not the case for the tertiary creep and the long term strength. Tubular specimen subjected to tension usually exhibit much shorter lifetime and lower ductility if compared to the case of pure torsion. This stress state effect has been observed for copper in Kowalewski (1995) and for austenitic steels in Niu et al. (2002); Trivaudey and Delobelle (1993), for example.

Tests under combined tension-torsion loading are useful to formulate and to verify constitutive models. Figure 12.17a shows the plot of the equation $\sigma^2 + 3\tau^2 = \sigma_{\mathrm{vM}}^2 = const$ with respect to coordinates σ and $\sqrt{3}\tau$. Different stress states leading to the same fixed value of the von Mises stress can be conveniently characterized by the angle α (stress state angle). The corresponding values for the normal and the shear stress can be then calculated as follows

$$\sigma = \sigma_{\mathrm{vM}} \cos\alpha, \quad \tau = \sigma_{\mathrm{vM}} \frac{\sin\alpha}{\sqrt{3}}$$

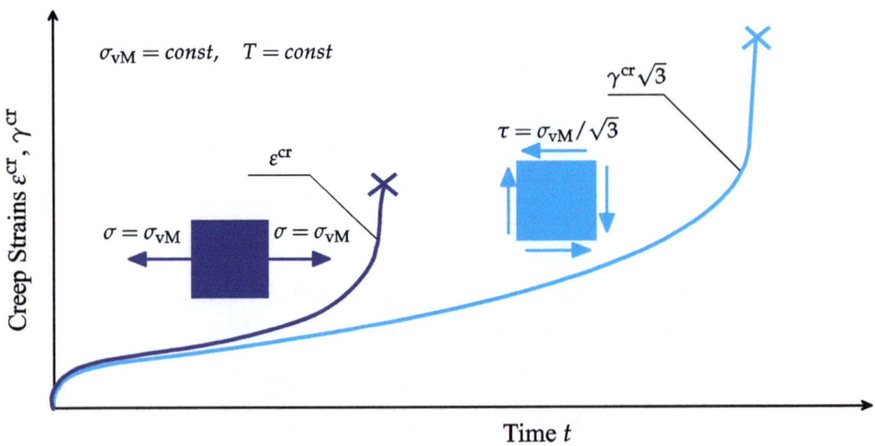

Fig. 12.16 Stress state effect of tertiary creep

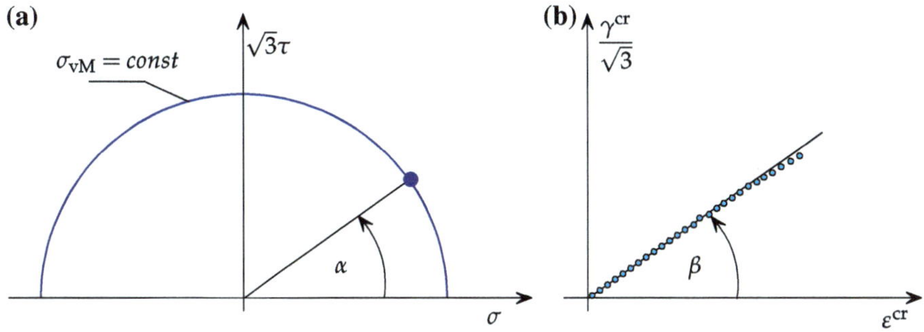

Fig. 12.17 Creep response under combined tension-torsion loading. **a** Plane stress state, **b** strain trajectory

For example, $\alpha = 0$ corresponds to the case of tension and $\alpha = \pi/2$ to the case of torsion. $0 < \alpha < \pi/2$ characterizes the combined loading case. The loading conditions realized in creep tests can be classified as follows

(a) stationary σ_{vM} and α,
(b) time-varying (e.g. stepwise or cyclic) σ_{vM} under fixed α,
(c) time-varying α under fixed σ_{vM} and
(d) both σ_{vM} and α are time-varying.

The loading cases (a) and (b) are called simple or proportional loadings, while the cases (c) and (d) are classified as non-proportional loadings. The results of creep tests under the combined loading can be conveniently presented as $\gamma^{cr}/\sqrt{3}$ versus ε^{cr} curves (so-called strain trajectories), e.g. Murakami and Sanomura (1985); Nikitenko (1984). A sketch of such a curve for the loading case (a) is presented in Fig. 12.17b. For many metals and alloys the direction of the strain trajectory characterized by the angel β, Fig. 12.17b, coincides with the direction of the applied stress state characterized by the angle α. Experimental data are discussed in Murakami and Sanomura (1985); Nikitenko (1984); Oytana et al. (1982). According to this one can assume that the creep rate tensor is coaxial and collinear with the stress deviator, i.e. $\dot{\boldsymbol{\varepsilon}} = \lambda \boldsymbol{s}$. Taking into account (12.6) and (12.8) the following relations can be obtained

$$\frac{3}{2}\dot{\varepsilon}_L = \lambda\sigma, \quad \frac{1}{2}\dot{\gamma} = \lambda\tau \quad \Rightarrow \quad \frac{\dot{\varepsilon}_L}{\dot{\gamma}/\sqrt{3}} = \frac{\sigma}{\sqrt{3}\tau}$$

For many materials, experimental results show that the above relations are well satisfied, e.g. Hayhurst and Leckie (1990); Murakami and Sanomura (1985); Nikitenko (1984); Oytana et al. (1982).

Non-coincidence of the strain-trajectory and the stress state angles indicates the anisotropy of inelastic behavior and/or dependency of the inelastic strain rate on the kind of the stress state. Anisotropic creep may be caused either by the initial anisotropy of the material microstructure as a result of material processing or by the anisotropy induced in the course of inelastic deformation. Examples for tension-torsion creep of initially anisotropic materials are presented for a directionally solidified nickel-based superalloy in Ohno and Takeuchi (1994) and for a fiber-reinforced material in Robinson et al. (2003a, b).

The trajectories of creep strains presented in Kawai (1989) for austenitic steel tubes demonstrate that initial small plastic pre-strain causes the stress state dependence of subsequent creep behavior. Different creep curves under tension, compression, torsion as well as combined tension-torsion and compression-torsion are obtained. This stress state effect of creep is related to the anisotropic hardening induced during the plastic pre-strain. Creep curves illustrating different behavior under tensile and compressive loadings are presented for several alloys in El-Magd and Nicolini (1999); Lucas and Pelloux (1981); Stouffer and Dame (1996); Zolochevskij (1988) and for ceramics in Pintschovius et al. (1989).

The origins of induced anisotropy for polycrystals may be related to hardening due to directed residual stress state in grains, crystallographic texture, formation and growth of cavities and microcracks on grain boundaries. For short fiber reinforced polymers fibers may change the orientation state as a result of creeping flow of the matrix material (Altenbach et al. 2003, 2007). Several methods exist to detect deformation and/or damage induced anisotropy. For example one may interrupt the test after a certain creep exposure, cut specimen in different directions and perform subsequent creep tests to establish the effect of the induced anisotropy. Such tests are discussed in Betten et al. (1995); El-Magd et al. (1996). Another approach is to subject the specimen to the non-proportional loading with varying principal directions, or in other words to rotate the loading with respect to the material without interrupting the test. To illustrate this consider again the stress state (12.5) for combined tension (compression) and torsion. The stress tensor $\boldsymbol{\sigma}$ can also be given in the spectral form as follows

$$\boldsymbol{\sigma} = \sigma_I \boldsymbol{n}_I \otimes \boldsymbol{n}_I + \sigma_{III} \boldsymbol{n}_{III} \otimes \boldsymbol{n}_{III},$$

where

$$\sigma_I = \frac{\sigma + \sqrt{\sigma^2 + \tau^2}}{2}, \quad \sigma_{III} = \frac{\sigma - \sqrt{\sigma^2 + \tau^2}}{2}$$

are principal stress values and the unit vectors

$$\boldsymbol{n}_I = \cos\vartheta \boldsymbol{k} + \sin\vartheta \boldsymbol{e}_\varphi, \quad \boldsymbol{n}_{III} = -\sin\vartheta \boldsymbol{k} + \cos\vartheta \boldsymbol{e}_\varphi, \quad \tan\vartheta = \frac{\tau}{\sigma_I}$$

define the principal directions, where ϑ is the angle between the first principal direction and the axial direction $\boldsymbol{k}$. By changing the values of σ and/or τ during the test, the angle ϑ can be manipulated. Results of creep tests under non-proportional loading with changing principal directions are published by Trampczynski et al. (1981) and by Murakami and Sanomura (1985). Tubular copper specimens were subjected to combined tension and torsion with varying loading amplitudes of σ and τ, correspondingly, Fig. 12.18a. The reversal of torque at time t_1 leads to the change of the angle θ characterizing the first principal direction of the stress tensor. As the absolute value of the shear stress remains the same, the value of the von Mises equivalent stress is not affected by the reversal. Therefore, one expects no changes in creep responses if the material is isotropic. In Murakami and Sanomura (1985) creep responses after shear stress reversals leading to different values of the angle ϑ are systematically analyzed. A sketch of a typical creep response is presented in Fig. 12.18b. Normal and shear creep strain versus time curves for stationary stress values as well as corresponding creep curves after the reversal of shear stress are given. As a result of shear stress reversal two effects can be recognized. After the shear stress reversal both the rate of the normal strain and the rate of the shear strain are significantly affected. This indicates on the anisotropic nature of hardening, induced during the first loading phase before the reversal. Furthermore, the creep life of the specimen after the shear stress reversal is significantly longer than the corresponding life under the stationary loading. This effect can be explained by the anisotropic damage evolution. The principal creep damage mechanism in copper is the nucleation, growth and coalescence of cavities at grain boundaries. The cavitation is found to take place mainly on those grain boundaries, which are orthogonal to the first principal direction of the stress tensor. After the shear stress reversal, former cavities stop to grow and new cavities nucleate and grow on new grain boundaries closely orthogonal to the rotated first principal direction. Cavitated grain boundaries before and after the stress reversal are shown schematically in Fig. 12.18b.

Examples discussed in this section are limited to force (torque) controlled tests. Engineering structures may be subjected to varying external loadings and thermal environment resulting in local non-proportional changes in stress and strain states. Stress and/or strain controlled tests on cruciform specimen with various bi-axial loading and temperature profiles are presented in (Zhang et al. 2007; Berger et al. 2008; Wang et al. 2014; Cui et al. 2013).

12.3 High-Temperature Inelasticity in Structures

Analysis of the structural behavior at high temperature is crucial for understanding time-dependent changes in stress and strain states as a result of constant or variable external loading. Local increase of strains, relaxation and redistribution of stresses are examples for such changes. For adequate experimental analysis of the material behavior a specimen should be subjected to realistic stress, strain and temperature

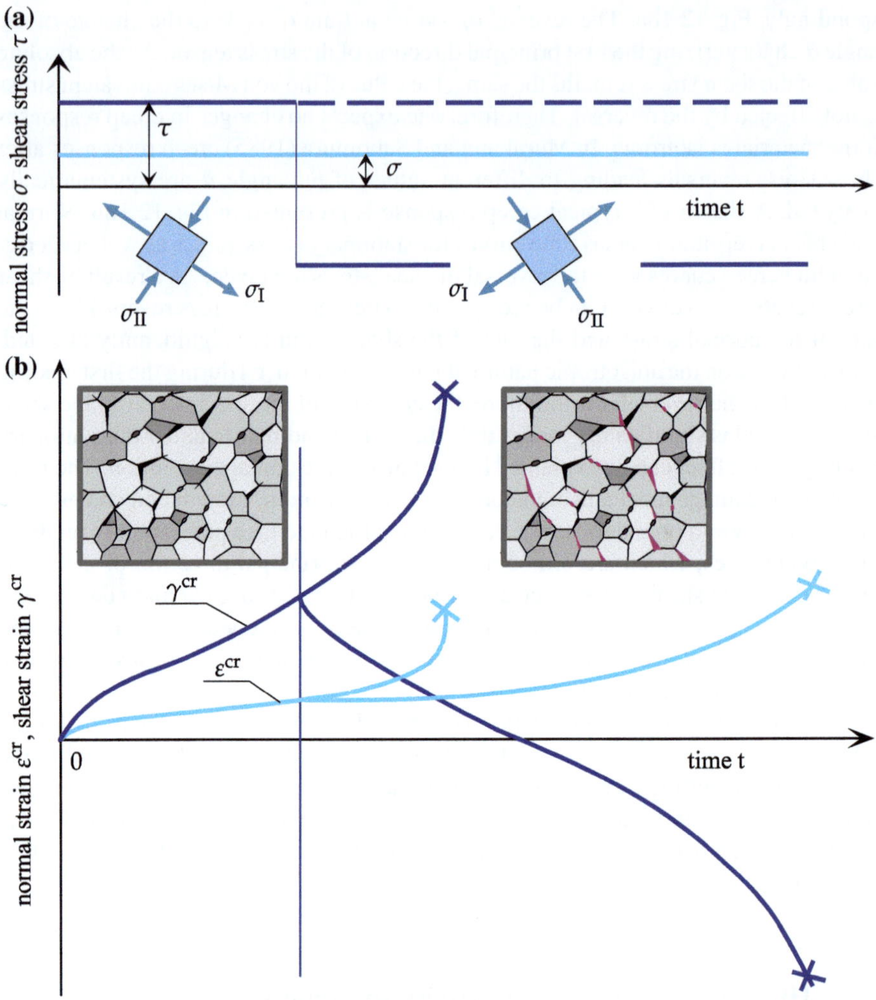

Fig. 12.18 Loading history and creep responses under combined tension and torsion (after Murakami and Sanomura (1985))

profiles. Such local profiles can be generated by heat transfer and structural analysis of a component. Examples of high-temperature applications are structural components of power plants, chemical refineries or heat engines, e.g. Gooch (2003). Design of pipework systems, rotors, turbine blades, etc. requires the consideration of inelastic processes. Inelasticity may cause excessive deformations, damage, buckling, crack initiation and growth.

Different types of failures are documented in the literature. Examples of critical structural members include pipe bends (May et al. 1994), welds (Shibli 2002), turbine blade root fixings (Gooch 2003), etc. The possibilities to analyze a structural prototype in the laboratory are limited by the long duration of tests and related costs. Furthermore, examinations of deformation and damage states in a structure during the service (e.g. replicas) can be only made at specific outer surface positions and after certain periods of time. The modeling of inelastic processes in structures is therefore an essential contribution to optimal design and residual life assessment. Furthermore it contributes to understanding and analysis of time-dependent deformations, stress redistributions and damage growth under given temperature and loading conditions. The aim of this section is to present simulation results for several structures.

12.3.1 Steam Transfer Line

An example for a steam transfer line between a header and a desuperheater of a boiler is presented in May et al. (1994). The pipeline from steel 1Cr0.5Mo (13CrMo4-5) had operated under the temperature in the range 500–550 °C and the internal pressure 11.8 MPa. After a service life of 77,000 h rupture occurred along the outer radius of a pipe bend. Metallographic analysis of a section cut from the bend close to the main crack has shown typical creep damage due to microvoids and microcracks on grain boundaries, Fig. 12.19.

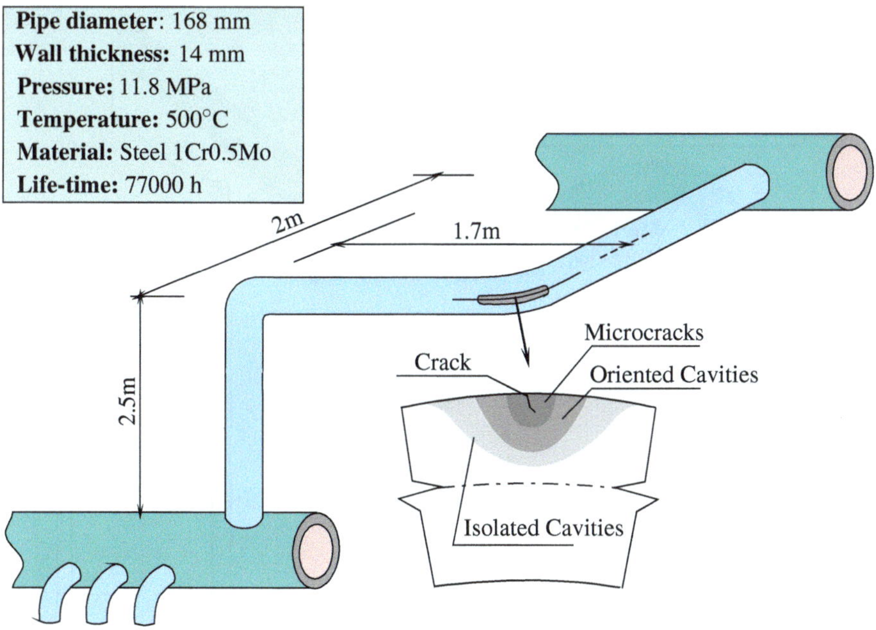

Fig. 12.19 Creep failure in a steam transfer line (after May et al. (1994))

Several incidents of pipe bend failures in different power plants are reported in Hald (1998). Inspection techniques were developed to examine the state of creep damage during the service. However, as noted in May et al. (1994) any inspection must be conducted at exactly the critical position, or the presence of damage may not be detected. Many investigations have addressed the analysis of mechanical behavior of pressurized curved tubes. Results of studies on elastic and elasto-plastic deformation and stability are reviewed in Bielski and Skrzypek (1989); Libai and Simmonds (1998). Creep and damage processes in curved tubes were discussed in Altenbach et al. (2001); Boyle and Spence (1983); Hyde et al. (2002). These studies were concerned with the analysis of a single pipe bend subjected to special loading conditions, i.e. in-plane bending moments and internal pressure. In the following example we analyze the behavior of pipe bends in a real spatial pipeline. Figure 12.20 shows the reference geometry of the structure which includes three straight pipe segments (I, III and V) and two pipe bends (II and IV). The lengths of the pipe segments, the mean diameter of the cross section and the wall thickness correspond to the data given in May et al. (1994). In addition, we take into account the non-uniformity of the wall thickness in the pipe bends as a result of processing by induction bending. The flanges of the pipeline are clamped. The internal pressure $p = 11.8\,\mathrm{MPa}$ and the temperature $T = 550\,^{\circ}\mathrm{C}$ are assumed to be constant. The following constitutive model for the creep-damage process is applied

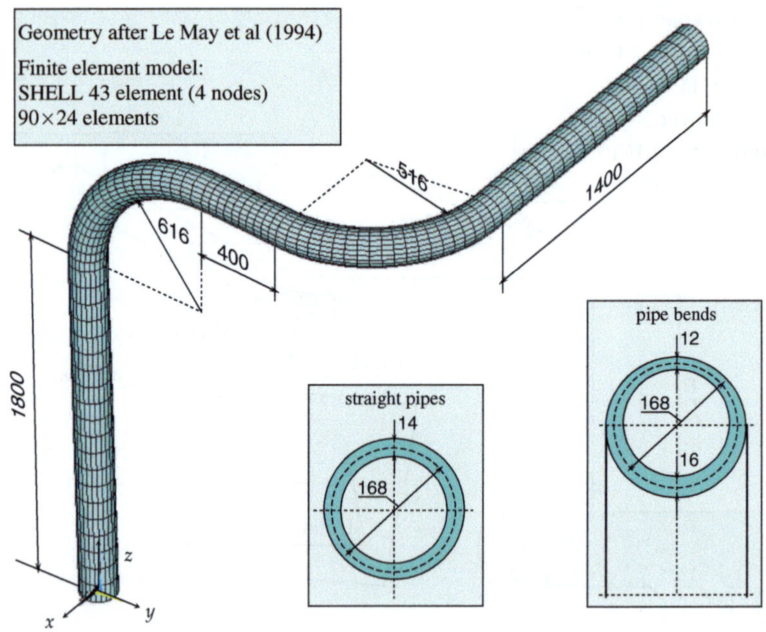

Fig. 12.20 *Pipeline* geometry and finite element mesh

$$\begin{aligned} &\dot{\boldsymbol{\varepsilon}}^{\mathrm{cr}} = \frac{3}{2} f_1(\sigma_{\mathrm{vM}}) g_1(\omega) \frac{\boldsymbol{s}}{\sigma_{\mathrm{vM}}}, \quad \dot{\omega} = f_2(\sigma_{\mathrm{eq}}) g_2(\omega), \\ &\boldsymbol{\varepsilon}^{\mathrm{cr}}|_{t=0} = \boldsymbol{0}, \quad \omega|_{t=0} = 0, \quad 0 \le \omega \le \omega_*, \\ &\boldsymbol{s} = \boldsymbol{\sigma} - \frac{1}{3} \mathrm{tr} \boldsymbol{\sigma} \boldsymbol{I}, \quad \sigma_{\mathrm{vM}} = \sqrt{\frac{3}{2} \boldsymbol{s} \cdot\cdot \boldsymbol{s}} \end{aligned} \tag{12.9}$$

Here $\boldsymbol{\varepsilon}^{\mathrm{cr}}$ is the creep strain tensor, $\boldsymbol{\sigma}$ is the stress tensor, ω is the scalar valued damage parameter and σ_{eq}^{ω} is the damage equivalent stress. The response functions f_1, f_2, g_1, and g_2 as well as the material constants are taken from Segle et al. (1996) for steel 1Cr0.5Mo at 550 °C as follows

$$\begin{aligned} &f_1(\sigma) = a\sigma^n, \quad g_1(\omega) = 1 - \rho + \rho(1-\omega)^{-n}, \\ &f_2(\sigma) = b\sigma^k, \quad g_2(\omega) = (1-\omega)^{-l} \\ &a = 1.94 \times 10^{-15}\,\mathrm{MPa}^{-n}/\mathrm{h}, \quad b = 3.302 \times 10^{-13}\,\mathrm{MPa}^{-k}/\mathrm{h}, \\ &n = 4.354, \quad k = 3.955, \quad l = 1.423, \quad \rho = 0.393, \quad \omega_* = 0.74 \end{aligned} \tag{12.10}$$

The damage equivalent stress is assumed in the form

$$\sigma_{\mathrm{eq}} = \alpha \frac{\sigma_I + |\sigma_I|}{2} + (1-\alpha)\sigma_{\mathrm{vM}},$$

where σ_I is the first principal stress and $\alpha = 0.43$. The elastic material constants are $E = 1.6 \times 10^5$ MPa and $\nu = 0.3$. Figure 12.21 illustrates the deformed shape and the distribution of the magnitude of the displacement vector in the reference state.

Figure 12.22a shows the distribution of the von Mises equivalent stress in the reference state. From the results we may conclude that both the pipe bends are subjected to complex spatial loading and deformation conditions as a result of internal pressure and uniform heating. In addition, the values of the von Mises equivalent stress in three points of the pipe bend IV are plotted as functions of time. According to the results the creep process of the pipeline may be divided into three stages. During the first stage (approximately 50 % of the total life) significant stress redistributions occur leading to quite different stress state in the pipeline (cp. Figs. 12.22 and 12.23). The second stage (approximately 45 % of the total life) is characterized by slow changes in the stress state. In the final stage (approximately 5 % of the total life) we observe additional stress redistributions, Fig. 12.23b. The distribution of damage parameter at the final time step is shown in Fig. 12.24. According to the results the critical position of possible creep failure is the point A of the pipe bend IV. This result agrees well with the data presented in May et al. (1994), where the creep failure has been detected at the same position. The processes of time dependent deformation, stress relaxation and redistribution have been illustrated in the literature based on different examples for beams, plates and shells (Altenbach et al. 1997; Altenbach and Naumenko 1997; Altenbach et al. 2000b, 2001, 2002; Altenbach and Naumenko 2002; Altenbach et al. 2004). One feature of the example considered here is that the final creep stage is not only the result of the local material deterioration but is additionally governed by the flattening (ovalisation) of the pipe bend cross section.

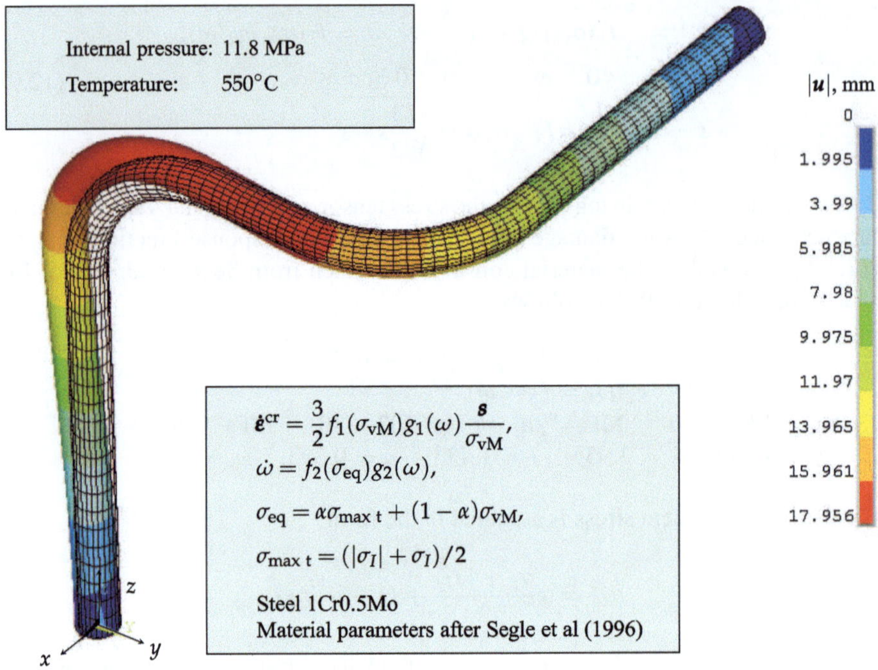

Fig. 12.21 Deformed shape and magnitude of the displacement vector in the reference elastic state

Let us note that some parameters of the reference pipe bend geometries were not given in May et al. (1994) and have been assumed in the presented calculation. Many additional details of geometry including the initial out of roundness of the cross section, inhomogeneous material properties as a result of processing, shutdowns and startups during the service, are not included in the presented model. Furthermore, the utilized material model (12.9) does not take into account primary creep. Tertiary creep is described by the single damage parameter and the corresponding kinetic equation does not distinguish between processes leading to the accelerated creep, for example coarsening of precipitates and cavitation. Therefore the obtained numerical result for the failure time (49,000 h) "slightly" differs from the value 77,000 h given

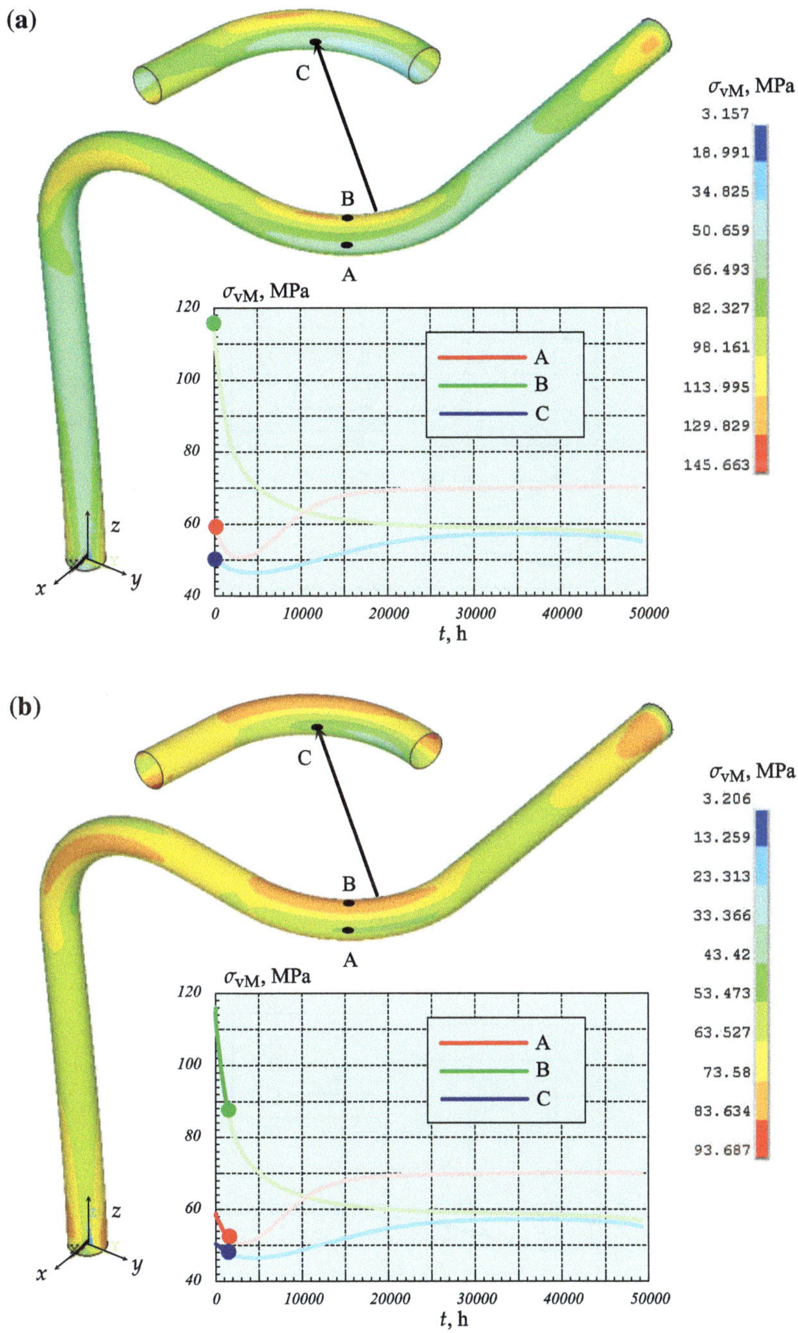

Fig. 12.22 Distribution of the von Mises equivalent stress and corresponding time variations in three points of the pipe bend. **a** Reference elastic state, **b** $t = 2000\,\text{h}$

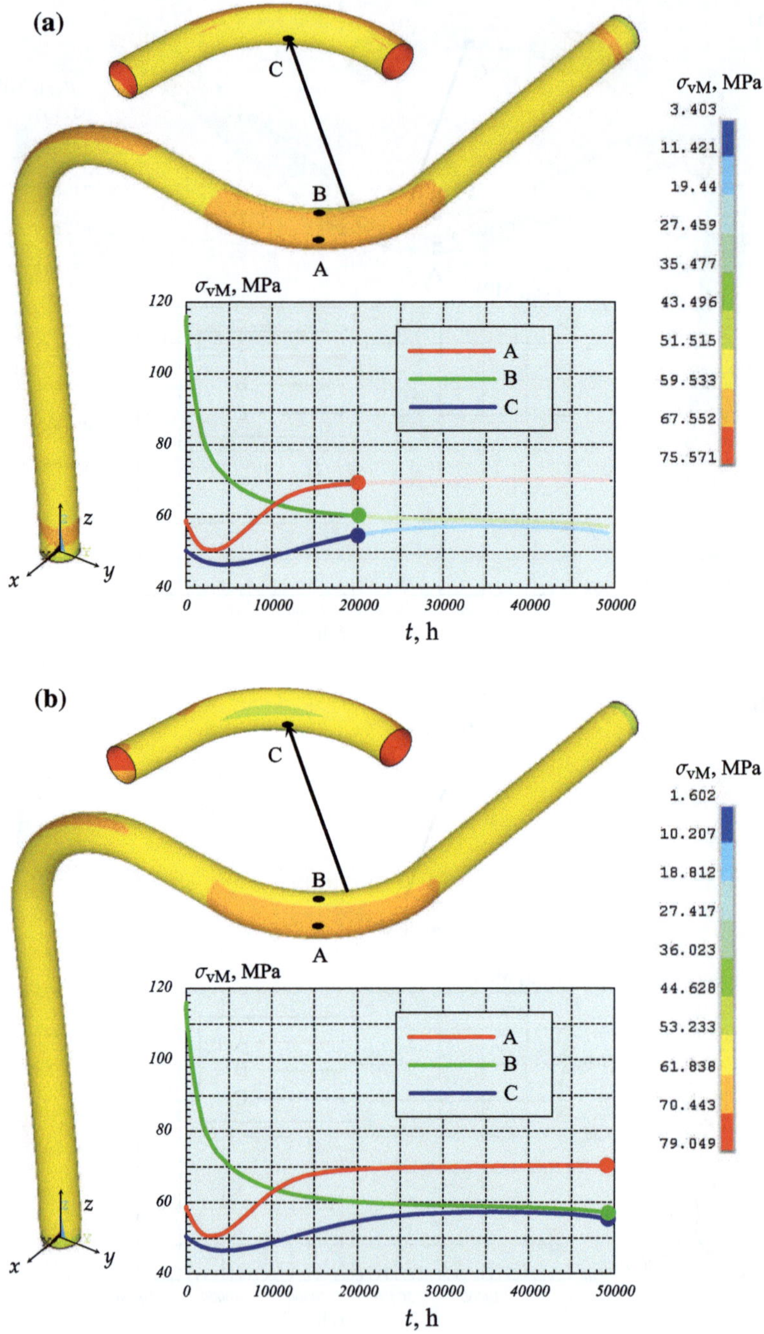

Fig. 12.23 Distribution of the von Mises equivalent stress and corresponding time variations in three points of the pipe bend. **a** $t = 20{,}000\,$h, **b** last time step

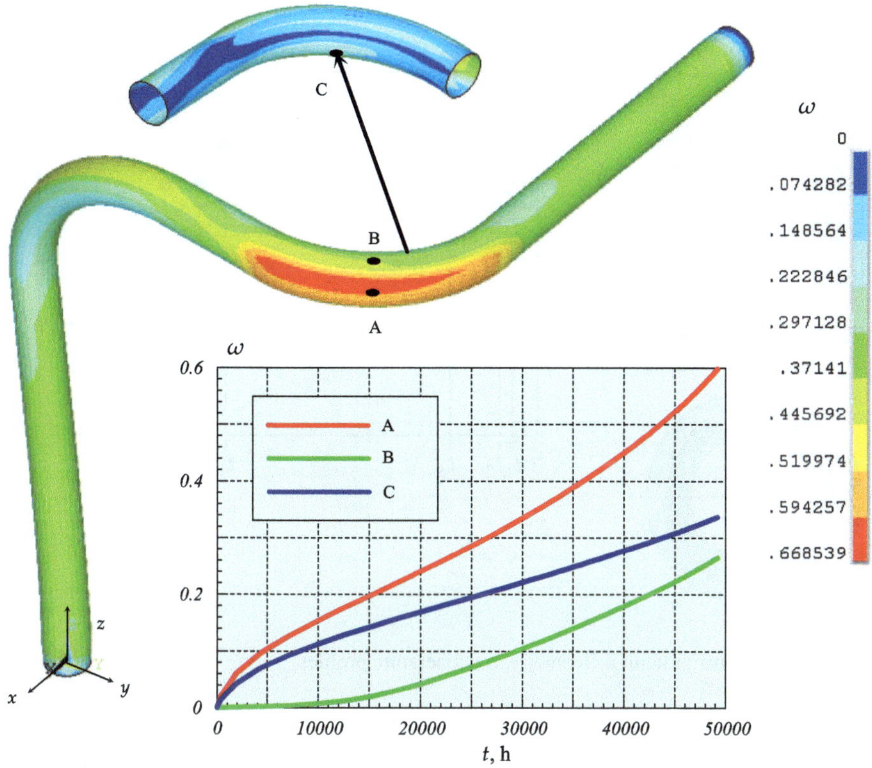

Fig. 12.24 Distribution of the damage parameter at the last time step and corresponding time variations

in May et al. (1994). Nevertheless, the results demonstrate the ability of the modeling to represent basic features of the creep process in a structure and to predict critical zones of possible creep failure.

12.3.2 Two-Bar System Under Thermo-Mechanical Loading

Engineering structures are often subjected to non-stationary and non-uniform thermal environment, for example during start-up and shut-down phases. The aim of this section is to present examples illustrating local changes in stress and strain states as a result of thermo-mechanical cycles.

To discuss basic features of the thermo-mechanical loading let us consider two pipes, rigidly connected as shown in Fig. 12.25a. For the sake of simplicity assume that the diameter of each pipe is much less than the length, such that the stress state is uni-axial and a two-bar model can be used to analyze the structural behavior.

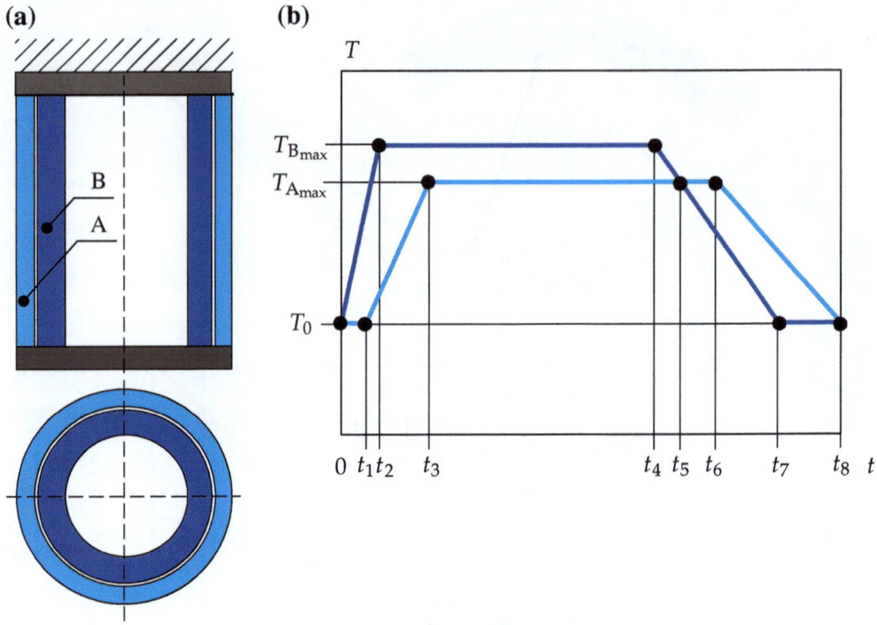

Fig. 12.25 Two bar system. **a** Geometry, **b** temperature profiles

Furthermore, assume that the pipes have the same cross section area and are made from the same material. First assume that the pipes are uniformly heated such that $\Delta T_A = \Delta T_B$, where $\Delta T_A = T_A - T_0$, $\Delta T_B = T_B - T_0$. T_A and T_B are absolute temperatures in the actual state and T_0 is the reference temperature. As a result of such a uniform heating the two-bar system will freely expand without stresses since the materials and cross section areas are the same. Now assume that the temperatures of the pipes are different such that $T_B > T_A$. Such a temperature difference may arise during the heat-up stage as a result of hot steam flow through the pipe B. The non-uniform temperature state will cause the non-uniform stress state in the system such that the pipe B will be subjected to compression while the pipe A to tension. To illustrate this remove the rigid connection as shown in Fig. 12.26b. Due to the assumed difference in temperatures, the pipe B is longer than the pipe A in the actual state as a result of thermal expansion.

To keep the elongation of the pipes the same, or in other words, to provide the strain compatibility of the structure the forces must be applied, as shown in Fig. 12.26c—the compressive force to the pipe B and tensile force to the bar A. This elementary example explains why the heated surface of a component is usually subjected to compressive stress state.

To illustrate changes in a stress state during a thermal cycle, consider an idealized temperature profile as shown in Fig. 12.25b. Assume that the temperature of the pipe B rapidly increases from T_0 up to $T_{B_{max}}$ over a time interval $0 - t_2$. As the heat flows towards the pipe A the temperature T_A increases with a delay over the

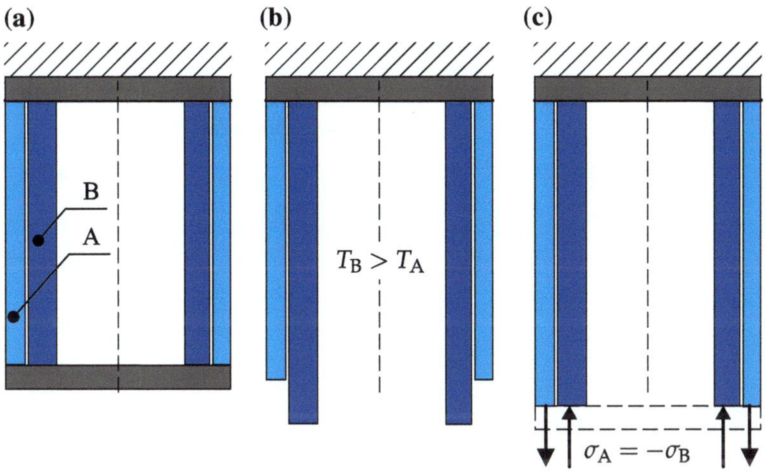

Fig. 12.26 Stress state in a two bar system during heat-up stage. **a** Reference state, **b** free thermal expansion during heat-up with $T_B > T_A$, **c** stresses (compressive for a and tensile for b) required to keep elongations the same

time interval $t_1 - t_2$. The greatest difference in temperatures is observed at the time point t_2. This is sometimes called "upshock" (Skelton 2003), the pipe B is subjected to the compression with the maximum stress magnitude. During the time interval $t_2 - t_3$ the temperature difference between the pipes decreases while the absolute temperature of the pipe A increases. The time interval $t_3 - t_4$ is the steady operation period. The time interval $t_4 - t_7$ is the cool-down stage, the steam temperature and the temperature of the pipe B decrease. During the time interval $t_5 - t_7$ the temperature difference increases again. However, the temperature of the pipe B is now lower that the temperature of the pipe A. During this "downshock" stage the pipe B is subjected to tension with the maximum stress value at the time point t_7.

Let us analyze stress and strain states in the pipes during the whole thermal cycle. For the sake of brevity assume that the Young's modulus E and the thermal expansion coefficient α_T are constant within the temperature interval $T_0 - T_{B_{max}}$. In general, the material properties are functions of the absolute temperature. However, the temperature dependence of E and α_T is usually much weaker in comparison to the temperature dependence of the inelastic strain rates. Let N_A and N_B be the internal forces in the bars. The obvious equilibrium condition for the forces yields $N_A = -N_B$. Since the cross section areas are assumed the same the relation between the stresses is $\sigma_A = -\sigma_B$. The constitutive equations for the bars can be formulated as follows

$$\varepsilon_A = \frac{\sigma_A}{E} + \varepsilon_A^{pl} + \alpha_T \Delta T_A, \quad \varepsilon_B = \frac{\sigma_B}{E} + \varepsilon_B^{pl} + \alpha_T \Delta T_B, \tag{12.11}$$

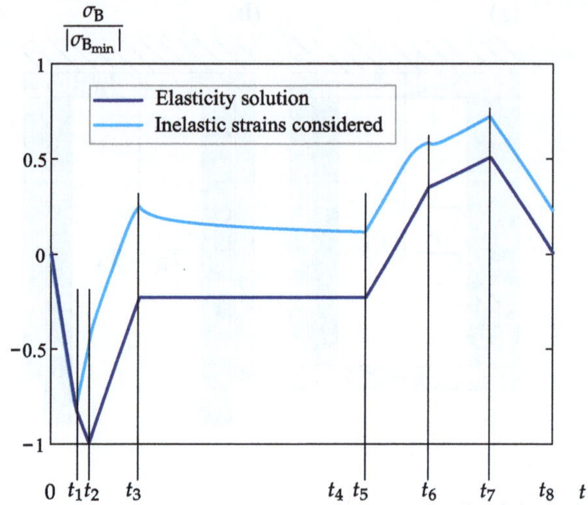

Fig. 12.27 Normalized stress in the pipe B versus time

where ε_A^{pl} and ε_B^{pl} are inelastic strain components in the bars A and B, respectively. The compatibility condition is $\varepsilon_A = \varepsilon_B = \varepsilon$. Subtracting Eq. $(12.11)_2$ from $(12.11)_1$ provides the stress in the bar B

$$\sigma_B = \frac{E}{2}(\varepsilon_A^{pl} - \varepsilon_B^{pl}) + \frac{E\alpha_T}{2}(T_A - T_B) \tag{12.12}$$

If the inelastic strains are negligible, then the stress in the bar B is related to the difference in absolute temperatures between the bars A and B. As a result the bar B is subjected to compression during warm-up stage and tension during the cool-down stage. After the adding Eq. (12.11) the mean strain of the bar system can be computed

$$\varepsilon = \frac{1}{2}(\varepsilon_A^{pl} + \varepsilon_B^{pl}) + \frac{\alpha_T}{2}(\Delta T_A + \Delta T_B) \tag{12.13}$$

Figure 12.27 shows the stress in the pipe B as a function of time. The results are normalized with the minimum stress value during the warm-up stage computed as follows

$$\sigma_{B_{min}} = \frac{E\alpha_T}{2}\left[T_A(t_2) - T_B(t_2)\right]$$

Two stress peaks are observed—the compressive one during the warm up stage with the maximum temperature difference $T_B - T_A$ at the time point t_2 and the tensile peak during the cool-down stage with the minimum temperature difference at the time point t_7. To analyze the inelastic response a constitutive model is required that is able to reflect the material behavior during the relatively fast start-up and shut-down stages and the slow regime during the stationary operation. Unified models of inelastic deformation were developed and utilized to capture both plasticity and creep deformation mechanisms under both rapid and slow loadings at high temperatures Chaboche (2008). A challenging problem is to describe complex interaction of damage processes including the cyclic fatigue damage accumulation and slow creep damage evolution. To simplify the analysis of structures it is suggested to distinguish between different thermo-mechanical loading profiles such as cold starts, warm starts and hot starts (Berger et al. 2008). Figure 12.27 illustrates qualitatively the stress variation for a hot start loading cycle. During the time period t_1–t_2 the stress distributions based on elasticity solution and the inelastic response coincide. With the increase of absolute temperature and in the stress magnitude compressive inelastic strain starts to accumulate. This leads to the decrease in the stress level for the inelastic solution and the residual tensile stress at the beginning of the stationary stage. This tensile stress relaxes down as a result of creep processes during the stationary operation period. During the cool-down stage an additional tensile inelastic strain accumulation has to be taken into account.

Variations in stress and strain states presented in Sect. 12.3.2 for a two-bar system are typical for many components subjected to variable thermal environment. Examples include turbine rotors (Naumenko et al. 2011b; Kostenko et al. 2013; Holdsworth et al. 2007; Colombo et al. 2008), turbocharger casings (Laengler et al. 2010; Nagode et al. 2011; Längler et al. 2014) and many other structures.

12.4 Microstructural Features and Length Scale Effects

Macroscopic inelastic properties of materials are strongly affected by microstructure. Limiting to the analysis of high temperature inelastic behavior and crystalline materials let us discuss basic microstructural features. For polycrystals the mean grain size is found to influence the macroscopic response. For low homologous temperatures and for a certain range of the mean grain size, the grain boundary strengthening (Hall-Petch) effect (Hall 1951; Petch 1953) is usually observed. The lattice structure of adjacent grains differs in orientation. An additional energy is required to change directions of moving dislocations. Impeding the dislocation movement through grain boundaries will hinder the onset of plasticity and hence increase the yield strength of the material. The decrease of mean grain size by an appropriate material processing leads to an increase in the yield strength. Based on the experimental data the following relation can be established see, for example, Roesler et al. (2007)

$$\sigma_{\mathrm{y}} = \sigma_0 + \frac{k}{\sqrt{d}},$$

where σ_0 and k are constants and d is the mean grain size.

In contrary, several mechanisms may operate leading to the weakening effect of grain boundaries, if the material is loaded at elevated temperature. For moderate stress levels and temperatures over 0.5 T_{m} diffusion of vacancies may control the deformation process. Different theories of diffusion creep provide the following relationship between the inelastic strain rate and the mean grain size

$$\dot{\varepsilon}^{\text{pl}} \sim \frac{1}{d^k}$$

The exponent k takes the value 2 according to the theory of lattice diffusion as discussed by Nabarro (1948, 2002) and Herring (1950) and 3 according to the theory of grain boundary diffusion as proposed by Coble (1963). An additional mechanism leading to the weakening effect of grain boundaries is the grain boundary sliding. Experimental works devoted to the analysis of grain boundary sliding are reviewed in Nabarro and Villiers (1995); Nørbygaard (2002); Langdon (2006). An elementary model of creep considering both the deformation of grains and sliding of grain boundaries can be based on the mixture rule (Illschner 1973). Here the total strain is a sum of strains due to grain interiors and grain boundaries weighted by the corresponding volume fractions. Materials with smaller grains exhibit higher volume fractions of grain boundaries and consequently have more essential contribution of grain boundary sliding to the overall deformation. Again the grain boundary sliding is one of the mechanisms leading to the weakening effect of grain boundaries. To establish the influence of mean grain size on the overall creep rate several additional effects should be considered.

Stress concentrations in the regions of grain intersections lead to the non-uniform creep deformation along grain boundaries. Furthermore, the effect of grain boundary deformation depends essentially on the stress level. For lower stress values grain boundary sliding and diffusion of vacancies have an essential influence on the deformation while for high stresses levels grain the deformation is primarily controlled by glide processes inside the grains. For many polycrystalline materials the dominant creep damage mechanism is the formation, growth and coalescence of cavities on grain boundaries. As an example, Fig. 12.28 shows a micrograph of copper speci-

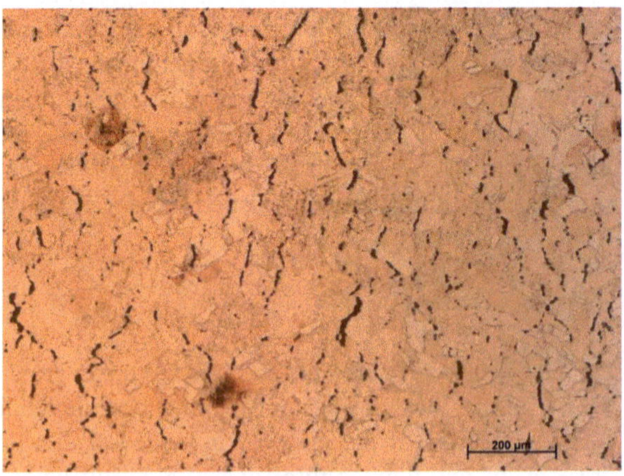

Fig. 12.28 Micrograph of copper specimen tested 20 h under constant stress of 10 MPa (stress direction *horizontal*) and temperature level of 550 °C (Ozhoga-Maslovskaja 2014)

men subjected to uni-axial load at high temperature. Cavities are formed on grain boundaries orthogonal to the stress axis. This kind of damage leads again to the weakening effect of grain boundaries. Therefore, one may expect that a material with finer grains would exhibit a shorter lifetime. Summing up, the grain boundaries may have hardening and/or weakening influence on the inelastic deformation and lifetime depending on the loading and temperature levels. Therefore the optimal grain size for a given material depends on the conditions under which it is to be used (Nabarro and Villiers 1995). The overlapping of several deformation and damage mechanisms makes it difficult to derive a relationship between the deformation rate and the grain size. Recently simulations of three-dimensional polycrystalline aggregates were performed illustrating several deformation and damage mechanisms like grain boundary sliding and cavitation as they influence the overall behavior (Ozhoga-Maslovskaja et al. 2015).

Apart from grains there are several microstructural features at lower length scales that affect inelastic deformation. Examples include cells and subgrains, that are dislocation substructures formed as a result of clustering of uniformly distributed dislocations (Raj et al. 1996). Cells consist of broad diffused walls containing dislocation tangles while boundaries of subgrains are narrow and formed by dislocation networks. Cell walls and subgrain boundaries separate the crystal into the regions with slightly different crystallographic orientations. Subgrain boundaries have larger misorientation than cell walls. Cells and subgrains may form in the course of inelastic deformation process for materials with relatively low dislocation density. At the macroscale the formation of substructures is observed as hardening, for example the decrease in the creep rate during the primary creep. Many high temperature materials, for example high-chromium steels possess fine subgrain structure. Lower mean subgrain size provides lower inelastic strain rate and higher creep resistance. To describe this mean subgrain size effect composite (phase mixture) model can be applied. Here a composite with two constituents having different initial dislocation densities and as a result different inelastic properties is assumed (Raj et al. 1996; Blum 2001, 2008). The inelastic soft constituent is the zone of cell or subgrain interiors with relatively low dislocation density. The inelastic hard regions include cell or subgrain boundaries with high dislocation density. The deformation process in such a composite is accompanied by a stress redistribution between the constituents - the stress level in the inelastic hard regions increases while the stress level within the inelastic soft regions increases. Lower size of subgrains leads to higher volume fraction of the hard regions and as a result to the lover overall deformation rate. The subgrain microstructure is not stable and subgrain coarsening usually takes place where the average size increases while the number of subgrains decreases. This leads to the overall softening under stationary or cyclic loadings (Fournier et al. 2011; Kimura et al. 2006; Qin et al. 2003).

Additional strengthening mechanisms for high temperature materials are precipitation and dispersion hardening. For example heat-resistant steels usually contain several kinds of precipitate particles in the matrix and at grain boundaries including carbonitrides and intermetallic compounds (Abe 2008). An important role in strengthening of aluminium alloys for high temperature applications plays the θ'

phase, in the form of rod-like precipitates aligned along the <001> crystallographic directions of the α-Al matrix (Naumenko and Gariboldi 2014). Dispersed fine precipitates are obstacles for mobile dislocations. Several mechanisms for interaction between mobile dislocations and particles are reviewed in Illschner (1973); Kassner and Pérez-Prado (2004); Roesler et al. (2007) among others. Both theoretical and experimental results show that the inelastic strain rate primarily depends on the mean spacing between particles. The mean particle spacing can also be related to the mean particle size. As an example let us consider regular arrangement of spheres with the diameter d, spacing between centers l, the volume fraction η_{p} and the number of particles N_{p} in a unit volume. In this case the following relation can be derived

$$l = \frac{1}{N_{\mathrm{p}}^{1/3}} = \frac{\pi}{6} \frac{d}{\eta_{\mathrm{p}}^{1/3}},$$

Experimental data suggest that strength of alloys is determined by spacing and diameter of precipitates. The greatest impedance to dislocation motion and hence the maximum potential for strengthening will occur when an alloy contains precipitates that are large enough to resist shearing by dislocations and too finely spaced to be bypassed by moving dislocations (Polmear 2004). Furthermore precipitates stabilize dislocations in the matrix and subgrain boundaries. This enhances strain hardening, for example primary creep. To capture these phenomena kinetic equations for dislocation density are developed, see for example Estrin (1996); Naumenko and Gariboldi (2014). One feature of the proposed constitutive and evolution equations is the dependence of the inelastic strain rate as well as rate of change of hardening variables on size parameters, such as mean spacing (or mean diameter) of precipitates (Estrin 1996; Roesler et al. 2007). At high temperature the microstructure of precipitates is not stable and evolves over time as a result of diffusion processes. Coarsening of precipitates in steels (Abe 2008) and aluminum alloys (Naumenko and Gariboldi 2014) leads to loss of the strength. To account for coarsening processes equations with respect to the size parameter—the mean particle diameter are required. Examples are presented by Kowalewski et al. (1994); Naumenko and Gariboldi (2014) for aluminum alloys and Dyson and McLean (2001); Blum (2008) for steels. Figure 12.29 provides a summary of several microstructural features discussed in this Section, where size parameters, length scales and associated deformation or damage mechanisms are presented.

Microstructural features and microstructural size parameters may have different kinds of influences on the behavior of components at high temperature. For large structural components, for example power plant components, the local inelastic behavior is usually determined by the local stress state and the current state of microstructure defined in terms of mean quantities like dislocation density, mean diameter of precipitates, damage parameter etc. Such a description is possible since a material point in the sense of continuum mechanics contains a huge number of microstructural elements, for example grains. An exception for large structural components is the behavior in the zones of stress concentrations, for example in the vicinity of notches. Here the macroscopic quantities like stress or strain may vary rapidly

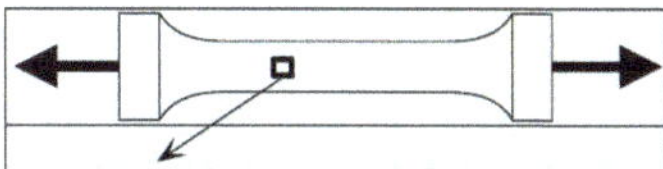	Size parameter	Mechanism	Length scales
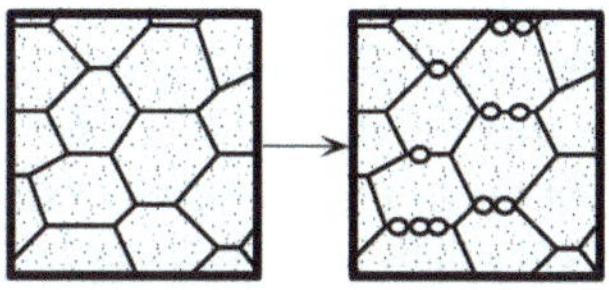	grain size	grain boundary sliding, diffusion, creep cavities	1 μm – 0.1 mm
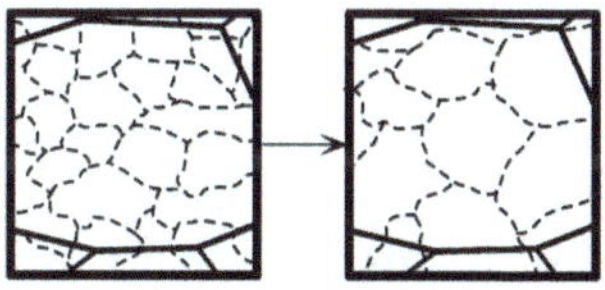	cell size, subgrain size	interaction of, moving dislocations with subgrains, substructure evolution	0.1 μm – 1 μm
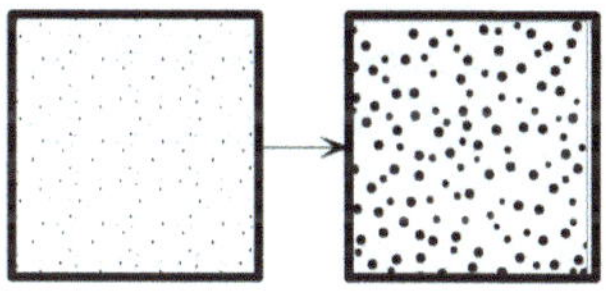	particle spacing	interaction of, moving dislocations with particles, coarsening of particles	10 nm – 0.1 μm

Fig. 12.29 Basic microstructure features of heat resistant materials, length scales, size parameters and mechanisms of changes in microstructure

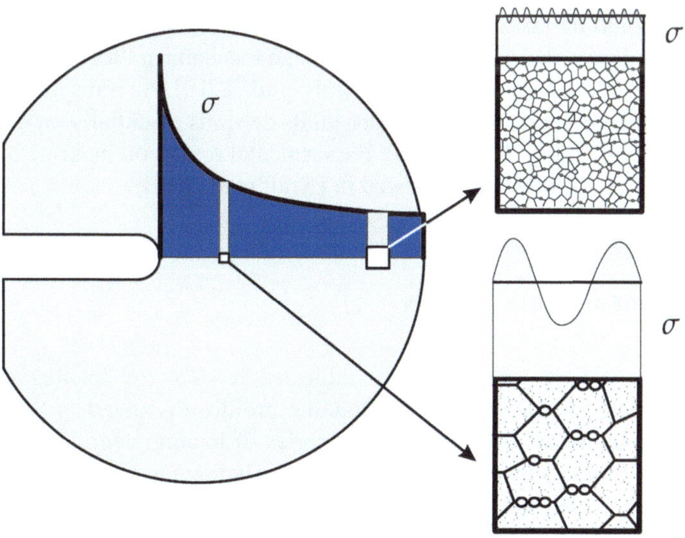

Fig. 12.30 Qualitative stress distribution in the vicinity of the notch root. To identify the behavior in zones with higher stress gradients smaller material samples are required

over a certain direction and the characteristic length of this change may be comparable with microstructural size quantities, for example the grain size. Figure 12.30 shows a sketch of a stress variation in the vicinity of the notch root.

Far from the notch root, where stresses vary slowly a sample with a large number of microstructural features, for example grains can be used to identify the material response. In zones with higher stress gradients smaller samples are required to analyze the material behavior. For smaller samples microstructure features must be taken into account to capture the material behavior. Such a smaller volume element with microstructure, if subjected to a uniform elongation on the boundary would exhibit highly non-uniform stress distribution such that higher moments of the stress tensor might be required to capture the averaged stress state. Furthermore the local response in a material point at the macroscopic scale may be affected not only by the local stress state and the local state of the microstructure but also by the neighborhood.

In general, the coupling between different size dependencies that determines the properties of a material should be analyzed (Arzt 1998). To this end one has to deal with the interaction of the following two length scales. The first one is responsible for the phenomenon involved, for example variation of the stress state in the vicinity of a notch, and is usually called the characteristic length. The second one is a microstructural dimension, denoted as the size parameter.

Analysis of different length scale effects ant their interactions is required in particular for small structures, such as components of microelectronics, thin films and coatings. Miniaturized components are frequently subjected to complex thermal and mechanical loading cycles. The dimension of such a part, for example, the diameter of a bondwire may be of several micrometers such that grain size and grain boundary effects should be taken into account in analysis of inelastic behavior. For thin films and thin layers the deformation or damage mechanism "feel" the presence of the surface or an interface (Arzt 1998; Kraft et al. 2010). As a result, the inelastic response of microcomponents and solder joints depends essentially on their dimensions (Wiese et al. 2008; Wiese 2010). Experimental results on inelastic behavior of thin films and micropillars are reviewed in (Kraft et al. 2010).

12.5 Temporal Scale Effects

Engineering structures are frequently subjected to complex loading conditions. Examples include thermo-mechanical loading profiles discussed in Sect. 12.2.1.4. Analysis of material behavior over many cycles of loading is crucial for life time estimations of components. Phenomena like cyclic hardening, cyclic softening, creep ratcheting, fatigue damage evolution etc. are usually observable with respect to the global time scale after a certain number of loading cycles. On the other hand, the type of loading and the response within one loading cycle can be related to the local time scale. At elevated temperature not only the amplitudes of stresses/strains within a cycle, but also loading rates, hold times and many other factors have an influence on the component life. Simulations of components in inelastic range for many cycles of loading is time consuming, if ever possible. For an efficient analysis it is convenient to introduce two or more time scales Altenbach et al. (2000a); Devulder et al. (2010); Fish et al. (2012). A "slow or macroscopic" time scale can be used to capture the

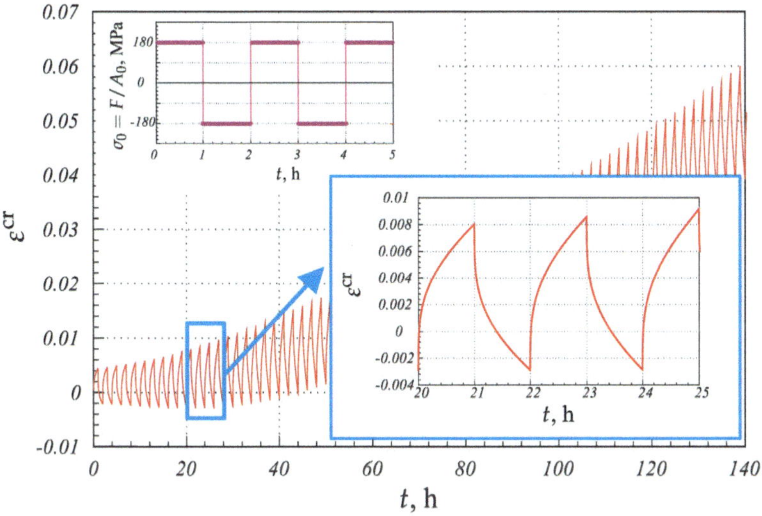

Fig. 12.31 Accumulation of inelastic strain for a 12 % Cr steel at 600 °C under cyclic force with hold times (simulation)

global cyclic behavior like cyclic hardening, softening or creep ratcheting. For the structural analysis within one loading cycle "fast or microscopic" time scales are useful. As an example, Fig. 12.31 illustrates the accumulation of the inelastic strain as a result of cyclic force with hold times. Two regimes are clearly seen, the global one with the growth in the strain amplitude as a function of the "slow" time or cycle number and the rapid change of the inelastic strain within several cycles of loading.

Inelastic behavior of structural materials at high temperature is controlled by several microstructural processes having different characteristic times. The inelastic deformation of crystalline materials can be explained by dislocation glide and dislocation climb (Frost and Ashby 1982; Nabarro and Villiers 1995). The glide motion of dislocations dominates at lower homologous temperatures and higher stress levels, while the climb of dislocations over obstacles—a nonconservative motion controlled by the diffusion of lattice vacancies, becomes important in high-temperature regimes and moderate stress levels. As a thermally activated process, the diffusion of vacancies occurs over time scales that are much longer than the times required for glide steps. The difference in the time scales may be of many orders of magnitude depending on the stress and temperature levels.

12.6 Modeling Approaches

The basic approaches to the description of inelastic behavior can be classified as follows. The *empirical modeling* is the study of correlations between the inelastic strain rate, stress, temperature and time. In addition, extrapolation methods are developed to predict time-dependent deformations and life time of a structure based on experimental data from short-term uni-axial creep tests. The aim of this approach is to derive simple formulae for an estimation of the structural behavior under creep conditions. An example is the Monkman-Grant relationship which states that the product of the minimum creep rate and the time to fracture is a constant (Monkman and Grant 1956).

$$\dot{\varepsilon}_{\min} t_{\mathrm{f}} = \mathrm{const}, \tag{12.14}$$

where $\dot{\varepsilon}_{\min}$ is the minimum creep rate and t_{f} is the time to fracture. Once the constant is identified from a short-term creep test, Eq. (12.14) can be used to estimate the time to fracture for lower stress levels. To this end only the minimum creep rate versus stress dependence is required.

Another example is the linear damage summation rule for the life time assessment of components. Following Taira (1962)

$$L = L_{\mathrm{f}} + L_{\mathrm{c}},$$

where the damage L in the creep-fatigue range is defined as a sum of pure fatigue damage L_{f} and creep damage L_{c}. A component failure is expected from this rule if the relative fatigue damage and the relative creep damage reach in the sum a critical value. Within the empirical approach the meaning of damage is related to the life time fraction. The fatigue damage is defined by the empirical rules of Palmgren (1924) and Miner (1945)

$$L_{\mathrm{f}} = \sum_{i=1}^{k} \frac{N_i}{N_{A_i}}$$

where N_i and N_{A_i} are the number of cycles experienced and the number of cycles to failure at the constant strain amplitude, respectively. The sum over k loading cycles provides the relative fatigue life L_{f}. A similar rule to estimate creep damage was proposed by Robinson (1952)

$$L_{\mathrm{c}} = \sum_{i=1}^{k} \frac{t_i}{t_{*_i}},$$

where t_i is the time spent at constant stress and temperature levels and t_{*_i} is the time to fracture for the same loading conditions.

Many different empirical relations of this type are reviewed in Penny and Mariott (1995); Viswanathan (1989). Empirical models are useful in early stages of design for a rapid estimation of the components operation life. It should be noted that the empirical approach provides one-dimensional relations. The dependencies of inelastic behavior on the type of stress state are not discussed. Furthermore, stress redistributions, creep-plasticity interactions, cyclic strain accumulations and many other effects cannot be considered.

Within the *materials science modeling*, the inelastic deformation process is characterized by a variety of microstructural rearrangements. According to assumed scenarios of transport processes in the microscale (diffusion of vacancies, climb and glide of dislocations, etc.) equations for the inelastic strain rate are derived. The form of the specific rate equation depends on the assumed deformation and damage mechanisms for specific stress and temperature regimes, e.g. (Frost and Ashby 1982). As an example consider the equation for the inelastic strain rate as proposed by Estrin (1996)

$$\dot{\varepsilon}^{\mathrm{p}} = \dot{\varepsilon}_0 \left(\frac{\sigma}{\hat{\sigma}}\right)^n, \tag{12.15}$$

where $\dot{\varepsilon}_0$ and n are constants, and the drag stress $\hat{\sigma}$ can be defined as a sum of dislocation and particle hardening contributions

$$\hat{\sigma} = M\alpha G b\sqrt{\rho} + M\chi G b\frac{1}{l}, \tag{12.16}$$

where G is the shear modulus, M is the Taylor factor, b is the magnitude of the Burgers vector, ρ is the dislocation density, l is the mean spacing between particles. α and χ are empirical constants. Equation (12.16) is based on the Taylor mechanism (dislocation-dislocation interactions) and the Orowan bowing mechanism (bypassing of dislocation over particles).

Mechanism based equations of this type are reviewed in François et al. (2012); Frost and Ashby (1982); Illschner (1973); Kassner and Pérez-Prado (2004); Nabarro and Villiers (1995), among others. In addition, kinetic equations for internal state variables are introduced. Examples for these variables include dislocation density (Estrin 1996) internal (back) stress, e.g. François et al. (2012), and various damage parameters associated with ageing and cavitation processes (Dyson and McLean 1998). The aim of the materials science models is to provide explicit correlations between quantities characterizing the type of microstructure and processing (grain size, types of alloying and hardening, etc.) and quantities characterizing the material behavior, e.g. the creep rate. Furthermore, the mechanisms based classification of different forms of creep equations including different stress and temperature functions is helpful in the structural analysis. However, the majority of models proposed within the materials science are one-dimensional and operate with scalar-valued quantities like magnitudes of stress and strain rates.

The objective of *continuum mechanics modeling* is to investigate inelastic behavior in idealized three-dimensional solids. The idealization is related to the hypothesis of a continuum, e.g. Haupt (2002). The approach is based on balance equations and assumptions regarding the kinematics of deformation and motion. Inelastic behavior is described by means of constitutive equations which relate multi-axial deformation and stress states. Topological details of microstructure are not considered. Processes associated with the microstructural changes like hardening, recovery, ageing and damage can be taken into account by means of hidden or internal state variables and corresponding evolution equations (Betten 2005; Chaboche 2008; Lemaitre and Desmorat 2005; Maugin 1992; Rabotnov 1969; Skrzypek and Ganczarski 1998). Various models and methods developed within the solid mechanics can be applied to the structural analysis in the inelastic range. Examples are theories of rods, plates shells and three-dimensional solids as well as direct variational methods, e.g. Altenbach et al. (1998); Betten (2005); Boyle and Spence (1983); Hyde et al. (2013); Malinin (1981); Podgorny et al. (1984); Skrzypek (1993). Numerical solution techniques, for example the finite element method can be combined with various time step integration techniques to simulate time dependent structural behavior up to critical state of failure.

The classical continuum mechanics of solids assumes only translational degrees of freedom for motion of material points. The local mechanical interactions between the material points are characterized by forces. Furthermore, it is assumed that the stress state at a point in the solid depends only on the deformations and state variables of a vanishingly small volume element surrounding the point. To account for the heterogeneous deformation various extensions to the classical continuum mechanics were proposed. Micropolar theories assume that a material point behaves like a rigid body, i.e. it has translation and rotation degrees of freedom. The mechanical interactions are due to forces and moments. Constitutive equations are formulated for force and moment stress tensors. An example, where the micropolar theory should be preferred over the classical one is the short-fiber reinforced material. Short fibers may rotate and align towards certain orientation states as a result of non-uniform deformation or flow during the processing. To account for the fiber orientation, rotational degrees of freedom are required (Altenbach et al. 2003, 2007; Eringen 2001). Micropolar theories of plasticity are presented in Forest et al. (1997); Altenbach and Eremeyev (2014), among others. Inelastic deformation process is highly heterogeneous at the microscale and several effects cannot be described by the classical continuum mechanics accurately. For example, the dependence of the yield strength on the mean grain size and on the mean size of precipitates, see Sect. 12.4, are not considered within the classical theories since they do not possess intrinsic length scales. To analyze such effects non-local continuum theories are developed. Examples include strain gradient (Fleck and Hutchinson 1997; Gao et al. 1999) and micromorphic theories (Forest 2009) where a gradient or the rotation (curl) of the inelastic strain are considered as additional degrees of freedom. Non-local and phase field theories of damage and fracture were recently advanced to capture initiation and propagation of cracks in solids (Miehe et al. 2010; Schmitt et al. 2013). One problem in the use of enhanced continuum theories is related to the forces thermodynamically conjugate to the introduced degrees of freedom. It is not easy to give a clear interpretation to the higher rank stress tensors associated with higher deformation gradients.

Continuum mechanics is widely used for the inelastic analysis of structures. Examples of high temperature applications are presented for circumferentially notched bars in (Hayhurst 1994), pipe weldments in Hayhurst et al. (2002) and thin-walled tubes in (Krieg 1999), where qualitative agreement between the theory and experiments carried out on model structures have been established. Constitutive equations with internal state variables have been found to be mostly suited for the inelastic analysis of structures (Hayhurst et al. 2002; Hyde et al. 2013; Naumenko and Altenbach 2007).

Classical or enhanced continuum mechanics approaches require experimental data to calibrate constitutive equations over a wide range of stress, strain rates and temperatures as well as for multi-axial stress states. Accurate experimental data, in particular data related to long term creep regime are rarely available.

The *micromechanical models* deal with discrete simulations of material behavior for a representative volume element with geometrically idealized microstructure. Simplifying assumptions are made for the behavior of constituents and their interactions, for the type of the representative volume element and for the exerted boundary conditions. Within the *continuum micromechanics* classical or enhanced continuum models are used to analyze constituents and interfaces. An example is the continuum crystal viscoplasticity model. Here the discrete dislocation substructure in a crystal is ignored, considering instead that plastic deformation occurs in the form of smooth shearing deformations on certain planes and in certain directions—the slip systems (Rice 1971; Hutchinson 1976). Pioneering works to the micromechanics modeling of high-temperature inelasticity assumed idealized, usually two-dimensional microstructures and simplified constitutive models. Examples include numerical simulations of void growth in a power law creeping matrix material, e.g. Tvergaard (1990); Giessen et al. (1995), crack propagation through a two-dimensional polycrystal Onck et al. (2000); Giessen and Tvergaard (1995). In the last two decades computational approaches were advanced to analyze realistic three-dimensional microstructures of polycrystals (Cailletaud et al. 2003; Roters et al. 2011). Figure 12.32 shows examples for microstructural polycrystal models generated by Voronoi tessellation. Based on such geometrical models several mechanisms of deformation and damage at the microscale can be analyzed. Examples include grain boundary sliding and boundary cavitation and—two important deformation

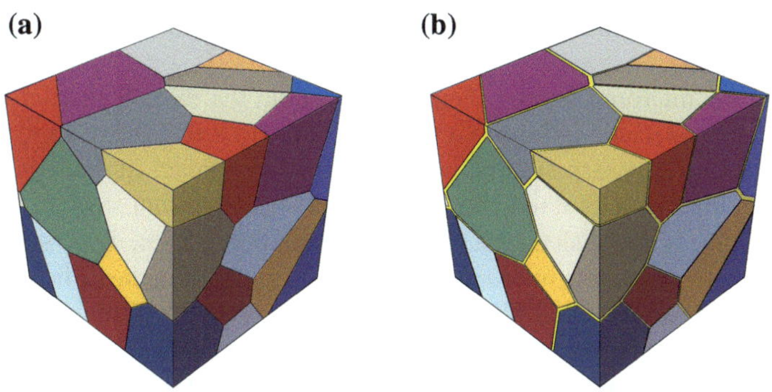

Fig. 12.32 Geometrical models of polycrystalline volume elements. **a** zero grain boundary thickness, **b** finite grain boundary thickness (Ozhoga-Maslovskaja 2014)

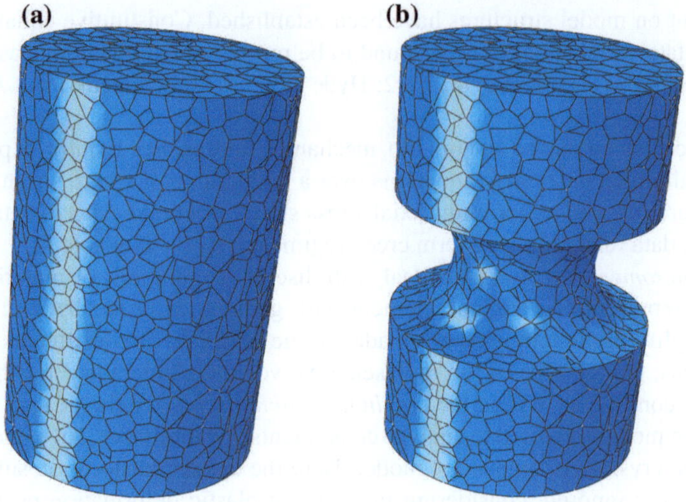

Fig. 12.33 Geometrical models of polycrystalline cylindrical specimen. **a** smooth specimen, **b** circumferentially notched specimen (after (Prygorniev and Naumenko 2013))

and damage mechanisms under high temperature (Ozhoga-Maslovskaja et al. 2015; Ozhoga-Maslovskaja 2014).

The representative volume (RVE) technique is widely used to analyze microstructural behavior in a bulk material. Indeed, a stress response of a RVE can be analyzed applying special types of deformation on the boundaries, for example periodic boundary conditions. By this approach gradient effects related to interfaces, free surfaces and notches cannot be captured. Recently polycrystal models with large numbers of grains are applied to analyze structures. As an example Fig. 12.33 shows geometrical polycrystal models for smooth and circumferentially notched cylindrical specimen generated by Voronoi tessellation. Simulations of inelastic behavior for such microstructural realizations and a subsequent statistical analysis contribute to understanding free-surface and notch stress effects in inelastic range (Prygorniev and Naumenko 2013).

Continuum micromechanics models are useful for the illustration of certain mechanisms of inelastic deformation. With respect to engineering applications the micromechanics approach suffers, however, from significant limitations. A typical high-temperature structural material, for example heat resistant steel, has a complex composition including dislocation structures, grain boundaries, dispersion particles, precipitates, etc. A reliable micromechanical description of inelastic behavior would therefore require a rather complex model of a multi-phase medium with many interacting constituents. Furthermore long term deformation at high temperature is usually accompanied by microstructural changes. For example subgrain boundaries may form, migrate and/or disappear. Voids form, grow and coalesce on grain boundaries leading to initiation of cracks and failure. Carbide precipitates coarsen during the high temperature exposure. Understanding of microstructural changes is crucial for design of high temperature materials. Direct simulations of topological changes in microstructure for real heat resistant alloys are time consuming, if ever possible.

Besides the continuum micromechanics several discrete methods were developed to analyze material behavior at lower length scales. The *discrete dislocation dynamics* (DD) method is suitable for the analysis of plastic deformation on the microscale and mesoscale (i.e. the size range of a fraction of a micrometer to tens of micrometers). Dislocations are explicitly represented as line defects embedded in an elastic medium. The main idea to derive and to solve equations of motion for dislocation loops. To this end driving forces (configurational forces) acting on dislocation line segments are defined. An example is the Peach-Köhler force, the energetic force work conjugate to the dislocation motion in an elastic continuum (Maugin 1993, 2011). In the early versions of DD, the collective behavior of dislocation ensembles was determined by interactions between infinitely long straight dislocations (Lepinoux and Kubin 1987). Simulations were two-dimensional and consisted of periodic cells containing multiple dislocations whose behavior was governed by a set of simplified rules. Recently, the DD methodology is extended to the more physical, three-dimensional simulations (Kubin 2013).

DD simulations give access to the dislocation patterning, interactions of dislocations with obstacles, subgrain boundaries, etc. but also to the mechanical response of a representative volume containing large number of dislocation lines. For thin films and small scale specimen DD simulations explain the size dependency of the observed flow stress (Kraft et al. 2010). DD is used for assessing the performance of enhanced continuum mechanics models in analysis of small scale structures and thin films (Aifantis et al. 2012) since it provides the knowledge of microscale stress and strain patterns, which are not easy to determine experimentally.

Several limitations exist when applying this method to the analysis of high-temperature phenomena. The inelastic strain at high temperature is mainly controlled by dislocation glide at a rate given by thermally-activated dislocation climb due to diffusion of vacancies (Frost and Ashby 1982). Numerical analysis of the phenomenon is challenging due to the complexity of incorporating both vacancies and dislocations in a single computational framework. Only recently two-dimensional DD simulations are performed illustrating dislocation glide and climb such that power law creep phenomenon can be reproduced (Keralavarma et al. 2012).

12.7 Conclusions and Recommendations

The objective of this review was to present current knowledge on modeling of high temperature material behavior for the structural analysis. Examples for high temperature applications include components of power plant, turbochargers, engines etc. Small scale components of microelectronics are further examples. The requirements for the modeling with respect to engineering applications are

- Ability to describe basic features of inelastic deformation, hardening, softening and damage processes for a wide range of stress, strain rate and temperature levels,
- Robustness and minimum number of functions and material properties to be identified from tests,
- Compatibility with structural mechanics methods, for example FEM

To met these requirements several theoretical and experimental approaches should be applied simultaneously. The continuum mechanics provides a rational framework for the analysis of real three-dimensional structures under complex thermo-mechanical loading paths. Appropriate stress and deformation measures are introduced to capture complex local multi-axial loadings. General forms of constitutive and evolution equations are defined such that invariance requirements with respect to the choice of reference frame, laws of continuum thermodynamics and other principles are fulfilled. To specialize the constitutive equation results of basic tests of the material behavior, such as tensile test, creep test, relaxation test, etc. should be systematically analyzed. On the other hand basic features of materials microstructure in the reference state and after a course of inelastic deformation process should be established. For example, to formulate a robust model it is not enough to say, that a material exhibits anisotropic properties. Even in the case of linear elasticity 21 material constants must be identified from tests if the kind of anisotropy is not specified. Microstructural analysis and appropriate assumptions with regard to symmetries of microstructure would reduce the identification effort essentially. Different types of material symmetries and appropriate forms of constitutive laws are defined within the continuum mechanics, while microstructural analysis is usually the subject of the materials science.

Once a first guess to the constitutive model, for example, elastoviscoplastic model with kinematic and isotropic hardening is selected, appropriate test programme to generate reliable databases, for example tensile tests for a range of temperatures and strain rates must be defined. The experimental data should be used to identify the constitutive functions, for example a function that captures the strain rate sensitivity. With the developed and identified constitutive model simulations of the material behavior under service-like loading conditions should be performed and compared with experimental data. For example, uni-axial tests under thermo-mechanical loading profile simulating start-up, running and shut-down stages of a component can be used for verification of the developed constitutive equation. Material testing under multi-axial stress states, for example tests on cruciform specimen as well as material testing under non-uniform stress states and stress concentrations, for example notched specimen, are useful to verify the modeling assumptions. The verification process may require to modify the model, for example to change the constitutive functions of stress and temperature. To this end additional information regarding deformation mechanisms is useful. Furthermore, loading profiles for laboratory testing must be extended to analyze the material behavior for wide ranges of stresses, strain rates and temperatures.

References

Abe F (2008) Strengthening mechanisms in steel for creep and creep rupture. In: Abe F, Kern TU, Viswanathan R (eds) Creep-esistant steels. Woodhead Publishing, Cambridge, pp 279–304
Abe F (2009) Analysis of creep rates of tempered martensitic 9 % Cr steel based on microstructure evolution. Mater Sci Eng A 510:64–69
Aifantis KE, Weygand D, Motz C, Nikitas N, Zaiser M (2012) Modeling microbending of thin films through discrete dislocation dynamics, continuum dislocation theory, and gradient plasticity. J Mat Res 27:612–618
Aktaa J, Petersen C (2009) Challenges in the constitutive modeling of the thermo-mechanical deformation and damage behavior of eurofer 97. Eng Fract Mech 76(10):1474–1484
Altenbach H, Eremeyev V (2014) Strain rate tensors and constitutive equations of inelastic micropolar materials. Int J Plast 63:3–17
Altenbach H, Naumenko K (1997) Creep bending of thin-walled shells and plates by consideration of finite deflections. Comput Mech 19:490–495
Altenbach H, Naumenko K (2002) Shear correction factors in creep-damage analysis of beams, plates and shells. JSME Int J Series A 45:77–83
Altenbach H, Morachkovsky O, Naumenko K, Sychov A (1997) Geometrically nonlinear bending of thin-walled shells and plates under creep-damage conditions. Arch Appl Mech 67:339–352
Altenbach H, Altenbach J, Naumenko K (1998) Ebene Flächentragwerke. Springer, Berlin
Altenbach H, Breslavsky D, Morachkovsky O, Naumenko K (2000a) Cyclic creep damage in thin-walled structures. J Strain Anal Eng Des 35(1):1–11
Altenbach H, Kolarow G, Morachkovsky O, Naumenko K (2000b) On the accuracy of creep-damage predictions in thinwalled structures using the finite element method. Comput Mech 25:87–98
Altenbach H, Kushnevsky V, Naumenko K (2001) On the use of solid- and shell-type finite elements in creep-damage predictions of thinwalled structures. Arch Appl Mech 71:164–181
Altenbach H, Huang C, Naumenko K (2002) Creep damage predictions in thin-walled structures by use of isotropic and anisotropic damage models. J Strain Anal Eng Des 37(3):265–275
Altenbach H, Naumenko K, Zhilin P (2003) A micro-polar theory for binary media with application to phase-transitional flow of fiber suspensions. Continuum Mech Thermodyn 15:539–570
Altenbach H, Naumenko K, Pylypenko S, Renner B (2007) Influence of rotary inertia on the fiber dynamics in homogeneous creeping flows. ZAMM J Appl Math Mechanics/Zeitschrift für Angewandte Mathematik und Mechanik 87(2):81–93
Altenbach H, Naumenko K, Gorash Y (2008) Creep analysis for a wide stress range based on stress relaxation experiments. Int J Mod Phys B 22(31n32):5413–5418
Altenbach H, Kozhar S, Naumenko K (2013) Modeling creep damage of an aluminum-silicon eutectic alloy. Int J Damage Mech 22(5):683–698
Altenbach J, Altenbach H, Naumenko K (2004) Egde effects in moderately thick plates under creep damage conditions. Tech Mech 24(3–4):254–263
Arzt E (1998) Size effects in materials due to microstructural and dimensional constraints: a comparative review. Acta Mater 46(16):5611–5626
Berger C, Scholz A, Mueller F, Schwienherr M (2008) Creep fatigue behaviour and crack growth of steels. In: Abe F, Kern TU, Viswanathan R (eds) Creep-resistant steels. Woodhead Publishing, Cambridge, pp 446–471
Bernhardt EO, Hanemann H (1938) Über den Kriechvorgang bei dynamischer Belastung und den Begriff der dynamischen Kriechfestigkeit. Z für Metallkunde 30(12):401–409
Betten J (1976) Plastic anisotropy and Bauschinger-effect—general formulation and comparison with experimental yield curves. Acta Mech 25(1–2):79–94
Betten J (2001) Kontinuumsmechanik. Springer, Berlin
Betten J (2005) Creep Mechanics. Springer, Berlin
Betten J, El-Magd E, Meydanli SC, Palmen P (1995) Untersuchnung des anisotropen Kriechverhaltens vorgeschädigter Werkstoffe am austenitischen Stahl X8CrNiMoNb 1616. Arch Appl Mech 65:121–132

Bielski J, Skrzypek J (1989) Failure modes of elastic-plastic curved tubes under external pressure with in-plane bending. Int J Mech Sci 31:435–458
Blum W (2001) Creep of crystalline materials: experimental basis, mechanisms and models. Mater Sci Eng A 319:8–15
Blum W (2008) Mechanisms of creep deformation in steel. In: Abe F, Kern TU, Viswanathan R (eds) Creep-resistant steels. Woodhead Publishing, Cambridge, pp 365–402
Boyle JT, Spence J (1983) Stress analysis for creep. Butterworth, London
Bunch JO, McEvily AJ (1987) On the behavior of ferritic steels subjected to load controlled cycling at elevated temperatures. In: Rie KT (ed) Cycle low fatigue and elasto-plastic behaviour of materials, Springer, Berlin, pp 252–257
da C Andrade EN (1910) On the viscous flow of metals, and allied phenomena. Proc R Soc Lond ALXXXIV:1–12
Cailletaud G, Forest S, Jeulin D, Feyel F, Galliet I, Mounoury V, Quilici S (2003) Some elements of microstructural mechanics. Comput Mater Sci 27(3):351–374
Chaboche JL (2008) A review of some plasticity and viscoplasticity constitutive equations. Int J Plast 24:1642–1693
Coble RL (1963) A model for boundary diffusion controlled creep in polycrystalline materials. J Appl Phys 34(6):1679–1682
Coffin LF (1954) A study of the effects of cyclic thermal stresses on a ductile metal. Trans ASME 76:931–950
Colombo F, Mazza E, Holdsworth SR, Skelton RP (2008) Thermo-mechanical fatigue tests on uniaxial and component-like 1CrMoV rotor steel specimens. Int J Fatigue 30:241–248
Cui L, Wang P (2014) Two lifetime estimation models for steam turbine components under thermomechanical creep-fatigue loading. Int J Fatigue 59:129–136
Cui L, Scholz A, von Hartrott P, Schlesinger M (2009) Entwicklung von Modellen zur Lebensdauervorhersage von Kraftwerksbauteilen unter thermisch-mechanischer Kriechermüdungsbeanspruchung. Abschlussbericht, AVIF Vorhaben Nr. 895, FVV, Heft 888, Frankfurt am Main
Cui L, Wang P, Hoche H, Scholz A, Berger C (2013) The influence of temperature transients on the lifetime of modern high-chromium rotor steel under service-type loading. Mater Sci Eng A 560:767–780
Devulder A, Aubry D, Puel G (2010) Two-time scale fatigue modelling: application to damage. Comput Mech 45(6):637–646
Dyson BF, McLean M (1998) Microstructural evolution and its effects on the creep performance of high temperature alloys. In: Strang A, Cawley J, Greenwood GW (eds) Microstructural stability of creep resistant alloys for high temperature plant applications. Cambridge University Press, Cambridge, pp 371–393
Dyson BF, McLean M (2001) Micromechanism-quantification for creep constitutive equations. In: Murakami S, Ohno N (eds) IUTAM symposium on creep in structures. Kluwer, Dordrecht, pp 3–16
El-Magd E, Nicolini G (1999) Creep behaviour and microstructure of dispersion-strengthened pm-aluminium materials at elevated temperatures. In: Mughrabi H, Gottstein G, Mecking H, Riedel H, Tobolski J (eds) Microstructure and mechanical properties of metallic high-temperature materials: research Report/DFG. Wiley-VCH, Weinheim, pp 445–464
El-Magd E, Betten J, Palmen P (1996) Auswirkung der Schädigungsanisotropie auf die Lebensdauer von Stählen bei Zeitstandbeanspruchung. Mat-wiss u Werkstofftechn 27:239–245
Eringen AC (2001) Microcontinuum field theories, vol II: Fluent Media. Springer, New York
Estrin Y (1996) Dislocation-density-related constitutive modelling. In: Krausz AS, Krausz K (eds) Unified constitutive laws of plastic deformation. Academic Press, San Diego, pp 69–104
Faruque MO, Zaman M, Hossain MI (1996) Creep constitutive modelling of an aluminium alloy under multiaxial and cyclic loading. Int J of Plast 12(6):761–780
Fish J, Bailakanavar M, Powers L, Cook T (2012) Multiscale fatigue life prediction model for heterogeneous materials. Int J Numer Meth Eng 91(10):1087–1104
Fleck NA, Hutchinson JW (1997) Strain gradient plasticity. Adv Appl Mech 33:295–361

Forest S (2009) Micromorphic approach for gradient elasticity, viscoplasticity, and damage. J Eng Mech 135(3):117–131

Forest S, Cailletaud G, Sievert R (1997) A cosserat theory for elastoviscoplastic single crystals at finite deformation. Arch Mech 49:705–736

Fournier B, Sauzay M, Caës C, Noblecourt M, Mottot M, Bougault A, Rabeau V, Pineau A (2008) Creep-fatigue-oxidation interactions in a 9Cr-1Mo martensitic steel. Part II: effect of compressive holding period on fatigue lifetime. Int J Fatigue 30(4):663–676

Fournier B, Dalle F, Sauzay M, Longour J, Salvi M, Caës C, Tournié I, Giroux PF, Kim SH (2011) Comparison of various 9–12 % Cr steels under fatigue and creep-fatigue loadings at high temperature. Mater Sci Eng A 528(22):6934–6945

François D, Pineau A, Zaoui A (2012) Mechanical behaviour of materials, mechanical behaviour of materials, Micro- and macroscopic constitutive behaviour, vol 1. Springer, Berlin

Frost HJ, Ashby MF (1982) Deformation-mechanism maps. Pergamon, Oxford

Gao H, Huang Y, Nix WD, Hutchinson JW (1999) Mechanism-based strain gradient plasticity-I. Theory. J Mech Phys Solids 47(6):1239–1263

Gariboldi E, Casaro F (2007) Intermediate temperature creep behaviour of a forged Al–Cu–Mg–Si–Mn alloy. Mater Sci Eng A 462(1):384–388

van der Giessen E, Tvergaard V (1995) Development of final creep failure in polycrystalline aggregates. Acta Metall Mater 42:959–973

van der Giessen E, van der Burg MWD, Needleman A, Tvergaard V (1995) Void growth due to creep and grain boundary diffusion at high triaxialities. J Mech Phys Solids 43:123–165

Gooch DJ (2003) Remnant creep life prediction in ferritic materials. In: Saxena A (ed) Comprehensive structural integrity, Creep and high-temperature failure, vol 5. Elsevier, Amsterdam, pp 309–359

Gorash Y, Altenbach H, Lvov G (2012) Modelling of high-temperature inelastic behaviour of the austenitic steel AISI type 316 using a continuum damage mechanics approach. J Strain Anal Eng Des 47(4):229–243

Hald J (1998) Service performance of a 12crmov steam pipe steel. In: Strang A, Cawley J, Greenwood GW (eds) Microstructural stability of creep resistant alloys for high temperature plant applications. Cambridge University Press, Cambridge, pp 173–184

Hall EO (1951) The deformation and ageing of mild steel: iii discussion of results. Proc Phys Soc Sect B 64(9):747

Haupt P (2002) Continuum mechanics and theory of materials. Springer, Berlin

Hayhurst DR (1994) The use of continuum damage mechanics in creep analysis for design. J Strain Anal Eng Des 25(3):233–241

Hayhurst DR, Leckie FA (1990) Yielding, damage and failure of anisotropic solids. In: Boehler JP (ed) High temperature creep continuum damage in metals. Mechanical Engineering Publ, London, pp 445–464

Hayhurst DR, Wong MT, Vakili-Tahami F (2002) The use of CDM analysis techniques in high temperature creep failure of welded structures. JSME Int J Ser A 45:90–97

Herring C (1950) Diffusional viscosity of a polycrystalline solid. J Appl Phys 21(5):437–445

Holdsworth S, Mazza E, Binda L, Ripamonti L (2007) Development of thermal fatigue damage in 1CrMoV rotor steel. Nucl Eng Des 237:2292–2301

Hutchinson JW (1976) Bounds and self-consistent estimates for creep of polycrystalline materials. Proc Roy Soc London Math Phys Sci 348(1652):101–127

Hyde T, Sun W, Hyde C (2013) Applied creep mechanics. McGraw-Hill Education, New York

Hyde TH, Xia L, Becker AA (1996) Prediction of creep failure in aeroengine materials under multi-axial stress states. Int J Mech Sci 38(4):385–403

Hyde TH, Sun W, Becker AA, Williams JA (1997) Creep continuum damage constitutive equations for the base, weld and heat-affected zone materials of a service-aged 1/2Cr1/2Mo1/4V:2 1/4Cr1Mo multipass weld at 640 °C. J Strain Anal Eng Des 32(4):273–285

Hyde TH, Sun W, Williams JA (1999) Creep behaviour of parent, weld and HAZ materials of new, service aged and repaired 1/2Cr1/2Mo1/4V: 21/4Cr1Mo pipe welds at 640 °C. Mater High Temp 16(3):117–129
Hyde TH, Yaghi A, Becker AA, Earl PG (2002) Finite element creep continuum damage mechanics analysis of pressurised pipe bends with ovality. JSME Int J Ser A 45(1):84–89
Hyde TH, Sun W, Agyakwa PA, Shipeay PH, Williams JA (2003) Anisotropic creep and fracture behaviour of a 9CrMoNbV weld metal at 650 °C. In: Skrzypek JJ, Ganczarski AW (eds) Anisotropic Behav Damaged Mater. Springer, Berlin, pp 295–316
Illschner B (1973) Hochtemperatur-Plastizität. Springer, Berlin
Inoue T (1988) Inelastic constitutive models under plasticity-creep interaction condition—theories and evaluations. JSME Int J Ser I 31(4):653–663
Kassner ME, Pérez-Prado MT (2004) Fundamentals of creep in melals and alloys. Elsevier, Amsterdam
Kawai M (1989) Creep and plasticity of austenitic stainless steel under multiaxial non-proportional loadings at elevated temperatures. In: Hui D, Kozik TJ (eds) Visco-plastic behavior of new materials, ASME, PVP-Vol 184, New York, pp 85–93
Keralavarma SM, Cagin T, Arsenlis A, Benzerga AA (2012) Power-law creep from discrete dislocation dynamics. Physical Review Letters 109(26):265,504
Kimura K, Kushima H, Sawada K (2009) Long-term creep deformation properties of 9Cr-1Mo steel. Mater Sci Eng A510–A511:58–63
Kimura M, Yamaguchi K, Hayakawa M, Kobayashi K, Kanazawa K (2006) Microstructures of creep-fatigued 9–12 % Cr ferritic heat-resisting steels. Int J Fatigue 28(3):300–308
Kloc L, Sklenička V (1997) Transition from power-law to viscous creep beahviour of P-91 type heat-resistant steel. Mater Sci Eng A234–A236:962–965
Kloc L, Sklenička V (2004) Confirmation of low stress creep regime in 9 % chromium steel by stress change creep experiments. Mater Sci Eng A387–A389:633–638
Kloc L, Sklenička V, Ventruba J (2001) Comparison of low creep properties of ferritic and austenitic creep resistant steels. Mater Sci Eng A319–A321:774–778
Kostenko Y, Almstedt H, Naumenko K, Linn S, Scholz A (2013) Robust methods for creep fatigue analysis of power plant components under cyclic transient thermal loading. In: ASME turbo Expo 2013: turbine technical conference and exposition, American Society of Mechanical Engineers, pp V05BT25A040–V05BT25A040
Kowalewski ZL (1995) Experimental evaluation of the influence of the stress state type on the creep characteristics of copper at 523 K. Arch Mech 47(1):13–26
Kowalewski ZL (2001) Assessment of the multiaxial creep data based on the isochronous creep surface concept. In: Murakami S, Ohno N (eds) IUTAM symposium on creep in structures. Kluwer, Dordrecht, pp 401–418
Kowalewski ZL, Hayhurst DR, Dyson BF (1994) Mechanisms-based creep constitutive equations for an aluminium alloy. J Strain Anal Eng Des 29(4):309–316
Kraft O, Gruber PA, Mönig R, Weygand D (2010) Plasticity in confined dimensions. Annu Rev Mater Res 40:293–317
Kraus H (1980) Creep Anal. Wiley, New York
Krausz AS, Krausz K (1996) Unified constitutive laws of plastic deformation. Academic Press, San Diego
Krempl E (1999) Creep-plasticity interaction. In: Altenbach H, Skrzypek J (eds) Creep and damage in materials and structures, Springer, New York, pp 285–348, CISM Lecture Notes No. 399
Krieg R (1999) Reactor pressure vessel under severe accident loading. Final Report of EU-Project Contract FI4S-CT95-0002. Technical report, Forschungszentrum Karlsruhe, Karlsruhe
Kubin L (2013) Dislocations, mesoscale simulations and plastic flow. Oxford Series on Materials Modelling, OUP Oxford
Laengler F, Mao T, Aleksanoglu H, Scholz A (2010) Phenomenological lifetime assessment for turbine housings of turbochargers. In: Proceedings of 9th international conference on multiaxial fatigue and fracture, 7th–9th June, 2010, Parma, Italy, (CD-ROM), Parma, pp 283–291

Langdon TG (2006) Grain boundary sliding revisited: developments in sliding over four decades. J Mater Sci 41:597–609
Längler F, Naumenko K, Altenbach H, Ievdokymov M (2014) A constitutive model for inelastic behavior of casting materials under thermo-mechanical loading. J Strain Anal Eng Des 49:421–428
Lazan BJ (1949) Dynamic creep and rupture properties of temperature-resistant materials under tensile fatigue stress. Proc ASTM 49:757–787
Le May I, da Silveria TL, Cheung-Mak SKP (1994) Uncertainties in the evaluations of high temperature damage in power stations and petrochemical plant. Int J of Press Vessels Pip 59:335–343
Lemaitre J, Desmorat R (2005) Engineering damage mechanics: ductile, creep, fatigue and brittle failures. Springer, Berlin
Lepinoux J, Kubin L (1987) The dynamic organization of dislocation structures: a simulation. Scr Metall 21(6):833–838
Libai A, Simmonds JG (1998) The nonlinear theory of elastic shells. Cambridge University Press, Cambridge
Lucas GE, Pelloux RMN (1981) Texture and stress state dependent creep in Zircaloy-2. Metall Trans A 12A:1321–1331
Malinin NN (1981) Raschet na polzuchest' konstrukcionnykh elementov (Creep calculations of structural elements, in Russ.). Mashinostroenie, Moskva
Manson SS (1953) Behavior of materials under conditions of thermal stress. NACA TN 2933
Maugin GA (1992) The thermomechanics of plasticity and fracture. Cambridge University Press, Cambridge
Maugin GA (1993) Material inhomogeneities in elasticity. Chapman Hall, London
Maugin GA (2011) Configurational forces: thermomechanics, physics, mathematics, and numerics. CRC Pressl, Boca Raton
Miehe C, Hofacker M, Welschinger F (2010) A phase field model for rate-independent crack propagation: robust algorithmic implementation based on operator splits. Comput Methods Appl Mech Eng 199(45–48):2765–2778
Miller AK (ed) (1987) Unified constitutive equations for creep and plasticity. Elsevier, London
Miner M (1945) Cumulative damage in fatigue. J Appl Mech 12:159–164
Monkman FC, Grant NJ (1956) An empirical relationship between rupture life and minimum creep rate in creep-rupture tests. In: Proc ASTM, vol 56, pp 593–620
Mughrabi H (2009) Cyclic slip irreversibilities and the evolution of fatigue damage. Metall Mater Trans B 40(4):431–453
Murakami S, Sanomura Y (1985) Creep and creep damage of copper under multiaxial states of stress. In: Sawczuk A, Bianchi B (eds) Plasticity today—modelling, methods and applications. Elsevier, London, pp 535–551
Nabarro FRN (1948) Report of a conference on the strength of solids. The Physical Society, London 75
Nabarro FRN (2002) Creep at very low rates. Metall Mater Trans A 33(2):213–218
Nabarro FRN, de Villiers HL (1995) The physics of creep, creep and creep-resistant alloys. Taylor & Francis, London
Nagode M, Längler F, Hack M (2011) Damage operator based lifetime calculation under thermo-mechanical fatigue for application on Ni-resist D-5S turbine housing of turbocharger. Eng Fail Anal 18(6):1565–1575
Naumenko K, Altenbach H (2007) Modelling of creep for structural analysis. Springer, Berlin
Naumenko K, Gariboldi E (2014) A phase mixture model for anisotropic creep of forged Al-Cu-Mg-Si alloy. Mater Sci Eng A 618:368–376
Naumenko K, Kostenko Y (2009) Structural analysis of a power plant component using a stress-range-dependent creep-damage constitutive model. Mater Sci Eng A510–A511:169–174
Naumenko K, Altenbach H, Gorash Y (2009) Creep analysis with a stress range dependent constitutive model. Arch Appl Mech 79:619–630

Naumenko K, Altenbach H, Kutschke A (2011a) A combined model for hardening, softening and damage processes in advanced heat resistant steels at elevated temperature. Int J Damage Mech 20:578–597

Naumenko K, Kutschke A, Kostenko Y, Rudolf T (2011b) Multi-axial thermo-mechanical analysis of power plant components from 9–12 % Cr steels at high temperature. Eng Fract Mech 78:1657–1668

Nikitenko AF (1984) Eksperimental'noe obosnovanie gipotezy suschestvovaniya poverkhnosti polzuchesti v usloviyakh slozhnogo nagruzheniya, (experimental justification of the hypothsis on existence of the creep surface under complex loading conditions, in russ.). Probl Prochn 8:3–8

Niu L, Kobayashi M, Takaku H (2002) Creep rupture properties of an austenitic steel with high ductility under multi-axial stresses. ISIJ Int 42:1156–1181

Nørbygaard T (2002) Studies of grain boundaries in materials subjected to diffusional creep. Ph.D. thesis, Risø National Laboratory, Roskilde

Odqvist FKG (1974) Mathematical theory of creep and creep rupture. Oxford University Press, Oxford

Odqvist FKG, Hult J (1962) Kriechfestigkeit metallischer Werkstoffe. Springer, Berlin

Ohno N (1998) Constitutive modeling of cyclic plasticity with emphasis on ratchetting. Int J Mech Sci 40(2):251–261

Ohno N, Takeuchi T (1994) Anisotropy in multiaxial creep of nickel-based single-crystal superalloy CMSX-2 (experiments and identification of active slip systems). JSME Int J Ser A 37:129–137

Ohno N, Kawabata M, Naganuma J (1990) Aging effects on monotonic, stress-paused, and alternating creep of type 304 stainless steel. Int J of Plast 6:315–327

Ohno N, Abdel-Karim M, Kobayashi M, Igari T (1998) Ratchetting characteristics of 316FR steel at high temperature, part I: strain-controlled ratchetting experiments and simulations. Int J Plast 14(4):355–372

Onck PR, Nguyen BN, Van der Giessen E (2000) Microstructural modelling of creep fracture in polycrystalline materials. In: Murakami S, Ohno N (eds) Creep in structures. Kluwer Academic Publishers, Dordrecht, pp 51–64

Oytana C, Delobelle P, Mermet A (1982) Constitutive equations study in biaxial stress experimants. Trans ASME J Eng Mat Techn 104(3):1–11

Ozhoga-Maslovskaja O (2014) Micro scale modeling grain boundary damage under creep conditions. Ph.D. thesis, Otto von Guericke University Magdeburg, Magdeburg

Ozhoga-Maslovskaja O, Naumenko K, Altenbach H, Prygorniev O (2015) Micromechanical simulation of grain boundary cavitation in copper considering non-proportional loading. Comput Mater Sci 96(Part A):178–184

Palmgren A (1924) Die Lebensdauer von Kugellagern. Z Ver Dtsch Ing 68(14):339–341

Penkalla HJ, Schubert F, Nickel H (1988) Torsional creep of alloy 617 tubes at elevated temperature. In: Reichman S, Duhl DN, Maurer G, Antolovich S, Lund C (eds) Superalloys 1988, the metallurgical society, pp 643–652

Penny RK, Mariott DL (1995) Design for creep. Chapman & Hall, London

Perrin IJ, Hayhurst DR (1994) Creep constitutive equations for a 0.5 Cr-0.5 Mo-0.25 V ferritic steel in the temperature range 600–675 °C. J Strain Anal Eng Des 31(4):299–314

Petch NJ (1953) The cleavage strength of polycrystals. J Iron Steel Inst 174:25–28

Pintschovius L, Gering E, Munz D, Fett T, Soubeyroux JL (1989) Determination of non-symmetric secondary creep behaviour of ceramics by residual stress measurements using neutron diffractometry. J Mater Sci Lett 8(7):811–813

Podgorny AN, Bortovoj VV, Gontarovsky PP, Kolomak VD, Lvov GI, Matyukhin YJ, Morachkovsky OK (1984) Polzuchest' elementov mashinostroitel'nykh konstrykcij (Creep of mashinery structural members, in Russ.). Naukova dumka, Kiev

Polák J (2003) 4.01—Cyclic deformation, crack initiation, and low-cycle fatigue. In: Karihaloo I, Milne R, Ritchie B (eds) Comprehensive structural integrity. Pergamon, Oxford, pp 1–39

Polmear IJ (2004) Aluminium alloys-a century of age hardening. Mater Forum 28:1–14

Prygorniev O, Naumenko K (2013) Surface layer effects in polycrystalline structures under cyclic viscoplasticity. In: CanCNSM 2013, 4th canadian conference on nonlinear solid mechanics, Montreal, Canada 2013.07.23-26, published on CD-ROM, 6 pp

Qin Y, Götz G, Blum W (2003) Subgrain structure during annealing and creep of the cast martensitic Cr-steel G-X12CrMoWVNbN 10-1-1. Mater Sci Eng A 341(1):211–215

Rabotnov YN (1969) Creep problems in structural members. North-Holland, Amsterdam

Raj SV, Iskovitz IS, Freed AD (1996) Modeling the role of dislocation substructure during class M and exponential creep. In: Krausz AS, Krausz K (eds) Unified constitutive laws of plastic deformation. Academic Press, San Diego, pp 343–439

Rice JR (1971) Inelastic constitutive relations for solids: an internal-variable theory and its application to metal plasticity. J Mech Phys Solids 19(6):433–455

Rieth M, Falkenstein A, Graf P, Heger S, Jäntsch U, Klimiankou M, Materna-Morris E, Zimmermann H (2004) Creep of the austenitic steel AISI 316 L(N). Experiments and models. Technical report, Forschungszentrum Karlsruhe, FZKA 7065, Karlsruhe

Robinson DN, Binienda WK, Ruggles MB (2003a) Creep of polymer matrix composites. I: Norton/Bailey Creep Law for transverse isotropy. Trans ASCE J Eng Mech 129(3):310–317

Robinson DN, Binienda WK, Ruggles MB (2003b) Creep of polymer matrix composites. II: Monkman-Grant failure relationship for transverse isotropy. Trans ASCE J Eng Mech 129(3):318–323

Robinson E (1952) Effect of temperature variation on the long-time rupture strength of steels. Trans ASME 74(5):777–781

Roesler J, Harders H, Baeker M (2007) Mechanical behaviour of engineering materials: metals, Ceramics, Polymers, and Composites. Springer, Berlin

Roters F, Eisenlohr P, Bieler T, Raabe D (2011) Crystal plasticity finite element methods: in materials science and engineering. Wiley, New York

Röttger D (1997) Untersuchungen zum Wechselverformungs- und Zeitstandverhalten der Stähle X20CrMoV121 und X10CrMoVNb91. Dissertation, Universität GH Essen, Fortschr.-Ber. VDI Reihe 5, Nr. 507, Düsseldorf

Sakane M, Hosokawa T (2001) Biaxial and triaxial creep testing of type 304 stainless steel at 923 k. In: Murakami S, Ohno N (eds) IUTAM symposium on creep in structures. Kluwer, Dordrecht, pp 411–418

Sakane M, Tokura H (2002) Experimental study of biaxial creep damage for type 304 stainless steel. Int J Damage Mech 11:247–262

Samir A, Simon A, Scholz A, Berger C (2005) Deformation and life assessment of high temperature materials under creep fatigue loading. Materialwiss Werkstofftech 36:722–730

Schmitt R, Müller R, Kuhn C, Urbassek H (2013) A phase field approach for multivariant martensitic transformations of stable and metastable phases. Arch Appl Mech 83:849–859

Segle P, Tu ST, Storesund J, Samuelson LA (1996) Some issues in life assessment of longitudinal seam welds based on creep tests with cross-weld specimens. Int J Press Vessels Pip 66:199–222

Shibli IA (2002) Performance of p91 thick section welds under steady and cyclic loading conditions: power plant and research experience. OMMI 1(3). http://www.ommi.co.uk

Simon A (2007) Zur Berechnung betriebsnah belasteter Hochtemperaturstähle mit einem konstitutiven Werkstoffmodell. Dissertation, Technische Universität Darmstadt, Berichte aus der Werkstofftechnik, Band 4/2007, Aachen

Skelton RP (2003) 5.02—Creep-fatigue interactions (Crack Initiation). In: Karihaloo I, Milne R, Ritchie B (eds) Comprehensive structural integrity. Pergamon, Oxford, pp 25–112

Skrzypek J, Ganczarski A (1998) Modelling of material damage and failure of structures. Foundation of engineering mechanics. Springer, Berlin

Skrzypek JJ (1993) Plasticity and creep. CRC Press, Boca Raton

Sosnin OV (1974) Energeticheskii variant teorii polzuchesti i dlitel'noi prochnosti. polzuchest' i razrushenie neuprochnyayushikhsya materialov (energetic variant of the creep and long-term strength theories. creep and fracture of nonhardening materials, in russ.). Probl Prochn 5:45–49

Sosnin OV, Gorev BV, Nikitenko AF (1986) Energeticheskii variant teorii polzuchesti (Energetic variant of the creep theory, in Russ.). Institut Gidrodinamiki, Novosibirsk
Stouffer DC, Dame LT (1996) Inelastic deformation of metals. Wiley, New York
Straub S (1995) Verformungsverhalten und Mikrostruktur warmfester martensitischer 12 %-Chromstähle. Dissertation, Universität Erlangen-Nürnberg, Fortschr.-Ber. VDI Reihe 5, Nr. 405, Düsseldorf
Taira S (1962) Lifetime of structures subjected to varying load and temperature. In: Hoff NJ (ed) Creep in structures. Springer, Berlin, pp 96–119
Taira S, Koterazawa R (1962) Dynamic creep and fatigue of an 18-8-Mo-Nb Steel. Bull JSME 5(17):15–20
Taira S, Ohtani R (1986) Teorija vysokotemperaturnoj prochnosti materialov (Theory of high-temperature strength of materials, in Russ.). Metallurgija, Moscow
Trampczynski WA, Hayhurst DR, Leckie FA (1981) Creep rupture of copper and aluminium under non-proportional loading. J Mech Phys Solids 29:353–374
Trivaudey F, Delobelle P (1993) Experimental study and modelization of creep damage under multi-axial loadings at high temperature. In: Wilshire B, Evans RW (eds) Creep and fracture of engineering materials and structures. Institute of Materials, London, pp 137–147
Tvergaard V (1990) Material failure by void growth to coalescence. Adv Appl Mech 27:83–51
Viswanathan R (1989) Damage mechanisms and life assessment of high temperature components. ASM international
Wang P, Cui L, Scholz A, Linn S, Oechsner M (2014) Multiaxial thermomechanical creep-fatigue analysis of heat-resistant steels with varying chromium contents. Int J Fatigue 67:220–227
Wiese S (2010) Verformung und Schädigung von Werkstoffen der Aufbau- und Verbindungstechnik: das Verhalten im Mikrobereich. Springer, Berlin
Wiese S, Roellig M, Mueller M, Wolter KJ (2008) The effect of downscaling the dimensions of solder interconnects on their creep properties. Microelectron Reliab 48(6):843–850
Winstone MR (1998) Microstructure and alloy developments in nickel-based superalloys. In: Strang A, Cawley J, Greenwood GW (eds) Microstructural stability of creep resistant alloys for high temperature plant applications. Cambridge University Press, Cambridge, pp 27–47
Yagi K, Merckling G, Kern TU, Irie H, Warlimont H (2004) Creep properties of heat resistant steels and superalloys. Landolt-Börnstein—Group VIII advanced materials and technologies: numerical data and functional relationships in science and technology. Springer, Berlin
Yaguchi M, Takahashi Y (2005) Ratchetting of viscoplastic material with cyclic softening, part 1: experiments on modified 9Cr–1Mo steel. Int J Plast 21(1):43–65
Zhang S, Harada M, Ozaki K, Sakane M (2007) Multiaxial creep-fatigue life using cruciform specimen. Int J Fatigue 29(5):852–859
Zolochevskij AA (1988) Kriechen von Konstruktionselementen aus Materialien mit von der Belastung abhängigen Charakteristiken. Tech Mech 9:177–184

Chapter 13
Onset of Matrix Cracking in Fiber Reinforced Polymer Composites: A Historical Review and a Comparison Between Periodic Unit Cell Analysis and Analytic Failure Criteria

Tomonaga Okabe, Yuta Kumagai, Ryo Higuchi and Masaaki Nishikawa

Abstract This paper explains previous studies addressing the onset crack or matrix crack in composite materials and presents a brief history of this field for the understanding of readers. Next, the analytic criterion and periodic unit cell analysis are compared for thermosetting or thermoplastic matrices. For both matrix resins, comparisons show that the Tsai-Hill criterion obviously cannot reproduce the results obtained from the periodic unit cell analysis, and the Hashin and Christensen criteria may give an appropriate failure envelope. Furthermore, macroscopic yielding and nonlinear deformation occur due to the plastic deformation of matrix resin before the failure. Thus it is appropriate to consider the elastoplastic or viscoplastic behavior of matrix resin. For thermoplastic resin, macroscopic yielding and nonlinear deformation occur due to the viscoplastic deformation of matrix resin much before its failure. Hence nonlinear deformation including creep may be more important than failure for thermoplastic resin.

Keywords Fiber reinforced polymer composites · Matrix cracking · Periodic unit cell analysis · Analytic failure criteria

T. Okabe (✉) · Y. Kumagai · R. Higuchi
Department of Aerospace Engineering, Tohoku University, 6-6-01 Aoba-yama, Aoba-ku, Sendai, Miyagi 980-8579, Japan
e-mail: okabe@plum.mech.tohoku.ac.jp

Y. Kumagai
e-mail: kumagai@plum.mech.tohoku.ac.jp

R. Higuchi
e-mail: higuchi@plum.mech.tohoku.ac.jp

M. Nishikawa
Department of Mechanical Engineering and Science, Kyoto University, C3 Kyoto Daigaku-Katsura, Nishikyo-ku, Kyoto 615-8540, Japan
e-mail: nishikawa@me.kyoto-u.ac.jp

H. Altenbach et al. (eds.), *From Creep Damage Mechanics to Homogenization Methods*, Advanced Structured Materials 64,
DOI 10.1007/978-3-319-19440-0_13

13.1 Introduction

The application of composite materials is essential in the field of transport aircraft to reduce weight. Boeing stated that the Dreamliner 787 utilizes more composite materials in its airframe and primary structure than any previous commercial airplane. Fifty weight percent of the Dreamliner is made of advanced composites, so that it weighs twenty percent less than the previous aluminium airplanes. Recently, Boeing stated that the fourth-generation 777X wing, which has a longer span than today's 777, will be made of advanced composite. The wings are thicker than the other primary structures, so this application contributes to weight reduction. Hereafter, it is hoped that advanced composite materials will be applied to small or medium airplanes.

Its application to automobiles is drastically expanding to improve fuel efficiency. Recently, BMW began selling the BMW i3 and i8, which are made of carbon fiber reinforced plastic (CFRP) composites. The primary structures of these cars are produced by resin transfer molding, and they reported that this application contributes to weight reduction. Other automobile companies are also trying to apply CFRP to their structures, and they are searching for efficient ways of using thermoplastic resin to reduce production time.

CFRP is widely applied in the transportation field because its strength and modulus per unit weight are superior to those of other engineering materials (Ashby 2011). Ceramics also have those excellent characteristics. However, Ceramics are quite brittle, so their application is limited to secondary structures. CFRP has high tensile and compressive strength and high ductility. These are the main reasons for the drastic expansion of its application.

CFRP has a characteristic hierarchy since it consists of fibers with micron-scale diameters and matrix. Therefore, that hierarchy induces multiple and complex failure modes. The final failure of unidirectional fiber reinforced composites under tension is controlled by the fiber strength, and its compressive strength is controlled by the matrix yielding stress. Its onset damage is controlled by the matrix properties and is called matrix crack. Delamination is major damage seen in composite laminates that are widely used in engineering applications. Thus, the modeling of damage and failure in composites is more complex than that of metal.

As mentioned above, composite materials have been applied to primary structures, so its damage modeling is critical for the design and reliability of composite transportation systems. From the viewpoint of design and reliability, the specific modeling of matrix cracks is essential and establishes the design criterion. In consequence, many studies have addressed matrix cracks for the last half century. These studies are classified as analytic and numerical. Analytical studies can be easily handled in the design, and numerical studies can provide specific and physical understanding of the damage process.

Boeing's researchers (Gosse and Christensen 2001) proposed a new unit cell analysis to predict the strain of matrix cracks. They called the approach "strain invariant failure theory" (SIFT). It was surprising that they tried to use the method in the realistic design of an airplane. Thereafter, many researchers have been trying to make a more specific and physical model since the original SIFT is too simplified for application to structural design. This paper explains previous studies addressing onset cracks or matrix cracks in composite materials and presents a brief history of this field for the understanding of readers. Next, we compare analytic criteria and our periodic unit cell analysis for thermosetting or thermoplastic matrices.

13.2 A Brief Historical Review of Matrix Crack Modeling

13.2.1 Analytic Failure Criterion

The failure criterion of composite materials has been studied for a long time. At the beginning of the studies, stress or strain criteria were utilized to predict failures. These criteria can be very useful for the uniaxial stress state but cannot be applied to the multiaxial stress states. The Tsai-Hill criterion (Azzi and Tsai 1965) is probably the first and most well-known criterion to overcome this difficulty. Several criteria have been proposed so far. Christensen described the analytic failure criterion in his book (Christensen 2005) and concluded that three analytic failure criteria (Tsai-Wu, Hashin and Christensen) (Tsai and Wu 1971; Hashin 1980; Christensen 1997) are well defined and others are not so good. Those are called "tensor form" and have a polynomial form. In this sub-section, these analytic criteria are briefly explained to illustrate their characteristics.

The strength parameters used in analytic failure criteria are defined as follows:

- T_{11} and C_{11}: Fiber-direction uniaxial tensile and compressive strengths
- T_{22} and C_{22}: Transverse uniaxial tensile and compressive strengths
- S_{12} and S_{23}: Fiber-direction and Transverse shear strengths

First, the Tsai-Wu criterion is given as

$$\begin{aligned}
&\left(\frac{1}{T_{11}}-\frac{1}{C_{11}}\right)\sigma_{11}+\left(\frac{1}{T_{22}}-\frac{1}{C_{22}}\right)(\sigma_{22}+\sigma_{33})+\frac{\sigma_{11}^2}{T_{11}C_{11}}\\
&+\frac{1}{T_{22}C_{22}}(\sigma_{22}+\sigma_{33})^2+2\left(\frac{1}{4S_{23}^2}-\frac{1}{T_{22}C_{22}}-\frac{1}{4T_{11}C_{11}}\right)\sigma_{11}(\sigma_{22}+\sigma_{33})\\
&+\frac{1}{S_{23}^2}(\sigma_{23}^2+\sigma_{22}\sigma_{33})+\frac{1}{S_{12}^2}(\sigma_{12}^2+\sigma_{31}^2)\leq 1,
\end{aligned} \tag{13.1}$$

where σ_{ij} is the Cauchy stress tensor. This criterion has a polynomial form and gives the interaction of stress tensors. In contrast, the Tsai-Hill criterion does not include the differences between tensile strength T_{ii} and compressive strength C_{ii} and is given by

$$\frac{\sigma_{11}^2}{T_{11}^2} + \frac{1}{T_{22}^2}(\sigma_{22} + \sigma_{33})^2 + 2\left(\frac{1}{4S_{23}^2} - \frac{1}{T_{22}^2} - \frac{1}{4T_{11}^2}\right)\sigma_{11}(\sigma_{22} + \sigma_{33}) + \frac{1}{S_{23}^2}(\sigma_{23}^2 + \sigma_{22}\sigma_{33}) + \frac{1}{S_{12}^2}(\sigma_{12}^2 + \sigma_{31}^2) \leq 1. \quad (13.2)$$

As stated later, the Tsai-Wu criterion is more appropriate for explaining the failure properties than the Tsai-Hill criterion since tensile strength and compressive strength in composites are generally different.

The Hashin criterion also has a polynomial form but decomposes failure modes into matrix failure groups and fiber failure groups. It is given as follows.

- Tensile matrix failure ($\sigma_{22} + \sigma_{33} > 0$)

$$\frac{1}{T_{22}^2}(\sigma_{22} + \sigma_{33})^2 + \frac{1}{S_{23}^2}(\sigma_{23}^2 - \sigma_{22}\sigma_{33}) + \frac{1}{S_{12}^2}(\sigma_{12}^2 + \sigma_{31}^2) \leq 1, \quad (13.3)$$

- Compressive matrix failure ($\sigma_{22} + \sigma_{33} < 0$)

$$\frac{1}{C_{22}}\left[\left(\frac{C_{22}}{2S_{23}}\right)^2 - 1\right](\sigma_{22} + \sigma_{33}) + \frac{1}{4S_{23}^2}(\sigma_{22} + \sigma_{33})^2 + \frac{1}{S_{23}^2}(\sigma_{23}^2 - \sigma_{22}\sigma_{33}) + \frac{1}{S_{12}^2}(\sigma_{12}^2 + \sigma_{31}^2) \leq 1, \quad (13.4)$$

- Tensile fiber failure ($\sigma_{11} > 0$)

$$\left(\frac{\sigma_{11}}{T_{11}}\right)^2 + \frac{1}{S_{12}^2}(\sigma_{12}^2 + \sigma_{31}^2) \leq 1, \quad (13.5)$$

- Compressive fiber failure ($\sigma_{11} < 0$)

$$\left(\frac{\sigma_{11}}{C_{11}}\right)^2 \leq 1 \quad (13.6)$$

Christensen specifically described in his famous book that failure mode decomposition is necessary and unavoidable, and there is no necessity to decompose failure modes into compressive groups and tensile groups. The third criterion proposed by Christensen is given by

- Matrix failure

$$\left(\frac{1}{T_{22}} - \frac{1}{C_{22}}\right)(\sigma_{22} + \sigma_{33}) + \frac{1}{T_{22}C_{22}}\left[(\sigma_{22} - \sigma_{33})^2 + 4\sigma_{23}^2\right] + \frac{\sigma_{12}^2 + \sigma_{31}^2}{S_{12}^2} \leq 1, \quad (13.7)$$

- Fiber failure

$$\left(\frac{1}{T_{11}} - \frac{1}{C_{11}}\right)\sigma_{11} + \frac{\sigma_{11}^2}{T_{11}C_{11}} - \frac{1}{4}\left(\frac{1}{T_{11}} + \frac{1}{C_{11}}\right)^2 \sigma_{11}(\sigma_{22} + \sigma_{33}) \leq 1 \quad (13.8)$$

The Christensen criterion is surprisingly simple, and there are five parameters to describe failure properties. As stated later, the Christensen criterion is almost the same as the Hashin criterion but has a simpler form.

Other criteria have been presented. However, they do not substantially differ from these three criteria. It should be noted that these criteria give the failure strength of unidirectional composites. Generally, unidirectional composites suddenly fail due to onset damage. Thus, if those analytic criteria are used to predict matrix cracks in composite laminates or structures, it gives its onset stress.

13.2.2 Micromechanics of Matrix Cracks

The micromechanics of matrix cracks has been studied experimentally and numerically for four decades. In the 1970s, the propagation of transverse matrix cracks generated when the load direction is perpendicular to the fiber direction, was the main topic in this field. As reviewed by Pagano et al. (1998), this problem is almost solved using the Griffith criterion. The initiation or onset of transverse matrix cracks has been still the unsolved problem, not its propagation.

Asp et al. (1996b) examined the onset of transverse matrix cracks using finite element analysis with the unit cell model. They used three fiber arrangements and examined the yielding criterion and the dilatation criterion for cavitation induced failure. They concluded that cavitation induced failure occurs at the fiber pole before matrix yielding. In any case, this dilatation failure follows the criterion based on the dilatational energy density U_{v} of a linear elastic material given by

$$U_{\mathrm{v}} = \frac{3(1-2\nu)}{2E}\sigma_{\mathrm{m}}^2, \quad (13.9)$$

where ν is Poisson's ratio, E is Young's modulus, and σ_{m} is the hydrostatic stress. In a following work, they did the Poker-Chip test to evaluate the above criterion and validated Eq. (13.9) experimentally (Asp et al. 1996a).

Hobbiebrunken et al. (2006) observed the onset of transverse matrix cracks with cross-ply laminates. He conducted three-point bending tests with 90/0/90 laminates and photographed transverse matrix cracks. In the photograph, the cracks occurred at the fiber poles as predicted by Asp et al. (1996a, b).

Pagano et al. (1998) published a holistic paper to summarize the modeling of transverse matrix cracks and presented detailed experiments. In this paper, they stated that the free-edge effect is not negligible for the onset of transverse matrix cracks.

Okabe et al. (2015) experimentally observed the site of onset of transverse matrix cracks in cross-ply laminates. They found that transverse matrix cracks initiate at the 0/90 interface of the free edge. This site differs from that modeled by Wang et al. (1985) and is strongly affected by the free-edge effect. In addition, they demonstrated that the triaxial stress state induces transverse matrix cracks.

These researches mainly focus on transverse matrix cracks. In contrast, matrix cracks occurring in the angle ply, especially their onset, has not been sufficiently examined. Henceforth, detailed experiments and observations will be necessary for specific modeling of such angle matrix cracks.

13.2.3 Unit Cell Analysis

Unit cell analysis is not so new; it has been used for a long time. However, SIFT is epoch-making in its engineering application. The most important thing is that it was proposed by Boeing engineers to design a realistic airplane. Generally, an aerospace engineer is conservative and does not want to use advanced technology. Nevertheless, their SIFT is a typical example of an advanced multiscale approach.

The unique point of SIFT is that the transfer matrix from the macroscopic strain to the microscopic strain of the selected points in the unit cell model is built beforehand. Each point has failure criteria given as follows.

If the dilatation failure is assumed to be at the corresponding point, the criterion is

$$J_1 = \varepsilon_1 + \varepsilon_2 + \varepsilon_3 \tag{13.10}$$

Here, J_1 is the dilatational strain invariant.

If the yielding failure is assumed to be at the corresponding point, the criterion is

$$J_2' = \sqrt{\frac{1}{6}[(\varepsilon_1 - \varepsilon_2)^2 + (\varepsilon_1 - \varepsilon_3)^2 + (\varepsilon_2 - \varepsilon_3)^2]} \tag{13.11}$$

Here, J_2' is the distortional strain invariant and ε_i is the principle strain of the corresponding point. Hence, the dilatation and yielding criteria are used to judge the failure.

This approach is sufficiently sophisticated and can be used for realistic airplane design. However, the following questions arise

1. Is it appropriate not to consider the elastoplastic or viscoplastic behavior of matrix resin?
2. Is it appropriate to consider that the initial yielding with Eq. (13.11) is used for the failure criterion?

To overcome this problem, Huang et al. (2012) developed the SIFT as follows.

1. The following criterion of matrix crack is used:

$$\frac{\bar{\sigma}^2}{C_{mi}T_{mi}} + \left(\frac{1}{T_{mi}} - \frac{1}{C_{mi}}\right) I_1 \geq 1 \tag{13.12}$$

   Here, $\bar{\sigma}$ is the von Mises stress, T_{mi} and C_{mi} are the initial tensile and compressive strengths of matrix, and I_1 is the first stress invariant.
2. A scalar damage variable is introduced, and the matrix stiffness is reduced with the variable.
3. The criterion of interface separation between fiber and matrix is introduced to predict the interfacial failure when the interface is weak.

They called this approach "Micromechanics of failure (MMF)."

Totry et al. (2008a, b); Canal et al. (2009) proposed a unit cell analysis considering the elastoplastic deformation of the matrix. They adopted the Mohr-Coulomb criterion instead of the von Mises criterion. Similar to Huang et al. (2012), interface separation was also analyzed with the cohesive zone model. Obtained results were compared with experiment results and analytic criteria.

Melro et al. (2013) proposed a unit cell analysis that considers viscoplastic deformation of matrix and incorporates continuum damage mechanics. In their paper, numerical integration is introduced in the algorithm, and experiments for pure resin reported by Fiedler et al. (2001) can be reproduced in their scheme. In a following work, Camanho et al. (2015) proposed analytic criteria and compared them with simulated results using a unit cell analysis proposed by Melro et al. (2013).

Okabe et al. (2011) also presented a periodic unit cell analysis that considers viscoplastic deformation of matrix and the damage growth model based on the Gurson-Tvergaard-Needleman (GTN) model. They succeeded in reproducing the experiments reported by Hobbiebrunken et al. (2006). Furthermore, they coupled the cavitation induced criterion presented by Asp et al. (1996a, b) with the GTN model and predicted the onset of transverse matrix cracking and the strength of off-axis laminates (Sato et al. 2014).

13.3 Comparison Between Analytic Criteria and Periodic Unit Cell Analysis

This section compares the relationship between analytic criteria and our periodic unit cell (PUC) analysis for thermosetting or thermoplastic matrices for the understanding of the readers.

Here, fiber is assumed to be an orthotropic elastic solid and matrix is assumed to be an elasto-viscoplastic solid. As mentioned above, this section analyses two types of matrix, thermosetting resin (epoxy resin) and thermoplastic resin (polypropylene resin).

The elasto-viscoplastic constitutive relationship of thermosetting resin, including variable D^*, is presented as follows (Kobayashi et al. 2004), based on the hypothesis of strain equivalence (Murakami 2012).

$$\overset{\circ}{\hat{\boldsymbol{\sigma}}} = (1 - D^*)\boldsymbol{C}^{\mathrm{e}}_{\mathrm{m}} : \boldsymbol{D} - (1 - D^*)\frac{3\mu\dot{\bar{\varepsilon}}^{\mathrm{p}}}{\bar{\sigma}}\boldsymbol{\sigma}' - \frac{\dot{D}^*}{1 - D^*}\boldsymbol{\sigma} \tag{13.13}$$

Here, $\hat{\boldsymbol{\sigma}}$ is the Kirchhoff stress tensor, $\boldsymbol{C}^{\mathrm{e}}_{\mathrm{m}}$ is the elastic constitutive tensor, $\boldsymbol{D}$ is the deformation rate tensor, μ is Lamé elastic constant, $\bar{\sigma}$ is the von Mises stress, and $\boldsymbol{\sigma}'$ is the deviatoric stress tensor. Superscript $\circ$ and $\cdot$ indicate Jaumann and time differentiation, respectively. The equivalent plastic strain rate $\dot{\bar{\varepsilon}}^{\mathrm{p}}$ is determined by the following equation, presented by Matsuda et al. (2002), which is a hardening rule involving the effect of hydrostatic stress on an epoxy resin (Okabe et al. 2011)

$$\dot{\bar{\varepsilon}}^{\mathrm{p}} = \dot{\varepsilon}_{\mathrm{r}} \left(\frac{\bar{\sigma} + \beta\sigma_{\mathrm{m}}}{g(\bar{\varepsilon}^{\mathrm{p}})} \right)^{\frac{1}{m}}, \tag{13.14}$$

where

$$g(\bar{\varepsilon}^{\mathrm{p}}) = g_1(\bar{\varepsilon}^{\mathrm{p}})^{g_2} + g_3 \tag{13.15}$$

In Eq. (13.14), m is an exponent regarding strain rate sensitivity, $\dot{\varepsilon}_{\mathrm{r}}$ is the reference strain rate, σ_{m} is the hydrostatic stress, and β is the hydrostatic stress sensitivity. This study assumes $m = 1/35$, $\dot{\varepsilon}_{\mathrm{r}} = 1 \times 10^{-5}$, and $\beta = 0.2$. In Eq. (13.15), g_1, g_2, and g_3 are material constants. In this analysis, $g_1 = 90\,\mathrm{MPa}$, $g_2 = 0.08$, and $g_3 = 20\,\mathrm{MPa}$ are used. This analysis assumes the linear limit of the epoxy resin is $\bar{\sigma}_{\mathrm{Y}} = 75\,\mathrm{MPa}$. The material constants regarding the elastic properties of fiber and matrix are listed in Table 13.1.

Table 13.1 Material properties of carbon fiber and epoxy resin used in microscopic 3D PUC analysis

Fiber longitudinal Young's modulus E_{L}	230 GPa
Fiber transverse Young's modulus E_{T}	17.5 GPa
Fiber longitudinal Poisson's ratio ν_{L}	0.17
Fiber transverse Poisson's ratio ν_{T}	0.46
Fiber radius r_{f}	3.5 μm
Fiber's coefficient of thermal expansion for the longitudinal direction α_{L}	-1.1×10^{-6}/K
Fiber's coefficient of thermal expansion for the transverse direction α_{T}	10×10^{-6}/K
Matrix Young's modulus E_{m}	3.2 GPa
Matrix Poisson's ratio ν_{m}	0.38
Matrix's coefficient of thermal expansion α_{m}	60×10^{-6}/K

This study employs two failure criteria for thermosetting resin. The first is the dilatational energy density criterion reported by Asp et al. (1996a, b). This failure criterion is valid under triaxial and elastic deformation that appears in matrix resin near the poles of the fibers in composites under transverse tension. The dilatational energy density U_{v} of a linear elastic material is given by Eq. (13.9). The matrix is assumed to fail when the dilatational energy density reaches a critical value $U_{\mathrm{v}}^{\mathrm{crit}}$ as

$$U_{\mathrm{v}} \geq U_{\mathrm{v}}^{\mathrm{crit}} \qquad \text{for} \qquad \bar{\sigma} \leq \bar{\sigma}_{\mathrm{Y}} \tag{13.16}$$

Here, $U_{\mathrm{v}}^{\mathrm{crit}}$ is a constant and is determined to be 0.9 MPa by comparing the cracking strain of the PUC model under uniaxial transverse tension with the failure strain of the experiment for unidirectional 90° laminates (Okabe et al. 2015).

The second failure criterion is based on damage variable D, calculated using a ductile damage growth law under plastic deformation. This study uses a damage growth model based on the Gurson-Tvergaard-Needleman (GTN) model (Gurson 1977; Tvergaard 1982; Tvergaard and Needleman 1984; Needleman and Tvergaard 1984), which was modified for epoxy resin by Nishikawa (2008). Damage variable D is calculated using the following equation.

$$\dot{D} = H_{\mathrm{e}}(\bar{\sigma} - \bar{\sigma}_{\mathrm{Y}})(1 - D)C < \dot{\varepsilon}_{\mathrm{m}}^{\mathrm{p}} > + (B_0 + B_1 D)\dot{\bar{\varepsilon}}^{\mathrm{p}}, \tag{13.17}$$

where

$$C < \dot{\varepsilon}_{\mathrm{m}}^{\mathrm{p}} > = A\left[D\left(\frac{< \sigma_{\mathrm{m}} >}{\sigma_0}\right)^2\right]. \tag{13.18}$$

The first term of the right-hand side in Eq. (13.17) represents void growth caused by the average plastic vertical strain. This term is activated when the von Mises stress exceeds the linear limit, as expressed by the Heaviside function $H_{\mathrm{e}}(\bullet)$. The second term indicates the damage evolution caused by plastic deformation. Here, σ_0 is the reference stress; A, B_0, and B_1 are non-dimensional constants; and $< \ >$ is the Macaulay bracket. This study uses $A = 1.5$, $B_0 = 0.6$, $B_1 = 0.6$ and $\sigma_0 = 73\,\mathrm{MPa}$. The matrix is assumed to fail when the damage variable reaches a critical value D^{crit} as

$$D \geq D^{\mathrm{crit}} \qquad \text{for} \qquad \bar{\sigma} > \bar{\sigma}_{\mathrm{Y}} \tag{13.19}$$

The damage variable is calculated at integration points using the damage growth law mentioned above and averaged within each element at each step. Furthermore, $D^*(D)$ is introduced to include the effect of sudden damage evolution due to coalescence of micro voids (Tvergaard and Needleman 1995).

$$D^* = \begin{cases} D & (D < D_{\mathrm{c}}) \\ D_{\mathrm{c}} + \dfrac{D^{*\mathrm{crit}} - D_{\mathrm{c}}}{D^{\mathrm{crit}} - D_{\mathrm{c}}}(D \quad D_{\mathrm{c}}) & (D \geq D_{\mathrm{c}}) \end{cases} \tag{13.20}$$

Here, D_c is D when starting the coalescence of micro voids, and D^{crit} is D when the element has failed. Equation (13.17) provides the damage evolution rate, and Eq. (13.20) yields D^*, which determines the stiffness matrix and stress of the matrix resin. At each element under plastic deformation ($\bar{\sigma} > \bar{\sigma}_Y$), we recognize that the element fails when D reaches D^{crit} and remove the stress. This paper uses $D_c = 0.08$, $D^{crit} = 0.25$, and $D^{*crit} = 1/1.5$. Furthermore, non-localization of variable D^* is performed to avoid mesh dependence of matrix damage (Tvergaard and Needleman 1995; Bazant and Pijaudier-Cabot 1988).

For the constitutive law of thermoplastic matrix resin, this study utilized the constitutive model of PP matrix proposed by Kobayashi et al. (2004). They formulated a one-parameter damage mechanics model based on the hypothesis of strain equivalence to address the initiation and propagation of craze. Thus this damage variable reproduces the craze inducing matrix crack. We explain this constitutive model below. Within the framework of thermodynamics, the non-coaxial constitutive equation of a viscoplastic polymer with damage evolution is

$$\overset{\circ}{\hat{\boldsymbol{\sigma}}} = (1 - D)\boldsymbol{C}_{\mathrm{m}}^{\mathrm{v}} : \boldsymbol{D} - (1 - D)\frac{3\mu\dot{\bar{\varepsilon}}^{\mathrm{p}}\cos\delta}{\bar{\sigma}}\boldsymbol{\sigma}' - \frac{\dot{D}}{1 - D}\boldsymbol{\sigma} \tag{13.21}$$

where

$$\boldsymbol{C}_{\mathrm{m}}^{\mathrm{v}} = \frac{H}{H + 3\mu}\left[\boldsymbol{C}_{\mathrm{m}}^{\mathrm{e}} + \frac{3\mu}{H}\left\{\frac{3\lambda + 2\mu}{3}\boldsymbol{I} \otimes \boldsymbol{I} + 3\mu\frac{\boldsymbol{\sigma}' \otimes \boldsymbol{\sigma}'}{\bar{\sigma}^2}\right\}\right],$$

$$H = \frac{1}{1 - D}\frac{\dot{\bar{\sigma}}}{\dot{\bar{\varepsilon}}^{\mathrm{p}}k}, \quad \bar{\sigma} = \sqrt{\frac{3}{2}\boldsymbol{\sigma}' : \boldsymbol{\sigma}'}, \quad \dot{\bar{\sigma}} = \sqrt{\frac{3}{2}\dot{\boldsymbol{\sigma}}' : \dot{\boldsymbol{\sigma}}'},$$

where λ denotes Lamé elastic constant. The equivalent strain rate $\dot{\bar{\varepsilon}}^{\mathrm{p}}$ is defined by the following hardening rule for the PP matrix (Murakami et al. 2002):

$$\dot{\bar{\varepsilon}}^{\mathrm{p}} = \dot{\varepsilon}_{\mathrm{r}}\left(\frac{\bar{\sigma}}{g(\bar{\varepsilon}^{\mathrm{p}})}\right)^{\frac{1}{m}}, \tag{13.22}$$

where

$$g(\bar{\varepsilon}^{\mathrm{p}}) = \sigma_{\mathrm{r}}\{\tanh(k_1\bar{\varepsilon}^{\mathrm{p}}) + k_2 + H_{\mathrm{e}}(\bar{\varepsilon}^{\mathrm{p}} - \varepsilon_{\mathrm{r}})k_3(\exp\bar{\varepsilon}^{\mathrm{p}} - \exp\varepsilon_{\mathrm{r}})\}.$$

Here, ε_r is the reference strain at rehardening, and σ_r is the initial yield stress. In addition to these parameters, k_1, k_2, and k_3 are determined as material constants, and H_e is the Heaviside function. The non-coaxial angle δ in Eq. (13.21) between deviatoric stress $\boldsymbol{\sigma}'$ and deviatoric plastic strain rate $\dot{\bar{\varepsilon}}^{\mathrm{p}}$, changes with the varying direction of the rate $\dot{\boldsymbol{\sigma}}'$ of deviatoric stress, according to the following empirical relationship.

$$\sin\delta = k\sin\alpha, \tag{13.23}$$

where

$$k = \begin{cases} p_1 \sin\left(\dfrac{\pi}{2}\dfrac{m}{p_2}\right) & (0 \leq m \leq p_2), \\ p_1\{\tanh[-p_3(m - p_2)] + 1\} & (p_2 \leq m) \end{cases}$$

Here, p_1, p_2, and p_3 are material constants; α is the angle between deviatoric stress $\boldsymbol{\sigma}'$ and the rate $\dot{\boldsymbol{\sigma}}'$ of deviatoric stress and satisfies the following relationship.

$$\cos\alpha = \frac{\dot{\bar{\sigma}}}{\bar{\sigma}} \tag{13.24}$$

Finally, D in Eq. (13.21) is calculated at integration points based on the following evolutionary equation (Kobayashi et al. 2004).

$$\dot{D} = A_{\mathrm{p}}(1 - D) < \dot{\varepsilon}_{\mathrm{m}}^{\mathrm{p}} > + B_{\mathrm{p}}\frac{\ddot{\bar{\varepsilon}}^{\mathrm{p}}}{\dot{\bar{\varepsilon}}^{\mathrm{p}}} \tag{13.25}$$

The first term represents craze growth due to volumetric plastic strain $\bar{\varepsilon}^{\mathrm{p}}$; the second term represents craze initiation and annihilation. Here, A_{p} and B_{p} are non-dimensional constants. Non-localization (Tvergaard and Needleman 1995; Bazant and Pijaudier-Cabot 1988) is performed to avoid mesh dependence of matrix damage. The rate of volumetric plastic strain $\dot{\bar{\varepsilon}}_{m}^{\mathrm{p}}$ is related to the hydrostatic pressure σ_{m} by Kobayashi et al. (2004), utilizing the preliminary FEM results:

$$\dot{\varepsilon}_{\mathrm{m}}^{\mathrm{p}} = \{(q_1 D)\cosh(q_2\sigma_{\mathrm{m}}/\sigma_{\mathrm{r}})\}^{\cdot}, \tag{13.26}$$

where q_1 and q_2 are material constants. The material constants of the matrix are listed in Table 13.2. The elastic properties of fiber used in the 3D PUC analysis of polypropylene are the same as listed in Table 13.1.

A craze inducing matrix crack is judged at the integration point of each finite element. When D approaches 1, the contribution to the stiffness matrix approaches zero, and sometimes numerical instability occurs. Therefore, we eliminated the corresponding element when the averaged D in the element reached D_{cr}. In this study, we used $D_{\mathrm{cr}} = 0.9$. The successive elimination process yielded free nodes. For stability, we searched for such nodes and excluded them from the equilibrium equations of the finite element analysis.

Henceforth, we briefly describe the finite element formulation of the PUC analysis. Assuming the analysed volume as V and the traction surface as S_{t}, the updated Lagrangian formulation of the principle of virtual work is given by Bathe (1996)

$$\Delta t \int_{V_{\mathrm{f}}+V_{\mathrm{m}}} ({}^{t}\underline{\dot{\boldsymbol{\sigma}}} : \delta\boldsymbol{\varepsilon} + {}^{t}\boldsymbol{\sigma} : \delta\dot{\boldsymbol{\varepsilon}})\mathrm{d}V = \int_{S_{\mathrm{t}}} {}^{t+\Delta t}\boldsymbol{f} \cdot \delta\boldsymbol{u}\mathrm{d}S - \left(\int_{V_{\mathrm{f}}} {}^{t}\boldsymbol{\sigma} : \delta\boldsymbol{\varepsilon}\mathrm{d}V + \int_{V_{\mathrm{m}}} {}^{t}\boldsymbol{\sigma} : \delta\boldsymbol{\varepsilon}\mathrm{d}V\right), \tag{13.27}$$

Table 13.2 Material properties of polypropylene resin used in microscopic 3D PUC analysis

Young's modulus E_{m}	1.8 GPa
Poisson's ratio ν_{m}	0.33
Coefficient of thermal expansion α_{m}	114×10^{-6}/K
Initial yield stress σ_{r}	11.0 MPa
Reference strain at rehardening ε_r	0.175
Reference strain rate $\dot{\varepsilon}_r$	0.893/s
Rate-sensitivity parameter m	0.041
Material constants	
Hardening rule k_1	100
k_2	1.85
k_3	2.37
Non-coaxial angle p_1	$\sin(\pi/8)$
p_2	0.0001
p_3	10
Damage evolution law A_{p}	1.0
B_{p}	0.0095
Volumetric plastic strain rate q_1	0.018
q_2	1.5

where subscript f is fiber, subscript m is matrix, $\boldsymbol{\varepsilon}$ is the Green strain tensor, $\underline{\boldsymbol{\sigma}}$ is the second Kirchhoff stress tensor, and $\boldsymbol{f}$ is the external force vector per unit area. To employ the PUC analysis based on the homogenization approach (Ohno et al. 2002; Okumura et al. 2004), the term of external force in Eq. (13.27) can be deleted. Substituting Eq. (13.13) or (13.21) into Eq. (13.27) leads to

$$\begin{aligned}
&\int_{V_{\mathrm{f}}} (\bar{\boldsymbol{C}}^{\mathrm{e}}_{\mathrm{f}} : \Delta\boldsymbol{\varepsilon}) : \delta\boldsymbol{\varepsilon} \mathrm{d}V + \int_{V_{\mathrm{m}}} \left[(1 - D^*)\bar{\boldsymbol{C}}^{\mathrm{v}}_{\mathrm{m}} : \Delta\boldsymbol{\varepsilon}\right] : \delta\boldsymbol{\varepsilon} \mathrm{d}V \\
&+ \int_{V_{\mathrm{f}}+V_{\mathrm{m}}} {}^{t}\boldsymbol{\sigma} : \delta\Delta\boldsymbol{\varepsilon} \mathrm{d}V = - \int_{V_{\mathrm{f}}+V_{\mathrm{m}}} {}^{t}\boldsymbol{\sigma} : \delta\boldsymbol{\varepsilon} \mathrm{d}V \\
&+ \int_{V_{\mathrm{m}}} (1 - D^*) \frac{3\mu\Delta\bar{\varepsilon}^{\mathrm{p}} \cos\delta}{\bar{\sigma}} ({}^{t}\boldsymbol{\sigma}' : \delta\boldsymbol{\varepsilon}) \mathrm{d}V + \int_{V_{\mathrm{m}}} \frac{\Delta D^*}{1 - D^*} ({}^{t}\boldsymbol{\sigma} : \delta\boldsymbol{\varepsilon}) \mathrm{d}V
\end{aligned} \tag{13.28}$$

where $\Delta\boldsymbol{\varepsilon}$ is the strain increment and $D^* = D$ is used in the case of thermoplastic since the sudden damage evolution is not considered. When considering the finite deformation (Okumura et al. 2004), the components of $\bar{\boldsymbol{C}}^{\mathrm{v}}_{\mathrm{m}}$ and $\bar{\boldsymbol{C}}^{\mathrm{e}}_{\mathrm{f}}$ are written as $\bar{C}^{\mathrm{v}}_{\mathrm{m},ijkl}$ and $\bar{C}^{\mathrm{e}}_{\mathrm{f},ijkl}$ given by

$$\begin{aligned}
\bar{C}^{\mathrm{v}}_{\mathrm{m},ijkl} &= C^{\mathrm{v}}_{\mathrm{m},ijkl} - \frac{1}{2}(\sigma_{ik}\delta_{jl} + \sigma_{il}\delta_{jk} + \delta_{ik}\sigma_{jl} + \delta_{il}\sigma_{jk}), \\
\bar{C}^{\mathrm{e}}_{\mathrm{f},ijkl} &= C^{\mathrm{e}}_{\mathrm{f},ijkl} - \frac{1}{2}(\sigma_{ik}\delta_{jl} + \sigma_{il}\delta_{jk} + \delta_{ik}\sigma_{jl} + \delta_{il}\sigma_{jk}),
\end{aligned} \tag{13.29}$$

where $C^{\rm v}_{{\rm m},ijkl}$ and $C^{\rm e}_{{\rm f},ijkl}$ indicate the components of $\boldsymbol{C}^{\rm v}_{\rm m}$ and $\boldsymbol{C}^{\rm e}_{\rm f}$, respectively. If the matrix is thermosetting resin, $\boldsymbol{C}^{\rm v}_{\rm m}$ is equal to $\boldsymbol{C}^{\rm e}_{\rm m}$. When incremental analysis using the periodic cell is performed, the displacement $\boldsymbol{u}$ and strain increment $\Delta\boldsymbol{\varepsilon}$ should be separated into a global (macro) component corresponding to the whole unit cell and a local (micro) component for inside the unit cell.

$$\begin{aligned} \boldsymbol{u} &= \boldsymbol{u}_{\rm G} + \boldsymbol{u}_{\rm L}, \\ \Delta\boldsymbol{\varepsilon} &= \Delta\boldsymbol{\varepsilon}_{\rm G} + \Delta\boldsymbol{\varepsilon}_{\rm L} \end{aligned} \tag{13.30}$$

Here, subscripts G and L indicate global and local. This separation is introduced in Eq. (13.28), providing the following equation

$$({}^t\boldsymbol{K}_{\rm f} + {}^t\boldsymbol{K}_{\rm m})\Delta\boldsymbol{U}_{\rm L} = -({}^t\boldsymbol{Q}_{\rm f} + {}^t\boldsymbol{Q}_{\rm m}) + {}^t\boldsymbol{Q}_{\rm v} + {}^t\boldsymbol{Q}_{\rm dam} - (\Delta\boldsymbol{Q}_{\rm f,G} + \Delta\boldsymbol{Q}_{\rm m,G}), \tag{13.31}$$

where

$$\begin{aligned}
\boldsymbol{K}_{\rm f} &= \sum_{\rm e}\int_{V^{\rm e}_{\rm f}} (\boldsymbol{B}^{\rm e\,T}\bar{\boldsymbol{D}}^{\rm e}_{\rm f}\boldsymbol{B}^{\rm e} + \boldsymbol{B}^{\rm e\;T}_{\rm NL}\boldsymbol{T}\boldsymbol{B}^{\rm e}_{NL})\,{\rm d}V, \\
\boldsymbol{K}_{\rm m} &= \sum_{\rm e}\int_{V^{\rm e}_{\rm m}} (\boldsymbol{B}^{\rm e\,T}\bar{\boldsymbol{D}}^{\rm e}_{\rm m}\boldsymbol{B}^{\rm e} + \boldsymbol{B}^{\rm e\;T}_{\rm NL}\boldsymbol{T}\boldsymbol{B}^{\rm e}_{NL})\,{\rm d}V, \\
\boldsymbol{Q}_{\rm f} &= \sum_{\rm e}\int_{V^{\rm e}_{\rm f}} \boldsymbol{B}^{\rm e\,T}\tilde{\boldsymbol{\sigma}}\,{\rm d}V, \\
\boldsymbol{Q}_{\rm m} &= \sum_{\rm e}\int_{V^{\rm e}_{\rm m}} \boldsymbol{B}^{\rm e\,T}\tilde{\boldsymbol{\sigma}}\,{\rm d}V, \\
\boldsymbol{Q}_{\rm v} &= \sum_{\rm e}\int_{V^{\rm e}_{\rm m}} \frac{3\mu\Delta\bar{\varepsilon}^{\rm p}\cos\delta}{\bar{\sigma}}\boldsymbol{B}^{\rm e\,T}\tilde{\boldsymbol{\sigma}}'\,{\rm d}V, \\
\boldsymbol{Q}_{\rm dam} &= \sum_{\rm e}\int_{V^{\rm e}_{\rm m}} \frac{\Delta D^*}{1-D^*}\boldsymbol{B}^{\rm e\,T}\tilde{\boldsymbol{\sigma}}\,{\rm d}V, \\
\Delta\boldsymbol{Q}_{\rm f,G} &= \sum_{\rm e}\int_{V^{\rm e}_{\rm f}} \boldsymbol{B}^{\rm e\,T}\bar{\boldsymbol{D}}^{\rm e}_{\rm f}\Delta\boldsymbol{\varepsilon}_{\rm G}\,{\rm d}V, \\
\Delta\boldsymbol{Q}_{\rm m,G} &= \sum_{\rm e}\int_{V^{\rm e}_{\rm m}} \boldsymbol{B}^{\rm e\,T}\bar{\boldsymbol{D}}^{\rm e}_{\rm m}\Delta\boldsymbol{\varepsilon}_{\rm G}\,{\rm d}V
\end{aligned}$$

Here, $\Delta\boldsymbol{U}$ is the increment of nodal displacement vector, $\boldsymbol{K}$ is the stiffness matrix of fiber and matrix, $\boldsymbol{T}$ is the matrix of Cauchy stress, $\tilde{\boldsymbol{\sigma}}$ is the vector of Cauchy stress, $\boldsymbol{B}$ is the compatibility matrix between strain and displacement, $\bar{\boldsymbol{D}}$ is the constitutive matrix, $\boldsymbol{Q}_{\rm v}$ is the internal force vector derived from the viscosity component, $\boldsymbol{Q}_{\rm dam}$ is the internal force vector derived from the damage component, and $\boldsymbol{Q}_{\rm f,G}$ and $\boldsymbol{Q}_{\rm m,G}$ are the internal force vectors generated by the global strain increment $\Delta\boldsymbol{\varepsilon}_{\rm G}$. Furthermore,

Fig. 13.1 Failure envelopes of in-plane shear stress versus transverse stress for CF/Epoxy

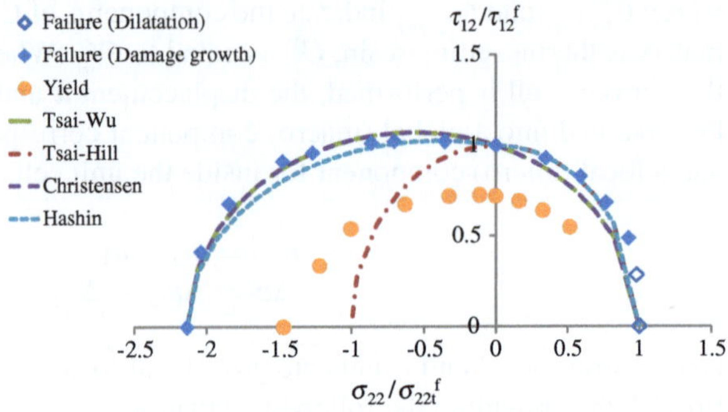

subscript e and NL denote element and nonlinear strain-displacement transformation. The present study adopts the tangent modulus method presented by Peirce et al. (1984) to make the time steps large and thereby reduce the calculation cost. The global stress $\boldsymbol{\sigma}^G$ is defined as the average stress in the periodic unit cell (Christensen 2005). In this calculation, we first calculated the thermal residual strain by decreasing the temperature (from curing temperature to room temperature, $\Delta T = -100$ K) of the unit cell. We then conducted the analysis by applying global strain incrementally to the unit cell and changing the strain ratios $\alpha_{yy} = \varepsilon_{yy}/\varepsilon_{xx}, \alpha_{zz} = \varepsilon_{zz}/\varepsilon_{xx}, \alpha_{xy} = \varepsilon_{xy}/\varepsilon_{xx}, \alpha_{yz} = \varepsilon_{yz}/\varepsilon_{xx}$, and $\alpha_{xz} = \varepsilon_{xz}/\varepsilon_{xx}$ to obtain the failure mode for the multiaxial stress state.

Figure 13.1 presents the failure envelope obtained from the PUC analysis with analytic failure criteria for thermosetting resin. The Tsai-Hill criterion obviously cannot reproduce the results obtained from the PUC analysis. It is quite natural that tensile and compressive onset strains differ because onset strain under tension is greatly affected by cavitation growth. Three criteria other than the Tsai-Hill criterion give almost the same results as the PUC analysis.

When the load condition can be regarded as almost pure mode I ($\sigma_{22}/\sigma_{22}^f \approx 1$), the dilatational energy density reaches a critical value. Thus, when the load condition is near that of pure mode I, the failure mode is brittle and its failure criterion is judged by Eq. (13.16). In contrast, in the other load conditions, the failure mode is ductile. In ductile failure, macroscopic yielding and nonlinear deformation occur due to the plastic deformation of matrix resin before the failure. Thus, it is appropriate to consider the elastoplastic or viscoplastic behavior of matrix resin. Huang et al. (2012) demonstrated similar results with nonlinear elasticity.

Figure 13.2 presents the failure envelope obtained from the PUC analysis with analytic failure criteria under biaxial loading. Two criteria other than Tsai-Wu criterion give almost the same result as the PUC analysis. In this case, since the matrix at the fiber surface deforms under the triaxial stress, the failure does not easily occur under biaxial compression. This is also shown in MMF. Therefore, Hashin and Christensen give an appropriate failure envelope.

Figure 13.3 depicts the failure sites seen in the PUC analysis. All failure sites are near the fiber surface. This implies that the improvement near the fiber surface,

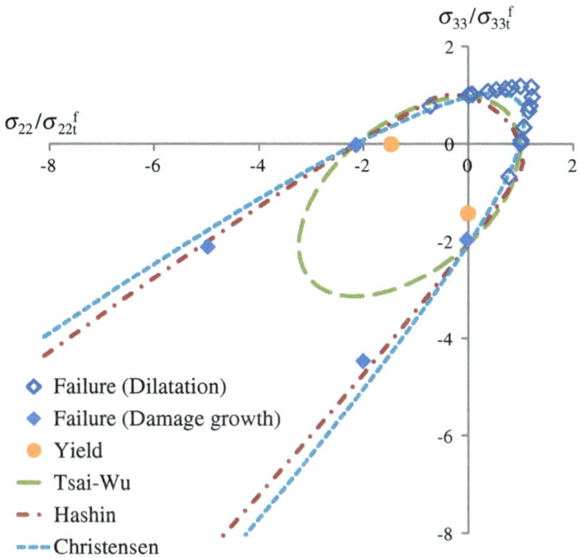

Fig. 13.2 Failure envelopes of biaxial loading for CF/Epoxy

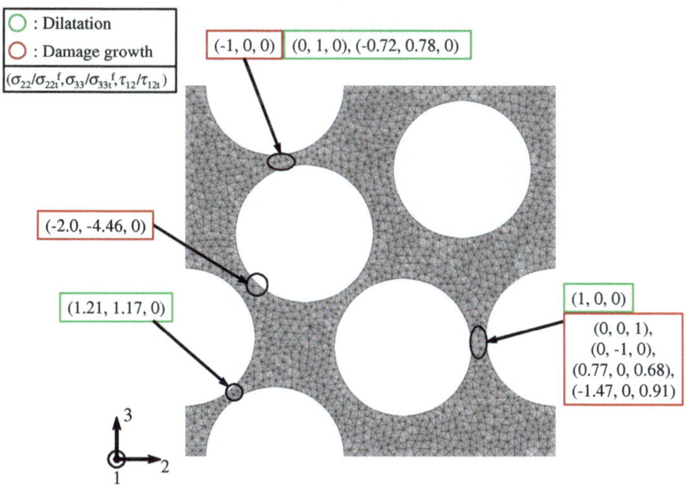

Fig. 13.3 Failure sites in PUC analysis for CF/Epoxy

including the interfacial treatment with nano-particles, may contribute to the suppression of matrix cracking.

Figure 13.4 plots the failure envelope obtained from the PUC analysis with analytic failure criteria for thermoplastic resin. The Tsai-Hill criterion obviously cannot reproduce the results obtained from the PUC analysis. It is quite natural that tensile and compressive onset strains are different because onset strain under tension is

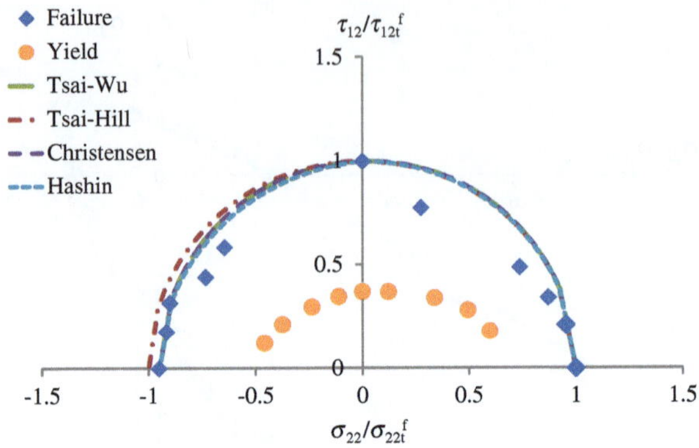

Fig. 13.4 Failure envelopes of in-plane shear stress versus transverse stress for CF/polypropylene

greatly affected by the craze growth. Three criteria other than the Tsai-Hill criterion give almost the same result as the PUC analysis. For thermoplastic resin, macroscopic yielding and nonlinear deformation occur due to the viscoplastic deformation of matrix resin much before its failure. Hence, for thermoplastic resin, nonlinear deformation including creep may be more important than failure. Figure 13.5 depicts the failure sites seen in the PUC analysis. Damage sites are similar to those for thermosetting resin.

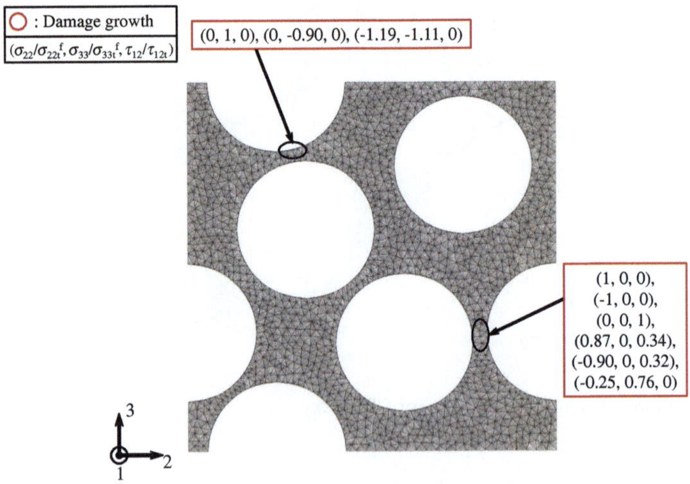

Fig. 13.5 Failure sites in PUC analysis for CF/polypropylene

13.4 Conclusions and Remarks

This paper explains previous studies addressing the onset crack or matrix crack in composite materials and presents a brief history of this field for the understanding of readers. Next, analytic criteria and our PUC analysis are compared for thermosetting or thermoplastic matrices. The comparisons supported the following conclusions.

The Tsai-Hill criterion obviously cannot reproduce the results obtained from the PUC analysis. The biaxial loading case shows that Hashin and Christensen may provide an appropriate failure envelope. Except for brittle failure, which occurs under load conditions similar to those of pure mode I, macroscopic yielding and nonlinear deformation occur due to the plastic deformation of matrix resin before the failure. Thus it is appropriate to consider the elastoplastic or viscoplastic behavior of matrix resin. For thermoplastic resin, macroscopic yielding and nonlinear deformation occur due to the viscoplastic deformation of the matrix resin much before its failure. Hence, in the case of thermoplastic resin, nonlinear deformation, including creep may be more important than failure.

Finally, we could not describe the multiscale modeling from the microscopic scale analysis stated above to the structural design. Readers interested in this problem should read LLorca et al. (2011); Okabe (2015).

References

Ashby MF (2011) Materials selection in mechanical design, 4th edn. Elsevier/Butterworth-Heinemann, Oxford

Asp LE, Berglund LA, Talreja R (1996a) A criterion for crack initiation in glassy polymers subjected to a composite-like stress state. Compos Sci Technol 56:1291–1301

Asp LE, Berglund LA, Talreja R (1996b) Prediction of matrix-initiated transverse failure in polymer composites. Compos Sci Technol 56:1089–1097

Azzi VD, Tsai SW (1965) Anisotropic strength of composites. Exp Mech 5:283–288

Bathe KJ (1996) Finite element procedures. Prentice Hall, New Jersey

Bazant ZP, Pijaudier-Cabot G (1988) Nonlocal continuum damage, localization instability and convergence. Trans ASME J Appl Mech 55:287–293

Camanho PP, Arteiro A, Melro AR, Catalanotti G, Vogler M (2015) Three-dimensional invariant-based failure criteria for fibre-reinforced composites. Int J Solids Struct 55:92–107

Canal LP, Segurado J, LLorca J, (2009) Failure surface of epoxy-modified fiber-reinforced composites under transverse tension and out-of-plane shear. Int J Solids Struct 46:2265–2274

Christensen RM (1997) Stress based yield/failure criteria for fiber composites. Int J Solids Struct 34(5):529–543

Christensen RM (2005) Mechanics of composite materials. Dover, New York

Fiedler B, Hojo M, Ochiai S, Schulte K, Ando M (2001) Failure behavior of an epoxy matrix under different kinds of static loading. Compos Sci Technol 61(11):1615–1624

Gosse JH, Christensen S (2001) Strain invariant failure criteria for polymers in composite materials. In: 19th AIAA applied aerodynamics conference, AIAA, 1184

Gurson AL (1977) Continuum theory of ductile rupture by void nucleation and growth: part i - yield criteria and flow rules for porous ductile media. Trans ASME J Eng Mater Technol 99:2–15

Hashin Z (1980) Failure criteria for unidirectional fiber composites. Trans ASME J Appl Mech 47:329–334

Hobbiebrunken T, Hojo M, Adachi T, de Jong C, Fiedler B (2006) Evaluation of interfacial strength in CF/epoxies using FEM and in-situ experiments. Compos A Appl Sci Manuf 37(12):2248–2256

Huang Y, Xu L, Ha SK (2012) Prediction of three-dimensional composite laminate response using micromechanics of failure. J Compos Mater 46:2431–2442

Kobayashi S, Tomii D, Shizawa K (2004) A modelling and simulation on failure prediction of ductile polymer based on craze evolution and annihilation. Trans Jpn Soc Mech Eng Ser A 70(694):810–817

LLorca J, González C, Molina-Aldareguía JM, Segurado J, Seltzer R, Sket F, Rodríguez M, Sádaba S, Muñoz R, Canal LP,(2011) Multiscale modeling of composite materials: a roadmap towards virtual testing. Adv Mater 23:5130–5147

Matsuda T, Ohno N, Tanaka H, Shimizu T (2002) Homogenized in-plane elastic-viscoplastic behavior of long fiber-reinforced laminates. JSME Int J Ser A Solid Mech Mater Eng 45(4):538–544

Melro AR, Camanho PP, Pires FMA, Pinho ST (2013) Micromechanical analysis of polymer composites reinforced by unidirectional fibres: Part I—Constitutive modelling. Int J Solids Struct 50:1897–1905

Murakami D, Kobayashi S, Torigaki T, Shizawa K (2002) A thermomechanical modeling and simulation of viscoplastic large deformation behavior for polymeric materials : 1st report, non-coaxiality of constitutive equation originated in strain rate dependence. Trans Jpn Soc Mech Eng Ser A 68(668):674–681

Murakami S (2012) Continuum damage mechanics: a continuum mechanics approach to the analysis of damage and fracture. Springer, New York

Needleman A, Tvergaard V (1984) An analysis of ductile rupture in notched bars. J Mech Phys Solids 32(6):461–490

Nishikawa M (2008) Multiscale modeling for the microscopic damage and fracture of fiber-reinforced plastic composites. Dr eng. thesis, The University of Tokyo, Tokyo

Ohno N, Okumura D, Noguchi H (2002) Microscopic symmetric bifurcation condition of cellular solids based on a homogenization theory of finite deformation. J Mech Phys Solids 50(5):1125–1153

Okabe T (2015) Recent studies on numerical modelling of damage progression in fibre-reinforced plastic composites. Mech Eng Rev 2(1):1–11

Okabe T, Nishikawa M, Toyoshima H (2011) A periodic unit-cell simulation of fiber arrangement dependence on the transverse tensile failure in unidirectional carbon fiber reinforced composites. Int J Solids Struct 48(20):2948–2959

Okabe T, Imamura H, Sato Y, Higuchi R, Koyanagi J, Talreja R (2015) Experimental and numerical studies of initial cracking in CFRP cross-ply laminates. Compos A Appl Sci Manuf 68:81–89

Okumura D, Ohno N, Noguchi H (2004) Elastoplastic microscopic bifurcation and post-bifurcation behavior of periodic cellular solids. J Mech Phys Solids 52(3):641–666

Pagano NJ, Schoeppner GA, Kim R, Abrams FL (1998) Steady-state cracking and edge effects in thermo-mechanical transverse cracking of cross-ply laminates. Compos Sci Technol 58(11):1811–1825

Peirce D, Shih CF, Needleman A (1984) A tangent modulus method for rate dependent solids. Comput Struct 18(5):875–887

Sato Y, Okabe T, Higuchi R, Yoshioka K (2014) Multiscale approach to predict crack initiation in unidirectional off-axis laminates. Adv Compos Mater 23(5–6):461–475

Totry E, González C, LLorca J (2008) Failure locus of fiber-reinforced composites under transverse compression and out-of-plane shear. Compos Sci Technol 68:829–839

Totry E, González C, LLorca J (2008) Prediction of the failure locus of C/PEEK composites under transverse compression and longitudinal shear through computational micromechanics. Compos Sci Technol 68:3128–3136

Tsai SW, Wu EM (1971) A general theory of strength for anisotropic materials. J Compos Mater 5:58–80

Tvergaard V (1982) On localization in ductile materials containing spherical voids. Int J Fract 18(4):237–252

Tvergaard V, Needleman A (1984) Analysis of the cup-cone fracture in a round tensile bar. Acta Metall 32(1):157–169

Tvergaard V, Needleman A (1995) Effects of nonlocal damage in porous plastic solids. Int J Solids Struct 32(8–9):1063–1077

Wang ASD, Kishore NN, Li CA (1985) Crack development in graphite-epoxy cross-ply laminates under uniaxial tension. Compos Sci Technol 24:1–31

Chapter 14
Swelling-Induced Buckling Patterns in Gel Films with a Square Lattice of Holes Subjected to In-Plane Uniaxial and Biaxial Pretensions

Dai Okumura, Akira Sasaki and Nobutada Ohno

Abstract In this study, we investigate swelling-induced buckling patterns in gel films containing a square lattice of holes subjected to in-plane pretensions. In accord with experiments, we simulate poly(dimethylsiloxane) (PDMS) films being prestrained and then swelled using toluene. Films are subjected to uniaxial and biaxial pretensions before swelling to investigate the potential ability of this system to generate complex buckling patterns. Finite element analysis is performed using an inhomogeneous field theory for polymeric gels. The resulting patterns are found to be highly diverse and depend sensitively on the type and magnitude of pretensions. The patterns arise from either transformation into diamond plate patterns (DPPs) or no pattern transformation. Diagrams of pattern transformation contain three regions of DPPs, transitional patterns, and monotonous patterns. Pretensions both distort the initial arrangement of the square lattice of holes and delay the onset of transformation into DPPs.

Keywords Gels · Swelling · Buckling · Films · FEA · Periodic structures

D. Okumura (✉)
Department of Mechanical Science and Engineering, Nagoya University,
Furo-cho, Chikusa-ku, Nagoya 464-8603, Japan
e-mail: okumura@mech.nagoya-u.ac.jp

A. Sasaki
Department of Computational Science and Engineering, Nagoya University,
Furo-cho, Chikusa-ku, Nagoya 464-8603, Japan
e-mail: sasaki@mml.mech.nagoya-u.ac.jp

N. Ohno
Department of Mechanical Science and Engineering, Nagoya University,
Chikusa-ku, Nagoya 464-8603, Japan
e-mail: ohno@mech.nagoya-u.ac.jp

H. Altenbach et al. (eds.), *From Creep Damage Mechanics to Homogenization Methods*, Advanced Structured Materials 64,
DOI 10.1007/978-3-319-19440-0_14

14.1 Introduction

Swelling-induced buckling generates a variety of complex and periodic wrinkle patterns in thin polymeric films (Tanaka et al. 1987; Breid and Crosby 2009) and causes more complicated pattern transformation in thin polymeric films with periodic arrangements of holes (Zhang et al. 2008; Wu et al. 2014). This buckling and the resulting pattern transformation are spontaneously induced by in-plane compressive stresses generated by the solvent swelling the thin polymeric films constrained on a substrate. Further, if the films are subjected to prestrain prior to swelling, the resulting buckling patterns depend on the type and magnitude of the prestrain (Zhang et al. 2008, 2009; Breid and Crosby 2011). The resulting periodic patterns have wavelengths in the order of 0.1–10 μm and can be formed homogeneously over large regions depending on the size of the film. This property has allowed researchers to generate complex patterns on nano- and microscales, control photonic and phononic properties, tune surface adhesion and wetting, and develop nanoprinting methods (Zhang et al. 2008, 2009; Yang et al. 2010).

When thin polymeric films with a square lattice of circular holes are exposed to a solvent, diamond plate patterns (DPPs) are typically observed in the absence of prestrain. The square array of circular holes buckles and transforms into a DPP, in which the circular holes are deformed into elliptical slits, and neighboring slits are arranged mutually perpendicular to each other. Zhang et al. (2008) observed DPPs in poly(dimethylsiloxane) (PDMS) films using toluene as a solvent, and investigated the effect of in-plane uniaxial pretension applied prior to solvent swelling on pattern transformation. Uniaxial pretension was applied along a lattice direction, and controlled by the nominal strain in this direction ($\phi = 0°$ and $\varepsilon_\phi \geq 0$ in Fig. 14.1). They

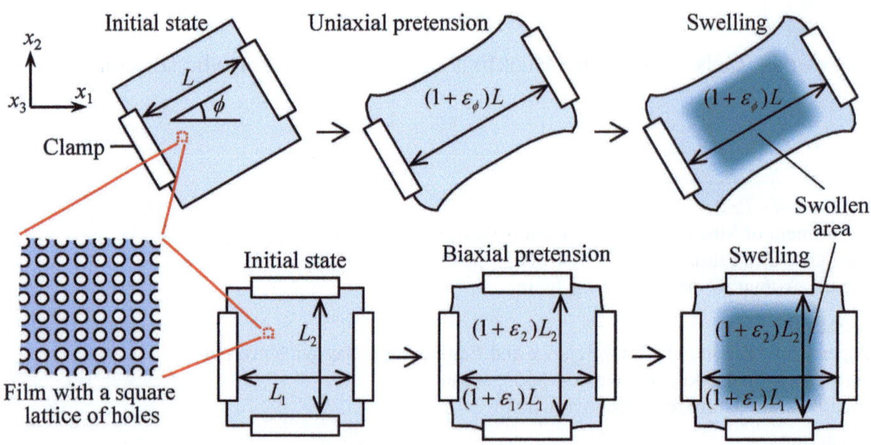

Fig. 14.1 Schematic diagrams of a polymeric film with a square lattice of holes being subjected to prestrain and then swelled by a solvent; uniaxial pretension is applied along a direction identified by the angle ϕ and biaxial pretension is identified by the biaxial strain ratio $\kappa\,(\varepsilon_2 = \kappa\varepsilon_1)$

observed that the resulting buckling pattern varied continuously as the magnitude of prestrain ε_ϕ was increased; the pattern evolved as a DPP ($\varepsilon_\phi = 0 - 0.1$), a slightly distorted DPP ($\varepsilon_\phi = 0.2$), a binary pattern of circles and lines ($\varepsilon_\phi = 0.3 - 0.5$) and a monotonous pattern of ellipses ($\varepsilon_\phi = 0.5 - 0.8$). In addition, when uniaxial pretension was applied along the direction 45° with respect to the lattice ($\phi = 45°$ and $\varepsilon_\phi = 0.64$ in Fig. 14.1), a monotonous pattern of slits formed with the slits aligned in the pretension direction (Zhang et al. 2009). Their experiments imply that by controlling the type and magnitude of prestrain, a rich variety of periodic patterns can be obtained from one polymeric film with a square lattice of holes and one solvent. More diverse types of periodic patterns are expected to form by imposing different types of prestrain, such as uniaxial pretension in other directions and biaxial pretension, but such investigations have not yet been reported.

DPPs formed in the absence of prestrain have been analyzed by Hong et al. (2009); Ding et al. (2013), and Okumura et al. (2014). They performed finite element analyses using the inhomogeneous field theory of polymeric gels in equilibrium. This theory was developed by Hong et al. (2009), in which the free-energy function of Flory and Rehner (1943) was applied to polymeric gels, and was implemented into the finite element software Abaqus using the user-defined material subroutine UHYPER. Okumura et al. (2015) also used this theory to analyze periodic units in the presence of uniaxial pretension in a lattice direction, and were able to successfully reproduce pattern changes as the pretension was increased of a DPP ($\varepsilon_\phi = 0$), a distorted DPP ($\varepsilon_\phi = 0.2$), a binary pattern of circles and lines ($\varepsilon_\phi = 0.4$), and a monotonous pattern of ellipses ($\varepsilon_\phi = 0.6$). They demonstrated that these different patterns appear continuously as transitional states during transformation into DPPs; the prestrain induced by uniaxial tension delays the onset of pattern transformation, while equilibrium swelling interrupts the progress of the transformation. This suggests that if uniaxial tension in other directions and biaxial tension are imposed as prestrains in finite element analysis, their effects on the resulting patterns can be analyzed using the approach of Okumura et al. (2015). It is interesting to investigate whether transformation into DPPs occurs regardless of the type of prestrain and whether more diverse patterns are obtained. Such an investigation may also allow us to attain a deeper understanding of the mechanism underlying swelling-induced buckling of polymeric films with a square lattice of holes subjected to prestrain.

The present study investigates swelling-induced buckling patterns in gel films with a square lattice of holes subjected to in-plane uniaxial and biaxial pretension. Finite element analysis is performed using the inhomogeneous field theory by Hong et al. (2009), which is briefly described in Sect. 14.2. Section 14.3 discusses numerical modeling based on the approach of Okumura et al. (2015). Periodic units including geometrical imperfections are analyzed under the assumption of generalized plane strain with the aid of artificial damping. A set of boundary and loading conditions and a set of material parameters are determined in accordance with experiments. Uniaxial and biaxial tensions are applied as pretensions prior to swelling. Section 14.4 presents and interprets the results of these simulations. Finally, conclusions are given in Sect. 14.5.

14.2 Inhomogeneous Field Theory

This section briefly describes the inhomogeneous field theory of polymeric gels in equilibrium, which was developed by Hong et al. (2009). This theory considers that a polymer network is in contact with a solvent and subjected to mechanical loads and geometric constraints at a constant temperature. If the stress-free, dry network is taken as a reference state, the deformation gradient of the network is defined as $F_{ij} = \mathrm{d}x_i(\boldsymbol{X})/\mathrm{d}X_j$, where X_j and $x_i(\boldsymbol{X})$ are the network coordinates of a gel system in reference and deformed states, respectively. When $C(\boldsymbol{X})$ is defined as the concentration of solvent molecules at a point in the gel system, the gel is in an equilibrium state characterized by the two fields $x_i(\boldsymbol{X})$ and $C(\boldsymbol{X})$. The free-energy density of the gel, W, is assumed to be a function of the deformation gradient, $\boldsymbol{F}$, and the concentration of solvent in the gel, C; i.e., $W(\boldsymbol{F}, C)$. The inhomogeneous field theory may be applied to various free-energy functions for swelling elastomers, but the present study uses the specific free-energy function of Flory and Rehner (1943). This is because this form is known to provide a basis for interpreting the swelling behavior of polymeric gels (Treloar 1975), and has been demonstrated to reproduce swelling-induced buckling patterns in gel films with a square lattice of holes, as mentioned in Sect. 14.1 (Hong et al. 2009; Ding et al. 2013; Okumura et al. 2014, 2015).

The free-energy function of Flory and Rehner for a polymeric gel consists of two terms associated with stretching and mixing of the free energies (Flory and Rehner 1943):

$$W = \frac{1}{2}NkT(I - 3 - 2\log J) - \frac{kT}{v}\left[vC\log\left(1 + \frac{1}{vC}\right) + \frac{\chi}{1 + vC}\right], \quad (14.1)$$

where $I = F_{ij}F_{ij}$ and $J = \det\boldsymbol{F}$ are invariants of the deformation gradient, N is the number of polymeric chains per reference volume, kT is the absolute temperature in the unit of energy, v is the volume per solvent molecule, and χ is the Flory-Huggins interaction parameter that characterizes the enthalpy of mixing.

Considering δx_i and δC to be arbitrary variations of x_i and C, respectively, from a state of equilibrium, the virtual work principle gives an equilibrium equation in which the change of the free energy of the gel equals the sum of the work done by the external mechanical force and external solvent:

$$\int_V \delta W \mathrm{d}V = \int_V B_i \delta x_i \mathrm{d}V + \int_A T_i \delta x_i \mathrm{d}A + \mu \int_V \delta C \mathrm{d}V, \quad (14.2)$$

where V is the reference volume and A is the reference surface. The first and second terms on the right-hand side of Eq. (14.2) are the mechanical work done by body forces and surface forces, respectively, and the third term represents the work done by the external solvent. Here, μ is the chemical potential of the external solvent, and is equivalent to that in the gel; i.e., $\mu = \partial W/\partial C$. A Legendre transformation allows the free-energy function $W(\boldsymbol{F}, C)$ to be transformed into another form

$$\hat{W} = W - \mu C, \tag{14.3}$$

which is defined as a function of $\boldsymbol{F}$ and μ; i.e., $\hat{W}(\boldsymbol{F}, \mu)$. Combination of Eqs. (14.2) and (14.3) leads to

$$\int_V \delta \hat{W} \mathrm{d}V = \int_V B_i \delta x_i \mathrm{d}V + \int_A T_i \delta x_i \mathrm{d}A \tag{14.4}$$

When the gel is in a state of equilibrium, the chemical potential of the solvent molecules in the gel is homogeneous and equals the chemical potential of the external solvent, μ. Consequently, μ is regarded as a state variable, and the equilibrium condition (14.4) takes the same form as that for a hyperelastic solid.

Assuming that the network of polymers and pure liquid solvent are incompressible, the volume of the gel can be expressed as the sum of the volume of the dry network and that of the swelling solvent. This assumption leads to

$$1 + vC = J \tag{14.5}$$

Using Eqs. (14.1), (14.3) and (14.5), the Flory-Rehner free-energy function can be rewritten as

$$\hat{W} = \frac{1}{2} NkT(I - 3 - 2\log J) - \frac{kT}{v}\left[(J-1)\log\frac{J}{J-1} + \frac{\chi}{J}\right] - \frac{\mu}{v}(J-1) \tag{14.6}$$

It is noted that the free-energy function in Eq. (14.6) takes an explicit form as a function of $\boldsymbol{F}$ and μ. Equation (14.6) acts as a free-energy function for a compressible hyperelastic material because of the volumetric change induced by solvent absorption. For example, the finite element package Abaqus offers a user-defined subroutine, UHYPER or UMAT, to implement the constitutive behavior of Eq. (14.6), and μ is passed into UHYPER or UMAT using the variable of temperature (Hong et al. 2009; Kang and Huang 2010).

The present study uses the user-defined subroutine UHYPER developed by Hong et al. (2009), in which the swelling process is simulated by increasing the chemical potential from a quasi-dry state ($\mu = \mu_0 < 0$) to an equilibrium swelling state ($\mu_0 = 0$). The quasi-dry state (Okumura et al. 2015) is a free swelling state, which is used to avoid a singularity in the dry state ($\mu = \infty$ because $vC = 0$), and is characterized by the homogeneous deformation gradient $F_{ij}^{\circ} = \lambda_0 \delta_{ij}$; i.e., $vC = \lambda_0^3 - 1 > 0$. The corresponding stress state is assumed to be homogeneously zero, and thus the finite value of μ_0 is calculated analytically using a small deviation of λ_0 from 1. Next, the equilibrium swelling state ($\mu = 0$) is defined as the state in which the network of polymers is in contact with the liquid solvent. The effect of atmospheric pressure is expected to be negligible, and thus can be ignored.

14.3 Numerical Modeling

Figure 14.1 shows schematic diagrams of polymeric films with a square lattice of holes subjected to in-plane pretension and then swelled by a solvent. The present study considers two types of pretension. Uniaxial pretension is applied by imposing a nominal strain $\varepsilon_\phi (\geq 0)$ along a direction identified by the angle ϕ, while biaxial pretension is applied by imposing nominal strains ε_1 and ε_2 along x_1 and x_2 directions, respectively ($\varepsilon_1 \geq 0$ and $\varepsilon_2 \geq 0$). The ratio of $\varepsilon_1 : \varepsilon_2$ is fixed in each simulation; i.e., $\varepsilon_2 = \kappa\varepsilon_1$, where $\kappa (\geq 0)$ is a constant. A controlled amount of nominal strain, ε_ϕ for uniaxial pretension or ε_1 and ε_2 for biaxial pretension, is retained during swelling. Swelling is induced by exposing the central region of the polymeric film surface to a liquid solvent. As a result, the swelling in this region of the polymeric film is constrained by the surrounding regions including its sides and bottom. The resulting in-plane compressive stress acts as a driving force to induce buckling and pattern transformation.

Zhang et al. (2008) experimented with PDMS films using toluene as a solvent, and observed diverse patterns, e.g., DPPs, distorted DPPs, binary patterns of circles and lines and monotonous patterns of ellipses, depending on the magnitude of uniaxial pretension applied along a lattice direction. These patterns were successfully analyzed by finite element analysis using generalized plane strain elements (Okumura et al. 2014, 2015). In this study, the approach developed by Okumura et al. (2015) is used to analyze the effect of uniaxial and biaxial pretensions on the pattern transformation of PDMS during swelling with toluene. A type of generalized plane strain element, CPEG4H, is used in the Abaqus analysis. The dimensions of the gel film are reduced to the hole diameter d and pitch l, where $d = 0.75\ \mu\text{m}$ and $l = 1.5\ \mu\text{m}$ (see Fig. 14.2). Material parameters for the system of PDMS and toluene are determined based on experimental data (Okumura et al. 2015); $kT = 4 \times 10^{-21}$ J (at room temperature), $NkT = 0.5$ MPa ($3NkT$ = Young's modulus for the dry network of PDMS), $v = 1.76 \times 10^{-28}$ m^3 and $\chi = 0.7$.

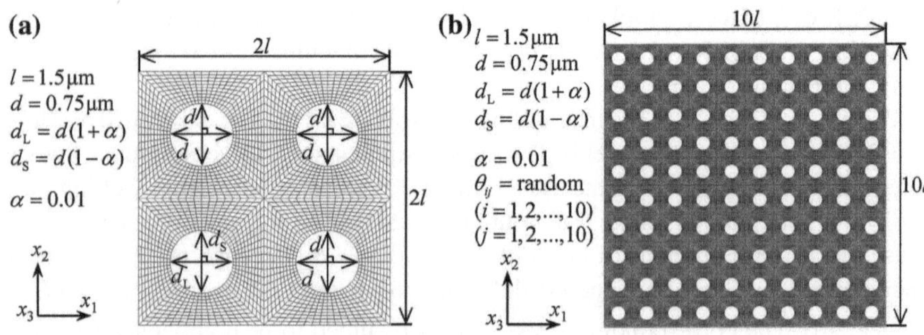

Fig. 14.2 Initial configuration and finite element meshes; (a) 2×2 unit cell with a simple imperfection and (b)10×10 unit cell with random imperfections

Periodic units consisting of 2 × 2 and 10 × 10 unit cells are analyzed. Figure 14.2 shows the periodic units and their finite element meshes. The numbers of nodes and elements are 2749 and 2560, respectively, for the 2 × 2 unit cell (Fig. 14.2a), and 67,421 and 64,000, respectively, for the 10 × 10 unit cell (Fig. 14.2b). Imperfections are introduced by considering each hole as a randomly oriented elliptical hole (Okumura et al. 2014, 2015). Elliptical holes are defined using two parameters, α and θ. The parameter α represents the magnitude of imperfections, so that the major and minor diameters of an elliptical hole are expressed as $d_L = d(1+\alpha)$ and $d_S = d(1-\alpha)$, respectively. The other parameter θ represents the angle between the major diameter and x_i direction. In the 10 × 10 unit cell (Fig. 14.2b), the angles $\theta_{ij}(i = 1-10,\ j = 1-10)$ are assigned to individual holes, and determined using random numbers. The set of θ_{ij} is not listed to avoid redundancy, and for simplicity, α is fixed at 0.01 for all holes. In contrast, the 2 × 2 unit cell (Fig. 14.2a) includes three perfect circular holes and one elliptical hole characterized by $\alpha = 0.01$ and $\theta = 0°$. The 2 × 2 unit cell is used for parametric studies to investigate the effect of in-plane pretensions on pattern transformation, while the 10 × 10 unit cell is used to verify the results obtained from the analysis of the 2 × 2 unit cell.

The periodic boundary conditions imposed on the boundary of each periodic unit can be expressed as (Bertoldi et al. 2008)

$$u_i^{(+)} - u_i^{(-)} = (\overline{F}_{ij} - \delta_{ij})(X_j^{(+)} - X_j^{(-)}) = \overline{H}_{ij}(X_j^{(+)} - X_j^{(-)}), \tag{14.7}$$

where $u_i^{(+)}$ and $u_i^{(-)}$ are the displacements at a point $X_j^{(+)}$ on the periodic unit boundary and the opposite point $X_j^{(-)}$, respectively, and the displacement is defined as $u_i = x_i - X_i$. In Eq. (14.7), $\overline{F}_{ij}$ and δ_{ij} denote the macroscopic deformation gradient and Kronecker delta, respectively, and thus $\overline{H}_{ij}$ is the macroscopic displacement gradient. The out-of-plane component $\overline{H}_{33}$ is determined using the generalized plane strain condition; i.e., using $\overline{S}_{33} = 0$, where $\overline{S}_{ij}$ denotes the macroscopic first Piola-Kirchhoff stress. Here, $\overline{F}_{ij}$, $\overline{H}_{ij}$ and $\overline{S}_{ij}$ are the macroscopic variables relative to the quasi-dry state, which is introduced using $\lambda_0 = 1.01$, resulting in $\mu_0 = -1.90\,kT$ (see Sect. 14.2). At this reference state ($\mu_0 = -1.90\,kT$), $\overline{F}_{ij}$, $\overline{H}_{ij}$ and $\overline{S}_{ij}$ are initialized as $\overline{F}_{ij} = \delta_{ij}$, $\overline{H}_{ij} = 0$ and $\overline{S}_{ij} = 0$. Uniaxial and biaxial pretensions are assumed to affect $\overline{F}_{ij}$, $\overline{H}_{ij}$ and $\overline{S}_{ij}$. Uniaxial pretension is expressed as $\overline{S}_{11} = \sigma_\phi \cos^2\phi$, $\overline{S}_{22} = \sigma_\phi \sin^2\phi$ and $\overline{S}_{12} = \sigma_\phi \sin\phi\cos\phi$. Here, σ_ϕ is the macroscopic nominal stress conjugate to $\varepsilon_\phi (\geq 0)$, and thus a set of $\overline{H}_{11}$, $\overline{H}_{22}$ and $\overline{H}_{12}$ is obtained when the magnitude of ε_ϕ is determined. In contrast, biaxial pretension is expressed as $\overline{H}_{11} = \varepsilon_1$, $\overline{H}_{22} = \varepsilon_2$ and $\overline{H}_{12} = 0$ ($\varepsilon_1 \geq 0, \varepsilon_2 \geq 0$), and the set is obtained when the magnitude of ε_1 and the biaxial strain ratio $\kappa(\varepsilon_2 = \kappa\varepsilon_1)$ are determined. Under the constraints of $\overline{H}_{11}$, $\overline{H}_{22}$ and $\overline{H}_{12}$, the swelling process is simulated by incrementally increasing μ from μ_0 to 0.

In the Abaqus analysis, automatic time incrementation is used, but the problem to be solved has snap-through instability just after the onset of pattern transformation, so iterative calculations for the next increment cannot be completed (Hong et al. 2009;

Okumura et al. 2014). To avoid this problem, the present study employs an automatic stabilization scheme with a constant damping factor, which is available in Abaqus using the STABILIZE option. The damping factor is automatically calculated from the dissipated energy fraction ω. The value of this fraction needs to be optimized by trial-and-error calculations, and is thus taken as $\omega = 10^{-12}$ for the 2×2 unit cell and $\omega = 10^{-10}$ for the 10×10 unit cell (see Okumura et al. 2015). Incidentally, the self-contact of individual surfaces included in periodic units can occur during analysis. Therefore, the CONTACT INCLUSIONS option is also used in the Abaqus analysis. Contacts are idealized as frictionless because the friction between contact surfaces is not expected to be a factor contributing to pattern transformation.

To quantify the progress of pattern transformation in a deformed state, a measure of the deviation from roundness for individual holes is used (Okumura et al. 2014, 2015). This scalar value is zero when the hole is a perfect circle, and increases as the hole deviates from a perfect circle. The deviation from roundness is defined as

$$\Delta_{ij} = \min_{\boldsymbol{x}_{ij} \in \boldsymbol{x}} (r_{\max}(\boldsymbol{x}_{ij}) - r_{\min}(\boldsymbol{x}_{ij})) \tag{14.8}$$

Here, $r_{\max}$ and $r_{\min}$ are the radii of circumscribed and inscribed circles of the hole identified by i and j (e.g., $i = 1-10$, $j = 1-10$ for 10×10 unit cells), respectively. Point $\boldsymbol{x}_{ij}$ is the center that minimizes the value $r_{\max} - r_{\min}$. To find the point $\boldsymbol{x}_{ij}$, trial calculations are performed at individual points on a square lattice with a 1 nm pitch. Consequently, $\overline{\Delta}$ and $\overline{\Delta}_{\rm sd}$ are calculated as the average of all holes and its standard deviation, respectively. As will be shown in the next section, the progress of pattern transformation is able to be clearly quantified using the sum of $\overline{\Delta}$ and $\overline{\Delta}_{\rm sd}$, because Okumura et al. (2015) reported that only the change of $\overline{\Delta}$ may miss the progress in the presence of prestrain.

14.4 Results and Discussion

14.4.1 Uniaxial Pretension

Figure 14.3 shows the deformed configurations at equilibrium swelling obtained for the 2×2 unit cell subjected to uniaxial pretension. Here, vC is a non-dimensional value of concentration C and represents the volumetric change induced by solvent swelling (Eq. (14.5)). Figure 14.3 indicates that the absence of prestrain ($\varepsilon_\phi = 0$) predicts well the transformation into a DPP (Okumura et al. 2014), and pretension along a lattice direction ($\phi = 0°$ and $\varepsilon_\phi = 0.2, 0.4$ and 0.6) successfully reproduces a distorted DPP ($\varepsilon_\phi = 0.2$), a binary pattern of circles and lines ($\varepsilon_\phi = 0.4$), and a monotonous pattern of ellipses ($\varepsilon_\phi = 0.6$), respectively. These predictions agree well with the experimental results reported by Zhang et al. (2008). Okumura et al. (2015) revealed that these different patterns appear continuously as transitional states during transformation into DPPs. Transformation into DPPs occurs despite the

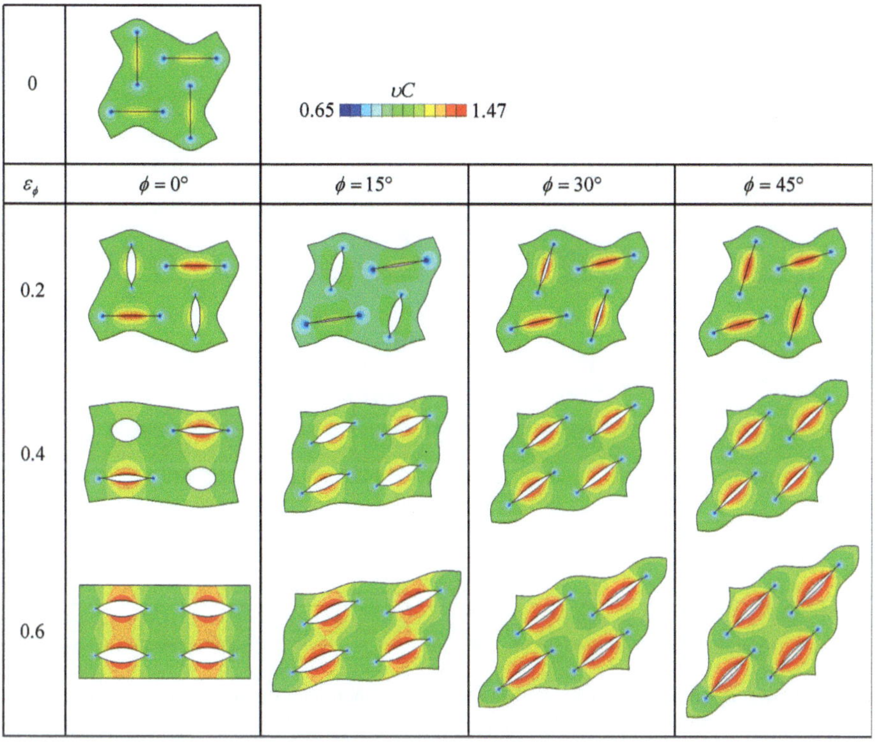

Fig. 14.3 Deformed configurations and normalized concentration distributions υC at equilibrium swelling for uniaxial pretensions determined by ϕ and ε_ϕ (2×2 unit cell)

presence of uniaxial pretension along a lattice direction, but an increase in the pretension delays the onset of pattern transformation. This delay limits further swelling during the transformation because the total amount of swelling is restricted by equilibrium swelling; i.e., a combination of polymer and solvent. As a result, transitional patterns, such as the distorted DPP and the binary pattern of circles and lines, are formed depending sensitively on the magnitude of pretension. Monotonous patterns are formed without pattern transformation when the magnitude of pretension is relatively larger. The resulting patterns can be categorized into three groups: DPPs, transitional patterns, and monotonous patterns.

Figure 14.3 also shows that uniaxial pretension along other directions ($\phi = 15°, 30°$ and $45°$) create diverse patterns that are sensitive to both ϕ and ε_ϕ. In the case of $\phi = 45°$ a DPP appears when $\varepsilon_\phi = 0.2$, but neighboring slits are not perpendicular to each other (cf. $\varepsilon_\phi = 0$) and have a specific angle arising from $\phi = 45°$ and $\varepsilon_\phi = 0.2$. When $\varepsilon_\phi = 0.4$ and 0.6, the resulting patterns are monotonous. Zhang et al. (2009) observed experimentally the corresponding pattern when $\phi = 45°$ and $\varepsilon_\phi = 0.64$. This agreement supports the validity of numerical modeling in the present study. In other cases ($\phi = 15°$ and $30°$), monotonous patterns are predicted

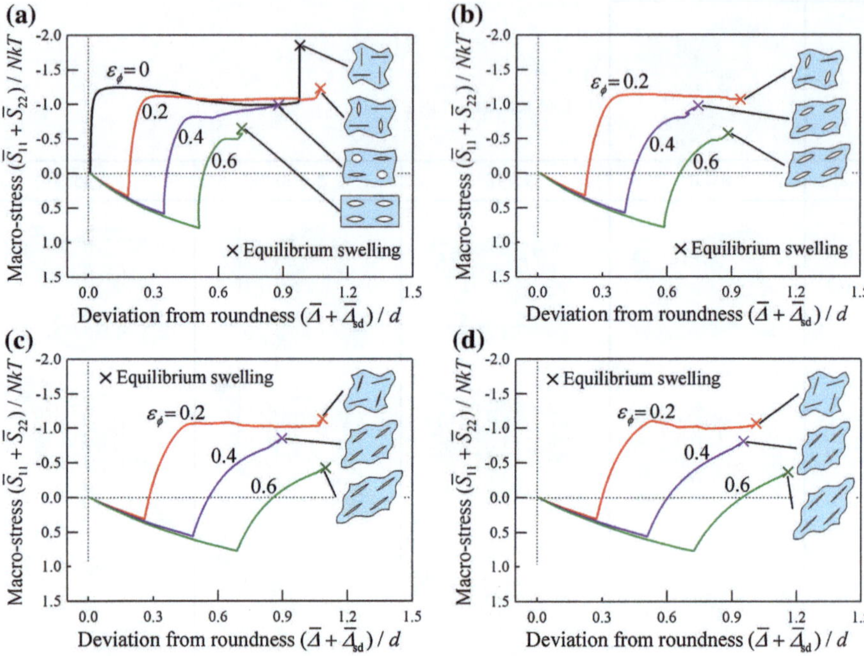

Fig. 14.4 Macroscopic stress, $(\overline{S}_{11}+\overline{S}_{22})/NkT$, as a function of the sum of the average deviation from roundness and its standard deviation, $(\overline{\Delta}+\overline{\Delta}_{sd})/d$; (**a**–**d**) for uniaxial pretensions along individual directions of $\phi = 0°$, $15°$, $30°$ and $45°$ (2×2 unit cell)

when the magnitude of pretension is large, i.e., $\varepsilon_\phi = 0.4$ and 0.6 ($\phi = 30°$) and $\varepsilon_\phi = 0.6(\phi = 15°)$, and elliptical slits align in the pretension direction. For moderate magnitudes of pretension, transitional patterns appear; for example, distorted DPPs for $\varepsilon_\phi = 0.2(\phi = 15°$ and $30°)$ and a slightly distorted monotonous pattern for $\varepsilon_\phi = 0.4(\phi = 15°)$. Although the resulting patterns are affected by both ϕ and ε_ϕ, a tendency is maintained regardless of the direction of pretension; the patterns evolve as DPPs, transitional patterns and monotonous patterns as the magnitude of pretension increases.

To investigate the progress of pattern transformation in greater detail, Fig. 14.4 plots the macroscopic stress as a function of the deviation from roundness. Swelling subsequent to pretension increases the sum of in-plane compressive stress components, $(\overline{S}_{11}+\overline{S}_{22})/NkT$. When this compressive stress attains a critical value to cause buckling, the resulting pattern transformation starts with a dramatic increase of the deviation from roundness for each hole. The sum of the average deviation from roundness and its standard deviation, $(\overline{\Delta}+\overline{\Delta}_{sd})/d$, which is used in the present study, estimates well the progress of pattern transformation in the presence of prestrains (cf. Okumura et al. 2015), as will be described below.

Figure 14.4 indicates that monotonous patterns ($\varepsilon_\phi = 0.6$ for $\phi = 0°$ and $15°$, and $\varepsilon_\phi = 0.4$ and 0.6 for $\phi = 30°$ and $45°$) form as a consequence of no pattern

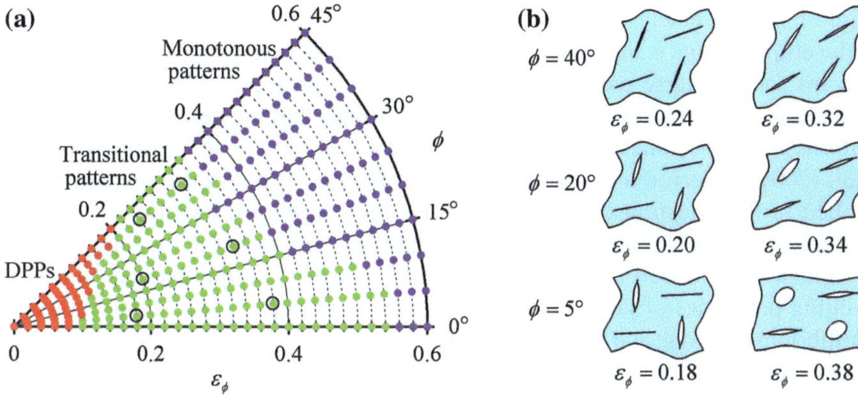

Fig. 14.5 Dependence of patterns at equilibrium swelling on uniaxial pretension; **a** diagram of pattern transformation of a PDMS film as a function of ε_ϕ and ϕ, and **b** transitional patterns predicted at marked points in the diagram (2 × 2 unit cell)

transformation because a dramatic increase of $(\overline{\Delta} + \overline{\Delta}_{\mathrm{sd}})/d$ does not appear except for $\varepsilon_\phi = 0.6(\phi = 0°)$. It is noted that in this particular case, the relatively small increase of $(\overline{\Delta} + \overline{\Delta}_{\mathrm{sd}})/d$ is caused by the appearance of creasing around the edges of individual holes (Okumura et al. 2015), so this increase does not lead to pattern transformation. In contrast, DPPs ($\varepsilon_\phi = 0$ and $\varepsilon_\phi = 0.2$ for $\phi = 45°$) and transitional patterns, such as distorted DPPs ($\varepsilon_\phi = 0.2$ for $\phi = 0°$, 15° and 30°) and binary patterns of circles and lines ($\varepsilon_\phi = 0.4$ for $\phi = 0°$), form as the value of $(\overline{\Delta} + \overline{\Delta}_{\mathrm{sd}})/d$ increases markedly under an almost constant stress. Transitional patterns are understood to form as a result of the transformation into DPPs being interrupted by swelling equilibrium (Okumura et al. 2015). This mechanism enables us to generate diverse patterns simply by controlling the magnitude of pretension. In the case of $\varepsilon_\phi = 0.4(\phi = 15°)$, a slightly distorted monotonous pattern is obtained, which originates from equilibrium swelling being reached immediately after the onset of the transformation into a DPP. A slight increase of $(\overline{\Delta} + \overline{\Delta}_{\mathrm{sd}})/d$ just before equilibrium swelling is caused by creasing, similar to that observed when $\varepsilon_\phi = 0.6(\phi = 0°)$, instead of pattern transformation. Figures 14.3 and 14.4 clearly show that transitional patterns appear in specific ranges of ε_ϕ and depend sensitively on ϕ. When $\phi = 0°$ and 15°, transitional patterns appear within a wide range of ε_ϕ of about $0.1 \le \varepsilon_\phi \le 0.5$. In contrast, when $\phi = 30°$ and 45°, transitional patterns appear within a narrow range of ε_ϕ of about $0.2 \le \varepsilon_\phi \le 0.3$.

Figure 14.5a depicts a diagram of pattern transformation of a PDMS film as a function of ε_ϕ and ϕ, which is obtained from analyses using a large number of combinations of ϕ and ε_ϕ. This diagram indicates three regions in which DPPs, transitional patterns and monotonous patterns appear at equilibrium swelling. In the region where DPPs form, the patterns are not all the same DPP, and have a specific angle between neighboring slits. The deviation of this angle from 90° increases as ϕ is increased from 0° to 45° and as the magnitude of ε_ϕ is increased from 0

(Fig. 14.3). Conversely, in the region where monotonous patterns form, the direction of the elliptical slits is identical to that of ϕ (Fig. 14.3). In the region where transient patterns form, diverse patterns appear that depend on the combination of ε_ϕ and ϕ (Figs. 14.3 and 14.5b). Figure 14.5a clearly shows that the range where transient patterns appear becomes narrower as ϕ is increased from 0° to 45°. Although Zhang et al. (2009) observed a monotonous pattern of slits when uniaxial pretension was applied in the direction of $\phi = 45°$, DPPs and transitional patterns can form in the range of $\varepsilon_\phi < 0.35$ (Fig. 14.5a). The resulting patterns strongly depend on the magnitude of ε_ϕ as well as on ϕ. This dependence originates from the mechanical and swelling properties of gels; swelling proceeds spontaneously but is interrupted by equilibrium swelling, and the pretension not only distorts the initial arrangements of a square lattice of holes, but also delays pattern transformation induced by compressive stresses from swelling.

14.4.2 Biaxial Pretension

Figure 14.6 shows the deformed configurations at equilibrium swelling obtained for the 2 × 2 unit cell subjected to biaxial pretension. The resulting patterns can

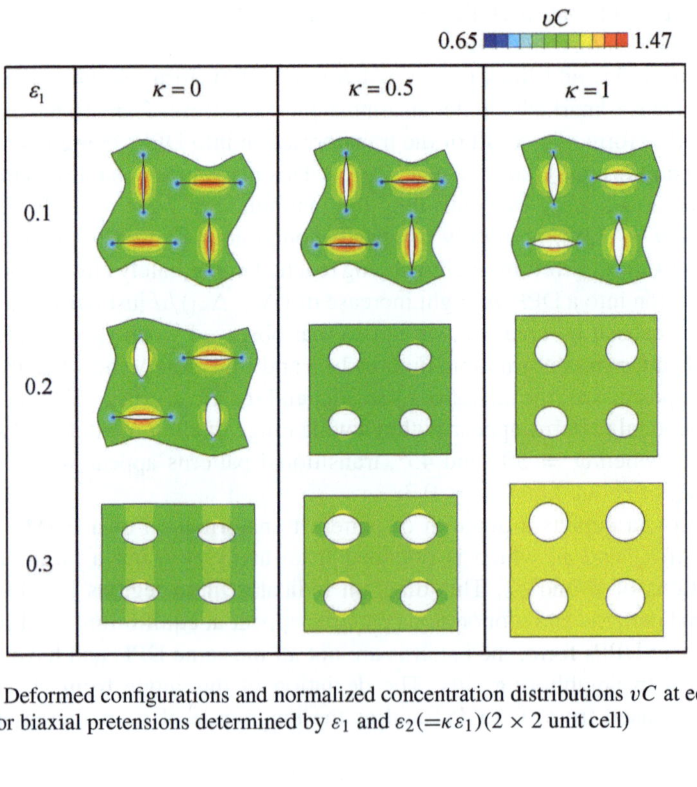

Fig. 14.6 Deformed configurations and normalized concentration distributions υC at equilibrium swelling for biaxial pretensions determined by ε_1 and $\varepsilon_2(=\kappa\varepsilon_1)(2 \times 2$ unit cell)

be categorized as DPPs, transient patterns and monotonous patterns. Like uniaxial pretension, biaxial pretension does not cause different buckling modes, and DPPs appear predominantly when pattern transformation occurs regardless of the presence of uniaxial or biaxial pretension. It is noted that biaxial pretension preserves a lattice direction perpendicular to another lattice direction, so that transformation into DPPs occurs under the condition where neighboring slits are arranged mutually perpendicular to each other (cf. uniaxial pretensions except for $\phi = 0°$ in Fig. 14.3). Although the selected values of $\varepsilon_1 = 0.1, 0.2$ and 0.3 ($\varepsilon_2 = \kappa\varepsilon_1$ with $\kappa = 0, 0.5$ and 1) are relatively smaller than the magnitude of prestrain in uniaxial pretension (i.e., $\varepsilon_\phi = 0.2, 0.4$ and 0.6), in the cases of $\varepsilon_1 = 0.3(\kappa = 0, 0.5$ and 1), and $\varepsilon_1 = 0.2(\kappa = 0.5$ and 1), pattern transformation does not occur and monotonous patterns form. Especially for $\kappa = 1$, i.e., equibiaxial pretension, the monotonous patterns consist of circular holes ($\varepsilon_1 = 0.2$ and 0.3) and the transient pattern ($\varepsilon_1 = 0.1$) is a DPP but does not consist of perfectly closed slits. This pattern is regarded as a transient pattern in the present study to distinguish it from DPPs with perfectly closed slits. Transient patterns appear only when $\varepsilon_1 = 0.1(\kappa = 0, 0.5$ and 1) and $\varepsilon_1 = 0.2(\kappa = 0)$. Biaxial pretension is found to have a stronger ability to prevent buckling than uniaxial tension (cf. Fig. 14.3).

Figure 14.7 shows the macroscopic stress as a function of the deviation from roundness for biaxial pretensions with biaxial strain ratios of $\kappa = 0, 0.5$ and 1.0.

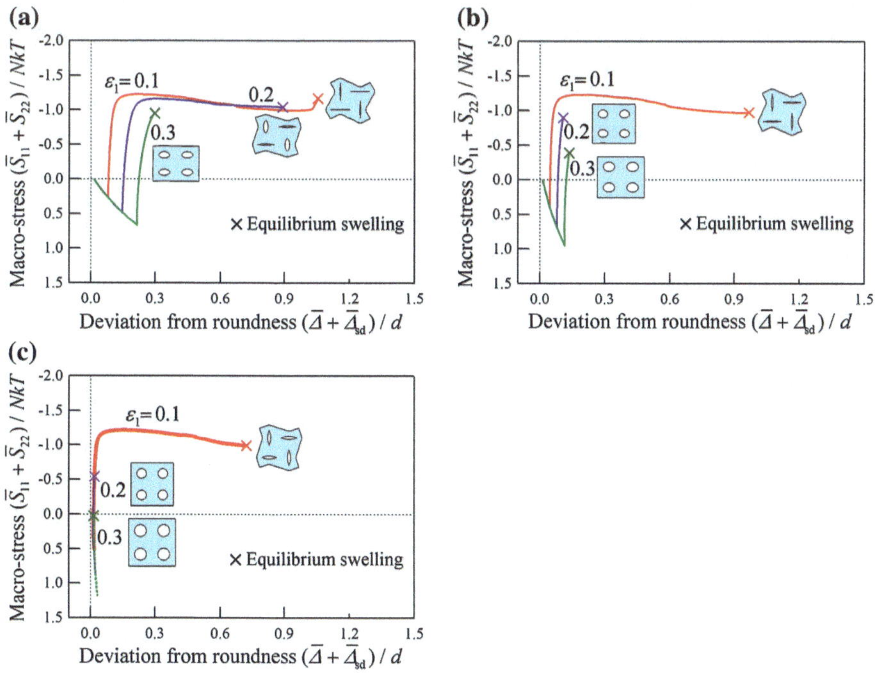

Fig. 14.7 Macroscopic stress, $(\overline{S}_{11} + \overline{S}_{22})/NkT$, as a function of the sum of the average deviation from roundness and its standard deviation, $(\overline{\Delta} + \overline{\Delta}_{sd})/d$; (**a**–**c**) for biaxial pretensions with biaxial strain ratios of $\kappa = 0, 0.5$ and 1 (2 × 2 unit cell)

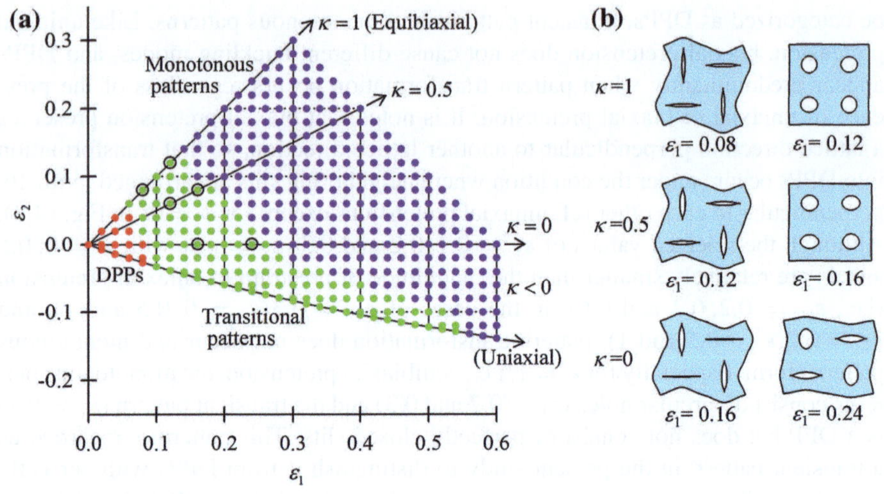

Fig. 14.8 Dependence of patterns at equilibrium swelling on biaxial pretension; **a** diagram of pattern transformation of a PDMS film as a function of ε_1 and $\varepsilon_2 (=\kappa\varepsilon_1)$, and **b** transitional patterns predicted at marked points in the diagram (2 × 2 unit cell)

Individual figures clearly show that when pretensions increase tensile stresses, this increase prevents compressive stresses induced by swelling from attaining a critical value to cause pattern transformation, so pattern transformation does not occur. If further swelling is allowed, monotonous and transitional patterns are expected to proceed to DPPs. However, the total amount of swelling is restricted by a combination of polymers and solvents, so a diverse range of transitional and monotonous patterns form. Figures 14.6 and 14.7 demonstrate that when biaxial pretension is applied to a PDMS film, transitional patterns appear within a narrow range of about $0.1 \leq \varepsilon_1 \leq 0.2$. This range becomes narrower as κ increases from 0 to 1.

Figure 14.8a shows a diagram of pattern transformation of a PDMS film as a function of ε_1 and ε_2. This diagram is obtained by also considering the region between the biaxial pretension defined by $\kappa = 0$ and the uniaxial pretension along the x_1 direction (i.e., the region where $\kappa < 0$ and $\overline{S}_{22} > 0$). Uniaxial pretension along the lattice direction has a wide region where transient patterns appear, while biaxial pretensions have narrower transitional regions as κ increases to 1. Figure 14.8b depicts transitional patterns at equilibrium swelling for $\kappa = 0$, 0.5 and 1. Distorted DPPs, binary patterns of circles and lines and distorted monotonous patterns are expected to appear in the narrow transitional region. The expected patterns form depending on ε_1 (Figs. 14.6 and 14.8b). When κ approaches 1, and especially in the case where $\kappa = 1$, the resulting patterns are not diverse and are mostly DPPs or monotonous patterns of circles. This is because equibiaxial pretension has a tendency to maintain a square array of circular holes.

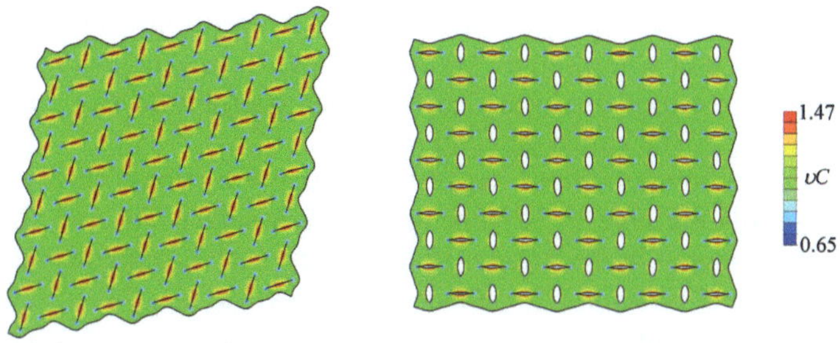

Fig. 14.9 Deformed configurations and normalized concentration distributions υC at equilibrium swelling for uniaxial and biaxial pretensions (10×10 unit cell)

14.4.3 Dependence on Unit Cell Size

The 2×2 unit cell analyzed in Sects. 14.4.1 and 14.4.2 may fail to predict the dominant buckling patterns of a larger unit cell because the resulting patterns are restricted by the periodicity of the 2×2 unit cell. Larger unit cells need to be analyzed to investigate long-wavelength buckling (Okumura et al. 2004; Ohno et al. 2004; Erami et al. 2006; Okumura et al. 2008; Takahashi et al. 2010) and verify the results obtained from the analysis of the 2×2 unit cell. Figure 14.9 presents the deformed configurations at equilibrium swelling obtained for the 10×10 unit cell. Deformed configurations illustrated in this figure are restricted to only two cases to avoid redundancy; Okumura et al. (2015) have reported them for the case of uniaxial pretension of $\phi = 0°$. These results indicate that long-wavelength buckling does not occur, and the resulting patterns have the periodicity of the 2×2 unit cells. The 2×2 unit cell used in the present study, which includes a simple imperfection (see Sect. 14.3), successfully reproduces the patterns predicted by the analysis of the 10×10 unit cell (see Figs. 14.3 and 14.6).

14.5 Conclusions

We investigated swelling-induced buckling patterns in gel films with a square lattice of holes subjected to in-plane uniaxial and biaxial pretensions. The inhomogeneous field theory (Hong et al. 2009) was used to analyze periodic units consisting of 2×2 and 10×10 unit cells. The main findings of this study are as follows.

The resulting patterns are highly diverse and strongly depend on the type and magnitude of pretensions. The patterns arise from either transformation into DPPs or no pattern transformation. Diagrams of pattern transformation consist of three regions: DPPs, transient patterns and monotonous patterns. Monotonous patterns form when there is no pattern transformation, while DPPs are generated by buckling and are the dominant buckling pattern, which is not affected by the presence of uniaxial or biaxial pretensions. Pretensions not only distort the initial arrangement of the square lattice of holes in the PDMS films, but also delay the onset of pattern transformation. Transitional patterns appear when equilibrium swelling interrupts the transformation of the lattice into DPPs.

The above-mentioned findings suggest that simply by controlling pretensions, numerous patterns are able to be generated from a single film and solvent. The approach demonstrated in the present study is very useful to predict formation of patterns that depend considerably on material properties, film dimensions, and the type and magnitude of pretensions.

References

Bertoldi K, Boyce M, Deschanel S, Prange S, Mullin T (2008) Mechanics of deformation-triggered pattern transformations and superelastic behavior in periodic elastomeric structures. J Mech Phys Solids 56:2642–2668

Breid D, Crosby A (2009) Surface wrinkling behavior of finite circular plates. Soft Matter 5:425–431

Breid D, Crosby A (2011) Effect of stress state on wrinkle morphology. Soft Matter 7:4490–4496

Ding Z, Liu Z, Hu J, Swaddiwudhipong S, Yang Z (2013) Inhomogeneous large deformation study of temperature-sensitive hydrogel. Int J Solids Struct 50:2610–2619

Erami K, Ohno N, Okumura D (2006) Long-wave in-plane buckling of elasto-plastic square honeycombs. Int J Plast 22:1569–1585

Flory P, Rehner J (1943) Statistical mechanics of cross-linked polymer networks. II swelling. J Chem Phys 11:521–526

Hong W, Liu Z, Suo Z (2009) Inhomogeneous swelling of a gel in equilibrium with a solvent and mechanical load. Int J Solids Struct 46:3282–3289

Kang M, Huang R (2010) A variational approach and finite element implementation for swelling of polymeric hydrogels under geometric constraints. Trans ASME J Appl Mech 77(061):004

Ohno N, Okumura D, Niikawa T (2004) Long-wave buckling of elastic square honeycombs subject to in-plane biaxial compression. Int J Mech Sci 46:1697–1713

Okumura D, Ohno N, Noguchi H (2004) Elastoplastic microscopic bifurcation and post-bifurcation behavior of periodic cellular solids. J Mech Phys Solids 52:641–666

Okumura D, Okada A, Ohno N (2008) Buckling behavior of Kelvin open-cell foams under [001], [011] and [111] compressive loads. Int J Solids Struct 45:3807–3820

Okumura D, Kuwayama T, Ohno N (2014) Effect of geometrical imperfections on swelling-induced buckling patterns in gel films with a square lattice of holes. Int J Solids Struct 51:154–163

Okumura D, Inagaki T, Ohno N (2015) Effect of prestrains on swelling-induced buckling patterns in gel films with a square lattice of holes. Int J Solids Struct 58:288–300

Takahashi Y, Okumura D, Ohno N (2010) Yield and buckling behavior of Kelvin open-cell foams subjected to uniaxial compression. Int J Mech Sci 52:377–385

Tanaka T, Sun S, Hirokawa Y, Katayama S, Kucera J, Hirose Y, Amiya T (1987) Mechanical instability of gels at the phase transition. Nature 325:796–798

Treloar L (1975) The physics of rubber elasticity. Oxford University Press, Oxford

Wu G, Xia Y, Yang S (2014) Buckling, symmetry breaking, and cavitation in periodically micro-structured hydrogel membranes. Soft Matter 10:1392–1399

Yang S, Khare K, Lin P (2010) Harnessing surface wrinkle patterns in soft matter. Adv Funct Mater 20:2550–2564

Zhang Y, Matsumoto E, Peter A, Lin P, Kamien R, Yang S (2008) One-step nanoscale assembly of complex structures via harnessing of an elastic instability. Nano Lett 8:1192–1196

Zhang Y, Reed J, Yang S (2009) Creating a library of complex metallic nanostructures via harnessing pattern transformation of a single PDMS membrane. ACS Nano 3:2412–2418

Chapter 15
A Method to Evaluate Creep Properties of Solder Alloys Using Micro Indentation

Katsuhiko Sasaki, Ken-ichi Ohguchi and Atsuko Takita

Abstract An indentation creep test is one of the effective methods to directly evaluate the creep deformation of solder joint. However, the indentation test does not give the same creep properties as those obtained by the uniaxial creep tests using a bulk specimen. In this paper, the authors proposes an indentation test conducted under the constant depth to determine a suitable reference area, which leads to the same creep characteristics as those obtained by the uniaxial creep test. A series of numerical micro indentation tests under a constant depth were conducted to construct a method to determine a new reference area without creep data obtained from the uniaxial creep test using bulk specimens. The numerical tests were conducted using finite element method (FEM). The numerical tests showed that the distribution of the principal stress plays an important role to determine the reference area of the indentation tests. Finally, it was found that the reference area obtained considering the distribution of the principal stress gives almost the same creep characteristic as those obtained by the uniaxial creep using bulk specimens.

Keywords Micro indentation · Creep · Solder alloys · FEM · Plasticity

15.1 Introduction

To evaluate the strength reliability of the electronic instruments, finite element method (FEM) analyses considering the creep deformation of the solder joints should be conducted. Usually the creep properties estimated by the tensile creep test using

---

K. Sasaki (✉) · A. Takita
Hokkaido University, Kita 13, Nishi 8, Kita-ku, Sapporo 060-8628, Hokkaido
e-mail: katsu@eng.hokudai.ac.jp

A. Takita
e-mail: atsuko_takita@ec.hokudai.ac.jp

K. Ohguchi
Akita University, 1-1 Tegata Gakuen-cho, Akita 010-8502, Japan
e-mail: ken@dips13.akita-u.ac.jp

H. Altenbach et al. (eds.), *From Creep Damage Mechanics to Homogenization Methods*, Advanced Structured Materials 64,
DOI 10.1007/978-3-319-19440-0_15

a bulk specimen are employed to express the creep behavior of the solder joint in the micron size order. However, there is a difference in the size between the bulk specimen used for the tensile creep test and the solder joint in-situ (Islam et al. 2005). The grain size of the bulk specimen is larger than that of the solder joint, because the cooling rate at center of the bulk specimen is slower than that of the solder joint. Therefore, the solder joint has a different creep deformation behavior from that of bulk specimen. The method to directly estimate the creep deformation of solder joint in-site is required to obtain the accuracy of FEM analysis for the evaluation of the strength reliability of electronic instruments.

The indentation creep test is one of effective tests to estimate the deformation in the micron size order specimens (Fujiwara and Otsuka 2001; Abell and Shen 2002). In this test, indenters such as Vickers, spherical, and conical are pushed into specimens with a constant loading rate until the load value reaches a predetermined constant value. Afterwards, the load is continuously kept at that value. In the constant load process, the indenter is continuously pushed into the specimen due to the development of the creep strain. Consequently, the indentation creep test can evaluate the steady-state creep deformation, which is expressed by Norton's law using the relationship between the indentation depth and time (Atkins et al. 1966). The deformation area due to the indentation test is too small: it is possible that the indentation creep test gives the characteristic of the micron size solder joint as conducted by Zhang et al. (2011) and Basaran et al. (2004). However, some researchers including the authors reported that the ordinary indentation creep test cannot give sufficient results for accurate analysis of the strength reliability calculation (Roebuck and Almond 1982; Ogawa et al. 2007; Miyamoto et al. 2002; Takagi et al. 2004).

There are some research works that verify the accuracy of the indentation creep test against the uniaxial creep tests using bulk specimens. For verification, a reference area was focused on to modify the evaluation method of the indentation creep test (Su et al. 2013; Takagi et al. 2005; Dorner et al. 2003). The indentation load is divided by three times the contact area between the indenter and specimen used in ordinary indentation creep test for calculating stress: the reference area of the indentation creep is three times the contact area between indenter and specimen. This method gives the same creep curve as that obtained from the uniaxial creep using bulk specimens. However, there is no physical meaning of why three times the constant area has to be used to obtain values that match those of the bulk specimens. There are also studies concerning the Vickers hardness test to obtain the same creep properties (Ogawa et al. 2007; Miyamoto et al. 2002). The method needs the numerical tests to determine the coefficient related to the contact area for the correction of the stress. The coefficient of this estimation method also lacks sufficient discussion for its physical meaning because it is determined by varying two parameters. The small punch (SP) creep test is also studied by many researchers (Nishioka et al. 2011; Kobayashi et al. 2009; Komazaki 2012; Zhao et al. 2013). SP creep test gives the uniaxial creep characteristics in the complex stress field.

Recently, the authors clarified that the indentation stress obtained from the indentation creep test gives larger values than the stress obtained from the uniaxial creep tests using bulk specimens (Takita et al. 2014). The authors also conducted the numerical indentation test with the constant depth process. The proposed method can easily obtain a suitable reference area to calculate the uniaxial stresses, which lead to the uniaxial creep strain. However, this method has a problem that the creep characteristics of the bulk specimen are needed to obtain the reference area. It should be noted that, in the case of the tensile creep test using a bar bulk specimen; the direction of the maximum principal stress coincides with the loading direction. This means that the direction of the maximum principal stress may give significant physical information for identifying the reference area of the indentation tests.

In this paper, the indentation tests with constant depth process are conducted by the numerical test using FEM to give the physical meaning to the reference area of the indentation tests. The principal stress plane in the indentation test is identified by the numerical test and the direction of the principal stress is detected. Considering the plane of the maximum principal stress, the reference area which gives reasonable stress can be obtained, thus, gives almost the same creep strain as that obtained by the uniaxial creep test using a bulk specimen. Four types of solder are chosen for the numerical test.

15.2 Experimental Discussions

15.2.1 Experimental Method

The specimen used for this study is made of Sn-3.0Ag-0.5Cu solder alloy, which is widely used in the solder joint. The geometry of the specimen is a disc having a diameter of 30 mm and thickness of 7 mm, respectively. The specimen is cut from an ingot of solder alloy casted at 250 °C. The cut specimen is grinded with mirrored surface and after then that the specimen is annealed at 155 °C for an hour.

Figure 15.1 shows the experimental equipment. Vickers indenter made of diamond was used in this study. Vickers indenter was fixed on the stepping motor (Lead

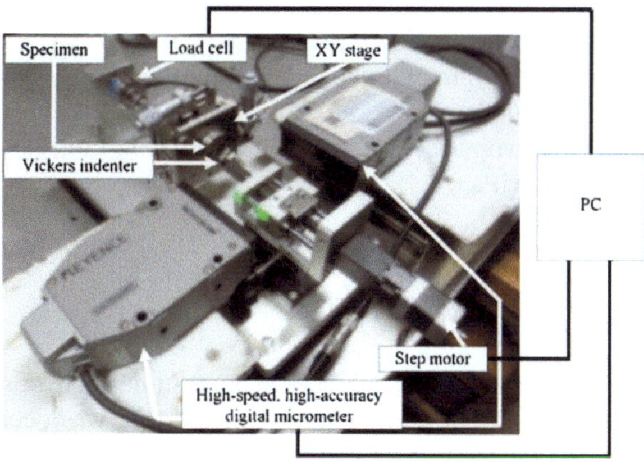

Fig. 15.1 Schematic drawing of the indentation machine

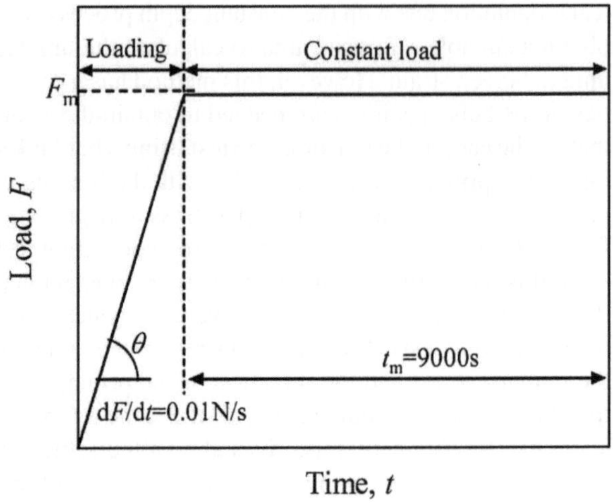

Fig. 15.2 Schematic outline of the loading for indentation creep test

engineering Co., Ltd.) and pushed into the specimen. The specimen was fixed on the XY stage using plate spring. The indentation load and displacement were measured continuously by using the load cell (KYOWA ELECTRONIC INSTRUMENTS Co., Ltd), and digital micrometer (KEYENCE Co., Ltd), respectively.

First, the indentation creep tests were conducted. The load profile for the test is shown as the schematic outline in Fig. 15.2. Indentation creep tests are conducted by pushing an indenter into a specimen in two steps. The first step is a loading process where the indenter is pushed into the specimen at a constant loading rate until a predetermined load value. The other step is a load maintenance process where the load value is maintained at the predetermined load value for an arbitrary time period. In this study, the indenter was pushed into the specimen until the load reached 1 N with a constant load rate of 0.01 N/s. After the load reached 1 N, the load was kept constant at the value of 1 N during 9000 s.

The tensile creep tests were also conducted to confirm that the creep properties obtained by the indentation creep test were not same as those obtained by the tensile creep test. The specimen made of Sn-3.0Ag-0.5Cu solder alloy is cylinder having a diameter of 8 mm and gauge length of 18 mm, respectively. The tensile specimens were annealed at 155 °C after the machining. The tensile creep tests were conducted using a Servopulser EHF-FBI (Shimazu Co., Ltd., Japan) at the three stresses of 20, 30 and 40 MPa.

15.2.2 Experimental Results

Figure 15.3 shows the indentation depth-time curve obtained by the indentation creep test. In Fig. 15.3, the indentation depth rate was decreases until about 2000 s just after the indentation creep test starts. After that, the indentation depth rate linearly

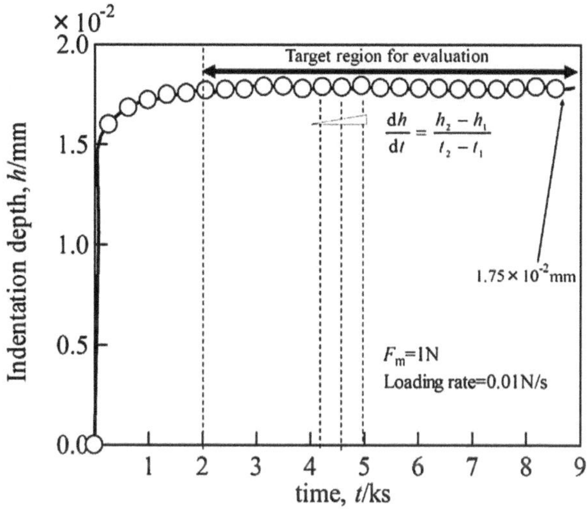

Fig. 15.3 *h-t* curve obtained by experiment and explanation of evaluation method

increases with time. In this paper, the initial time range until 2000 s was omitted, because the increase value of contact area between indenter and specimen is too large to identify the reference area. This means that the deformation after 2000 s is corresponding to the steady state creep curves expressed by the following Norton's law,

$$\dot{\epsilon}^{c} = A\sigma^{n} \tag{15.1}$$

where $\dot{\epsilon}^{c}$ and σ are uniaxial creep strain rate and stress, respectively. A and n are the material constants. The steady-state creep deformation during indentation creep tests is ordinarily evaluated by using the relationship between the equivalent creep strain rate and the equivalent stress during the load maintenance process. To evaluate the steady-state creep deformation, the uniaxial creep strain rate $\dot{\epsilon}^{c}$ and the uniaxial stress σ are ordinarily evaluated by the following procedure in the indentation creep test. The uniaxial creep strain rate $\dot{\epsilon}^{c}$ occurring in the microscopic region under indenter can be calculated by the following equation,

$$\dot{\epsilon}^{c} = \frac{1}{h}\frac{\mathrm{d}h}{\mathrm{d}t} \tag{15.2}$$

where h is the depth of indenter, and $\mathrm{d}h/\mathrm{d}t$ the slope as shown in Fig. 15.3 (Takagi et al. 2005). The uniaxial stress σ under the indenter can be calculated by dividing the constant load F by the contact area S_c between the specimen and the indenter as follows,

$$\sigma = \frac{F}{S_c} = \frac{F}{26.47h^2} \tag{15.3}$$

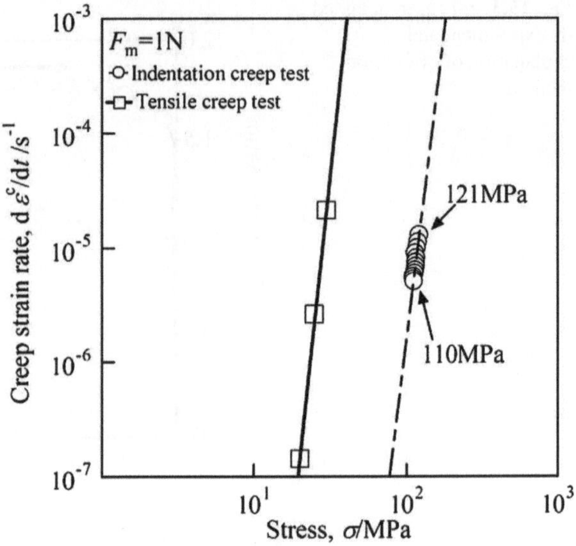

Fig. 15.4 Relationship between stress and creep strain rate obtained by the tensile creep test and the indentation creep test

Equation (15.3) means that the contact area S_c coincides with the reference area. Using indentation depth changing at the load maintenance process, indentation rate dh/dt and the contact area S_c are calculated in a series of point on the h-t curve except the initial region.

Figure 15.4 shows the relationship between the uniaxial creep strain rate and stress calculated by Eqs. (15.2) and (15.3). Circles in Fig. 15.4 show the relationship obtained by the indentation creep test. The solid line shows the relationship obtained by the tensile creep test using the bulk specimen of Sn-3.0Ag-0.5Cu. Comparing the result obtained by the indentation creep test with that obtained by the tensile creep test, it is found that the result obtained by the indentation creep test is not equal to that obtained by the tensile creep test.

The reason of the difference is that the definition of the reference area to determine the stress for the indentation tests cannot be explained just by the experimental results, because the indentation creep test is affected by the microstructures. To clarify the problem of the indentation creep test, the creep deformation of the specimen without the effect of the microstructures, such as that measured by tensile creep test, should be evaluated by the indentation creep test. Therefore, in this paper, we attempted to conduct all indentation tests by using the numerical test. Using the numerical test, the specimen is considered as uniform removing the effect of microstructures such as grain boundary, grain size, and inclusion.

15.3 The Indentation Test Including the Constant Depth Process

The evaluation method proposed in this paper is established by modifying the indentation test with the constant depth process (Takita et al. 2014). The indentation test with the constant depth process consists of three processes as shown in Fig. 15.5: the first process involves that the indenter being pushed into the specimen with a constant indentation rate until the indenter arrives at a predetermined depth. The second process involves the indenter depth being kept at that depth for an arbitrary duration. The third process involves that the indenter being removed from the specimen by a constant indentation rate. In the second process, the load relaxation curves are obtained because the load relaxes due to the creep deformation as shown in Fig. 15.6. In the indentation test with the constant depth process, the steady-state creep deformation is evaluated from the load relaxation curve (Sargent and Ashdy 1992), and it can be expressed by Norton's law (15.1). The uniaxial creep strain rate $\dot{\epsilon}^{\mathrm{c}}$ is obtained by using following equation from the stress relaxation curved (Kobayashi et al. 2009).

$$\dot{\epsilon}^{\mathrm{c}} = -\frac{1}{E}\frac{\mathrm{d}\sigma}{\mathrm{d}t} \tag{15.4}$$

where E is Young's modulus, $\mathrm{d}\sigma/\mathrm{d}t$ is the stress relaxation rate.

The uniaxial stress is obtained by dividing the load F by the new reference area S_{r} as follows,

$$\sigma = -\frac{F}{S_{\mathrm{r}}} \tag{15.5}$$

where the new reference area S_{r} is defined as the effective area to calculate the uniaxial stress σ.

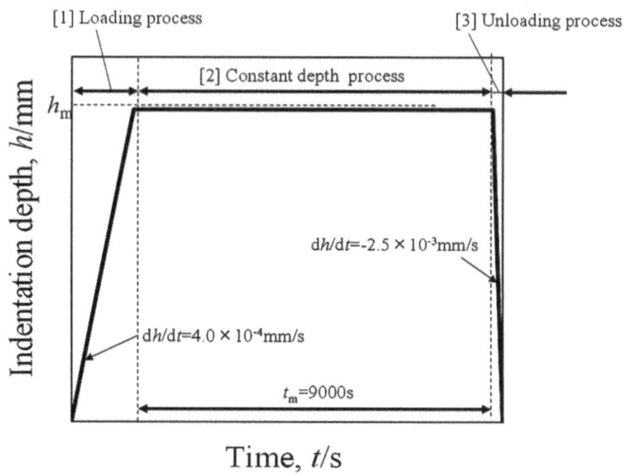

Fig. 15.5 Schematic outline of the loading for indentation test with the constant depth process

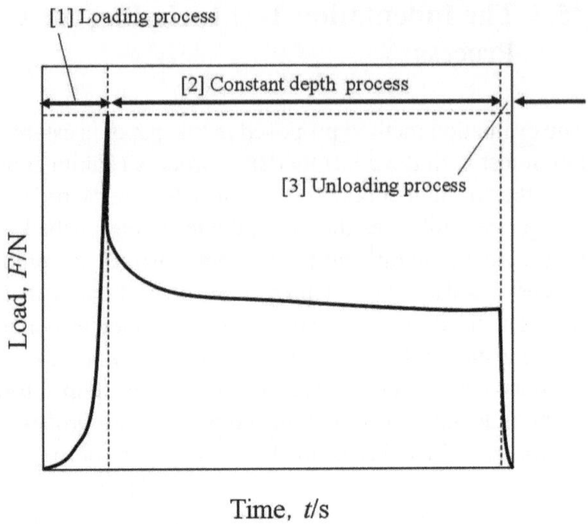

Fig. 15.6 The relationship between the load and the time obtained by the indentation test with constant depth process

The uniaxial stress of the indentation test is commonly obtained by dividing the load F by the contact area S_c between the indenter and the specimen (Sargent and Ashdy 1992). However, the authors has already clarified that the contact area S_c is not suitable for determining the stress (Takita et al. 2014). Therefore, the new reference area S_r is employed to obtain the stress.

The uniaxial creep strain rate $\dot{\varepsilon}^c$ in Eq. (15.5), which obtained by the indentation test with the constant depth process, should be equal to that obtained from Norton's law in Eq. (15.1). Therefore, the new reference area S_r is calculated by the following equation,

$$S_r = \left(\frac{-\dot{F}}{AEF^n} \right)^{\frac{1}{1-n}} \tag{15.6}$$

Since the reference area is assumed to be a constant value as the indenter depth is kept a constant value, the indentation test with the constant depth process has a benefit to determine the new reference area. However, to determine the new reference area, additional data is required such as creep curves obtained by the uniaxial creep tests using bulk specimens (Takita et al. 2014). This means that the creep characteristic cannot be determined from the indentation test with the constant depth process alone.

In this paper, the method to identify the creep properties using the indentation test without the additional test, such as uniaxial creep tests, is proposed. There are two methods to identify the new reference area S_r from the relationship between time, load and indentation depth. One of them uses a concentric circle centered on the vertex of the indenter (Johnson 1985). The other method uses a surface normal to the principal stresses i.e. the principal stress plane. The reference area of the tensile creep test is the cross section of a specimen, i.e., the principal stress plane of the uniaxial tensile test. Therefore, we focus on the principal stress to obtain the new reference area.

15.4 Method-FE Model

In this study, all indentation tests were conducted by numerical test to assume that the indentation specimen has the same mechanical properties as those of the bulk specimen used for the tensile creep test. This means that the effect of the microstructure on the mechanical properties is removed from the result obtained by the indentation test.

All numerical tests were conducted using commercial FEM analysis software Ansys 14.5 (ANSYS, Inc., Pennsylvania, USA). The process where the Vickers indenter was pushed into cuboid specimen was simulated. The Vickers indenter has a height of 0.7 mm, while the edge length of the specimen is 0.6 × 0.6 × 0.3 mm. Figure 15.7 shows the finite element model used for the numerical tests. Considering the symmetry of the model geometry, the FE model was described as a quarter three-dimensional symmetry model (0.3 × 0.3 × 0.3 mm). To simulate the contact between the indenter and the specimen, contact elements were used for boundary area between the indenter and specimen. Nodes on axial planes were fixed for the normal directions. The load was transferred to the specimen through the indenter by giving

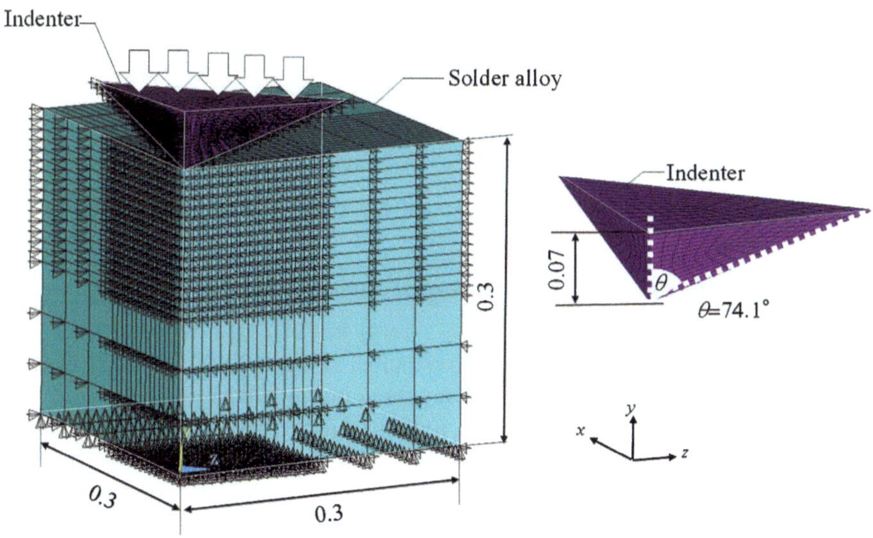

Fig. 15.7 FE model used in the indentation test. Nodes on axial plane were fixed for the normal direction of plane

Table 15.1 Material properties of solders; E Young's Modulus; ν Poisson's ratio; σ_Y Yield stress; H Tangent modulus; A and n Material constants in Norton's law

Solder	E GPa	ν	σ_Y MPa	H MPa	A	n
Sn-3.0Ag-0.5Cu	21	0.35	38	162	6.10×10^{-24}	12.6
Sn-3.5Ag	36.9	0.36	20	141	3.21×10^{-26}	14.7
Sn-37Pb	33.6	0.36	23	152	2.38×10^{-11}	4.38
Sn-58Bi	24	0.43	35	147	1.60×10^{-16}	6.01

the load for the top area of the indenter as shown in Fig. 15.7. The Vickers indenter is assumed as an elastic material by using the elastic properties of diamond; Young's modulus and Poisson's ration were 1141 GPa and 0.07, respectively. In this study, the diamond indenter was assumed as isotropic material to reduce load for the FE model. The creep deformation is simulated by using the Norton's law in Eq. (15.1). The solder specimen was assumed as an elastic-plastic-creep material. To express the elastic-plastic deformation, Hooke's law and the bilinear stress-strain relation with the isotropic hardening law properties of the solder alloys shown in Table 1 were used for solder model (Ohguchi et al. 2006; Sakane 2004, 2000; Matsumura and Yamamoto 2005). The mechanical properties in Table 15.1 were obtained by the tensile creep test and the tensile test.

In this study, the numerical tests were conducted by using four kinds of solder. All of the mechanical properties used in the numerical tests were the value obtained by tensile creep tests (Takita et al. 2014; Sakane 2004).

The load conditions of the indentation test contains the constant depth process with the constant indentation rate of 4.0×10^{-4} mm/s, six different predetermined constant depths, and the duration time for the constant indentation depths of 9000 s. The indenter was removed from the specimen at a rate of 2.5×10^{-3} mm/s.

15.5 Principal Stress Plane Caused by the Indentation Test

15.5.1 Identification of the Principal Stress Plane

Figure 15.8 shows the distribution of the principal stress at 2000 s just after the constant depth process starts: (a), (b) and (c) show the distributions of the first principal stress, the second principal stress, and the third principal stress, respectively. Figure 15.8 shows the results of Sn-3.0Ag-0.5Cu, for example. For the first and second principal stresses, both the compressive and tensile parts clearly exist: the compressive stress appears in specimen beneath the indenter while the outside is tensile. The third principal stress shown in Fig. 15.8c has only a small tensile component; therefore, the third principal stress mostly affects the creep characteristic of the specimen than other principal stresses (Shibutani et al. 2005). This suggests that

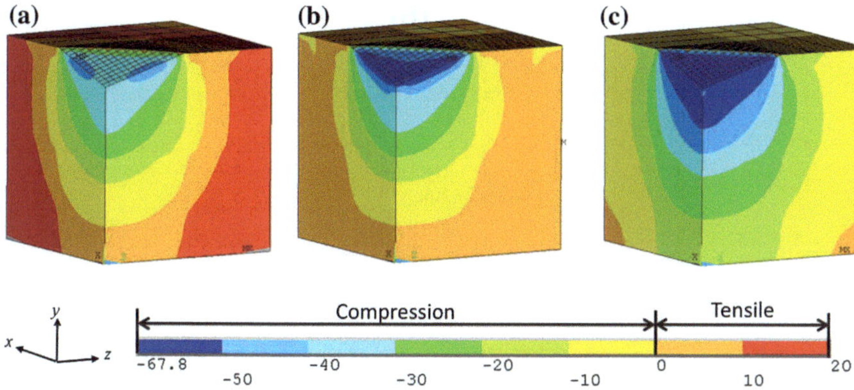

Fig. 15.8 Distribution of principal stress in the specimen of Sn-3.0Ag-0.5Cu (100 s). **a** First principal stress. **b** Second principal stress. **c** Third principal stress

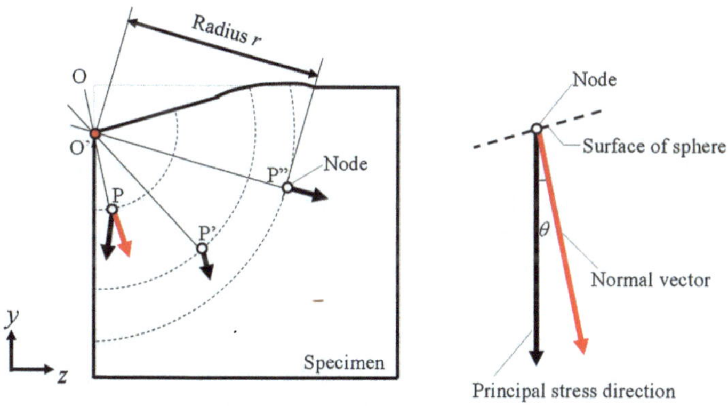

Fig. 15.9 Schematic outline the angle between the normal stress vector and the principal stress direction

there is a possibility that the third principal stress could be used for the identification of the reference area.

Figure 15.9 shows the schematic outline of the sphere centered on the vertex of the indenter. In Fig. 15.9, θ is the angle between the unit vector of the principal stress **m** and the normal vector **n** of the sphere. θ can be obtained by Eq. (15.7)

$$\cos\theta = \frac{m_x n_x + m_y n_y + m_z n_z}{\sqrt{m_x^2 + m_y^2 + m_z^2}\sqrt{n_x^2 + n_y^2 + n_z^2}} \tag{15.7}$$

where $\mathbf{m} = (m_x, m_y, m_z)$ and $\mathbf{n} = (n_x, n_y, n_z)$.

Figure 15.10a shows the FE model of the solder specimen before deformation. The angle of the principal stress in the nodes on the broken line were calculated by using Eq. (15.7) when the indenter is pushed into the specimen until 4.0×10^{-2} mm.

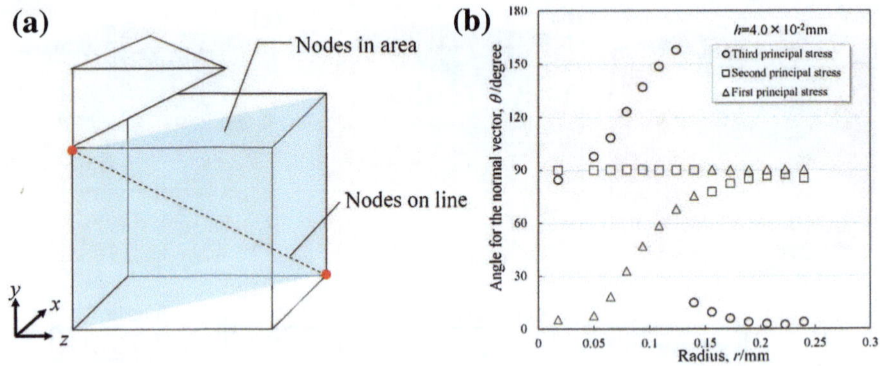

Fig. 15.10 Principal stress vector change due to radius from *top* of indenter. **a** Line which nodes located. **b** Relationship between radius and angle consisted with the principal stress vectors and the normal stress

Figure 15.10b shows the relationship between the angle θ and the radius from the vertex of the indenter to node on the broken line as shown in Fig. 15.10a. Circles, triangles and squares in Fig. 15.10b show the first principal stress, the second principal stress and the third principal stress, respectively. The direction of the first principal stress coincides with the normal direction of the sphere until the radius of 0.06 mm. The direction changes after the radius of 0.06 mm and the direction coincides with the tangent direction of the sphere after the radius of 0.15 mm.The direction of the second principal stress also coincides with the tangential direction of the sphere at any radius.

On the other hands, the angle θ between the direction of the third principal stress and the normal direction increases with increase in the radius until the radius of 0.14 mm. However, after the radius of 0.14 mm, the direction of the third principal stress converges at 0°. Namely, the direction of the third principal stress coincides with the normal direction of the sphere after the radius of 0.14 mm. This suggests that the indentation creep is dominated by the third principal stress.

It is well known that the stress occurring due to the indentation decreases with increase in the distance from the indenter (Johnson 1985). Therefore, the surface of sphere whose radius is about 0.14 mm has the largest principal stress (Ochiai et al. 2008; Anstis et al. 1980). However, experiments cannot figure out the distribution of the principal stress in the specimen. This paper proposes a method to identify the principal stress plane of the indentation test by using the relationship between time and load, and indentation depth obtained by experiments.

15.5.2 Method to Determine the New Reference Area

The principal stresses are distributed within the specimen due to the loading through the contact area (Johnson 1985). The existence of the contact part and non-contact

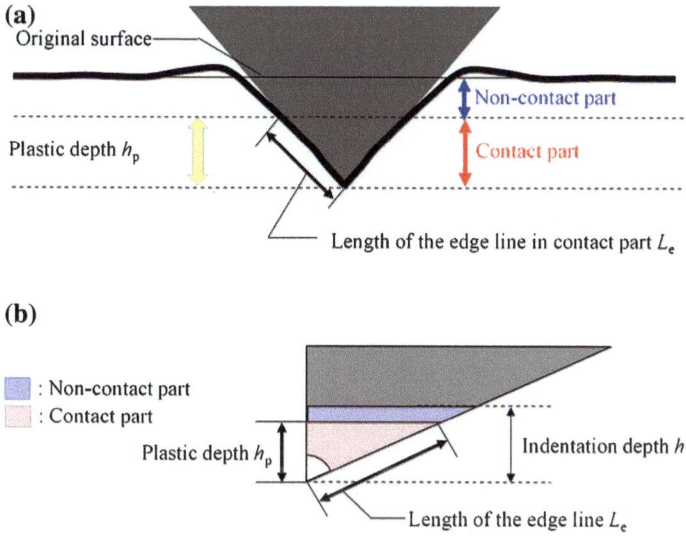

Fig. 15.11 Schematic outline of the contact state in the indentation. **a** Contact state when the indenter pushed into the specimen.**b** Relationship between the plastic depth h_p and the length of the *edge line* L_e

part is also known (Kanazawa and Yosizawa 1997; Pearce et al. 1997). Figure 15.11a shows the schematic outline of the contact state in the indentation. The contact part of indenter penetrates into the specimen. The surface of the non-contact part bends when the indenter is pushed into the specimen. Therefore the effective parameter to determine the principal stress plane may be affected by the contact conditions between the specimen and the indenter. Here, the effect of the contact conditions on the principal stress plane is first clarified.

The relationship between the plastic depth h_p and the boundary of the contact part and the non-contact part is shown in Fig. 15.11b. The distance from the vertex of the indenter to the boundary between the contact part and non-contact part, i.e. the length of the edge line L_e is calculated from the geometry of the indenter and the plastic depth h_p, which is the depth in the contact part of the indentation. The relationship between the plastic depth h_p and the length of the edge line L_e is expressed by the following equation

$$h_p = L_e \cos 74.1^\circ, \tag{15.8}$$

where 74.1° is the angle of the indenter as shows in Fig. 15.7.

The plastic depth h_p is the parameter, which is usually used to calculate Young's modulus of specimens based on the elastic contact theory (Oliver and Pharr 1992). The plastic depth h_p is obtained from the relationship between load and indentation depth during the unloading process as shown in Fig. 15.12. The ratio of the load to the indentation depth dF/dh is a constant value just after the unloading starts as shown by a letter A in Fig. 15.12. The broken line AB is the line drawn using the constant

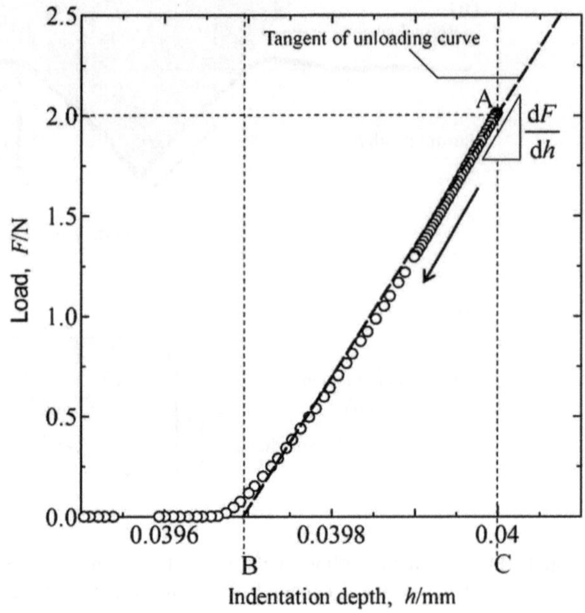

Fig. 15.12 Unload curve obtained by numerical test

ratio at point A. Since BC is the depth deformed by the elastic recovery, the depth at point B is the plastic depth h_p, which is calculated by the follows equation (Sargent and Ashdy 1992; Johnson 1985).

$$h_p = h_m - \frac{F_{us}}{dF/dh}, \tag{15.9}$$

where h_m, dF/dh, and F_{us} are the constant indentation depth, the ratio of the load to the depth at point A, and the load at point A, respectively.

The principal stress plane caused by the indentation test, which was normal for the third principal stress, was identified by the relationship between time, load, and indentation depth considering the relationship between the plastic depth h_p and the radius r of the principal stress plane. Using the numerical result of the indentation test with the constant depth process, the plastic depth h_p is calculated by Eq. (15.9). The radius of the principal stress plane r is assumed as the distance from the vertex of the indenter to the location where the angle of the third principal stress for the normal vector of sphere centered on the vertex of the indenter is 0° (±5°). Figure 15.13 shows the relationship between the plastic depth h_p and radius r. The plastic depth h_p is proportional to the location where the principal stress plane is formed as the spherical surface. The relationship between the plastic depth h_p and the radius r is rewritten as the following cosine function.

$$h_p = r \cos 73.1° \tag{15.10}$$

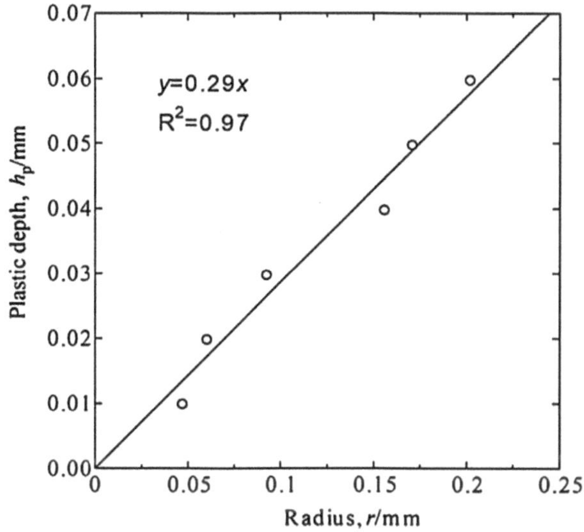

Fig. 15.13 The relationship between convergent radius r and plastic depth h_{p}

Comparing Eq. (15.10) with (15.8), it is found that the relationship between the plastic depth h_{p} and the radius of the principal stress plane r coincides well with that between the plastic depth h_{p} and the length of the edge line in the contact part L_{e}. Therefore, the principal stress plane caused by the indentation test is assumed as the spherical surface whose radius is the length of the edge line in the contact part of indentation. This means that the radius of the principal stress plane can be identified from the indentation test since the plastic depth h_{p} can be measured in the unloading process.

15.6 Evaluation of the Creep Deformation Considering the Principal Stress Plane

15.6.1 Behavior of the Plastic Depth in the Constant Depth Process

The total strain increment $d\epsilon$ is assumed as the sum of elastic strain increment ϵ^{e}, plastic strain increment $d\epsilon^{\mathrm{p}}$, and creep strain increment $d\epsilon^{\mathrm{c}}$ as shown in Ohguchi et al. (2006)

$$d\epsilon = d\epsilon^{\mathrm{e}} + d\epsilon^{\mathrm{p}} + d\epsilon^{\mathrm{c}} \tag{15.11}$$

When the strain of specimen is a constant value, the total strain increment and the plastic strain increment become zero. Therefore, Eq. (15.11) is rewritten as the following equation,

$$d\epsilon^{\mathrm{e}} = -d\epsilon^{\mathrm{c}} \tag{15.12}$$

Using Hooke's law, the creep strain increment is expressed in following equation,

$$d\epsilon^{\mathrm{c}} = -\frac{1}{E}d\sigma \tag{15.13}$$

The steady-state creep strain rate in Eq. (15.4) is derived by differentiating Eq. (15.13) with time t. Equation (15.13) means that the creep strain increases with decrease in the elastic strain when the indentation depth is kept constant. There is a possibility that the decrease in the elastic strain leads to the decrease in the non-contact part of the indenter during the constant depth process. And also the decrease in the non-contact area leads to the increase in the plastic depth h_{p} (Kanazawa and Yosizawa 1997; Pearce et al. 1997). Therefore, it is necessary to clarify the behavior of the plastic depth in the constant depth process.

15.6.2 The Method to Evaluate the Steady-State Creep

The indentation tests were conducted without the constant depth process. The plastic depth h_{p} obtained by the indentation test without the constant depth process was compared with that with the constant depth process. Figure 15.14 shows the schematic outline of the indentation test without the constant depth process. The indentation rate

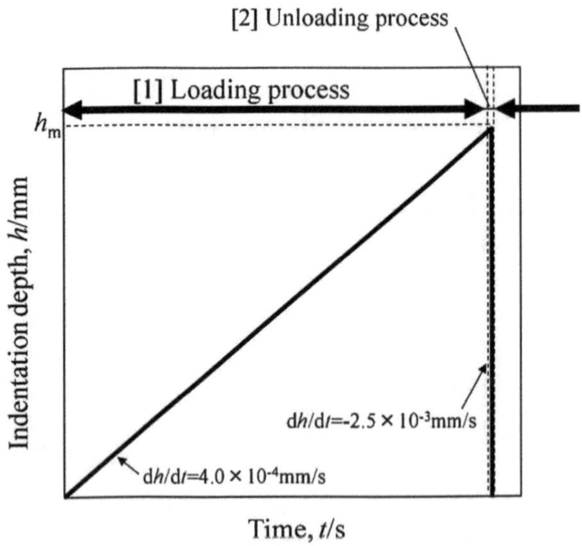

Fig. 15.14 Schematic outline of the loading for indentation unloading test

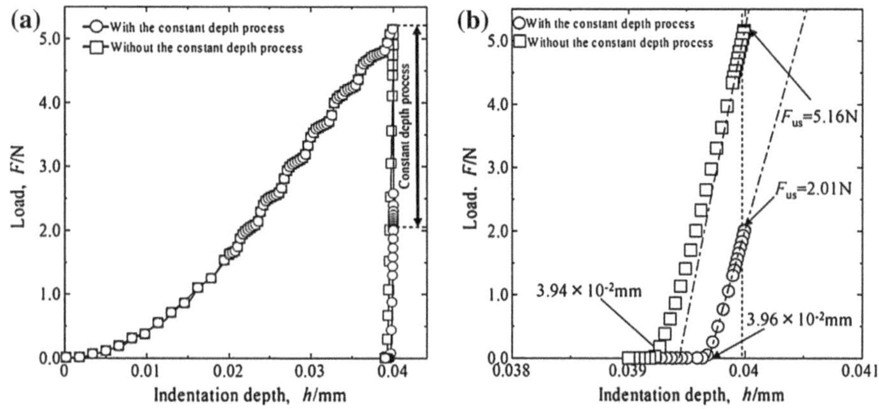

Fig. 15.15 The unloading curves obtained from the unloading process

of both loading and unloading processes were the same as those of the indentation test with the constant depth process as shown in Fig. 15.5.

Figure 15.15a shows the relationship between the load and the indentation depth (the $F-h$ curve) obtained by both the indentation tests with and without the constant depth processes. In case of the indentation test with the constant depth process, the load relaxes during the constant depth process as shown by the circles in Fig. 15.15a: the unloading process starts at 2.01 N. On the other hand, the unloading process starts at 5.16 N in the case of the indentation test without the constant depth process as shown by squares in Fig. 15.15a. Therefore, the load relaxation due to the recovery of the elastic strain without the constant depth process is larger than that with the constant depth process as shown in Fig. 15.15b. However, comparing the plastic depths h_p obtained by both the indentation tests, it is found that the plastic depth of the indentation test with the constant depth process is almost same as that without the constant depth process: h_p is about 0.039 mm in the both cases.

This indicates that the creep strain increases with decrease in the elastic strain on the contact part during the constant depth process. The plastic depth does not change while the indentation depth is kept constant. Therefore, the new reference area can be determined from the data obtained by unloading process after the constant depth process.

Figure 15.16 shows the procedure to evaluate the steady-state creep deformation from the indentation test with the constant depth process:

(1) The indentation test with the constant depth process is conducted at arbitrary depth h_m.
(2) The plastic depth h_p is obtained from the unloading curve using the ratio of load to depth dF/dh just after the unloading starts.
(3) The new reference area S_r and Young's modulus E are calculated by using the unloading curve. It has been clear that the reference area S_r is the half surface area of the sphere whose radius is the length of the edge line in the contact part

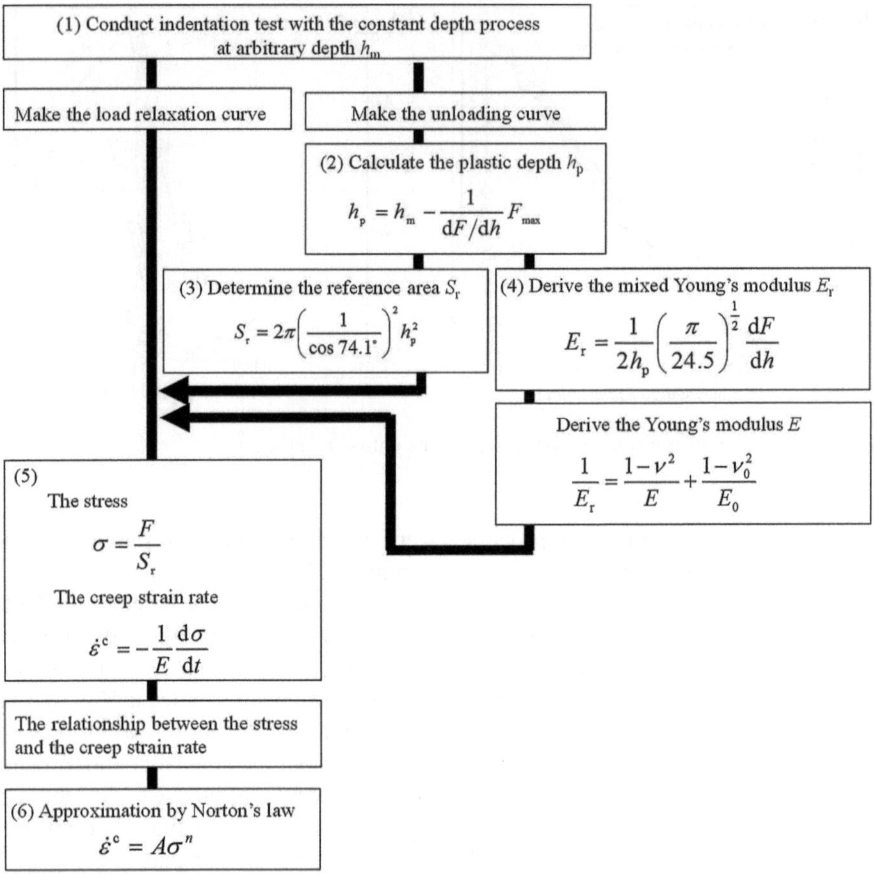

Fig. 15.16 The flowchart of evaluation process using new reference area

of indentation and whose center is the vertex of the indenter. Namely, the new reference area S_r is expressed by a function of the plastic depth h_p,

$$S_r = 2\pi \left(\frac{1}{\cos 74.1^\circ}\right)^{0.5} \frac{dF}{dh} = 83.7 h_p^2 \tag{15.14}$$

(4) Young's modulus is obtained by the following equation applying the non-contact part (Oliver and Pharr 1992; Seddon 1965; Doerner and Nix 1986).

$$E_r = \frac{1}{2h_p}\left(\frac{\pi}{24.5}\right)^{0.5} \frac{dF}{dh} \tag{15.15}$$

where E_r means Young's modulus mixed the indenter and specimen. Young's modulus of the specimen E is calculated by using the following equation,

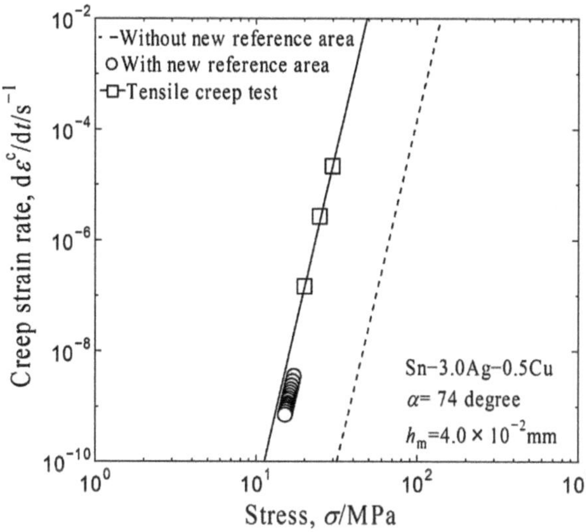

Fig. 15.17 The relationships between the stress and the creep strain rate in Sn-3.0Ag-0.5Cu

$$\frac{1}{E_r} = \frac{1 - \nu^2}{E} + \frac{1 - \nu_0^2}{E_0} \tag{15.16}$$

where ν_0, and ν are the Poisson ratio of the indenter and specimen, E_0 Young's modulus of the indenter, respectively.

(5) After the new reference area S_r is determined and the Young's modulus E is evaluated, the steady-state creep deformation is obtained from the load relaxation curve caused in the constant depth process. Namely, substituting the new reference area S_r and Young's modulus E to Eqs. (15.4) and (15.5), stress σ and the creep strain rate $\dot{\epsilon}^c$ are obtained.

(6) Finally, the steady-state creep curves can be determined from the indentation test with the constant depth process and the curves are expressed by Norton's law.

Figure 15.17 shows the relationship between stress and the creep strain rate obtained by the above procedure (1)–(6) for the solder alloy of Sn-3.0Ag-0.5Cu. The solid line shows the experimental result obtained by the tensile creep test using bulk specimens, while broken line shows the result obtained from the indentation creep test not using the new reference area proposed here. Squares show the result obtained from the indentation tests with the constant depth process using the proposed new reference area expressed by Eq. (15.16). It is clear that the new reference area determined here gives almost the same result as the result obtained from the tensile creep test.

The procedure is applied to the other three solder alloys; Sn-3.5Ag, Sn-37Pb, and Sn-58Bi. The steady-state creep properties in the three solder alloys are shown in Table 15.1. Figure 15.18 shows the relationship between the stress and the creep strain rate of each solder alloy. Circles, squares, and triangles show the results of Sn-3.5Ag, Sn-37Pb, and Sn-58Bi, respectively: these results are obtained using the

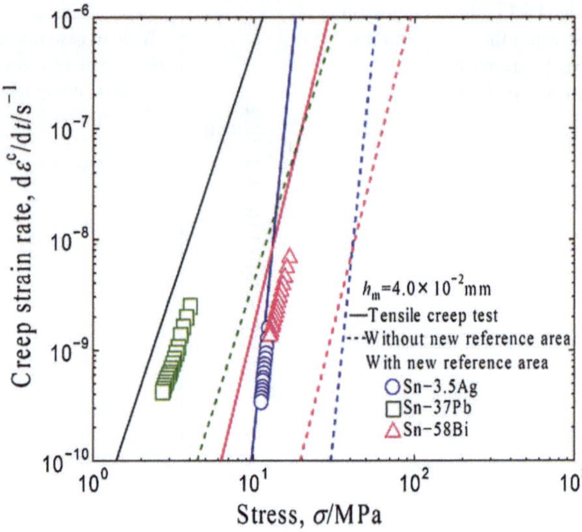

Fig. 15.18 The comparison between the creep properties obtained by the modified method and the tensile creep test in solder alloy other than Sn-3.0Ag-0.5Cu; Sn-3.5Ag, Sn-37Pb and Sn-58Bi solder alloy

new reference area. The broken lines show the result obtained without the new reference area, and solid lines the result obtained from the tensile creep test using bulk specimens, respectively. It is clear from Fig. 15.18 that the new reference area gives the almost same results as the tensile creep test for the all solder alloys.

15.7 Conclusion

In this paper, a series of numerical indentation tests with a constant depth were conducted to construct a method to determine a new reference area of the indentation for the estimation of the steady-state creep of small specimens. The numerical tests were conducted using finite element method (FEM). As a result, the following conclusions were obtained:

1. An indentation creep test using a small specimen and a pure tensile test using a bulk specimen made of Pb-3.0Ag-0.5Cu were conducted. The test results showed that the steady-state creep deformation obtained from the indentation creep tests was different from that obtained from the pure tensile test. This result suggests that it is necessary to clarify the correctness of the method of the indentation creep tests under the condition where the effect of microstructure of the specimen can be neglected.
2. An indentation test with a constant depth was proposed to obtain a new reference area of the indentation test. The new reference area is effective to obtain the steady-state creep strain from the indentation test. However the experimental method to identify the new reference area is required.

3. It is found that the direction vector of the maximum principal stress is normal to the spherical surface centered vertex of the indenter. Therefore, the plane of maximum principal stress of the indentation test is assumed to be the reference area.
4. The radius of the maximum principal stress plane can be determined as the length of edge line L_e in the contact surface between the indenter and specimen. The plastic depth h_p is also considered to determine the accurate length of edge line L_e in the contact part between the indenter and specimen.
5. Based on the conclusion (1) and (2), the method to evaluate the steady-state creep deformation by the indentation test was proposed. The proposed method can identify the new reference area S_r of the indentation tests and gives the steady-state creep curves from only an indentation tests with a constant depth process.
6. The method was applied to four solder alloys; Sn-3.0Ag-0.5Cu, Sn-3.5Ag, Sn-37Pb, and Sn-58Bi. The steady-state creep of each solder alloy can be well estimated by the proposed method.

References

Abell K, Shen Y (2002) Deformation induced phase rearrangement in near eutectic tin-lead alloy. Acta Mater 50(12):3191–3202

Anstis G, Chantikul P, Lawn B, Marshall D (1980) A critical evaluation of indentation techniques for measuring fracture toughness: i. direct crack measurements. J Am Ceram Soc 64:533–538

Atkins AG, Silverio A, Tabor D (1966) Indentation hardness and the creep of solids. J Inst Metals 94:369–378

Basaran C, Tang H, Nie S (2004) Experimental damage mechanics of microelectronic solder joints under fatigue loading. Mech Mater 36:1111–1121

Doerner M, Nix W (1986) A method for interpreting the data from depth-sensing indentation instruments. J Mater Res 1:601–609

Dorner D, Roller K, Skrotzki B, Stockhert B, Eggeler G (2003) Creep of a TiAl alloy: a comparison of indentation and tensile testing. Mat Sci Eng A 357:346–534

Fujiwara M, Otsuka M (2001) Indentation creep of β-Sn and Sn-Pb eutectic alloy. Mat Sci Eng A 319–321:929–933

Islam R, Wu B, Alam M, Chan Y, Jillek W (2005) Investigations on microhardness of Sn-Zn based lead-free solder alloys as replacement of Sn-Pb solder. J Alloy Compd 392:149–158

Johnson K (1985) Contact mechanics. Cambridge University Press, Cambridge, MA

Kanazawa K, Yosizawa Hirokawa TH (1997) Analysis of ultramicrohardness by conical indenter (1st report, analysis of indentation process using fem and correction of indentation process using fem and correction of indenter tip radius). J Soc Mecanical A 63:567–576

Kobayashi K, Kajihara I, Koyama H, Stratford G (2009) Deformstion and fracture mode during small punch creep tests. Jpn Soc Mech Eng A 75:669–676

Komazaki S (2012) The frontier of technology development in remaining life assessment for high temperature components. J Soc Mat Sci 61:980–986

Matsumura T, Yamamoto T (2005) Assembly technology using lead-free solder. FUJITSU 56:545–551

Miyamoto J, Ogawa T, Ohsawa T (2002) Prediction of mechanical properties of lead-free solders using indentation method. J Soc Mat Sci 51:445–450

Nishioka T, Uemura H, Ohsawa T, Sawaragi Y (2011) Effects of specimen surface condition and atmosphere on small punch creep property of low alloy steel. J Soc Mat Sci 60:146–152

Ochiai S, Ikeda S, Iwamoto S, Sha J, Okuda H, Waku Y, Nakagawa N, Mitani A, Sato M, Ishikawa T (2008) Residual stresses in YAG phase of melt growth Al2O3/YAG eutectic composite estimated by indentation fracture test and finite element analysis. J Eur Ceram Soc 28:2309–2317
Ogawa T, Kohketsu H, Takahashi K, Ohsawa T (2007) Influence of microstructure and specimen size on creep properties of solders. J Solid Mech Mater Eng 1:254–260
Ohguchi K, Sasaki K, Ishibashi M (2006) A quantitative evaluation of time-independent and time-dependent deformations of lead-free and lead-containing solder alloys. J Electron Mat 35:132–139
Oliver W, Pharr G (1992) An imporoved technique for determining hardness and elastic modulus using load and displacement sensing indentation experiments. J Mater Res 7:1564–1583
Pearce S, King J, Holdsworth M (1997) Axisymmetric indentation of curved elastic membranes by a convex rigid indenter. Int J Nonlin Mech 46:1128–1138
Roebuck B, Almond E (1982) Equivalent of indentation and compressive creep tests on WC/Co hardmetal. J Mater Sci Lett 1:519–521
Sakane M (ed) (2000) Factual database on creep and creep-fatigue properties of Sn-37Pb and Sn-3.5Ag solders. Jpn Soc Mater Sci, Japan
Sakane M (ed) (2004) Factual database on creep and creep-fatigue properties of Sn-37Pb and Sn-3.5Ag solders. Jpn Soc Mater Sci, Japan
Sargent P, Ashdy M (1992) Indentation creep Mater Sci Tech 8:594–600
Seddon I (1965) The relation between load and penetration in the axisymmetric boussinesq problem for a punch of arbitrary profile. Int J Eng Sci 3:47–57
Shibutani T, Yu Q, Shiratori M (2005) Evaluation of diffusion creep in low melting point materials by nanoindentation creep test. Jpn Soc Mech Eng A 71:1285–1291
Su C, Herbert E, Sohn S, LaManna J, Oliver W, Pharr G (2013) Measurement of power-law creep parameters by instrumented indentation methods. J Mech Phys Solids 61:517–536
Takagi H, Fujiwara M, Kakehi K (2004) Measuring Young's modulus of Ni-based superalloy single crystals at elevated temperatures through microindentation. Mater Sci Eng A 348:387–389
Takagi H, Mand Dao M, Fujiwara Otsuka M (2005) Detecting the transition of creep rate-controlling process in Al-Mg solid-solution alloy through instrumented indentation. J Japan Inst Metals 69(4):348–355
Takita A, Sasaki K, Ohguchi K (2014) A method to identify steady creep strain from indentation creep using a new reference area of indentation. J Electron Mater 43:2530–2539
Zhang N, Yang F, Shi Y, Guo F (2011) Compression creep of 63Sn37Pb solder balls. Acta Mater 59:3156–3163
Zhao L, Jing H, Xu L, Han Y, Xiu J, Qiao Y (2013) Evaluating of creep property of distinct zones in P92 steel welded joint by small punch creep test. Mater Design 47:677–686

Chapter 16
The Behavior of the Graded Cellular Material Under Impact

Changjian Shen and Guoxing Lu

Abstract Due to their excellent properties, including low weight, high stiffness and strength and heat insulation, cellular materials are widely used in engineering applications, especially in the aerospace and defense industries, as energy absorption devices. To improve the performance of the cellular material, optimization of the material property has been attracted considerable research interest. Introducing a gradient in material parameters into the cellular material will significantly influence the behavior of the cellular material under impact loadings. Current paper summarizes our research progress on the investigation of the influence of the gradient on the behavior of the graded cellular material under impact conditions, including the deformation mode, underlying mechanism, energy absorption capacity and the transmitted force on the protected structure. Finite element simulation, analytical study and experimental investigation are involved in this research.

Keywords Graded cellular material · Impact · Deformation mode · Energy absorption capacity

16.1 Introduction

Cellular solids, both natural and synthetic, have attracted numerous research interests due to their outstanding properties, including high relative stiffness and strength, heat insulation and efficient energy absorption (Gibson and Ashby 1997). In nature, cellular structures (e.g. in woods and bones) vary in cell wall thickness, size and shape. For cellular solids, a gradual variation in the cell size distribution or material

C. Shen
School of Mechanical and Aerospace Engineering, Nanyang Technological University, Nanyang, Singapore
e-mail: cjshen@ntu.edu.sg

G. Lu (✉)
Faculty of Science, Engineering and Technology, Swinburne University of Technology, Hawthorn Vic 3122, Australia
e-mail: glu@swin.edu.au

H. Altenbach et al. (eds.), *From Creep Damage Mechanics to Homogenization Methods*, Advanced Structured Materials 64,
DOI 10.1007/978-3-319-19440-0_16

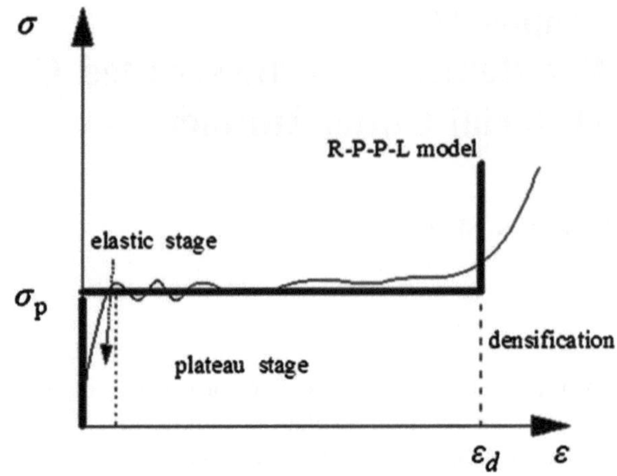

Fig. 16.1 General stress-strain relationship for the cellular material and the R-PP-L material model

strength can influence many properties such as mechanical shock resistance and thermal insulation. Up to now, few experiments and numerical simulations on non-homogenous cellular structures have been reported. Therefore, it is worthwhile to further explore the mechanical properties of graded cellular structures under dynamic loading.

In general, the stress-strain curve of cellular material possesses the feature sketched by the fine line in Fig. 16.1 and it is characterized by three stages, including a short elastic stage, a long plateau stage and finally a densification stage. Under dynamic loading, compaction shock waves show up, resulting in the dynamic stress enhancement. To explain such feature, Reid and Peng (1997) investigated the behavior of the wood under high speed impact and proposed a rigid-perfectly plastic-locking (R-PP-L) material model. After their pioneering work, improvements of the model were achieved by a number of researchers (Karagiozova et al. 2012; Lopatnikov et al. 2003, 2004; Pattofatto et al. 2007; Tan et al. 2005a, b). However, these studies only focused on the uniform cellular material. Most recently, more research interest has been attracted to the graded cellular material. The concept of such material is to change the foam property (i.e. parent material property, cell size/volume etc.) in one direction to improve its performance.

Most of the related researches in literature are based on finite element (FE) simulations. Due to its simply structure, early work employed regular (Ali et al. 2008) or Voronoi honeycomb (Ajdari et al. 2009, 2011; Wang et al. 2011) to investigate the behavior of the structure under impact, introducing a gradient in the density/thickness of the cell wall. For the foam material, most studies (Cui et al. 2009; Kiernan et al. 2009; Li et al. 2014) employed uniform foam with several layers with different material properties in each layer, instead of continuous material. These studies focused on the final energy absorption capacity, neglecting the influence of the impact velocity on the deformation process of the material.

Using the finite element simulation, Shen et al. (2013a) started the problem from the simplest case, in which a gradient only in the materials' yielding stress was introduced to the honeycomb structure, in order to reveal any new phenomena and define some parameters first, such as gradient and some dimension-less parameters. Two types of deformation modes were observed when the gradient was positive while three types of deformation modes existed when the gradient was negative. The deformation modes were nearly the same when the velocity was either small or sufficiently high, no matter whether the gradient was positive or negative. Due to the boundary and the two-dimensional effect, the compressed band was "V" shaped when velocity was low and it tended to be transverse when the velocity was higher. It was found that two shock fronts appeared when the strongest cell was placed at the impinged end while only one shock front appeared when the weakest cell was placed at the impinged end.

Based on the deformation modes, analytical models (Shen et al. 2013b) were proposed using the one-dimensional shock theory, in which the material is assumed to be continuous and rigid-perfect-perfect plastic (R-PP-L). Then, the analytical model was also used to study a more practical case, in which density gradient is involved (Shen et al. 2014b). For both cases, the analytical models were able to predict reasonable results. Some deviations were found at the densification length and the stresses at the proximal/impinged end of the specimen. The main reason is that the R-PP-L material model only involved two parameters (plateau stress and locking strain) and was not accurate enough for good predictions.

Afterwards, the authors conducted an experiment, in which a foam block with varying cross-section together with a back mass impinged onto a rigid target (Shen et al. 2014b, 2015). The proposed two deformation modes were identified in relation to the influence of the gradient. To better give the prediction, the analytical model employed the rigid-power-law densification material model, which is obtained from the real stress-strain curve. Comparison between the analytical and experimental results showed that the results from the power-law material model were more reasonable.

Current paper summarizes our research progress on the graded cellular material under impact. The deformation modes, analytical modeling, energy ab-sorption capacity and experimental studies are presented.

16.2 Finite Element Simulation of the Graded Honeycomb Structure

The Finite Element (FE) model is similar to that previously proposed by Ruan et al. (2003), but a gradient in the material's yielding stress is introduced to represent the graded hexagonal honeycomb structure (GHS) (Fig. 16.2a). The structure is composed of 16 cells in the X_1 direction and 15 in the X_2 direction. The material property of all the layers is the same except that the yielding stress varies layer by layer. It

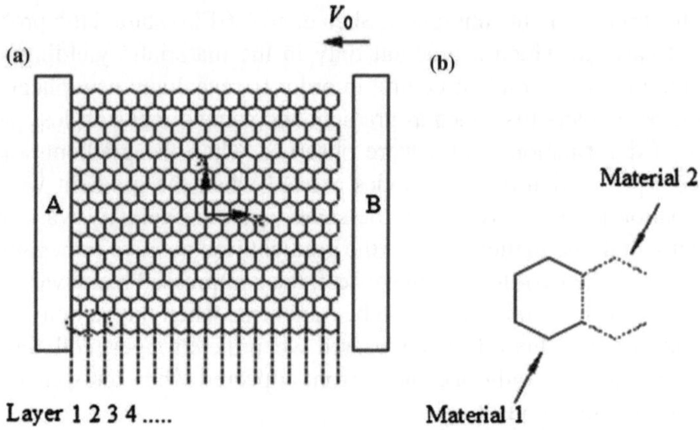

Fig. 16.2 **a** Finite element model composed by 16 hexagonal cells in X_1 direction and 15 cells in X_2 direction. **b** Within a cell, five edges have the same material property while the *left* one is assigned with different yielding stress. Plate A is fixed and Plate B is the impinging plate

is noted that the neighboring layers have cells sharing the same side. Within a cell, five edges are assigned the same material while the left one edge has a different material property (Fig. 16.2b). The parent material is assumed to be elastic, perfectly plastic with $E = 69\,\text{GPa}$ and $\nu = 0.3$, where E and ν are Young's modulus and Poisson's ratio of the material, respectively. The cell is regular with the same value of edge length, which is $l = 2.7\,\text{mm}$, and the corresponding angle between the edges is 120 °. The wall-thickness of the cells t and the density of the parent material ρ_S are 0.2 mm and 2700 kg/m^3, respectively. Hence, the density of the honeycomb is $\rho_0 = 2/\sqrt{3}(t/l)230$ kg/m^3.

The gradient is defined as follows,

$$\Theta = n\frac{\Delta\sigma_{\text{ys}}}{\sigma_{\text{ave}}}, \tag{16.1}$$

where n is the number of the layers, $\Delta\sigma_{\text{ys}}$ is the difference of the yielding strength of the material and σ_{ave} is the average strength of the whole model. For clarity, it is defined that the stress gradient is negative when the strongest layer is placed at the impact end while it is positive when the weakest layer is placed at the impact end.

Both plates A and B are defined as rigid body. As shown in Fig. 16.2, plate A is fixed by constraining all the degrees of freedom at the reference point; plate B only can move along the X_1 direction. Constant velocities are applied to the impinging plate (i.e. plate B). The deformation profiles are obtained to investigate the deformation modes of the GHS under dynamic loading.

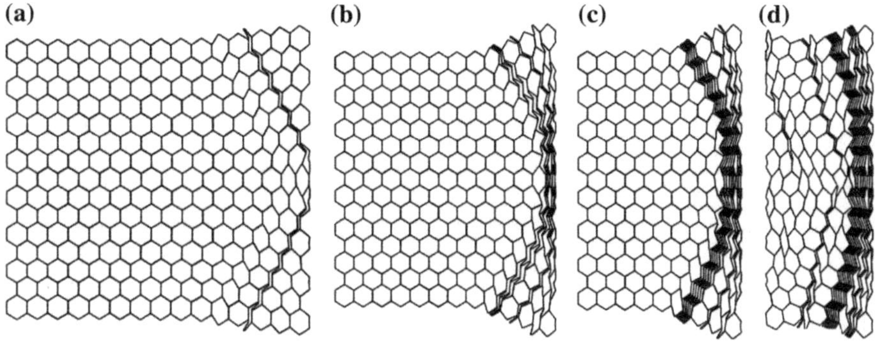

Fig. 16.3 Deformation profile of the graded honeycomb in the case of $\Theta = 0, 42$, $V = 5\,\text{m/s}$. **a** u = 7 mm, **b** u = 21 mm, **c** u = 35 mm, **d** u = 49 mm. u is the displacement of the impinging plate

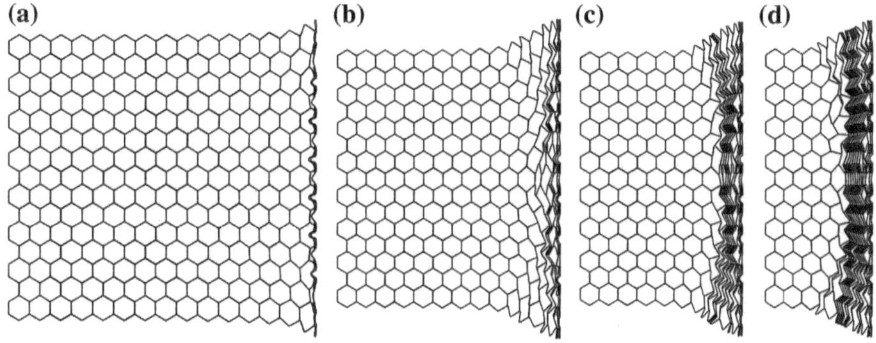

Fig. 16.4 Deformation profile of the graded honeycomb in the case of $\Theta = 0, 42$, $V = 35\,\text{m/s}$. **a** u = 7 mm, **b** u = 21 mm, **c** u = 35 mm, **d** u = 49 mm. u is the displacement of the impinging plate

Figures 16.3 and 16.4 show the deformation profiles of the GHS with a positive gradient of 0.42 under different velocities. The deformation profile was obtained at 4 instant when the displacement of the impinged plate is 7, 21, 35 and 49 mm. It was noted that only one compaction band imitated from the impinged end of the structure and the compaction band tended to be transverse when the velocity was high.

Figures 16.5, 16.6 and 16.7of the GHS with a negative gradient of −0.42 under different velocities. Similarly, the deformation profiles were obtained at 4 instants. Three deformation modes were found. When the velocity was low, the deformation occurred from the weakest end, forming a "V" shaped com-paction band, and propagated to the other end (Fig. 16.5). When the velocity was high enough (e.g. 50 m/s), only one compaction band initiated from the loading end, forming a "I" shaped compaction band (Fig. 16.7). A transitional mode was found when the velocity is intermediate, in which compaction zones showed up from both the proximal and impinged end (Fig. 16.6).

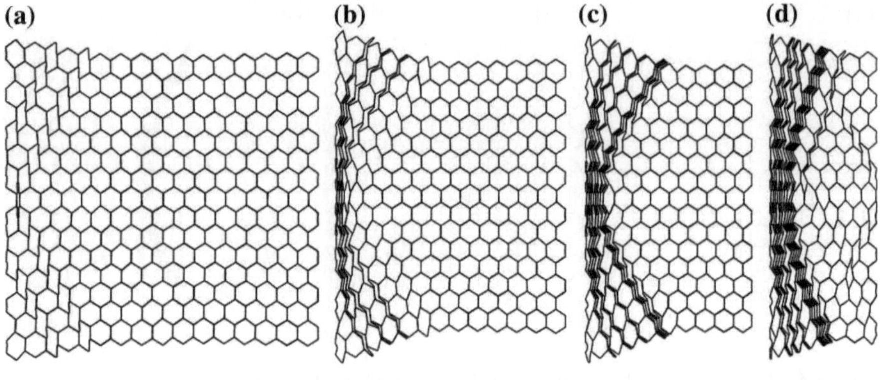

Fig. 16.5 Deformation profile of the graded honeycomb in the case of $\Theta = -0, 42$, $V = 5\,\text{m/s}$. **a** u = 7 mm, **b** u = 21 mm, **c** u = 35 mm, **d** u = 49 mm. u is the displacement of the impinging plate

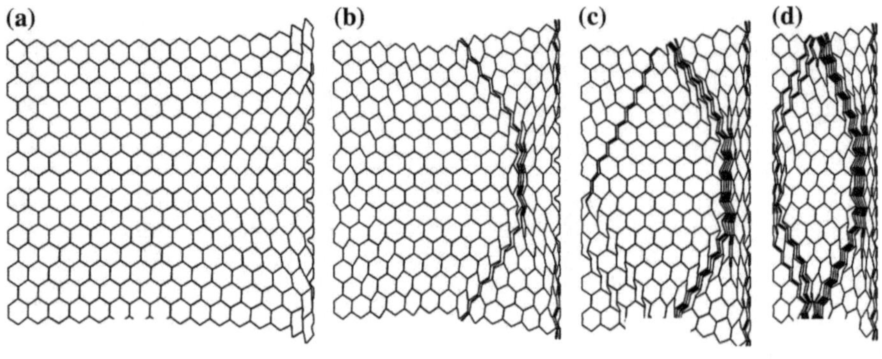

Fig. 16.6 Deformation profile of the graded honeycomb in the case of $\Theta = -0, 42$, $V = 20\,\text{m/s}$. **a** u = 7 mm, **b** u = 21 mm, **c** u = 35 mm, **d** u = 49 mm. u is the displacement of the impinging plate

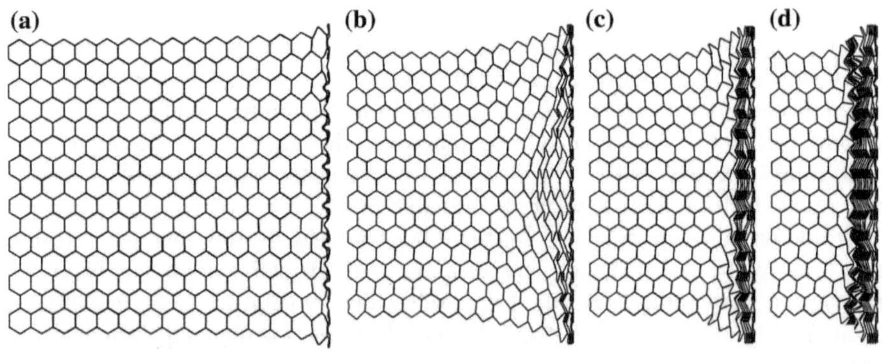

Fig. 16.7 Deformation profile of the graded honeycomb in the case of $\Theta = -0, 42$, $V = 50\,\text{m/s}$. **a** u = 7 mm, **b** u = 21 mm, **c** u = 35 mm, **d** u = 49 mm. u is the displacement of the impinging plate

16.3 Analytical Modeling of the Deformation Process in the Graded Cellular Rod

Based on the observation of the FE simulations, it is remarked that two compaction fronts appear when the strongest cell is placed at the impinged end while only one compaction front propagates if the weakest cell is placed at the impinged end. An analytical model was established for a graded continuum rod by using the shock theory. Figure 16.8 shows the impact scenario, in which a graded cellular rod is impinged by a rigid mass G with an initial velocity V_0. The deformation process of the cellular rod with positive gradient is similar to the one of the uniform rod, which has been investigated comprehensively in literature (Reid and Peng 1997; Tan et al. 2005a, b).

Figure 16.9a shows a DS mode which occurs when the gradient is negative. Crushing occurs initially from both the distal and impinged end. In this DS mode, a forward shock and a backward shock propagate simultaneously before they finally meet, if the input energy is sufficiently large. V_1 is the velocity of the impinging mass and V_2 is the velocity of the undeformed zone between the two shock fronts. The dynamic crushing stress in the graded cellular rod for the DS mode is depicted in Fig. 16.9b at an instant $t = t$.

From the geometry, the following relationship can be easily obtained,

$$\frac{\mathrm{d}u}{\mathrm{d}t} = V_1, \frac{\mathrm{d}y}{\mathrm{d}t} = V_{s2}, \frac{\mathrm{d}l}{\mathrm{d}t} = -\frac{1}{\varepsilon_\mathrm{d}}V_1, \frac{\mathrm{d}y}{\mathrm{d}t} = \frac{1-\varepsilon_\mathrm{d}}{\varepsilon_\mathrm{d}}V_2, \frac{\mathrm{d}x}{\mathrm{d}t} = \frac{1-\varepsilon_\mathrm{d}}{\varepsilon_\mathrm{d}}(V_1 - V_2) \tag{16.2}$$

Since the undeformed zone has velocity V_2, which varies with time, by taking the first shock front as the reference frame and considering the first shock front over the period from t to $(t + \mathrm{d}t)$, the conservation of moment gives,

$$-\sigma_\mathrm{p}^\mathrm{f}\mathrm{d}t = \left[\frac{G}{A_0} + \rho_\mathrm{s}(x + \mathrm{d}x)\right](V_1 + \mathrm{d}V_1) - \left(\frac{G}{A_0} + \rho_\mathrm{s}x\right)V_1 - \frac{\rho_0}{1-\varepsilon_\mathrm{d}}, \tag{16.3}$$

where $\sigma_\mathrm{p}^\mathrm{f} = \sigma_0 - \left(\frac{1}{2} - \frac{x}{1-\varepsilon_\mathrm{d}}\right)\frac{\Theta}{l_0}\sigma_0$ is the quasi-static plateau stress ahead of the forward shock front and $\rho_\mathrm{s} = \rho_0/(1-\varepsilon_\mathrm{d})$ is the density of the densified material.

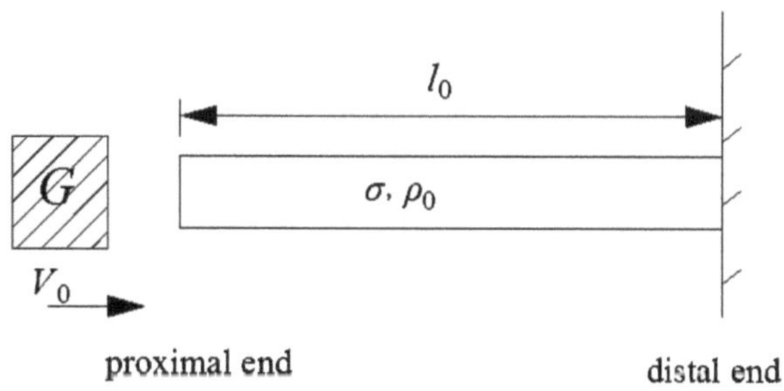

Fig. 16.8 A stationary cellular rod directly impinged by a rigid mass G

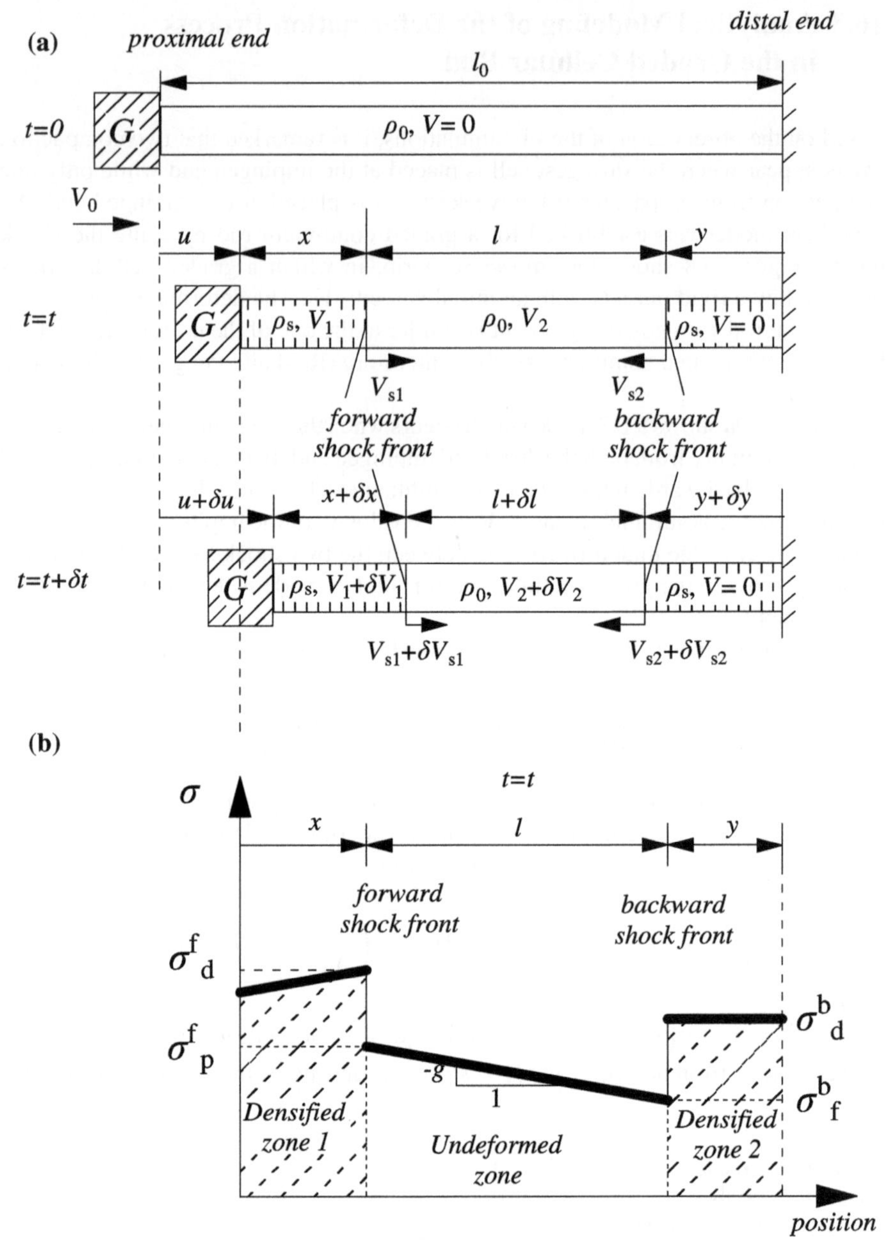

Fig. 16.9 Schematic of the DS mode and the stress distribution in the crushing foam rod **a** the DS mode in the graded cellular rod impinged by a rigid mass (negative gradient, double shock fronts); **b** corresponding stress distribution in the graded cellular rod at instant *t*

For the backward shock front, the reference frame can be fixed at the distal end. The conservation of the moment gives,

$$-\sigma_{\rm d}^{\rm b}{\rm d}t = {\rm d}\left[\left(\frac{G}{A_0} + \rho_{\rm s}x\right)V_1\right] + {\rm d}(\rho_0 l V_2), \tag{16.4}$$

where $\sigma_{\rm d}^{\rm b}$ is the reaction stress at the distal end. Since the densified zone adjacent to the distal end is in equilibrium, the reaction stress $\sigma_{\rm d}^{\rm b}$ equals the dynamic plateau stress behind the backward shock front

$$\sigma_{\rm d}^{\rm b} = \sigma_0 + \left(\frac{l_0}{2} - \frac{y}{1-\varepsilon_{\rm d}}\right)g\sigma_0 + \frac{\rho_0}{\varepsilon_{\rm d}}V_2^2 \tag{16.5}$$

Equations (16.3) and (16.4) are the governing equations for the double shock mode. Then, combing with the geometry relationships in Eq. (16.2) gives,

$$\frac{{\rm d}V_2}{{\rm d}t} = -\frac{\Theta\sigma_0}{\rho_0 l_0} \tag{16.6}$$

Initially, $V_2 = 0$ at $t = 0$. From Eq. (16.6), it is found that ${\rm d}V_2/{\rm d}t$ varies linearly with the gradient Θ. It is evident that when the gradient $\Theta > 0$ (i.e. the impact occurs from the weakest part), V_2 cannot decrease so that $V_2 \equiv 0$, implying that only a single shock occurs from the impinged end. On the other hand, when the gradient $\Theta < 0$ (i.e. the impact occurs from the strongest part), ${\rm d}V_2/{\rm d}t > 0$, indicating that V_2 increases with time. In the latter case, double shocks initiate and propagate from both the distal and proximal ends. However, Eq. (16.2) also requires $V_1 > V_2$ to ensure ${\rm d}x/{\rm d}t > 0$. Thus, two successive phases should exist with the DS mode. In Phase I, V_2 increases with time and V_1 decreases until they reach the same value. In Phase II, immediately following Phase I, $V_1 = V_2$ and the length of the densified zone near the proximal end, x, remains unchanged (i.e. ${\rm d}x/{\rm d}t = 0$), while only the densified zone adjacent to the distal end continues to expand.

Typical time histories of the non-dimensionalized velocities, v_1 and v_2 are plotted in Fig. 16.10a with mass ratio $\Gamma = 0.5$, gradient $\Theta = -1.0$ and initial velocity $v_0 = 2$. The details of the non-dimensionalization can be found in Shen et al. (2013b). Correspondingly, the time histories of the positions of the two shock fronts are shown Fig. 16.10b. Figure 16.10a shows that v_1 decreases, while v_2 increases in Phase I; Phase II starts at the end of Phase I (i.e. at instant $\tau = \tau^*$) with the condition of $v_1 = v_2$. The whole response ends either when v decreases to zero or when the undeformed length λ decreases to zero. Note that at the phase transition, ${\rm d}v/{\rm d}\tau$ exhibits discontinuity, deviating from ${\rm d}v_1/{\rm d}\tau$ and ${\rm d}v_2/{\rm d}\tau$, which is caused by a sudden drop of the stress at the forward shock front. When v_1 just decreases to v_2, no further deformation occurs at the forward shock front. In fact, in the R-PP-L material model, the stress at the forward shock front suddenly drops to a value lower than $\sigma_{\rm d}^{\rm f}$ at the instant at which the compaction ceases. In such a condition, the stresses can be determined by the Newton's second law. In the case that the initial kinetic

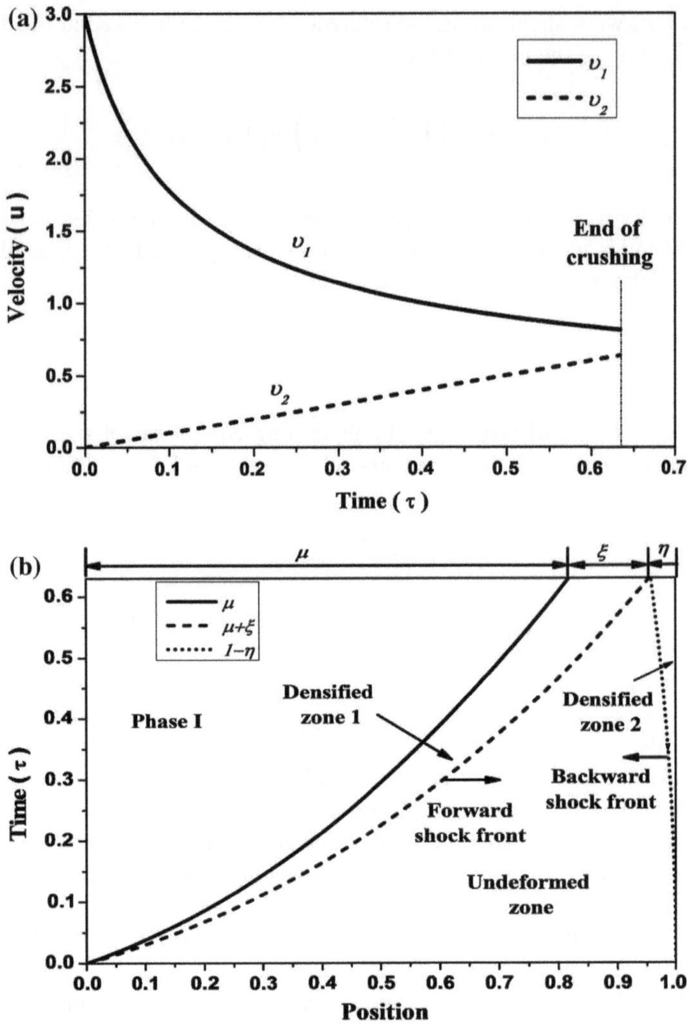

Fig. 16.10 A typical crushing process in DS mode **a** time histories of the non-dimensional velocity, υ_1 and υ_2; **b** typical time histories of the position of the shock front. $\Gamma = 0.5$, $\Theta = -1.0$ and $\upsilon_0 = 2$

energy is so large that the two shock fronts meet each other (i.e. the length of the undeformed zone λ decreases to zero) before υ_1 decreases to υ_2, Phase II does not appear.

Consider the critical case that the velocity of the mass decreases to zero exactly at the moment when the rod is just fully crushed. Figure 16.11 plots the energy ratio between the dynamic and static situations against the gradient with different mass ratios. It is noted that the positive gradient enhance the energy absorption capacity while the negative gradient weakens the capacity. The effect becomes more significant when the mass ratio is small.

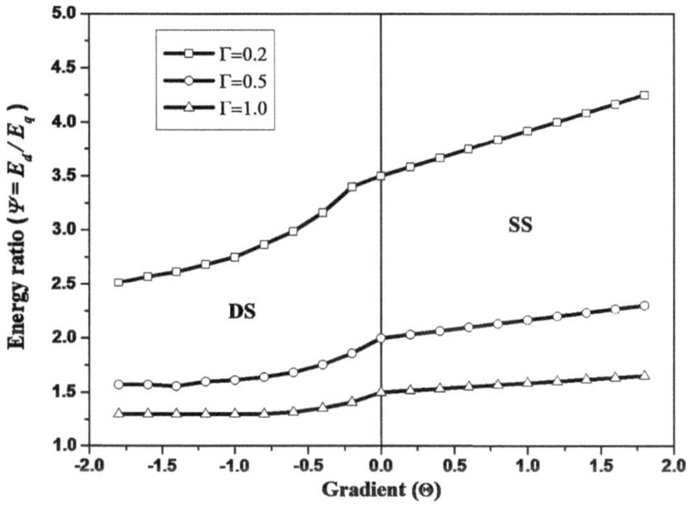

Fig. 16.11 Energy ratio $\psi = E_d/E_q$ versus gradient Θ

As a supplementary work, a more practical case was considered, in which the density gradient is introduced to a foam block (Shen et al. 2014b). The same approach was adopted in the analytical modeling using the R-PP-L material model. Similar conclusion was achieved. For both the cases, comparison with the finite element results showed that deviations existed at the densification length and the dynamic stresses at the proximal/impinged end.

16.4 Experimental Work on the Foam Block with Varying Cross-Section

The ALPORAS foam, a type of closed cell aluminum foam was used in the experiment. The chemical composition of the cell wall material is AL-0.45 wt.%Mg-0.52 wt.%Ca-0.21 wt.%Ni. The weight of each specimen was weighted to calculate the relative density, which is defined as the density of the foam divided by the density of the parent material (i.e. ρ_0/ρ_s). The density of the foam used in this study is in the range of 8–11.5 %. To obtain the stress-strain relationship of the foam, quasi-static uniaxial compression tests were carried out on a servo-hydraulic MTS machine, using uniform and cubic foam specimens with the dimension $50 \times 50 \times 50\,\text{mm}^3$. The nominal stress-strain relationship is shown in Fig. 16.12. Due to the inhomogeneity of the foam specimen, the densities of the tested specimen in the dynamic tests may be different from the one in the quasi-static compression test (10 %). Hence, the scaling method (Shen et al. 2014b) was implemented to obtain the stress-strain relationship for the foam specimen with different densities. The locking strain for

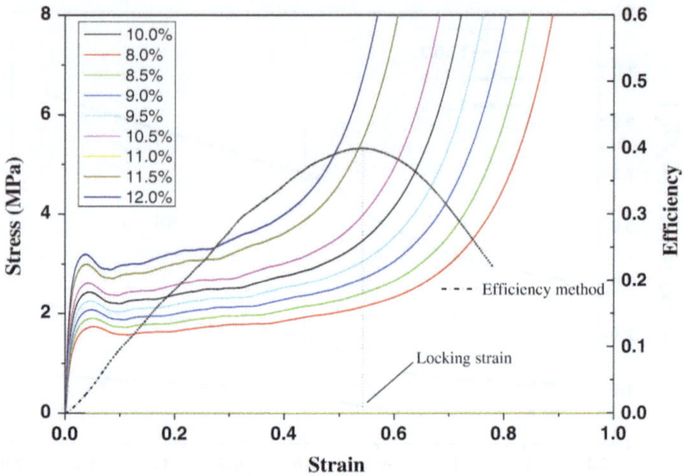

Fig. 16.12 The quasi-static compression stress-strain curve and the determination of the densified strain by using the efficiency method

(a)

(b)

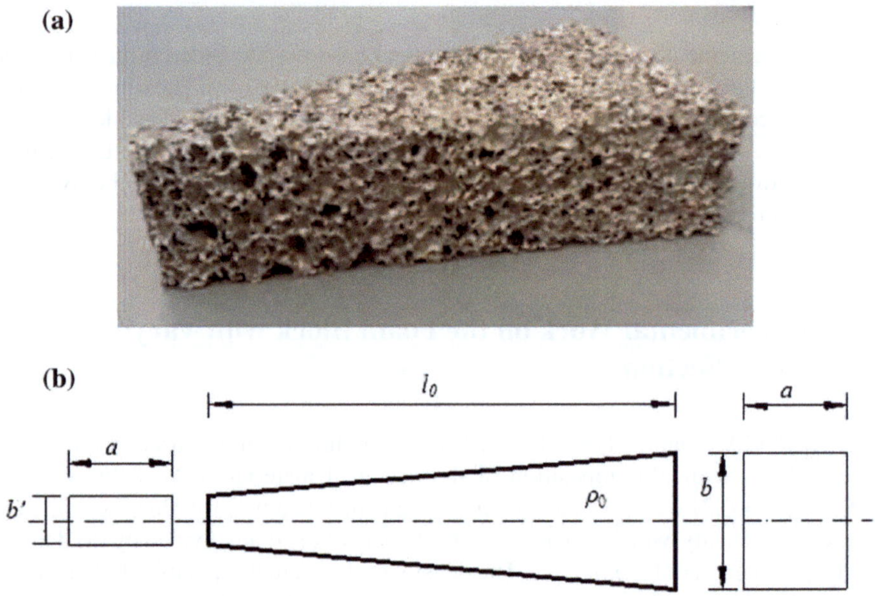

Fig. 16.13 Specimen and its sketch **a** the trapezoid foam specimen; **b** a sketch of the foam specimen

the foam (10 %) is determined as 0.48 by using the energy efficiency method (Tan et al. 2005b).

The foam specimen was a trapezoid block with varying cross-section along the loading direction, shown in Fig. 16.13. The initial length was l_0, the depth a is

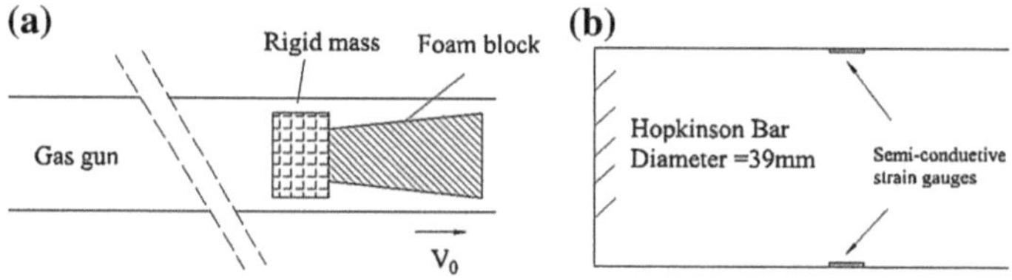

Fig. 16.14 Experimental set up: the specimen together with a rigid mass fired by a gas gun impinged onto a Hopkinson bar

a constant; and the width linearly decreased or increased along the loading direction. The gradient is defined as,

$$\Theta_A = \frac{\Delta A}{A_0} = 2\frac{b - b'}{b + b'}, \tag{16.7}$$

where ΔA is the difference between the cross-sectional areas of the two ends of the specimen, and A_0 is the average cross-sectional area.

The specimen together with a back mass was fired by using a gas gun, shown in Fig. 16.14. The sensitivity of the pressure bar is determined by the property of the material of the bar and the type of the strain gauges. To eliminate the influence of the reflected wave, a 4 m long aluminum cylindrical pressure bar with 39 mm diameter was employed to serve as the output bar, measuring the stress/strain signals generated by impact. The Young's modulus of the aluminum is 70 GPa and the density is 2700 kg/m^3. Compared with the foam specimen, the bar can be regarded as a rigid target. Two sets of semi-conductive strain gauges were mounted diametrically oppositely on the Hopkinson bar's cylindrical surface at a location of 300 mm away from the proximal end. The voltage signals were collected by a strain amplifier and then recorded by a Tektronix oscilloscope. A high speed camera was placed to capture the crushing process of the foam specimen. The camera was set to record images at a rate of 45,000 frames per second, which was sufficient to capture the details of the foam deformation. Auto-exposure was set before recording to avoid the blur effect.

Figure 16.15 shows a typical double shock mode. The force and velocity histories are plotted in Fig. 16.15a and the corresponding deformation profiles are shown in Fig. 16.15b. The force curve can be divided into four phases. In Phase I, an initial peak force appeared immediately after the impact. The peak value was much higher than those of the uniform/reversed tapered specimens since the impinged end had the largest cross-section. In this phase, the foam material at the distal end remained uncompressed because the elastic wave still had not reached the distal end. Phase II started when the force gradually decreased. Looking at the corresponding velocity history in Fig. 16.15a, the velocity only dropped a little from instant (1) to (5),

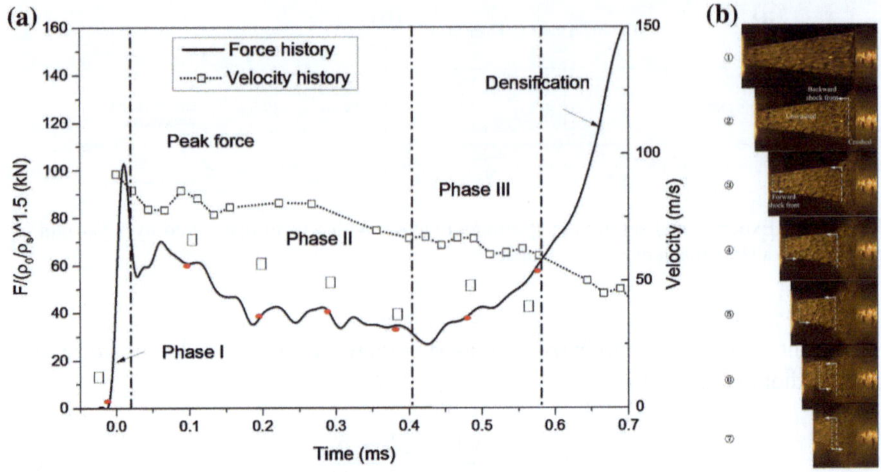

Fig. 16.15 Force and velocity histories and deformation profiles at 7 instant. **a** Force and velocity histories of the tapered specimen; **b** the corresponding deformation process. The impact velocity is 88 m/s, the relative density is 10.8 % and the time interval is 0.133 ms

indicating that the inertia effect was not the main reason for the decrease of the force. Referring to the deformation profile (Fig. 16.15b), it is noted a compaction front (backward shock front) was formed and propagated to the distal end due to the collapse of the foam at the proximal end. During the process, the cross-sectional area at the backward shock front was decreasing in the crushing, which may be the main reason for the force dropping. Meanwhile, a new compaction zone showed up at the distal end of the specimen due to the reflection of the elastic wave, forming another forward shock front. Then, it propagated to the other end. Phase II ended when the force dropped to the lowest value. Afterwards, Phase III started. In this phase, the force began to increase to higher values. No further collapse was found at the proximal end (see instants ⑥ and ⑦), indicating that the backward shock front at the proximal end ceased. The force at the proximal end was the same as the one at the forward shock front to maintain the equilibrium state. With the propagation of the forward shock wave, the cross-sectional area at the forward shock front increased, resulting in the raise of the force. After instant ⑦, the two shock fronts collided with each other and the residual velocity was around 60 m/s, bringing the whole specimen into the densification stage.

To give better prediction from the analytical result, the power-law material model proposed by Pattofatto et al. (2007) was used to fit the stress-strain relationship of ALPORAS foam

$$\sigma = \sigma_0 + k\varepsilon^m, \tag{16.8}$$

where k and m are constants determined to fitting the stress-strain curves. When $m = 1$, k is the linear hardening plastic (LHP) modulus for R-LHP model in Zheng et al. (2011). The stress change across the shock front, is expressed as,

$$[\sigma] = \frac{\rho_0[V]^2}{[\varepsilon_{[V]}]} \tag{16.9}$$

where [...] denotes the quantity change across the shock front.

Neglecting the elastic property of the foam leads to ε_Y. Consequently, the localized strain just behind the shock front, $\varepsilon_{[V]}$, is obtained, as follows,

$$\varepsilon_{[V]} = \left(\frac{\rho_0[V]^2}{k}\right)^{1/(m+1)} \tag{16.10}$$

The schematic of the impact of the tapered specimen is depicted in Fig. 16.16a and the corresponding stress distribution is shown in Fig. 16.16b. Similarly, the specimen is divided into three regimes, namely forward densification zone, undeformed zone, and the back densification zone.

The original length of the forward and backward densified zones are X and Y, respectively.

$$X = \int_0^x \frac{1}{1 - \varepsilon V_1 - V_2} \mathrm{d}x, \quad Y = \int_0^y \frac{1}{1 - \varepsilon V_1 - V_2} \mathrm{d}y \tag{16.11}$$

The cross-sectional areas at characteristic positions are expressed as,

$$\begin{aligned} &\frac{A_{\mathrm{bs}}}{A_0} = 1 - \left(\frac{1}{2} - \frac{Y}{l_0}\right)\Theta_A, \quad \frac{A_{\mathrm{b}}}{A_0} = 1 - \frac{1}{2}\left(1 - \frac{Y}{l_0}\right)\Theta_A, \\ &\frac{A_{\mathrm{m}}}{A_0} = 1 - \frac{1}{2}\left(\frac{X - Y}{l_0}\right)\Theta_A, \quad \frac{A_{\mathrm{f}}}{A_0} = 1 - \frac{1}{2}\left(\frac{X}{l_0} - 1\right)\Theta_A, \\ &\frac{A_{\mathrm{fs}}}{A_0} = 1 - \left(\frac{X}{l_0} - \frac{1}{2}\right), \end{aligned} \tag{16.12}$$

where A_{bs} and A_{fs} are the cross-sectional areas at the backward and forward shock front, respectively; A_{b}, A_{m} and A_{f} are the average cross-sectional areas in the backward, undeformed zone and forward densified zone, respectively.

The following geometry relationships are obtained,

$$\begin{aligned}
&\frac{du}{dt} = V_1,\ \frac{dx}{dt} = \frac{1-\varepsilon_{V_1-V_2}}{\varepsilon_{V_1-V_2}}(V_1 - V_2),\ \frac{dy}{dt} = \frac{1-\varepsilon_{V_1-V_2}}{\varepsilon_{V_2}}V_2,\\
&\frac{dl}{dt} = -\frac{dx}{dt} - \frac{dy}{dt} - \frac{du}{dt},\\
&\frac{dX}{dt} = \frac{V_1 - V_2}{\varepsilon_{V_1-V_2}},\ \frac{dY}{dt} = \frac{V_2}{\varepsilon_{V_2}},\ \frac{dl}{dt} = -\frac{dX}{dt} - \frac{dY}{dt}
\end{aligned} \tag{16.13}$$

Applying the conservation of momentum for the whole system over a small time interval dt gives,

$$-A_{\mathrm{bs}}\sigma_{\mathrm{d}}^{\mathrm{b}}\mathrm{d}t = \mathrm{d}\left[\left(G + \int_0^t \rho(x)xA_x\mathrm{d}x\right)V_1\right] + \mathrm{d}(\rho_0 l V_2 A_{\mathrm{m}}) \tag{16.14}$$

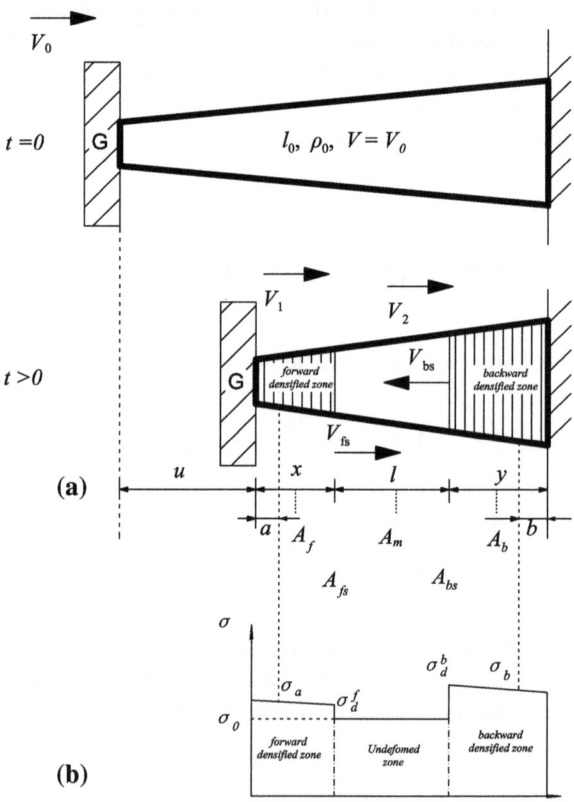

Fig. 16.16 Schematic of the DS mode for the tapered specimen and the stress distribution. **a** The impact of the graded foam on a rigid surface for the tapered specimen; **b** corresponding stress distribution at instant t

Also, at the forward shock front, the conservation of momentum gives,

$$-A_{fs}\sigma_0 dt = d\left[\left(G + \int_0^t \rho(x) x A_x dx\right) V_1\right] + \frac{\rho_0}{1 - \varepsilon_{V_1 - V_2}} l V_2 A_{fs} dx \quad (16.15)$$

Solving Eqs. (16.14) and (16.15) gives,

$$\begin{aligned} \frac{dV_1}{dt} &= -\frac{\left[1 - \left(\frac{X}{l_0} - \frac{1}{2}\right)\Theta_A\right]\left[\sigma_0 + \rho_0 (V_1 - V_2)^2/(1 - \varepsilon_{V_1 - V_2})\right] A_0}{G + \rho_0 X \left[1 - \frac{1}{2}\left(\frac{X}{l_0} - 1\right)\Theta_A\right] A_0} \\ \frac{dV_2}{dt} &= \frac{\sigma_0 \Theta_A}{\rho_0 l_0 \left[1 - \frac{1}{2}\left(\frac{X - Y}{l_0}\right)\Theta_A\right]} \end{aligned} \quad (16.16)$$

The initial conditions are

$$V_1\big|_{t=0} = V_2\big|_{t=0} = V_0, \quad X\big|_{t=0} = 0, \quad Y\big|_{t=0} = 0, \quad o\big|_{t=0} = 0, \quad l\big|_{t=0} = 0 \quad (16.17)$$

To ensure the existence of the forward densification zone, it is necessary that

$$\frac{dV_2}{dt} < \frac{dV_1}{dt} \quad (16.18)$$

Thus, substituting the initial conditions into Eq. (16.18) gives,

$$\Theta_A < -\frac{2}{1 + 2\Gamma} \quad (16.19)$$

here Γ is the mass ratio of the back mass over the mass of the foam block. Equation (16.19) is the same as that given in Shen et al. (2014a). When there is no back mass (i.e. $\Gamma = 0$), $\Theta_A < -2$, indicating that forward densification zone does not exist.

The deformation process may be divided into two phases. In Phase I, two compaction shock fronts imitate from both the proximal and distal ends. Phase II starts when V_2 decreases to zero. In Phase II, the backward densification zone ceases and the forward shock front continues to propagate to the proximal end until the kinetic energy is exhausted. The governing equations in Phase II can be obtained by applying the conditions, $V_1\big|_{t=t^*} = V_1$, $V_2 \equiv 0$ to Eq. (16.16). If the input kinetic energy is sufficient high, V_2 may be still larger than zero at the end of the crushing. In such case, Phase II does not exist.

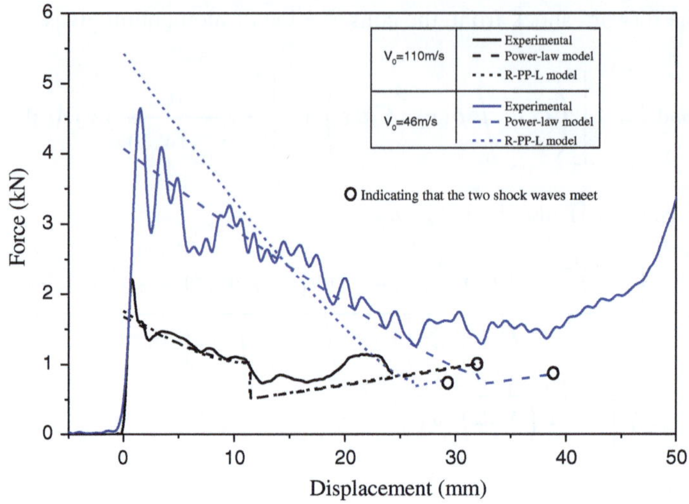

Fig. 16.17 Comparison of the force-displacement curves for the tapered specimens between the analytical and experimental results in Scenario I

Figure 16.17 shows the comparison of the force-displacement curves between the analytical and experimental results for tapered specimens. Under a low impact velocity of 46 m/s, both the power-law and the R-PP-L material models agree well with the experimental result. When the impact velocity is high (e.g. 110 m/s), the R-PP-L material model predicts higher initial value and drops much faster while the predictions by the power-law material model give good agreement. Similarly, the final displacement is different from the experiment because further crushing is not considered when the backward and forward shock waves meet. Figure 16.18a shows shock front positions for a typical case in Scenario I. The corresponding strain distribution and velocity histories are shown in Fig. 16.18b. The relative density of the foam is 10.0 %, the back mass and the specimen have an impact velocity of 65 m/s, the gradient is $\Theta_A = -1$ and back mass is 18.5 g. Different from the R-PP-L material model, the strain distribution in the foam specimen is non-uniform. In Phase I, V_2 decreases faster than V_1. Thus, the difference of the velocities across the forward shock front (i.e. $V_1 - V_2$) increases, resulting in the rise of the localized strain. It reaches the maximum when Phase I ends. Meanwhile, the localized strain at the backward shock front drops from the beginning and it reaches minimum when Phase I ends since V_2 keeps deceasing. In Phase II, the backward densification zone ceases (i.e. $V_2 \equiv 0$) while the forward densification zone continues to expand. The localized strain at the forward shock front starts to drop in Phase II.

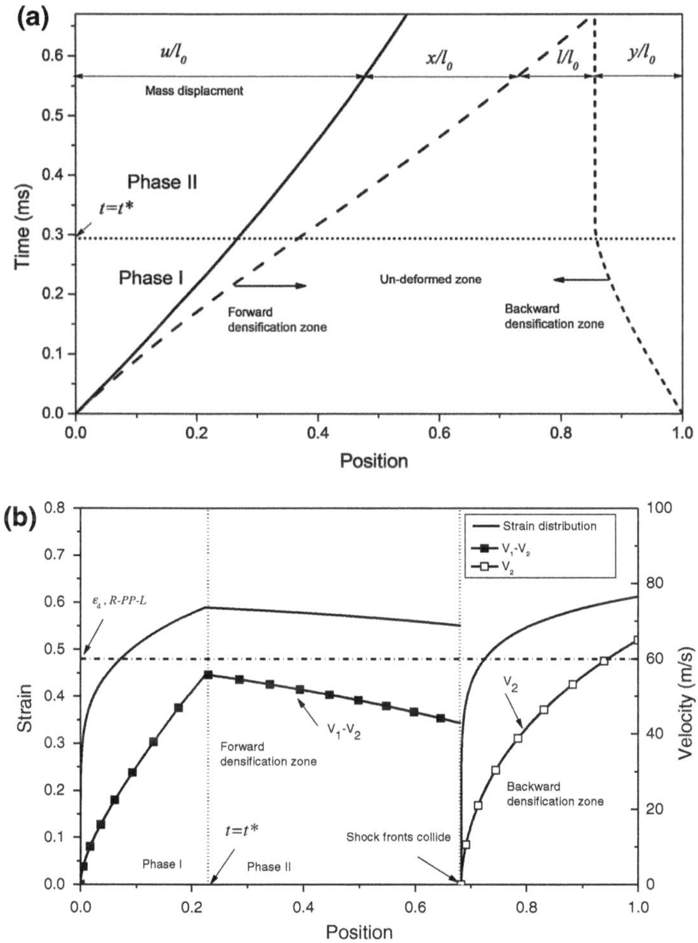

Fig. 16.18 Typical crushing process of the DS mode for the tapered specimen. **a** The shock front positions in the DS mode; **b** the corresponding strain distribution and instant velocity at the shock fronts

16.5 Conclusion

This chapter focuses on the investigation into the dynamic behavior of the graded cellular material under impact from three aspects, namely the numerical simulation (graded honeycomb structure), analytical modeling (graded foam rod) and the experimental study (graded foam block).

Two deformation modes were identified from these studies. When the gradient is positive, only one shock initiates from the proximal end while two shock fronts appear when the gradient is negative.

Both R-PP-L and Power-law material model can capture the features of the double shock mode. Comparison with the experimental result indicates that the power-law model gives better prediction. Investigation on the energy absorption capacity shows that the negative strength gradient weakens the capacity while the positive enhance the capacity.

Acknowledgments The authors would like to acknowledge Prof. Yu Tongxi at Hong Kong University of Science and Technology and Prof. Ruan Dong at Department of Mechanical and Product Design Engineering, Swinburne University of Technology, for their valuable support throughout the work.

References

Ajdari A, Canavan P, Nayeb-Hashemi H, Warner G (2009) Mechanical properties of functionally graded 2-D cellular structures: a finite element simulation. Mat Sci Eng A-Struct 499:434–439

Ajdari A, Babaee S, Vaziri A (2011) Mechanical properties and energy absorption of heterogeneous and functionally graded cellular structures. Procedia Eng 10:219–223

Ali M, Qamhiyah A, Flugrad D, Shakoor M (2008) Theoretical and finite element study of a compact energy absorber. Adv Eng Softw 39:95–106

Cui L, Kiernan S, Gilchrist MD (2009) Designing the energy absorption capacity of functionally graded foam materials. Mat Sci Eng A-Struct 507:215–225

Gibson L, Ashby M (1997) Cellular solids: structure and properties, 2nd edn. Cambridge University Press, Cambridge

Karagiozova D, Langdon GS, Nurick GN (2012) Propagation of compaction waves in metal foams exhibiting strain hardening. Int J Solids Struct 49:2763–2777

Kiernan S, Cui L, Gilchrist MD (2009) Propagation of a stress wave through a virtual functionally graded foam. Int J Nonlin Mech 44:456–468

Li S, Wang Z, Wu G, Zhao L, Li X (2014) Dynamic response of sandwich spherical shell with graded metallic foam cores subjected to blast loading. Compos A Appl Sci Manuf 56:262–271

Lopatnikov SL, Gama BA, Haque MJ, Krauthauser C, Gillespie JW, Guden M, Hall IW (2003) Dynamics of metal foam deformation during Taylor cylinder-Hopkinson bar impact experiment. Compos Struct 61:61–71

Lopatnikov SL, Gama BA, Haque MJ, Krauthauser C, Gillespie JW, Guden M, Hall IW (2004) High-velocity plate impact of metal foams. Int J Impact Eng 30:421–445

Pattofatto S, Elnasri I, Zhao H, Tsitsiris H, Hild F, Girard Y (2007) Shock enhancement of cellular structures under impact loading: part II analysis. J Mech Phys Solids 55:2672–2686

Reid SR, Peng C (1997) Dynamic uniaxial crushing of wood. Int J Impact Eng 19:531–570

Ruan D, Lu G, Wang B, Yu T (2003) In-plane dynamic crushing of honeycombs—a finite element study. Int J Impact Eng 28:161–182

Shen CJ, Lu G, Yu TX (2013a) Dynamic behavior of graded honeycombs—a finite element study. Compos Struct 98:282–293

Shen CJ, Yu TX, Lu G (2013b) Double shock mode in graded cellular rod under impact. Int J Solids Struct 50:217–233

Shen CJ, Lu G, Yu TX (2014a) Dynamic response of a cellular block with varying cross-section. Int J Impact Eng p (accepted)

Shen CJ, Lu G, Yu TX (2014b) Investigation into the behavior of a graded cellular rod under impact. Int J Impact Eng 74:92–106

Shen CJ, Lu G, Ruan D, Yu TX (2015) Propagation of the compaction waves in a cellular block with varying cross-section. Int J Solids Struct p (submitted)

Tan P, Reid S, Harrigan J, Zou Z, Li S (2005a) Dynamic compressive strength properties of aluminium foams. Part I—experimental data and observations. J Mech Phys Solids 53:2174–2205
Tan P, Reid S, Harrigan J, Zou Z, Li S (2005b) Dynamic compressive strength properties of aluminium foams. Part II—'shock' theory and comparison with experimental data and numerical models. J Mech Phys Solids 53:2206–2230
Wang XK, Zheng ZJ, Yu JL, Wang CF (2011) Impact resistance and energy absorption of functionally graded cellular structures. Appl Mech Mater 69:73–78
Zheng ZJ, Liu YD, Yu JL, Reid SR (2011) Dynamic crushing of cellular materials: continuum-based wave models for the transitional and shock modes. Int J Impact Eng 42:66–79

Chapter 17
Fracture Mechanics at Atomic Scales

Takahiro Shimada and Takayuki Kitamura

Abstract Fracture and strength in atomic components are governed by mechanical instabilities at atomic scales associated with irreversible deformations through bond breaking/switching, such as cleavage, dislocation nucleation, and phase transformations of crystal lattices. It is therefore of central importance to determine the critical conditions where atomic structures becomes mechanically unstable. Here we review the state-of-the-art theory for "fracture mechanics of atomic structures" that provides a rigorous description of mechanical instabilities in arbitrary atomic structures under any external loading/constraint. The theory gives the critical instability condition by positivity of the minimum eigenvalue of the Hessian matrix of the total energy with respect to degrees of freedom of the system (i.e., instability criterion), and it successfully provides atomistic insights into fracture in various atomic/nanoscale structures. The review also covers the recent development of theory extended to advanced systems including large-scale, finite temperature, and "multi-physics" instabilities in (ferro-)electric and magnetic materials as functional fracture.

Keywords Mechanical instabilities · Atomic components · Fracture criterion · Nanoscales · Molecular dynamics

17.1 Introduction

Understanding the nature of fracture, a catastrophic failure of materials, remains a major challenge in a wide range of fields including not only mechanical engineering but also physics, materials science, biology, and geophysics, because fracture is both a physically essential phenomenon and a practically inevitable issue that all

T. Shimada (✉) · T. Kitamura
Department of Mechanical Engineering and Science, Kyoto University,
Nishikyo-ku, Kyoto 615-8540, Japan
e-mail: shimada@me.kyoto-u.ac.jp

T. Kitamura
e-mail: kitamura@kues.kyoto-u.ac.jp

H. Altenbach et al. (eds.), *From Creep Damage Mechanics to Homogenization Methods*, Advanced Structured Materials 64,
DOI 10.1007/978-3-319-19440-0_17

materials in all range of scales intrinsically possess and commonly suffer from (Liebowitz 1968; Zhang et al. 2014; Holland and Marder 1999; Kermode et al. 2008; Pons and Karma 2010; Warner et al. 2007; Buehler and Gao 2006; Song and Curtin 2013; Livne et al. 2010; Nalla et al. 2003; Bažant 2002; Rubin 1995; Dombard 2007). Failure of materials is ultimately governed by bond breaking/switching at the atomic level of crystals. Brittle fracture through cleavage of crystal lattice is the simplest and most typical fracture that is associated with bond breaking. Bond switching often induces the slip of atomic planes resulting in nucleation and multiplication of dislocations and phase transformation of crystal lattices, which are closely related to strength of materials. Such a bond breaking/switching and resulting dramatic change in (global or local) atomic structures can be regarded as "mechanical instabilities" of atomic structures, which often accompany irreversible deformations of materials with a sudden drop of the external load. Such mechanical instabilities can be critical in nanoscale/atomic components such as nanofilms, nanotubes, nanowires and nanodots that have recently been developed due to remarkable advancement in manufacturing technology. In such atomic components only a single bond-breaking/switching often leads to immediate fracture or the fatal malfunction due to their ultimately small dimensions. Therefore, it is of central importance to investigate the nature of mechanical instabilities for thorough understanding of characteristic strength and fracture of atomic components.

Born and Huang (1954) and Milstein (1971); Hill and Milstein (1997); Milstein (1980) demonstrated that stability of a perfect crystal can be evaluated on the basis of the elastic constants. These criteria were extended by Wang et al. (1993, 1997, 1995) for finite deformations, where the crystal stability can be expressed as functions of the elastic stiffness coefficients (i.e., B-criterion). Wang's theorem can be successfully applied to, for example, bifurcation from the fcc to bcc structure in metals under tension (Luo et al. 2002; Černý et al. 2004). However, this criterion is primarily effective only for perfect crystals subject to uniform deformation. As an early attempt toward mechanical instability in inhomogeneous structures, Kitamura et al. (1997, 1998); Yashiro and Tomita (2010) have shown that the dislocation nucleation in a Ni nano-wire under tension by applying the criterion to the local crystal lattice. On the other hand, Li et al. (2003, 2002); van Vliet et al. (2003) employed the concept of phonon soft modes to a local site (Λ-criterion). Dmitriev et al. (2005a, b) applied the concept to the low-dimensional components. Thanks to numerous efforts, the mechanical instabilities in atomic structures have been understood to a certain extent. However, these criteria do not maintain rigorousness, because only the energy balance in a local region (i.e., limited degrees of freedom) is taken into account. For the precise evaluation of mechanical instability in an arbitrary atomic system, it is necessary to consider all the degrees of freedom of the entire system.

A theory that considers the energy balance with respect to all the degrees of freedom of atoms has been proposed by Kitamura et al. (2004a, b). This theory rigorously evaluates the criterion of mechanical instability in heterogeneous atomic systems under non-uniform loading or constraint (Kitamura et al. 2004a, b). Here we review the rigorous criterion that describes the mechanical instabilities in arbitrary atomic structures and its recent advances and applications. In Sect. 17.2, we provide a detailed theory to describe the mechanical instabilities in atomic structures and its applications. In Sect. 17.3, we discuss problems of this original theory for practical situations and introduce an improved approach that resolves the problems. Section 17.4 contributes to the review of recent advances and extension of the method toward complicated system such as finite temperature, electric, and magnetic systems. In Sect. 17.5 we summarize the present review with some future directions of mechanical instability issues.

17.2 Fracture Criterion for Mechanical Instability in Arbitrary Atomic Structures

To investigate the onset and nature of mechanical instabilities in atomic components, a rigorous and universal criterion has been proposed for an arbitrary atomic structure under any loading/constraint condition (Kitamura et al. 2004a, b). Here, let us consider an atomic system consisting of N atoms. The potential energy of the atomic system, Φ, can be represented as a function of atomic coordinate,

$$\Phi = \Phi(\boldsymbol{R}), \tag{17.1}$$

where $\boldsymbol{R}$ denotes the configuration vector consisting of atomic positions,

$$\begin{aligned}\boldsymbol{R} &\equiv^t (r_z^2, r_y^3, r_z^3, r_x^4, r_y^4, r_z^4, \ldots, r_i^\alpha, \ldots, r_x^N, r_y^N, r_z^N) \\ &=^t (R_1, R_2, \ldots, R_m, \ldots, R_M)\end{aligned} \tag{17.2}$$

Here, r_i^α denotes the coordinate of atom α in the $i(= x, y, z)$ direction. This is the general form to express the potential energy of atomic system, in any type of po-tential functions and force fields including first-principles (quantum-mechanics) approaches. The irreducible number of structural degrees of freedom (DOFs) in the atomic system without any constraint is $M = 3N - 6$ because the DOFs of rigid-body translations (3) and rotations (3) is subtracted from the total DOFs of atoms $(3N)$. Under a displacement constraint where some of atoms are fixed, the number of DOFs is $M = 3N - 3n_c - 6$, where n_c is the number of constrained atoms. Here, let us consider the atomic system under static external load, i.e., the atoms are located at their own optimal sites and are in balance with the external load and/or constraint. The total energy Π at the equilibrated atomic configuration ($\boldsymbol{R}_0$) under external load consists of the potential energy, Φ, and the work done by external load, W,

$$\Pi = \Phi + W \tag{17.3}$$

Assuming an infinitesimal deformation, $\delta\boldsymbol{R}$, to the equilibrated system, the total energy of slightly deformed system, $\Pi(\boldsymbol{R}_0 + \delta\boldsymbol{R})$, can be described by the Taylor's series expansion of total energy $\Pi(\boldsymbol{R}_0)$ with respect to $\delta\boldsymbol{R}$,

$$\Pi(\boldsymbol{R}_0+\delta\boldsymbol{R}) = \Pi(\boldsymbol{R}_0) + \sum_{m=1}^{M}\left(\left.\frac{\partial\Phi}{\partial R_m}\right|_{\boldsymbol{R}=\boldsymbol{R}_0} + \left.\frac{\partial W}{\partial R_m}\right|_{\boldsymbol{R}=\boldsymbol{R}_0}\right)\delta R_m$$
$$+\frac{1}{2}\sum_{m=1}^{M}\sum_{n=1}^{M}\left.\frac{\partial^2\Phi}{\partial R_m\partial R_n}\right|_{\boldsymbol{R}=\boldsymbol{R}_0}\delta R_m\delta R_n + \cdots \tag{17.4}$$
$$+\frac{1}{2}\sum_{m=1}^{M}\sum_{n=1}^{M}\left.\frac{\partial^2 W}{\partial R_m\partial R_n}\right|_{\boldsymbol{R}=\boldsymbol{R}_0}\delta R_m\delta R_n + \cdots,$$

where R_m denotes a component of configuration vector $\boldsymbol{R}$ included in the DOFs of system, i.e., $R_m = r_i^\alpha$. Since the first derivative of total energy (i.e., force acting on atoms) is zero due to the system at equilibrium, the second term on the right-hand side can be eliminated

$$\frac{\partial\Pi}{\partial R_m}\delta R_m = \left(\left.\frac{\partial\phi}{\partial R_m}\right|_{\boldsymbol{R}=\boldsymbol{R}_0} + \left.\frac{\partial W}{\partial R_m}\right|_{\boldsymbol{R}=\boldsymbol{R}_0}\right)\delta R_m \tag{17.5}$$

Considering that the external load is constant due to the static loading, the work is proportional to the displacement of atoms on which the external load is applied. Thus, we obtain

$$\frac{\partial^2 W}{\partial R_m\partial R_n}\delta R_m\delta R_n = 0 \tag{17.6}$$

Using Eqs. (17.4)–(17.6) and ignoring the higher-order terms, the change in total energy $\delta\Pi$ with respect to the infinitesimal deformation $\delta\boldsymbol{R}$ is given by

$$\delta\Pi(\boldsymbol{R}_0) = \Pi(\boldsymbol{R}_0+\delta\boldsymbol{R}) - \Pi(\boldsymbol{R}_0) = \frac{1}{2}\sum_{m=1}^{M}\sum_{n=1}^{M}\frac{\partial^2\Phi}{\partial R_m\partial R_n}\delta R_m\delta R_n = \frac{1}{2}\delta\boldsymbol{R}^T\boldsymbol{H}\delta\boldsymbol{R}, \tag{17.7}$$

where $\boldsymbol{H}$ is the $M\times M$ Hessian matrix of potential energy Π with respect to $\boldsymbol{R}$, and the superscript ^{T} means transposition. The component of Hessian matrix, H_{mn}, is

$$H_{mn} \equiv \left.\frac{\partial^2\Phi}{\partial R_m\partial R_n}\right|_{\boldsymbol{R}=\boldsymbol{R}_0} \tag{17.8}$$

By solving the eigenvalue problem of the Hessian matrix $\boldsymbol{H}$, we obtain

$$\boldsymbol{H}\boldsymbol{p}_m = \eta_m\boldsymbol{p}_m, \tag{17.9}$$

where η_m is the eigenvalue of the Hessian matrix $\boldsymbol{H}$, and $\boldsymbol{p}_m$ is the corresponding eigenvector. Using the eigenvector $\boldsymbol{p}_m$, the Hessian matrix is diagonalized as

$$P^{-1}HP = P^T HP = \begin{pmatrix} \eta_1 & & 0 \\ & \ddots & \\ 0 & & \eta_M \end{pmatrix}, \tag{17.10}$$

where $\boldsymbol{P} = (\boldsymbol{p}_1 \ldots \boldsymbol{p}_M)$. Since the eigenvectors are an orthogonal basis set of M-dimensional vector space, any infinitesimal deformation $\delta\boldsymbol{R}$ can be expressed as a linear combination of the eigenvectors as

$$\delta\boldsymbol{R} = \sum_{m=1}^{M} u_m \boldsymbol{p}_m = \boldsymbol{Pu}, \tag{17.11}$$

where u_m is the component of $\delta\boldsymbol{R}$ in the $\boldsymbol{p}_m$ direction, $\boldsymbol{u} = (u_1 \ldots u_M)$. Therefore, the total energy change with $\delta\boldsymbol{R}$ in Eq. (17.7) becomes

$$\delta\Pi(\boldsymbol{R}_0) = \frac{1}{2}(\boldsymbol{Pu})^T \boldsymbol{H}(\boldsymbol{Pu}) = \frac{1}{2}\boldsymbol{u}(\boldsymbol{P}^T \boldsymbol{HP})\boldsymbol{u} = \sum_{m=1}^{M} \eta_m u_m^2 \tag{17.12}$$

In this equation, the eigenvalue, η_m, signifies the potential energy curvature against the deformation along the corresponding eigenvector, $\boldsymbol{p}_m$. Therefore, if the minimum eigenvalue is positive ($\eta_1 > 0$), the change in the total energy, $\delta\Pi$, is always positive with respect to any infinitesimal deformation, indicating that the system is mechanically stable. On the other hand, when the minimum eigenvalue becomes zero or negative ($\eta_1 \leq 0$), the total energy can decrease along the corresponding eigenvector, $\boldsymbol{p}_1$. This indicates that the system is mechanically unstable and the corresponding eigenvector $\boldsymbol{p}_1$ represents the atomic motion at the beginning of instability (unstable mode). Therefore, the onset of mechanical instabilities in atomic structures can be determined by the positivity of the eigenvalue of Hessian matrix $\boldsymbol{H}$, i.e., $\eta_1 = 0$ (Kitamura et al. 2004a, b).

The validity of the theory has been verified by applying the criterion to a nanoscale cracked atomic body under external load (Kitamura et al. 2004b), as shown in Fig. 17.1. Two different loading conditions that lead to different fracture modes were applied to the model: the external load-controlled (Case A) and the displacement-controlled (Case B) conditions. Figure 17.2a plots the stress-strain curves of Cases A and B. The two stress-strain curves exhibit the different behavior, especially, the different critical strain where fracture occurs. Note that these bring about different morphology of mechanical instability at the critical strain: the cleavage crack propagation is observed for Case A (i.e., brittle fracture), while the dislocation emission from the crack tip (i.e., ductile fracture) occurs for Case B. The above criterion was applied to these two deformations and fracture, and the minimum eigenvalue as a criterion is evaluated for each case, as shown in Fig. 17.2b. Although the mechanical behavior and fracture mode are totally different between Cases A and B, the minimum eigenvalues become zero at each critical strain where stress dramatically drops (or the strain diverges) and fracture occurs. This indicates that the instability

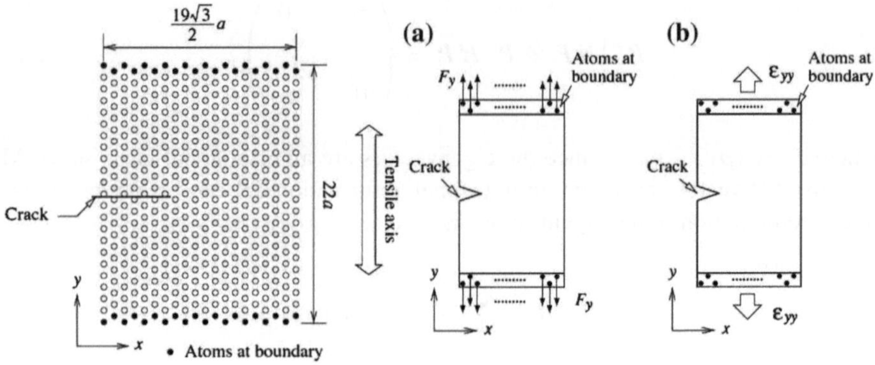

Fig. 17.1 An atomic structure with a crack. **a** Case A: load control simulation. **b** Case B: displacement control simulation (Kitamura et al. 2004b)

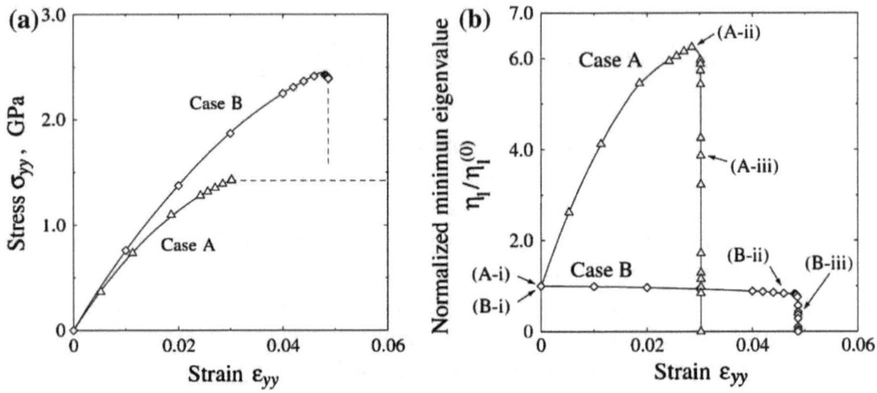

Fig. 17.2 Results of computational tensile tests shown in Fig. 17.1. **a** Stress versus strain curves of cracked atomic structures, **b** minimum eigenvalue normalized by that of unstrained one as a function of strain (Kitamura et al. 2004b)

criterion, $\eta_1 = 0$, exactly captures the critical strain where fracture occurs regardless of the fracture modes.

Figure 17.3 shows the unstable deformation mode, $\boldsymbol{p}_1$, corresponding the minimum eigenvalue of $\eta_1 = 0$. The dominant mode vector for Case A represents the opening mode of crack, i.e., cleavage fracture. On the other hand, the unstable mode of Case B shows the sliding mode of neighboring atomic planes, i.e., the dislocation emission from the crack tip. These modes are perfectly consistent with the fracture behaviors directly obtained by the molecular dynamics simulations. Therefore, the method can rigorously evaluate the instability criterion as well as the deformation mode of the mechanical instabilities.

Owing to the capability of the criterion as a powerful tool to explore the nature of fracture in atomic components, the method has been applied to various kinds of

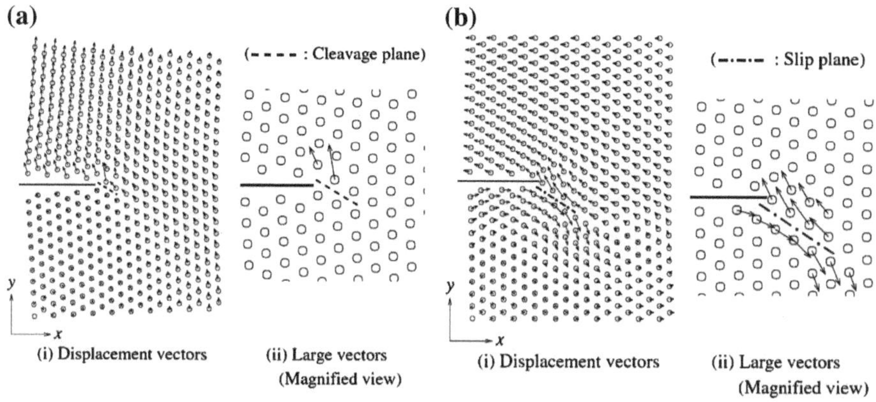

Fig. 17.3 Unstable deformation modes (eigenvectors corresponding to minimum eigenvalues) for **a** Case A and **b** Case B (Kitamura et al. 2004b)

systems (Kitamura et al. 2004a; Umeno et al. 2007; Shimada et al. 2008, 2009a; Kubo et al. 2013), including metallic glasses and amorphous metals (Umeno et al. 2007; Shimada et al. 2008, 2009a), interface cracking of nanofilms (Kitamura et al. 2004a), and sliding and/or switching of domain/grain boundaries (Umeno et al. 2009). For example, the method clarified the fracture mode of bi-material interfaces and its dominant region near the interface edge (Kitamura et al. 2004a), as shown in Fig. 17.4a. The method was also applied to more complicated atomic structures, such as amorphous metals, where atoms are arranged almost randomly in contrast to crystal lattices. Although it was difficult to extract detailed deformation processes that contribute to macroscopic plasticity in amorphous metals due to the random atomic arrangements, the method successfully captured the localized atomic motions dominated by several tens of atoms (see Fig. 17.4b) as an elementary process of plasticity in amorphous metals (Umeno et al. 2007; Shimada et al. 2008, 2009a).

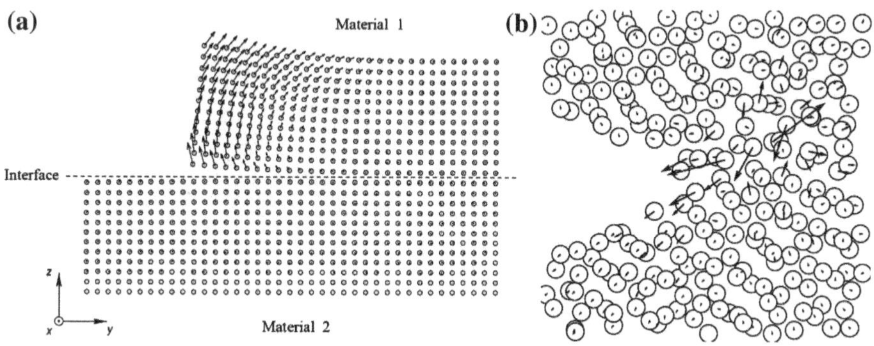

Fig. 17.4 Unstable deformation modes of **a** delamination of bi-material interfaces (Kitamura et al. 2004a) and **b** amorphous metals near a crack-tip (Shimada et al. 2009a)

Thus, the method has successfully revealed the nature of mechanical instabilities at the atomic scales and the elementary process of complicated deformation and fracture in atomic structures.

17.3 Simplified Evaluation of Fracture Criterion for Large-Scale Atomic Structures

Although the applicability of the instability criterion is quite large and a variety of mechanical instabilities and fracture issues can be solved by the theory, there are still some problems. One of the difficulties is to apply the criterion to a large-scale atomic structure. When the system contains huge number of atoms, the DOFs of the system and the resulting Hessian matrix, $\boldsymbol{H}$, in Eq. (17.8) becomes extremely large so that the eigenvalue problem of Hessian matrix cannot be computationally solved due to its huge memory requirements.

In this section, we introduce a simplified method to evaluate mechanical instabilities even in a large-scale atomic structure using a reasonable concept to reduce the effective degrees of freedom in the system (Shimada et al. 2009b, 2010b). In an attempt to re-duce the degrees of freedom in the computation, a part of the atomic system is divided into elements and the displacement of atoms in the element is described using a linear function of the nodal coordinate, like the well-established concept of finite element analysis. Figure 17.5 schematizes the concept applied to a three-dimensional atomic structure with a crack (Shimada et al. 2010b). In the area

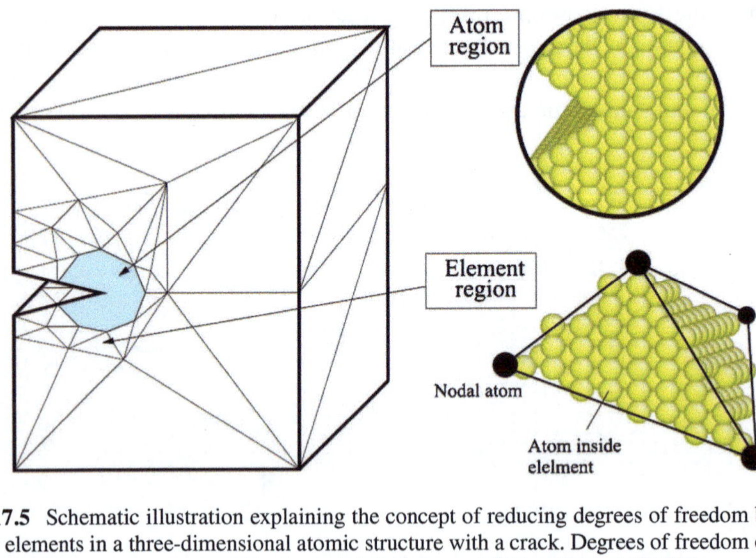

Fig. 17.5 Schematic illustration explaining the concept of reducing degrees of freedom by using linear elements in a three-dimensional atomic structure with a crack. Degrees of freedom of atoms inside an element are represented by its nodal displacement (Shimada et al. 2010b)

far from the crack, the element includes many atoms because the strain gradient is moderate and the deformation of the atoms can be well-described by the nodal displacements of element. On the other hand, the number of atoms in the element should decrease in the region near the crack tip where the strain concentrates, and the DOFs of each atom is fully taken into account in the proximity of the crack-tip where the unstable deformation preferentially occurs.

Denoting the number of atoms and nodes in the system as N'_{atom} and N'_{node}, respectively, we can re-describe the deformation of the system by the vector $\boldsymbol{R}'$,

$$\begin{aligned}\boldsymbol{R}' \equiv^t \Big(& r_x^1, r_y^1, r_z^1, \ldots, r_i^\alpha, \ldots, r_x^{N'_{\text{atom}}}, r_y^{N'_{\text{atom}}}, r_z^{N'_{\text{atom}}}, \\ & S_x^1, S_y^1, S_z^1, \ldots, S_i^\xi, \ldots, S_x^{N'_{\text{node}}}, S_y^{N'_{\text{node}}}, S_z^{N'_{\text{node}}}\Big)\end{aligned} \tag{17.13}$$

where S_i^ξ denotes the i-th coordinate of a node, ξ. Following the concept of the finite element analyses, it is assumed that the displacements in the x, y, and z direction of an atom α in an element, u_x^α, u_y^α, and u_z^α, respectively, can be represented using the linear function as

$$\begin{aligned} u_x^\alpha &= a_1 + a_2 r_x^\alpha + a_3 r_y^\alpha + a_4 r_z^\alpha, \\ u_y^\alpha &= a_5 + a_6 r_x^\alpha + a_7 r_y^\alpha + a_8 r_z^\alpha, \\ u_z^\alpha &= a_9 + a_{10} r_x^\alpha + a_{11} r_y^\alpha + a_{12} r_z^\alpha, \end{aligned} \tag{17.14}$$

For the node , ξ, we obtain

$$\begin{aligned} U_x^\xi &= a_1 + a_2 S_x^\xi + a_3 S_y^\xi + a_4 S_z^\xi, \\ U_y^\xi &= a_5 + a_6 S_x^\xi + a_7 S_y^\xi + a_8 S_z^\xi, \\ U_z^\xi &= a_9 + a_{10} S_x^\xi + a_{11} S_y^\xi + a_{12} S_z^\xi, \end{aligned} \tag{17.15}$$

where U_i^ξ $(i = x, y, z)$ denotes the nodal displacement. Thus, the displacement of atoms in the element can be described by the displacements of 4 nodes that consist of the tetrahedral element as follows

$$\begin{Bmatrix} u_x^\alpha \\ u_y^\alpha \\ u_z^\alpha \end{Bmatrix} = C_\alpha^1 \begin{Bmatrix} U_x^1 \\ U_y^1 \\ U_z^1 \end{Bmatrix} + C_\alpha^2 \begin{Bmatrix} U_x^2 \\ U_y^2 \\ U_z^2 \end{Bmatrix} + C_\alpha^3 \begin{Bmatrix} U_x^3 \\ U_y^3 \\ U_z^3 \end{Bmatrix} + C_\alpha^4 \begin{Bmatrix} U_x^4 \\ U_y^4 \\ U_z^4 \end{Bmatrix}, \tag{17.16}$$

where the coefficient of node ξ, C_α^ξ, is represented as a function of the coordinate of atom and node, r_i^α and S_i^α (Shimada et al. 2010b).

The number of reduced degrees of freedom can now be $M' = 3(N'_{\text{atom}} + N'_{\text{node}}) - 6$ (or, $M' = 2(N'_{\text{atom}} + N'_{\text{node}}) - 3$ for a two-dimensional system). It should be noted that the reduced DOFs of M' is quite smaller than the original DOFs of atoms M because large number of atoms are included in the element regions (i.e., $M' \ll M$). The Hessian matrix of this atom-element system, $\boldsymbol{H}'$, to be solved is now rewritten as

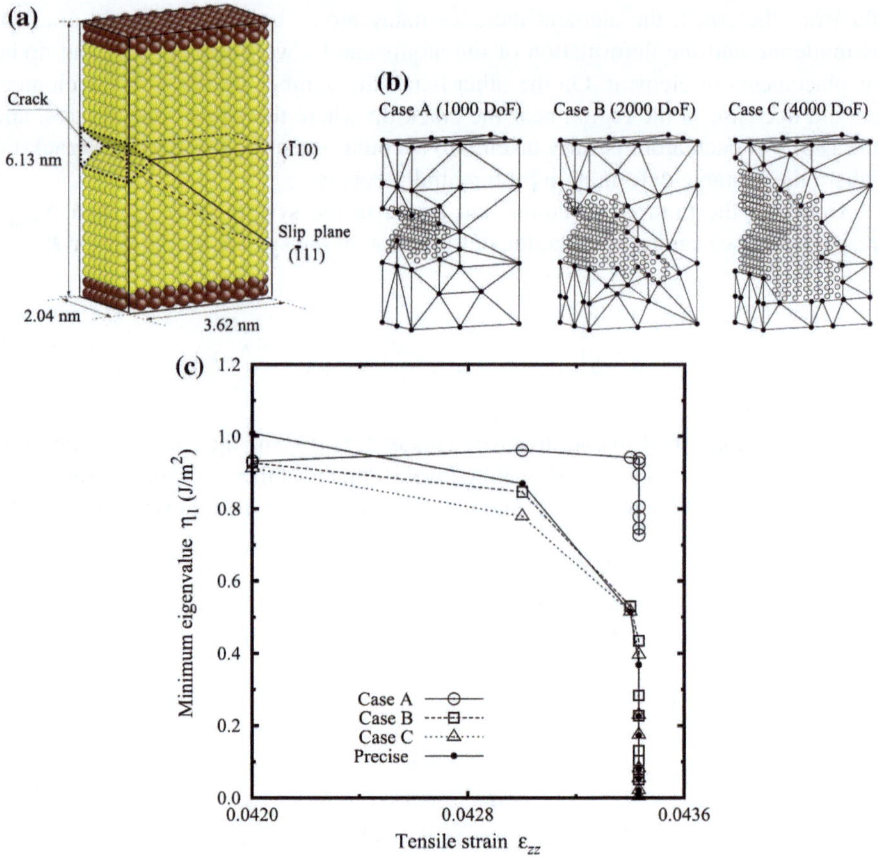

Fig. 17.6 **a** Three-dimensional atomic structure of Cu with a crack. **b** Elements and atoms for 1000 (Case A), 2000 (Case B), and 4000 (Case C) degrees of freedom generated according to the deformation gradient. Note that the number of degrees of freedom of all atoms is 11127. **b** Minimum eigenvalues normalized by those of the unstrained state as a function of strain obtained by Cases A, B, and C and exact solution of original method as "Precise" (Shimada et al. 2010b)

$$H'_{mn} = \frac{\partial^2 U}{\partial R'_m \partial R'_n}, \tag{17.17}$$

where the size $(M' \times M')$ is significantly reduced from the original size $(M \times M)$ because $M' \ll M$ (Shimada et al. 2010b). Therefore, by applying the atom-element concept, the eigenvalues (instability criterion) as well as the corresponding eigenvectors (unstable mode) can be computed even in a large-scale atomic structure.

The atom-element concept for the mechanical instability analysis of large-scale atomic structures has been validated by applying the concept to the three-dimensional Cu atomic structure with a crack under tension (Shimada et al. 2010b), as shown in Fig. 17.6a. During tension tests, a dislocation is emitted from the crack tip at a critical strain. The stress-strain curve and the resulting dislocation emission are similar to those shown in Fig. 17.1 (Case B), although we do not show the details of tensile deformations here. Figure 17.6b shows the mesh division of a three-dimensional atomic component with a crack. The atomic structure is divided into elements according to the strain gradient (Shimada et al. 2009b, 2010b). Here, atomic displacement in an element is linearly related with the displacement at the nodes. The models of Cases A, B, and C have approximately 1000, 2000, and 4000 degrees of freedom, respectively. Note that the total number of degrees of freedom (atomic degrees of freedom) is 11127. The atomic region appears near the crack tip and elements become rough far from the crack. Figure 17.6c plots the minimum eigenvalues of the Hessian matrix for Cases A, B and C. The model with the lowest degrees of freedom (Case A) significantly differs from the exact numerical solution that considers all the degrees of freedom shown as "Precise." On the other hand, by increasing the degrees of freedom, Cases B and C give almost the same curves of the exact solution. Cases B and C are therefore a quite good approximation of the exact solution so that we are able to determine the onset of mechanical instability from the minimum eigenvalue with reduced DOFs and Hessian matrix $\boldsymbol{H}'$.

Figure 17.7 shows the unstable eigenvector modes corresponding to the minimum eigenvalues for Cases A, B, and C. Case A gives a mode that is remarkably deviated from the exact solution. This is because the Case A does not afford enough DOFs to

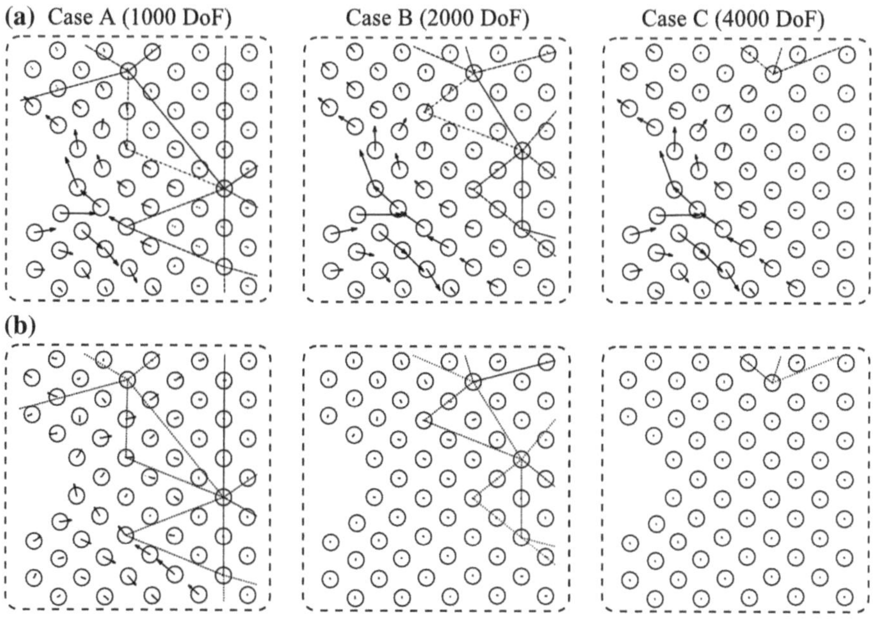

Fig. 17.7 Simplified instability analysis results for Cases A, B, and C. **a** Unstable mode vectors. **b** Error vectors representing the difference from the exact unstable mode. Only the region near the crack tip is shown (Shimada et al. 2010b)

represent the unstable mode of discretized atomic motion near the crack tip. On the other hand, the unstable deformation mode is correctly described when the degrees of freedom is more than 2000 (Case B). Here, the size of matrix, $\boldsymbol{H}'$, is reduced to about 1/25 compared with that of the original approach, $\boldsymbol{H}$. Here, it should be noted that the model introduced here is relatively small for the purpose of the validation of the concept. When we deal with larger atomic structures that contain millions of atoms as molecular dynamics simulations usually treat, a huge number of atoms are included in the element region and the size of Hessian matrix can be reduced dramatically. We also remark that, although the present model has a crystalline structure which is more likely to apply the atom-element concept due to its regular atomic arrangement, it has been confirmed that the concept can be applied to even highly-disordered structures such as amorphous metals where the atomic displacement exhibits quite nonlinear behavior (Shimada et al. 2009a; Kubo et al. 2013). Therefore, the atom-element concept introduced here is quite successful and therefore enables us to evaluate the mechanical instability criterion even for large-scale atomic systems.

17.4 Recent Advances of Instability Criteria for Complicated Systems

Very recently, numerous attempts and efforts have been made to extend and apply the instability criterion to more complicated systems and phenomena (Umeno et al. 2009, 2010; Shimada et al. 2010a; Yan et al. 2012; Shimada et al. 2012, 2015), such as finite temperature, dislocation structures, ferroelectrics, magnetic materials, etc. Especially, it has recently attracted much attention that mechanical stress/strain strongly affects and interacts with ferroelectric and magnetic properties, i.e., "multi-physics" coupling. Nanostructures where the discreteness of atoms becomes dominant exhibit novel multi-physics properties that is distinct from the bulk counterpart, as reviewed in (Shimada and Kitamura 2014). Here, we thus focus on advanced instability criteria related to these hot "multi-physics" phenomena.

For ferroelectric materials where a spontaneous ionic displacement brings about electric dipoles in perovskite lattices, i.e., spontaneous polarization, the spontaneous polarization interacts with external electric field. When the external field exceeds the critical value, the spontaneous polarization and internal ionic displacement becomes unstable and finally switches its directions, namely, do-main switching. Thus, the domain switching is one of instabilities of atoms arising from not only mechanical but also electric loading, i.e., multi-physics instabilities. The criterion for such multi-physics instabilities in ferroelectrics has been proposed by extending the original mechanical instability analysis (Shimada et al. 2012). In addition to the original framework, here the total energy of the system Π consists of not only the potential energy Φ and the work done by external load W, but also the electro-static energy due to external electric field V_{elec},

$$\Pi = \Phi + W + V_{\mathrm{elec}} \tag{17.18}$$

By following the same manner as the original theory, the Hessian matrix to be solved $\boldsymbol{H}$ is now re-formulated as

$$H_{mn} = \left.\frac{\partial^2 \Phi}{\partial R_m \partial R_n}\right|_{\boldsymbol{R}=\boldsymbol{R}_0} + \left.\frac{\partial^2 V_{\mathrm{elec}}}{\partial R_m \partial R_n}\right|_{\boldsymbol{R}=\boldsymbol{R}_0} \tag{17.19}$$

Therefore, the eigenvalue of the re-formulated Hessian matrix can be a criterion for ferroelectric systems that essentially include a strong interaction with both mechanical and electric loading.

This criterion was applied to the switching of 180° domain walls (DWs) in ferroelectric $PbTiO_3$ under external electric field (Shimada et al. 2012), as shown in Fig. 17.8a. This simulation model with periodic boundary conditions contains a 180° DW at the center of cell. The left side of DW shows spontaneous polarization in the +z direction while that of the right side is the opposite $-z$ direction. When the external electric field is applied in the $+z$ direction, spontaneous polarization (referred as ferro-electric distortion δ_{FE} here) of the right side of DW (denoted as plane 1–4) smoothly increases, as shown in Fig. 17.8b. When the electric field reaches the critical value of $E_{\mathrm{c}} = 4.225$ MV/cm, δ_{FE} changes from the negative to positive value, indicating that spontaneous polarization switches from $-z$ to $+z$, i.e., the domain switching occurs. The minimum eigenvalue, as a criterion based on the above extended the-

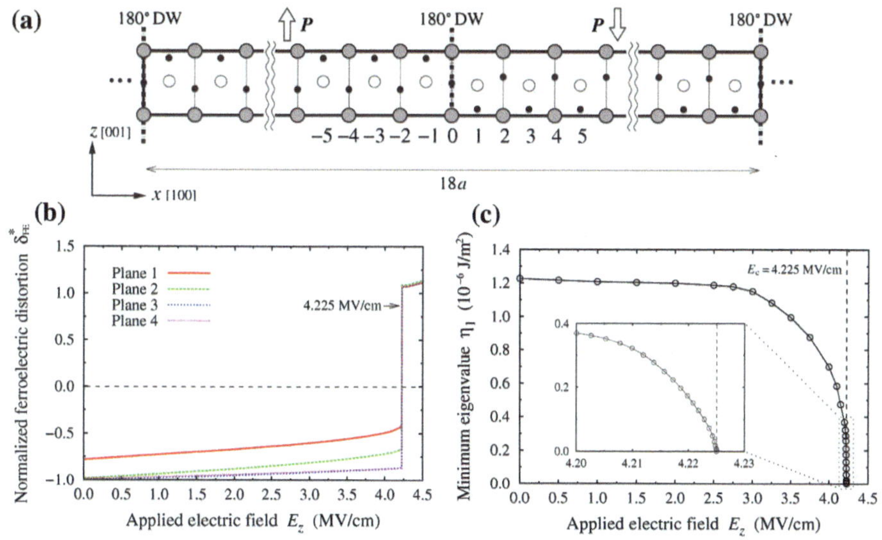

Fig. 17.8 Computational results for domain switching in ferroelectric $PbTiO_3$. **a** Simulation supercell for 180° domain walls in ferroelectric $PbTiO_3$. **b** Ferroelectric distortion (spontaneous polarization) as a function of external electric field. **c** Minimum eigenvalue as a function of applied electric field (Shimada et al. 2012)

ory, smoothly decreases with increasing electric field, as shown in Fig. 17.8c. The minimum eigenvalue reaches zero (criterion) at the field of 4.225 MV/cm, which corresponds to the point where domain switching starts. This consistency therefore demonstrates that the criterion based on the extended theory can successfully capture (ferro-)electric instabilities in such ionic systems.

The instability criterion has also been extended to magnetic (spin-lattice) systems (Shimada et al. 2015). The potential energy of the system with magnetic moments, such as ferromagnetic Fe, is described by not only the atomic coordinates, $\boldsymbol{R}$, but also magnetic moments, $\boldsymbol{m}$, and is given by

$$\Phi = \Phi(\boldsymbol{R}, \boldsymbol{m}), \tag{17.20}$$

where

$$\boldsymbol{m} \equiv^t (m_x^1, m_y^1, m_z^1, m_x^2, m_y^2, m_z^2, \ldots, m_x^N, m_y^N, m_z^N) \tag{17.21}$$

Here, m_i^α denotes the magnetic moment of atom α in the i direction. Since the DOFs of magnetic moment in the system is $3N$, the total number of DOFs including atomic and spin DOFs is now $M = 6N - 6$. Here, an arbitrary deformation and/or perturbation of the magnetic moment of the system can be represented by a change in the following M-dimensional vector $\boldsymbol{X}$ consisting of all DOFs,

$$\begin{aligned} \boldsymbol{X} \equiv^t\, & (r_z^2, r_y^3, r_z^3, r_x^4, r_y^4, r_z^4, \ldots, r_x^N, r_y^N, r_z^N) \\ & m_x^1, m_y^1, m_z^1, m_x^2, m_y^2, m_z^2, \ldots, m_x^N, m_y^N, m_z^N) \end{aligned} \tag{17.22}$$

The total energy of the spin-lattice system Π now consists of not only the potential energy Φ and the work done by external load W, but also the magnetostatic energy due to external magnetic field V_{mag},

$$\Pi = \Phi + W + V_{\mathrm{mag}} \tag{17.23}$$

By following the same manner as the original concept, the Hessian matrix to be solved $\boldsymbol{H}$ is now re-written as

$$H_{mn} = \left.\frac{\partial^2 \Phi}{\partial X_m \partial X_n}\right|_{\boldsymbol{X}=\boldsymbol{X}_0} + \left.\frac{\partial^2 V_{\mathrm{mag}}}{\partial X_m \partial X_n}\right|_{\boldsymbol{X}=\boldsymbol{X}_0} \tag{17.24}$$

Again, the eigenvalue of the re-formulated Hessian matrix can be a criterion for a magnetic system. This extended criterion was applied to magnetization switching in ferromagnetic Fe under external magnetic field, as shown in Fig. 17.9. The magnetic moment in Fe is initially along $+z$ direction ($\theta = 0\,\mathrm{rad}$; point A in Fig. 17.9a). When the external magnetic field is applied in $+x$ direction, the magnetic moment slightly rotates (point B).When the magnetic field reaches at the coercive field H_c, the magnetic moment suddenly rotates from $+z$ to $+x$ direction (points C-D; see

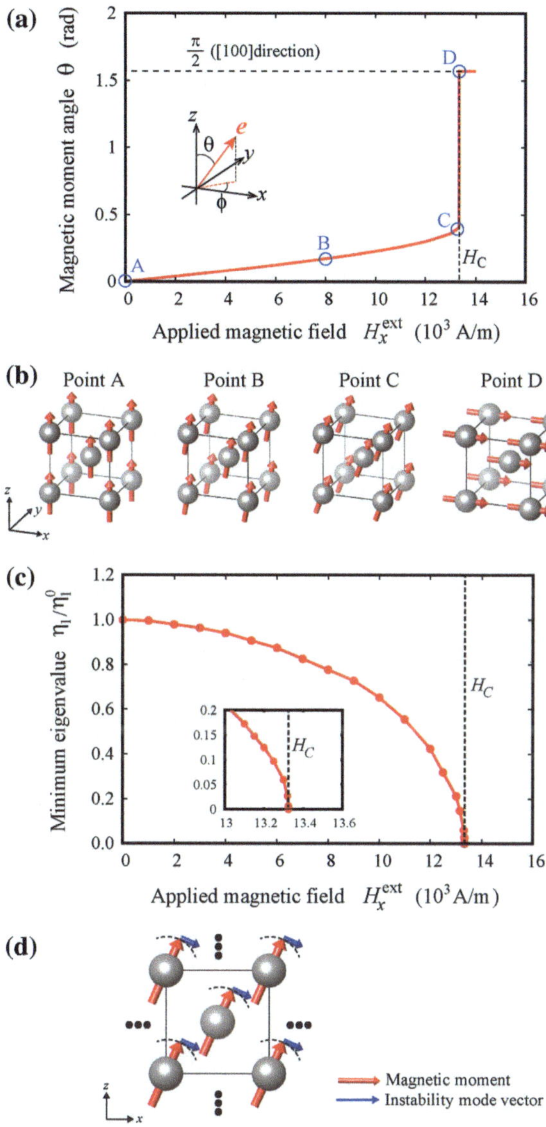

Fig. 17.9 Computational results for magnetization switching in ferromagnetic iron. **a** Magnetic moment as a function of external magnetic field. **b** Magnetic configurations under applied magnetic fields. **c** Minimum eigenvalue of criterion as a function of magnetic field. **d** Unstable mode of magnetic moments under the coercive field (Shimada et al. 2015)

also Fig. 17.9b), i.e., magnetization switching occurs. The minimum eigenvalue, as a magnetic criterion based on the extended theory, smoothly decreases with increasing magnetic field (see Fig. 17.9c). The minimum eigenvalue finally becomes zero (criterion) at the coercive field of H_c, which corresponds to the points C-D where magnetization switching occurs. This agreement shows that the criterion based on the extended theory can successfully describe the magnetic instabilities in spin-lattice systems. In addition, the unstable mode of magnetic moments shown in Fig. 17.9d demonstrates the rotation of spins toward the $+x$ direction, which correctly represents the behavior of magnetization switching shown in Fig. 17.9b.

17.5 Conclusions

This review covered the state-of-the-art theory of "fracture mechanics of atomic structures" that provides a rigorous description of mechanical instabilities in arbitrary atomic structures under any external loading/constraint. The theory gives the critical instability condition by positivity of the minimum eigenvalue of the Hessian matrix of the total energy with respect to degrees of freedom of the system as well as the fracture mode at the onset of instability. The theory successfully provides atomistic insights into fracture in various atomic/nanoscale structures including nanocracks, interface edges, defects, amorphous and non-crystalline structures. The review concludes with recent advances of the theory extended to complicated and multi-functional systems including large-scale, finite temperature, and multi-physics instabilities in (ferro-)electric and magnetic materials as functional fracture. Such extension of the instability theory for multi-physics phenomena can open up a new discipline underlying between material strength and physical properties.

Acknowledgments The authors acknowledge financial support of this work by the Grant-in-Aid for Specially Promoted Research (Grant No. 25000012) from the Japan Society of Promotion of Science (JSPS).

References

Bažant ZP (2002) Concrete fracture models: testing and practice. Eng Frac Mech 69:165–205
Born M, Huang K (1954) Dynamical theory of crystal lattices. Oxford University Press, Oxford
Buehler MJ, Gao H (2006) Dynamical fracture instabilities due to local hyperelasticity at crack tips. Nature 439:307–310
Černý M, Šob M, Pokluda J, Šandra P (2004) Ab initio calculation of ideal tensile strength and mechanical stability in copper. J Phys Condens Matter 16:1045–1052
Dmitriev SV, Kitamura T, Li J, Umeno Y, Yashiro K, Yoshikawa N (2005a) Near-surface lattice instability in 2D fiber and half-space. Acta Mater Sci 53:1215–1224
Dmitriev SV, Li J, Yoshikawa N, Shibutani Y (2005b) Theoretical strength of 2D hexagonal crystals: application to bubble raft indentation. Philos Mag 85:2177–2195
Dombard AJ (2007) Planetary science crack under stress. Nature 447:276–277
Hill R, Milstein F (1997) Principles of stability analysis of ideal crystals. Phys Rev B 15:3087–3096
Holland D, Marder M (1999) Crack and atoms. Adv Mater 11:793–806
Kermode JR, Albaret T, Sherman D, Bernstein N, Gumbsch P, Payne MC, Csányi G, de Vita A (2008) Low-speed fracture instabilities in a brittle crystal. Nature 455:1224–1227
Kitamura T, Yashiro K, Ohtani R (1997) Atomic simulation on deformation and fracture of nano-single crystal of nickel in tension. JSME Int J Ser A 40:430–435

Kitamura T, Umeno Y, Fushino R (2004a) Instability criterion of inhomogeneous atomic system. Mater Sci Eng A 379:229–233
Kitamura T, Umeno Y, Tsuji N (2004b) Analytical evaluation of unstable deformation criterion of atomic structure and its application to nanostructure. Comp Mater Sci 29:499–510
Kitamura T, Yashiro K, Ohtani R (1998) Mesoscopic dynamics of fracture. Computational materials design. Springer, Berlin, pp 120–130
Kubo A, Albina JM, Umeno Y (2013) Atomistic study of stress-induced switch-ing of 90° ferroelectric domain walls in PbTiO3: size, temperature and structural effect. Model Simul Mater Sci Eng 21(065):019
Li J, van Vliet KJ, Zhu T, Yip S, Suresh S (2002) Atomistic mechanisms governing elastic limit and incipient plasticity in crystals. Nature 418:307–310
Li J, Ngan AHW, Gumbsch P (2003) Atomic modeling of mechanical behavior. Acta Mater 51:5711–5742
Liebowitz H (1968) Fracture—an advanced treatise. Academic Press, New York
Livne A, Bouchbinder E, Svetlizky I, Fineberg J (2010) The near-tip fields of fast cracks. Science 327:1359–1363
Luo W, Roundy D, Cohen ML, Morris JW (2002) Ideal strength of bcc molybdenum and niobium. Phys Rev B 66(094):110
Milstein F (1971) Theoretical strength of a perfect crystal. Phys Rev B 3:1130–1141
Milstein F (1980) Theoretical elastic behaviour of crystals at large strains. J Mater Sci 15:1071–1084
Nalla RK, Kinney JH, Richie RO (2003) Mechanistic fracture criteria for the failure of human cortical bone. Nat Mater 2:164–168
Pons AJ, Karma A (2010) Helical crack front instability in mixed-mode fracture. Nature 464:85–89
Rubin AM (1995) Propagation of magma-field cracks. Annu Rev Earth Planet Sci 23:287–336
Shimada T, Okawa S, Minami S, Kitamura T (2008) Development of efficient instability analysis method for atomic structures using linear elements and its application to amorphous metal (in Japanese). Trans Jpn Soc Mech Eng A 74:1328–1335
Shimada T, Okawa S, Minami S, Kitamura T (2009a) Development of efficient instability analysis method for atomic structures using linear elements and its application to amorphous metal. J Solid Mech Mater Eng 3(5):807–818
Shimada T, Okawa S, Minami S, Kitamura T (2009b) Simplified evaluation of mechanical instability in large-scale atomic structures. Mater Sci Eng A 513–514:166–171
Shimada T, Kondo T, Sumigawa T, Kitamura T (2010a) Mechanical instability criterion of dislocation structures based on discrete dislocation dynamics. Trans Jpn Soc Mech Eng A 76:1721–1728
Shimada T, Okawa S, Kitamura T (2010b) Simplified analysis of mechanical in-stability three-dimensional atomic components and its application to nanoscale crack. J Solid Mech Mater Eng 4(7):1071–1082
Shimada T, Sakamoto H, Kitamura T (2012) Development of multi-physics in-stability criterion for atomic structures and application to domain switching in ferroelectrics under external electric field. J Soc Mater Sci 61:155–161
Shimada T, Kitamura T (2014) Multi-physics properties in ferroelectric nanostructure. Bull JSME 1(2):SMM0009-SMM0009
Shimada T, Ouchi K, Ikeda I, Ishii Y, Kitamura T (2015) Magnetic instability criterion for spin-lattice systems. Comp Mater Sci 97:216–221
Song J, Curtin WA (2013) Atomic mechanism and prediction of hydrogen embrittlement in iron. Nat Mater 12:145–151
Umeno Y, Kitamura T, Tagawa M (2007) Mechanical instability in non-uniform atomic structure: application to amorphous metal. Mater Sci Eng A 462:450–455
Umeno Y, Shimada T, Kitamura T (2009) Instability mode analysis of dislocation nucleation from notch based on atomistic model (instability activation mechanism under finite temperature). Trans Jpn Soc Mech Eng A 75:1247–1254
Umeno Y, Shimada T, Kitamura T (2010) Dislocation nucleation in a thin Cu film from molecular dynamics simulations: instability activation by thermal fluctuations. Phys Rev B 82(104):108

van Vliet KJ, Li J, Zhu T, Yip S, Suresh S (2003) Quantifying the early stages of plasticity through nanoscale experiments and simulations. Phys Rev B 67(104):105

Wang J, Yip S, Phillpot SR, Wolf D (1993) Crystal instabilities at finite strain. Phys Rev Lett 71:4182–4185

Wang J, Li J, Yip S, Wolf D, Phillpot SR (1997) Unifying two criteria of born: elastic instability and melting of homogeneous crystals. Physica A 240:396–403

Wang J, Li J, Yip S, Phillpot SR, Wolf D (1995) Mechanical instabilities of homogeneous crystals. Phys Rev B 52:12,627–12,635

Warner DH, Curtin WA, Qu S (2007) Rate dependence of crack-tip processes predicts twinning trend in f.c.c. metals. Nat Mater 6:876–881

Yan Y, Kondo T, Shimada T, Sumigawa T, Kitamura T (2012) Criterion of mechanical instabilities for dislocation structures. Mater Sci Eng A 534:681–687

Yashiro K, Tomita Y (2010) Local lattice instability at a dislocation nucleation and motion. J Phys IV 11:Pr5-3-Pr5-10

Zhang P, Ma L, Fan F, Zeng Z, Peng C, Loya PE, Liu Z, Gong Y, Zhang J, Zhang X, Ajayan PM, Zhu T, Lou J (2014) Fracture toughness of graphene. Nat Commun 5:3782

Chapter 18
Radiation Damage Evolution in Ductile Materials

Błażej Skoczeń and Aneta Ustrzycka

Abstract The problem investigated in the present work concerns the physical processes involved in radiation damage and the way they affect the mechanical properties of ductile materials. Multiscale modeling of evolution of radiation induced micro-damage in ductile materials subjected to periodic stress states in the inelastic range is presented. The resulting micro-structural and damage evolution causes profound changes of the macroscopic properties and severely degrades the lifetime of the components subjected to irradiation. The evolution of radiation induced damage is combined with the evolution of classical micro-damage of mechanical origin (micro-cracks and micro-voids), within the common framework of Continuum Damage Mechanics (CDM). An additive formulation with respect to damage parameters (tensors) has been used. The Rice & Tracey kinetic law may be conveniently applied to describe the evolution of radiation induced damage in the form of clusters of voids embedded in the metallic matrix. Closed form analytical solutions for the problem of periodic irradiation combined with cyclic axial loads, corresponding to R&T law was obtained.

Keywords Radiation induced damage · Frenkel pairs · Vacancies · Interstitial atoms · Displacement per atom · Evolution of clusters of voids · Continuum damage mechanics (CDM) framework

18.1 Introduction

Exposure to high energy radiation (flux of particles) degrades the microstructure of materials. Energetic particles penetrating a solid displace the lattice atoms from their original positions. In the elastic collisions (initiated when a given atom is struck by

---

B. Skoczeń · A. Ustrzycka (✉)
Cracow University of Technology, Warszawska 24, 31-155 Kraków, Poland
e-mail: blazej.skoczen@pk.edu.pl

A. Ustrzycka
e-mail: anetaustrzycka@mech.pk.edu.pl

H. Altenbach et al. (eds.), *From Creep Damage Mechanics to Homogenization Methods*, Advanced Structured Materials 64,
DOI 10.1007/978-3-319-19440-0_18

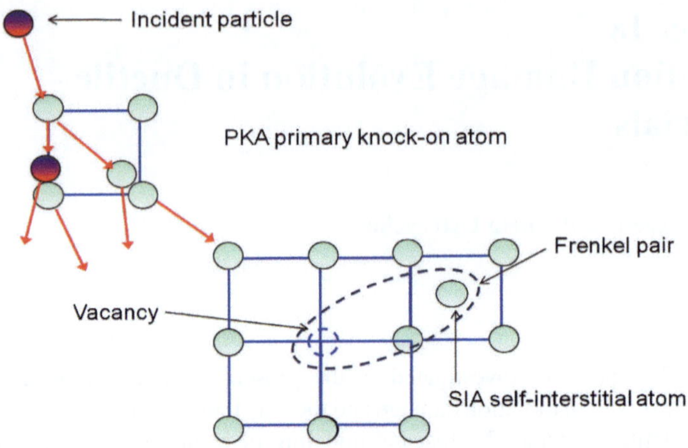

Fig. 18.1 Displacement cascade and formation of Frenkel pairs

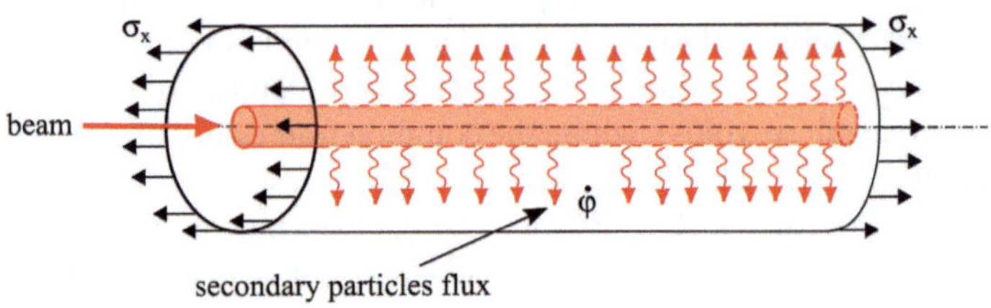

Fig. 18.2 Coaxial target—detector configuration (round bar and cylindrical shell)

a high energy particle), incident particles transfer recoil energy to the lattice atoms. The initial primary knock-on atoms will recoil with a given amount of kinetic energy that is dissipated in a sequence of collisions with the other lattice atoms. As a result of cascade process, atoms in the solid can be displaced from their equilibrium lattice positions, creating vacancies and interstitials (see Zinkle and Singh 1993). These vacancy-interstitial pairs are called Frenkel pairs.The vacancies of the Frenkel pairs often form clusters by means of diffusion Fig. 18.1.

As an example of typical problem related to evolution of radiation induced damage, degradation of material properties in a thin-walled irradiated cylindrical part of detector of particles has been investigated. The coaxial target—detector configuration is shown in Fig. 18.2.

The target is hit by high-energy particles beam. The process of beam absorption is associated with emission of secondary particles flux in the radial direction. The secondary particles induce micro-damage in the thin-walled cylinder surrounding the target. As the cylinder is simultaneously subjected to mechanical loads, the fields of radiation and mechanically induced damage occur in the same lattice. This problem is used as an illustration of different nature of both damage types. The nature of mechanically induced micro-damage, comprising micro-voids and micro-cracks, is

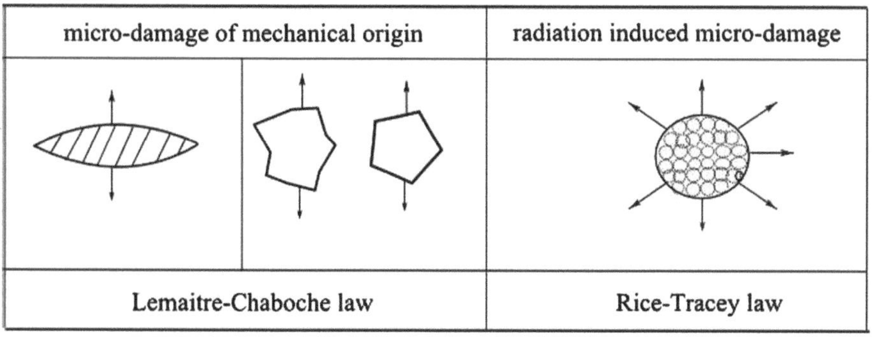

Fig. 18.3 Type of micro-damage and the corresponding laws of damage evolution

different than the nature of irradiation induced micro-damage, comprising clusters of micro-voids resulting from elastic interactions of particles with lattice atoms. The clusters shape in general has been assumed in the form of spherical micro-void.The mechanically induced micro-damage shows usually directional response and anisotropic features. Both types of micro-damage: radiation induced and of mechanical origin are presented in Fig. 18.3.

Thus, the present work constitutes an attempt to fill in the gap and offers consistent methodology to trace the evolution of micro-damage of different nature and to predict the time to rupture of the component. Based on the expected displacement per atom *dpa*, an attempt to identify the critical radiation damage effects for structural materials is made. Thus, the ultimate goal is lifetime prediction expressed in terms of the number of beam cycles for components of particle detectors (Baussan et al. 2014; Edgecock et al. 2013).

18.2 Constitutive Relations

The present constitutive model of anisotropic damage evolution in ductile materials is based on the assumptions of small strains. The main mechanism of damage onset and further evolution is related to formation of plastic strain fields. In order to describe the evolution of anisotropic damage, second order damage tensor $\underline{\underline{D}}$ is introduced (Garion and Skoczeń 2003; Egner and Skoczeń 2010):

$$\underline{\underline{D}} = \sum_{i=1,3} D_i \underline{n}_i \otimes \underline{n}_i \tag{18.1}$$

Here, $\otimes$ denotes the dyadic product and $\underline{n}_i$ stands for the base vector associated with the principal direction i. D_i denotes the component of damage tensor related to direction i, defined by:

$$D_i = \frac{\mathrm{d}S_{D\underline{n}_i}}{\mathrm{d}S_{\underline{n}_i}}, \tag{18.2}$$

where $\mathrm{d}S_{D\underline{n}_i}$ is the area of damage in the section $\mathrm{d}S_{\underline{n}_i}$, represented by the normal $\underline{n}_i$. The effective stress obeys the strain equivalence principle (Lemaitre 1999):

$$\underline{\underline{\widetilde{\sigma}}} = \underline{\underline{\underline{\underline{E}}}} : \underline{\underline{\epsilon}}, \tag{18.3}$$

where $\underline{\underline{\underline{\underline{E}}}}$ denotes the fourth rank elastic stiffness tensor. The general relationship between the stress and the effective stress tensors reads:

$$\underline{\underline{\widetilde{\sigma}}} = \underline{\underline{\underline{\underline{M}}}}^{-1} : \underline{\underline{\epsilon}}, \tag{18.4}$$

where $\underline{\underline{\widetilde{\sigma}}}$ is the effective stress and $\underline{\underline{\underline{\underline{M}}}}^{-1}$ stands for the fourth rank symmetric damage effect tensor (Chaboche 1983), that depends on the damage state:

$$\underline{\underline{\underline{\underline{M}}}} = \frac{1}{2}\underline{\underline{\underline{\underline{I}}}} - \frac{1}{4}\left(\underline{\underline{D}}\otimes\underline{\underline{I}} + \underline{\underline{D}}\overline{\otimes}\underline{\underline{I}} + \underline{\underline{I}}\otimes\underline{\underline{D}} + \underline{\underline{I}}\overline{\otimes}\underline{\underline{D}}\right) \tag{18.5}$$

The yield surface is defined in the following way:

$$f_{\mathrm{y}}\left(\underline{\underline{\widetilde{\sigma}}}, \underline{\underline{\widetilde{X}}}, \widetilde{R}\right) = J_2\left(\underline{\underline{\widetilde{\sigma}}} - \underline{\underline{\widetilde{X}}}\right) - \sigma_{\mathrm{y}} - \widetilde{R}, \tag{18.6}$$

where σ_{y} denotes the yield stress of undamaged material. The second invariant of the stress tensor $J_2\left(\underline{\underline{\widetilde{\sigma}}} - \underline{\underline{\widetilde{X}}}\right)$ is defined:

$$J_2\left(\underline{\underline{\widetilde{\sigma}}} - \underline{\underline{\widetilde{X}}}\right) = \sqrt{\frac{3}{2}\left(\underline{\underline{\widetilde{s}}} - \underline{\underline{\widetilde{X}}}\right) : \left(\underline{\underline{\widetilde{s}}} - \underline{\underline{\widetilde{X}}}\right)}, \tag{18.7}$$

where, $\underline{\underline{\widetilde{s}}}$ denotes the effective deviatoric stress and:

$$\underline{\underline{\widetilde{X}}} = \underline{\underline{\underline{\underline{M}}}}^{-1} : \underline{\underline{X}}; \quad \widetilde{R} = \frac{R}{\left(1 - \sqrt{\underline{\underline{D}} : \underline{\underline{D}}}\right)}, \tag{18.8}$$

stand for the effective kinematic and the effective isotropic hardening variables, respectively. Here, $\underline{\underline{X}}$ denotes the back stress coupled to the back strain, and R denotes the isotropic hardening parameter coupled to the plastic strain. The associated flow rule takes the form:

$$\mathrm{d}\underline{\underline{\epsilon}}^{\mathrm{p}} = \frac{\partial f_{\mathrm{y}}\left(\underline{\underline{\widetilde{\sigma}}}, \underline{\underline{\widetilde{X}}}, \underline{\underline{\widetilde{R}}}\right)}{\partial \underline{\underline{\widetilde{\sigma}}}}\mathrm{d}\lambda, \tag{18.9}$$

where $\mathrm{d}\underline{\underline{\epsilon}}^{\mathrm{p}}$ denotes the plastic strain tensor. The evolution laws for hardening variables are represented by the following equations:

$$\mathrm{d}\underline{\underline{\widetilde{X}}} = \frac{2}{3} C_X \mathrm{d}\underline{\underline{\epsilon}}^{\mathrm{p}}; \quad \mathrm{d}\widetilde{R} = C_R \mathrm{d}p, \tag{18.10}$$

where C_X denotes the kinematic and C_R the isotropic hardening moduli, respectively. Here, $\mathrm{d}p$ stands for the accumulated plastic strain increment:

$$\mathrm{d}p = \sqrt{\frac{2}{3}\underline{\underline{\epsilon}}^{\mathrm{p}} : \underline{\underline{\epsilon}}^{\mathrm{p}}} \tag{18.11}$$

Finally, the kinetic evolution law for mechanically induced micro-damage is postulated in the following form (Garion and Skoczeń 2003):

$$\underline{\underline{\dot{D}}} = \underline{\underline{C}}\,\underline{\underline{Y}}\,\underline{\underline{C}}^{\mathrm{T}} \dot{p} H\left(p - p_{\mathrm{D}}\right), \tag{18.12}$$

where $H\left(p - p_{\mathrm{D}}\right)$ is the Heaviside function, p_{D} denotes the damage threshold and $\underline{\underline{Y}}$ stands for the strain energy density release rate tensor (Chaboche 1983):

$$\underline{\underline{Y}} = \frac{1}{4}\left[\underline{\underline{\epsilon}}^{\mathrm{e}}\left(\underline{\underline{\underline{\underline{E}}}} : \underline{\underline{\epsilon}}^{\mathrm{e}}\right) + \left(\underline{\underline{\underline{\underline{E}}}} : \underline{\underline{\epsilon}}^{\mathrm{e}}\right)\underline{\underline{\epsilon}}^{\mathrm{e}}\right], \tag{18.13}$$

and $\underline{\underline{\epsilon}}^{\mathrm{e}}$ denotes the elastic strain tensor.

18.3 Kinetics of Evolution of Radiation Induced Damage

It is assumed that the flux of secondary particles emitted from the target as well as the related distribution of NIEL (non-ionizing energy loss) (Bacon et al. 2000) in the cylindrical shell reflect the profile of beam deposited power in the target. Thus, typical *dpa* distribution in the target-detector coaxial configuration is illustrated in Fig. 18.4.

In order to carry out the analysis, typical *dpa* distribution has been described by the following function:

$$dpa\left(x\right) = a x^{b} \mathrm{e}^{cx} \tag{18.14}$$

Based on the known *dpa* in the representative volume element (RVE), the density of defects (clusters of voids) caused by irradiation is computed. The dependence between the cluster density q_{c} , the average cluster radius r_{c} and the dpa value is of power type (Nita et al. 2004):

$$q_{\mathrm{c}} = \begin{cases} C_{qI}\left(dpa\right)^{n_{qI}} & \text{for} \quad dpa \geq D_{\mathrm{S}} \\ C_{qII}\left(dpa\right)^{n_{qII}} & \text{for} \quad dpa < D_{\mathrm{S}} \end{cases} \tag{18.15}$$

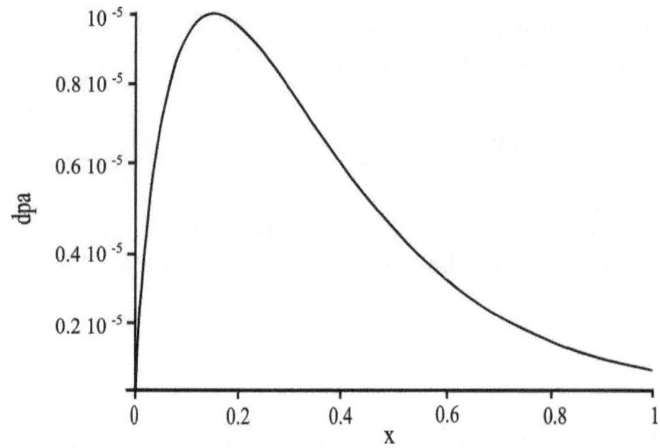

Fig. 18.4 Typical distribution of *dpa* in the detector wall (along the generator of the cylindrical shell)

In a similar way, the radius of clusters is computed. Here, again type power law is applied:

$$r_c = \begin{cases} C_r\,(dpa)^{n_r} & \text{for}\quad dpa \geq D_S \\ r_{cr} & \text{for}\quad dpa < D_S \end{cases} \tag{18.16}$$

where C_{qI}, C_{qII}, n_{qI}, n_{qII}, C_r, r_{cr} denote the material parameters, D_S denotes the saturation irradiation dose (in *dpa* units) and r_{cr} is the mean cluster radius with respect to the irradiation dose equal to D_S. Interstitials show a strong tendency to cluster. For smaller values of *dpa* interstitials and di-interstitials dominate. For higher *dpa* values (above 10^{-2}) the interstitial clusters prevail. Most of the vacancies, however, survive as single vacancies. For the values of *dpa* exceeding 10^{-3} the number of stable vacancy clusters increases substantially, however, the average size of the vacancy clusters stabilizes (Fig. 18.5) around *dpa* equal to 10^{-1}. In order to determine the material parameters, the clusters density and the average cluster size expressed by the number of atoms per cluster are plotted as a function of the irradiation dose in Fig. 18.5. Here, the irradiation dose is expressed in *dpa* units.

After the critical *dpa* value has been reached the clusters size evolution stops and for Al corresponds typically to some 55 atoms. Also, the number of clusters per unit volume is saturated and evolves slowly for higher doses. For Al the *dpa* threshold value stays in the range of 10^{-2}–10^{-1}. The surface density of clusters of voids is recalculated $q_c^{(2/3)} = q_A$ in order to obtain the damage parameter, according to the following equation:

$$D_{r0} = q_A \pi r_{c0}^2 \tag{18.17}$$

Further evolution of post-irradiation damage is related to the growth of clusters of voids under mechanical loads. The clusters shape in general has been assumed in the form of spherical micro-void. In order to compute the evolution of radiation

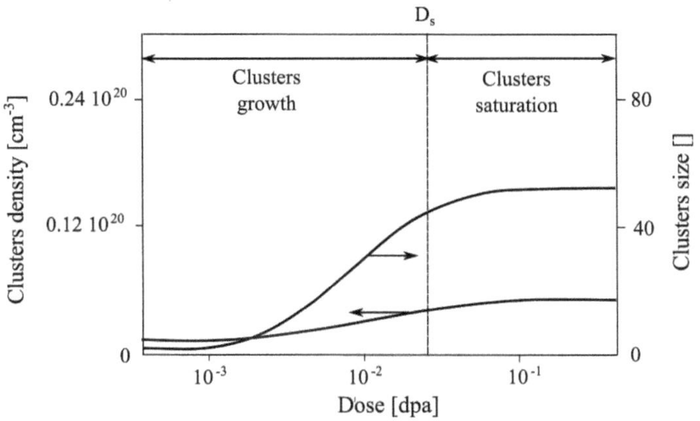

Fig. 18.5 Clusters density and average cluster size as a function of irradiation dose (for Al, computed based on Verbiest and Pattyn (2004))

induced micro-damage under mechanical loads, the evolution law proposed by Rice and Tracey (2004) has been implemented. The Rice & Tracey model predicts the growth of an initially spherical void in an infinite, rigid—perfectly plastic material subjected to a uniform remote strain field. Under the assumption of spherical void representation, the average rate of growth is given by:

$$\mathrm{d}r_{\mathrm{c}} = r_{\mathrm{c}}\alpha_r \exp\left(\frac{3\sigma_{\mathrm{m}}}{2\sigma_{\mathrm{eq}}}\right)\mathrm{d}p \tag{18.18}$$

The radius increment $\mathrm{d}r_{\mathrm{c}}$ is derived as a function of the equivalent plastic strain $\mathrm{d}p$, the stress triaxiality factor $\frac{3\sigma_{\mathrm{m}}}{2\sigma_{\mathrm{eq}}}$, the current radius r_{c} and a scalar multiplier α_r.

The increment $\mathrm{d}A_{rm}$ of clusters cross-section can be now expressed as a difference between the post-irradiation and the initial clusters cross-sections:

$$\mathrm{d}A_{rm} = \pi\left((r_{c0} + \mathrm{d}r_{\mathrm{c}})^2 - r_{c0}^2\right) = \pi\left(2r_{c0}\mathrm{d}r_{\mathrm{c}} + \mathrm{d}r_{\mathrm{c}}^2\right) \tag{18.19}$$

The post-irradiation damage increment $\mathrm{d}D_{rm}$ is defined as a function of the mean size of clusters cross-section and the surface density of clusters:

$$D_{rm} = q_A d A_{rm} = q_A 2\pi r_{c0}\mathrm{d}r_{\mathrm{c}} \tag{18.20}$$

Integrating the Rice & Trice evolution law:

$$\int_{r_{c0}}^{\widetilde{r}_c} \frac{\mathrm{d}r_{\mathrm{c}}}{r_{\mathrm{c}}} = \alpha_r \int_0^{\widetilde{p}} \exp\left(\frac{3\sigma_{\mathrm{m}}}{2\sigma_{\mathrm{eq}}}\right)\mathrm{d}p, \tag{18.21}$$

one obtains:

$$r_c = r_{c0}\alpha_r \exp\left[\alpha_r \int_0^{\tilde{p}} \exp\left(\frac{3\sigma_m}{2\sigma_{eq}}\right) dp\right] \tag{18.22}$$

Applying Eq. (18.22) the post-irradiation damage evolution law is derived as a function of the mean radius of cluster after irradiation and the accumulated plastic strain increment dp:

$$dD_{rm} = q_A 2\pi r_{c0}^2 \exp\left[\alpha_r \int_0^{\tilde{p}} \exp\left(\frac{3\sigma_m}{2\sigma_{eq}}\right) dp\right] \exp\left(\frac{3\sigma_m}{2\sigma_{eq}}\right) dp \tag{18.23}$$

Integrating the above equation for the uniaxial stress:

$$\int_0^{\tilde{D}_{rm}} dD_{rm} = q_A 2\pi r_{c0}^2 \int_0^{\tilde{p}} \exp\left[\alpha_r \exp\left(\frac{1}{2}\right) p\right] \exp\left(\frac{1}{2}\right) dp \tag{18.24}$$

Finally, the post-irradiation damage law is described by the exponential function of plastic strain:

$$\tilde{D}_{rm} = q_A 2\pi r_{c0}^2 e^{\alpha_r C \tilde{p}}, \tag{18.25}$$

$$\tilde{D}_{rm} = C_1 e^{C_2 \tilde{p}}, \tag{18.26}$$

where

$$C_1 = q_A 2\pi r_{c0}^2, \quad C_2 = \alpha_r C \tag{18.27}$$

The dependence between the damage parameter and the plastic strain for the Rice & Tracey model is illustrated in Fig. 18.6.

The Rice & Tracey law is exponential and D_{rm} reaches 1 for the accumulated plastic strain p_{cr}. Thus, the radiation induced damage is represented by a scalar parameter reflecting its isotropic nature. For the sake of simplicity, it is assumed that the evolution of radiation induced damage is also isotropic (at least as long as the contrary has not been proved). Therefore, the tensorial representation of the radiation induced damage reads:

$$\underline{\underline{D}}_{rm} = \frac{1}{3} D_{rm} \underline{\underline{I}}, \tag{18.28}$$

where $\underline{\underline{I}}$ denotes the identity tensor. In order to combine both types of damage: radiation induced and of mechanical origin, additive rule is postulated. Applying the

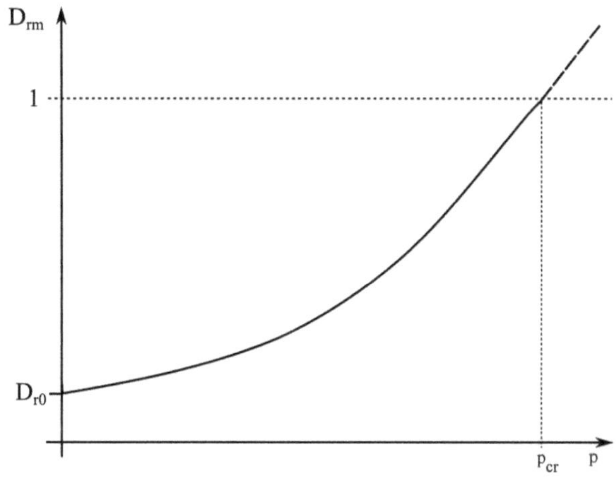

Fig. 18.6 The post-irradiation damage as a function of plastic strain

additive rule, one obtains the total damage tensor as a sum of the mechanically and the radiation induced damage tensors:

$$\underline{\underline{D}} = \underline{\underline{D}}_m + \underline{\underline{D}}_{rm} = \underline{\underline{D}}_m + \frac{1}{3} D_{rm} \underline{\underline{I}} \tag{18.29}$$

It can be further expressed in terms of damage rates:

$$\mathrm{d}\underline{\underline{D}} = \mathrm{d}\underline{\underline{D}}_m + \mathrm{d}\underline{\underline{D}}_{rm} = \mathrm{d}\underline{\underline{D}}_m + \frac{1}{3} \mathrm{d}D_{rm} \underline{\underline{I}}, \tag{18.30}$$

where:

$$\mathrm{d}\underline{\underline{D}}_m = \underline{\underline{C}}\,\underline{\underline{Y}}\,\underline{\underline{C}}^{\mathrm{T}} \mathrm{d}p, \quad \mathrm{d}D_{rm} = q_A 2\pi r_{c0} \mathrm{d}r_{\mathrm{c}} \tag{18.31}$$

The model can be easily reduced to its scalar form by means of contraction operation:

$$\mathrm{d}D = \mathrm{d}D_m + \mathrm{d}D_{rm}, \tag{18.32}$$

where (cf. Lemaitre 1999):

$$\mathrm{d}D_m = \frac{Y}{S} \mathrm{d}p, \quad Y = \frac{\sigma_{\mathrm{eq}}^2}{2E\,(1-D)^2} \left[\frac{2}{3}(1+\nu) + 3\,(1-2\nu) \left(\frac{\sigma_{\mathrm{m}}}{\sigma_{\mathrm{eq}}} \right)^2 \right] \tag{18.33}$$

Here, Y stands for the strain energy density release rate and S denotes the strength energy of damage. The driving force of damage evolution remains the accumulated plastic strain p. The strain energy density release rate Y has been expressed here as a function of the von Mises equivalent stress σ_{eq}, the hydrostatic stress σ_m and the Poisson coefficient ν.

18.4 Analytical Solutions for the Problem of Periodic Irradiation Combined with Cyclic Axial Loads

The mechanism of damage evolution for multiple irradiation cycles is illustrated in Fig. 18.7. Single cycle is composed of emission of the particles flux, production of clusters of voids and further mechanical loading. The number of clusters of voids increases at each subsequent cycle. Given the fact, that increment of damage is for

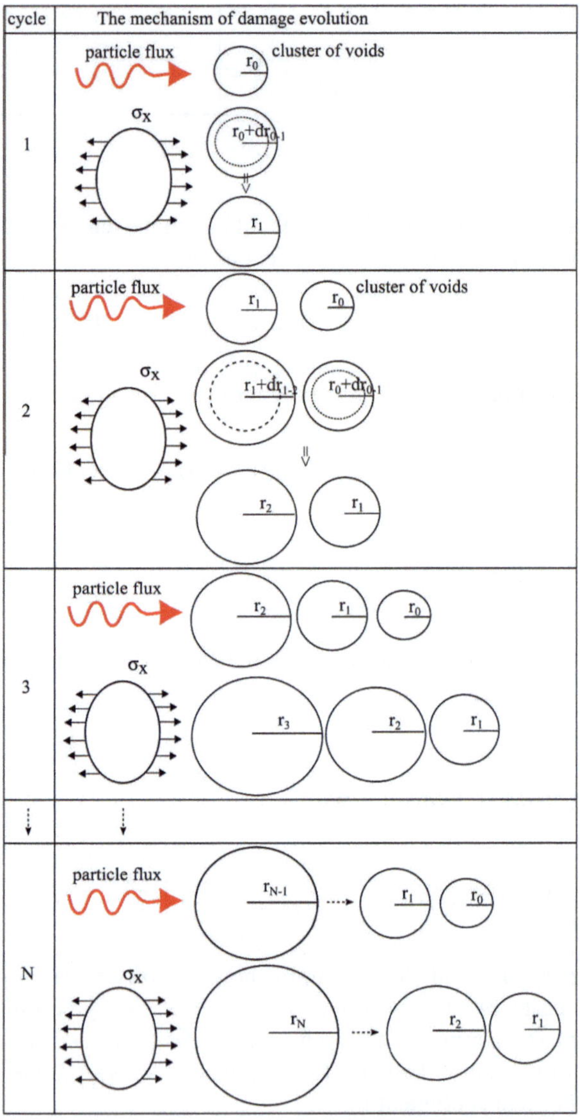

Fig. 18.7 The mechanism of damage evolution corresponding to cycles comprising radiation (particle flux) and mechanical loading (σ_x)

real applications rather small ($dpa_{\max} \cong 10^{-5}$), the effect of existing voids, forming effective sinks for new radiation induced vacancies, is negligible and it slightly affects formation of new clusters. Therefore, the constant increment of damage induced by each subsequent irradiation cycle is assumed. The radii of the newly formed clusters, as well as these created in the previous cycles, increase due to the mechanical loads. To find a closed-form analytical solution for one dimensional problem presented in Fig. 18.7 the recurrence relations are postulated.

To solve the problem, the recurrence relations are postulated. Integrating the post-irradiation damage evolution equation:

$$\int_{D_i}^{D_{i+1}} \mathrm{d}D = q_A 2\pi \int_{r_i}^{r_{i+1}} r\mathrm{d}r, \tag{18.34}$$

one obtains the following recurrence relation for damage parameter:

$$\Delta D_{i\to i+1} = q_A \pi \left(r_{i+1}^2 - r_i^2\right) \tag{18.35}$$

Integrating the Rice & Tracey law:

$$\int_{r_i}^{r_{i+1}} \frac{\mathrm{d}r_{\mathrm{c}}}{r_{\mathrm{c}}} = \alpha_r \exp\left(\frac{3\sigma_{\mathrm{m}}}{2\sigma_{\mathrm{eq}}}\right) \int_0^{\widetilde{p}} \mathrm{d}p, \tag{18.36}$$

the increase of cluster of voids radius reads:

$$r_{i+1} = r_i \mathrm{e}^{A\widetilde{p}} \tag{18.37}$$

The recurrence relation for damage parameter takes the form:

$$\Delta D_{i\to i+1} = q_A \pi r_i^2 \left(\mathrm{e}^{2A\widetilde{p}} - 1\right), \tag{18.38}$$

where:

$$A = \alpha_r \exp\left(\frac{3\sigma_{\mathrm{m}}}{2\sigma_{\mathrm{eq}}}\right) \tag{18.39}$$

Each further term of the sequence is defined as a function of the preceding terms, where the basic formula takes the form:

$$D_{r0} = q_A \pi r_{c0}^2 \tag{18.40}$$

Below a sequence of expressions describing the damage parameter using the recurrence relations Eqs. (18.37) and (18.38) is derived:

$$\begin{aligned} D_{rm1} &= D_{r0} + \Delta D_{rm(0\to 1)} = D_{r0} + q_A \pi r_{c0}^2 \left(e^{2A\tilde{p}} - 1 \right), \\ D_{rm2} &= D_{rm1} + \Delta D_{rm(1\to 2)} + D_{r0} + \Delta D_{rm(0\to 1)}, \\ &= 2D_{r0} + q_A \pi r_{c0}^2 e^{2A\tilde{p}} + q_A \pi r_{c0}^2 e^{4A\tilde{p}} - 2q_A \pi r_{c0}^2, \\ D_{rm_{i+1}} &= D_{rm_i} + D_{r0} + \Delta D_{rm(i\to i+1)} + \Delta D_{rm(i-1\to i)} + \cdots + \Delta D_{rm(0\to 1)} \end{aligned} \tag{18.41}$$

Here, both mechanisms of creation of new clusters of voids and evolution of size of existing clusters of voids are accounted for Fig. 18.7. Finding the Nth term in the sequence indicates that all terms create a geometric series. Each successive term can be obtained by multiplying the previous one by the expression $e^{2A\tilde{p}}$:

$$\begin{aligned} D_{rmN} = {} & N D_{r0} + q_A \pi r_0^2 e^{2A\tilde{p}} + q_A \pi r_0^2 e^{4A\tilde{p}} + q_A \pi r_0^2 e^{6A\tilde{p}} + \cdots \\ & + q_A \pi r_0^2 e^{2NA\tilde{p}} - N q_A \pi r_0^2, \end{aligned} \tag{18.42}$$

$$D_{rmN} = q_A \pi r_{c0}^2 \sum_{n=1}^{N} e^{2nA\tilde{p}} \tag{18.43}$$

In what follows, a_1 is a scaling factor, equal to:

$$a_1 = q_A \pi r_{c0}^2 e^{2nA\tilde{p}}, \tag{18.44}$$

and $q \neq 0$ is the common ratio:

$$q = e^{2nA\tilde{p}} \tag{18.45}$$

The following sum of geometric series is obtained:

$$S_N = a_1 \frac{1 - q_N}{1 - q}, \quad S_N = q_A \pi r_{c0}^2 e^{2nA\tilde{p}} \frac{1 - e^{2ApN}}{1 - e^{2Ap}} \tag{18.46}$$

Using Eq. (18.45) the $N-th$ term of damage evolution sequence can be rewritten as:

$$D_{rmN} = q_A \pi r_{c0}^2 e^{2nA\tilde{p}} \frac{1 - e^{2ApN}}{1 - e^2 Ap} \tag{18.47}$$

In the present paper, the following criterion has been used:

$$D_{rmN_f} = q_A \pi r_{c0}^2 e^{2nA\tilde{p}} \frac{1 - e^{2ApN}}{1 - e^{2Ap}} = D_{cr}, \tag{18.48}$$

where D_{cr} denotes a critical value of damage parameter, corresponding to lattice failure ($D_{cr} = 0.1$). The value $\tilde{p}$ denotes a constant increment of p on each cycle.

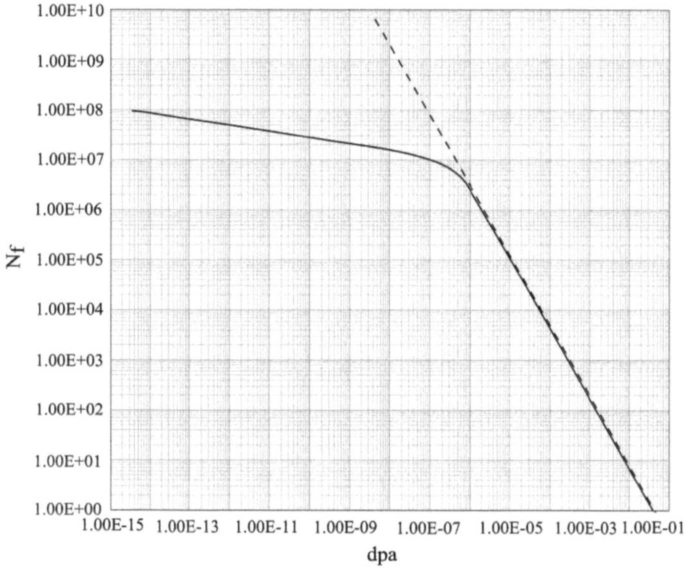

Fig. 18.8 Number of cycles to failure N_{f} as a function of dpa corresponding to Rice & Tracey model

Using this formula one can calculate damage parameter for any number of cycles N_{f} . The relation between the number of cycles to failure N_f as a function of dpa on single cycle, according to Eq. (18.47) is presented in Fig. 18.8.

The solid line represents the performance of Rice & Tracey model. For comparison, the dashed line represents the number of cycles to failure corresponding to irradiation of the material not subjected to mechanical stress. For the values of dpa exceeding 10^{-6}, the Rice & Tracey model shows in double logarithmic scale linear response. For smaller values, well below 10^{-6}, the evolution of mechanically induced damage significantly reduces the number of cycles to failure (Fig. 18.8). The solid line represents the performance of Rice & Tracey model. Based on the profile of presented results, an attempt to derive bilinear approximation for Rice & Tracey model was made (Fig. 18.9).

The following simple analytical formula was obtained:

$$\log(N_{\mathrm{c}}) = a + b\ \log(dpa_{\mathrm{max}}), \tag{18.49}$$

and

$$N_{\mathrm{c}} = 10^{a} dpa_{\mathrm{max}}^{b} \tag{18.50}$$

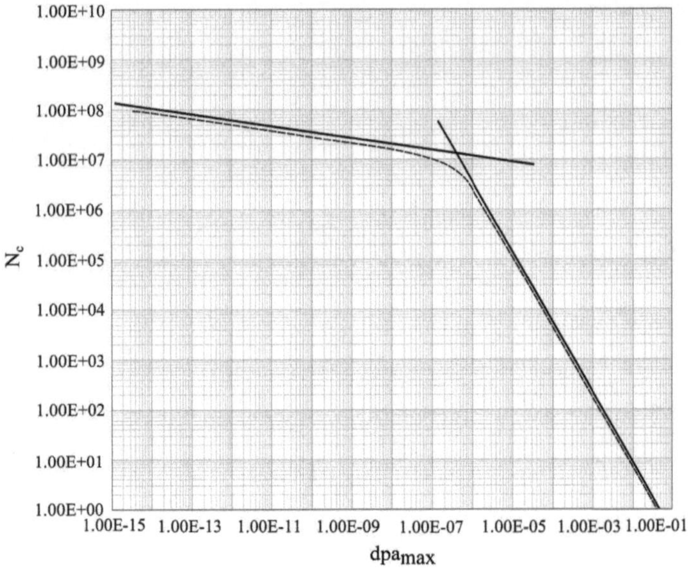

Fig. 18.9 Bilinear approximations for Rice & Tracey model

Finally, for the Rice & Tracey model the following analytical formula has been derived:

$$N_c = \begin{cases} 10^{1.9} dpa_{\max}^{-1.4} & \text{for} \quad dpa_{\max} \geq 10^{-6} \\ 10^{6.1} dpa_{\max}^{-0.13} & \text{for} \quad dpa_{\max} < 10^{-6} \end{cases} \tag{18.51}$$

18.5 Concluding Remarks

In the present work a new constitutive model has been built in order to include the evolution of radiation induced damage in the lifetime estimation of critical components subjected to irradiation and mechanical loads. Both types of damage fields, radiation induced and mechanical, are functions of the accumulated plastic strain and start evolving beyond the yield stress. The model has been developed in the framework of continuum damage mechanics and contains strong physical background related to the mechanism of generation of clusters of voids in the irradiated solids. It has been demonstrated in the course of the present work that the number of cycles to failure depends strongly on the accumulation of micro-damage due to irradiation. The lifetime of irradiated components has been expressed as a function of two parameters: maximum *dpa* and maximum axial stress on cycle. In order to make the model fully efficient it is fundamental to establish with enough certitude maximum *dpa* on single cycle by means of suitable experiments. The model remains relatively simple and its calibration is fully feasible due to the limited number of parameters.

References

Bacon D, Gao F, Osetsky YN (2000) The primary damage state in fcc, bcc and hcp metals as seen in molecular dynamics simulations. J Nucl Mater 276:1–12

Baussan E, Bielski J, Bobeth C, Bouquerel E, Oand Caretta P, Cupial Davenne T, Densham C, Dracos M, Fitton M, Gaudiot G, Kozien M, Lacny L, Lepers B, Longhin A, Loveridge P, Osswald F, Poussot P, Rooney M, Skoczen B, Szybinski B, Ustrzycka A, Vassilopoulos N, Wilcox D, Wroblewski A, Wurtz J, Zeter V, Zito M (2014) The neutrino super beam based on the superconducting proton linac. Phys Rev ST Accel Beams 17(031):001

Chaboche JL (1983) Anisotropic creep damage in the framework of continuum damage mechanics. Nucl Eng Des 79:309–319

Edgecock TR, Caretta O, Davenne T, Densam C, Fitton M, Kelliher D, Loveridge P, Machida S, Prior C, Rogers C, Rooney M, Thomason J, Wilcox D, Wildner E, Efthymiopoulos I, Garoby R, Gilardoni S, Hansen C, Benedetto E, Jensen E, Kosmicki A, Martini M, Osborne J, Prior G, Stora T, Melo Mendonca T, Vlachoudis V, Waaijer C, Cupial P, Chancé A, Longhin A, Payet J, Zito M, Baussan E, Bobeth C, Bouquerel E, Dracos M, Gaudiot G, Lepers B, Osswald F, Poussot P, Vassilopoulos N, Wurtz J, Zeter V, Bielski J, Kozien M, Lacny L, Skoczen B, Szybinski B, Ustrycka A, Wroblewski A, Marie-Jeanne M, Balint P, Fourel C, Giraud J, Jacob J, Lamy T, Latrasse L, Sortais P, Thuillier T, Mitrofanov S, Loiselet M, Keutgen T, Delbar T, Debray F, Trophine C, Veys S, Daversin C, Zorin V, Izotov I, Skalyga V, Burt G, Dexter AC, Kravchuk VL, Marchi T, Cinausero M, Gramegna F, De Angelis G, Prete G, Collazuol G, Laveder M, Mazzocco M, Mezzetto M, Signorini C, Vardaci E, Di Nitto A, Brondi A, La Rana G, Migliozzi P, Moro R, Palladino V, Gelli N, Berkovits D, Hass M, Hirsh TY, Schaumann M, Stahl A, Wehner J, Bross A, Kopp J, Neuffer D, Wands R, Bayes R, Laing A, Soler P, Agarwalla SK, Cervera Villanueva A, Donini A, Ghosh T, Gómez Cadenas JJ, Hernández P, Martín-Albo J, Mena O, Burguet-Castell J, Agostino L, Buizza-Avanzini M, Marafini M, Patzak T, Tonazzo A, Duchesneau D, Mosca L, Bogomilov M, Karadzhov Y, Matev R, Tsenov R, Akhmedov E, Blennow M, Lindner M, Schwetz T, Fernández Martinez E, Maltoni M, Menéndez J, Giunti C, González García MC, Salvado J, Coloma P, Huber P, Li T, López Pavón J, Orme C, Pascoli S, Meloni D, Tang J, Winter W, Ohlsson T, Zhang H, Scotto-Lavina L, Terranova F, Bonesini M, Tortora L, Alekou A, Aslaninejad M, Bontoiu C, Kurup A, Jenner LJ, Long K, Pasternak J, Pozimski J, Back JJ, Harrison P, Beard K, Bogacz A, Berg JS, Stratakis D, Witte H, Snopok P, Bliss N, Cordwell M, Moss A, Pattalwar S, Apollonio M (2013) High intensity neutrino oscillation facilities in europe. Phys Rev ST Accel Beams 16(021):002

Egner H, Skoczeń B (2010) Ductile damage development in two-phase metallic materials applied at cryogenic temperatures. Int J Plast 26:488–506

Garion H, Skoczeń B (2003) Combined model of strain-induced phase transformation and orthotropic damage in ductile materials at criogenic temperatures. Int J Damage Mech 12:331–356

Lemaitre J (1999) A course on damage mechanics. Springer, Berlin

Nita N, Schaeublin R, Victoria M (2004) Impact of irradiation on the microstructure of nanocrystalline materials. J Nucl Mater 1:329–333

Rice JR, Tracey DM (2004) On the ductile enlargement of voids in triaxial stress fields. J Mech Phys Solids 17:201–217

Verbiest E, Pattyn H (2004) Study of radiation damage in aluminum. Phys Rev B 8:5097–5121

Zinkle S, Singh B (1993) Analysis of displacement damage and defect production under cascade damage conditions. J Nucl Mater 199:173–191

Chapter 19
Capabilities of the Multi-mechanism Model in the Prediction of the Cyclic Behavior of Various Classes of Metals

Lakhdar Taleb, Kacem Saï and Georges Cailletaud

Abstract The paper deals with an evaluation of the multi-mechanism (MM) approach capabilities in the prediction of the cyclic behavior of different classes of metallic materials. For this objective, the tests detailed in (Taleb, Int J Plast 43:1–19, 2013a) have been simulated here by the MM model. In these tests, six alloys were considered: two ferritic steels (35NCD16 and XC18), two austenitic stainless steels (304L and 316L), one "extruded" aluminum alloy (2017A) and one copper-zinc alloy (CuZn27). The specimens have been subjected to proportional and non-proportional stress as well as the combination of stress and strain control at room temperature. The identification of the material parameters has been carried out using exclusively strain controlled experiments under proportional and non-proportional loading paths performed in the present study for each material. The model may describe a large number of phenomena with twenty five parameters in total but, it appears that for a given material under the adopted conditions, the activation of all parameters may be not necessary. Our attention was focused mainly on the capabilities to predict correctly the cyclic accumulation of the inelastic strain including the shape of the hysteresis loops. The comparison between test responses and their predictions by the MM model are generally satisfactory with relatively small number of material parameters (between eight and thirteen according to the material). One can also highlight the capability of the MM model to describe a transient ratcheting without activation of the dynamic recovery term in the kinematic variables. Finally, the MM model deserves improvement for a better description of the cyclic behavior of anisotropic materials.

L. Taleb (✉)
INSA, GPM, CNRS UMR 6634, BP 08 Avenue de L'université,
76801 St. Etienne du Rouvray Cedex, France
e-mail: lakhdar.taleb@insa-rouen.fr

K. Saï
UGPMM Ecole Nationale d'Ingénieurs de Sfax, B.P. 1173 Sfax, Tunisia
e-mail: kacemsai@yahoo.fr

G. Cailletaud
MINES ParisTech, Centre des Matériaux, CNRS UMR 7633,
BP 87, 91003 Evry Cedex, France
e-mail: Georges.Cailletaud@ensmp.fr

H. Altenbach et al. (eds.), *From Creep Damage Mechanics to Homogenization Methods*, Advanced Structured Materials 64,
DOI 10.1007/978-3-319-19440-0_19

Keywords Cyclic loading · Metallic materials · Mechanical testing · Ratcheting · Constitutive equations · Multi-mechanism model

19.1 Introduction

Phenomenological modeling of the cyclic behavior of metallic materials is one of the most active subjects in mechanical engineering for the last decades. For that purpose one can find generally two distinct approaches: the one where all mechanisms of the inelastic strain are expressed through the evolution of one variable, in contrast with the second approach (called multi-mechanism) which enables the possibility to describe the inelastic strain through more than one variable representing different mechanisms. In addition to the pioneering papers (Chaboche 1986, 1989, 1991, 1994; Chaboche et al. 1979; Ohno 1990; Ohno and Wang 1991, 1993, 1994), one can note that during the last 20 years period, the published works according to the first approach are numerous (Abdel-Karim and Ohno 2000; Abdel-Karim 2009, 2010, 2011; Bari and Hassan 2000, 2001, 2002; Chaboche 2008; Chaboche et al. 2012; Chen and Jiao 2004; Chen and Kim 2003; Chen et al. 2003; Dafalias and Feigenbaum 2011; Feigenbaum et al. 2012; Hassan et al. 2008; Jiang and Zhang 2008; Kang 2008; Kang et al. 2002, 2004, 2005; Kang and Gao 2002, 2004; Kang and Kan 2007; Kobayashi and Ohno 2002; Krishna et al. 2009; Ohno and Abdel-Karim 2000; Portier et al. 2000; Vincent et al. 2002, 2004; Yaguchi and Takahashi 2005; Yu et al. 2012; Yoshida 2000; Yoshida and Uemori 2002). For the same period, without forgetting the pioneering works (Cailletaud and Saï 1995; Contesti and Cailletaud 1989; Pugh 1978), different papers have been published in literature about the multi-mechanism approach (Saï and Cailletaud 2007; Saï 2011; Saï et al. 2012, 2014; Taleb et al. 2006, 2014; Taleb and Cailletaud 2010, 2011; Wolff and Taleb 2008). Regarding the quantity of literature works dealing with both approaches, it appears clearly that the first category (mono-mechanism) has received more interest compared to the multi-mechanism approach. This imbalance is may be due to the apparent complexity of the MM models. Indeed, one can think for instance that have more than one inelastic strain variable leads to very large number of material parameters! However, this is not true as the number of parameters remains similar between the two approaches as we will demonstrate in the present study. Two main differences between both approaches may be pointed out:

- If the thermodynamic consistency is generally ensured for the models of the first approach whatever the (positive) values of the material parameters, this is not necessarily true for certain versions of the MM model where restrictive conditions have to be verified (Wolff and Taleb 2008).
- Contrary to the first approach, the MM model may lead to ratcheting even in the absence of dynamic recovery term in the kinematic hardening variables (Taleb et al. 2014).

In this paper, the capabilities of the multi-mechanism (MM) model will be evaluated considering the experimental responses of different classes of alloys subjected to proportional and non-proportional stress and strain control: two ferritic steels (35NCD16 and XC18), one austenitic stainless steel (316L), one aluminum alloy (2017A) and, one copper-zinc alloy (CuZn27). The choice of these materials may be justified by two main considerations:

- They are largely used in current industrial applications: building, nuclear, aeronautics, automotive…
- They may exhibit various inelastic phenomena: rate-dependency of the behavior, cyclic hardening, extra-hardening in presence of non-proportional loading path, cyclic softening, memory of the strain hardening, cyclic accumulation of the inelastic strain under unsymmetrical stress control, anisotropy …

Qualitatively, all these phenomena may be described by the constitutive equations of the MM model; the accuracy of such description will be studied here considering the experimental responses performed earlier in (Taleb 2013a) under stress and combined stress-strain control following proportional as well as non-proportional loading paths. The calibration of the material parameters of the model has been carried out through additional strain controlled experiments performed in the present work. Two versions of the MM model have been considered here: the one described in Taleb and Cailletaud (2010) and its extension to anisotropic materials Saï et al. (2012). After a brief review of the experiments carried out by Taleb (2013a), their numerical simulation by the MM model is detailed and the comparison experiment-modeling discussed.

19.2 Experimental Procedure

19.2.1 Materials, Specimens and Experimental Device

In Taleb (2013a), six different metallic materials have been considered: two austenitic stainless steels (304L and 316L), two ferritic steels (35NCD16 and XC18), one aluminum alloy 2017A and finally one copper-zinc alloy (CuZn27). In the present paper, the responses of 304L SS will not be considered as they were extensively studied and simulated in Hassan et al. (2008); Saï et al. (2014); Taleb and Cailletaud (2011). The average chemical compositions of the rest of the considered materials are given in Table 19.1a–e.

The same specimen geometry has been considered for all materials and all tests: tubular shape with a gage length of 46 mm where a central part of 25 mm is used for the extensometry. In this zone, the outer and inner diameters are respectively equal to 20 and 17 mm that enables to have relatively thin tubes.

The steel specimens have been machined from solid bars and then heat treated in order to have almost the same metallurgical and mechanical initial states. The

Table 19.1 Average chemical composition (in % mass) of the different materials used in (Taleb 2013a)

(a) 316L SS

C	Si	Mn	Ni	Cr	Mo	S	P	Fe
0.02	0.36	1.32	10.13	16.57	2.03	0.029	0.031	Balance

(b) 35NCD16

C	Si	Mn	P	S	Cr	Ni	Mo	Cu	Fe
0.35	0.29	0.48	0.21	0.011	1.81	3.78	0.27	0.17	Balance

(c) XC18

C	Si	Mn	P	S	Cr	Ni	Mo	Cu	Fe
0.20	0.22	0.57	0.022	0.021	0.11	0.06	0.01	0.18	Balance

(d) 2017A

Cu	Mg	Mn	Si	Fe	Al
3.95	0.67	0.53	0.76	0.90	Balance

(e) CuZn27

Cu	Pb	Sn	Fe	Ni	Al	Mn	Si	Zn
63.0	1.50	1.00	1.25	2.50	1.75	2.25	1.00	Balance

applied thermal cycle is composed of heating up to 1050 °C followed by a hold time of 1 h at this temperature and, finally a slow cooling up to room temperature. The aluminum alloy specimens have been machined from extruded solid bars parallel to the extrusion axis; they have been also heat treated in order to ensure almost the same initial state. The treatment includes a hold time of 1 h at 500 °C followed by a quenching up to room temperature. Finally, the copper-zinc alloy specimens have been obtained by casting process performed by FAVI Company.[1]

The tests were carried out on a MTS axial-torsional servo-hydraulic machine with a capacity of ±250 KN/±75 mm in tension-compression and ±2200 Nm/±45° in torsion. The axial displacement and the rotation in the central zone of the specimen were measured by means of a MTS axial-torsional extensometer with 25 mm gage length and a capacity of ±2.5 mm/±5°. Hydraulic collet grips enable to rigidly maintain the heads of the specimen under adjustable pressure control.

The stress and strain tensors are assumed constant in the gage length; they have the following general form in a cylindrical coordinate system (r, z, θ) with z oriented along the specimen axis:

$$\underline{\sigma} = \begin{pmatrix} 0 & 0 & 0 \\ 0 & 0 & \tau \\ 0 & \tau & \sigma \end{pmatrix} \quad \text{and} \quad \underline{\varepsilon} = \begin{pmatrix} \varepsilon_{rr} & 0 & 0 \\ 0 & \varepsilon_{\theta\theta} & \gamma/2 \\ 0 & \gamma/2 & \varepsilon_{zz} \end{pmatrix}$$

[1] http://www.favi.com/ang/index.php.

Table 19.2 Main characteristics of the tests performed in Taleb (2013a) and considered in the present paper

Material	Tension-compression $\sigma_{max}/\sigma_{min}$	Triangle $\sigma_{max} = \gamma_{max}/\sqrt{3}$	Reference
316L	250/−150		316L_50-200
		250/0.4	316L_250-0.4
35NCD16	1250/−850		35NCD16_200-1050
		900/0.4	35NCD16_900-0.4
XC18	370/−270		XC18_50-320
		250/0.4	XC18_250-0.4
2017A	300/−200		2017A_50-250
		300/0.4	2017A_300-0.4
CuZn27	300/−200		CuZn27_50-250
		300/0.4	CuZn27_250-0.4

Stresses (MPa) and strains (%)

19.2.2 Tests Performed in Earlier Work

Two cyclic loading paths, one proportional and the other non-proportional, have been considered in Taleb (2013a). The proportional loading path was a non-symmetric tension-compression under load control while the non-proportional one was a combination of axial force and torsional rotation following a triangular shape. The choice of the latter path is motivated by its "complexity" as several earlier studies show that different constitutive equations fail in simulating such a test type (Abdel-Karim 2010; Hassan et al. 2008).

The list providing the main characteristics of all tests performed in Taleb (2013a) is given in Table 19.2. In tension-compression tests, 100 cycles have been applied (Fig. 19.1a). In order to minimize the contribution of creep in the cyclic accumulation of the inelastic strain, "creep-ratcheting" tests have been performed where the

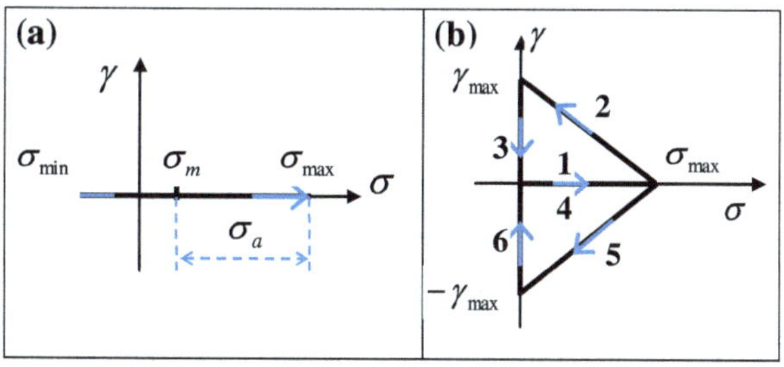

Fig. 19.1 Loading paths applied in the tests performed: **a** Cyclic tension-compression under stress control with tensile mean stress (σ_m); **b** Path "triangle" combining prescribed stress between 0 and σ_{max} and torsional strain between γ_{max} and $-\gamma_{max}$ following *triangular* shape

maximum force of the first cycle was maintained constant during a certain time. In the multiaxial tests, 100 cycles following a "triangular" path have been applied (Fig. 19.1b). For the particular case of 316L SS for which the behavior is time-dependent, the maximum force of the first cycle was maintained constant during 2 h in order to minimize the contribution of creep in the cyclic accumulation of the inelastic strain. All tests have been carried out at room temperature.

Due to the presence of the Lüders phenomenon in XC18 steel, the tension-compression test started by a first pre-loading under strain control ensuring to go over the plateau. However, the stress range of such pre-loading remains lower than the stress range of the subsequent cycles performed at force control.

19.3 Numerical Simulations

19.3.1 Constitutive Equations of the MM Model

We are interested here by the MM model based on two mechanisms and one criterion (called 2M1C). The first version of this model has been proposed in Cailletaud and Saï (1995). Several improvements of this model have been performed successively in Saï and Cailletaud (2007); Saï et al. (2012); Taleb et al. (2006); Taleb and Cailletaud (2010). This section shows the version of the 2M1C model described in Taleb and Cailletaud (2010) and its extension in Saï et al. (2012) without detailing the full equation set that can be found in the different original papers.

A mechanism is defined by a local stress, a kinematic variable, and a flow rule. Two mechanisms are combined in one loading function f to define one criterion.

The inelastic strain is composed of two components expressing two different mechanisms depending on the same viscoplastic potential (or the same multiplier in the time-independent plasticity):

$$\underline{\varepsilon}^{in} = \underline{\varepsilon}_1^{in} + \underline{\varepsilon}_2^{in} \tag{19.1}$$

Two local stresses are associated to these components. Here they are equal to the macroscopic stress,

$$\underline{\sigma}_1 = \underline{\sigma}_2 = \underline{\sigma} \tag{19.2}$$

The kinematic hardening is described by two state variables $\underline{\alpha}_1$ and $\underline{\alpha}_2$ which vary according to the Armstrong-Frederick model (Armstrong and Frederick 1966) combined with radial fading memory proposed in Burlet and Cailletaud (1987). The associated thermodynamical forces $\underline{X}_1$ and $\underline{X}_2$ respectively are coupled through the material parameter C_{12},

$$\underline{X}_1 = \frac{2}{3} C_{11} \underline{\alpha}_1 + \frac{2}{3} C_{12} \underline{\alpha}_2, \tag{19.3}$$

$$\underline{X}_2 = \frac{2}{3} C_{12} \underline{\alpha}_1 + \frac{2}{3} C_{22} \underline{\alpha}_2, \tag{19.4}$$

$$\dot{\underline{\alpha}}_1 = \underline{n}_1 \dot{v} - D_1[(1-\eta)\underline{\alpha}_1 + \eta(\underline{\alpha}_1 : \underline{m}_1)\underline{m}_1]\dot{v}, \tag{19.5}$$

$$\dot{\underline{\alpha}}_2 = \underline{n}_2 \dot{v} - D_2[(1-\eta)\underline{\alpha}_2 + \eta(\underline{\alpha}_2 : \underline{m}_2)\underline{m}_2]\dot{v}, \tag{19.6}$$

where the tensor $\underline{m}_j$ is defined by

$$\underline{m}_j = \frac{\underline{n}_j}{||\underline{n}_j||} = \frac{\underline{\sigma}'_j - \underline{X}'_j}{||\underline{\sigma}'_j - \underline{X}'_j||} \tag{19.7}$$

Note that the parameter η is only active for the case of multiaxial loading path. $\underline{\sigma}'$ is the deviatoric part of $\underline{\sigma}$.

C_{11}, C_{22}, C_{12}, D_1, D_2 and η are material parameters while $\dot{v}$ is a positive scalar defined below and, $\underline{n}_i$ is the gradient of the function f in the stress space,

$$\underline{n}_i = \frac{\partial f}{\partial \underline{\sigma}_i} = \frac{3}{2} \frac{\underline{\sigma}'_i - \underline{X}'_i}{J} \tag{19.8}$$

The yield function is given by,

$$f = J - R \tag{19.9}$$

R is a function of the size of the elastic domain:

$$R = R_0 + Q_1(1 - \mathrm{e}^{-b_1 v}) + Q_2(1 - \mathrm{e}^{-b_2 v}) + Q_3(1 - \mathrm{e}^{-b_3 v}), \tag{19.10}$$

where R_0 is the initial value of R (see below its relation with the initial yield stress) and b_i some material parameters.

Q_2 is a material parameter while Q_1 and Q_3 are variables defined as follows. Q_1 takes into account the strain memory effect following the concept of memory surface in the plastic strain space summarized below (Chaboche et al. 1979; Nouailhas et al. 1985; Ohno 1982),

$$Q_1 = Q_{1_0} + (Q_{1_m} - Q_{1_0})(1 - \mathrm{e}^{-2\mu q}) \tag{19.11}$$

Let $\underline{n}^*$ the gradient of the function F in the plastic strain space and $\underline{n} = \underline{n}_1 + \underline{n}_2$ If $\underline{n} : \underline{n}^* < 0$,

$$\mathrm{d}q = 0 \quad \text{and} \quad \mathrm{d}\underline{\xi} = \underline{0} \tag{19.12}$$

If $\underline{n} : \underline{n}^* > 0$ and $F = \sqrt{\frac{2}{3}(\underline{\varepsilon}^{\text{in}} - \underline{\xi}) : (\underline{\varepsilon}^{\text{in}} - \underline{\xi})} - q > 0$,

$$\mathrm{d}q = \eta_{\text{mem}}(\underline{n} : \underline{n}^*)\mathrm{d}v \quad \text{and} \quad \mathrm{d}\underline{\xi} = (1 - \eta_{\text{mem}})(\underline{n} : \underline{n}^*)\underline{n}^*\mathrm{d}v \tag{19.13}$$

If $\underline{n} : \underline{n}^* > 0$ and $F < 0$,

$$\mathrm{d}q = -0.1\eta_{\text{mem}}(\underline{n} : \underline{n}^*)\mathrm{d}v \quad \text{and} \quad \mathrm{d}\underline{\xi} = \underline{0} \tag{19.14}$$

Q_{1_0}, Q_{1_m}, μ and η_{mem} are material parameters.

Q_3 is a variable that introduces extra-hardening due to non-proportional loading paths. Benallal and Marquis (1987) expressed the non-proportionality through the angle α between the plastic strain rate and the deviatoric stress rate: we chose here a simplified version of this model:

$$\mathrm{d}Q_3 = d_3 A(Q_{3_s} - Q_3)\mathrm{d}v \tag{19.15}$$

where if, $\sqrt{\mathrm{d}\sigma'_{ij}\mathrm{d}\sigma'_{ij}} \neq 0$ and $\sqrt{\mathrm{d}\varepsilon^{\text{in}}_{ij}\mathrm{d}\varepsilon^{\text{in}}_{ij}} \neq 0$,

$$Q_{3_s} = AQ_{3_m}, \tag{19.16}$$

$$A = 1 - \cos\alpha\cos\alpha, \tag{19.17}$$

$$\cos\alpha = (\mathrm{d}\sigma'_{ij}\mathrm{d}\varepsilon^{\text{in}}_{ij})/(\sqrt{\mathrm{d}\sigma'_{ij}\mathrm{d}\sigma'_{ij}}\sqrt{\mathrm{d}\varepsilon^{\text{in}}_{ij}\mathrm{d}\varepsilon^{\text{in}}_{ij}}) \tag{19.18}$$

d_3 and Q_{3_m} are material parameters.

It is important to know that the evaluation of the non-proportionality by means of the angle between plastic strain rate and the deviatoric stress rate has two main drawbacks related to certain specific situations. This angle will be left undefined each time the deviatoric stress rate is zero ($\underline{\dot{\sigma}}' = \underline{0}$); this case corresponds for instance to classical creep tests. The second drawback can be exhibited in situations where non-proportional effects may exist despite $\alpha = 0$. Such cases may be encountered in multistep tests like cyclic tension compression (sequence 1) followed by cyclic torsion (sequence 2). In such configuration, if after the first sequence, an adequate path is applied in order to come back to zero axial stress and strain (Jiang and Zhang 2008; Murakami et al. 1989), the application of the second sequence may be accompanied by extra hardening despite the fact that $\alpha = 0$. In this work, when the MM model is used for simulating creep tests, we admit $A = 0$ which corresponds to proportional loading conditions.

The expression of J is given by:

$$J = \sqrt{J_1^2 + J_2^2} \tag{19.19}$$

with

$$J_i = \sqrt{\frac{3}{2}(\underline{\sigma}'_i - \underline{X}'_i) : (\underline{\sigma}'_i - \underline{X}'_i)} \tag{19.20}$$

As a consequence, the initial yield stress σ_y is given by,

$$R_0 = \sqrt{2}\sigma_y \tag{19.21}$$

The model can be written in the viscoplastic framework by introducing the material parameters K and n,

$$\underline{\dot{\varepsilon}}_i^{\text{in}} = \dot{v}\underline{n}_i = \langle \frac{f}{K} \rangle^n \underline{n}_i \tag{19.22}$$

Note that v is not exactly the accumulated inelastic strain, since:

$$\underline{\dot{\varepsilon}}_i^{\text{in}} : \underline{\dot{\varepsilon}}_i^{\text{in}} = (\dot{v})^2 (\underline{n}_1 + \underline{n}_2) : (\underline{n}_1 + \underline{n}_2) \tag{19.23}$$

For anisotropic materials like aluminum alloy 2017A considered here, J_i is replaced by J_{Li} defined by,

$$J_{Li} = \sqrt{\frac{3}{2}(\underline{\sigma}'_i - \underline{X}'_i) : \underline{\underline{L}} : (\underline{\sigma}'_i - \underline{X}'_i)}, \tag{19.24}$$

where the second order tensors are represented by a six-component vector such as:

$$\underline{T} = \begin{pmatrix} T_{11} & T_{12} & T_{13} \\ T_{21} & T_{22} & T_{23} \\ T_{31} & T_{32} & T_{33} \end{pmatrix} \Rightarrow \vec{T} = \begin{pmatrix} T_{11} \\ T_{22} \\ T_{33} \\ \sqrt{2}T_{12} \\ \sqrt{2}T_{23} \\ \sqrt{2}T_{31} \end{pmatrix} \tag{19.25}$$

$\underline{\underline{L}}$ is a fourth order tensor given by

$$\underline{\underline{L}} = \begin{pmatrix} L_a & 0 & 0 & 0 & 0 & 0 \\ 0 & L_b & 0 & 0 & 0 & 0 \\ 0 & 0 & L_c & 0 & 0 & 0 \\ 0 & 0 & 0 & L_d & 0 & 0 \\ 0 & 0 & 0 & 0 & L_e & 0 \\ 0 & 0 & 0 & 0 & 0 & L_f \end{pmatrix} \tag{19.26}$$

where L_a, L_b, L_c, L_d, L_e and L_f are some material parameters.

Note that in the last conditions, n_i will be given by,

$$\underline{n}_i = \frac{\partial f}{\partial \underline{\sigma}_i} = \frac{3}{2} \frac{\underline{\underline{L}} : (\underline{\sigma}_i - \underline{X}_i)}{J} \tag{19.27}$$

19.3.2 Identification of the Material Parameters

19.3.2.1 Overall Strategy

The identification process has been conducted following three main steps.

* *Step 1: Manual identification.* In this step, an approximation of the saturation values of the isotropic variables was performed. For this goal, strain controlled experiments have been considered: Q_{1_0}, Q_{1_m} and Q_2 have been evaluated through tension-compression responses while an approximation of Q_{3_0} and Q_{3_m} have been deducted from non-proportional responses. It is worth noting that according to the cyclic behavior of the material, the identified values may represent a minimum (cyclic hardening) or a maximum (cyclic softening) of the parameters under question. Figure 19.2 gives an example considering the responses of the copper-zinc alloy CuZn27.
* *Step 2: Optimization process.* Using the values estimated in step 1, the parameters Q_{1_0}, Q_{1_m}, Q_2, Q_{3_0} and Q_{3_m} were supposed constants in this step. The evaluation of the remaining parameters is then performed through an optimization module provided with ZSeT code included in the Finite Elements code ZeBuLoN developed at MINES ParisTech and ONERA, France (see Besson et al. (1998), or http://www.zset-software.com/products/z-opt for more details).

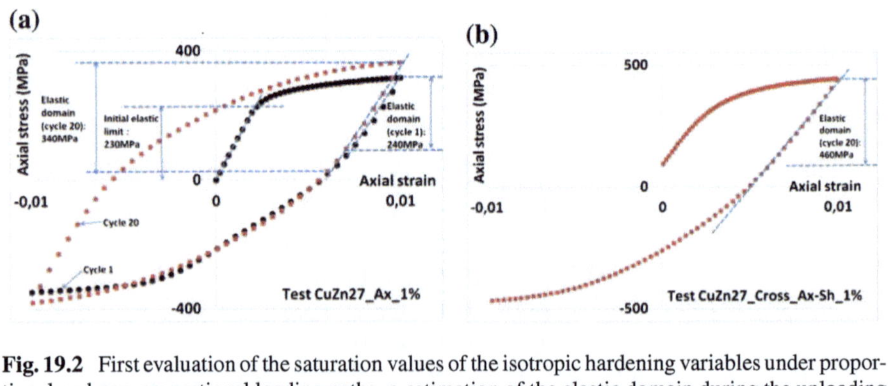

Fig. 19.2 First evaluation of the saturation values of the isotropic hardening variables under proportional and non-proportional loading paths. **a** estimation of the elastic domain during the unloading part of the first cycle (evaluation of Q_{1_0}) and the 20th cycle (evaluation of Q_{1_m}); **b** Estimation of the elastic domain after 20 cycles in the test "CuZn27_Cross_Ax-Sh_1%" (evaluation of Q_{3_m})

Only strain controlled experiments in addition to creep and/or relaxation tests when these phenomena are important were used for the identification of the material parameters. When the optimization process leads to very small values of certain parameters, the latter were assumed equal to zero in order to have the simplest version of the model.

* *Step 3: Manual improvement of the parameters.* In this last step, when it is necessary, some of the parameters estimated "manually" in the first step were adjusted once again "manually" in order to enable better representation of the strain controlled experiments.

Note that except for 316L SS, time-dependent phenomena were found not significant.

19.3.2.2 Tests Performed in the Current Work

As advocated above, strain controlled experiments have been performed for each material under proportional and non-proportional loading paths in order to identify the material parameters of the model. For the proportional loading, each material has been subjected to 20 cycles of tension-compression with amplitude of ε_a (Fig. 19.3a). In order to check the importance of the rate-dependency of the behavior, the peak strain of the first cycle is maintained constant for 15 min (relaxation test); no hold time is applied in the subsequent 19 cycles.

For the non-proportional tests, a cross path is chosen where the specimen is subjected successively to: one cycle of tension-compression with a given amplitude $\varepsilon_a \rightarrow$ one cycle of fully reversed torsion with the same equivalent amplitude $\rightarrow$ one cycle of tension-compression with ε_a and so on. 20 cycles have been applied in each direction (Fig. 19.3b). No hold time is applied during the cycles.

The main characteristics of the tests performed under strain control are summarized in Table 19.3.

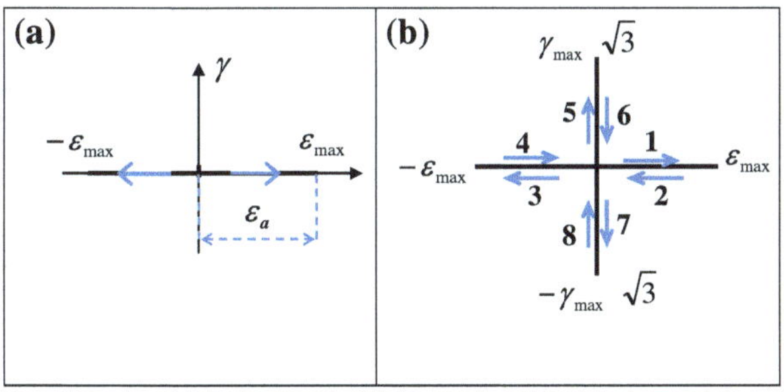

Fig. 19.3 Loading paths applied in the tests performed: **a** Cyclic tension-compression under strain control; **b** Path "cross" where each cycle is composed of one cycle of tension-compression with given amplitude ε_a followed by one cycle of fully reversed torsion with the same equivalent (von Mises) amplitude

Table 19.3 Main characteristics of the tests performed under strain control

Material	Loading path	Reference
316L	20 cycles: $\varepsilon_{zz} = \pm 1\,\%$	316_Ax_1%
35NCD16	20 cycles: $\varepsilon_{zz} = \pm 1\,\%$	35NCD16_Ax_1%
	20 cycles: $\varepsilon_{zz} = \pm 1\,\% \rightarrow \gamma/\sqrt{3} = \pm 1\,\%$	35NCD16_Cross_1%
XC18	20 cycles: $\varepsilon_{zz} = \pm 1.5\,\%$	XC18_Ax_1.5%
	20 cycles: $\varepsilon_{zz} = \pm 1.5\,\% \rightarrow \gamma/\sqrt{3} = \pm 1.5\,\%$	XC18_Cross_1.5%
2017A	20 cycles: $\varepsilon_{zz} = \pm 0.8\,\%$	2017A_Ax_0.8%
	20 cycles: $\gamma/\sqrt{3} = \pm 0.8\,\%$	2017A_Sh_0.8%
	20 cycles: $\varepsilon_{zz} = \pm 0.8\,\% \rightarrow \gamma/\sqrt{3} = \pm 0.8\,\%$	2017A_Cross_0.8%
CuZn27	20 cycles: $\varepsilon_{zz} = \pm 1\,\%$	CuZn27_Ax_1%
	20 cycles: $\varepsilon_{zz} = \pm 1\,\% \rightarrow \gamma/\sqrt{3} = \pm 1\,\%$	CuZn27_Cross_1%

19.3.2.3 Results of the Parameters Identification and Comparison with Experiments

35NCD16

According to the test conditions considered here, the following assumptions were adopted for 35NCD16 steel in the identification process:

- No extra hardening is significant under non-proportional loading: $Q_3 = 0$.
- No significant memory of the strain hardening: $\mu = 0$.
- Time-dependent phenomena have been neglected: $K = 10$ and $n = 20$.

Furthermore, the optimization leads to very small values of D_1 and D_2, therefore we choose: $D_1 = D_2 = 0$. The rest of the identified parameters are given in Table 19.4. The simulations are generally satisfactory even if a slight overestimation of the cyclic hardening may be noticed for tension-compression (Fig. 19.4a). Under cross path one can remark the good representation of the cyclic hardening followed by a moderate cyclic softening exhibited by the material (Fig. 19.4b).

Table 19.4 Identified parameters for the 35NCD16 steel

R_0	Q_{1_0}	b_1	Q_2	b_2	C_{11}	C_{22}	C_{12}
849	700	100	−211	5.025	1,419,347	77	51

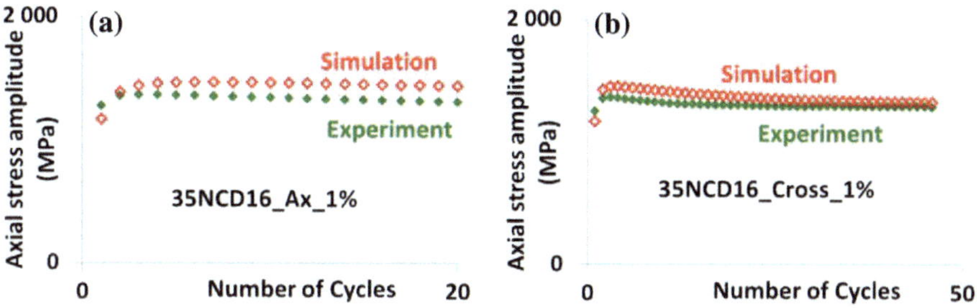

Fig. 19.4 Comparison between experiments and simulations of the tests used in the optimization process for 35NCD16 steel: **a** Axial stress amplitude versus the number of cycles for the test 35NCD16_Ax_1%; **b** Axial stress amplitude versus the number of cycles for the test 35NCD16_Cross_1%

Table 19.5 Identified parameters for the XC18 steel

R_0	Q_{1_0}	b_1	Q_2	b_2	Q_{3_m}	b_3	d_3	η	C_{11}	C_{22}	D_1	D_2
424	−110	312	80	12.1	200	10,000	100	0.9	53,395	3358	143	0

XC18

The test results obtained here suggest to neglect the memory of the strain hardening ($\mu = 0$) and the time-dependent phenomena ($K = 10$ and $n = 20$). Furthermore, the optimization leads to very small value of C_{12}, therefore we neglect the coupling between the kinematic hardening variables: $C_{12} = 0$. In addition we choose $C_{3_0} = 0$. The rest of the identified parameters are given in Table 19.5.

Figure 19.5 shows the quasi-perfect agreement between the experimental responses and their simulations in both cases of the loading proportional (Fig. 19.5a) and non-proportional (Fig. 19.5b).

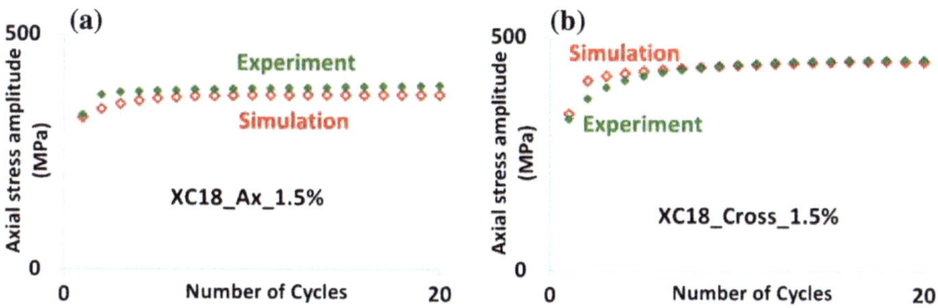

Fig. 19.5 Comparison between experiments and simulations of the tests used in the optimization process for XC18 steel: **a** Axial stress amplitude versus the number of cycles for the test XC18_Ax_1.5%; **b** Axial stress amplitude versus the number of cycles for the test XC18_Cross_1.5%

Table 19.6 Identified parameters for the 2017A steel

R_0	Q_2	b_2	Q_{3m}	b_3	d_3	C_{11}	C_{22}	C_{12}	L_a	L_b	L_c	L_d	L_e	L_f
283	92	12.2	500	18.5	15	5026	607,427	−35,000	1	1	1	1	2.56	1

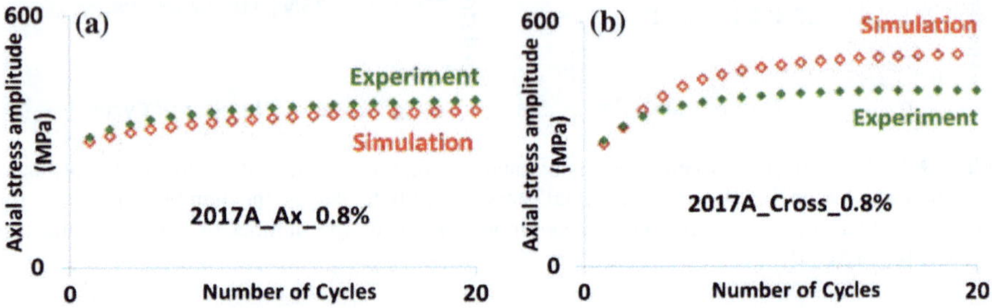

Fig. 19.6 Comparison between experiments and simulations of the tests used in the optimization process for 2017A alloy: **a** Axial stress amplitude versus the number of cycles for the test 2017A_Ax_0.8 %; **b** Axial stress amplitude versus the number of cycles for the test 2017A_Cross_0.8 %

2017A

Under the test conditions considered here, we assume that memory of the strain hardening as well as time-dependent phenomena are not significant: $\mu = 0$, $K = 31$ and $n = 11$. Furthermore, the optimization leads to very small values of D_1 and D_2, therefore we choose: $D_1 = D_2 = 0$ in addition to, $Q_{3_0} = 0$. The rest of the identified parameters are given in Table 19.6.

The predictions of the MM model are very good under tension-compression (Fig. 19.6a) while one can notice an overestimation of the extra-hardening under non-proportional loading path after the third cycle. This difference may be due to the anisotropic character of the material which is taken into account here only through the yield stress. This “defect” deserves improvement in the future by incorporating the possibility to take into account the difference in the cyclic hardening according to the direction considered. More investigations about the cyclic behavior of 2017A alloy may be found in May et al. (2013).

CuZn27

The test results obtained here suggest to neglect memory of the strain hardening ($\mu = 0$) and time-dependent phenomena ($K = 10$ and $n = 20$). Furthermore, we choose: $D_1 = D_2 = 0$ and $Q_{3_0} = 0$. The rest of the identified parameters are given in Table 19.7.

Figure 19.7 shows general good agreement between experiments and the predictions of the MM model despite the slight difference in the representation of the cyclic

Table 19.7 Identified parameters for the CuZn27

R_0	Q_{1_0}	b_1	Q_2	b_2	Q_{3_m}	b_3	d_3	C_{11}	C_{22}	C_{12}
255	-40	301	120	21	150	300	202	186,233	7710	82

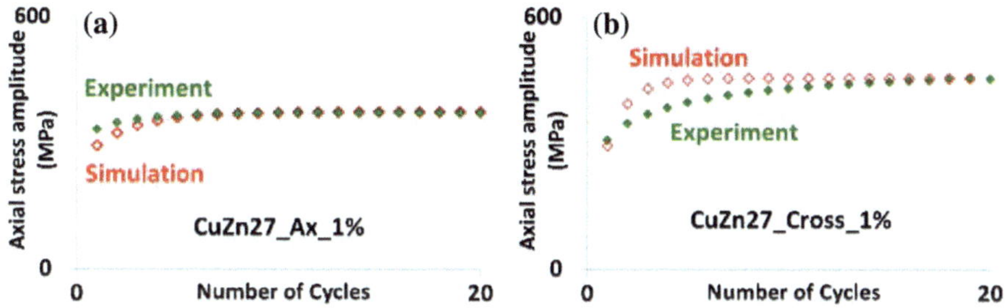

Fig. 19.7 Comparison between experiments and simulations of the tests used in the optimization process for CuZn27 alloy: **a** Axial stress amplitude versus the number of cycles for the test CuZn27_Ax_1%; **b** Axial stress amplitude versus the number of cycles for the test CuZn27_Cross_1%

hardening rate during the first cycles. Such a difference is more important for the non-proportional loading path (Fig. 19.7b).

316L

In the absence of specific tests, and in order to simplify the identification procedure, the importance of the following phenomena was assumed the same as for the 304 L stainless steel at room temperature (Taleb and Cailletaud 2011)

- Extra hardening due to non-proportional loading path (see May et al. (2013)).
- Memory of the strain hardening: investigations about this phenomenon may be found in Belattar et al. (2012); Taheri et al. (2011).
- Time-dependent phenomena. One can notice that 316L SS is rate-dependent at room temperature but rate-independent at 350 °C (Taleb 2013b).

Furthermore, we choose: $D_1 = D_2 = 0$ and $Q_{1_0} = Q_2 = Q_{3_0} = 0$. The rest of the identified parameters are given in Table 19.8.

Table 19.8 Identified parameters for the 316L SS

R_0	Q_{m_0}	b_1	η_{mem}	μ	Q_{3_m}	b_3	d_3	n	K	C_{11}	C_{22}	C_{12}
212	180	11.1	0.15	117	450	14	51	3.7	650	103,776	5840	−20,000

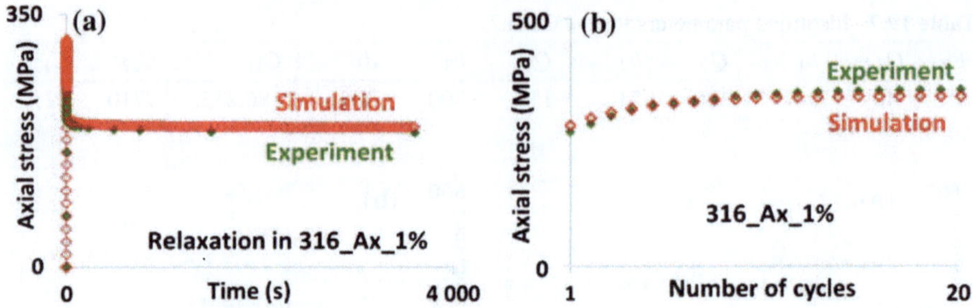

Fig. 19.8 Comparison between experiments and simulations of the test 316_Ax_1% used in the optimization process: **a** Relaxation of the axial stress under constant axial strain control of 1 % (peak strain in the first cycle); **b** Axial stress amplitude versus the number of cycles

The relaxation of the axial stress under constant axial strain control of 1 % is well represented by the model (Fig. 19.8a). The same observation was performed about the prediction of the cyclic hardening under tension-compression (Fig. 19.8b).

19.3.3 Summary of the Qualitative Capabilities of the MM Model

Table 19.9 summarizes the phenomena which may be taken into account by the MM model with the associated material parameters. In its full version, the model may necessitate the identification of up to 25 parameters but the case where the material exhibits all the listed phenomena is rather scarce in current engineering applications as we will see below.

19.3.4 Thermodynamic Consistency of the Model with the Identified Parameters

Let us now check if the parameters identified above ensure the thermodynamic consistency of the model established by Wolff and Taleb (2008). It is demonstrated that the free energy is non-negative if and only if the following relation is fulfilled:

$$C_{12}^2 \leq C_{11}C_{22} \tag{19.28}$$

This condition is verified with the identified values of the parameters for all materials under question.

Table 19.9 Phenomena taken into account by the MM model and associated material parameters

Phenomenon and associated material parameters		Number of parameters
Viscosity (K, n)		2
Anisotropy about the yield stress $(L_a, L_b, \ldots, L_f)$		6
Kinematic hardening (6)	Linear term (C_1, C_2)	2
	Dynamic recovery term (D_1, D_2, η)	3
	Coupling between both mechanisms (C_{12})	1
	Initial yield stress (R_0)	1
Isotropic hardening (11)	"Classical" term (Q_2, b_2)	2
	Memory of the strain hardening $(Q_1 : Q_{1_0}, Q_{1_m}, b_1, \eta_{\text{mem}}, \mu)$	5
	Extra hardening under non-proportional loading path $(Q_3 : Q_{3_m}, b_3, d_3)$	3
Total		25

A sufficient condition to ensure the positivity of the dissipation is also established (Wolff and Taleb 2008) about the parameters of the model:

$$C_{12}^2(D_1 - D_2)^2 \leq 4D_1D_2(C_{11}C_{22}(1-\eta)^2 - C_{12}^2) \tag{19.29}$$

This relation is automatically verified for 35NCD16, 316L, 2017A and CuZn27 since $D_1 = D_2 = 0$ for these materials. For XC18 steel, the coupling in the kinematic hardening is neglected ($C_{12} = 0$) which makes the condition (19.29) also satisfied.

19.3.5 Simulation of the Stress Controlled Experiments of Taleb (2013a)

19.3.5.1 35NCD16

Figure 19.9 gives the results of the tests performed on 35NCD16 steel and their simulations with the MM model. The material exhibits moderate ratcheting under tension-compression which is well predicted by the simulation (Fig. 19.9a–c) especially after the first few cycles. Under non-proportional loading path, one can observe a transient cyclic accumulation of the axial inelastic strain which stabilizes after 10 cycles: this observation is also very well reproduced by the MM model (Fig. 19.9d).

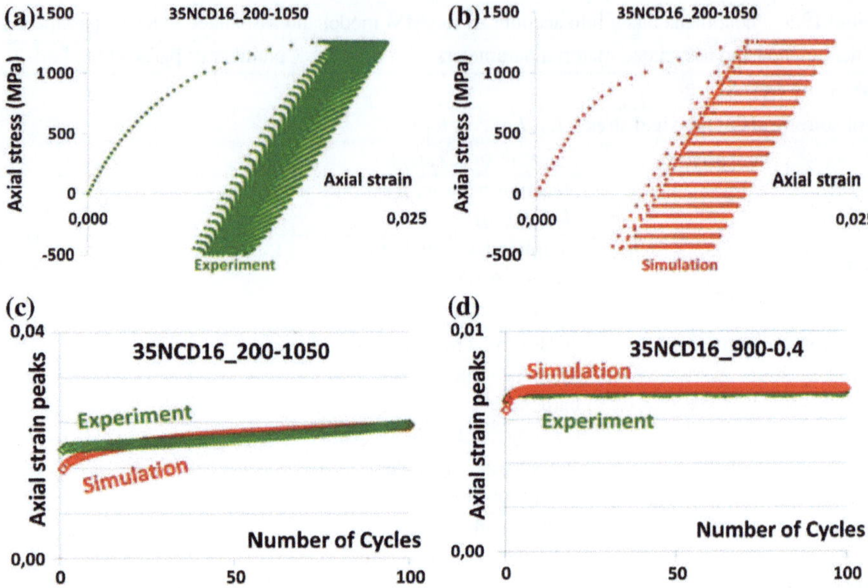

Fig. 19.9 Simulations of the tests performed on 35NCD16 steel under stress control. Axial stress-strain diagrams for the test 35NCD16_200-1050: **a** experiment; **b** simulation. Axial strain peaks versus the number of cycles, comparison between experiments and their simulations: **c** test 35NCD16_200-1050; **d** test 35NCD16_900-0.4

19.3.5.2 XC18

The results of the tests performed on XC18 steel and their numerical simulations are given in Fig. 19.10. During the first loading in plasticity, the material exhibits Lüders phenomenon characterized by a plateau with a quasi-constant axial stress while the axial strain evolves (Fig. 19.10a). This phenomenon is very well predicted by the MM model (Fig. 19.10b). After the first cycle in tension-compression, the material shows regular cyclic accumulation of the axial inelastic strain having the characteristics of the ratcheting phenomenon which is perfectly predicted by the model (Fig. 19.10c). Under triangle path (non-proportional loading), one can also observe a cyclic accumulation of the inelastic axial strain (ratcheting) but with decreasing rate which stabilizes after about 50 cycles. The simulation of this observation is satisfactory especially for the second half of the test (Fig. 19.10d)).

19.3.5.3 2017A

Figure 19.11 gives the results of the tests carried out on 2017A alloy and their simulations by the MM model. Under tension-compression, no significant cyclic accumulation of the inelastic strain is observed (Fig. 19.11a, c) while the MM model

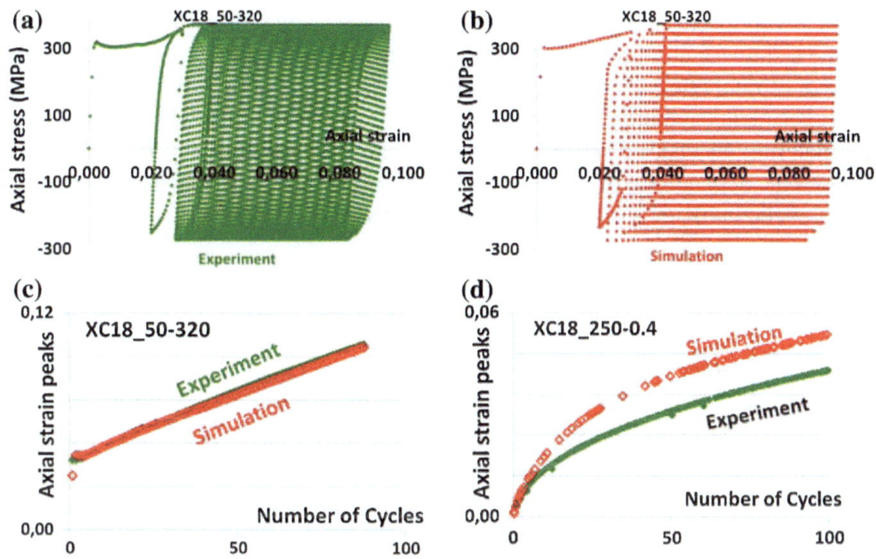

Fig. 19.10 Simulations of the tests performed on XC18 steel under stress control. Axial stress-strain diagrams for the test XC18_50-320: **a** experiment; **b** simulation. Axial strain peaks versus the number of cycles, comparison between experiments and their simulations: **c** test XC18_50-320; **d** test XC18_250-0.4

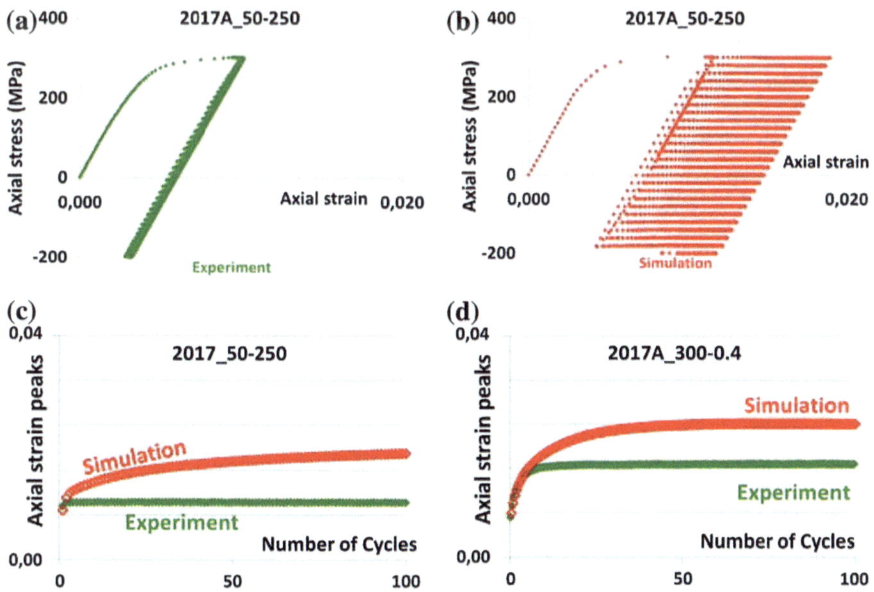

Fig. 19.11 Simulations of the tests performed on 2017A alloy under stress control. Axial stress-strain diagrams for the test 2017A_50-250: **a** experiment; **b** simulation. Axial strain peaks versus the number of cycles, comparison between experiments and their simulations: **c** test 2017A_50-250; **d** test 2017A_300-0.4

predicts a transient ratcheting (Fig. 19.11b, c). The same remark may be advocated for the test under non-proportional loading path where the simulation overestimates the rate of ratcheting for the first few cycles (Fig. 19.11d). This difference between the experiments and their simulations may be linked to the anisotropic character of the material. Indeed, this character is taken into account by the MM model only through the dependence of the yield stress on the direction; the anisotropic character of the cyclic hardening is also important to consider.

19.3.5.4 CuZn27

The results of the tests performed on CuZn27 alloy and their simulations by the MM model are given in Fig. 19.12. The behavior of this material is unusual compared to the other metals considered in this study. Indeed, the unloading part of the first cycle shows a significant decrease of the elastic domain which is well represented by the model (Fig. 19.12a). Considering the cyclic evolution of the axial strain peaks under tension-compression, one can notice negative ratcheting which is also well predicted by the model (Fig. 19.12c). In order to understand this observation, we have represented the axial stress-strain loops of the first and the last cycles (Fig. 19.12b). As it can be remarked, due to the evolution of the cyclic hardening, the axial strain peaks decrease while the axial strain valleys increase during the cycles. This evolution is

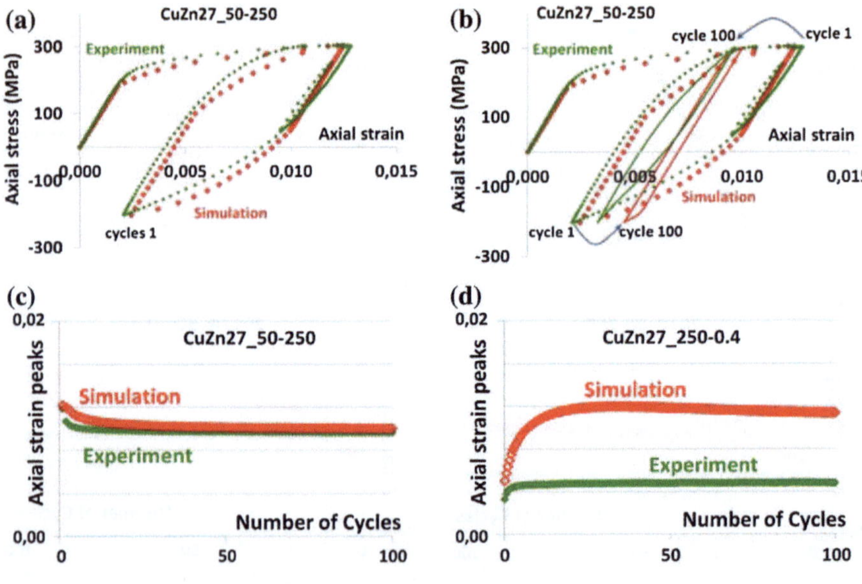

Fig. 19.12 Simulations of the tests performed on CuZn27 alloy under stress control. Axial stress-strain diagrams for the test CuZn27_50-250, comparison between experiments and their simulations: **a** first cycle; **b** first cycle with cycle 100. Axial strain peaks versus the number of cycles, comparison between experiments and their simulations: **c** test CuZn27_50-250; **d** test CuZn27_250-0.4

qualitatively well represented by the model despite the overestimation of the isotropic hardening for the stress-strain loop of the steady state (cycle 100, Fig. 19.12b). Under non-proportional loading, the material shows a transient ratcheting in the axial direction which is overestimated by the model for the first 20 cycles (Fig. 19.12d).

19.3.5.5 316L

The results of the tests performed on 316 L stainless steel as well as their simulations by the MM model are given in Fig. 19.13. The material exhibits rate-dependent behavior at room temperature; under constant axial stress, significant creep is observed (Fig. 19.13a). The phenomenon is generally well predicted by the model (Fig. 19.13c–d up to $t = 7200$ s) despite the bad representation of the transition between the elastic and the inelastic behavior during the loading in the first cycle (Fig. 19.13b). This discrepancy may be linked with the assumptions adopted in order to simplify the identification of the material parameters. After the creep sequence no significant cyclic accumulation of the inelastic strain is observed during the following cycles

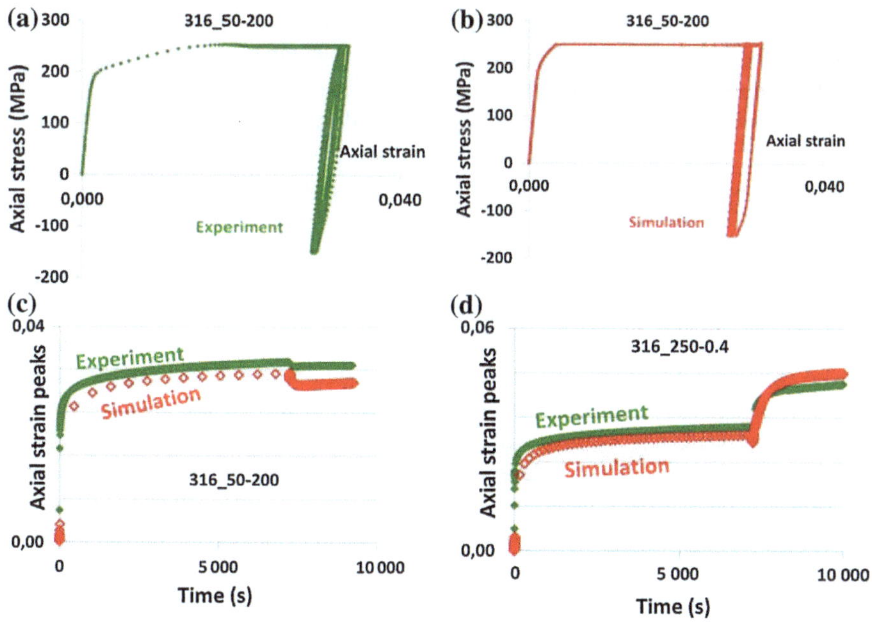

Fig. 19.13 Simulations of the tests performed on 316L SS under stress control. Axial stress-strain diagrams for the test 316L_50-200 composed of a creep sequence with $\sigma_{zz} = 250$ MPa up to $t = 7200$ s followed by 100 cycles between 250 and -150 MPa: **a** experiment; **b** simulation. Axial strain peaks versus time, comparison between experiments and their simulations: **c** test 316L_50-200; **d** test 316L_250-0.4 composed of creep sequence with $\sigma_{zz} = 250$ MPa up to $t = 7200$ s followed by 100 cycles "triangle"

(Fig. 19.13a, c) which are well predicted by the model (Fig. 19.13b, c). For the tests performed under non-proportional loading path, one can notice a clear cyclic accumulation of the axial inelastic strain after the creep sequence which is also very well predicted by the model.

19.4 Discussion

The predictions of the MM model appear as qualitatively satisfactory in general. This quality is associated with relatively reduced number of parameters and simplified experimental identification. Indeed, as advocated above (see Table 19.9), the MM model has several capabilities in the prediction of the cyclic inelastic behavior of materials; in its full version, twenty five material parameters are needed for isothermal conditions. This number of parameters may appear large. Nevertheless, considering current industrial applications, it is rather scarce to find a metal which exhibits all phenomena listed in Table 19.9 and, therefore the number of "active" parameters is generally smaller. Based on the present study, Table 19.10 summarizes for each material the significant phenomena and associated material parameters. One can remark that if we consider for instance 35NCD16 steel, only eight parameters were necessary for the simulations. This reduced number is due to the fact that, in the test conditions considered here, this material may be characterized by: rate-independent behavior, isotropy, kinematic hardening without dynamic recovery term, no memory of the strain hardening and finally, no extra-hardening due to non-proportional loading paths. Despite this relatively small number of parameters, the predictions are very satisfactory for proportional as well as non-proportional loading paths used in the present study. On the other hand, 316L and XC18 steels were the two materials for which the maximum number of parameters was necessary for the simulations. However, this maximum remains relatively reasonable as it is equal to 13.

Besides the number of material parameters, their experimental calibration may also be an essential factor in the choice of the model. Note that here for all considered materials under the considered conditions, only classical tests performed under strain control were necessary for the identification of the MM model: monotonic tensile curve, cyclic behavior under tension-compression, cyclic behavior under non-proportional loading path. Note that no test under cyclic stress control is necessary for the identification. Of course, this property is valid for the test conditions considered in the present study; its generalization needs further works. Another specificity of the MM model is related to the dynamic recovery term in the kinematic hardening. Indeed, this term was activated only one time (case of XC18 steel) in presence of significant ratcheting. The activation of this term is not always necessary as the description of ratcheting by the MM model is possible without the presence of this term. Indeed, a transient uniaxial ratcheting may be described by the model with a rate linked to the difference $C_{11}C_{22} - C_{12}^2$ as illustrated in Taleb et al. (2014).

Among the weaknesses of the model, one can notice the relatively bad prediction in the case of 2017A alloy, an improvement of the model in order to take into account

Table 19.10 Phenomena exhibited by each material and related parameters

		35NCD16	2017A	CuZn27	XC18	316L
Viscosity (K, n)		No	No	No	No	Yes
Anisotropy about the yield stress $(L_a, L_b, \ldots, L_f)$		No	Yes	No	No	No
Kinematic hardening (6)	Linear term (C_{11}, C_{22})	Yes	Yes	Yes	Yes	Yes
	Dynamic recovery term (D_1, D_2, η)	No	No	No	Yes	No
	Coupling between both mechanisms (C_{12})	Yes (low)	Yes (high)	Yes (low)	No	Yes (high)
Isotropic hardening (11)	Initial yield stress (R_0)	Yes	Yes	Yes	Yes	Yes
	"Classical" term (Q_2, b_2)	Yes	Yes	Yes	Yes	No
	Memory of the strain hardening $(Q_1 : Q_{1_0}, Q_{1_m}, b_1, \eta_{\text{mem}}, \mu)$	No Only Q_{1_0}, b_1	No	No Only Q_{1_0}, b_1	No Only Q_{1_0}, b_1	Yes with $Q_{1_0} = 0$
	Extra hardening under non-proportional loading path $(Q_3 : Q_{3_m}, b_3, d_3)$	No	Yes	Yes	Yes	Yes
Total		8	10	11	13	13

the anisotropy linked to the cyclic behavior in addition to the yield stress is necessary. Furthermore, the model deserves improvement about the simulation of the materials like CuZn27 under non-proportional loading paths.

19.5 Concluding Remarks

The comparison between the test results and their predictions by the MM model enables to make the following concluding remarks:

- The predictions of the MM model are satisfactory for the ferritic steels 35NCD16 and XC18 under both proportional and non-proportional loading paths considered here.
- For the extruded (anisotropic) alloy 2017A, the model overestimates the cyclic accumulation of the inelastic strain during the first cycles in general but it stabilizes quickly producing very small stain rate in accordance with experiments.
- For CuZn27 alloy, the predictions are satisfactory in tension-compression contrarily to the non-proportional case where the cyclic accumulation of the inelastic strain is overestimated.
- For 316L SS in the considered test conditions, no significant cyclic accumulation of the inelastic strain is observed in tension-compression after creep sequence; this observation is very well predicted by the model. The simulations are also satisfactory for the test with non-proportional loading path.

Among the strengths of the model, one can put forward its large capabilities associated with relatively small number of parameters and simplified experimental identification. Among the weaknesses of the model, we notice the need to improve its capabilities in the prediction of the anisotropic cyclic behavior. Of course these concluding remarks are valid in the conditions of loading and environment adopted in the present study. Further works are necessary in order to make these properties more general for the MM model.

Acknowledgments The authors acknowledge the "Région Haute Normandie" and the European Community, through the FEDER program for their grateful financial support.

References

Abdel-Karim M (2009) Modified kinematic hardening rules for simulations of ratchetting. Int J Plast 25:1560–1587

Abdel-Karim M (2010) An evaluation for several kinematic hardening rules on prediction of multiaxial stress-controlled ratchetting. Int J Plast 26:711–730

Abdel-Karim M (2011) Effect of elastic modulus variation during plastic deformation on uniaxial and multiaxial ratchetting simulations. Eur J Mech A/Solids 30:11–21

Abdel-Karim M, Ohno N (2000) Kinematic hardening model suitable for ratcheting with steady-state. Int J Plast 16:225–240

Armstrong PJ, Frederick CO (1966) A mathematical representation of the multiaxial Bauschinger effect. CEGB Report RD/B/N 731, Berkely Nuclear Laboratories, Berkely, UK

Bari S, Hassan T (2000) Anatomy of coupled constitutive models for ratchetting simulation. Int J Plast 16:381–409

Bari S, Hassan T (2001) Kinematic hardening rules in uncoupled modeling for multiaxial ratchetting simulation. Int J Plast 17:885–905

Bari S, Hassan T (2002) An advancement in cyclic plasticity modeling for multiaxial ratcheting simulation. Int J Plast 18:873–894

Belattar A, Taleb L, Hauet A, Taheri S (2012) Dependence of the cyclic stress-strain curve on the loading history and interaction with the fatigue of the 304L stainless steel. Mater Sci Eng A 563:170–180

Benallal A, Marquis D (1987) Constitutive equations for nonproportional cyclic elasto-viscoplasticity. Trans ASME J Eng Mater Tech 109:326–336

Besson J, Leriche R, Foerch R, Cailletaud G (1998) Object-oriented programming applied to the finite element method. Part II. Application to material behaviors. Revue Européenne des Éléments Finis 7(5):567–588

Burlet H, Cailletaud G (1987) Modeling of cyclic plasticity in finite element codes. In: Desai CS (ed) 2nd International conference on constitutive laws for engineering materials: theory and applications. Elsevier, Tuscon, pp 1157–1164

Cailletaud G, Saï K (1995) Study of plastic/viscoplastic models with various inelastic mechanisms. Int J Plast 11:991–1005

Chaboche JL (1986) Time-independent constitutive theories for cyclic plasticity. Int J Plast 2:149–188

Chaboche JL (1989) Constitutive equations for cyclic plasticity and cyclic viscoplasticity. Int J Plast 5:247–302

Chaboche JL (1991) On some modifications of kinematic hardening to improve the description of ratcheting effects. Int J Plast 7:661–678

Chaboche JL (1994) Modeling of ratcheting: evaluation of various approaches. Eur J Mech A/Solids 13(4):501–518

Chaboche JL (2008) A review of some plasticity and viscoplasticity constitutive theories. Int J Plast 24:1642–1693

Chaboche JL, Dang-Van K, Cordier G (1979) Modelization of the strain memory effect on the cyclic hardening of 316 stainless steel. In: Proceedings of the 5th International conference on SMiRT, Berlin, Paper L 11/3

Chaboche JL, Kanouté P, Azzouz F (2012) Cyclic inelastic constitutive equations and their impact in the fatigue life predictions. Int J Plast 35:44–66

Chen X, Jiao R (2004) Modified kinematic hardening rule for multiaxial ratchetting prediction. Int J Plast 20:871–898

Chen X, Kim KS (2003) Modeling of ratcheting behavior under multiaxial cyclic loading. Acta Mech 163:9–23

Chen X, Jiao R, Kim KS (2003) Simulation of ratcheting strain to a high number of cycles under biaxial loading. Int J Solids Struct 40:7449–7461

Contesti E, Cailletaud G (1989) Description of creep-plasticity interaction with non-unified constitutive equations: application to an austenitic stainless steel. Nucl Eng Des 116(3):265–280

Dafalias YF, Feigenbaum HP (2011) Biaxial ratchetting with novel variations of kinematic hardening. Int J Plast 27:479–491

Feigenbaum H, Dugdale J, Dafalias YF, Kourousis K, Plesek J (2012) Multiaxial ratcheting with advanced kinematic and directional distortional hardening rules. Int J Sol Struct 49:3063–3076

Hassan T, Taleb L, Krishna S (2008) Influences of non proportional loading paths on ratcheting responses and simulations by two recent cyclic plasticity models. Int J Plast 24:1863–1889

Jiang Y, Zhang J (2008) Benchmark experiments and characteristic cyclic plasticity deformation. Int J Plast 24:1481–1515
Kang GZ (2008) Ratchetting: recent progresses in phenomenon observation, constitutive modeling and application. Int J Fatigue 30:1448–1472
Kang GZ, Gao Q (2002) Uniaxial and non-proportionally multiaxial ratcheting of U71Mn rail steel: experiments and simulations. Mech Mater 34:809–820
Kang GZ, Gao Q (2004) Temperature-dependent cyclic deformation of S304 stainless steel under non-proportionally multiaxial load and its constitutive modeling. Key Eng Mater 274:247–252
Kang GZ, Kan Q (2007) Constitutive modeling for uniaxial time-dependent ratcheting of SS304 stainless steel. Mech Mater 39:488–497
Kang GZ, Gao Q, Yang XJ (2002) A visco-plastic constitutive model incorporated with cyclic hardening for uniaxial/multiaxial ratchetting of SS304 stainless steel at room temperature. Mech Mater 34:521–531
Kang GZ, Gao Q, Yang XJ (2004) Uniaxial and nonproportional multiaxial ratchetting of SS304 stainless steel at room temperature: experiments and simulations. Int J Non-linear Mech 39:843–857
Kang GZ, Li YG, Gao Q (2005) Non-proportionally multiaxial ratcheting of cyclic hardening materials at elevated temperatures and its constitutive modeling. Mech Mater 37:1101–1118
Kobayashi M, Ohno N (2002) Implementation of cyclic plasticity models based on a general form of kinematic hardening. Int J Numer Meth Eng 53:2217–2238
Krishna S, Hassan T, Ben Naceur I, Saï K, Cailletaud G (2009) Macro versus micro-scale constitutive models in simulating proportional and non proportional cyclic and ratcheting responses of stainless steel 304. Int J Plast 25:1910–1949
May A, Taleb L, Belouchrani M (2013) Analysis of the cyclic behavior and fatigue damage of extruded 2017 aluminum alloy. Mater Sci Eng A 571:123–136
Murakami S, Kawai M, Ohmi YJ (1989) Effects of amplitude-history and temperature-history on multiaxial cyclic behavior of type 316 stainless steel. Trans ASME J Eng Mater Tech 111:278–285
Nouailhas D, Cailletaud G, Policella H, Marquis D, Dufailly J, Lieurade HP, Ribes A, Bollinger E (1985) On the description of cyclic hardening and initial cold working. Eng Fract Mech 21:887–895
Ohno N (1982) A constitutive model of cyclic plasticity with a nonhardening strain region. Trans ASME J Appl Mech 49:721–727
Ohno N (1990) Recent topics in constitutive modeling of cyclic plasticity and viscoplasticity. Appl Mech Rev 43:283–295
Ohno N, Abdel-Karim M (2000) Uniaxial ratcheting of 316FR steel at room temperature. Part II: Constitutive modeling and simulation. Trans ASME J Eng Mater Tech 122:35–41
Ohno N, Wang JD (1991) Transformation of a nonlinear kinematic hardening rule to a multisurface form under isothermal and nonisothermal conditions. Int J Plast 7:879–891
Ohno N, Wang JD (1993) Kinematic hardening rules with critical state of dynamic recovery. Part I: Formulations and basic features for ratcheting behavior. Int J Plast 9:375–403
Ohno N, Wang JD (1994) Kinematic hardening rules for simulation of ratchetting behavior. Eur J Mech A/Solids 13:519–531
Portier L, Calloch S, Marquis D, Geyer P (2000) Ratchetting under tension-torsion loadings: experiments and modelling. Int J Plast 16:303–335
Pugh CE (1978) On establishing constitutive equations for use in design of high temperature fast-reactor structure. Nucl Eng Des 51(1):23–27
Saï K (2011) Multi-mechanism models: present state and future trends. Int J Plast 27:250–281
Saï K, Cailletaud G (2007) Multi-mechanism models for the description of ratchetting: effect of the scale transition rule and of the coupling between hardening variables. Int J Plast 23:1589–1617
Saï K, Taleb L, Cailletaud G (2012) Numerical simulation of an anisotropic behavior of 2017 aluminum alloy. Comput Mater Sci 65:48–57
Saï K, Taleb L, Guesmi F, Cailletaud G (2014) Multi-mechanism modeling of proportional and non-proportional ratchetting of stainless steel 304. Acta Mech 225:3265–3283

Taheri S, Hauet A, Taleb L, Kpodekon C (2011) Micro-macro investigations about the fatigue behaviour of pre-hardened 304L steel. Int J Plast 27:1981–2004
Taleb L (2013a) About the cyclic accumulation of the inelastic strain observed in metals subjected to cyclic stress control. Int J Plast 43:1–19
Taleb L (2013b) On the cyclic behavior of 304L and 316L stainless steels. Key Eng Mater 535–536:201–204
Taleb L, Cailletaud G (2010) An updated version of the multimechanism model for cyclic plasticity. Int J Plast 26:859–874
Taleb L, Cailletaud G (2011) Cyclic accumulation of the inelastic strain in the 304L SS under stress control at room temperature: ratcheting or creep? Int J Plast 27:1936–1958
Taleb L, Cailletaud G, Blaj L (2006) Numerical simulation of complex ratcheting tests with a multi-mechanism model type. Int J Plast 22:724–753
Taleb L, Cailletaud G, Saï K (2014) Experimental and numerical analysis about the cyclic behavior of the 304L and 316L stainless steels at 350°c. Int J Plast 61:32–48
Vincent L, Calloch S, Kurtyka T, Marquis D (2002) An improvement of multiaxial ratcheting modeling via yield surface distortion. J Eng Mater Technol 124:402–411
Vincent L, Calloch S, Marquis D (2004) A general cyclic plasticity model taking into account yield surface distortion for multiaxial ratcheting. Int J Plast 20:1817–1850
Wolff M, Taleb L (2008) Consistency of two multi-mechanism models in isothermal plasticity. Int J Plast 24:2059–2083
Yaguchi M, Takahashi Y (2005) Ratchetting of viscoplastic material with cyclic softening: II. Application of constitutive models. Int J Plast 21:835–860
Yoshida F (2000) A constitutive model of cyclic plasticity. Int J Plast 16:359–380
Yoshida F, Uemori T (2002) A model for large strain cyclic plasticity describing the bauschinger effect and work hardening stagnation. Int J Plast 18:661–686
Yu D, Chen X, Yu W, Chen G (2012) Thermo-viscoplastic modeling incorporating dyamic strain aging effect on the uniaxial behavior of Z2CND18.12N stainless steel. Int J Plast 37:119–139

Chapter 20
Phase-Field Modeling for Dynamic Recrystallization

Tomohiro Takaki, Akinori Yamanaka and Yoshihiro Tomita

Abstract Hot working is a process in which metallic materials are worked at the elevated temperatures above the recrystallization temperature. During the hot working of the low-to-medium stacking fault energy metals, the dynamic recrystallization (DRX) occurs. The mechanical properties of the DRX materials during the hot working are largely affected by the nucleation and growth of the dynamic recrystallized grains. In this article, the application of a phase-field method, which has emerged as a powerful numerical tool to simulate the material microstructure evolutions, to the simulations of the deformation and microstructure during the DRX is reviewed. First, the multi-phase-field dynamic recrystallization (MPF-DRX) model, which can simulate the mechanical behaviors of a computational domain based on the DRX microstructure evolutions simulated by the multi-phase-field (MPF) method, is introduced. Next, a hot-working multi-scale model, where the macro deformation is simulated by the finite element (FE) method and the microstructure evolution is simulated by the MPF-DRX method, is presented.

Keywords Phase-field method · Dynamic recrystallization · Hot working · Multiscale modeling · Finite element method

---

T. Takaki (✉)
Faculty of Mechanical Engineering, Kyoto Institute of Technology, Matsugasaki, Sakyo, Kyoto 606-8585, Japan
e-mail: takaki@kit.ac.jp

A. Yamanaka
Mechanical Systems Engineering, Tokyo University of Agriculture and Technology, Naka-cho,Koganei, Tokyo184-8588, Japan
e-mail: a-yamana@cc.tuat.ac.jp

Y. Tomita
Kobe University, Kobe, Japan
e-mail: tomita@kobe-u.com

H. Altenbach et al. (eds.), *From Creep Damage Mechanics to Homogenization Methods*, Advanced Structured Materials 64,
DOI 10.1007/978-3-319-19440-0_20

20.1 Introduction

Hot working is a process in which metallic materials are worked at elevated temperatures above than recrystallization temperature, which is almost half of the melting temperature. During the hot working of low-to-medium stacking fault energy metals, such as copper and austenitic iron, the dynamic recrystallization (DRX) occurs (Doherty et al. 2004; Humphreys and Hatherly 1995; Sakai et al. 2014; Sakai and Jonas 1984). The DRX treated here is a conventional discontinuous dynamic recrystallization (Sakai and Jonas 1984) where the recrystallized grains nucleate and grow; however, there is also a continuous dynamic recrystallization where the dislocation subgrain structures gradually transform into grains with high-angle boundaries at heavy deformation (Sakai et al. 2014). The DRX is similar to the primary static recrystallization (SRX) with nucleation and following grain growth during post-deformation annealing (Humphreys and Hatherly 1995). The main difference is that the DRX occurs during straining in continuous deformation. Therefore, the dislocation densities change not only in the deformation grains, but also in the recrystallized grains. Figure 20.1 schematically illustrates the stress-strain curves during the hot deformation. For materials without DRX, the stress increases monotonically and becomes a steady-state flow stress (Fig. 20.1a). This type of curve can be seen in high stacking fault energy materials, such as aluminum, magnesium, and ferritic iron, where the dislocation density cannot reach the critical value at which the recrystallization occurs, as the dislocations can move easily or dynamic recovery (DRV) is the main restoration process. On the other hand, as shown in Fig. 20.1b, the DRX materials show characteristic stress-strain curves depending on the strain rate, temperature, and initial grain size (Blaz et al. 1983; Sakai and Jonas 1984). When the dislocation density reaches the critical value to initiate the recrystallization, the recrystallized grains are nucleated and begin to grow. As the dislocation density inside the recrystallized grains is low compared to that in the deformed grains, the work hardening gradually decreases, and the stress reaches the maximum peak stress where the work

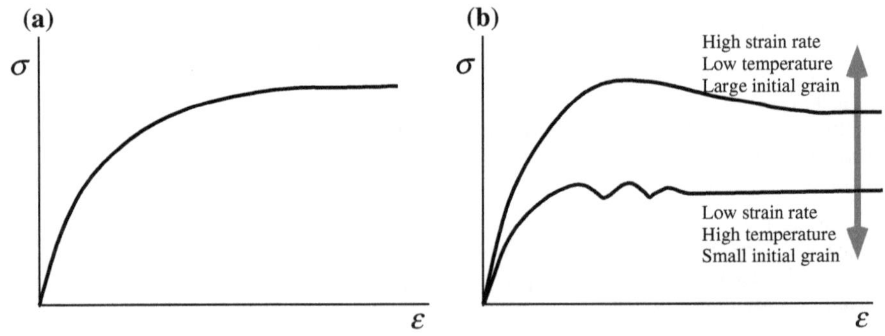

Fig. 20.1 Stress-strain curves for **a** Dynamic recovery (DRV) and **b** Dynamic recrystallization (DRX) materials during hot deformation

hardening and softening balance each other. The maximum peak stress is higher at higher strain rate, lower temperature, and larger initial grain size. After the maximum stress point, the work softening can be observed. Here, for a high strain rate, low temperature, and large initial grain size, a single peak curve can be observed. On the other hand, for a low strain rate, high temperature, and small initial grain size, a multiple peak curve is present. These mechanical properties are dominated by the microstructure evolution, or the nucleation and growth of recrystallized grains.

Finite element (FE) simulations for hot working, such as hot rolling, fogging etc., have been widely conducted to predict the thermomechanical behavior during the process (Dawson 1984; Galantucci and Tricarico 1999; Li et al. 2001; Montmitonnet 2006; Oh 1982; Oh et al. 1991; Song et al. 2002; Yamada et al. 1991). In those conventional simulations, constitutive equations depending on the strain, strain rate, or temperature have been used (Anand 1985; Anand et al. 1990; Gronostajski 2000). To predict the microstructure and introduce the effects of the microstructure evolution on the mechanical behavior into the simulations, constitutive equations introducing the information of the volume fraction of the recrystallization, average grain size, etc., as internal state variables have been presented (Adebanjo and Miller 1989; Brown et al. 1989; Busso 1998; Cho et al. 2005; Davenport et al. 2000; Dunne et al. 1997; Fan and Yang 2011; Yanagimoto et al. 1998; Yeom et al. 2005). Recently, detailed constitutive studies have been conducted (Brown and Bammann 2012; McQueen and Ryan 2002; Momeni et al. 2012), confirming that such models are in very good agreement with the experimentally observed microstructure (Li et al. 2015; Qu et al. 2005; Quan et al. 2015; Yanagimoto et al. 2014; Yin et al. 2014). As shown here, finite element simulations have been successfully used to evaluate the macroscopic mechanical behaviors during hot working. On the other hand, it might be difficult to develop accurately the constitutive equations for the complicated thermal and working histories. In addition, we cannot obtain the microstructures directly from those simulations.

Several microstructure evolution models are available, such as vertex (Anderson et al. 1984), Monte Carlo (MC) (Hassold et al. 1990; Holm and Battaile 2001), cellular automaton (CA) (Wolfram 1983), level-set (Bernacki et al. 2011), and phase-field (PF) (Chen 2014b) methods. For the DRX modeling, the CA method is the most widely used. Goetz and Seetharaman (1998) related the DRX microstructure evolutions computed by CA with the mechanical behavior of the system. Ding and Guo (1984, 2002, 2004); Qian and Guo (1993) developed the cellular automaton dynamic recrystallization (CA-DRX) model, in which the DRX microstructure evolutions are computed by the CA model, the evolution of the dislocation density is modeled using the Kocks-Mecking model (Kocks 1976; Mecking and Kocks 1981), and the average stress of the system is computed by the Bailey-Hirsch equation (Bailey and Hirsch 1960). Although the CA-DRX model does not solve the inhomogeneous material deformation, it has succeeded to express the characteristic properties observed in the DRX, such as the single and multiple peaks stress-strain curves, as shown in Fig. 20.1b, depending on the microstructure evolutions. The CA-DRX model is applied to various materials and phenomena with DRX (Goetz 2005; Hallberg et al. 2010; Jin and Cui 2010; Kugler and Turk 2004; Madison et al. 2012; Svyetlichnyy

2010; Xiao et al. 2008; Yazdipour et al. 2008; Zheng et al. 2009, 2008). The MC (Peczak 1995; Peczak and Luton 1993; Rollett et al. 1992) and PF methods (Takaki et al. 2008, 2009) are also used for the DRX modeling in a similar framework to that of the CA-DRX model. However, the number of published papers is considerably lower compared to that of the CA-DRX model. Presumably, in the CA method, the operation to express the grain boundary migration is easier and more flexible than that required in the MC and PF methods. However, the CA method presents some difficulties in expressing accurately the curvature effect.

In the DRX model, it is essential to solve the inhomogeneous deformations and the microstructures simultaneously. Although some models (Abrivard et al. 2012; Chuan et al. 2013; Li et al. 2013) are presented, those are thought to be not practical in the present due to the high computational costs and the insufficient in the coupling of deformation and microstructure. Lee et al. computed the microstructures that distribute nonuniformly in the hot-worked specimen by using the CA and FE methods (Lee and Im 2010; Lee et al. 2014). This study is promising to relate the macro mechanical behaviors with the microstructures. On the other hand, the method is a one-way coupling from the FE method to the CA method. The FE simulations for the macro field need a constitutive equation.

The PF method has emerged as the most powerful numerical model to simulate the material microstructure evolutions. There are many good review papers on the PF method (Asta et al. 2009; Chen 2014a, b; Moelans et al. 2008; Singer-Loginova and Singer 2008; Steinbach 2009; Takaki 2014). The PF method introduces an order parameter, or phase-field variable, to discriminate the different conditions such as phases and grains. The microstructure evolutions are expressed by solving the time evolution equations of the phase field. To express multi phases and multi grains, the multi-phase-field (MPF) model is typically adopted (Fan and Chen 1997; Steinbach and Pezzolla 1999), and the multiple phase-field variables are employed to express the multiple different conditions. For the normal grain growth and the problems of the constant chemical driving force, the PF and MPF methods give accurate microstructure evolutions by taking into account the curvature effects.

We have developed a multi-phase-field dynamic recrystallization (MPF-DRX) model (Takaki et al. 2008, 2009), where the MPF model (Steinbach and Pezzolla 1999) is used instead of the CA method utilized in the CA-DRX model. Compared to the CA-DRX model, the DRX microstructure evolutions can be simulated more accurately by using the MPF-DRX model (Takaki et al. 2008). Recently, this model was applied to the deformation-induced ferrite transformation (Yamanaka and Takaki 2014). For the problems concerning the non-uniform macroscopic deformation, the multi-phase-field finite element dynamic recrystallization (MPFFE-DRX) model was developed (Takaki 2014; Yoshimoto and Takaki 2014), after we confirmed that the MPF-DRX model can be applied to the transient deformations (Takaki et al. 2011). In the MPFFE-DRX model, the macro deformations and the DRX microstructures are computed by the conventional FE method and the MPF-DRX model, respectively, and those two models are coupled dynamically. In the remaining of this article, the MPF-DRX and MPFFE-DRX models are reviewed with some computational examples.

20.2 MPF-DRX Model

To express the DRX phenomenon, the deformation and the microstructure change should be considered simultaneously. In the MPF-DRX model, however, the deformation is not solved directly. Instead, the time evolution of the dislocation density, due to the continuous deformation, is expressed by simple equations. The DRX grain growth is expressed by the MPF model. Here, for the nucleation of the DRX grain, a critical dislocation density necessary to initiate the DRX grain and the nucleation rate model are employed.

20.2.1 *Deformation*

The recrystallized grains grow by eliminating the dislocations introduced during the plastic deformation. The driving force of the recrystallized grain growth is the energy of the dislocation $\tau \approx 0.5Gb^2$, where G is the shear modulus and b is the amount of the Burgers vector. Considering the dislocation density ρ, the stored energy f_{stor} can be expressed as $f_{\text{stor}} \approx 0.5\rho Gb^2$ (Humphreys and Hatherly 1995). In the case of the static recrystallization (SRX), the driving force for the grain boundary (GB) migration can be approximated by f_{stor}, as the dislocation density in the recrystallized grain is kept at a very low value. On the other hand, in the case of the DRX, the dislocation density in the recrystallized grain increases with the continuous deformation. Figure 20.2a shows the schematic illustration of the dislocation density profile near the GB between the deformed grain (DG) and the circular recrystallized grain (RG) with radius R, which migrates with velocity V_{GB} (Roberts and Ahlblom 1978). In the DG, the dislocation density DG is high and its increasing rate is relatively slow. In the RG, the dislocation density is very low just behind the GB, ρ_0, it increases with approaching to the center of the RG, and reaches ρ_{RG} at the center of the circular RG. Owing to the recovery effect, the rate of the dislocation accumulation decreases in the high dislocation density region, as shown by the blue arrows in Fig. 20.2a. Therefore, the GB migration rate V_{GB} decreases with the deformation because the driving force for the GB migration is the difference of stored energy between the RG and the DG.

The time evolution of the dislocation density is modeled by the Kocks-Mecking model (Kocks 1976; Mecking and Kocks 1981)

$$\frac{d\rho}{d\varepsilon} = k_1\sqrt{\rho} - k_2\rho, \tag{20.1}$$

where ε is the true strain, and k_1 and k_2 are the coefficients. The first term on the right hand side of Eq. (20.1) is the dislocation accumulation term, whereas the second term is the recovery term. These terms contribute to the hardening and softening, respectively. The true stress σ is related to the dislocation density ρ by the Bailey-Hirsch equation (Bailey and Hirsch 1960)

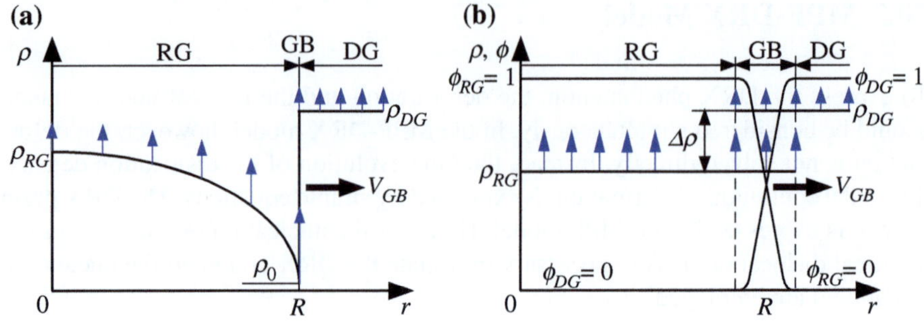

Fig. 20.2 Image of the dislocation density profiles (*solid lines*) and the rate of the dislocation density (*blue arrows*) around the grain boundary (GB) between the deformed grain (DG) and the recrystallized grain (RG) with radius R. **a** Actual material and **b** The MPF-DRX model

$$\sigma = \alpha G b \sqrt{\rho}, \tag{20.2}$$

where α is a constant of the order of 0.1. From Eqs. (20.1) and (20.2), the following equation can be derived:

$$\frac{d\sigma}{d\varepsilon} = \frac{1}{2}\alpha G b k_1 \left(1 - \frac{1}{\alpha G b}\frac{k_2}{k_1}\sigma\right) \tag{20.3}$$

k_1 and k_2 can be obtained by setting $\sigma = 0$ and $d\sigma/d\varepsilon = 0$, respectively, as

$$k_1 = \frac{2}{\alpha G b}\left.\frac{d\sigma}{d\varepsilon}\right|_{\sigma=0} \quad \text{and} \quad k_2 = \frac{\alpha G b}{\sigma_s}k_1 = \frac{2}{\sigma_s}\left.\frac{d\sigma}{d\varepsilon}\right|_{\sigma=0}, \tag{20.4}$$

where σ_s is the steady state stress. From Eq. (20.4), it can be seen that $k_1 = k_1(T)$ and $k_2 = k_2(T, \dot{\varepsilon})$. Figure 20.3 shows a schematic illustration of the stress-strain curves for the actual DRX material and as calculated from Eq. (20.3). The steady state stress σ_s should be determined by the uniaxial tests. Ding and Guo (1984) used the following equation to determine σ_s (Humphreys and Hatherly 1995):

$$\dot{\varepsilon} = a_1 \sigma_s^{a_2} \exp\left(-\frac{Q_{\text{act}}}{RT}\right) \tag{20.5}$$

where Q_{act} is the activation energy, and a_1 and a_2 are the constants determined by the uniaxial tests. As shown in Fig. 20.3, Eq. (20.3) depicts the stress-strain curve by the accumulation and recovery of dislocations, as for the DRV material shown in Fig. 20.1a. To express the softening caused by the nucleation and growth of the recrystallized grains, the DRX microstructures need to be computed.

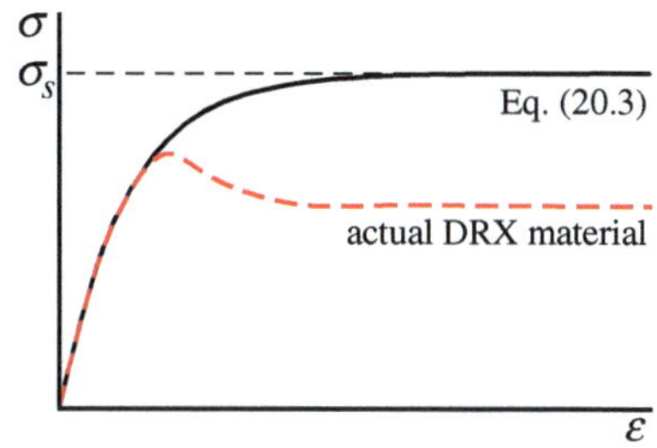

Fig. 20.3 Stress-strain curves of the DRX material and of the DRV material expressed by Eq. (20.3)

20.2.2 DRX Grain Growth

As the dislocation density in the DRX grains is lower than that in the deformed grains, the material becomes soft by growing the DRX grains. The discrepancy between Eq. (20.3) and the actual material, shown in Fig. 20.3, is expressed by the nucleation and the growth of the DRX grains. In this model, the microstructure evolutions, or the DRX grain growth, are expressed by the MPF model. We adopted the Steinbach's MPF model (Steinbach and Pezzolla 1999). The MPF model does not employ the normal double-well potential, but the double-obstacle potential. Therefore, the equivalent profile of the phase-field variable becomes a sine function, and the interface region is strictly determined. Thus, by applying the active parameter tracking algorithm (Kim et al. 2006; Ohno et al. 2011; Suwa et al. 2007), the polycrystal grain growth can be efficiently and accurately simulated.

To indicate the multiple grains, multiple phase-field variables ϕ_i are used. As shown in Fig. 20.4, ϕ_i takes the value of 1 in the ith grain and of 0 in the other grains, and it changes smoothly from 1 to 0 across the GB. The time evolution of ϕ_i is expressed by solving the following phase-field equation

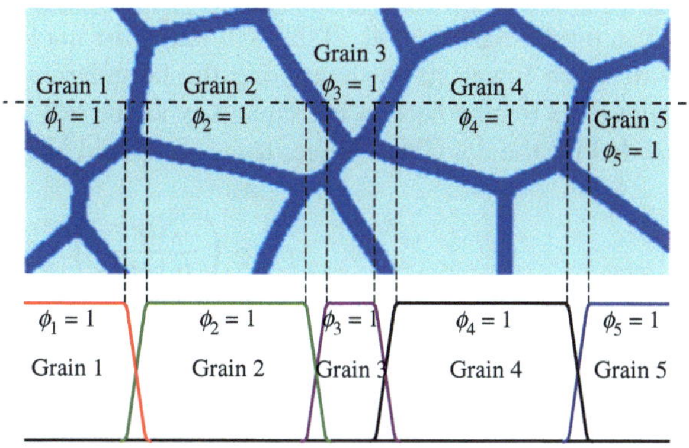

Fig. 20.4 Definition of phase-field variables, ϕ_i, in the multi-phase-field (MPF) model to express the polycrystal. The *light blue* regions are in the grains ($\phi_i = 1$) and the *thick blue* regions are the grain boundary regions ($0 < \phi_i < 1$)

$$\frac{\partial \phi_i}{\partial t} = \sum_{j=1}^{n} \frac{2M_{ij}^{\phi}}{n} \left[\sum_{k=1}^{n} \left\{ (W_{ik} - W_{jk})\phi_k + \frac{1}{2}(a_{ik}^2 - a_{jk}^2)\nabla^2 \phi_k \right\} - \frac{8}{\pi}\sqrt{\phi_i \phi_j}\Delta f_{ij} \right], \tag{20.6}$$

where n is the number of the phase-field variables having non-zero value at a space point; a_{ij}, W_{ij}, and M_{ij}^{ϕ} are the gradient coefficient, height of double-obstacle potential, and phase-field mobility, respectively, between the neighboring two grains expressed by ϕ_i and ϕ_j. These parameters are related by the following equations to the GB thickness δ_{ij}, GB energy γ_{ij}, and GB mobility M_{ij}.

$$a_{ij} = \frac{2}{\pi}\sqrt{2\delta_{ij}\gamma_{ij}}, \quad W_{ij} = \frac{4\gamma_{ij}}{\delta_{ij}} \quad \text{and} \quad M_{ij}^{\phi} = \frac{\pi^2}{8\delta_{ij}} M_{ij} \tag{20.7}$$

The Δf_{ij} is the driving force of the GB between the ith and the jth grains, and is the stored energy difference. Although the dislocation density in the actual grain is inhomogeneous, as shown in Fig. 20.2a, we assumed for simplicity a constant dislocation density in a grain, as shown in Fig. 20.2b. Accordingly, Δf_{ij} can be expressed as

$$\Delta f_{ij} = \frac{1}{2} G b^2 (\rho_i - \rho_j), \tag{20.8}$$

for the dislocation density ρ_i and ρ_j in the ith and jth grains, respectively.

20.2.3 DRX Grain Nucleation

In the conventional DRX, the bulging of the pre-existing GB by deformation is the main nucleation mechanism (Miura et al. 1994, 2004, 2007; Sakai et al. 2014; Wusatowska-Sarnek et al. 2002). Although the nucleation occurs at triple junctions with much lower strain than along the boundaries (Miura et al. 2005), the triple junction is treated here in the same way as the GB. The critical dislocation density needed for the flat GB to bulge, ρ_c, is computed as (Roberts and Ahlblom 1978)

$$\rho_c = \left(\frac{20\gamma\dot{\varepsilon}}{3blM\tau^2} \right)^{1/3}, \tag{20.9}$$

where γ is the GB energy, M is the GB mobility, and l is the mean free path of the dislocation. This equation is derived based on the classical nucleation theory assuming very ideal conditions. Accordingly, it is needed to model and identify the nucleation rate to fill up the gap shown in Fig. 20.3. Ding and Guo (1984) used the following equation for the nucleation rate:

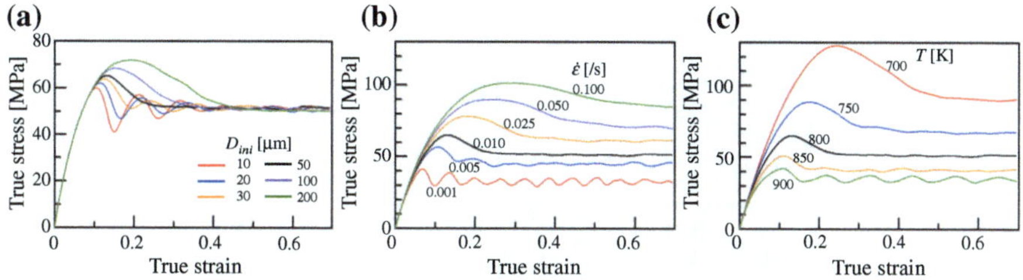

Fig. 20.5 Stress-strain curves of the system when changing the **a** initial average grain size, **b** strain rate, and **c** temperature in uniaxial loading

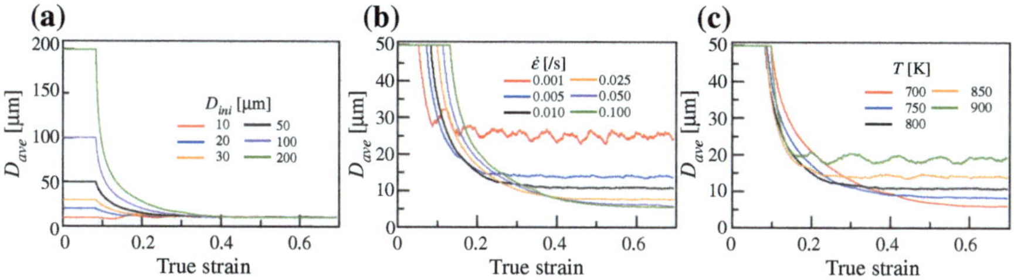

Fig. 20.6 Variations of the average grain size when changing the **a** initial average grain size, **b** strain rate, and **c** temperature in uniaxial loading

$$\dot{n} = C\dot{\varepsilon}^m \exp\left(-\frac{Q_{act}}{RT}\right), \tag{20.10}$$

where C and m are constants that are determined by comparison with the experiments.

20.2.4 Simulations

Some uniaxial simulations are shown here. Figures 20.5 and 20.6 show the stress-strain curves of the system and the changes of the average grain size, D_{ave}, when changing the initial grain size, D_{ini}, the strain rate, $\dot{\varepsilon}$, and the temperature, T, in a standard condition of $D_{ini} = 50\,\text{mm}$, $\dot{\varepsilon} = 0.01\,/\text{s}$, and $T = 800\,\text{K}$. The computational domain is fixed to $140 \times 140\,\mu\text{m}^2$ (560×560 meshes) except for $D_{ini} = 100$ µm, which uses $280.25 \times 280.25\,\mu\text{m}^2$ (1121×1121 meshes), and $200\,\mu\text{m}$, which uses $560.5 \times 560.5\,\mu\text{m}^2$ (2242×2242 meshes). These computational domains are prepared so as to include at least 10 grains in the system. The following parameters are used in the present simulations: $Q_{act} = 275{,}000$ J/mol, $a_1 = 5 \times 10^{-45}$, $a_2 = 7.60$, $b = 0.256\,\text{nm}$, $\alpha = 0.5$, $\mu = 42.1\,\text{GPa}$, $k_1 = 4 \times 10^8\,/\text{m}$, $\delta_{ij} = \delta = 5\Delta x = 1.25\,\mu\text{m}$, $\gamma_{ij} = \gamma = 0.21$ J/m^2, $M_{ij} = M = bM_0/kT \exp(-Q_b/RT)$, $M_0 = 7.5 \times 10^{-15}$ m^3/s, $Q_b = 110{,}000$ J/mol, and $l = 6.325\,\mu\text{m}$.

From Fig. 20.5, multiple peak stress-strain curves are observed for the small D_{ini}, low $\dot{\varepsilon}$, and high T. On the other hand, the single peak curves are seen for the large

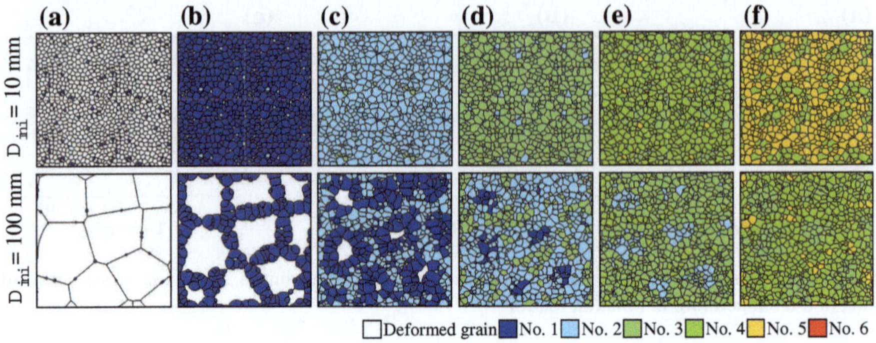

Fig. 20.7 Dynamic recrystallization (DRX) microstructure changes at the true strain ε = **a** 0.1, **b** 0.2, **c** 0.3, **d** 0.4, **e** 0.5, and **f** 0.6 for $D_{\text{ini}} = 10\,\mu\text{m}$ and $100\,\mu\text{m}$. The color indicates the DRX No

D_{ini}, high $\dot{\varepsilon}$, and low T. Besides, the steady state stresses are independent of the size of D_{ini}, whereas they increase for higher $\dot{\varepsilon}$ and lower T. From Fig. 20.6a, the average grain size, D_{ave}, reaches a constant value independent of the initial grain size. Furthermore, from Fig. 20.6b, c, the steady state sizes of D_{ave} become smaller as the strain rate increases and the temperature decreases. As shown in Figs. 20.5 and 20.6, the results computed using the MPF-DRX model accurately reveal the typical stress-strain curves and the variations of the average grain size observed in the uniaxial compression tests (Blaz et al. 1983; Sakai et al. 2014; Sakai and Jonas 1984).

Figure 20.7 shows the DRX microstructure evolutions for $D_{\text{ini}} = 10\,\mu\text{m}$ and $100\,\mu\text{m}$ at $T = 800\,\text{K}$ and $\dot{\varepsilon} = 0.01/\text{s}$. As the computation for $D_{\text{ini}} = 100\,\mu\text{m}$ ($1121\Delta x \times 1121\Delta x$) uses a four times larger domain compared to that for $D_{\text{ini}} = 10\,\mu\text{m}$ ($560\Delta x \times 560\Delta x$), four identical figures are arranged for $D_{\text{ini}} = 10\,\mu\text{m}$ to fix the visualization regions. The white grains indicate the initial deformed grains, whereas the colored grains indicate the DRX grains. The DRX grains are numbered and colored at every DRX cycle (Sakai and Jonas 1984; Takaki et al. 2009). Figure 20.8 shows the variations of the DRX fractions in every DRX cycle, where the line colors correspond to those of Fig. 20.7. The black lines in Fig. 20.8 are the stress-strain curves shown in Fig. 20.5. As shown in Fig. 20.7a, some DRX grains nucleate on the GB. Here, the number of the DRX grains is larger for $D_{\text{ini}} = 10\,\mu\text{m}$ than for $D_{\text{ini}} = 100\,\mu\text{m}$ because of the larger GB fraction. Therefore, the softening starts already at $\varepsilon = 0.1$ for $D_{\text{ini}} = 10\,\mu\text{m}$. In $\varepsilon = 0.2$ shown in Fig. 20.7b, for $D_{\text{ini}} = 10\,\mu\text{m}$, the whole region is filled by the DRX grains of the first DRX cycle, whereas, for $D_{\text{ini}} = 100\,\mu\text{m}$, the initial grains remain and are surrounded by the first DRX grains. This is very typical necklace structure. Increasing the deformation to $\varepsilon = 0.3$, as shown in Fig. 20.7c, for $D_{\text{ini}} = 10\,\mu\text{m}$, the whole region is almost completely filled with second DRX grains. On the other hand, for $D_{\text{ini}} = 100\,\mu\text{m}$, the second DRX grains are observed in the first DRX grains, but the initial deformed grains still remain. As shown in Fig. 20.8a for $D_{\text{ini}} = 10\,\mu\text{m}$, the overlaps of the

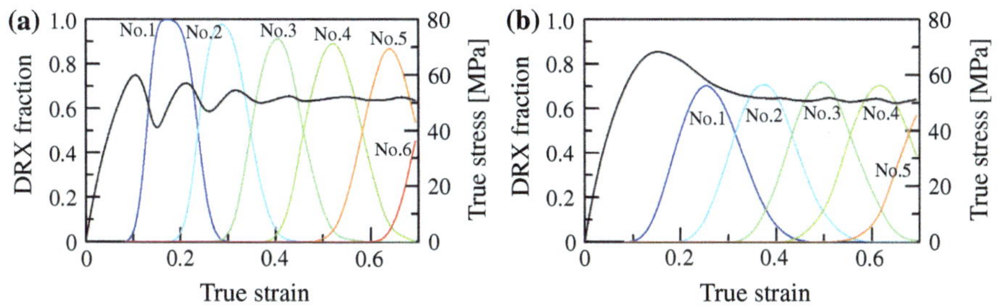

Fig. 20.8 Dynamic recrystallization (DRX) cycles for D_{ini} = **a** 10 μm and **b** 100 μm. The colors of the DRX fraction curves correspond to those used in Fig. 20.7. The black lines indicate the stress-strain curves of the system shown in Fig. 20.5

DRX cycles are relatively small and increase with the deformation. By increasing the overlap of the DRX cycle or decreasing the peak of the DRX fraction, the amplitude of the fluctuated stress-strain curve decreases. Meanwhile, in Fig. 20.8b for $D_{\text{ini}} = 100\,\mu\text{m}$, it can be observed that the overlaps and peaks of the DRX fraction curves are respectively larger and smaller than those of $D_{\text{ini}} = 10\,\mu\text{m}$. Since the DRX grains with different DRX number contribute to the softening and hardening, in a large initial grain as $D_{\text{ini}} = 100\,\mu\text{m}$, a steady state condition is achieved from smaller deformation than the case of the small initial grain due to the large overlap of the DRX cycles.

20.3 MPFFE-DRX Model

As shown in the previous section, the MPF-DRX model can express the stress-strain curves of the system based on the DRX microstructure evolutions depending on the temperature and strain rate. It has also been confirmed that the MPF-DRX model can express the transient deformations (Takaki et al. 2011), where the temperature and strain rate change during the deformation (Frommert and Gottstein 2009; Sakai et al. 1983; Sakai and Jonas 1984), although the MPF-DRX simulations shown in the previous section were performed under constant temperature and strain rate. Therefore, the stress-strain relations in the uniaxial condition depending on the temperature and strain rate can be obtained by using the MPF-DRX model. Merely, the stress-strain curve obtained in the MPF-DRX computation can be used in the uniaxial constitutive equation in the finite element simulations during the hot working, such as hot rolling, fogging, etc. Consequently, we have developed the MPFFE-DRX model (Takaki 2014; Yoshimoto and Takaki 2014), in which the macroscopic mechanical behaviors during the hot working are simulated by the conventional finite element method, and the DRX microstructures are simulated by the MPF-DRX model. In this model, we do not need the complicated constitutive equations depending on the temperature, strain rate, and internal variables such as the DRX fraction and the DRX grain size.

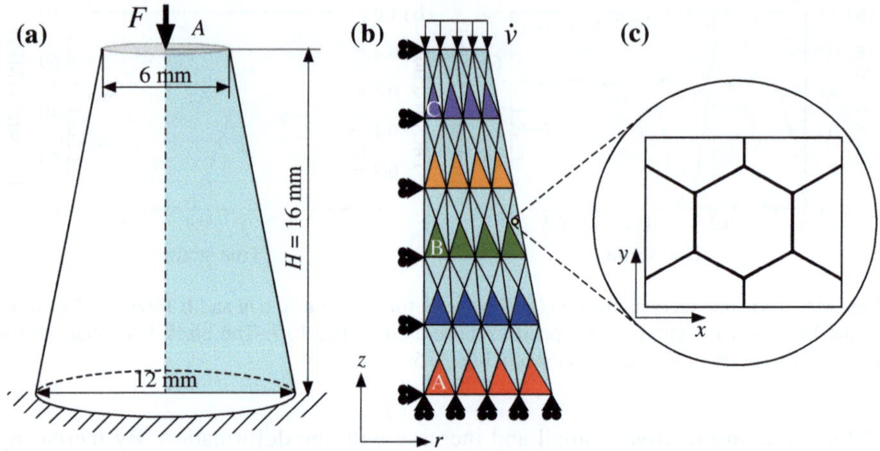

Fig. 20.9 Computational conditions for the compression simulation of a circular truncated cone. **a** Size of the circular truncated cone, **b** Finite element meshes and boundary conditions in macro computation, and **c** Micro computational domain and initial grain structure. The 80 micro domains (**c**) is prepared for all the 80 triangular elements in (**b**)

In the actual hot working, the temperatures of the specimen change. Here, for simplicity, we assume that the temperature is constant during the working. Figure 20.9 shows the computational conditions for the example under investigation with the MPFFE-DRX model. Here, we perform the uniaxial compression simulation of the circular truncated cone shown in Fig. 20.9a; the top and bottom diameters are 6 and 12 mm, respectively, and the height is $H = 16$ mm. This cone is modeled as an axisymmetric body and is meshed by the 5×4 crossed-triangular elements or 80 triangular elements, as shown in Fig. 20.9b. The displacements in the z-direction of the bottom are fixed, and a constant velocity $\dot{v}$ is applied to the top surface, as $\dot{v}/H = 0.01$/s. The compression is applied under the friction-free assumption. The DRX microstructures are computed using the MPF-DRX model, and the two-dimensional computational domains shown in Fig. 20.9c are prepared. The same 80 micro domains are prepared for all the 80 triangular elements shown in Fig. 20.9b. The micro domain size is set to $380\Delta x \times 329\Delta x$ with a lattice size $\Delta x = 0.25\,\mu$m. The four hexagonal regular grains are arranged under the periodic boundary conditions. The temperature is fixed to 800 K and the other parameters are the same as those used in the previous section. The time increment Δt is set to $\Delta t = \Delta x^2/(16M_\phi a^2)$, which is four times smaller than the stability condition in the first order forward difference scheme when solving Eq. (20.9) explicitly. The macro finite element equation is solved every ten steps of the computation of the micro domain.

In the macroscopic deformation simulations, we use a simple elastic-plastic constitutive equation derived from Hook's law and J_2 flow theory, that is independent of the strain rate and temperature because the effects of the strain rate and temperature

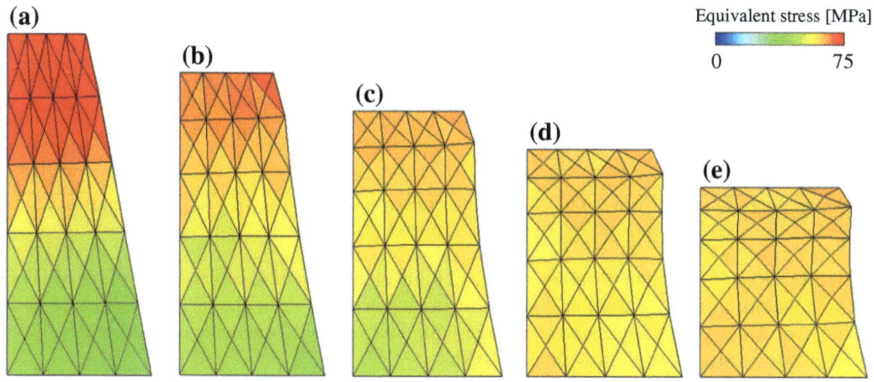

Fig. 20.10 Macro deformations and equivalent stress distributions at v/H = **a** −0.1, **b** −0.2, **c** −0.3, **d** −0.4, and **e** −0.5

are incorporated into the MPF-DRX model. In this case, the relation between the Jaumann rate of the Kirchhoff stress, $\overset{\nabla}{S}_{ij}$, and strain rate $\dot{\varepsilon}_{ij}$ is indicated:

$$\overset{\nabla}{S}_{ij} \left(D^{\mathrm{e}}_{ijkl} - \frac{2G}{g} \sigma'_{ij} \sigma'_{ij} \right) \dot{\varepsilon}_{kl}, \tag{20.11}$$

where σ'_{ij} is the deviatoric stress and D^{e}_{ijkl} is the elastic coefficient tensor. The subscripts indicate the tensor components. In addition, g is expressed as

$$g = \frac{2}{3}\bar{\sigma}^2 \left(1 + \frac{h}{2G} \right), \frac{1}{h} = \frac{3}{2} \left(\frac{1}{E_t} - \frac{1}{E} \right), \tag{20.12}$$

where, E is the Young's modulus and E_t is the tangent modulus of the stress-strain curve computed in the MPF-DRX simulation. The equivalent strain rates and temperatures (constant here) computed in the macro simulations, which are different in all the elements, are transferred to the micro field computations, and the tangent moduli of the stress-strain curves, E_t, computed in the micro fields are transferred to the macro fields. Therefore, the MPFFE-DRX model is an actual dynamic coupling model between the macro mechanical behaviors and the DRX microstructures.

Figure 20.10 shows the deformations and equivalent stress distributions of the macro field shown in Fig. 20.9b. Figure 20.11a shows the F/A versus v/H curve, where F is the force applied to the top surface, A is the initial area of the top surface, v is the displacement of top surface, and H is the initial height as shown in Fig. 20.9a. Figure 20.11b, c shows the changes of the equivalent stress and average grain size of the 20 elements colored in Fig. 20.9b, in which the line colors correspond to those in Fig. 20.9b. In Fig. 20.10a, at $v/H = -0.1$, the stress concentrates around the top surface because of the small area. As shown in Fig. 20.11b, the equivalent stresses around the top surface achieve a peak point at about $v/H = -0.1$. Therefore,

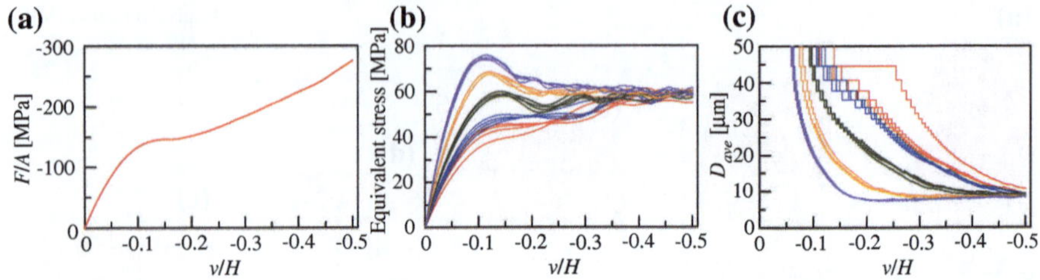

Fig. 20.11 Results of the multi-phase-field and finite element dynamic recrystallization (MPFFE-DRX) simulations. **a** Macroscopic load and displacement curve of the circular truncated cone. Changes of **b** equivalent stress and **c** average grain size of the 20 elements colored in Fig. 20.9b; the line colors correspond to those in Fig. 20.9b

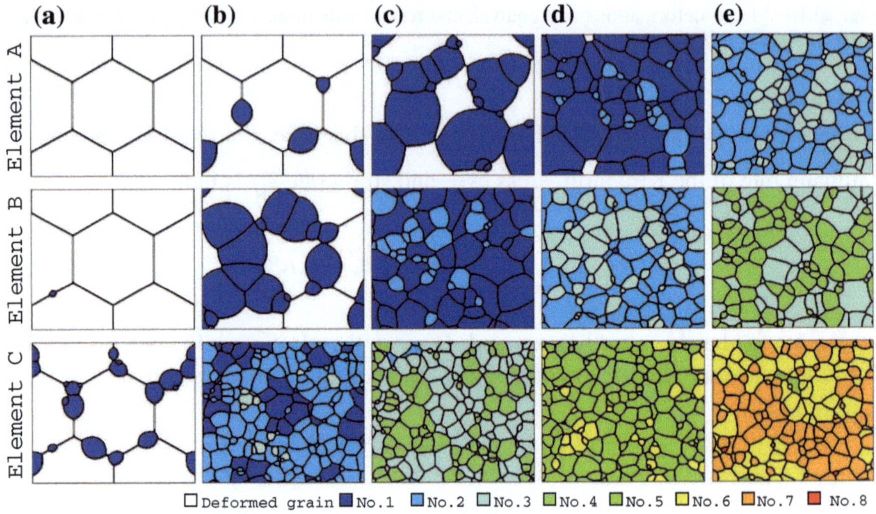

Fig. 20.12 Dynamic recrystallization (DRX) microstructure evolutions of the elements A, B, and C shown in Fig. 20.9b at v/H = **a** −0.1, **b** −0.2, **c** −0.3, **d** −0.4, and **e** −0.5. The color indicates the DRX number

the equivalent stresses around the top surface decrease in the deformation from $v/H = -0.1$ to -0.2. With the compression, the high stress region around the middle and bottom regions propagates from top to bottom, as shown in Fig. 20.10. This is also confirmed by the shift of the peak points of the equivalent stress to the high v/H side, as shown in Fig. 20.11b. Finally at $v/H = -0.5$, the circle truncated cone changes its shape, almost becoming a cylinder, as shown in Fig. 20.10e. Then, almost uniform stress distributions are achieved, as shown in Figs. 20.10e and 20.11b. These results agree well with the experimental observation (Manonukul and Dunne 1999). Figure 20.12 shows the DRX microstructure changes of the elements A, B, and C shown in Fig. 20.9b. The color indicates the DRX number as in Fig. 20.7. From Fig. 20.12, the DRX nucleation occurs by following the element order C, B,

A, which is from top to bottom of the circular truncated cone. With increasing the deformation, the grain size of all the elements saturates to a constant value. This is also observed in Fig. 20.11c. These results confirm that the MPFFE-DRX model can simulate the macro deformation based on the DRX microstructures as well as the DRX microstructures based on the macro deformation.

20.4 Conclusions

We reviewed our application of the phase-field method to the hot working with the DRX. First, the MPF-DRX model was introduced. In the MPF-DRX model, the DRX microstructures are simulated by the MPF method and the deformations are expressed by the Kocks-Mecking model and the Bailey-Hirsch equation. The results of the example simulations performed by changing the initial grain size, temperature, and strain rate, showed that the MPF-DRX model can express the characteristic behaviors observed in the DRX, such as the transit from multiple peaks to single peak of the stress-strain relations. Furthermore, the MPFFE-DRX model enabling the hot-working multiscale simulation was introduced. In the model, the macro hot-working simulations are performed by the normal finite element simulations, and the DRX microstructures are simulated by the MPF-DRX model. As an example, the compression simulation of a circle truncated cone was performed. It was shown that the MPFFE-DRX model can simulate the macro deformation based on the DRX microstructures as well as the DRX microstructures based on the macro deformation. Although this is not the goal of the multiscale simulation during the DRX, however, it provides a promising model for the macro mechanical behaviors and the microstructure.

References

Abrivard G, Busso EP, Forest S, Appolaire B (2012) Phase field modelling of grain boundary motion driven by curvature and stored energy gradients. Part II: application to recrystallisation. Philos Mag 92(28–30):3643–3664

Adebanjo RO, Miller AK (1989) Modelling the effects of recrystallization on the flow behavior during hot deformation by modifying an existing constitutive model I: conceptual development of the MATMOD-ReX equations. Mater Sci Eng A 119:87–94

Anand L (1985) Constitutive equations for hot-working of metals. Int J Plast 1(3):213–231

Anand L, Zavaliangos A, von Turkovich BF (1990) Hot working—constitutive equations and computational procedures. CIRP Ann Manuf Technol 39(1):235–238

Anderson MP, Srolovitz DJ, Grest G, Sahni PS (1984) Computer simulation of grain growth—I Kinetics. Acta Metall 32(5):783–791

Asta M, Beckermann C, Karma A, Kurz W, Napolitano R, Plapp M, Purdy G, Rappaz M, Trivedi R (2009) Solidification microstructures and solid-state parallels: recent developments, future directions. Acta Metall 57(4):941–971

Bailey JE, Hirsch PB (1960) The dislocation distribution, flow stress, and stored energy in cold-worked polycrystalline silver. Philos Mag 53(5):485–497

Bernacki M, Logé RE, Coupez T (2011) Level set framework for the finite-element modelling of recrystallization and grain growth in polycrystalline materials. Scripta Mater 64(6):525–528

Blaz L, Sakai T, Jonas JJ (1983) Effect of initial grain size on dynamic recrystallization of copper. Met Sci 17(12):609–616

Brown AA, Bammann DJ (2012) Validation of a model for static and dynamic recrystallization in metals. Int J Plast 32–33:17–35

Brown SB, Kim KH, Anand L (1989) An internal variable constitutive model for hot working of metals. Int J Plast 5(2):95–130

Busso EP (1998) A continuum theory for dynamic recrystallization with microstructure-related length scales. Int J Plast 14(4–5):319–353

Chen LQ (2014a) Phase-field method and materials genome initiative (MGI). Chin Sci Bull 59(15):1641–1645

Chen LQ (2014b) Phase-field models for microstructure evolution. Annu Rev Mater Sci 32:113–140

Cho JR, Jeong HS, Cha DJ, Bae WB, Lee JW (2005) Prediction of microstructural evolution and recrystallization behaviors of a hot working die steel by FEM. J Mater Process Technol 160(1):1–8

Chuan W, He Y, Wei LH (2013) Modeling of discontinuous dynamic recrystallization of a near-α titanium alloy IMI834 during isothermal hot compression by combining a cellular automaton model with a crystal plasticity finite element method. Comput Mater Sci 79:944–959

Davenport SB, Silk NJ, Sparks CN, Sellars CM (2000) Development of constitutive equations for modelling of hot rolling. Mater Sci Technol 16(5):539–546

Dawson PR (1984) A model for the hot or warm forming of metals with special use of deformation mechanism maps. Int J Mech Sci 26(4):227–244

Ding R, Guo Z (2002) Microstructural modelling of dynamic recrystallisation using an extended cellular automaton approach. Comput Mater Sci 23(1–4):209–218

Ding R, Guo ZX (1984) Coupled quantitative simulation of microstructural evolution and plastic flow during dynamic recrystallization. Acta Mater 49(16):3163–3175

Ding R, Guo ZX (2004) Microstructural evolution of a Ti-6Al-4V alloy during β-phase processing: experimental and simulative investigations. Mater Sci Eng A 365(1–2):172–179

Doherty RD, Hughes DA, Humphreys FJ, Jonas JJ, Jensen DJ, Kassner ME, King WE, McNelley TR, McQueen HJ, Rollett AD (2004) Current issues in recrystallization: a review. Mater Sci Eng A 238(2):219–274

Dunne FPE, Nanneh MM, Zhou M (1997) Anisothermal large deformation constitutive equations and their application to modelling titanium alloy in forging. Philos Mag A 75(3):587–610

Fan D, Chen LQ (1997) Anisothermal large deformation constitutive equations and their application to modelling titanium alloy in forging. Acta Mater 45(2):611–622

Fan XG, Yang H (2011) Internal-state-variable based self-consistent constitutive modeling for hot working of two-phase titanium alloys coupling microstructure evolution. Int J Plast 27(11):1833–1852

Frommert M, Gottstein G (2009) Mechanical behavior and microstructure evolution during steady-state dynamic recrystallization in the austenitic steel 800H. Mater Sci Eng A 506(1–2):101–110

Galantucci LM, Tricarico L (1999) Thermo-mechanical simulation of a rolling process with an FEM approach. J Mater Process Technol 92–93:494–501

Goetz RL (2005) Particle stimulated nucleation during dynamic recrystallization using a cellular automata model. Scripta Mater 52(9):851–856

Goetz RL, Seetharaman V (1998) Modeling dynamic recrystallization using cellular automata. Scripta Mater 38(3):405–413

Gronostajski Z (2000) The constitutive equations for FEM analysis. J Mater Process Technol 106(1–3):40–44

Hallberg H, Wallin M, Ristinmaa M (2010) Simulation of discontinuous dynamic recrystallization in pure Cu using a probabilistic cellular automaton. Comput Mater Sci 49(1):25–34

Hassold GN, Holm EA, Srolovitz DJ (1990) Effects of particle size on inhibited grain growth. Scr Metall Mater 24(1):101–106
Holm E, Battaile C (2001) The computer simulation of microstructural evolution. JOM J Min Met Mater Soc 53(9):20–23
Humphreys F, Hatherly M (1995) Recrystallization and related annealing phenomena. Pergamon, Oxford
Jin Z, Cui Z (2010) Investigation on strain dependence of dynamic recrystallization behavior using an inverse analysis method. Mater Sci Eng A 527(13–14):3111–3119
Kim S, Kim D, Kim W, Park Y (2006) Computer simulations of two-dimensional and three-dimensional ideal grain growth. Phys Rev E Stat Nonlinear Soft Matter Phys 74(6):061605
Kocks UF (1976) Laws for work-hardening and low-temperature creep. Trans ASME J Eng Mater Technol 98(1):76–85
Kugler G, Turk R (2004) Modeling the dynamic recrystallization under multi-stage hot deformation. Acta Mater 52(15):4659–4668
Lee HW, Im YT (2010) Numerical modeling of dynamic recrystallization during nonisothermal hot compression by cellular automata and finite element analysis. Int J Mech Sci 52(10):1277–1289
Lee HW, Kang SH, Lee Y (2014) Prediction of microstructure evolution during hot forging using grain aggregate model for dynamic recrystallization. Int J Precis Eng Manuf 15(6):1055–1062
Li G, Jinn JT, Wu WT, Oh SI (2001) Recent development and applications of three-dimensional finite element modeling in bulk forming processes. J Mater Process Technol 113(1–3):40–45
Li H, Wu C, Yang H (2013) Crystal plasticity modeling of the dynamic recrystallization of two-phase titanium alloys during isothermal processing. Int J Plast 51:271–291
Li X, Duan L, Li J, Wu X (2015) Experimental study and numerical simulation of dynamic recrystallization behavior of a micro-alloyed plastic mold steel. Mater Des A 66:309–320
Madison JD, Tikare V, Holm EA (2012) A hybrid simulation methodology for modeling dynamic recrystallization in UO2 LWR nuclear fuels. J Nucl Mater 425(1–3):173–180
Manonukul A, Dunne FPE (1999) Initiation of dynamic recrystallization under inhomogeneous stress states in pure copper. Acta Mater 47(17):4339–4354
McQueen HJ, Ryan ND (2002) Constitutive analysis in hot working. Mater Sci Eng A 322(1–2):43–63
Mecking H, Kocks UF (1981) Kinetics of flow and strain-hardening. Acta Metall 29(11):1865–1875
Miura H, Aoyama H, Sakai T (1994) Effect of grain-boundary misorientation on dynamic recrystallization of Cu-Si bicrystals. J Jpn Inst Met 58(3):267–275
Miura H, Sakai T, Mogawa R, Gottstein G (2004) Nucleation of dynamic recrystallization at grain boundaries in copper bicrystals. Philos Mag 51(7):671–675
Miura H, Sakai T, Andiarwanto S, Jonas JJ (2005) Nucleation of dynamic recrystallization at triple junctions in polycrystalline copper. Philos Mag 85(23):2653–2669
Miura H, Sakai T, Mogawa R, Jonas JJ (2007) Nucleation of dynamic recrystallization and variant selection in copper bicrystals. Philos Mag 87(27):4197–4209
Moelans N, Blanpain B, Wollants P (2008) An introduction to phase-field modeling of microstructure evolution. Calphad: Comput Coupling Phase Diagrams Thermochem 32(2):268–294
Momeni A, Abbasi S, Badri H (2012) Hot deformation behavior and constitutive modeling of VCN200 low alloy steel. Calphad: Comput Coupling Phase Diagrams Thermochem 36(11):5624–5632
Montmitonnet P (2006) Hot and cold strip rolling processes. Comput Methods Appl Mech Eng 195(48–49):6604–6625
Oh SI (1982) Finite element analysis of metal forming processes with arbitrarily shaped dies. Int J Mech Sci 24(8):479–493
Oh SI, Wu WT, Tang JP, Vedhanayagam A (1991) Capabilities and applications of FEM code deform: the perspective of the developer. J Mater Process Technol 27(1–3):25–42
Ohno M, Tsuchiya S, Matsuura K (2011) Formation conditions of coarse columnar austenite grain structure in peritectic carbon steels by the discontinuous grain growth mechanism. Acta Mater 59(14):5700–5709

Peczak P (1995) A Monte Carlo study of influence of deformation temperature on dynamic recrystallization. Acta Metall Mater 43(3):1279–1291
Peczak P, Luton MJ (1993) A Monte Carlo study of the influence of dynamic recovery on dynamic recrystallization. Acta Metall Mater 41(1):59–71
Qian M, Guo ZX (1993) Cellular automata simulation of microstructural evolution during dynamic recrystallization of an HY-100 steel. Mater Sci Eng A 365(1–2):180–185
Qu J, Jin QL, Xu BY (2005) Parameter identification for improved viscoplastic model considering dynamic recrystallization. Int J Plast 21(7):1267–1302
Quan G, Luo G, Liang J, Wu D, Mao A, Liu Q (2015) Modelling for the dynamic recrystallization evolution of Ti-6Al-4V alloy in two-phase temperature range and a wide strain rate range. Comput Mater Sci 97:136–147
Roberts W, Ahlblom B (1978) A nucleation criterion for dynamic recrystallization during hot working. Acta Metall 26(5):801–813
Rollett AD, Luton MJ, Srolovitz DJ (1992) Microstructural simulation of dynamic recrystallization. Acta Metall Mater 40(1):43–55
Sakai T, Jonas JJ (1984) Dynamic recrystallization: mechanical and microstructural considerations. Acta Metall 32(2):189–209
Sakai T, Akben MG, Jonas JJ (1983) Dynamic recrystallization during the transient deformation of a vanadium microalloyed steel. Acta Metall 31(4):631–641
Sakai T, Belyakov A, Kaibyshev R, Miura H, Jonas JJ (2014) Dynamic and post-dynamic recrystallization under hot, cold and severe plastic deformation conditions. Prog Mater Sci 60(1):130–207
Singer-Loginova I, Singer HM (2008) The phase field technique for modeling multiphase materials. Rep Prog Phys 71(10):106501
Song JL, Dowson AL, Jacobs MH, Brooks J, Beden I (2002) Coupled thermo-mechanical finite-element modelling of hot ring rolling process. J Mater Process Technol 121(2–3):332–340
Steinbach I (2009) Phase-field models in materials science. Modell Simul Mater Sci Eng 17(7):073001
Steinbach I, Pezzolla F (1999) A generalized field method for multiphase transformations using interface fields. Physica D Nonlinear Phenom 134(4):385–393
Suwa Y, Saito Y, Onodera H (2007) Phase-field simulation of abnormal grain growth due to inverse pinning. Acta Mater 55(20):6881–6894
Svyetlichnyy DS (2010) Modelling of the microstructure: from classical cellular automata approach to the frontal one. Comput Mater Sci 50(1):92–97
Takaki T (2014) Phase-field modeling and simulations of dendrite growth. ISIJ Int 54(2):437–444
Takaki T, Hirouchi T, Hisakuni Y, Yamanaka A, Tomita Y (2008) Multi-phase-field model to simulate microstructure evolutions during dynamic recrystallization. Mater Trans 49(11):2559–2565
Takaki T, Hisakuni Y, Hirouchi T, Yamanaka A, Tomita Y (2009) Multi-phase-field simulations for dynamic recrystallization. Comput Mater Sci 45(4):881–888
Takaki T, Yamanaka A, Tomita Y (2011) Multi-phase-field simulations of dynamic recrystallization during transient deformation. ISIJ Int 51(10):1717–1723
Wolfram S (1983) Statistical mechanics of cellular automata. Rev Mod Phys 55(3):601–644
Wusatowska-Sarnek AM, Miura H, Sakai T (2002) Nucleation and microtexture development under dynamic recrystallization of copper. Mater Sci Eng A 323(1–2):177–186
Xiao N, Zheng C, Li D, Li Y (2008) A simulation of dynamic recrystallization by coupling a cellular automaton method with a topology deformation technique. Comput Mater Sci 41(3):366–374
Yamada K, Ogawa S, Hamauzu S (1991) Two-dimensional thermo-mechanical analysis of flat rolling using rigid-plastic finite element method. ISIJ Int 31(36):566–570
Yamanaka A, Takaki T (2014) Multi-phase-field simulation of flow stress and microstructural evolution during deformation-induced ferrite transformation in a Fe-C alloy. ISIJ Int 52(12):2917–2925

Yanagimoto J, Karhausen K, Brand AJ, Kopp R (1998) Incremental formulation for the prediction of flow stress and microstructural change in hot forming. Trans ASME J Manuf Sci Eng 120(2):316–322

Yanagimoto J, Dupin E, Liu JS, Yanagida A (2014) Numerical analysis for microstructure control in hot forming process. Procedia Eng 81:38–43

Yazdipour N, Davies CHJ, Hodgson PD (2008) Microstructural modeling of dynamic recrystallization using irregular cellular automata. Comput Mater Sci 44(2):566–576

Yeom JT, Lee CS, Kim JH, Park NK (2005) Finite-element analysis of microstructure evolution in the cogging of an alloy 718 ingot. Mater Sci Eng A 449–451:722–726

Yin F, Hua L, Mao H, Han X, Qian D, Zhang R (2014) Microstructural modeling and simulation for GCr15 steel during elevated temperature deformation. Mater Des 55:560–573

Yoshimoto C, Takaki T (2014) Multiscale hot-working simulations using multi-phase-field and finite element dynamic recrystallization model. ISIJ Int 54(2):452–459

Zheng C, Xiao N, Li D, Li Y (2008) Numerical simulation of dynamic strain-induced austenite-ferrite transformation in a low carbon steel. Comput Mater Sci 44(2):507–514

Zheng C, Xiao N, Hao L, Li D, Li Y (2009) Numerical simulation of dynamic strain-induced austenite-ferrite transformation in a low carbon steel. Acta Mater 57(10):2956–2968

Chapter 21
Mechanical Properties of Shape Memory Alloy and Polymer

Hisaaki Tobushi, Ryosuke Matsui and Kohei Takeda

Abstract In order to describe the shape memory effect, superelasticity and recovery stress due to the martensitic transformation and the R-phase transformation of TiNi shape memory alloy, a thermomechanical constitutive equation considering the volume fractions of induced phases associated with both transformations is developed. The conditions for progress of phase transformation and subloop-deformation behavior are discussed. A nonlinear thermomechanical constitutive model of polyurethane-shape memory polymer is developed to describe the thermomechanical properties such as shape fixity, shape recovery and recovery stress, by modifying a linear model. The coefficients in the model are expressed by the single exponential functions of temperature in order to describe the variation in mechanical properties of the material due to the glass transition.

Keywords Shape memory alloy · Shape memory polymer · Constitutive modelling · Thermomechanical property · Phase transformation

21.1 Introduction

In shape memory alloy (SMA), the residual strain of several percent disappears by heating which is called the shape memory effect (SME) and the transformed strain recovers by unloading which is called the superelasticity (SE) or pseudoelasticity (Funakubo 1987; Otsuka and Wayman 1998; Tobushi et al. 2013a). These deformation properties occur due to the martensitic transformation (MT). In TiNi SMA, the properties occur not only due to the MT, but also due to the rhombohedral phase

H. Tobushi (✉) · R. Matsui · K. Takeda
Department of Mechanical Engineering, Aichi Institute of Technology,
1247 Yachigusa, Yakusa-cho, Toyota 470-0392, Japan
e-mail: tobushi@aitech.ac.jp

R. Matsui
e-mail: r_matsui@aitech.ac.jp

K. Takeda
e-mail: k-takeda@aitech.ac.jp

H. Altenbach et al. (eds.), *From Creep Damage Mechanics to Homogenization Methods*, Advanced Structured Materials 64,
DOI 10.1007/978-3-319-19440-0_21

transformation (RPT) (Miyazaki and Otsuka 1984, 1986; Miyazaki et al. 1988). Therefore, the deformation behavior of TiNi SMA due to the MT and the RPT is very complex. In order to design SMA elements, since the stress-strain-temperature relationship is necessary for this purpose, the authors proposed the theory which was useful for the practical application and described well the deformation behavior of TiNi SMA due to the MT (Tobushi et al. 2013a; Miyazaki and Otsuka 1984, 1986; Miyazaki et al. 1988; Tanaka 1986; Tanaka et al. 1986, 1988, 1992). In the theory, a stress-strain-temperature equation was used considering a volume fraction of induced phase. Furthermore they proposed a constitutive equation which described the deformation properties due to the RPT (Tobushi et al. 1992; Sawada et al. 1993; Lexcellent et al. 1994; Lin et al. 1995a).

The development of applications of shape memory polymer (SMP) is currently of great interest in wide fields (Irie 1998; Liang et al. 1991). In order to design SMP elements rationally, the constitutive equation which expresses the thermomechanical properties of the material is necessary. The characteristics such as shape recovery and shape fixity of SMP exist due to the glass transition (Takahashi et al. 1996; Hayashi et al. 1993). Because the thermomechanical properties of SMP vary markedly in the glass transition region, the properties are complex and therefore it is not so easy to express them. From the practical viewpoint, both the stress-strain-temperature-time relationship which is the most base to evaluate SMP and the simple constitutive equation which expresses the relationship are important. For this purpose, the authors proposed a model in which a standard linear viscoelastic model is modified by taking account of a slip mechanism due to internal friction and furthermore thermal expansion is considered (Tobushi et al. 1997). The coefficients in the model are expressed by the single exponential functions of temperature. They confirmed that the proposed model is useful to evaluate the thermomechanical properties of SMP. Recoverable strain of SMP is several hundred percent. Most of SMP elements are practically used in the range of strain below 20 % (Hayashi 1993; Tobushi et al. 1998). This is due to the fact that bending deformation is widely employed in order to deform SMP elements and that large deflection is easily obtained in the range of small strain through bending. In SMP, nonlinear strain appears above 3 % of strain. Therefore, even below 20 % of strain, it is important to take account of nonlinear strain in practical applications.

In the present chapter, based on the previous studies, a constitutive equation which describes the deformation behavior associated with the MT and the RPT in SMA is proposed. Based on the experimental data, the material parameters are determined and the calculated results are compared with the experimental results. In order to understand the reason why the pseudoviscoelastic behaviors of creep, stress relaxation, a decrease in strain during unloading and an increases in strain during reloading which appear under the stress-controlled condition, the conditions for progress of the phase transformation and the corresponding subloop-deformation behavior are discussed. A thermomechanical constitutive model of SMP is proposed in which nonlinear terms of stress are considered for both an elastic term and a viscous term in the linear model (Tobushi et al. 2001). In order to evaluate validity of the model, several kinds of thermomechanical tests on SMP of polyurethane series are carried out.

21.2 Thermomechanical Properties and Modeling of Shape Memory Alloy

21.2.1 Constitutive Relationship for Martensitic Transformation

The stress-strain-temperature relationship is discussed in the case of uniaxial tension. For a constitutive equation which describes the deformation behavior associated with the MT, the constitutive equation which is developed based on the relationship in the previous studies (Tanaka 1986; Tanaka et al. 1986, 1988, 1992; Tobushi et al. 1992; Sawada et al. 1993) is proposed.

We assume the following constitutive equation which describes the thermomechanical behavior of the material.

$$\dot{\sigma} = D\dot{\varepsilon} + \Theta\dot{T} + \Omega\dot{\xi} \tag{21.1}$$

We assume the following transformation kinetics which governs the progress of the transformation.

$$\xi = \Xi(\sigma, T) \tag{21.2}$$

In these equations, σ, ε and T represent the stress, strain and temperature, respectively. The internal state variable $\xi (0 \leq \xi \leq 1)$ represents the volume fraction of the martensitic phase. The volume fraction ξ is prescribed by the transformation kinetics (21.2).

The dot on the letter in Eq. (21.1) denotes the time derivative. The coefficients D and Θ represent the modulus of elasticity and the thermoelastic constant, respectively. The quantity $|\Omega/D|$ represents the transformation strain range due to the MT, which denotes the maximum recoverable strain and can be evaluated from the crystallographic difference between the parent (austenite) phase and the martensitic phase, respectively.

21.2.2 Transformation Kinetics of Martensitic Transformation

We assume the concrete form of the transformation kinetics (21.2) which describes the progress of the transformation. With respect to the MT, we employ the following form, the same as the previous paper (Tanaka 1986; Tanaka et al. 1986, 1988, 1992)

$$\frac{\dot{\xi}}{1-\xi} = b_{\mathrm{M}} C_{\mathrm{M}} \dot{T} - b_{\mathrm{M}} \dot{\sigma} \geq 0, \tag{21.3}$$

$$-\frac{\dot{\xi}}{\xi} = b_A C_A \dot{T} - b_A \dot{\sigma} \geq 0 \tag{21.4}$$

Equation (21.3) is applied to the MT and Eq. (21.4) to its reverse transformation. The coefficients b_M, C_M, b_A and C_A are the material constants. If we assume that these material parameters are constant and integrate Eqs. (21.3) and (21.4), we obtain the following equations

$$\xi = 1 - \exp\{b_M C_M (M_s - T) + b_M \sigma\}, \tag{21.5}$$

$$\xi = \exp\{b_A C_A (A_s - T) + b_A \sigma\} \tag{21.6}$$

M_s and A_s respectively represent the temperatures at which the MT and its reverse transformation start under stress-free conditions. The MT starts or completes when $\xi = 0$ or $\xi = 1$ in Eq. (21.2), respectively. These conditions are represented by the curves on the stress-temperature plane and are referred to as the transformation starting line and the transformation completing line. With respect to Eqs. (21.5) and (21.6), the equations for these lines are obtained as follows. The transformation starting line and the transformation completing line are expressed as

$$\sigma = C_M (T - M_s), \tag{21.7}$$

$$\sigma = C_M (T - M_s) - 2 \ln 10 / b_M, \tag{21.8}$$

for the MT and

$$\sigma = C_A (T - A_s), \tag{21.9}$$

$$\sigma = C_A (T - A_s) - 2 \ln 10 / b_A \tag{21.10}$$

for its reverse transformation, respectively. In deriving Eq. (21.8), as usual in metallurgy, the transformation is understood to complete when ξ reaches 0.99, instead of 1, and 0.01 in deriving Eq. (21.10). Each transformation progresses in the respective region between these two lines, which is called the transformation zone.

21.2.3 Constitutive Relationship Containing Martensitic and R-Phase Transformations

We assume the following constitutive equation which describes the thermomechanical behavior of the material

$$\dot{\sigma} = D\dot{\varepsilon} + \Theta \dot{T} + \Omega \dot{\xi} + \Psi \dot{\eta} \tag{21.11}$$

We assume the following transformation kinetics which governs the progress of the R-phase transformation.

$$\eta = H(\sigma, T) \tag{21.12}$$

The internal state variable $\eta (0 \leq \eta \leq 1)$ represents the volume fraction of the rhombohedral phase (R-phase). The volume fractions ξ and η satisfy the conditions: $0 \leq \xi + \eta \leq 1$. The volume fractions ξ and η are respectively prescribed by the transformation kinetics (21.2) and (21.12).

The quantity $|\Psi/D|$ represents the transformation strain range due to the RPT, which denotes the maximum recoverable strain and can be evaluated from the crystallographic difference between the parent phase and the R-phase, respectively.

21.2.4 Transformation Kinetics of R-Phase Transformation

With respect to the RPT, we employ the following form, the same as the previous paper (Tobushi et al. 1992; Sawada et al. 1993)

$$\eta = b'_{\mathrm{M}} C'_{\mathrm{M}} ((T - M'_{\mathrm{s}}) - b'_{\mathrm{M}} \sigma, \tag{21.13}$$

$$\eta = 1 + b'_{\mathrm{A}} C'_{\mathrm{A}} ((A'_{\mathrm{s}} - T) + b'_{\mathrm{A}} \sigma \tag{21.14}$$

Equation (21.13) is applied to the RPT and Eq. (21.14) to its reverse transformation. The coefficients b'_{M}, C'_{M}, b'_{A} and C'_{A} are the material constants. M'_{s} and A'_{s} respectively represent the temperatures at which the RPT and its reverse transformation start under stress-free conditions. The RPT starts or completes when $\eta = 0$ or $\eta = 1$ in Eq. (21.12), respectively. These conditions are represented by the curves on the stress-temperature plane. With respect to Eqs. (21.13) and (21.14), the equations are obtained as follows. The transformation starting line and transformation completing line are expressed as

$$\sigma = C'_{\mathrm{M}} (T - M'_{\mathrm{s}}), \tag{21.15}$$

$$\sigma = C'_{\mathrm{M}} (T - M'_{\mathrm{s}}) - \frac{1}{b'_{\mathrm{M}}} \tag{21.16}$$

for the RPT and

$$\sigma = C'_{\mathrm{A}} (T - A'_{\mathrm{s}}), \tag{21.17}$$

$$\sigma = C'_{\mathrm{A}} (T - A'_{\mathrm{s}}) - \frac{1}{b'_{\mathrm{A}}} \tag{21.18}$$

for its reverse transformation, respectively. At the low temperatures of $T \le M_s'$ and $T \le A_s'$, the starting and completing stresses of the rearrangement of the R-phase are almost constant. Thus we assume that the respective starting and completing lines of the transformation are expressed as

$$\sigma = \sigma_{M0}', \tag{21.19}$$

$$\sigma = \sigma_{M0}' - \frac{1}{b_M'} \tag{21.20}$$

for the rearrangement of the R-phase. Therefore the starting and completing lines of the RPT are represented by the respective bilinear lines. The RPT progresses in the region between the two bilinear lines.

21.2.5 Stress-Strain Curves and Transformed Region

The stress-strain relationship and the transformed region due to the MT and the RPT are schematically explained as follows. For simplicity, we consider uniaxial tension at constant temperature. The stress-strain curves are shown in Fig. 21.1. The relationship between the transformed region and the volume fraction corresponding to the deformation state in the gauge length of a specimen is shown in Fig. 21.2. When we draw up Figs. 21.1 and 21.2, we consider the following points. If the specimen of TiNi SMA is loaded in tension at constant temperature, the RPT appears at first and the MT follows. During the progress of each transformation, the interface between the induced phase and the parent phase moves like Lüder's bands in the yield stage of mild steel.

As shown in Figs. 21.1 and 21.2, in the case that the RPT starts at first under the stress σ_{Ms}', if the volume fraction of the R-phase is η, the volume fraction of the parent phase is $1 - \eta$. If the RPT completes under the stress σ_{Mf}', $\eta = 1$. In the case that stress increases after completion of the RPT and the MT starts under the stress σ_{Ms}, if the volume fraction of the martensitic phase is ξ, the volume fraction of the R-phase η is $1 - \xi$. If the MT completes under the stress σ_{Mf}, $\xi = 1$ and $\eta = 0$.

21.2.6 Results and Discussion of Modeling

21.2.6.1 Coefficients and Transformation Lines

We carried out the tension tests at a constant temperature and the heating tests under a constant strain for Ti-55.3 wt% Ni SMA wire, 0.75 mm in diameter. The respective values of coefficients for the MT and the RPT determined based on the experimental results are shown in Tables 21.1 and 21.2. Because the thermoelastic effect is small

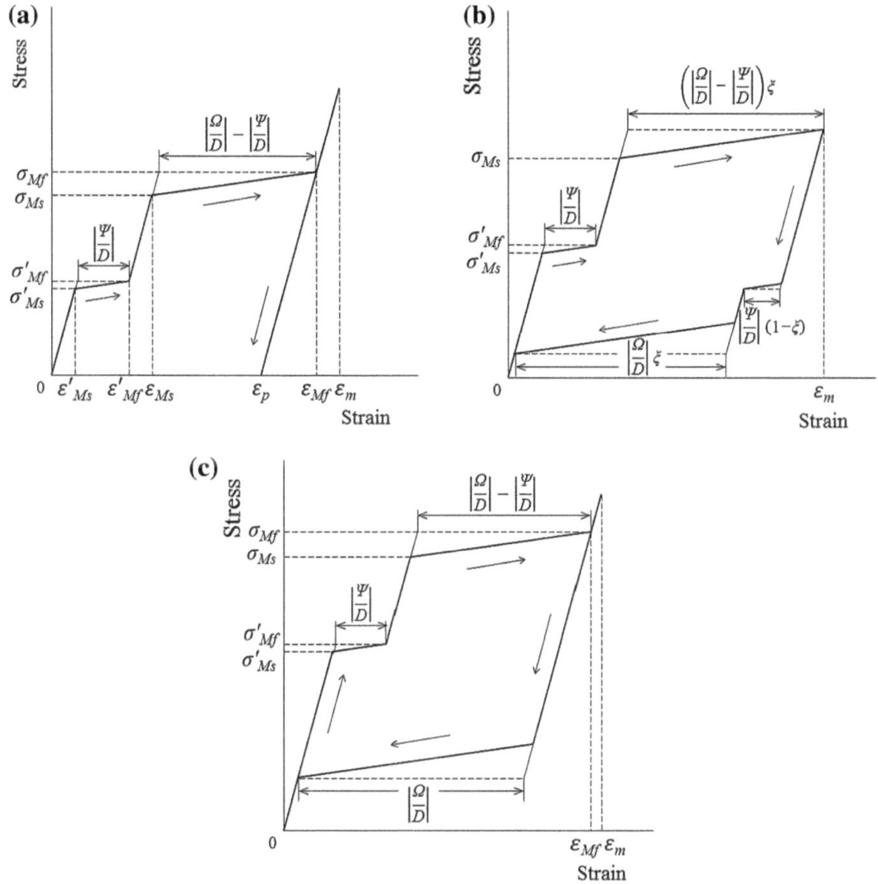

Fig. 21.1 Stress-strain curves at constant temperature T and transformation strain ranges. **a** $T < A_s, \varepsilon_m > \varepsilon_{Mf}$. **b** $T > A_f, \varepsilon_m < \varepsilon_{Mf}$. **c** $T > A_f, \varepsilon_m > \varepsilon_{Mf}$

in the range treated in the present chapter, we assume $\Theta = 0$. Although the modulus of elasticity D depends on T, ξ and η, we assume that D is constant because we are describing the overall behavior of SMA using a simple model and are proposing the model useful for design of SMA elements. The transformation strain range $|\Omega/D|$ due to the MT increases in proportion to T and the range $|\Psi/D|$ due to the RPT decreases. Therefore if D is assumed constant, the coefficients Ω and Ψ are expressed as follows

$$\Omega = AT + B, \tag{21.21}$$

$$\Psi = A'T + B' \tag{21.22}$$

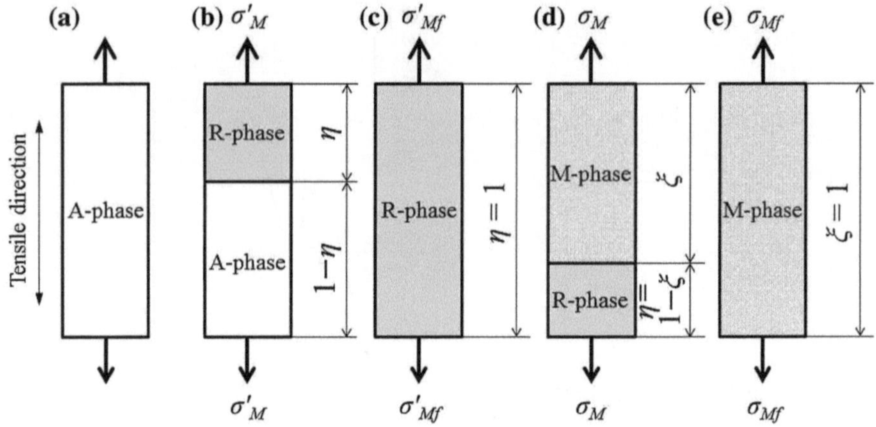

Fig. 21.2 Relationship between transformed region and volume fractions due to the MT and RPT. **a** $\sigma = 0, \varepsilon = 0$. **b** $\sigma = \sigma'_M, \varepsilon'_{Ms} < \varepsilon < \varepsilon'_{Mf}$. **c** $\sigma = \sigma'_{Mf}, \varepsilon = \varepsilon'_{Mf}$. **d** $\sigma = \sigma_M, \varepsilon_{Ms} < \varepsilon < \varepsilon_{Mf}$. **e** $\sigma = \sigma_{Mf}, \varepsilon = \varepsilon_{Mf}$

Table 21.1 Coefficients for the MT

D (GPa)	A (MPa/K)	B (MPa)	C_M (MPa/K)	C_A (MPa/K)	b_M (1/MPa)	b_A (1/MPa)	M_s (K)	A_s (K)
60	−6	−1962	6.125	8.8	−0.132	0.132	272.5	318

The starting lines and the completing lines for the MT and its reverse transformation S_M, F_M, S_A and F_A and the starting lines and the completing lines for the RPT and its reverse transformation S'_M, F'_M, S'_A and F'_A determined by using these coefficients are shown in Fig. 21.3. As seen in Fig. 21.3, the MT zone is located on the low temperature side by about 40 K from its reverse transformation zone. The RPT zone at $T \geq A'_s$ overlaps with its reverse transformation zone, and the slopes of each transformation line are steep.

21.2.6.2 Stress-Strain Relationship

The stress-strain curves calculated with various maximum strains ε_m for the tension tests at $T = 303$ and 333 K are shown in Fig. 21.4. In the case of $T = 303$ K, residual strain appears after unloading and diminishes by heating under no-stress, showing the SME. In the case of $T = 333$ K, the transformed strain recovers by unloading, showing the SE.

Comparing the curves shown in Fig. 21.4 with the experimental results (Tobushi et al. 1992; Sawada et al. 1993), we find that the calculated results quantitatively describe the overall stress-strain relationship due to the RPT and the MT.

Table 21.2 Coefficients for the RPT

D (GPa)	A' (MPa/K)	B' (MPa)	C'_{M} (MPa/K)	C'_{A} (MPa/K)	b'_{M0} (1/MPa)	b'_{M}		b'_{A} (1/MPa)	M'_{s} (K)	A'_{s} (K)
						$T < M'_{\mathrm{s}}$ (1/MPa)	$T \geq M'_{\mathrm{s}}$ (1/MPa)			
60	6.49	−2386	15.2	17.4	25	−0.094	−0.028	0.025	317	318

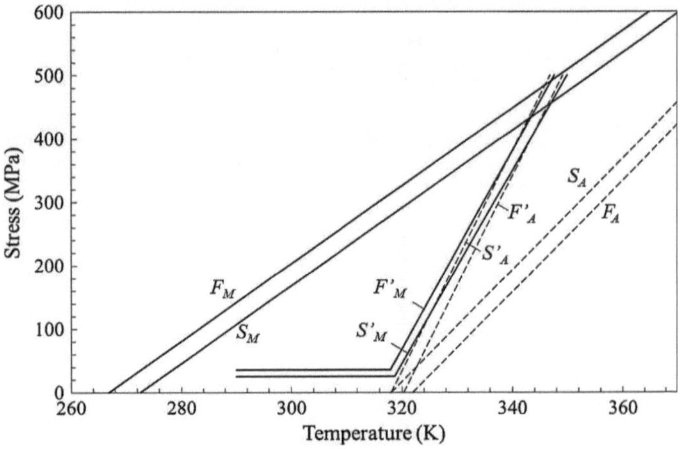

Fig. 21.3 Transformation starting and completing lines for the MT and RPT on the stress-temperature plane

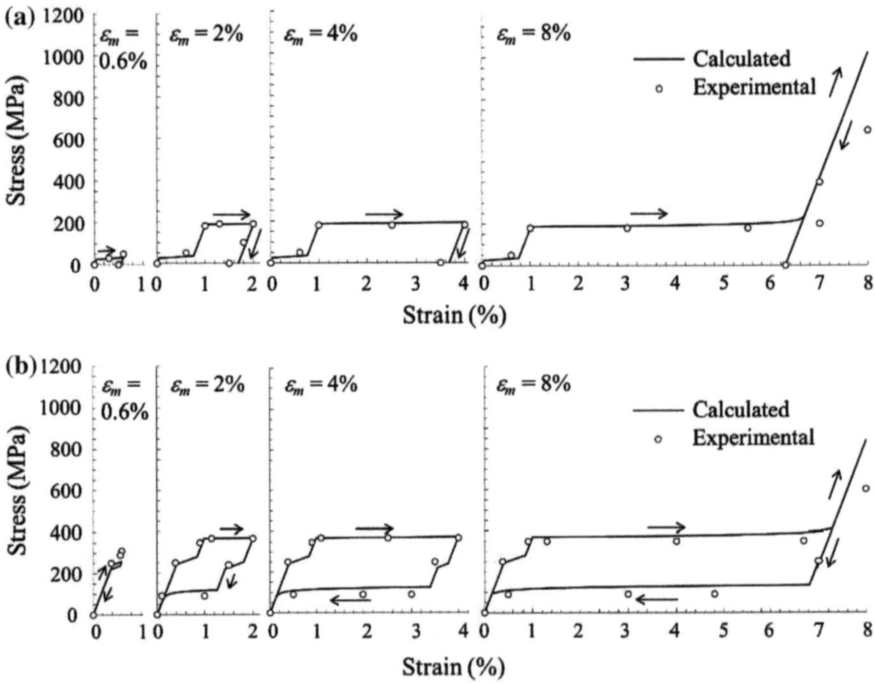

Fig. 21.4 Stress-strain curves for the tension tests at constant temperature. **a** $T = 303$ K. **b** $T = 333$ K

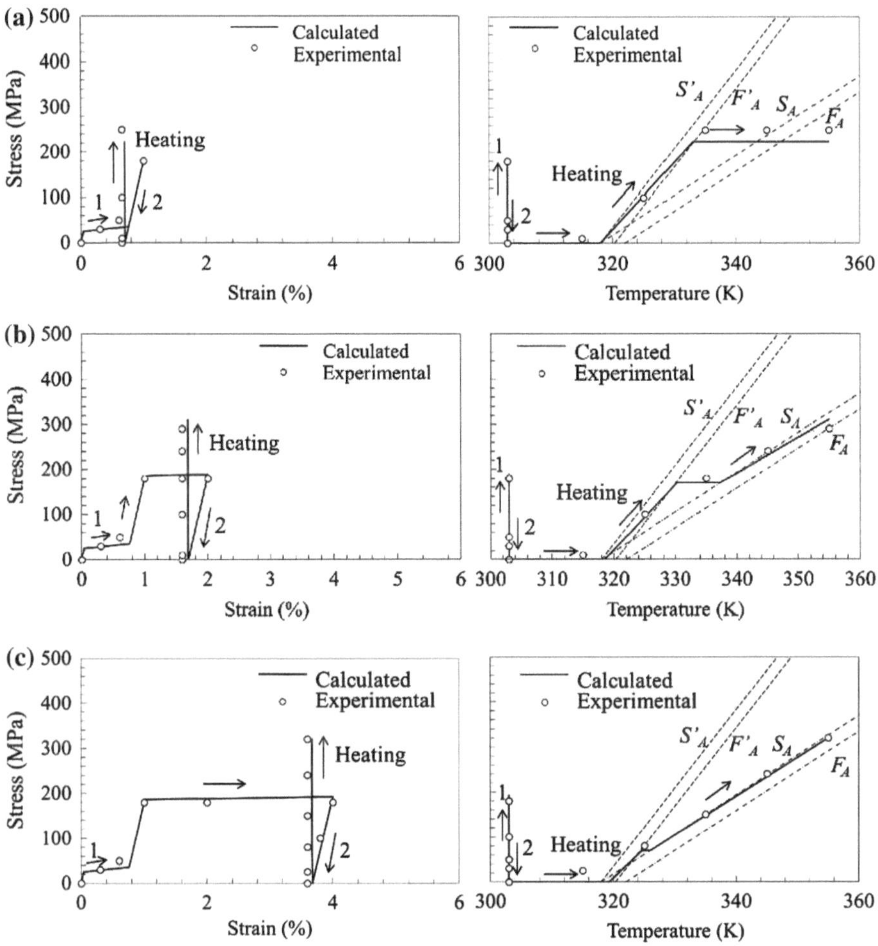

Fig. 21.5 Relationships of stress versus strain and stress versus temperature for the heating test: Loading (1) and unloading (2) were followed by heating under constant residual strain. **a** $\varepsilon_m = 1\,\%$. **b** $\varepsilon_m = 2\,\%$. **c** $\varepsilon_m = 4\,\%$

21.2.6.3 Recovery Stress

Stress-strain curves and stress-temperature curves calculated for the heating tests under constant residual strain which appears during loading and unloading processes at $T = 303$ K and these under constant maximum strain are shown in Figs. 21.5 and 21.6, respectively. In the stress-temperature diagram in Figs. 21.5 and 21.6, the reverse transformation lines S_{A} and F_{A} for the MT and these S'_{A} and F'_{A} for the RPT are plotted by chain lines and dashed lines, respectively.

As seen from the stress-temperature curves during the heating process in Figs. 21.5 and 21.6, in the case of $\varepsilon_m \leq 1\,\%$, recovery stress increases along the reverse

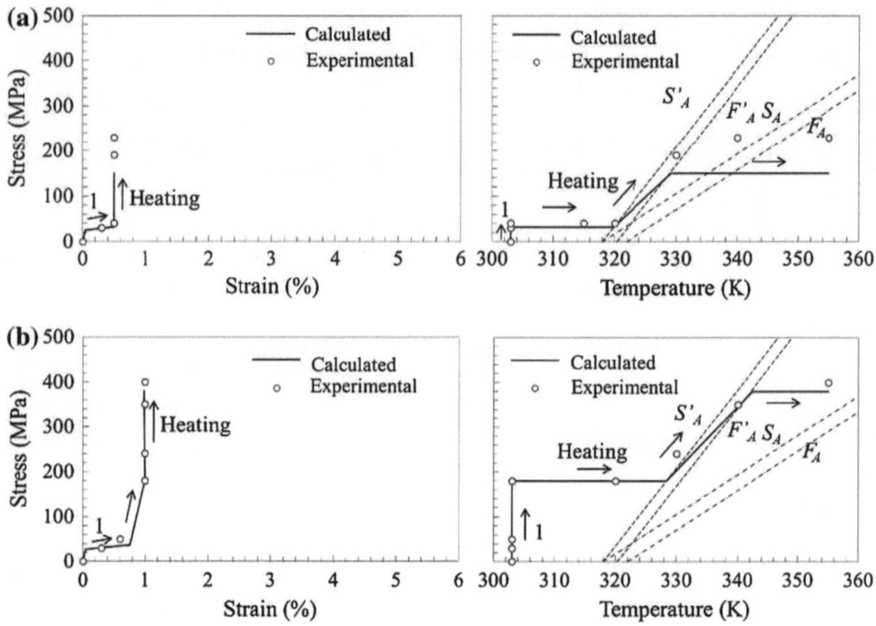

Fig. 21.6 Relationships of stress versus strain and stress versus temperature for the heating test: Loading (1) was followed by heating under constant maximum strain. **a** $\varepsilon_m = 0.5\,\%$. **b** $\varepsilon_m = 1\,\%$

transformation line for the RPT. If ε_m is in the MT region, recovery stress increases at first along the reverse transformation line for the RPT and follows along that for the MT. In this case, if ε_m is large, recovery stress due to the RPT is small and that due to MT is large.

Comparing the curves shown in Figs. 21.5 and 21.6 with the experimental results (Tobushi et al. 1992; Sawada et al. 1993), we find that the calculated results quantitatively describe the overall behavior of recovery stress.

21.2.7 Conditions for Progress of Phase Transformation and Subloop-Deformation Behavior

21.2.7.1 Conditions for Progress of Phase Transformation in the Subloop Loading

From Eq. (21.3), the condition for progress of the MT becomes as follows since $b_M < 0$

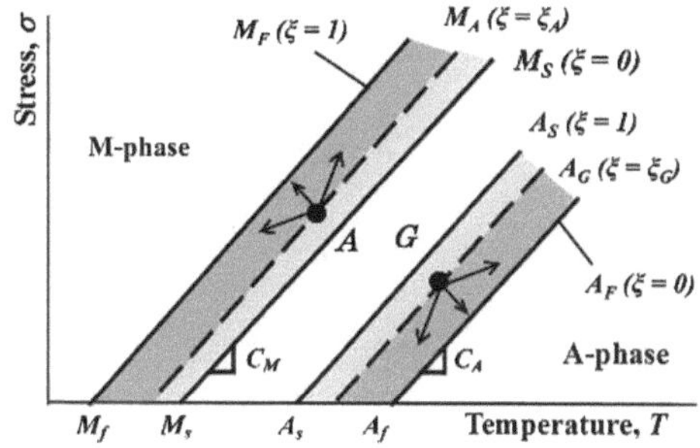

Fig. 21.7 Conditions for progress of the MT and the reverse transformation in the subloop loading on the stress-temperature phase diagram

$$\begin{aligned} \frac{\mathrm{d}\sigma}{\mathrm{d}T} > C_{\mathrm{M}} &: \quad \text{for} \quad \mathrm{d}T > 0, \\ \frac{\mathrm{d}\sigma}{\mathrm{d}T} < C_{\mathrm{M}} &: \quad \text{for} \quad \mathrm{d}T < 0 \end{aligned} \tag{21.23}$$

From Eq. (21.4), the condition for progress of the reverse transformation becomes as follows since $b_{\mathrm{A}} > 0$

$$\begin{aligned} \frac{\mathrm{d}\sigma}{\mathrm{d}T} < C_{\mathrm{A}} &: \quad \text{for} \quad \mathrm{d}T > 0, \\ \frac{\mathrm{d}\sigma}{\mathrm{d}T} > C_{\mathrm{A}} &: \quad \text{for} \quad \mathrm{d}T < 0 \end{aligned} \tag{21.24}$$

The conditions for progress of the phase transformation in the subloop loading during the phase transformation are shown on the stress-temperature phase diagram in Fig. 21.7. In Fig. 21.7, $M_{\mathrm{S}}(\xi = 0)$ and $M_{\mathrm{F}}(\xi = 1)$ denote the MT start and finish lines with a slope of C_{M}, respectively, and $A_{\mathrm{S}}(\xi = 1)$ and $A_F(\xi = 0)$ denote the reverse transformation start and finish lines with a slope of C_{A}, respectively. Points A and G in Fig. 21.7 represent respectively the state of progress of the MT and the reverse transformation and the volume fractions of the M-phase at each point are ξ_{A} and ξ_{G}. The broken lines M_{A} and A_{G} denote the states with the volume fractions ξ_A and ξ_G, respectively. The conditions prescribed by Eqs. (21.23) and (21.24) for progress of the phase transformation from the points A and G mean that stress and temperature vary in the directions shown by the arrows in Fig. 21.7.

21.2.7.2 Subloop-Deformation Behavior During the Loading Process

The stress-strain diagram and stress-temperature path subjected to various thermomechanical subloop loadings from the point A during the loading process under the

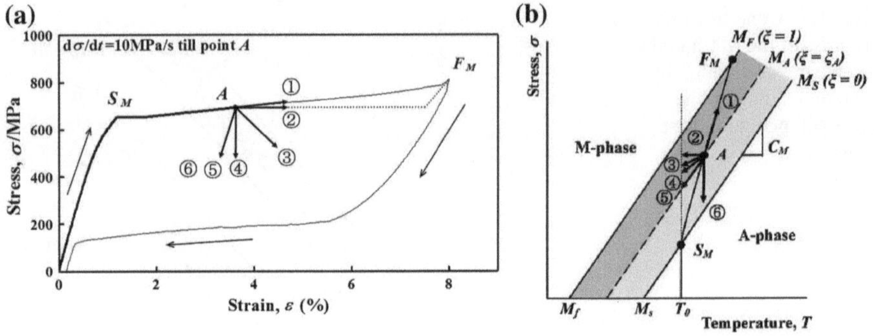

Fig. 21.8 Stress-strain diagram and various thermomechanical-subloop loading paths during the loading process under the stress-controlled condition with constant stress rate till the point *A*. **a** Stress-strain diagram for loading paths (1)–(6). **b** Thermomechanical-subloop loading paths (1)–(6)

stress-controlled condition are schematically shown in Fig. 21.8. The stress-strain relationships (1)–(6) shown by the arrows for the subloop process from the point *A* in Fig. 21.8a correspond respectively to the thermomechanical-subloop loading paths (1)–(6) shown by the arrows in Fig. 21.8b.

Tensile load is applied at temperature T_0 under a constant stress rate up to the point *A*. The MT starts at the point S_M during loading. The upper stress plateau starts from the point S_M in the stress-strain diagram and temperature increases due to the MT. The stress-strain curve and variation in temperature depend on the loading rate (Tobushi et al. 1999; Pieczyska et al. 2006; Gadaj et al. 1999). The thermomechanical paths (1)–(6) from the point *A* and the corresponding deformation behaviors can be explained as follows. The path (1) corresponds to the continuous loading under a constant stress rate. In this case, the MT progresses with the condition $\mathrm{d}\sigma/\mathrm{d}T > C_M$ and the upper stress plateau finishes at the point F_M. The path (2) expresses the condition with holding stress constant. In this case, the MT progresses due to decrease in temperature under the constant stress and therefore creep deformation is induced due to the MT as observed in (Takeda et al. 2013a, b). The path (3) corresponds to unloading under a constant stress rate. In this case, both stress and temperature decrease with the condition $\mathrm{d}\sigma/\mathrm{d}T < C_M$ and the MT progresses, resulting in increase in strain. The path (4) expresses the condition with holding strain constant. In this case, the MT progresses due to decrease in temperature under the constant strain and stress relaxation is induced due to the MT as observed in Takeda et al. (2014). The path (5) represents the thermomechanical neutral loading. In this case, the MT stops with the condition $\mathrm{d}\sigma/\mathrm{d}T = C_M$. The path (6) is the case of elastic unloading. In this case, the MT does not appear with the condition $\mathrm{d}\sigma/\mathrm{d}T > C_M$.

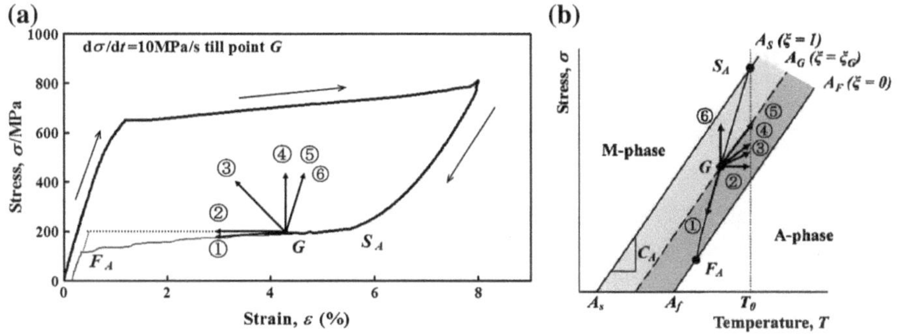

Fig. 21.9 Stress-strain diagram and various thermomechanical-subloop loading paths during the unloading process under the stress-controlled condition with constant stress rate till the point G. **a** Stress-strain diagram for loading paths ①-⑥. **b** Thermomechanical-subloop loading paths ①-⑥

21.2.7.3 Subloop-Deformation Behavior During the Unloading Process

The stress-strain diagram and stress-temperature path subjected to various thermomechanical subloop loadings from the point G during the unloading process under the stress-controlled condition are schematically shown in Fig. 21.9. Tensile loading and unloading are applied continuously under a constant stress rate till the point G. The stress-strain relationships ①-⑥ shown by the arrows for the subloop process from the point G in Fig. 21.9a correspond respectively to the thermomechanical-subloop loading paths ①-⑥ shown by the arrows in Fig. 21.9b.

The reverse transformation starts at the point S_A during unloading at temperature T_0. The lower stress plateau starts from the point S_A in the stress-strain diagram and temperature decreases due to the reverse transformation. The paths ①-⑥ from the point G and the corresponding deformation behaviors can be explained as follows. The path ① corresponds to the continuous unloading under a constant stress rate. In this case, the reverse transformation progresses with $d\sigma/dT > C_A$ and the lower stress plateau finishes at the point F_A. The path ② expresses the condition with holding stress constant. In this case, the reverse transformation progresses with increasing temperature under the constant stress and creep recovery is induced due to the reverse transformation as observed in Takeda et al. 2013a, b. The path ③ corresponds to reloading under a constant stress rate. In this case, both stress and temperature increases with $d\sigma/dT < C_A$ and the reverse transformation progresses, resulting in decrease in strain. The path ④ expresses the condition with holding strain constant. In this case, the reverse transformation progresses with increasing temperature under the constant strain and stress recovery is induced due to the reverse transformation as observed in Takeda et al. (2014). The path ⑤ represents the thermomechanical neutral reloading. In this case, $d\sigma/dT = C_A$ and the reverse transformation stops. The path ⑥ is the case of elastic reloading. In this case $d\sigma/dT > C_A$ and the reverse transformation does not appear.

21.3 Thermomechanical Properties and Modeling of Shape Memory Polymer

21.3.1 Linear Constitutive Equation

In Tobushi et al. (1997), the authors proposed a linear constitutive model which was developed by modifying a linear viscoelastic theory. The modified linear constitutive model involves a slip element due to internal friction in a three-element standard linear viscoelastic model. The model expresses the fact that, in the process of creep recovery under no-load which follows loading with creep strain ε_c, a certain part of ε_c remains as irrecoverable strain ε_s. Thermal expansion is also considered in the model. Thermal expansion is assumed to be independent on the mechanical behavior. The relationship in the modified linear constitutive equation is expressed as follows

$$\dot{\varepsilon} = \frac{\dot{\sigma}}{E} + \frac{\sigma}{\mu} - \frac{\varepsilon - \varepsilon_s}{\lambda} + \alpha \dot{T}, \tag{21.25}$$

$$\varepsilon_s = S\varepsilon_c \tag{21.26}$$

where σ, ε and T denote stress, strain and temperature, respectively. The dot denotes time derivative. E, μ, λ and α represent modulus of elasticity, viscosity, retardation time and coefficient of thermal expansion. E, μ and λ depend on temperature. In the case of a standard linear elastic model, $\varepsilon_s = \alpha = 0$ in Eq. (21.25). In the creep recovery process, $\sigma = 0$ and therefore Eq. (21.25) becomes $\dot{\varepsilon} + \varepsilon/\lambda = 0$. Based on this equation, strain is recovered with lapse of time and finally is recovered perfectly. S in Eq. (21.26) is a proportional coefficient. As expressed by Eq. (21.26), a certain ratio S of creep strain ε_c appears as irrecoverable strain ε_s. At high temperature T_h above the glass transition temperature T_g, large ε_c appears under low stress and ε_c is recovered after unloading. Therefore S is small at T_h. On the contrary, at low temperature T_l below T_g, ε_c appears under high stress and large amount of ε_c remains after unloading. Therefore S is large at T_l.

21.3.2 Nonlinear Constitutive Equation

Large strain is recovered by heating in SMP. The behavior of large strain cannot be expressed by a linear model. In order to express the behavior of large strain, the linear constitutive equation is modified. The clear yielding stage appears in the stress-strain curve and strain is recovered elastically with a steep slope during unloading, resulting in large amount of residual strain at temperatures below T_g. In order to express nonlinear time-independent strain, with respect to a linearly elastic term $\dot{\varepsilon}/E$ in Eq. (21.25), let us consider a nonlinear term expressed by a power function of stress. Because the rate of increase in creep strain increases nonlinearly with an increase in stress (Tobushi et al. 1996), a nonlinear term expressed by a power function of stress is added to a linearly viscous term σ/μ. By considering these nonlinear terms, Eq. (21.25) becomes

$$\dot{\varepsilon} = \frac{\dot{\sigma}}{E} + m\left(\frac{\sigma - \sigma_y}{k}\right)^{m-1}\frac{\dot{\sigma}}{k} + \frac{\sigma}{\mu} + \frac{1}{b}\left(\frac{\sigma}{\sigma_c} - 1\right)^n - \frac{\varepsilon - \varepsilon_s}{\lambda} + \alpha\dot{T}, \quad (21.27)$$

where σ_y and σ_c denote proportional limits of stress in the time-independent term and the viscous term, respectively, and correspond to yield stress and creep limit.

If large strain is applied at a constant strain rate in the loading process, seemingly time-independent plastic strain ε_p remains after unloading. Let us consider ε_p corresponding to time-dependent ε_c in the right-hand side of Eq. (21.26). In order to express irrecoverable strain ε_s, Eq. (21.26) becomes

$$\varepsilon_s = S(\varepsilon_c + \varepsilon_p), \quad (21.28)$$

where S is a proportional coefficient.

The material parameters involved in Eqs. (21.27) and (21.28) vary depending on temperature.

21.3.3 Dependence of Coefficients on Temperature

The mechanical properties of SMP vary markedly in the glass transition region. This appears due to the fact that SMP is composed of soft segments and hard segments and that micro-Brownian motion of soft segments is activated at temperatures above T_g but is frozen at temperatures below T_g (Tobushi et al. 1996). In order to express the properties in the glass transition region, let us consider the dependence of coefficients in Eqs. (21.27) and (21.28) on temperature.

The relationship between modulus of elasticity E and temperature T obtained by the dynamic mechanical tests on SMP is shown in Fig. 21.10a. In Fig. 21.10a, E is expressed in logarithmic scale and T in a reciprocal normalized by T_g. As seen in Fig. 21.10a, E varies markedly in the glass transition region but is almost constant above and below the glass transition region ($T_W = T_g \pm 15$ K). As shown in Fig. 21.10b, the relationship between $\log E$ and T_g/T is approximately represented by a straight line with a slope of a. The relationship in the glass transition region $T_g - T_w \leq T \leq T_g + T_w$ is expressed as follows:

$$\log E - \log E_g = a\left(\frac{T_g}{T} - 1\right)$$

Thus E becomes

$$E = E_g \exp\left[a\left(\frac{T_g}{T} - 1\right)\right] \quad (21.29)$$

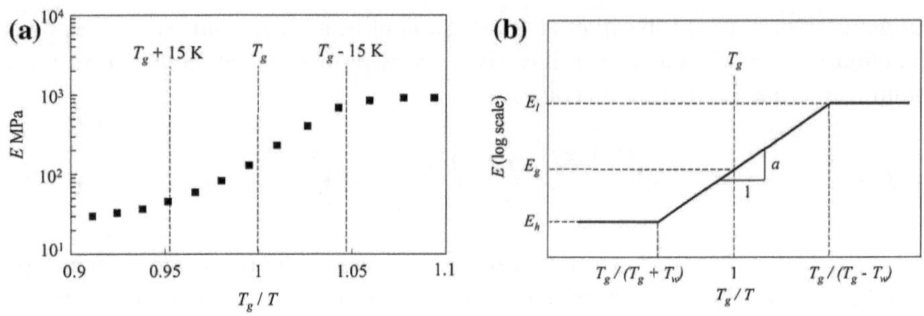

Fig. 21.10 Relationship between elastic modulus and temperature. **a** Experimental result. **b** Approximate curve

where E_g is the value of E at $T = T_g$. As seen in Fig. 21.10b, E_h and E_l denote the values of E above and below the glass transition region, respectively. The exponential function (21.29) is similar to the equation for viscosity which was derived theoretically by Eyring and empirically by Arrhenius.

In the same manner as E, in order to express the deformation properties in the glass transition region, the material parameters $k, \sigma_y, \mu, \sigma_c, \lambda$, and S mentioned in Sect. 21.3.2 are expressed by an exponential function of temperature T. These parameters are denoted by x and are expressed as follows

$$x = x_g \exp\left[a\left(\frac{T_g}{T} - 1\right)\right], \tag{21.30}$$

where x_g is the value of x at $T = T_g$. The boundaries of the glass transition region are $T_g \pm T_w$. Each coefficient is constant above and below the glass transition region. As observed in Tobushi et al. (1997), values of the exponent a in Eq. (21.30) are 34–45 for the large majority of coefficients x. In the present study, for simplicity of the theory, a is assumed to be constant for each coefficient. Therefore, by taking into a constant a, each coefficient x can be evaluated by the characteristic value x_g at T_g. Therefore, coefficients are determined easily based on the tests at T_g.

21.3.4 Results and Discussion of Modeling

21.3.4.1 Material Parameters

The material was an SMP thin film of polyurethane series (Diary MS5510; produced by SMP Technologies Inc.). The glass transition temperature T_g was about 328 K. Based on the experimental results, coefficients in Eqs. (21.27)–(21.30) were determined as follows. At first, the coefficients of time-dependent terms were determined based on the creep test under constant stress and the stress relaxation test

Table 21.3 Values of coefficients at T_g

E_g (MPa)	k_g (MPa)	σ_{yg} (MPa)	μ_g (GPa s)	σ_{cg} (MPa)	λ_g (s)	S_g (−)
146	400	4.0	35.0	0.5	390	0.1

Table 21.4 Values of coefficients above and below T_g calculated by Eq. (21.30)

Coefficients	$T \geq T_g + 15$ (K)	$T \leq T_g - 15$ (K)
E (MPa)	27.7	902
k (MPa)	75.9	2470
σ_y (MPa)	0.759	24.7
μ (GPa s)	6.64	216
σ_c (MPa)	0.0949	3.09
λ (s)	74	2410
S (-)	0.019	0.618

under constant strain. Next to the tests, the coefficients of time-independent terms were determined based on the tension tests. In this procedure, the time-independent stress-strain curves were obtained by removing the time-dependent strain from the stress-strain curves obtained through the tension tests. Finally, the values of coefficients were revised to express the overall characteristics of the thermomechanical properties and recovery stress. From the results of the experiment, it was found that the behavior of stress and strain varied markedly in the temperature region $T_g - 15$ K $\leq T \leq T_g + 15$ K but varied slightly above and below the temperature region. The temperature width T_w of the glass transition region was determined to be 15 K. Based on the tests, the exponent a in Eqs. (21.29) and (21.30) was decided to be 38. The values of x_g at T_g for the coefficients in Eqs. (21.29) and (21.30) are shown in Table 21.3. The values of the coefficients x above and below the glass transition region calculated by Eq. (21.30) are shown in Table 21.4. The dependence of the material parameters m, n, b and α on temperature was slight compared with other parameters x. Therefore these parameters were assumed to be independent on temperature for simplicity of the theory. The parameters m, n, b and α were determined to be 2, 1.01, 10^4 s and 11.6×10^{-5} K^{-1}, respectively. These values were average values at T_g, above T_g and below T_g.

21.3.4.2 Shape Fixity and Shape Recovery

Loading Path and Deformation Property

The three-dimensional stress-strain-temperature diagram in the thermomechanical tests is shown in Fig. 21.11. As shown in Fig. 21.11, the maximum strain ε_m was applied at high temperature T_h in the process ①. Maintaining ε_m constant, the specimen was cooled to T_l in the process ②. The specimen was unloaded at T_l in

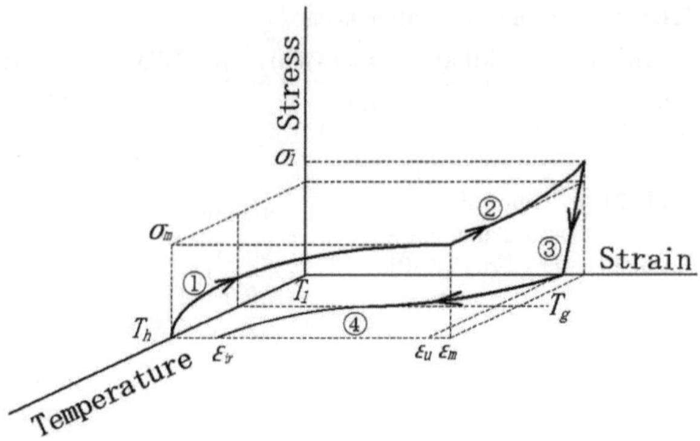

Fig. 21.11 Three-dimensional stress-strain-temperature diagram showing the loading path in the thermomechanical test

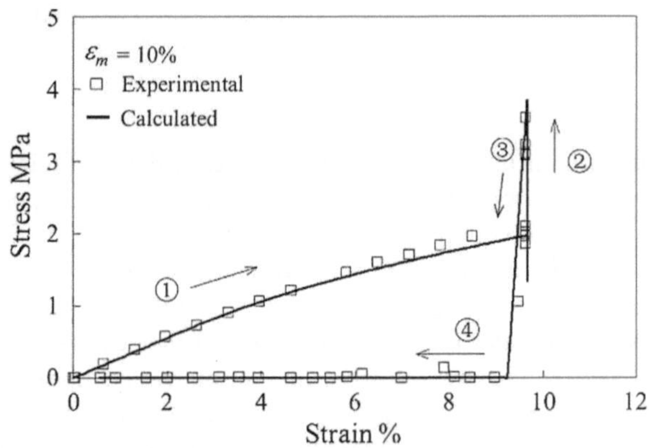

Fig. 21.12 Relationship between stress and strain in the thermomechanical test; $\varepsilon_m = 10\,\%$

the process ③. It was heated from T_l to T_h under no-load condition in the process ④. T_h was $T_g + 20$ K, $T_l = T_g - 20$ K and $\varepsilon_m = 10\,\%$. Strain rate was $8.33 \times 10^{-3}\ s^{-1}$. Average heating rate was 0.0667 K/s but average cooling rate 0.133 K/s. The stress-strain curves, stress-temperature curves and strain-temperature curves obtained from the test are shown in Figs. 21.12, 21.13 and 21.14, respectively. In these figures, symbols ①–④ denote the loading process. The solid curves denote the calculated results. In the following sections, let us discuss the experimental results and calculated results for each figure.

Stress-Strain Relationship

As seen in Fig. 21.12, during the loading process ① up to ε_m at T_h, strain increases nonlinearly when stress becomes large. In the cooling process ② from T_h to T_l under constant ε_m, stress increases. In the unloading process ③, the slope of the curve is steep and the strain ε_u at a termination point of unloading is close to ε_m. The ratio $\varepsilon_u/\varepsilon_m$ represents the rate of shape fixity and approaches 1 if ε_m becomes large.

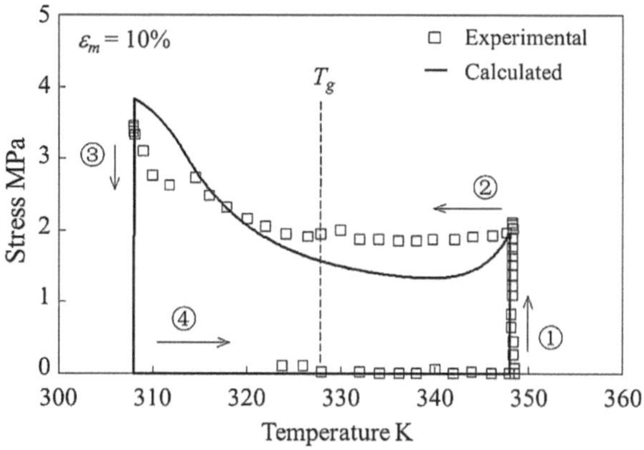

Fig. 21.13 Relationship between stress and temperature in the thermomechanical test; $\varepsilon_m = 10\,\%$

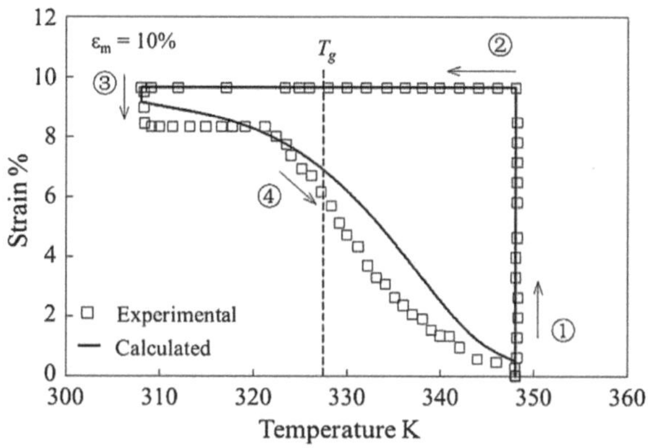

Fig. 21.14 Relationship between strain and temperature in the thermomechanical test; $\varepsilon_m = 10\,\%$

Modulus of elasticity of SMP is large at T_l and small at T_h. We can use excellent shape fixity of SMP elements by applying the difference of modulus of elasticity between T_l and T_h. The calculated results express the inclination of the experimental results.

Stress-Temperature Relationship

As seen in Fig. 21.13, during the cooling process ② from T_h to T_l under constant ε_m, stress decreases a little in the beginning, is constant till the vicinity of T_g and increases below T_g. This phenomenon appears due to the following reason. In the beginning of cooling, because stress relaxation occurs by maintaining ε_m constant at high temperature, stress decreases. At temperatures below T_g, modulus of elasticity increases by cooling. As the results, thermal stress against thermal contraction increases and stress therefor increases. In the vicinity of T_g, the decrease in stress due to stress relaxation is balanced with the increase in stress due to thermal stress and stress is therefore constant. In the test, average cooling rate $\dot{T}$ was 0.133 K/s but cooling rate was large in the beginning. If cooling rate is large, stress relaxation is small. The calculated results express the overall inclination of the experimental results.

Strain-Temperature Relationship

As seen in Fig. 21.14, during the heating process (4) under no-load, strain is recovered. Strain is markedly recovered in the vicinity of T_g. SMP is composed of soft segments and hard segments. Micro-Brownian motion of soft segments is frozen at temperatures below T_g but is activated at temperatures above T_g. Therefore the amount of strain recovery is small at low temperature, but becomes large in the vicinity of T_g due to activated micro-Brownian motion. The calculated results express the experimental results.

Influence of Heating Rate on Recovery Strain

Because recovery strain depends on time, let us discuss the dependence of recovery strain on heating rate. The same loading path as the thermomechanical test is considered. The calculated results of strain recovery at various heating rates $\dot{T}$ during the heating process (4) under no-load for $\varepsilon_m = 20\,\%$ are shown in Fig. 21.15. The relationship between irrecoverable strain ε_{ir} at high temperature T_h and $\dot{T}$ is shown in Fig. 21.16. As seen in Fig. 21.15, recovery of strain delays in the case of large $\dot{T}$. As seen in Fig. 21.16, if $\dot{T}$ is larger than 0.1 K/s, response of recovery strain delays markedly. In the case of large SMP elements subjected to large heating rate, even if surface elements reach high temperature, the increase in temperature of internal elements delays. Therefore in the design of large SMP elements, it is important to consider response speed of recovery deformation.

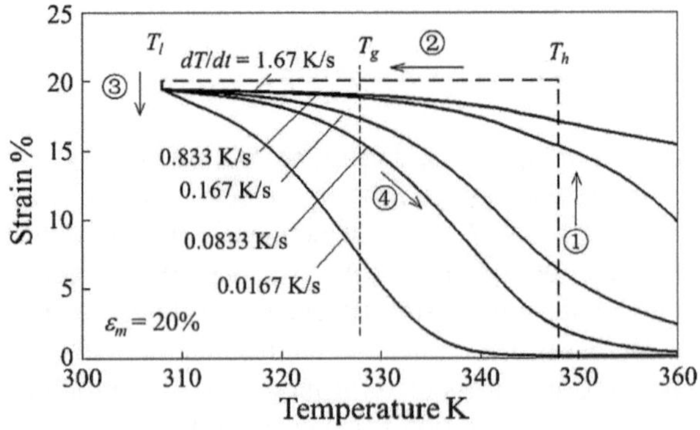

Fig. 21.15 Strain recovery during the heating process under no-load at various heating rates

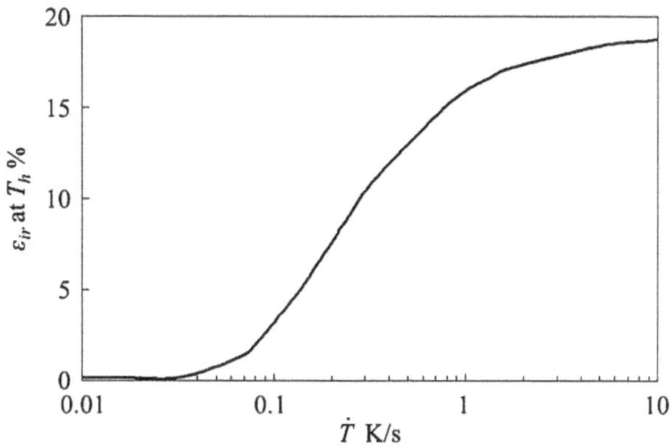

Fig. 21.16 Dependence of irrecoverable strain at T_h on heating rate $\dot{T}$

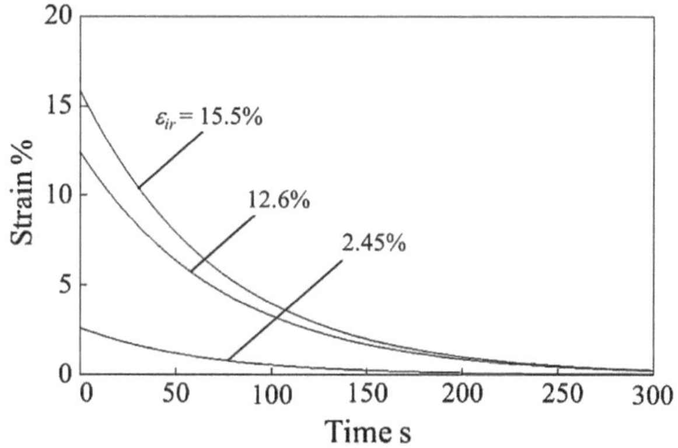

Fig. 21.17 Strain recovery with lapse of time under no-load at T_h

Although irrecoverable strain is ε_{ir} when temperature reaches T_h, strain is recovered gradually with lapse of time under no-load at T_h. With respect to various ε_{ir}, the calculated results of strain recovery with lapse of time under no-load at T_h are shown in Fig. 21.17. The rate of strain recovery is large in the early stage but decreases with lapse of time. As seen in Figs. 21.15, 21.16 and 21.17, in the case of large heating rate, recovery strain delays during the heating process but appears gradually after reaching high temperature. In applications of SMP elements, heating rate differs depending on objectives. Therefore it is necessary to design SMP elements by evaluating the behavior of both strain recovered depending on heating rate and strain recovered with lapse of time after reaching T_h.

21.3.4.3 Recovery Stress

The three-dimensional stress-strain-temperature diagram in the recovery-stress tests is shown in Fig. 21.18. As shown in Fig. 21.18, the initial loading paths (1)–(3) were the same as those in the thermomechanical tests. In the process (4), no-load condition

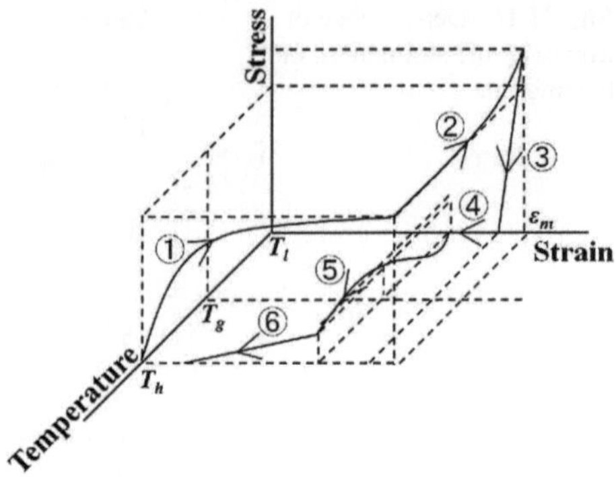

Fig. 21.18 Three-dimensional stress-strain-temperature diagram showing the loading path in the recovery-stress test

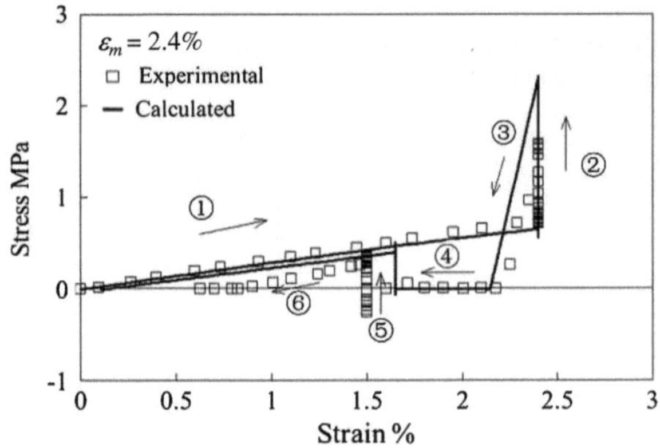

Fig. 21.19 Relationship between stress and strain in the recovery-stress test; $\varepsilon_m = 2.4\,\%$

was maintained at low temperature T_l for 10 min. In the process (5), the specimen was heated from T_l to high temperature T_h under constant strain. In the process (6), it was unloaded at T_h. The maximum strain ε_m was 2.4 %. Loading and unloading rates and heating and cooling rates were the same as those in the thermomechanical test. T_h and T_l were $T_g \pm 20$ K, respectively.

Stress-Strain Relationship

The stress-strain curves obtained by the recovery-stress tests are shown in Fig. 21.19. As seen in Fig. 21.19, the relationship in the processes (1)–(3) is the same as that in the processes (1)–(3) in Fig. 21.12. In the process (4) under no-load, strain is recovered by about 1 % for 10 min. Strain is recovered due to the delay of response for the unloading process (3). The decrease in strain is the same as creep recovery and retarded elasticity after unloading. The recovered strain is saturated for 10 min.

In the heating process (5), stress increases. The slope of the curve in the unloading process (6) is almost the same as that in the loading process (1) because strain is

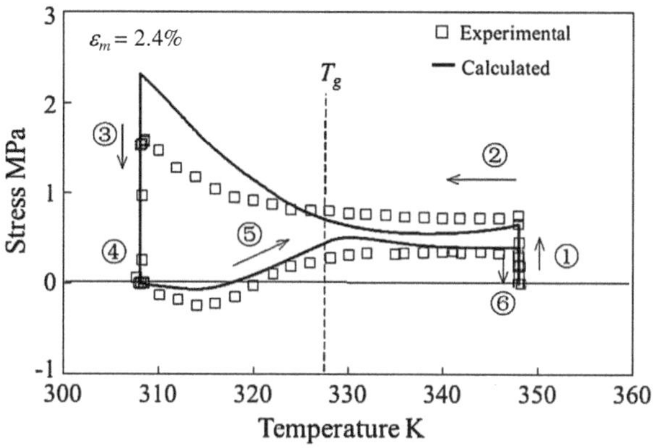

Fig. 21.20 Relationship between stress and temperature in the recovery-stress test; $\varepsilon_m = 2.4\,\%$

small in this case and main deformation is elastic. The calculated results shown by the solid curves express the inclination of the experimental results.

Stress-Temperature Relationship

The stress-temperature curves obtained by the recovery-stress tests are shown in Fig. 21.20. As seen in Fig. 21.20, the relationship in the processes (1)–(3) is the same as that in the processes (1)–(3) in Fig. 21.13. In the heating process (5) under constant strain which follows the process (4), stress decreases a little in the vicinity of T_l, increases around T_g and is almost constant above T_g. The decrease in stress which is observed in the early stage of heating process (5) appears due to compressive stress against thermal expansion. The increase in stress around T_g appears against the constraint of strain which has the inclination to be recovered due to the micro-Brownian motion of molecular chain activated by heating. If strain is not constrained, stress does not increase by heating but strain is only recovered as observed in Fig. 21.13. The increase in stress is the same as recovery stress which appears against shape memory effect due to the reverse transformation in normal shape memory alloys (Lin et al. 1995b). Although recovery stress increases around T_g and is almost constant thereafter, stress decreases a little in the vicinity of T_h. The decrease in stress appears due to stress relaxation at high temperature. Although the amount of variation in the estimated stress shown by the solid curves is larger than that in stress obtained by the test, the calculated results express the overall inclination of the experimental results.

In the design of SMP elements, the development of the theory to evaluate secondary shape forming of SMP (Tobushi et al. 2006) is necessary. The theory to evaluate the deformation properties of shape memory composite with SMP and shape memory alloy (Tobushi et al. 2013b, 2011) is also the future subject.

21.4 Conclusion

The thermomechanical constitutive equation considering the volume fractions of induced phase due to the MT and the RPT of TiNi SMA was developed. The proposed constitutive equation expresses well the properties of the SME, SE and recovery stress. The proposed theory is useful for design of SMA elements. The conditions for progress of phase transformation and subloop-deformation were discussed.

In order to describe the thermomechanical properties of SMP by a simple model from the practical viewpoint, a nonlinear constitutive equation was developed by modifying a linear constitutive model. The calculated results were discussed by performing several kinds of thermomechanical tests. The proposed theory expresses well the thermomechanical properties of SMP, such as shape fixity, shape recovery and recovery stress in the strain range of practical use. In the design of SMP elements, it is necessary to prescribe the working characteristics of the elements, such as the working start and completion temperatures, the amount of shape recovery and shape fixity, the tightening force and their response speed. The proposed theory is useful for these objectives.

References

Funakubo H (ed) (1987) Shape memory alloys. Gordon and Breach Science Publishers, New York

Gadaj SP, Nowacki WK, Tobushi H (1999) Temperature evolution during tensile test of TiNi shape memory alloy. Arch Mech 51(6):649–663

Hayashi S (1993) Properties and applications of polyurethane series shape memory polymer. Int Prog Urethanes 6:90–115

Hayashi S, Ishikawa N, Jiordano C (1993) High moisture permeability polyurethane for textile applications. J Coat Fabr 23:74–83

Irie M (1998) Shape memory polymers. In: Otsuka K, Wayman CM (eds) Shape memory materials. Cambridge University Press, Cambridge, pp 203–219

Lexcellent C, Tobushi H, Ziolkowski A, Tanaka K (1994) Thermomechanical model of reversible R-phase transformation in TiNi shape memory alloy. Int J Press Ves Pip 58:51–57

Liang C, Rogers CA, Malafeew E (1991) Preliminary investigation of shape memory polymers and their hybrid composites. In: Smart structures and materials, pp 97–105

Lin PH, Tobushi H, Ikai A, Tanaka K (1995a) Deformation properties associated with the martensitic and R-phase transformations in TiNi shape memory alloy. J Appl Biomech 10(2):4–11

Lin PH, Tobushi H, Tanaka K, Lexcellent C, Ikai A (1995b) Recovery stress of TiNi shape memory alloy under constant strain. Arch Mech 47(2):281–293

Miyazaki S, Otsuka K (1984) Mechanical behaviour associated with the premartensitic rhombohedral-phase transition in a Ti50Ni47Fe3 alloy. Philos Mag A 50(3):393–408

Miyazaki S, Otsuka K (1986) Deformation and transition behavior associated with the R-phase in Ti-Ni alloys. Metall Trans 17A:53–63

Miyazaki S, Kimura S, Otsuka K (1988) Shape-memory effect and pseudoelasticity associated with the R-phase transition in Ti-50.5 % Ni single crystals. Philos Mag A 57(3):467–478

Otsuka K, Wayman CM (eds) (1998) Shape memory materials. Cambridge University Press, Cambridge

Pieczyska EA, Tobushi H, Gadaj SP, Nowacki WK (2006) Superelastic deformation behaviors based on phase transformation bands in TiNi shape memory alloy. Mater Trans 47(3):670–676

Sawada T, Tobushi H, Kimura K, Hattori T, Tanaka K, Lin PH (1993) Stress-strain-temperature relationship associated with the R-phase transformation in TiNi shape memory alloy. JSME Int J A 35(3):395–401
Takahashi T, Hayashi N, Hayashi S (1996) Structure and properties of shape-memory polyurethane block copolymers. J Appl Polym Sci 60:1061–1069
Takeda K, Matsui R, Tobushi H, Pieczyska EA (2013a) Creep and creep recovery under stress-controlled subloop loading in TiNi shape memory alloy. Arch Mech 65(5):429–444
Takeda K, Tobushi H, Pieczyska EA (2013b) Transformation-induced creep and creep recovery of shape memory alloy. Materials 5:909–921
Takeda K, Matsui R, Tobushi H, Pieczyska EA (2014) Transformation-induced relaxation and stress recovery of TiNi shape memory alloy. Materials 7:1912–1926
Tanaka K (1986) A thermomechanical sketch of shape memory effect: one-dimensional tensile behavior. Res Mech 18:251–263
Tanaka K, Kobayashi S, Sato Y (1986) Thermomechanics of transformation pseudoelasticity and shape memory effect in alloys. Int J Plast 2:59–72
Tanaka K, Tobushi H, Iwanaga H (1988) Continuum mechanical approach to thermomechanical behavior of TiNi alloys. In: Proceedings of 31st Japan congress on materials research, pp 51–56
Tanaka K, Hayashi T, Itoh Y, Tobushi H (1992) Analysis of thermomechanical behavior of shape memory alloys. Mech Mater 13:207–215
Tobushi H, Tanaka K, Kimura K, Hori T, Sawada T (1992) Stress-strain-temperature relationship associated with the R-phase transformation in TiNi shape memory alloy. JSME Int J Ser I 35(3):278–284
Tobushi H, Hara H, Yamada E, Hayashi S (1996) Thermomechanical properties in a thin film of shape memory polymer of polyurethane series. Smart Mater Struct 5:483–491
Tobushi H, Hashimoto T, Hayashi S, Yamada E (1997) Thermomechanical constitutive modeling in shape memory polymer of polyurethane series. J Intell Mater Syst Struct 8:711–718
Tobushi H, Hashimoto T, Ito N, Hayashi S, Yamada E (1998) Shape fixity and shape recovery in a film of shape memory polymer of polyurethane series. J Intell Mater Syst Struct 9:127–136
Tobushi H, Tanaka K, Shimeno Y, Nowacki WK, Gadaj SP (1999) Influence of strain rate on superelastic behaviour of TiNi shape memory alloy. Proc Inst Mech Eng Part L J Mater Des Appl 213:93–102
Tobushi H, Okumura K, Hayashi S, Ito N (2001) Thermomechanical constitutive model of shape memory polymer. Mech Mater 33:545–554
Tobushi H, Hayashi S, Hoshio K, Miwa N (2006) Influence of strain-holding conditions on shape recovery and secondary-shape forming in polyurethane-shape memory polymer. Smart Mater Struct 15:1033–1038
Tobushi H, Hayashi S, Pieczyska EA, Date K, Nishimura Y (2011) Three-way actuation of shape memory composite. Arch Mech 63(5–6):443–457
Tobushi H, Matsui R, Takeda K, Pieczyska EA (eds) (2013a) Mechanical properties of shape memory materials. Nova Science Publishers, New York
Tobushi H, Takeda K, Matsui R, Hayashi S (2013b) Shape-memory composite actuator with SMA and SMP. In: Proceedings of 19th international conference on composite materials, pp 9285–9295

Chapter 22
Constitutive Model of Discontinuously-Reinforced Composites Taking Account of Reinforcement Damage and Size Effect and Its Application

Keiichiro Tohgo

Abstract In discontinuously-reinforced composites, their mechanical properties are affected by cracking or debonding damage and size of reinforcements such as particles or short-fibers. This article deals with a constitutive model of discontinuously-reinforced composites which can describe the matrix plasticity, evolution of cracking or debonding damage of reinforcements, and reinforcement size effects on deformation and damage. The model is developed based on the Eshelby equivalent inclusion method and Mori-Tanaka mean field concept, and can be applied to particulate- or short-fiber-reinforced composites with progressive cracking or debonding damage. Influence of progressive damage and size of reinforcements on the stress-strain relations of the composites are demonstrated based on the numerical results by the model. Furthermore, a finite element method (FEM) is developed based on the model for particulate-reinforced composites with progressive debonding damage. As an example, FEM analysis is carried out on a crack-tip field in the composites, and the influence of particle size and debonding damage on an elastic-plastic singular field around a crack-tip is discussed based on the numerical results.

Keywords Constitutive model · Discontinuously-reinforced composites · Cracking damage · Debonding damage · Reinforcement size effect · Crack-tip field

22.1 Introduction

The technique to improve mechanical performance of materials by dispersing particles or short-fibers in a matrix has been applied to ceramic-matrix, metal-matrix and polymer-matrix composites, and these materials are called particulate-reinforced

---

K. Tohgo (✉)
Department of Mechanical Engineering, Shizuoka University, 3-5-1, Johoku, Nakaku, Hamamatsu 432-8561, Japan
e-mail: tohgo.keiichiro@shizuoka.ac.jp

H. Altenbach et al. (eds.), *From Creep Damage Mechanics to Homogenization Methods*, Advanced Structured Materials 64,
DOI 10.1007/978-3-319-19440-0_22

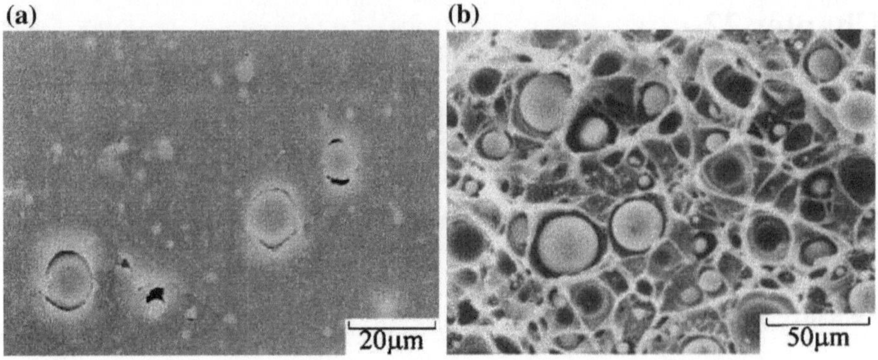

Fig. 22.1 Debonding damage of particles observed on the side surface (**a**) and fracture surface (**b**) in glass-particle-reinforced nylon 66 composites under uniaxial tension (Tohgo et al. 2001)

composites or short-fiber-reinforced composites and discontinuously-reinforced composites for both. Such composites exhibit high performance in terms of stiffness, strength and resistance to fatigue and wear, while they generally exhibit poor ductility, low fracture toughness, and strong dependency on processing (Bayha et al. 1992; Llorca et al. 1993; Whitehouse and Clyne 1993; Corbin and Wilkinson 1994; Lloyd 1994; Caceres and Griffiths 1996; Kiser et al. 1996; Kouzeli and Mortensen 2002; Hartingsveldt and Aartsen 1989; Tohgo et al. 1998, 2001; Cho et al. 2006; Adachi et al. 2008; Fu et al. 2008). In the composites, a variety of damage modes such as fracture of reinforcements, interfacial debonding between reinforcements and matrix, and cracking in matrix adjacent to hard reinforcements develop from an early stage of deformation under monotonic and cyclic loads. Observed damage modes depend on the combination of the mechanical properties of constituents and the in-situ interfacial strength between them. Figure 22.1 shows an example of debonding damage of particles and fracture surface in glass-particle-reinforced nylon 66 composites under uniaxial tension (Tohgo et al. 2001). These damage modes strongly affect mechanical performances such as stress-strain relation, tensile strength and fracture toughness (Bayha et al. 1992; Llorca et al. 1993; Whitehouse and Clyne 1993; Corbin and Wilkinson 1994; Lloyd 1994; Caceres and Griffiths 1996; Kiser et al. 1996; Kouzeli and Mortensen 2002; Hartingsveldt and Aartsen 1989; Tohgo et al. 1998, 2001; Cho et al. 2006; Adachi et al. 2008; Fu et al. 2008). It is well known that in discontinuously-reinforced composites mechanical performances also depend on reinforcement size; the reinforcement size is smaller, the yield and flow stresses are more enhanced, while the fracture toughness and ductility are often more reduced (Lloyd 1994; Kouzeli and Mortensen 2002). On the development of high-performance composites, it is important to make clear the influence of the size and damage progress of reinforcements on mechanical performances of discontinuously-reinforced composites, and constitutive models to describe the deformation of such composites is necessary.

In this article, the development of a constitutive model of discontinuously-reinforced composites taking account of the progressive reinforcement damage and reinforcement size effect is shown along the research works carried out by author and his group (Tohgo and Chou 1996; Tohgo and Weng 1994; Cho et al. 1997; Tohgo and Cho 1999; Tohgo and Itoh 2005; Tohgo et al. 2010, 2014). Furthermore, for the case of particulate-reinforced composites with progressive debonding damage, the development of a finite element method (FEM) based on the model and FEM analysis of a crack-tip field in composites are also shown (Tohgo et al. 2014).

22.2 Load Carrying Capacity of a Broken Ellipsoidal Inhomogeneity

The Eshelby equivalent inclusion method has been applied to derive constitutive models of composites by using the solution for an ellipsoidal inclusion in an infinite body (Eshelby 1957; Mori and Tanaka 1973; Mura 1982; Tandon and Weng 1988). When we consider a cracked ellipsoidal inhomogeneity in an infinite body, since there is no solution for such problem, it is difficult to derive constitutive models of the composites containing cracked reinforcements in a matrix. However, if we introduce a concept of load carrying capacity of an inhomogeneity into this problem, the constitutive models of the composites with reinforcement damage can be derived in the scheme of Eshelby equivalent inclusion method.

Load carrying capacity of a reinforcement in a composite can be defined by its average stress. High average stress of the reinforcement comparing with overall composite stress means high load carrying capacity of the reinforcement. On the other hand, when the average stress of the reinforcement is reduced by debonding or cracking damage, the load carrying capacity is also reduced, and the stress free in a void means that the load carrying capacity of the void is equal to zero.

In this section, the load carrying capacity of an intact or broken ellipsoidal in-homogeneity in an infinite body subjected to the remote applied stress as shown in Fig. 22.2 is discussed (Cho et al. 1997).

22.2.1 Intact Ellipsoidal Inhomogeneity

Figure 22.2a shows an intact ellipsoidal inhomogeneity embedded in an infinite body under applied stress $\boldsymbol{\sigma}$. The elastic stiffness tensors of the infinite body (matrix) and the inhomogeneity are denoted by $\boldsymbol{D}_0$ and $\boldsymbol{D}_\mathrm{r}$, respectively. The stress of the ellipsoidal inhomogeneity $\boldsymbol{\sigma}^\mathrm{r}$ is uniform and is given by the Eshelby equivalent inclusion method as well known (Eshelby 1957; Mura 1982)

$$\boldsymbol{\sigma}^\mathrm{r} = \boldsymbol{D}_0(\boldsymbol{S}-\boldsymbol{I})\{(\boldsymbol{D}_\mathrm{r}-\boldsymbol{D}_0)\boldsymbol{S}+\boldsymbol{D}_0\}^{-1}\boldsymbol{D}_0(\boldsymbol{S}-\boldsymbol{I})^{-1}\boldsymbol{D}_0^{-1}\boldsymbol{\sigma}, \tag{22.1}$$

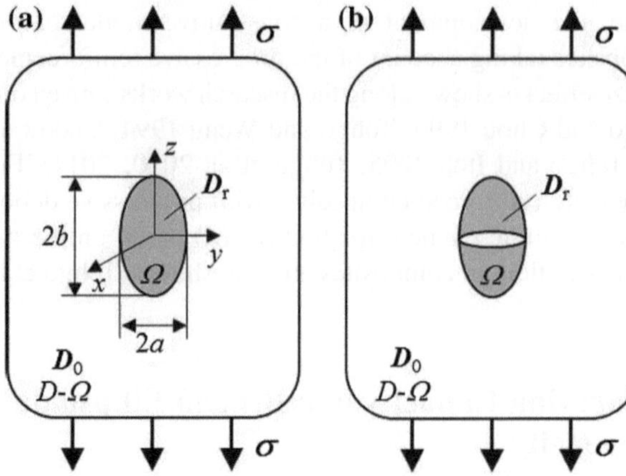

Fig. 22.2 An intact (**a**) or broken (**b**) ellipsoidal inhomogeneity in an infinite body

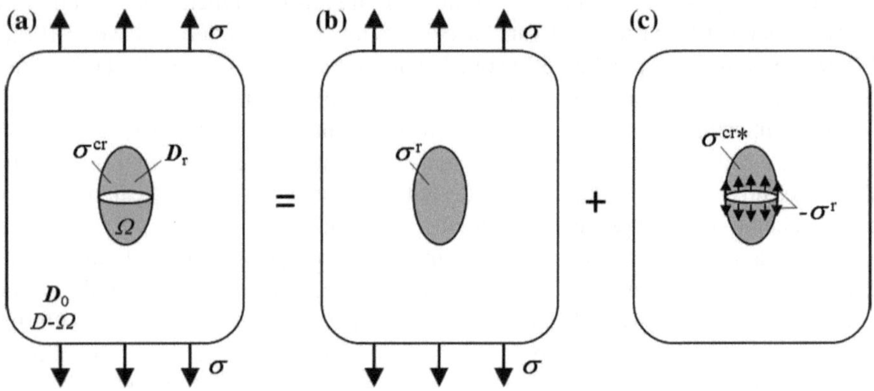

Fig. 22.3 Principle of superposition for a broken ellipsoidal inhomogeneity in an infinite body

where $\boldsymbol{S}$ is the Eshelby tensor which is expressed as a function of shape of the inhomogeneity and Poisson's ratio of the matrix (Eshelby 1957; Mura 1982), and $\boldsymbol{I}$ is the unit tensor.

22.2.2 Broken Ellipsoidal Inhomogeneity

For an ellipsoidal inhomogeneity cracked in the cross section of xy-plane as shown in Fig. 22.2b, stress distribution in the inhomogeneity seems to be complex. Figure 22.3 shows the principle of superposition for a broken ellipsoidal inhomogeneity in an infinite body. The stress state in the broken inhomogeneity $\boldsymbol{\sigma}^{cr}$ is given by the sum of the stresses $\boldsymbol{\sigma}^{r}$ and $\boldsymbol{\sigma}^{cr^*}$, where $\boldsymbol{\sigma}^{r}$ is the stress in the intact inhomogeneity under the applied stress $\boldsymbol{\sigma}$ and $\boldsymbol{\sigma}^{\mathrm{cr}^*}$ is the stress in the broken inhomogeneity subjected to internal stress $-\boldsymbol{\sigma}^{\mathrm{r}}$ on the crack surface. Therefore, the average stress of the broken inhomogeneity is expressed by

$$\bar{\boldsymbol{\sigma}}^{\mathrm{cr}} = \boldsymbol{\sigma}^{\mathrm{r}} + \bar{\boldsymbol{\sigma}}^{\mathrm{cr}^*} \tag{22.2}$$

Since $\bar{\boldsymbol{\sigma}}^{\mathrm{cr}^*}$ is given as a function of the stress of the intact inhomogeneity $\boldsymbol{\sigma}^{\mathrm{r}}$, the Eq. (22.2) can be written as

$$\bar{\boldsymbol{\sigma}}^{\mathrm{cr}} = \boldsymbol{\sigma}^{\mathrm{r}} + \boldsymbol{h}\boldsymbol{\sigma}^{\mathrm{r}} = (\boldsymbol{I} + \boldsymbol{h})\boldsymbol{\sigma}^{\mathrm{r}} = \boldsymbol{k}\boldsymbol{\sigma}^{\mathrm{r}}, \tag{22.3}$$

where $\boldsymbol{h}$ is a coefficient expressing reduction of the average stress due to the cracking damage of an ellipsoidal inhomogeneity and $\boldsymbol{k}$ is a ratio of the average stresses of the broken and intact inhomogeneities. The components of the average stresses are given by

$$\boldsymbol{\sigma}^{\mathrm{r}} = [\sigma_x^{\mathrm{r}}, \sigma_y^{\mathrm{r}}, \sigma_z^{\mathrm{r}}, \tau_{yz}^{\mathrm{r}}, \tau_{zx}^{\mathrm{r}}, \tau_{xy}^{\mathrm{r}}], \tag{22.4}$$

$$\bar{\boldsymbol{\sigma}}^{\mathrm{cr}} = [\bar{\sigma}_x^{\mathrm{cr}}, \bar{\sigma}_y^{\mathrm{cr}}, \bar{\sigma}_z^{\mathrm{cr}}, \bar{\tau}_{yz}^{\mathrm{cr}}, \bar{\tau}_{zx}^{\mathrm{cr}}, \bar{\tau}_{xy}^{\mathrm{cr}}], \tag{22.5}$$

$$\bar{\boldsymbol{\sigma}}^{\mathrm{cr}^*} = [\bar{\sigma}_x^{\mathrm{cr}^*}, \bar{\sigma}_y^{\mathrm{cr}^*}, \bar{\sigma}_z^{\mathrm{cr}^*}, \bar{\tau}_{yz}^{\mathrm{cr}^*}, \bar{\tau}_{zx}^{\mathrm{cr}^*}, \bar{\tau}_{xy}^{\mathrm{cr}^*}], \tag{22.6}$$

In Fig. 22.3c, the components of traction on the crack surface are $-\sigma_z^{\mathrm{r}}$, $-\tau_{yz}^{\mathrm{r}}$ and $-\tau_{zx}^{\mathrm{r}}$. Taking account of the symmetry of the stress distribution, the average stress components in the broken inhomogeneity are $\bar{\sigma}_z^{\mathrm{cr}^*}$, $\bar{\sigma}_x^{\mathrm{cr}^*}$ and $\bar{\sigma}_y^{\mathrm{cr}^*}$ due to $-\sigma_z^{\mathrm{r}}$, $\tau_{yz}^{\mathrm{cr}^*}$ due to $-\tau_{yz}^{\mathrm{r}}$, and $\tau_{zx}^{\mathrm{cr}^*}$ due to $-\tau_{zx}^{\mathrm{r}}$. Therefore, we have

$$\boldsymbol{h} = \begin{bmatrix} 0 & 0 & h_{13} & 0 & 0 & 0 \\ 0 & 0 & h_{23} & 0 & 0 & 0 \\ 0 & 0 & h_{33} & 0 & 0 & 0 \\ 0 & 0 & 0 & h_{44} & 0 & 0 \\ 0 & 0 & 0 & 0 & h_{55} & 0 \\ 0 & 0 & 0 & 0 & 0 & 0 \end{bmatrix} \tag{22.7}$$

and

$$\boldsymbol{k} = \begin{bmatrix} 1 & 0 & h_{13} & 0 & 0 & 0 \\ 0 & 1 & h_{23} & 0 & 0 & 0 \\ 0 & 0 & 1 + h_{33} & 0 & 0 & 0 \\ 0 & 0 & 0 & 1 + h_{44} & 0 & 0 \\ 0 & 0 & 0 & 0 & 1 + h_{55} & 0 \\ 0 & 0 & 0 & 0 & 0 & 1 \end{bmatrix} \tag{22.8}$$

Furthermore, in the case of the axisymmetric ellipsoidal inhomogeneity, the next relations are obtained

$$h_{13} = h_{23}, \qquad h_{44} = h_{55} \tag{22.9}$$

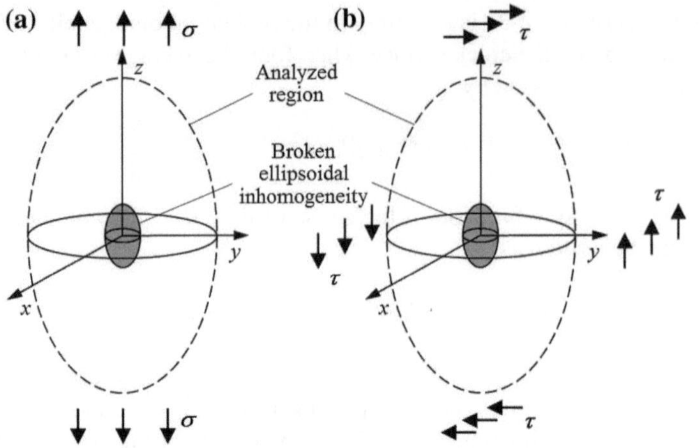

Fig. 22.4 A broken ellipsoidal inhomogeneity in an infinite body under uniaxial tension and pure shear. **a** an inhomogeneity under uniaxial tension; **b** an inhomogeneity under pure shear

Fig. 22.5 Coefficient (h_{33}) and load carrying capacity ratio on tensile stress as a function of an aspect ratio (Cho et al. 1997)

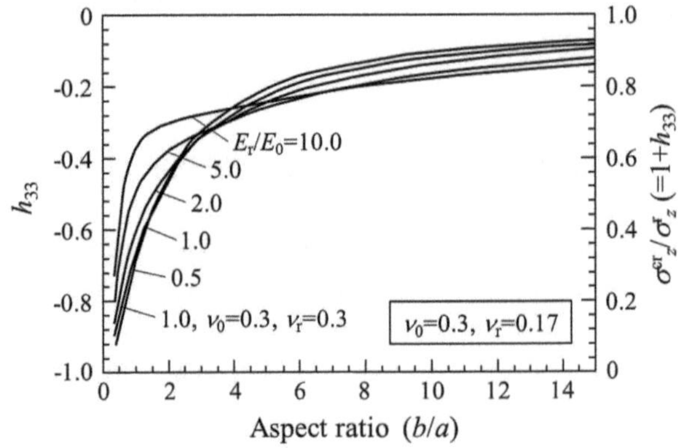

As a result, once the three components h_{33}, h_{13} and h_{44} are obtained, the average stress, i.e. the load carrying capacity, of the broken ellipsoidal inhomogeneity can be evaluated by the average stress of the intact inhomogeneity given by Eq. (22.1). Out of the three components, h_{33} and h_{13} are determined by the analysis under uniaxial tension, and h_{44} by the analysis under pure shear.

Finite element analysis of the broken ellipsoidal inhomogeneity in an infinite body under uniaxial tension and under pure shear as shown in Fig. 22.4 was carried out, and then the three components h_{33}, h_{13} and h_{44} were obtained (Cho et al. 1997). h_{33}, h_{13} and h_{44} were given as functions of an aspect ratio for a variety of combinations of the elastic moduli of inhomogeneity (E_r, ν_r) and matrix (E_0, ν_0) as shown in Figs. 22.5, 22.6 and 22.7.

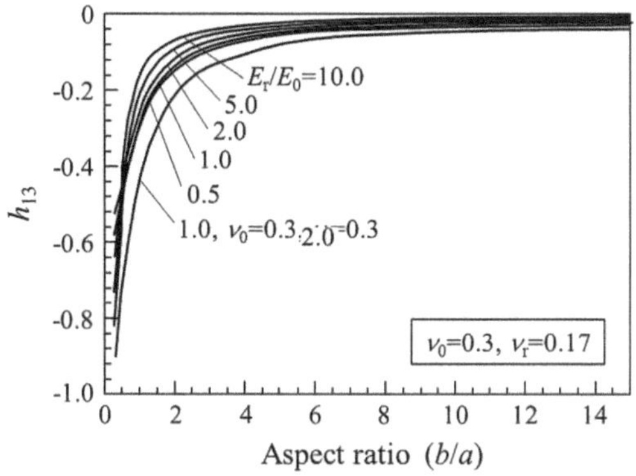

Fig. 22.6 Coefficient (h_{33}) as a function of an aspect ratio (Cho et al. 1997)

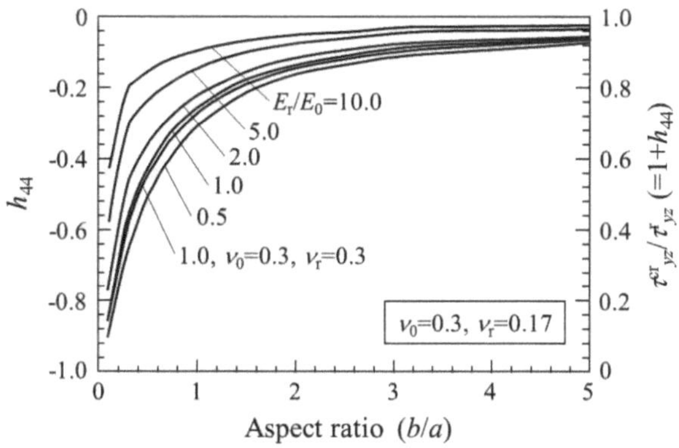

Fig. 22.7 Coefficient (h_{44}) and load carrying capacity ratio on shear stress as a function of an aspect ratio (Cho et al. 1997)

22.3 An Incremental Damage Model of Discontinuously-Reinforced Composites

In discontinuously-reinforced composites in which hard particles or short-fibers are dispersed in a ductile matrix, cracking damage of reinforcements or debonding damage between reinforcements and a matrix progresses as shown in Fig. 22.8a. Influence of progressive damage on the stress-strain relation of discontinuously-reinforced composites was studied with two schemes. Finite element analysis for a unit cell containing one reinforcement in a matrix was widely applied to the fracture or debonding of reinforcements as in Fig. 22.8b (Needleman 1987; Bao 1992; Finot et al. 1994; Brockenbrough and Zok 1995; Llorca and Segurado 2004; Eckschlager et al. 2002; Segurado and Llorca 2005, 2006). The unit cell analysis has an advantage to provide details of damage process in one reinforcement. However, because it is assumed in the unit-cell analysis that all reinforcements are in the same stage of damage process, we cannot obtain the overall stress-strain response of the realistic

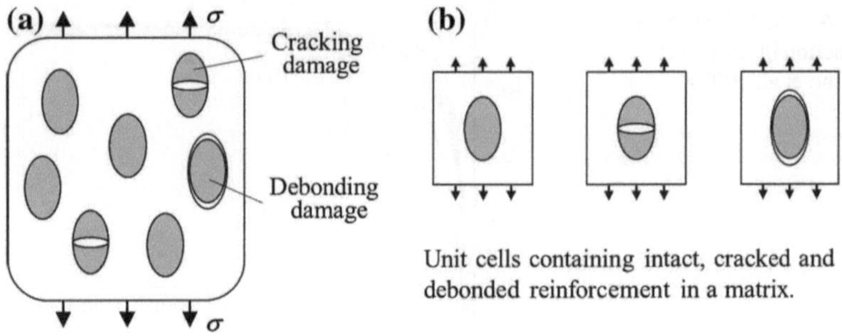

Fig. 22.8 Cracking damage and debonding damage in discontinuously-reinforced composites (**a**) and unit-cells containing intact or damaged reinforcement in a matrix (**b**)

composites containing intact reinforcements and damaged reinforcements. To overcome this problem, some modifications were proposed; for examples, the damaged cell model (Bao 1992; Brockenbrough and Zok 1995) and multi-particles cell model (Llorca and Segurado 2004; Eckschlager et al. 2002; Segurado and Llorca 2005, 2006). On the other hand, some micromechanics-based models were developed to describe the overall stress-strain relation of discontinuously-reinforced composites. In the micromechanics-based models, the damage evolution in composites can be described by transition of a volume fraction from intact reinforcements to damaged reinforcements. Tohgo and Chou (1996) and Tohgo and Weng (1994) proposed an incremental damage model of particulate-reinforced composites taking into account plasticity of a matrix and progressive debonding damage of particles based on the Eshelby equivalent inclusion method (Eshelby 1957) and Mori-Tanaka mean field concept (Mori and Tanaka 1973). This model was also extended to progressive cracking damage of reinforcements in discontinuously-reinforced composites (Cho et al. 1997; Tohgo and Cho 1999). Furthermore, constitutive models of particulate-reinforced composites with progressive debonding damage of particles were developed by Chen et al. (2003); Matous (2003); Ju and Lee (2001) and Sun et al. (2003).

In this section, an incremental damage model of discontinuously-reinforced composites with progressive damage is formulated by using the concept of load carrying capacity of a broken ellipsoidal inhomogeneity in the Eshelby equivalent inclusion method and Mori-Tanaka mean field concept (Tohgo and Chou 1996; Tohgo and Weng 1994; Cho et al. 1997; Tohgo and Cho 1999).

22.3.1 Properties of Constituent Materials

In the composite system, the reinforcements are elastic and the matrix is elastic-plastic. An elastic incremental stress-strain relation of the reinforcements is given in the isotropic form:

$$d\boldsymbol{\sigma}^{\mathrm{r}} = \boldsymbol{D}_{\mathrm{r}}(E_{\mathrm{r}}, \nu_{\mathrm{r}})d\boldsymbol{\varepsilon}^{\mathrm{r}}, \tag{22.10}$$

where $\boldsymbol{D}_{\mathrm{r}}$ is the elastic stiffness described by the Young's modulus E_r and Poisson's ratio ν_{r} of the reinforcements. The elastic behavior of the matrix is also given in the isotropic form with the elastic stiffness $\boldsymbol{D}_0$ described by the Young's modulus E_0 and Poisson's ratio ν_0 of the matrix as follows:

$$d\boldsymbol{\sigma}^{0} = \boldsymbol{D}_0(E_0, \nu_0)d\boldsymbol{\varepsilon}^{0} \tag{22.11}$$

The Prandtl-Reuss equation (the J2-flow theory) to describe the elastic-plastic deformation of the matrix is approximated by the same form as the elastic relation (Tohgo and Chou 1996):

$$d\boldsymbol{\sigma}^{0} = \boldsymbol{D}_0(E_0', \nu_0')d\boldsymbol{\varepsilon}^{0}, \tag{22.12}$$

where $\boldsymbol{D}_0(E_0', \nu_0')$ is equivalent tangential modulus obtained by replacing E_0 and ν_0 with E_0' and ν_0' in $\boldsymbol{D}_0(E_0, \nu_0)$. E_0' and ν_0' are the tangential Young's modulus and Poisson's ratio for elastic-plastic deformation and given by Berveiller and Zaoui (1979)

$$E_0' = \frac{E_0}{1 + \frac{E_0}{H'}}, \tag{22.13}$$

$$\nu_0' = \frac{\nu_0 + \frac{E_0}{2H'}}{1 + \frac{E_0}{H'}}, \tag{22.14}$$

where H' is the work-hardening ratio of the matrix:

$$H' = \frac{d\sigma_{\mathrm{e}}^{0}}{d\varepsilon_{\mathrm{e}}^{0\mathrm{p}}} \tag{22.15}$$

σ_{e}^{0} and $d\varepsilon_{\mathrm{e}}^{0\mathrm{p}}$ are the von Mises equivalent stress and incremental equivalent plastic strain, respectively,

$$\sigma_{\mathrm{e}} = \left(\frac{3}{2}\sigma_{ij}^{0\prime}\sigma_{ij}^{0\prime}\right)^{1/2}, \tag{22.16}$$

$$d\varepsilon_{\mathrm{e}}^{0\mathrm{p}} = \left(\frac{2}{3}d\varepsilon_{ij}^{0\mathrm{p}}d\varepsilon_{ij}^{0\mathrm{p}}\right)^{1/2} \tag{22.17}$$

Equation (22.12) is strictly consistent with the Prandtl-Reuss equation in the case of monotonic proportional loading.

In the composite system the stress and strain of the intact and damaged reinforcements and matrix are represented by symbols with the superscripts r, d and 0, respectively, and the overall composite stress and strain are shown by symbols without superscript.

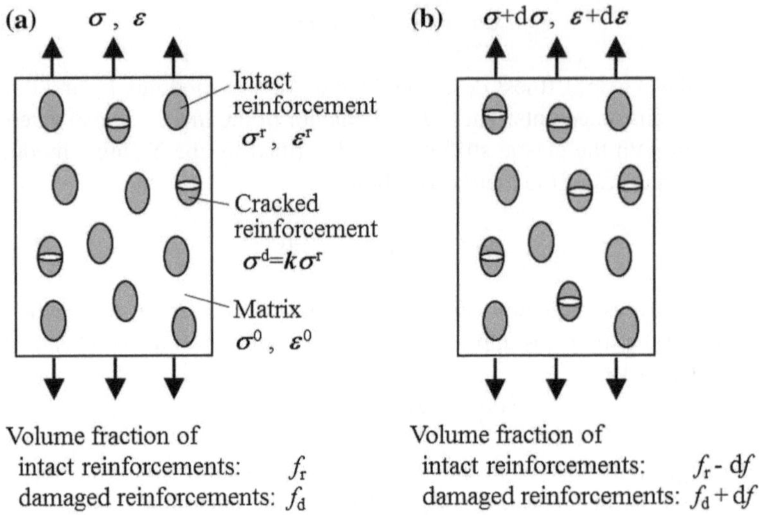

Fig. 22.9 The states of composite undergoing damage process before and after incremental deformation. $\mathrm{d}f$ is a volume fraction of the reinforcements damaged in the incremental deformation, **a** before incremental deformation, **b** after incremental deformation

22.3.2 Modeling of Progressive Damage of Reinforcements

Figure 22.9 schematically illustrates the states before and after incremental deformation of particle or aligned short-fiber-reinforced composites in damage process. As shown in Fig. 22.9, the composites contain intact and cracked reinforcements in a matrix. In composites, the microscopic stresses and strains in the reinforcements and matrix are created due to material heterogeneity, in addition to the overall composite stress and strain. Furthermore, the cracking of reinforcement leads to the stress release and reduction of load carrying capacity. To describe the progressive cracking damage of the reinforcements in the composites under incremental deformation, the damage process can be simulated as follows:

(1) The cracking damage of reinforcements is controlled by the in-situ stress of the reinforcements and the statistical behavior of strength of the reinforcements.
(2) The stress of the reinforcement is released during cracking and its load carrying capacity after cracking is described by the average stress of the broken reinforcement.
(3) The progressive damage of the composite is described by a decrease in intact reinforcement volume fraction and an increase in cracked reinforcement volume fraction.

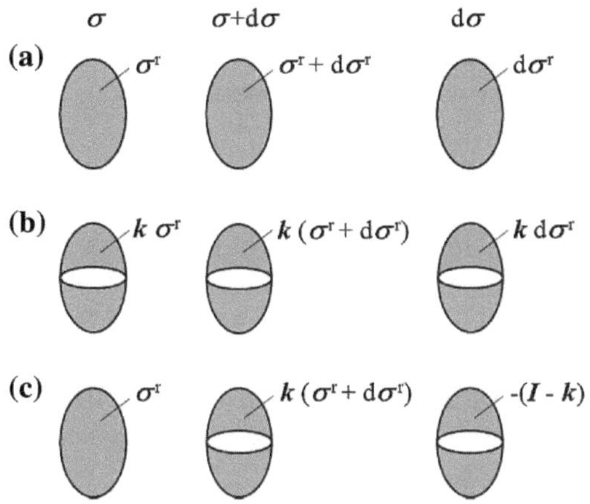

Fig. 22.10 Average stresses before and after incremental deformation and their increments for the intact and cracked reinforcements and the reinforcements to be cracked, **a** intact reinforcements, **b** cracked reinforcements, **c** reinforcements to be cracked

22.3.3 Formulation

The state before deformation shown in Fig. 22.9a is described in terms of volume fractions of the intact and damaged reinforcements f_r and f_d. If the volume fraction of the reinforcements cracked during incremental deformation is denoted by $\mathrm{d}f$, the state after deformation shown in Fig. 22.9b is described in terms of the volume fractions of $f_\mathrm{r}-\mathrm{d}f$ and $f_\mathrm{d}+\mathrm{d}f$. The composite undergoing damage process contains the intact and cracked reinforcements and the reinforcements to be cracked during the incremental deformation in the matrix.

Figure 22.10 shows the average stresses before and after incremental deformation and their increments for the intact and cracked reinforcements and the reinforcements to be cracked during incremental deformation. In the figure, the stress of the intact reinforcements $\boldsymbol{\sigma}^\mathrm{r}$ is given by the Eshelby equivalent inclusion method, and the stress of the broken reinforcements is described by $\boldsymbol{k}\boldsymbol{\sigma}^\mathrm{r}$ as in Eq. (22.3).

Following the Eshelby equivalence principle combined with Mori-Tanaka mean field concept, the incremental stress in the intact reinforcements shown in Fig. 22.10a is given by

$$\begin{aligned} \mathrm{d}\boldsymbol{\sigma}^\mathrm{r} &= \mathrm{d}\boldsymbol{\sigma} + \mathrm{d}\tilde{\boldsymbol{\sigma}} + \mathrm{d}\boldsymbol{\sigma}_1^\mathrm{pt} \\ &= \boldsymbol{D}_\mathrm{r}(\mathrm{d}\hat{\boldsymbol{\varepsilon}} + \mathrm{d}\tilde{\boldsymbol{\varepsilon}} + \mathrm{d}\boldsymbol{\varepsilon}_1^\mathrm{pt}) = \boldsymbol{D}_0(\mathrm{d}\hat{\boldsymbol{\varepsilon}} + \mathrm{d}\tilde{\boldsymbol{\varepsilon}} + \mathrm{d}\boldsymbol{\varepsilon}_1^\mathrm{pt} - \mathrm{d}\boldsymbol{\varepsilon}_1^*) \end{aligned} \tag{22.18}$$

The Eshelby equivalence principle for the cracked reinforcements shown in Fig. 22.10b can be written in the form of

$$\begin{aligned} \boldsymbol{k}\mathrm{d}\boldsymbol{\sigma}^\mathrm{r} &= \mathrm{d}\boldsymbol{\sigma} + \mathrm{d}\tilde{\boldsymbol{\sigma}} + \mathrm{d}\boldsymbol{\sigma}_2^\mathrm{pt} \\ &= \boldsymbol{k}\boldsymbol{D}_\mathrm{r}(\mathrm{d}\hat{\boldsymbol{\varepsilon}} + \mathrm{d}\tilde{\boldsymbol{\varepsilon}} + \mathrm{d}\boldsymbol{\varepsilon}_1^\mathrm{pt}) = \boldsymbol{D}_0(\mathrm{d}\hat{\boldsymbol{\varepsilon}} + \mathrm{d}\tilde{\boldsymbol{\varepsilon}} + \mathrm{d}\boldsymbol{\varepsilon}_2^\mathrm{pt} - \mathrm{d}\boldsymbol{\varepsilon}_2^*) \end{aligned} \tag{22.19}$$

Furthermore, for the reinforcements to be cracked shown in Fig. 22.10c, since the current reinforcement stress $\boldsymbol{\sigma}^{\mathrm{r}}$ should be released up to $\boldsymbol{k\sigma}^{\mathrm{r}}$ in incremental deformation, the next equation is obtained:

$$-(\boldsymbol{I}-\boldsymbol{k})\boldsymbol{\sigma}^{\mathrm{r}} = \mathrm{d}\boldsymbol{\sigma} + \mathrm{d}\tilde{\boldsymbol{\sigma}} + \boldsymbol{\sigma}_3^{\mathrm{pt}} = \boldsymbol{D}_0(\mathrm{d}\hat{\boldsymbol{\varepsilon}} + \mathrm{d}\tilde{\boldsymbol{\varepsilon}} + \boldsymbol{\varepsilon}_3^{\mathrm{pt}} - \boldsymbol{\varepsilon}_3^*) \tag{22.20}$$

Neglecting high-order terms of increments, the above equation becomes

$$-(\boldsymbol{I}-\boldsymbol{k})\boldsymbol{\sigma}^{\mathrm{r}} = \boldsymbol{\sigma}_3^{\mathrm{pt}} = \boldsymbol{D}_0(\boldsymbol{\varepsilon}_3^{\mathrm{pt}} - \boldsymbol{\varepsilon}_3^*) \tag{22.21}$$

In the above equations, $\mathrm{d}\boldsymbol{\sigma}$ and $\mathrm{d}\tilde{\boldsymbol{\sigma}}$ are the incremental applied stress and incremental average stress based on the Mori-Tanaka mean field concept, and they are related to $\mathrm{d}\hat{\boldsymbol{\varepsilon}}$ and $\mathrm{d}\tilde{\boldsymbol{\varepsilon}}$ by

$$\mathrm{d}\boldsymbol{\sigma} = \boldsymbol{D}_0\mathrm{d}\hat{\boldsymbol{\varepsilon}}, \tag{22.22}$$

$$\mathrm{d}\tilde{\boldsymbol{\sigma}} = \boldsymbol{D}_0\mathrm{d}\tilde{\boldsymbol{\varepsilon}} \tag{22.23}$$

$\mathrm{d}\sigma_1^{\mathrm{pt}}$, $\mathrm{d}\sigma_2^{\mathrm{pt}}$, σ_3^{pt} and $\mathrm{d}\varepsilon_1^{\mathrm{pt}}$, $\mathrm{d}\varepsilon_2^{\mathrm{pt}}$, $\varepsilon_3^{\mathrm{pt}}$ represent the perturbed parts of the stresses and strains in the intact and cracked reinforcements and reinforcements to be cracked, respectively. $\mathrm{d}\varepsilon_1^*$, $\mathrm{d}\varepsilon_2^*$ and ε_3^* are the Eshelby equivalent transformation strains. The perturbed strains are related to the transformation strains:

$$\mathrm{d}\varepsilon_1^{\mathrm{pt}} = \boldsymbol{S}\mathrm{d}\varepsilon_1^* \tag{22.24}$$

$$\mathrm{d}\varepsilon_2^{\mathrm{pt}} = \boldsymbol{S}\mathrm{d}\varepsilon_2^* \tag{22.25}$$

$$\varepsilon_3^{\mathrm{pt}} = \boldsymbol{S}\varepsilon_3^* \tag{22.26}$$

$\boldsymbol{S}$ is the Eshelby tensor for an ellipsoidal inclusion (Eshelby 1957; Mura 1982). $\mathrm{d}\sigma_1^{\mathrm{pt}}$, $\mathrm{d}\sigma_2^{\mathrm{pt}}$ and σ_3^{pt} are described by

$$\mathrm{d}\sigma_1^{\mathrm{pt}} = \boldsymbol{D}_0(\boldsymbol{S}-\boldsymbol{I})\mathrm{d}\varepsilon_1^* \tag{22.27}$$

$$\mathrm{d}\sigma_2^{\mathrm{pt}} = \boldsymbol{D}_0(\boldsymbol{S}-\boldsymbol{I})\mathrm{d}\varepsilon_2^* \tag{22.28}$$

$$\sigma_3^{\mathrm{pt}} = \boldsymbol{D}_0(\boldsymbol{S}-\boldsymbol{I})\varepsilon_3^* \tag{22.29}$$

Since the incremental overall composite stress $\mathrm{d}\boldsymbol{\sigma}$ is represented by the incremental average stress of the composite as follows:

$$\mathrm{d}\boldsymbol{\sigma} = (f_{\mathrm{r}} - \mathrm{d}f)\mathrm{d}\boldsymbol{\sigma}^{\mathrm{r}} + f_{\mathrm{d}}\boldsymbol{k}\mathrm{d}\boldsymbol{\sigma}^{\mathrm{r}} - \mathrm{d}f(\boldsymbol{I}-\boldsymbol{k})\boldsymbol{\sigma}^{\mathrm{r}} + (1 - f_{\mathrm{r}} - f_{\mathrm{d}})(\mathrm{d}\boldsymbol{\sigma} + \mathrm{d}\tilde{\boldsymbol{\sigma}}) \tag{22.30}$$

Neglecting high-order terms of increments and substituting Eqs. (22.18), (22.19) and (22.21) in the above equation, the incremental average stress $\mathrm{d}\tilde{\boldsymbol{\sigma}}$ is given by

$$\mathrm{d}\tilde{\boldsymbol{\sigma}} = -f_{\mathrm{r}}\mathrm{d}\boldsymbol{\sigma}_1^{\mathrm{pt}} - f_{\mathrm{d}}\mathrm{d}\boldsymbol{\sigma}_2^{\mathrm{pt}} - \mathrm{d}f\boldsymbol{\sigma}_3^{\mathrm{pt}} \tag{22.31}$$

Substituting Eqs. (22.23), (22.27), (22.28) and (22.29) into the above equation, the next equation is obtained for the incremental average strain:

$$\mathrm{d}\tilde{\boldsymbol{\varepsilon}} = -(\boldsymbol{S} - \boldsymbol{I})(f_{\mathrm{r}}\mathrm{d}\boldsymbol{\varepsilon}_1^* + f_{\mathrm{d}}\mathrm{d}\boldsymbol{\varepsilon}_2^* + \mathrm{d}f\boldsymbol{\varepsilon}_3^*) \tag{22.32}$$

The incremental overall strain $\mathrm{d}\boldsymbol{\varepsilon}$ of the composite is expressed by the incremental average strain of the composite as follows:

$$\begin{aligned}\mathrm{d}\boldsymbol{\varepsilon} = {} & (f_{\mathrm{r}} - \mathrm{d}f)(\mathrm{d}\hat{\boldsymbol{\varepsilon}} + \mathrm{d}\tilde{\boldsymbol{\varepsilon}} + \mathrm{d}\boldsymbol{\varepsilon}_1^{\mathrm{pt}}) + f_{\mathrm{d}}(\mathrm{d}\hat{\boldsymbol{\varepsilon}} + \mathrm{d}\tilde{\boldsymbol{\varepsilon}} + \mathrm{d}\boldsymbol{\varepsilon}_2^{\mathrm{pt}}) \\ & + \mathrm{d}f(\mathrm{d}\hat{\boldsymbol{\varepsilon}} + \mathrm{d}\tilde{\boldsymbol{\varepsilon}} + \boldsymbol{\varepsilon}_3^{\mathrm{pt}}) + (1 - f_{\mathrm{r}} - f_{\mathrm{d}})(\mathrm{d}\hat{\boldsymbol{\varepsilon}} + \mathrm{d}\tilde{\boldsymbol{\varepsilon}})\end{aligned} \tag{22.33}$$

Considering Eqs. (22.24), (22.25), (22.26) and (22.32) and neglecting high-order terms of increments, the above equation becomes

$$\mathrm{d}\boldsymbol{\varepsilon} = \mathrm{d}\hat{\boldsymbol{\varepsilon}} + f_{\mathrm{r}}\mathrm{d}\boldsymbol{\varepsilon}_1^* + f_{\mathrm{d}}\mathrm{d}\boldsymbol{\varepsilon}_2^* + \mathrm{d}f\boldsymbol{\varepsilon}_3^* \tag{22.34}$$

By solving the above equations, the Eshelby equivalent transformation strains $\mathrm{d}\boldsymbol{\varepsilon}_1^*$, $\mathrm{d}\boldsymbol{\varepsilon}_2^*$ and $\boldsymbol{\varepsilon}_3^*$ can be described as functions of the incremental overall composite stress $\mathrm{d}\boldsymbol{\sigma}$ and the volume fraction of reinforcements to be cracked $\mathrm{d}f$. Finally, the incremental overall strain $\mathrm{d}\boldsymbol{\varepsilon}$ versus stress $\mathrm{d}\boldsymbol{\sigma}$ relation of the composite is obtained as

$$\begin{aligned}\mathrm{d}\boldsymbol{\varepsilon} &= (\boldsymbol{I} + f_{\mathrm{r}}\boldsymbol{A}_1^{-1}\boldsymbol{B}_1 + f_{\mathrm{d}}\boldsymbol{A}_2^{-1}\boldsymbol{B}_2)\boldsymbol{D}_0^{-1}\mathrm{d}\boldsymbol{\sigma} \\ &\quad + \{(\boldsymbol{I} - \boldsymbol{S})^{-1} + f_{\mathrm{r}}\boldsymbol{A}_1^{-1}\boldsymbol{B}_1 + f_{\mathrm{d}}\boldsymbol{A}_2^{-1}\boldsymbol{B}_2\}\boldsymbol{D}_0^{-1}(\boldsymbol{I} - \boldsymbol{k})\boldsymbol{\sigma}^{\mathrm{r}}\mathrm{d}f \\ &= \mathrm{d}\boldsymbol{\varepsilon}_{\mathrm{st}} + \mathrm{d}\boldsymbol{\varepsilon}_{\mathrm{dam}},\end{aligned} \tag{22.35}$$

where

$$\begin{aligned}\boldsymbol{A}_1 = {} & (\boldsymbol{D}_{\mathrm{r}} - \boldsymbol{D}_0)^{-1}\{\boldsymbol{D}_0 + (\boldsymbol{D}_{\mathrm{r}} - \boldsymbol{D}_0)\boldsymbol{S} + (\boldsymbol{D}_{\mathrm{r}} - \boldsymbol{D}_0)(\boldsymbol{I} - \boldsymbol{S})f_{\mathrm{r}}\} \\ & - f_{\mathrm{d}}\{\boldsymbol{D}_0 + (\boldsymbol{k}\boldsymbol{D}_{\mathrm{r}} - \boldsymbol{D}_0)f_{\mathrm{d}}\}^{-1}\{\boldsymbol{k}\boldsymbol{D}_{\mathrm{r}}\boldsymbol{S} + (\boldsymbol{k}\boldsymbol{D}_{\mathrm{r}} - \boldsymbol{D}_0)(\boldsymbol{I} - \boldsymbol{S})f_{\mathrm{r}}\},\end{aligned} \tag{22.36}$$

$$\boldsymbol{B}_1 = f_{\mathrm{d}}\{\boldsymbol{D}_0 + (\boldsymbol{k}\boldsymbol{D}_{\mathrm{r}} - \boldsymbol{D}_0)f_{\mathrm{d}}\}^{-1}(\boldsymbol{k}\boldsymbol{D}_{\mathrm{r}} - \boldsymbol{D}_0) - \boldsymbol{I}, \tag{22.37}$$

$$\begin{aligned}\boldsymbol{A}_2 = {} & \{\boldsymbol{D}_0 + (\boldsymbol{D}_{\mathrm{r}} - \boldsymbol{D}_0)\boldsymbol{S} + (\boldsymbol{D}_{\mathrm{r}} - \boldsymbol{D}_0)(\boldsymbol{I} - \boldsymbol{S})f_{\mathrm{r}}\}^{-1}(\boldsymbol{D}_{\mathrm{r}} - \boldsymbol{D}_0)(\boldsymbol{I} - \boldsymbol{S})f_{\mathrm{d}} \\ & - \{\boldsymbol{k}\boldsymbol{D}_{\mathrm{r}}\boldsymbol{S} + (\boldsymbol{k}\boldsymbol{D}_{\mathrm{r}} - \boldsymbol{D}_0)(\boldsymbol{I} - \boldsymbol{S})f_{\mathrm{r}}\}^{-1}\{\boldsymbol{D}_0 + (\boldsymbol{k}\boldsymbol{D}_{\mathrm{r}} - \boldsymbol{D}_0)f_{\mathrm{d}}\}(\boldsymbol{I} - \boldsymbol{S}),\end{aligned} \tag{22.38}$$

$$\begin{aligned}\boldsymbol{B}_2 = {} & \{\boldsymbol{k}\boldsymbol{D}_{\mathrm{r}}\boldsymbol{S} + (\boldsymbol{k}\boldsymbol{D}_{\mathrm{r}} - \boldsymbol{D}_0)(\boldsymbol{I} - \boldsymbol{S})f_{\mathrm{r}}\}^{-1}(\boldsymbol{k}\boldsymbol{D}_{\mathrm{r}} - \boldsymbol{D}_0) \\ & - \{\boldsymbol{D}_0 + (\boldsymbol{D}_{\mathrm{r}} - \boldsymbol{D}_0)\boldsymbol{S} + (\boldsymbol{D}_{\mathrm{r}} - \boldsymbol{D}_0)(\boldsymbol{I} - \boldsymbol{S})f_{\mathrm{r}}\}^{-1}(\boldsymbol{D}_{\mathrm{r}} - \boldsymbol{D}_0)\end{aligned} \tag{22.39}$$

In Eq. (22.35), the first and second terms, $\mathrm{d}\boldsymbol{\varepsilon}_{\mathrm{st}}$ and $\mathrm{d}\boldsymbol{\varepsilon}_{\mathrm{dam}}$, represent the strain increment due to the stress increment and the strain increment due to the cracking damage, respectively. The incremental average stresses of the matrix and the intact and cracked reinforcements are given by

$$\begin{aligned} \mathrm{d}\boldsymbol{\sigma}^0 &= \boldsymbol{D}_0(\boldsymbol{I}-\boldsymbol{S})\{(\boldsymbol{I}-\boldsymbol{S})^{-1} + f_\mathrm{r}\boldsymbol{A}_1^{-1}\boldsymbol{B}_1 + f_\mathrm{d}\boldsymbol{A}_2^{-1}\boldsymbol{B}_2\}\boldsymbol{D}_0^{-1} \\ &\times \{\mathrm{d}\boldsymbol{\sigma} + (\boldsymbol{I}-\boldsymbol{k})\boldsymbol{\sigma}^\mathrm{r}\mathrm{d}f\}, \end{aligned} \tag{22.40}$$

$$\begin{aligned} \mathrm{d}\boldsymbol{\sigma}^\mathrm{r} &= \boldsymbol{D}_0(\boldsymbol{I}-\boldsymbol{S})\{(\boldsymbol{I}-\boldsymbol{S})^{-1} - (1-f_\mathrm{r})\boldsymbol{A}_1^{-1}\boldsymbol{B}_1 + f_\mathrm{d}\boldsymbol{A}_2^{-1}\boldsymbol{B}_2\}\boldsymbol{D}_0^{-1} \\ &\times \{\mathrm{d}\boldsymbol{\sigma} + (\boldsymbol{I}-\boldsymbol{k})\boldsymbol{\sigma}^\mathrm{r}\mathrm{d}f\}, \end{aligned} \tag{22.41}$$

$$\mathrm{d}\boldsymbol{\sigma}^\mathrm{d} = \boldsymbol{k}\mathrm{d}\boldsymbol{\sigma}^\mathrm{r} \tag{22.42}$$

Furthermore, the incremental average strains of the matrix and the intact and cracked reinforcements are given by

$$\mathrm{d}\boldsymbol{\varepsilon}^0 = \boldsymbol{D}_0^{-1}\mathrm{d}\boldsymbol{\sigma}^0, \tag{22.43}$$

$$\mathrm{d}\boldsymbol{\varepsilon}^\mathrm{r} = \boldsymbol{D}_\mathrm{r}^{-1}\mathrm{d}\boldsymbol{\sigma}^\mathrm{r}, \tag{22.44}$$

$$\begin{aligned} \mathrm{d}\boldsymbol{\varepsilon}^\mathrm{d} &= [(\boldsymbol{I}-\boldsymbol{S})\{(\boldsymbol{I}-\boldsymbol{S})^{-1} + f_\mathrm{r}\boldsymbol{A}_1^{-1}\boldsymbol{B}_1 + f_\mathrm{d}\boldsymbol{A}_2^{-1}\boldsymbol{B}_2\} + \boldsymbol{S}\boldsymbol{A}_2^{-1}\boldsymbol{B}_2]\boldsymbol{D}_0^{-1} \\ &\times \{\mathrm{d}\boldsymbol{\sigma} + (\boldsymbol{I}-\boldsymbol{k})\boldsymbol{\sigma}^\mathrm{r}\mathrm{d}f\} \end{aligned} \tag{22.45}$$

22.3.4 Cumulative Probability of the Cracked Reinforcements

The following Weibull distribution is adopted for the cumulative probability of the cracked reinforcements:

$$P_\mathrm{d}(\sigma_{\max}^\mathrm{r}) = 1 - \exp\left\{-\left(\frac{\sigma_{\max}^\mathrm{r}}{S_0}\right)^m\right\}, \tag{22.46}$$

where $\sigma_{\max}^\mathrm{r}$ is the maximum tensile stress of the intact reinforcements, and S_0 and m are the scale and shape parameters, respectively. The average strength of the reinforcements is given by

$$\bar{\sigma}_{\max}^\mathrm{r} = S_0\Gamma\left(1 + \frac{1}{m}\right), \tag{22.47}$$

where $\Gamma(\ldots)$ is the Gamma-function. If a reinforcement volume fraction is denoted by $f_{\mathrm{r}0}$, as a cumulative volume fraction of the cracked reinforcements is represented by $f_{\mathrm{r}0}P_\mathrm{d}$, the volume fraction $\mathrm{d}f$ of the reinforcements to be cracked in incremental deformation is given by

$$\mathrm{d}f = f_{\mathrm{r}0}\frac{\mathrm{d}P_\mathrm{d}}{\mathrm{d}\sigma_{\max}^r}\mathrm{d}\sigma_{\max}^r, \tag{22.48}$$

22.3.5 Equivalent Stress of the Matrix in Composite

In order to describe the matrix plasticity, we need to estimate the von Mises equivalent stress of the matrix in a composite. In the composite, as the matrix deforms heterogeneously, the microscopic stress and equivalent stress are not uniform but distributed in the matrix. On the other hand, the present constitutive model gives the average microscopic stress of the matrix. Tohgo and Weng (1994) proposed expressions to describe the equivalent stress of the matrix in the incremental damage model of the composites with debonding damage. This approach is applicable to the composite with reinforcement cracking damage.

The equivalent stress σ_{e}^{0} of the matrix in the composite before plastic deformation and damage is given by

$$\left(\sigma_{\mathrm{e}}^{0}\right)^{2}=\frac{3E_{0}}{2(1-f_{\mathrm{r}0})(1+\nu_{0})}(2U-f_{\mathrm{r}0}\boldsymbol{\sigma}^{\mathrm{r}}\boldsymbol{\varepsilon}^{\mathrm{r}})-\frac{9(1-2\nu_{0})}{2(1+\nu_{0})}\left(\sigma_{\mathrm{m}}^{0}\right)^{2}, \tag{22.49}$$

where σ_{m}^{0} is the average hydrostatic stress of the matrix, and $\boldsymbol{\sigma}^{\mathrm{r}}$ and $\boldsymbol{\varepsilon}^{\mathrm{r}}$ are the stress and strain of the reinforcements. U is the energy of a unit volume of the composite

$$U=\frac{1}{2}\boldsymbol{\sigma}\boldsymbol{\varepsilon} \tag{22.50}$$

After incremental deformation the equivalent stress of the matrix is estimated by $\sigma_{\mathrm{e}}^{0}+\mathrm{d}\sigma_{\mathrm{e}}^{0}$, where σ_{e}^{0} and $\mathrm{d}\sigma_{\mathrm{e}}^{0}$ denote the current equivalent stress before incremental deformation and its increment, respectively. In the numerical analysis σ_{e}^{0} is known and $\mathrm{d}\sigma_{\mathrm{e}}^{0}$ is given by

$$\begin{aligned}\mathrm{d}\sigma_{\mathrm{e}}^{0}=&\frac{3E_{0}}{2\sigma_{\mathrm{e}}^{0}(1-f_{\mathrm{r}}-f_{\mathrm{d}})(1+\nu_{0})}\\&\times\left\{\mathrm{d}U-\mathrm{d}R-f_{\mathrm{r}}\boldsymbol{\sigma}^{\mathrm{r}}\mathrm{d}\boldsymbol{\varepsilon}^{\mathrm{r}}-f_{\mathrm{d}}\boldsymbol{\sigma}^{\mathrm{d}}\mathrm{d}\boldsymbol{\varepsilon}^{\mathrm{d}}+\frac{1}{2}\mathrm{d}f(\boldsymbol{\sigma}^{\mathrm{r}}\mathrm{d}\boldsymbol{\varepsilon}^{\mathrm{r}}-\boldsymbol{\sigma}^{\mathrm{d}}\mathrm{d}\boldsymbol{\varepsilon}^{\mathrm{d}})\right\}\\&-\frac{9(1-2\nu_{0})}{2\sigma_{\mathrm{e}}^{0}(1+\nu_{0})}\sigma_{\mathrm{m}}^{0}\mathrm{d}\sigma_{\mathrm{m}}^{0}\end{aligned} \tag{22.51}$$

where $\boldsymbol{\sigma}^{\mathrm{d}}$ and $\boldsymbol{\varepsilon}^{\mathrm{d}}$ are the stress and strain of the cracked reinforcements. $\mathrm{d}U$ is the incremental energy of the composite and $\mathrm{d}R$ is the energy released by the cracking damage

$$\mathrm{d}U=\boldsymbol{\sigma}\,\mathrm{d}\boldsymbol{\varepsilon}, \tag{22.52}$$

$$\mathrm{d}R=\frac{1}{2}\boldsymbol{\sigma}\,\mathrm{d}\boldsymbol{\varepsilon}_{\mathrm{dam}}, \tag{22.53}$$

where $\mathrm{d}\boldsymbol{\varepsilon}_{\mathrm{dam}}$ is the strain increment due to damage corresponding to the second term in Eq. (22.35).

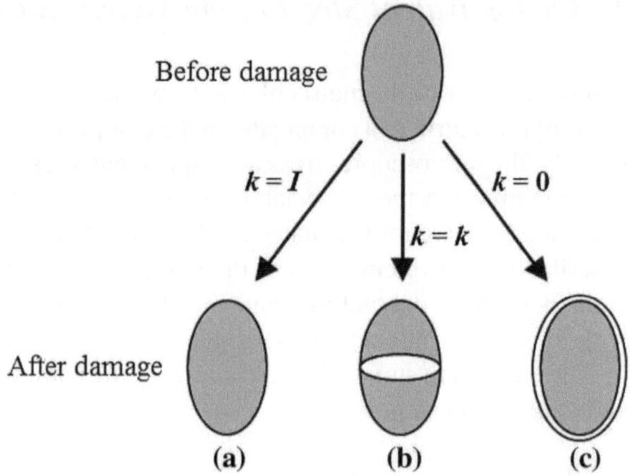

Fig. 22.11 Relationship between the type of damage and the ***k***-matrix, **a** no damage, **b** cracking damage, **c** debonding damage

If the matrix is in elastic-plastic state, the elastic moduli E_0 and ν_0 in all equations in Sect. 22.3 reduce to their elastic-plastic counterparts E_0' and ν_0', respectively.

22.3.6 Cracking Damage and Debonding Damage of Reinforcements

Although the above incremental damage model have been developed for the cracking damage of the reinforcements, this model also describes the progressive interfacial debonding damage by modifying the $\boldsymbol{k}$-matrix in Eq. (22.3). The load carrying capacity of the damaged reinforcements is described by the $\boldsymbol{k}$-matrix as shown in Fig. 22.11. The model with the $\boldsymbol{k}$-matrix for the broken inhomogeneity describes the progressive cracking damage as in Fig. 22.11b. If the unit matrix is set for the $\boldsymbol{k}$-matrix ($\boldsymbol{k} = \boldsymbol{I}$) as in Fig. 22.11a, the model describe the incremental deformation without damage, because the reinforcements maintain the whole load carrying capacity after damage. If null matrix is set for the $\boldsymbol{k}$-matrix ($\boldsymbol{k} = \boldsymbol{0}$) as in Fig. 22.11c, the model describes the progressive debonding damage, because the reinforcements lose the whole load carrying capacity after damage.

Furthermore, if $\boldsymbol{k} = \boldsymbol{0}$ and a reinforcement volume fraction f_{r0} and void volume fraction f_{d0} are set arbitrarily in the model, the model describes the deformation of the composites containing reinforcements and voids and also the porous material containing voids.

22.4 Influence of Reinforcement Damage on Stress-Strain Response of Composites

As an example of application of the incremental damage model developed in Sect. 22.3, the stress-strain response of particle or aligned short-fiber-reinforced composites under uniaxial tension was analyzed taking account of the progressive damage of reinforcements (Tohgo and Cho 1999).

The Young's modulus and Poisson's ratio were assumed as $E_0 = 500\sigma_0$, $\nu_0 = 0.3$ for a matrix, and as $E_r = 2500\sigma_0$, $\nu_r = 0.17$ for reinforcements. The equivalent stress-equivalent plastic strain relation of the matrix was given by

$$\sigma_e^0 = \sigma_0\left(1 + \frac{\varepsilon_e^{0p}}{\varepsilon_0}\right)^{0.1}, \qquad \varepsilon_0 = \frac{\sigma_0}{E_0}, \tag{22.54}$$

where σ_0 is the reference stress or the yield stress of the matrix. For damage of the reinforcements, Weibull distribution Eq. (22.46) with $m = 5.0$ and $\sigma_{max}^r = 3.0\sigma_0$ was adopted. A volume fraction of the reinforcements contained in the composites was 20 %.

22.4.1 Elastic Stress-Strain Response

Figures 22.12 and 22.13 show the influence of reinforcement damage on the stress-strain relations of elastic composites in which both reinforcements and matrix are elastic. The volume fraction of the damaged reinforcements is also shown in these figures.

Hereafter, the composite in which any damage does not develop during deformation is referred to as "perfect composite", and the composite in which the cracking

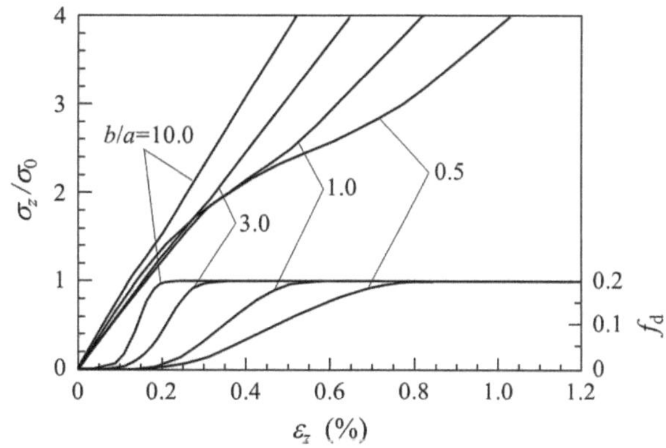

Fig. 22.12 Influence of an aspect ratio of reinforcements on the stress-strain relations of the elastic composites with progressive cracking damage (Tohgo and Cho 1999)

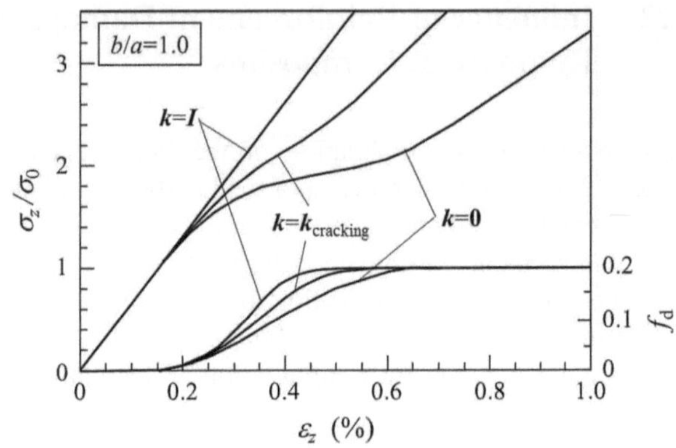

Fig. 22.13 Influence of the ***k***-matrix on the stress-strain relations of the elastic composites (Tohgo and Cho 1999)

or debonding damage occurs with deformation is referred to as "composite with damage" or "damaged composite". On the composite with progressive damage, the stress-strain relations deviate from linear relations for the perfect composite with increasing damage of the reinforcements. After all reinforcements are subjected to the damage, the stress-strain relations become linear again. Figure 22.12 shows the influence of an aspect ratio of the reinforcements on the stress-strain relations of the composites with progressive cracking damage. When the aspect ratio is lower, the influence of the cracking damage on the stress-strain relations is more drastic because the reinforcements lose larger amount of load carrying capacity (see Fig. 22.5).

The present model can describe not only the cracking damage of the reinforcements but also the interfacial debonding damage by modifying the ***k***-matrix as mentioned in Sect. 22.3.6. Figure 22.13 shows the influence of the ***k***-matrix on the stress-strain relations of particulate-reinforced composites (aspect ratio: $b/a = 1.0$). When ***k*** is the unit matrix, the stress-strain relation is linear and corresponds to the perfect composite because of no loss of the load carrying capacity of reinforcements. On the other hand, in the case of null matrix for ***k***, the stress-strain relation corresponds to the composite with progressive full-debonding damage because of complete loss of the load carrying capacity of reinforcements. It is noted from Fig. 22.13 that the full-debonding damage described by $\boldsymbol{k} = \boldsymbol{0}$ gives the lower limit of the stress-strain relation of the composite with reinforcement damage.

22.4.2 Elastic-Plastic Stress-Strain Response

Figure 22.14 shows the elastic-plastic stress-strain relations of the perfect composites and the composites with progressive cracking or debonding damage. The stress-strain relations of the composites are affected by the aspect ratio and progressive damage of the reinforcements. On comparison of the stress-strain relations under constant aspect ratio, the perfect composite gives the highest stress-strain relation, the

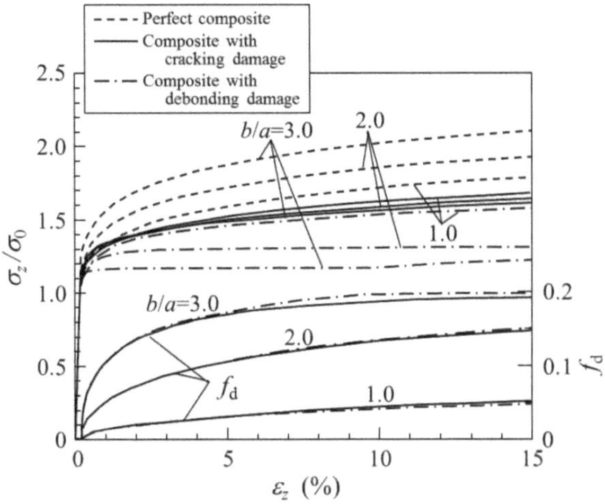

Fig. 22.14 Influence of an aspect ratio of reinforcements on the elastic-plastic stress-strain relations of the perfect composites and the composites with progressive cracking or debonding damage (Tohgo and Cho 1999)

stress-strain relation shifts downward from the relation of perfect composite by progressive damage and then the composite with debonding damage gives the lowest stress-strain relation. The influence of the aspect ratio is more complicated and is different in each type of composites. On the perfect composites, the aspect ratio of reinforcement is higher; the stress-strain relation is higher because the reinforcements carry the higher stress. On the other hand, on the composites with damage, the influence of aspect ratio comes out through the damage progress and load carrying capacity of the reinforcements after damage. The damage progress is enhanced due to high stress of the reinforcements with high aspect ratio as shown by f_d in Fig. 22.14, while the load carrying capacity is maintained in high level on the reinforcements with high aspect ratio after the cracking damage and is completely lost irrespective of the aspect ratio after the debonding damage. As a result, the stress-strain relations of the composites with debonding damage are strongly affected by the aspect ratio and shift downward with high aspect ratio, while the stress-strain relation of the composites with cracking damage are not so much affected by the aspect ratio.

22.5 Consideration of Particle Size Effects

In discontinuously-reinforced composites the mechanical performances depend on reinforcement size; the reinforcement size is smaller, the yield and flow stresses are more enhanced, while the fracture toughness and ductility are often more reduced (Lloyd 1994; Kouzeli and Mortensen 2002).

The reinforcement size effects on overall deformation behavior of the composites come from the reinforcement size effects on deformation and on damage. The reinforcement size effect on deformation is explained by an interphase between reinforcements and matrix for polymer-matrix and metal-matrix composites and by strain gradient plasticity of a matrix for metal-matrix composites. In some composite systems, an interphase with thickness of a characteristic length scale is created between reinforcements and matrix. If the mechanical properties of the interphase are different from those of the reinforcements and matrix, the overall deformation behavior of the composite depends on the reinforcement size due to surface effect of reinforcements. Based on this concept, Liu and Sun (2005, 2008), Jiang et al. (2009) and Yang et al. (2011) developed micromechanics-based models of particulate-reinforced composites. When we consider reinforcements in a matrix under plastic deformation, strain gradient around reinforcements is more enhanced by smaller reinforcements, and then geometrically necessary dislocations (GND) are stored like an interphase around reinforcements and the yield and flow stresses in the matrix especially around reinforcements become high. By introducing the influence of strain gradient plasticity, Nan and Clarke (1996) extended the Nan-Yuan effective medium approach (Nan and Yuan 1993), and Taupin et al. (2010) developed an internal length mean field approach.

The reinforcement size effect on damage should be also considered because it is well known that the fracture and debonding of reinforcements in composites is harder to occur on smaller reinforcements. For the reinforcement size effect on damage, the strain energy release rate criterion and the Weibull probability given in terms of reinforcement stress and size were used to describe nucleation and progress of the reinforcement damage as in references (Cho et al. 2006; Brockenbrough and Zok 1995; Segurado and Llorca 2006; Chen et al. 2003).

In this section, for the particulate-reinforced composites with progressive debonding damage, the incremental damage model is extended to consider the particle size effects (Tohgo et al. 2010).

22.5.1 Particle Size Effect on Deformation

In order to introduce particle size effects into the incremental damage model developed in Sect. 22.3, the Nan-Clarke simple method (Nan and Clarke 1996) has been used.

Particle size effect on deformation of the matrix is realized by introducing the dislocation plasticity for the stress-strain relation of the in-situ matrix in composites. The stress-strain relation is given by Ramberg-Osgood relation as follows:

$$\varepsilon_{\mathrm{e}}^{0} = \frac{\sigma_{\mathrm{e}}^{0}}{E_0} + \lambda \frac{\sigma_0^0}{E_0}\left(\frac{\sigma_{\mathrm{e}}^{0}}{\sigma_0^0}\right)^{1/N}, \tag{22.55}$$

where σ_{e}^{0} and $\varepsilon_{\mathrm{e}}^{0}$ are the equivalent stress and strain, E_0, σ_0^0 and N are the Young's modulus, yield stress and work-hardening exponent, respectively, and λ is a material constant. These material constants are determined for Eq. (22.55) to fit the uniaxial stress-strain relation of a bulk matrix material. Nan and Clarke (1996) assumed that σ_0^0 is affected by particles in composites, and then modified as follows:

$$\sigma_0^0 = (\sigma_0^0)_{\text{Bulk}} + \Delta\sigma_0^0, \tag{22.56}$$

where $(\sigma_0^0)_{\text{Bulk}}$ is the yield stress for a bulk of the matrix material. $\Delta\sigma_0^0$ expresses hardening of the matrix caused by particles in composites and is given by the dislocation plasticity as follows:

$$\begin{aligned}(\Delta\sigma_0^0)^2 &= (\Delta\sigma_{\text{OR}}^0 + \Delta\sigma_{\text{KIN}}^0)^2 + (\Delta\sigma_{\text{ISO}}^0)^2 + (\Delta\sigma_{\text{CTE}}^0)^2 \\ &= \left(\xi\mu_0 b\sqrt{\frac{4f_{\text{r}}}{\pi d^2}} + \zeta\mu_0 f_{\text{r}}\sqrt{\frac{\varepsilon_{\text{e}}^{0\text{p}} b}{d}}\right)^2 \\ &\quad + \left(\eta\mu_0\sqrt{\frac{f_{\text{r}}\varepsilon_{\text{e}}^{0\text{p}} b}{d}}\right)^2 + \left(\gamma\mu_0 b\sqrt{\frac{6\Delta T\Delta\theta f_{\text{r}}}{bd(1-f_{\text{r}})}}\right)^2,\end{aligned} \tag{22.57}$$

where μ_0, b, $\varepsilon_{\text{e}}^{0\text{p}}$ are the shear modulus, Burgers vector and equivalent plastic strain of the matrix, respectively, and d is particle diameter. ξ (=~ 1.0), η (=~ 0.4), ζ (=~ 2.0) and γ (=~ 1.0) are constants. $\Delta\sigma_{\text{OR}}^0$ is the Orowan stress for dislocation to pass through aligned particles, $\Delta\sigma_{\text{ISO}}^0$ and $\Delta\sigma_{\text{KIN}}^0$ are isotropic and kinematic contributions due to the effects of strain gradient plasticity related to the geometrically necessary dislocations for heterogeneous deformation around a particle in composites (Brown and Stobbs 1976), and $\Delta\sigma_{\text{CTE}}^0$ is a contribution of the dislocations stored in fabrication due to thermal expansion mismatch $\Delta\theta$ and temperature change ΔT (Arsenault and Shi 1986; Arsenault et al. 1991). $\Delta\theta$ is the difference in thermal expansion coefficient between particles and matrix, and ΔT is the difference between fabrication or heat treatment temperature and room temperature.

Contributions of the four kinds of hardening mechanisms seem to depend on composites. In this article, for the contributions of each hardening mechanism, a nonlinear form as in Eq. (22.57) was assumed (Clyne and Withers 1993).

22.5.2 *Particle Size Effect on Debonding Damage*

To consider the particle size effect on damage, it is assumed that debonding damage is controlled by a critical energy criterion for particle-matrix interfacial debonding. Let us consider a particle with diameter of d in a composite. The particle encounters debonding damage when the microscopic tensile stress of the particle reaches a critical value $\sigma_{\text{cr}}^{\text{r}}$. Since elastic strain energy stored in the particle is released and the void surface is created by full debonding damage, the following relation is obtained from energy balance during debonding process of a particle:

$$\frac{(\sigma_{\mathrm{cr}}^{\mathrm{r}})^2}{E_{\mathrm{r}}} d^3 \propto \Gamma d^2, \tag{22.58}$$

where Γ is the specific interface energy and E_{r} is the Young's modulus of the particle. Furthermore Γ means the critical strain energy release rate for particle-matrix interfacial debonding from a viewpoint of fracture mechanics. By introducing a critical stress intensity factor K_C defined by $\Gamma = K_C^2/E_{\mathrm{r}}$, Eq. (22.58) gives a proportional relationship between $\sigma_{\mathrm{cr}}^{\mathrm{r}}$ and $K_C/\sqrt{d}$ (Chen et al. 2003). In this article, the following relation is used for the sake of simplicity

$$\sigma_{\mathrm{cr}}^{\mathrm{r}} = \frac{K_C}{\sqrt{d}} \tag{22.59}$$

As Γ or K_C is uniquely given for the combination of constituent materials in composites, Eq. (22.59) represents the particle size effect on debonding damage.

22.5.3 Composites Containing Various Sized Particles

For the composites containing various sized particles, it is assumed that distribution of particle size in a composite is given by probabilistic distribution $p(d)$ represented by number frequency. Since the overall macroscopic stress and strain of the composites are given by volume average of the microscopic stress and strain, the influence of particles on macroscopic mechanical properties of the composite is not described by the number frequency $p(d)$, but volume frequency $p_{\mathrm{v}}(d)$. The volume frequency of particle $p_{\mathrm{v}}(d)$ is given by

$$p_{\mathrm{v}}(d) = \frac{1}{\int d^3 p(d)\mathrm{d}d} d^3 p(d) \tag{22.60}$$

Consequently, some values A, such as the stress, strain, intact particle volume fraction, damaged particle volume fraction and so on, of the composite containing various sized particles can be obtained by the following stochastic averaging,

$$A = \int A(d) p_{\mathrm{v}}(d)\mathrm{d}d, \tag{22.61}$$

where $A(d)$ exhibits the values for a composite with constant sized particles of d.

22.6 Influence of Debonding Damage and Particle Size in Particulate-Reinforced Composites

In this section, numerical analysis of particulate-reinforced composites under uniaxial tension was carried out based on the incremental damage model and the influence of the debonding damage and particle size on the stress-strain response was demonstrated (Tohgo et al. 2010).

22.6.1 Composites Containing Constant Sized Particles

Numerical analysis was carried out on composites containing constant sized particles under uniaxial tension to examine the influence of the particle size effects on deformation and debonding damage. The composites were assumed as a SiC particle reinforced aluminum (Al) alloy composite (SiC/A356-T4). The material properties of an Al-alloy matrix were $E_0 = 70$ GPa, $\nu_0 = 0.33$, $(\sigma_0^0)_{\text{Bulk}} = 86$ MPa, $N = 0.212$, $\lambda = 3/7$ and $b = 0.283$ nm. The material properties of SiC particles were $E_{\text{r}} = 427$ GPa and $\nu_{\text{r}} = 0.17$. It was assumed that the composites contained constant sized particles with 15 % of a particle volume fraction and then the particle size was changed in the analysis. Thermal expansion mismatch $\Delta\theta$ and temperature change ΔT for T4 heat treatment were $\Delta\theta = 19.3 \times 10^{-6}$/°C and $\Delta T = 520$ °C. In Eq. (22.57), the values of ξ, η, ζ and γ were assigned as 1.0, 0.4, 2.0 and 1.0, respectively, as in the Nan-Clarke model (Nan and Clarke 1996). In the composites with debonding damage, $\Gamma = 15.8$ N/m ($K_C = 2.6$ MPa$\sqrt{\text{m}}$) was used for interfacial debonding between particles and matrix.

Figure 22.15 shows the particle size effects on stress-strain relations of the composites without debonding damage and comparison between the present model and Nan-Clarke model (Nan and Clarke 1996). If the particle size is larger than 50 μm, the stress-strain relation converges to the result by the incremental damage model

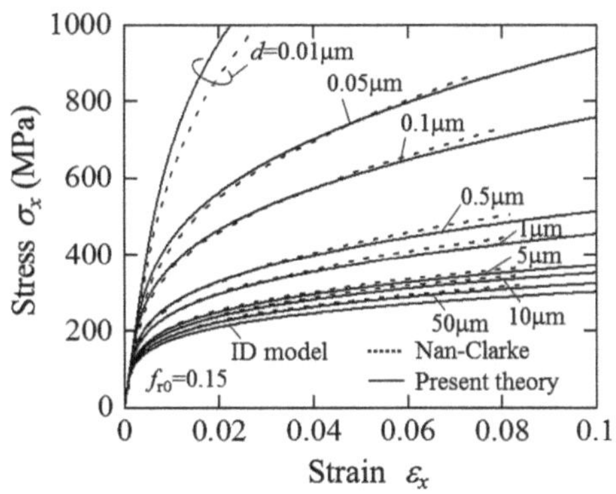

Fig. 22.15 Particle size effects on stress-strain relations of the composites and comparison between the present model and Nan-Clarke model. ID model shows the model without particle size effect (Tohgo et al. 2010)

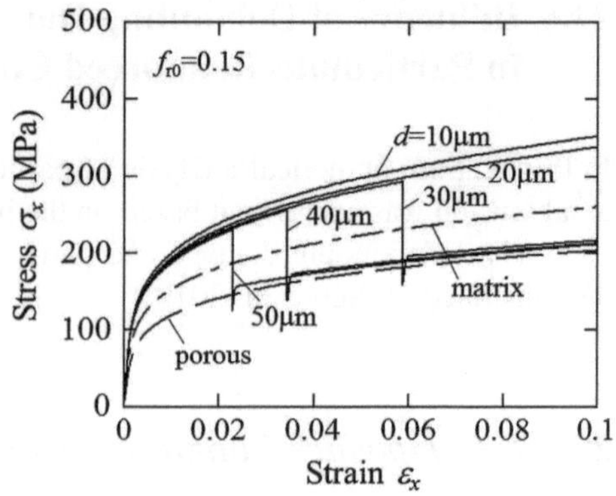

Fig. 22.16 Particle size effects on stress-strain relations of the composites with debonding damage (Tohgo et al. 2010)

without particle size effects (ID model). However, if the particle size is smaller than 10 μm, the dislocation strengthening becomes dominant with decreasing particle size as shown in Fig. 22.15. Figure 22.15 also shows that the present model predicts almost the same stress-strain relations as the Nan-Clarke model. The present model is based on the Eshelby equivalent inclusion method and Mori-Tanaka mean field concept as a composite model and adopts the tangent modulus for matrix plasticity, while the Nan-Clarke model is based on the effective medium approach (Nan and Yuan 1993) and the secant modulus for matrix plasticity.

In the composites with constant sized particles, Eq. (22.59) is spontaneously satisfied and debonding damage occurs on all particles. To describe this debonding behavior, the following numerical procedure was taken. When Eq. (22.59) was satisfied at a certain deformation stage in the numerical analysis for strain increments, the progress of debonding damage was analyzed for damage increments under the strain increment of zero ($d\boldsymbol{\varepsilon} = \mathbf{0}$) in Eq. (22.35). After all particles were debonded as $f_r = 0$ and $f_d = f_{r0}$, the numerical analysis was continued for strain increments. This numerical analysis corresponds to the finite element analysis of debonding damage process in a unit cell containing one particle.

Figure 22.16 shows the influence of particle size on the stress-strain relations of the composites with debonding damage. The stress-strain relations of the matrix material and porous material containing voids of 15 % are also shown in Fig. 22.16. The composite stress decreases down with debonding damage, and then after debonding damage the stress-strain relation of the composite is almost consistent with that of the porous material because the composite behaves as a porous material. The following two points are noticeable in Fig. 22.16. The particle size is smaller, the debonding damage is more delayed; this is obvious from Eq. (22.59). The stress-strain relation after debonding damage is slightly higher than that of the porous material; this might be caused by hardening of matrix material due to the particle size effect derived by the plastic deformation before debonding damage.

The step-wise stress decrease and pop-in on the stress-strain relations in Fig. 22.16 are the result of strict application of the deterministic criterion Eq. (22.59) to debonding damage in the composites with constant sized particles, and the similar results are also obtained in the finite element analysis of a unit cell analysis (Needleman 1987; Bao 1992; Finot et al. 1994). If the probabilistic Weibull criterion is used for the nucleation and progress of debonding damage, the smooth stress-strain relation can be obtained as shown in Fig. 22.14. The step-wise stress-strain relation due to the deterministic criterion may suggest that the nucleation of debonding damage can be a trigger of final fracture in such composites, or that the probabilistic criterion might be more realistic than the deterministic criterion.

22.6.2 Composites Containing Various Sized Particles

Numerical analysis was carried out on the composites containing various sized particles under uniaxial tension to describe the Lloyd's experimental result for a composite (15vol%SiC/A356-T4) (Lloyd 1994). The material properties used in the analysis were the same as those in Sect. 22.6.1. To describe the particle size effects, $\xi = 1.8$, $\eta = 0.4$, $\zeta = 2.0$, $\gamma = 1.8$ and $\Gamma = 15.8$ N/m ($K_C = 2.6$ MPa$\sqrt{\text{m}}$) were used. Number frequency of particles was assumed to follow the lognormal distribution,

$$p(d) = \frac{1}{\sqrt{2\pi}\,\delta d} \exp\left[-\frac{(\ln d - \phi)^2}{2\delta^2}\right], \tag{22.62}$$

where δ and ϕ are constants and the mean particle diameter $\bar{d}$ is given by

$$\bar{d} = \exp\left(\phi + \frac{\delta^2}{2}\right) \tag{22.63}$$

In Eq. (22.62), δ was set as 0.55 and ϕ were determined as 1.864 so as to be $\bar{d} = 7.5\,\mu\text{m}$. For the volume frequency of particles $p_V(d)$ obtained by Eq. (22.60), the mean particle diameter was given as $\bar{d}_V = 18.3\,\mu\text{m}$. In the numerical analysis, it was assumed that the composite contained the particles discreetly distributed in size from 0.175 to 69.825 μm with an interval of 0.35 μm according to the lognormal distribution. Incremental analysis of the composites with constant sized particles was simultaneously carried out to obtain $A(d)$, and the values A of the composite containing various sized particles was obtained by substituting the numerical results $A(d)$ for each deformation stage into Eq. (22.61).

Figure 22.17 shows the Lloyd's experimental results (Lloyd 1994) and numerical results for the stress-strain relations of the composite and matrix material under uniaxial tension. Firstly, from the comparison between the experimental result and the result without particle size effects and debonding damage (ID model), the experimental result shows higher stress-strain relation; this means that the particle size effects should be considered. Secondly, as compared the experimental result with the numerical result taking account of particle size distribution and no debonding damage (ID-PSE-SD), both results show a good agreement with each other in an

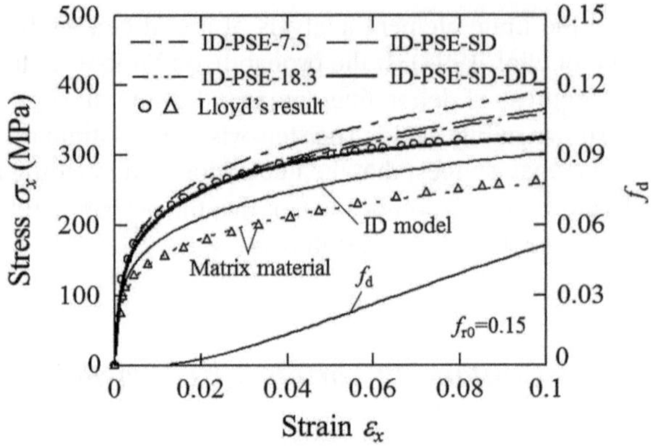

Fig. 22.17 Comparison of stress-strain relations between the predictions and Lloyd's experimental result for a composite (15vol%SiC/A356-T4). ID model is the model without particle size effect. ID-PSE-7.5 and ID-PSE-18.3 are the models considering particle size effects for composites with particle size of 7.5 and 18.3 mm, respectively. ID-PSE-SD is the model considering particle size effects and size distribution. ID-PSE-SD-DD is a model considering particle size effects, size distribution and debonding damage (Tohgo et al. 2010)

early stage of deformation, but the stress in the numerical result becomes higher than that in the experimental result as the deformation progresses. In Fig. 22.17, the numerical results for the perfect composites with constant sized particles of number average diameter 7.5 μm and volume average diameter 18.3 μm (ID-PSE-7.5 and ID-PSE-18.3) are also shown. When these results are compared with the result for the composite with particle size distribution, it is found that the composite with particle size distribution are well described by the composite with constant sized particles of volume average diameter.

As mentioned in Sect. 22.6.1, in the composite containing constant sized particles the debonding damage controlled by Eq. (22.59) occurs simultaneously. On the other hand, in the composite containing various sized particles the debonding damage occurs in turn from larger particles to smaller particles. For the composite with particle size distribution and debonding damage (ID-PSE-SD-DD) the stress-strain relation and progress of debonding damage given by void volume fraction are shown in Fig. 22.17. Finally, on the numerical result of ID-PSE-SD-DD, as the debonding damage progresses, the stress-strain relation shifts from the result of ID-PSE-SD to lower side of the stress and then describes well the experimental result.

Consequently, it is found that the experimental stress-strain relation of the SiC/A356-T4 can be described by the present model taking account of the particle size effects, particle size distribution and debonding damage.

22.7 Finite Element Method Based on the Constitutive Model

A finite element method (FEM) was developed based on the incremental damage model of particulate-reinforced composites containing various sized particles (Tohgo et al. 2014). The EFM was formulated on the basis of quadrilateral 8-noded isoparametric elements with integration at four Gauss points. Figure 22.18 shows a schematic illustration of material image at the Gauss points in finite element mesh and derivation of the tangential modulus and stress reduction due to progressive damage. The material image at the Gauss points is a composite containing various sized particles; small intact particles, already debonded large particles and particles to be damaged in the next incremental deformation. This composite is referred to as "mother composite" here. It is considered that the mother composite consists of n elements of composites in which a particle size is constant and a particle volume fraction is the same as the mother composite ($f = f_{r0}$). Their particle sizes are discretely distributed ($d = d_1$ to d_n) with an interval of Δd according to the particle size distribution. The stress-strain

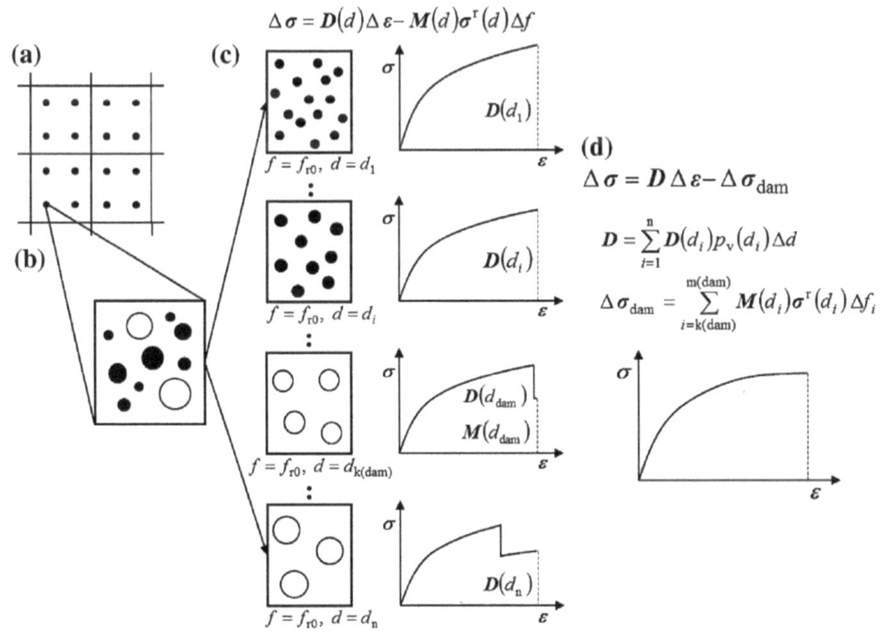

Fig. 22.18 Schematic illustration of material image at the Gauss points in finite element mesh and derivation of the tangential modulus and stress reduction due to progressive damage (Tohgo et al. 2014), **a** finite element mesh, **b** material image on Gauss points, **c** composite elements with constant particle size and their stress-strain relations and tangential moduli, **d** stress-strain relation and tangential modulus of composite with particle size distribution

relation of each composite element is described by Eq. (22.35) which is rewritten in the form of

$$d\boldsymbol{\sigma} = \boldsymbol{D}(d)d\boldsymbol{\varepsilon} - \boldsymbol{M}(d)\boldsymbol{\sigma}^{\mathrm{r}}(d)df \tag{22.64}$$

The first and second terms of right hand side in Eq. (22.64) are a stress increment due to strain increment and stress reduction due to damage progress respectively, and $\boldsymbol{D}(d)$ is the tangential modulus. The stress increment $\Delta\boldsymbol{\sigma}$ of each composite element is calculated under the same strain increment $\Delta\boldsymbol{\varepsilon}$ as the mother composite, and then the stress-strain relations up to current state and tangential moduli are obtained as in Fig. 22.18c. For the composite elements (i=k(dam) to m(dam)) in critical stage of debonding damage (see Eq. (22.59)), the debonding damage occurs in the next incremental deformation and the stress is abruptly reduced. The composite elements (i=m(dam)+1 to n) with large particles behave as porous materials because the debonding damage already occurs. By applying the stochastic averaging Eq. (22.61) to Eq. (22.64), the stress-strain relation, tangential modulus and stress reduction due to damage for the mother composite in Fig. 22.18d are given as follows:

$$\sum_{i=1}^{n} \Delta\boldsymbol{\sigma}(d_i)p_{\mathrm{v}}(d_i)\Delta d = \sum_{i=1}^{n} \boldsymbol{D}(d_i)p_{\mathrm{v}}(d_i)\Delta d\Delta\boldsymbol{\varepsilon} - \sum_{i=\mathrm{k(dam)}}^{\mathrm{m(dam)}} \boldsymbol{M}(d_i)\boldsymbol{\sigma}^{\mathrm{r}}(d_i)f_{\mathrm{r0}}p_{\mathrm{v}}(d_i)\Delta d \tag{22.65}$$

and

$$\Delta\boldsymbol{\sigma} = \boldsymbol{D}\Delta\boldsymbol{\varepsilon} - \Delta\boldsymbol{\sigma}_{\mathrm{dam}} \tag{22.66}$$

where

$$\Delta\boldsymbol{\sigma} = \sum_{i=1}^{n} \Delta\boldsymbol{\sigma}(d_i)p_{\mathrm{v}}(d_i)\Delta d, \tag{22.67}$$

$$\boldsymbol{D} = \sum_{i=1}^{n} \boldsymbol{D}(d_i)p_{\mathrm{v}}(d_i)\Delta d, \tag{22.68}$$

$$\Delta\boldsymbol{\sigma}_{\mathrm{dam}} = \sum_{i=\mathrm{k(dam)}}^{\mathrm{m(dam)}} \boldsymbol{M}(d_i)\boldsymbol{\sigma}^{\mathrm{r}}(d_i)\Delta f_i, \tag{22.69}$$

$$\Delta f_i = f_{\mathrm{r0}}p_{\mathrm{v}}(d_i)\Delta d \tag{22.70}$$

In the formulation of the FEM, the tangential modulus Eq. (22.68) and stress reduction Eq. (22.69) are used. And, after a stiffness equation of the FEM is solved for incremental deformation, the stress and strain increments of the mother composite at each Gauss point are obtained. Then the values $A(d_i)$ such as the stress, strain, intact particle volume fraction, damaged particle (void) volume fraction and so on, for each composite element are calculated and these values A for the mother composite are also obtained as

$$A = \sum_{i=1}^{n} A(d_i) p_{\mathrm{v}}(d_i) \Delta d \tag{22.71}$$

22.8 An Elastic-Plastic Singular Field Around a Crack-Tip in Particulate-Reinforced Composites

A singular stress/strain field around a crack-tip in particulate-reinforced composites is very complicated, because microscopic stress/strain fields of particles and matrix exist in addition to macroscopic stress/strain field and furthermore these stress/strain fields are affected by particle content, particle size and progressive damage. The fracture toughness is determined by the energy dissipation par unit area of crack growth when the fracture occurred at such a crack-tip. Then, a lot of mechanisms were proposed for toughening in particulate-reinforced composites; crack front bowing (or crack pinning), crack-tip blunting, matrix plastic yielding, particle-matrix interfacial debonding, micro-cracking, crack deflection by hard particles, and so on (Cho et al. 2006; Adachi et al. 2008; Fu et al. 2008; Kotoul and Vrbka 2003; Qiao 2003; Sun et al. 2011; Salviato et al. 2013). The transition of fracture morphology from ductile manner to brittle manner is also important. For example, the crack extension in brittle manner is likely induced by an increase in content of hard and small particles while is likely restrained by debonding and cracking damage of particles. As the specific fracture toughness of a particulate-reinforced composite depends on its fracture process and on which mechanisms are acting as toughening, it is difficult to derive the unified conclusions on the influence of particle content, particle size and particle damage on fracture toughness. To discuss the fracture toughness of particulate-reinforced composites, we need the details of a singular stress/strain field around a crack-tip in particulate-reinforced composites.

In this section, analysis of a crack-tip field in particulate-reinforced composites with debonding damage and containing various sized particles has been carried out by the FEM developed in Sect. 22.7. The influence of particle size and debonding damage on an elastic-plastic singular field around a crack-tip is discussed based on the numerical results (Tohgo et al. 2014).

22.8.1 Numerical Procedure

22.8.1.1 Material Properties and Debonding Properties

Numerical analysis was carried out on SiC particle reinforced aluminum (Al) alloy composites (15vol%SiC/A356-T4). For an Al-alloy matrix, the Young's modulus and Poisson's ratio were $E_0 = 70\,\mathrm{GPa}$, $\nu_0 = 0.33$, the yield stress, work hardening

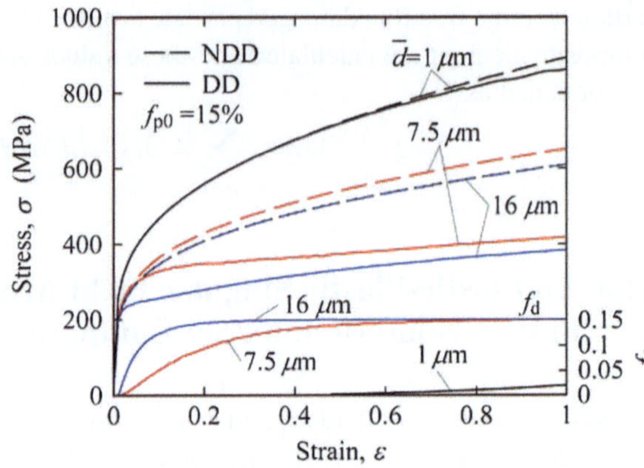

Fig. 22.19 Stress-strain relations and void volume fractions of composites under uniaxial tension (Tohgo et al. 2014)

exponent and material constant in Eq. (22.55) were $(\sigma_0^0)_{\text{Bulk}} = 86$ MPa, $N = 0.212$ and $\lambda = 3/7$, and the Burgers vector was $b = 0.283$ nm. For SiC particles, the Young's modulus and Poisson's ratio were $E_r = 427$ GPa and $\nu_r = 0.17$. The thermal expansion mismatch $\Delta\theta$ and temperature change ΔT for T4 heat treatment were $\Delta\theta = 19.3 \times 10^{-6}/°\text{C}$ and $\Delta T = 520\,°\text{C}$. To describe the particle size effects, $\xi = 1.0$, $\eta = 0.4$, $\zeta = 2.0$, $\gamma = 1.0$ in Eq. (22.57) and $K_C = 2.6\ \text{MPa}\sqrt{\text{m}}$ in Eq. (22.59) were used.

A particle volume fraction was 15vol%. Number frequency of particles was assumed to follow the lognormal distribution Eq. (22.62). In Eq. (22.62), δ was set as 0.55 and ϕ were changed as −0.151, 1.864 and 2.621 so as to be $\bar{d} = 1, 7.5$ and 16 μm. The average particle diameters in $p_v(d)$ are $\bar{d}_v = 2.5$, 18.3 and 43.8 μm for the distributions with $\bar{d} = 1, 7.5$ and 16 μm, respectively.

FEM analysis was carried out for the three kinds of composites in the cases of no debonding damage (*perfect composite*) and with debonding damage (*composite with damage*). Figure 22.19 shows stress-strain relations and damaged particle (void) volume fractions of the composites under uniaxial tension obtained by the present constitutive model. Although the fracture strain of 15vol%SiC/A356-T4 is less than 10 %, the stress-strain relations are shown up to 100 % in Fig. 22.19. On the perfect composites denoted by NDD, the stress-strain relation shifts upward with decreasing particle size. On the composites with damage denoted by DD, the stress-strain relations of the composites with $\bar{d} = 7.5$ and 16 μm drastically go down from those of perfect composites with increasing debonding damage, while in the composite with $\bar{d} = 1$ μm the debonding damage is depressed and the stress-strain relation is almost the same as the perfect composite.

22.8.1.2 Analysis of Edge-Cracked Three-Point Bending Specimen

Numerical analysis of an edge-cracked three-point bending specimen was carried out under plane strain condition by using the FEM. Figure 22.20 shows the three-point

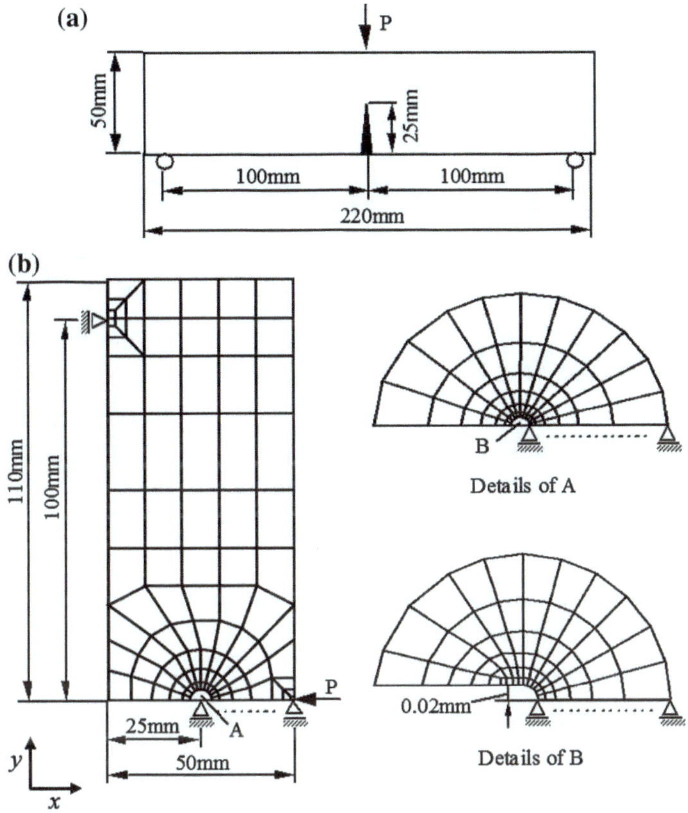

Fig. 22.20 Three-point bending specimen and its finite element mesh (Tohgo et al. 2014), **a** three-point-bending specimen, **b** finite element mesh

bending specimen and a finite element mesh for a half region of the specimen. In order to avoid the numerical problem for crack-tip singularity and to describe the crack-tip blunting due to the plastic deformation, a very thin notch with notch radius of 0.02 mm is regarded as a crack. The finite element mesh contains 228 elements and 759 nodes. The smallest element size is about 0.005 mm × 0.007 mm around a notch tip. The first step of incremental analysis was solved for linear problem up to the applied displacement which corresponds to the damage initiation or plastic deformation at only one Gauss point in the whole region. In the following steps, the incremental analysis was repeated for an incremental bending deformation up to large scale yielding of the specimen. The elastic-plastic fracture mechanics parameter J-integral was calculated for each stage of deformation from the load-displacement relation by the simple estimation method of three-point bending specimen (Rice et al. 1973).

The present FEM analysis provides us with distributions of overall macroscopic stress/strain, distributions of average microscopic stresses/strains of the matrix and

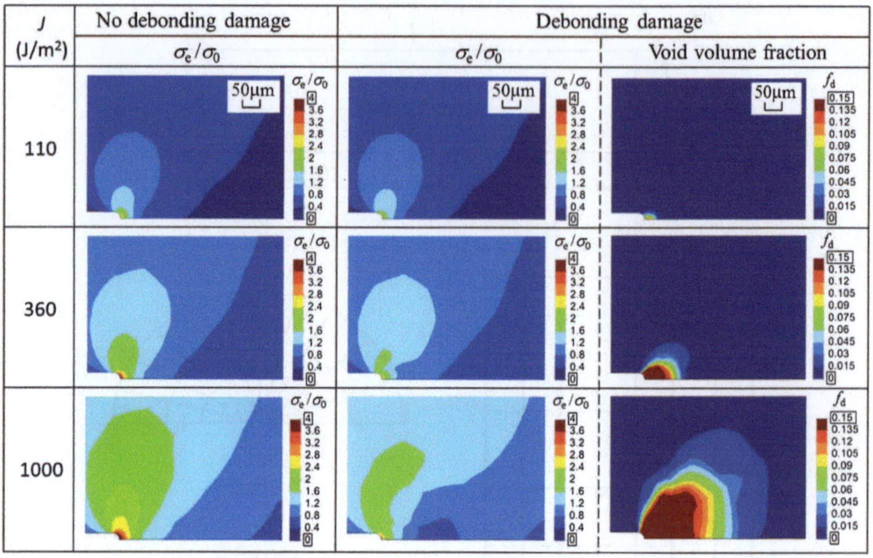

Fig. 22.21 Macroscopic equivalent stress and void volume fraction around a crack-tip (Average particle size: 16 μm) (Tohgo et al. 2014)

particles, plastic zone due to matrix plasticity and damage zone due to debonding damage of particles.

22.8.2 Elastic-Plastic Singular Field Around a Crack-Tip

22.8.2.1 Equivalent Stress Distribution and Damage Zone Around a Crack-Tip

Figure 22.21 shows the macroscopic equivalent stress and void volume fraction (damage evolution) around a crack-tip in the composites with average particle size of 16 μm for the cases of no damage and with damage. Under the low J value, the macroscopic equivalent stress is almost the same in both cases because the debonding damage is limited in the vicinity of the crack-tip. However, with increasing J value, the debonding damage develops and spreads out from a crack-tip in the composite with damage, and the macroscopic equivalent stress in the composite with damage is much lower than that in the perfect composite. This is caused by the stress reduction due to debonding damage of particles. Figure 22.22 shows the influence of particle size on the macroscopic equivalent stress and void volume fraction of the perfect composite and composite with damage under $J = 1000\,\mathrm{J/m^2}$. On the perfect composites, the macroscopic equivalent stress around a crack-tip becomes high with decreasing particle size. On the composites with damage, with increasing particle

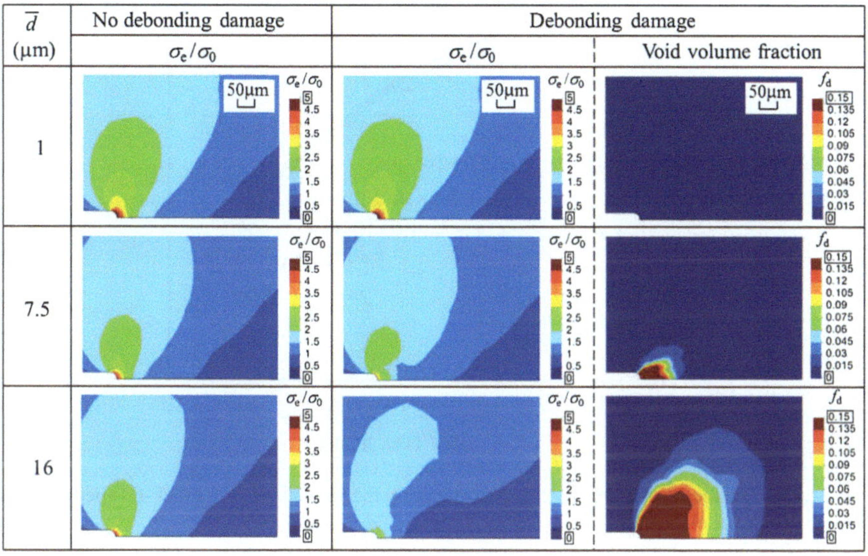

Fig. 22.22 Influence of particle size on macroscopic equivalent stress and void volume fraction around a crack-tip ($J = 1000\ \mathrm{J/m^2}$) (Tohgo et al. 2014)

size, the debonding damage becomes easy to occur and damage zone spreads out widely and as a result the macroscopic equivalent stress is drastically reduced. In the present composites, the influence of particle size on the macroscopic equivalent stress around a crack-tip is more attributed to the particle size effect on damage than the particle size effect on deformation.

22.8.2.2 Stress and Strain Distributions Ahead of a Crack-Tip

Distributions of the macroscopic stress along the line ahead of a crack-tip under five J levels are shown in Figs. 22.23 and 22.24. In these figures, the macroscopic stress is presented as a function of the normalized distance from a crack-tip in a log-log diagram, and distributions of void volume fraction are also shown for the right hand linear scale. For the elastic-plastic materials, the elastic region is described by a straight line of gradient $-1/2$, i.e. the elastic singular stress field, and the plastic region is described by a line of gradient $-N/(N+1)$, i.e. the elastic-plastic singular stress field (HRR field), where N is the work-hardening exponent of each material (Hutchinson 1968; Rice and Rosengren 1968). Figure 22.23 exhibits the influence of particle size on the stress distributions in the perfect composites. The stress distribution shifts upward in the HRR field with decreasing particle size.

Figure 22.24a, b and c show the influence of debonding damage on the stress distributions in the composites with average particle size of 1, 7.5 and 16 μm, respectively. In these figures, the stress distributions of the perfect composites are also shown for

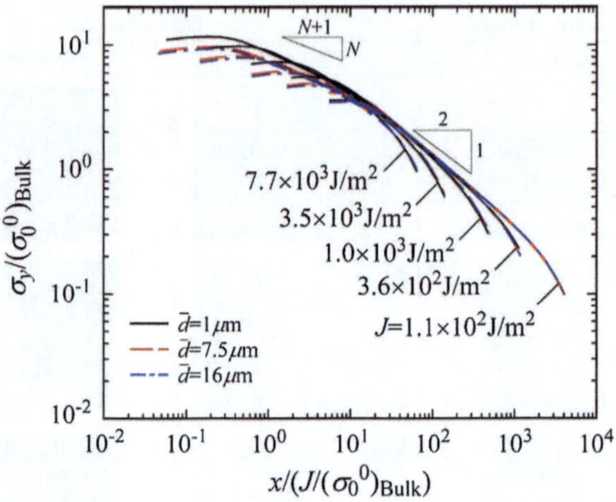

Fig. 22.23 Distributions of macroscopic stress along the line ahead of a crack-tip (perfect composites) (Tohgo et al. 2014)

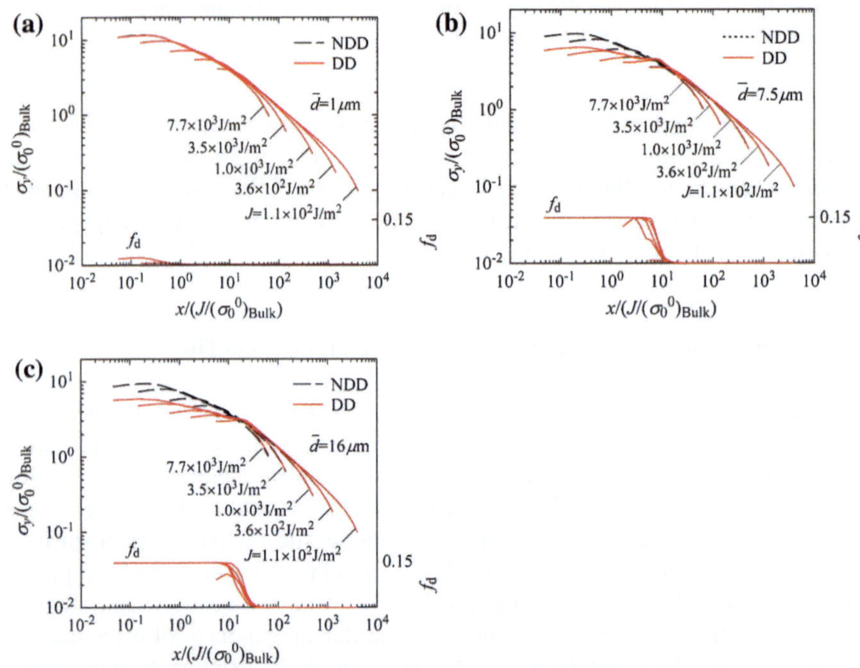

Fig. 22.24 Distributions of macroscopic stress and void volume fraction along the line ahead of a crack-tip (composites with damage) (Tohgo et al. 2014), **a** average particle size is 1 μm, **b** average particle size is 7.5 μm, **c** average particle size is 16 μm

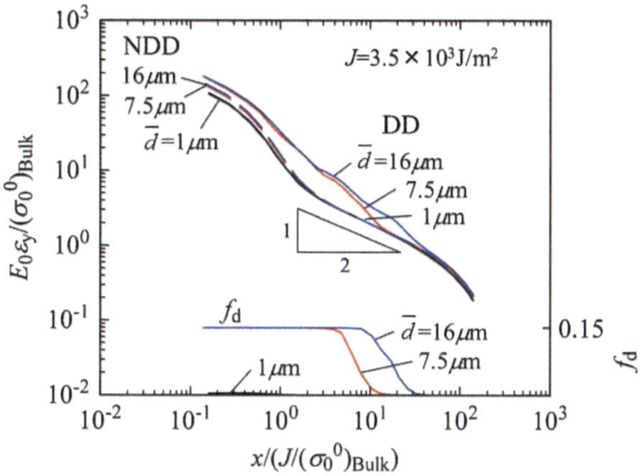

Fig. 22.25 Distributions of macroscopic strain along the line ahead of a crack-tip in perfect composites (NDD) and composites with damage (DD) under $J = 3.5 \times 103$ J/m^2 (Tohgo et al. 2014)

comparison. In Fig. 22.24a for average particle size of 1 μm, because of less debonding damage the stress distribution of the composite with damage is almost the same as that of the perfect composite. On the other hand, in Fig. 22.24b and c for average particle size of 7.5 and 16 μm, the debonding damage develops well and damage zone is created, and then the stress distribution shifts downward in the damage zone from that of the perfect composite. The particle size is larger, the damage zone is larger and the stress reduction is more remarkable.

Distributions of the macroscopic strain along the line ahead of a crack-tip under $J = 3.5 \times 10^3$ J/m^2 for the all particulate-reinforced composites are shown in Fig. 22.25. According to the HRR field for the elastic-plastic materials, as the plastic strain ε_y^p along the line ahead of a crack-tip vanishes (Hutchinson 1968; Rice and Rosengren 1968), the total strain ε_y is described by a straight line of gradient $-1/2$ due to elastic strain ε_y^e in a log-log diagram. In Fig. 22.25, the upward shifts of strain from the distribution of gradient $-1/2$ are due to the plastic deformation around a rounded crack-tip for the perfect composites (NDD) and due to the debonding damage for the composites with damage (DD). It is found from Fig. 22.25 that the influence of particle size and debonding damage on the strain distribution ahead of a crack-tip appears contrary to the stress distribution. In the perfect composites the strain distribution shifts downward with decreasing particle size, while in the composites with damage the strain distributions within the damage zone shift upward from the strain distributions for the perfect composites.

The influence of particle size and debonding damage on an elastic-plastic singular field around a crack-tip is summarized as in Fig. 22.26. In the perfect composites without debonding damage, the particle size is smaller, the stress distribution shifts upward in the HRR field. This is caused by the particle size effect on deformation. In the composites with debonding damage, the debonding damage spreads out from a crack-tip and the stress reduction occurs in damaged zone from the stress distributions for the perfect composites. The particle size is larger, the damaged zone spreads out wider and the stress reduction due to debonding damage is more remarkable. This

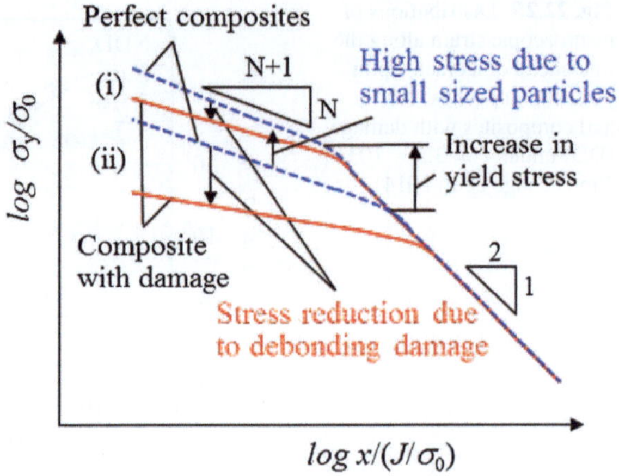

Fig. 22.26 Schematic illustration of the influence of particle size and debonding damage on the stress field around a crack-tip in elastic-plastic matrix composites (Tohgo et al. 2014), *i* composites with small sized particles, *ii* composites with large sized particles

is caused by the particle size effect on debonding damage. The influence of particle size and debonding damage on the strain distribution ahead of a crack-tip appears contrary to the stress distribution.

22.9 Summary

In this article, the constitutive model of discontinuously-reinforced composites such as particulate-reinforced composites and short-fiber-reinforced composites has been developed based on the Eshelby equivalent inclusion method and Mori-Tanaka mean field concept. The model can describe the matrix plasticity, evolution of cracking or debonding damage of reinforcements, and reinforcement size effects on deformation and damage. The influence of progressive damage and size of the reinforcements on the stress-strain relations of the composites were demonstrated based on the numerical results by the model. The experimental results of SiC/Al composites by Lloyd (1994) was described by the model taking account of the particle size effects, particle size distribution and debonding damage.

Furthermore, a finite element method (FEM) has been developed based on the model for particulate-reinforced composites with progressive debonding damage. As an example, FEM analysis was carried out on a crack-tip field in the composites, and the influence of particle size and debonding damage in an elastic-plastic singular field around a crack-tip was discussed based on the numerical results. The influence of particle size and debonding damage on an elastic-plastic singular field around a crack-tip is summarized, and this might be useful in interpretation of fracture behavior of particulate-reinforced composites.

References

Adachi T, Osaki M, Araki W, Kwon S (2008) Fracture toughness of nano- and micro-spherical silica-particle-filled epoxy composites. Acta mater 56:2101–2109

Arsenault R, Shi N (1986) Dislocation generation due to differences between the coefficients of thermal expansion. Mater Sci Eng 81:175–187

Arsenault R, Wang L, Feng C (1991) Strengthening of composites due to microstructural changes in the matrix. Acta Metall Mater 39:47–57

Bao G (1992) Damage due to fracture of brittle reinforcements in a ductile matrix. Acta Metall Mater 40:2547–2555

Bayha T, Kilmer R, Wawner F (1992) The fracture characteristics of Al-9Ti/SiCp metal matrix composites. Metall Trans A 23A:1653–1665

Berveiller M, Zaoui A (1979) An extension of the self-consistent scheme to plastically-flowing polycrystals. J Mech Phys Solids 26:325–344

Brockenbrough J, Zok F (1995) On the role of particle cracking in flow and fracture of metal matrix composites. Acta Metall Mater 43:11–20

Brown L, Stobbs W (1976) The work-hardening of copper-silica: V. Equilibrium plastic relaxation by secondary dislocations. Phil Mag 34:351–372

Caceres C, Griffiths J (1996) Damage by cracking of silicon particles in an Al-7Si-0.4Mg casting alloy. Acta Mater 44:25–33

Chen J, Huang Z, Mai Y (2003) Constitutive relation of particulate-reinforced viscoelastic composite materials with debonded microvoids. Acta Mater 51:3375–3384

Cho J, Joshi M, Sun C (2006) Effect of inclusion size on mechanical properties of polymeric composites with micro and nano particles. Compos Sci Technol 66:1941–1952

Cho Y, Tohgo K, Ishii H (1997) Load carrying capacity of a broken ellipsoidal inhomogeneity. Acta Mater 45:4787–4795

Clyne T, Withers P (1993) An introduction to metal matrix composites. Cambridge University Press, Cambridge

Corbin S, Wilkinson D (1994) Influence of matrix strength and damage accumulation on the mechanical response of a particulate metal matrix composite. Acta Metall Mater 42:1329–1335

Eckschlager A, Han W, Böhm H (2002) A unit cell model for brittle fracture of particles embedded in a ductile matrix. Comput Mater Sci 25:85–91

Eshelby J (1957) The determination of the elastic field of an ellipsoidal inclusion, and related problems. Proc R Soc Lond A 241:376–396

Finot M, Shen Y, Needleman A, Suresh S (1994) Micromechanical modeling of reinforcement fracture in particle-reinforced metal-matrix composites. Metall Mater Trans A 25A:2403–2420

Fu S, Feng X, Lauke B, Mai Y (2008) Effects of particle size, particle/matrix interface adhesion and particle loading on mechanical properties of particulate-polymer composites. Compos Part B 39:933–961

Hartingsveldt E, Aartsen J (1989) Interfacial debonding in polyamide-6/glass bead composites. Polymer 30:1984–1991

Hutchinson J (1968) Singular behaviour at the end of a tensile crack in a hardening material. J Mech Phys Solids 16:13–31

Jiang Y, Tohgo K, Shimamura Y (2009) A micro-mechanics model for composites reinforced by regularly distributed particles with an inhomogeneous interphase. Comput Mater Sci 46:507–515

Ju J, Lee H (2001) A micromechanical damage model for effective elastoplastic behavior of partially debonded ductile matrix composites. Int J Solids Struct 38:6307–6332

Kiser M, Zok F, Wilkinson D (1996) Plastic flow and fracture of a particulate metal matrix composite. Acta Mater 44:3465–3476

Kotoul M, Vrbka J (2003) Crack bridging and trapping mechanisms used to toughen brittle matrix composite. Theor Appl Fract Mech 40:23–44

Kouzeli M, Mortensen A (2002) Size dependent strengthening in particle reinforced aluminum. Acta Mater 50:39–51

Liu H, Sun L (2005) Multi-scale modeling of elastoplastic deformation and strenthening mechanism in aluminum-based amorphous nanocomposites. Acta Mater 53:2693–2701

Liu H, Sun L (2008) A micromechanics-based elastoplastic model for amorphous composites with nanoparticle interactions. Trans ASME J Appl Mech 75:031, 009-1-10

Llorca J, Segurado J (2004) Three-dimensional multiparticle cell simulation of deformation and damage in sphere-reinforced composites. Mater Sci Eng A 365:267–274

Llorca J, Martin A, Ruiz J, Elices M (1993) Particulate fracture during deformation of a spray formed metal-matrix composite. Metall Trans A 24A:1575–1588

Lloyd D (1994) Particle reinforced aluminium and magnesium matrix composites. Int Mater Rev 39(1):1–23

Matous K (2003) Damage evolution in particulate composite materials. Int J Solids Struct 40:1489–1503

Mura T (1982) Micromechanics of defects in solids. Martinus Nijhoff, The Hague

Mori T, Tanaka K (1973) Average stress in matrix and average elastic energy of materials with misfitting inclusions. Acta Metall 21:571–574

Nan C, Clarke D (1996) The influence of particle size and particle fracture on the elastic/plastic deformation of metal matrix composites. Acta mater 44:3801–3811

Nan C, Yuan R (1993) Multiple-scattering solution to nonlinear mechanical properties of binary elastic-plastic composite media. Phys Rev B 48:3042–3047

Needleman A (1987) Continuum model for void nucleation by inclusion debonding. Trans ASME J Appl Mech 54:525–531

Qiao Y (2003) Fracture toughness of composite materials reinforced by debondable particulates. Scripta Mater 49:491–496

Rice J, Rosengren G (1968) Plane strain deformation near a crack tip in a power-law hardening material. J Mech Phys Solids 16:1–12

Rice J, Paris P, Merkle J (1973) Some further results on J-integral analysis and estimates. ASTP STP 536:231–245

Salviato M, Zappalorto M, Quaresimin M (2013) Plastic shear bands and fracture toughness improvements of nanoparticle filled polymer: a multiscale analytical model. Compos Part A 48:144–152

Segurado J, Llorca J (2005) A computational micromechanics study of the effect of interface decohesion on the mechanical behavior of composites. Acta Mater 53:4931–4942

Segurado J, Llorca J (2006) Computational micromechanics of composites: the effect of particle spatial distribution. Mech Mater 38:873–883

Sun L, Ju J, Liu H (2003) Elasticplastic modeling of metal matrix composites with evolutionary particle debonding. Mech Mater 35:559–569

Sun L, Gibson R, Gordaninejad F (2011) Multiscale analysis of stiffness and fracture of nanoparticle-reinforced composites using micromechanics and global-local finite element models. Eng Frac Mech 78:2645–2662

Tandon G, Weng G (1988) Theory of particle-reinforced plasticity. Trans ASME J Appl Mech 55:126–135

Taupin V, Berbenni S, Fressengeas C, Bouaziz O (2010) On particle size effects: an internal length mean field approach using field dislocation mechanics. Acta Mater 58:5532–5544

Tohgo K, Cho Y (1999) Theory of reinforcement damage in discontinuously-reinforced composites and its application. JSME Int J 42:521–529

Tohgo K, Chou T (1996) Incremental theory of particulate-reinforced composites including debonding damage. JSME Int J 39:389–397

Tohgo K, Itoh T (2005) Elastic and elastic-plastic singular fields around a crack-tip in particulate-reinforced composites with progressive debonding damage. Int J Solids Struct 42:6566–6585

Tohgo K, Weng G (1994) A progressive damage mechanics in particle-reinforced metal-matrix composites under high triaxial tension. Trans ASME J Eng Mater Technol 116:414–420

Tohgo K, Mochizuki M, Ishii H (1998) Incremental damage theory and its application to glass-particle-reinforced Nylon 66 composites. Int J Mech Sci 40:199–213

Tohgo K, Fukuhara D, Hadano A (2001) The influence of debonding damage on fracture toughness and crack-tip field in glass-particle-reinforced Nylon 66 composites. Compos Sci Technol 61:1005–1016

Tohgo K, Itoh T, Shimamura Y (2010) A constitutive model of particulate-reinforced composites taking account of particle size effects and damage evolution. Compos Part A 41:313–321

Tohgo K, Fujii T, Kato D, Shimamura Y (2014) Influence of particle size and debonding damage on an elastic-plastic singular field around a crack-tip in particulate-reinforced composites. Acta Mech 225:1373–1389

Whitehouse A, Clyne T (1993) Cavity formation during tensile straining of particulate and short fibre metal matrix composites. Acta Metall Mater 41:1701–1711

Yang H, Chen P, Jiang Y, Tohgo K (2011) Incremental damage theory of particulate-reinforced composites with a ductile interphase. Compos Struct 93:2655–2662

Chapter 23
A Study of Metal Fatigue Failure as Inherent Features of Elastoplastic Constitutive Equations

Zhao-Ling Wang and Heng Xiao

Abstract Elastoplastic constitutive equations in a novel sense are proposed toward a direct simulation of metal fatigue failure as inherent features of elastoplastic behavior from a fresh standpoint. Such equations are established with consideration of realizing a smooth transition from the elastic to the plastic state and, in particular, characterizing asymptotic loss of the stress-bearing capacity up to failure. It is demonstrated that certain salient features of both high and low cycle fatigue failure of metals under cyclic loadings may be derived as direct consequences from these simple equations.

Keywords Metal · Fatigue · Elastoplasticity · Direct simulation

23.1 Introduction

It is of wide interest to develop an effective methodology to assess reliability and safety problems of materials and structures associated with fatigue, fracture and failure. Background of metal fatigue may be found in related monographs, e.g., Pook (2007). Usually, the fatigue failure behavior of metals under repeated loading conditions may be studied from various standpoints. Numerous results have been obtained in the past decades. For certain representatives of most recent results, reference

---

Z.-L.Wang
Shanghai Institute of Applied Mathematics and Mechanics, Shanghai University,
Yanchang Road 149, Shanghai 200072, China

Z.-L. Wang
School of Mathematics and Information Sciences, Weifang University,
Weifang, Shandong, China
e-mail: wangzhaoling555@163.com

H. Xiao (✉)
State Key Laboratory of Advanced Special Steels and Shanghai Institute of Applied
Mathematics and Mechanics, Shanghai University,
Yanchang Road 149, Shanghai 200072, China
e-mail: xiaoheng@shu.edu.cn

H. Altenbach et al. (eds.), *From Creep Damage Mechanics to Homogenization Methods*, Advanced Structured Materials 64,
DOI 10.1007/978-3-319-19440-0_23

may be made to Stoughton and Zhu (2004) for a review of failure criteria for sheet metals and, moreover, to Wu et al. (2005) for forming limit analysis; Bronkhorst et al. (2006) for a study of the localization behavior of tantalum and stainless steel; Shyam et al. (2007) for a model of small fatigue crack growth in metallic materials; Yoshida et al. (2007) and Benzerga et al. (2012) for the path-dependence property of the forming limit and the fracture locus; Stoughton and Yoon (2005, 2011, 2012) for anisotropic materials under non-proportional loading and for path-independent forming limits etc.; Kim et al. (2011) for a study of the shear fracture; Jansen et al. (2013) for an anisotropic stress-based criterion to predict the fracture mechanism etc.; Khan and Liu (Khan and Liu 2012a, b) for latest advances concerning strain rate and temperature effects, and many others. On the other hand, a widely used approach is based on continuum damage mechanics. For some recent results in this respect, refer to, e.g., Brünig (Brünig 2003a, b) for a ductile damage model with irreversible thermodynamics and for numerical treatment; Bonora et al. (2005) for ductile damage evolution under multi-axial stress states; Brünig and Ricci (2005) for a nonlocal model for anisotropically damaged metals; Brünig et al. (2008) for a ductile damage criterion under multi-axial stress states; Brünig and Gerke (2011) for damage evolution simulation under dynamic loading; Chung et al. (2011) for a modified damage model; Malcher et al. (2012) for a study of ductile fracture under various cases of triaxial stress; Brünig et al. (2013) for micro-mechanical studies in triaxial stress cases; Shojaei et al. (2013) for a study of brittle to ductile damage based on viscoplastic models; and many others under cyclic loadings. In particular, significant studies in the active area of cyclic plasticity have been made in direct relation to fatigue behavior of metals under various cyclic loadings. For representees of original contributions and developments in this respect, refer to the review articles earlier by Ohno (1990) and later by Chaboche (2008) as well as references therein.

In the existing studies, results have been derived based on either additional criteria for failure or augmented constitutive structures. In a most recent study (Xiao 2014; Xiao et al. 2014), a direct, alternative approach has been suggested to model fatigue, fracture and failure behaviors of metals and alloys etc. from a fresh standpoint. The central idea is to establish new elastoplasticity models into which the fatigue failure behavior is incorporated as inherent constitutive features. Toward this objective and others, it has been indicated that the usual notion of yielding should be abandoned toward a realistic, smooth transition from the elastic to the plastic state. Moreover, asymptotic loss of the stress-bearing capacity up to failure should be incorporated. In this contribution, we are going to propose a simple form of elastoplasticity model of the nature just indicated and, then, demonstrate that the fatigue failure behavior may be derived as direct consequences from this simple model.

The main content of this contribution is arranged as follows. In Sect. 23.2, a simple form of elastoplasticity model is proposed in the aforementioned novel sense; predictions of this model under uniaxial cyclic loadings up to failure are then studied in Sect. 23.3 and numerical examples are presented in Sect. 23.4; and, finally, some remarks are given in Sect. 23.5.

23.2 Smooth Elastoplastic Equations with Asymptotic Loss of the Stress-Bearing Capacity Upto Failure

Consider an elastoplastic body undergoing finite deformation. Let $\boldsymbol{F}$ and $\boldsymbol{L}$ be the deformation gradient and the velocity gradient and, moreover, let $\boldsymbol{D}$, $\boldsymbol{W}$ and $\boldsymbol{\tau}$ be the stretching tensor, the vorticity tensor and the Kirchhoff stress, namely,

$$\boldsymbol{D} = \frac{1}{2}(\boldsymbol{L} + \boldsymbol{L}^T), \quad \boldsymbol{W} = \frac{1}{2}(\boldsymbol{L} - \boldsymbol{L}^T), \quad \boldsymbol{L} = \dot{\boldsymbol{F}} \cdot \boldsymbol{F}^{-1},$$

$$\boldsymbol{\tau} = J\boldsymbol{\sigma}, \quad J = \det\boldsymbol{F}.$$

In the above, J and $\boldsymbol{\sigma}$ are the volumetric ratio and the Cauchy stress (true stress), respectively.

Various formulations of finite elastoplastic deformations have been developed in the past decades (e.g. Xiao et al. 2006). In the subsequent development, we direct attention to the Eulerian rate formulation based on the additive separation:

$$\boldsymbol{D} = \boldsymbol{D}^e + \boldsymbol{D}^p \tag{23.1}$$

with the elastic stretching $\boldsymbol{D}^e$ and the plastic stretching $\boldsymbol{D}^p$. Constitutive formulations for these two parts will be presented in two steps below.

The first step is to give an elastic rate equation of hypo-elastic type. This is done by relating the elastic stretching $\boldsymbol{D}^e$ with an objective rate of the Kirchhoff stress $\boldsymbol{\tau}$. In a self-consistent sense of characterizing the elastic behavior, it is required that, for every process of pure elastic deformation, this Eulerian rate equation should be exactly integrable to really deliver a finite hyperelastic relation based on an elastic potential. For metals experiencing small elastic strains, a self-consistent Eulerian rate formulation in this sense may be given below (e.g. Bruhns et al. 1999, 2003, 2005; Xiao et al. 2000, 2006, 2007):

$$\boldsymbol{D}^e = \frac{1}{2G}\overset{\circ}{\boldsymbol{\tau}}{}^{\log} - \frac{\nu}{E}(\mathrm{tr}\dot{\boldsymbol{\tau}})\boldsymbol{I}\,. \tag{23.2}$$

In the above, G, E and ν are the shear modulus, Young's modulus and the Poisson ratio with the relation $E = 2G(1+\nu)$ and, moreover, $\overset{\circ}{\boldsymbol{\tau}}{}^{\log}$ is the co-rotational logarithmic rate of the Kirchhoff stress $\boldsymbol{\tau}$, viz.

$$\overset{\circ}{\boldsymbol{\tau}}{}^{\log} = \dot{\boldsymbol{\tau}} + \boldsymbol{\tau} \cdot \boldsymbol{\Omega}^{\log} - \boldsymbol{\Omega}^{\log} \cdot \boldsymbol{\tau}\,, \tag{23.3}$$

with the logarithmic spin

$$\boldsymbol{\Omega}^{\log} = \boldsymbol{W} + \sum_{r\neq s=1}^{n}\left(\frac{1+\frac{b_r}{b_s}}{1-\frac{b_r}{b_s}} + \frac{2}{\ln\frac{b_r}{b_s}}\right)\boldsymbol{B}_r \cdot \boldsymbol{D} \cdot \boldsymbol{B}_s\,, \tag{23.4}$$

where the b_t and $\boldsymbol{B}_t$ are the n distinct eigenvalues of the left Cauchy-Green tensor $\boldsymbol{B} = \boldsymbol{F} \cdot \boldsymbol{F}^T$ and the corresponding eigenprojections of $\boldsymbol{B}$, respectively. Details may be found in Xiao et al. (1997).

The next step is to formulate a flow rule for the plastic stretching $\boldsymbol{D}^p$. The normality flow rule is adopted and of the form (Bruhns et al. 1999; Xiao et al. 2000; Bruhns et al. 2003, 2005; Xiao et al. 2007):

$$\boldsymbol{D}^p = \varrho \frac{\breve{f}}{\breve{h}} \frac{\partial f}{\partial \boldsymbol{\tau}}, \tag{23.5}$$

where the yield function f and the factor $\xi = \breve{f}/\breve{h}$ will be given slightly later. According to the classical theory of elastoplasticity, the plastic indicator ϱ in Eq. (23.5) is associated with the concept of yielding and represents the very feature of plastic behavior in an idealized sense. In fact, it is assumed that no plastic deformation would be induced, namely, $\varrho = 0$ prior to yielding, whereas plastic deformation would be induced, namely, $\varrho = 1$ at yielding. Specifically, the loading-unloading conditions associated with the yield condition should be introduced to specify whether the plastic indicator ϱ takes value 0 or takes value 1.

The representation of the plastic behavior in the above idealized or simplified sense implies that there always emerges a non-smooth transition from the elastic to the plastic state, with a strong discontinuity in the tangent modulus. For consequences in a few respects, refer to Xiao et al. (2014); Xiao (2014) for detail.

For realistic elastoplastic behavior, the plastic deformation may be induced at any stress level with a non-vanishing plastic indicator $0 \leq \varrho \leq 1$, albeit the latter is close to 1 merely in the case when the stress level is such that the value of the yield function f is close to zero. With this in mind, a more realistic representation may be presented by specifying $0 \leq \varrho \leq 1$ in the following manner: The value of ϱ is close to 1 whenever the stress point $\boldsymbol{\tau}$ stays in the vincinity of the yield surface $f = 0$, whereas the value of ϱ becomes vanishingly small whenever the stress point stays far away from the yield surface $f = 0$.

With the above idea in mind, we may introduce a plastic indicator ϱ of the above property and then replace the idealized plastic indicator in the usual flow rule. As a result, a new flow rule may be given as follows (Xiao et al. 2014; Xiao 2014):

$$\boldsymbol{D}^p = \frac{g}{r} \exp\left[-m\left(1 - \frac{g}{r}\right)\right] \frac{\xi + |\xi|}{2} \frac{\partial f}{\partial \boldsymbol{\tau}} \tag{23.6}$$

with $m \geq 0$ a non-negative dimensionless parameter. In the above, f is the von Mises yield function of the form:

$$\begin{cases} f = g - r\,, \\ g = \dfrac{1}{2} J_2 = \dfrac{1}{2} \mathrm{tr}\tilde{\boldsymbol{\tau}}^2\,, \\ r = \dfrac{1}{3} q^2\,, \end{cases} \tag{23.7}$$

where the $\tilde{\boldsymbol{\tau}}$ is the deviatoric part of $\boldsymbol{\tau}$ and the q, called the stress limit, relies on the plastic work κ, namely, $q = q(\kappa)$ with

$$\dot{\kappa} = \boldsymbol{\tau} : \boldsymbol{D}^p . \tag{23.8}$$

Moreover, the ξ in Eq. (23.6) is given by

$$\xi = \frac{\breve{f}}{\breve{h}} , \tag{23.9}$$

where

$$\breve{f} = 2G\tilde{\boldsymbol{\tau}} : \boldsymbol{D} , \tag{23.10}$$

$$\breve{h} = \frac{2}{3} J_2(3G + qq'), \quad q' = \frac{dq}{d\kappa} . \tag{23.11}$$

Equations (23.1), (23.2) and (23.6)–(23.11) together produce

$$\frac{\overset{\circ}{\boldsymbol{\tau}}{}^{\log}}{2G} = \frac{\nu}{1-2\nu}(\mathrm{tr}\boldsymbol{D})\boldsymbol{I} + \boldsymbol{D} - \frac{1.5GJ_2^{-1}}{3G+qq'} \exp\left[-m\left(1 - \frac{3J_2}{2q^2}\right)\right] (\tilde{\boldsymbol{\tau}} : \boldsymbol{D} + |\tilde{\boldsymbol{\tau}} : \boldsymbol{D}|)\tilde{\boldsymbol{\tau}} \tag{23.12}$$

with the plastic work determined by

$$\dot{\kappa} = \frac{1.5G}{3G+qq'} \exp\left[-m\left(1 - \frac{3J_2}{2q^2}\right)\right] (\tilde{\boldsymbol{\tau}} : \boldsymbol{D} + |\tilde{\boldsymbol{\tau}} : \boldsymbol{D}|) , \tag{23.13}$$

Unlike the classical formulation for rate-independent elastoplasticity, now the elastoplastic constitutive equations proposed are fully free in a sense without involving the yield condition as well as the loading-unloading conditions and, moreover, give rise to plastic flow at any non-zero stress level whenever $\tilde{\boldsymbol{\tau}} : \boldsymbol{D} > 0$. For metals etc., it is observed that plastic strain will become dominant whenever the stress reaches such a level that a certain condition (yield condition) is met, whereas it may be negligibly small prior to this stress level. Now this feature is incorporated into the new elastoplastic equations with a smooth transition from the elastic to the plastic state. Certain significant consequences implied by such new equations, such as relations to the smooth transition and the fatigue behavior etc., may be found in Xiao et al. (2014); Xiao (2014).

This contribution aims to present an initial study of the fatigue behavior for metals etc. Here, the essential point lies in the fact that, from a phenomenological standpoint, the physical essence of material failure would be just the loss of stress-bearing capability attendant with fully developed plastic flows. The quantitative characterization of this fact in a broad sense has been suggested in a newest study (Xiao 2014). The necessary and sufficient conditions for the stress limit etc. have been suggested in the just mentioned reference. Toward our purpose, a simple form of the stress limit $q = q(\kappa)$ is taken into consideration and given below:

$$q = \frac{1}{2} q_0 \left[1 + \tanh \beta \left(\kappa_c - \kappa\right)\right] . \tag{23.14}$$

In the above, q_0, β and κ_c are three positive material parameters. The stress limit given above exhibits asymptotic loss of the stress-bearing capacity, since it tends to vanish as the plastic work κ goes to infinity, namely, $q|_{\kappa \to +\infty} = 0$. Detail in this respect may be found in Xiao (2014).

The main feature of the stress limit q given by Eq. (23.14) is as follows: For a fairly large β, the stress limit q given by Eq. (23.14) actually yields a constant value, i.e., q_0, before the plastic work κ reaches a critical value slightly smaller than κ_c, whereas it goes rapidly to vanish after κ exceeds a critical value slightly greater than κ_c.

In addition to the elastic constants E and ν, the proposed model is characterized by the parameters m, q_0, β and κ_c for the plastic behavior. In the next section, it will be shown that certain salient features of both high and low cycle fatigue behaviors of metals may be derived from the new elastoplastic equations proposed here.

23.3 Fatigue Failure Under Uniaxial Cyclic Loadings

Consider a cylindrical metal bar undergoing uniaxial deformations. Let τ be the axial Kirchhoff stress and let a and l be the stretch ratios in the axial direction and the lateral direction, respectively. Then, from Eqs. (23.12) and (23.13) the rate equations for the axial Hencky strain $h = \ln a$ and the plastic work κ may be derived as follows:

$$\dot{h} = \frac{q_0}{E} \frac{1 + \frac{qq'}{3G} - [\bar{\tau}\dot{h}]\frac{1-2\nu}{3} \exp\left(-m\left(1 - \bar{\tau}^2\right)\right)}{1 + \frac{qq'}{3G} - [\bar{\tau}\dot{h}] \exp\left(-m\left(1 - \bar{\tau}^2\right)\right)} \dot{\bar{\tau}} , \tag{23.15}$$

$$\dot{\kappa} = \frac{2}{3}(1+\nu)\frac{q_0^2}{E} \frac{[\bar{\tau}\dot{h}] \exp\left(-m\left(1 - \bar{\tau}^2\right)\right)}{1 + \frac{qq'}{3G} - [\bar{\tau}\dot{h}] \exp\left(-m\left(1 - \bar{\tau}^2\right)\right)} \bar{\tau}\dot{\bar{\tau}} , \tag{23.16}$$

where

$$qq' = -\frac{1}{4}\beta q_0^2 \frac{1 + \tanh \beta(\kappa - \kappa_c)}{\cosh^2 \beta(\kappa_c - \kappa)} . \tag{23.17}$$

Moreover, the lateral stretch ratio l is given by

$$\ln l = -\frac{1}{2}h + \frac{1-2\nu}{2}\frac{q_0}{E}\bar{\tau} . \tag{23.18}$$

Here and henceforth we denote

$$\bar{\tau} = \frac{\tau}{q_0} ,$$

$$[x] = \frac{1}{2}(1 + \mathrm{sgn}(x)) = \begin{cases} 1 \text{ for } x > 0 , \\ 0 \text{ for } x \leq 0 . \end{cases}$$

Given any process of the axial loading, namely, $\tau = \tau(t)$, the response of the axial Hencky strain $h = h(t)$ may be derived from Eqs. (23.15) and (23.16). Here, two processes of cyclic loading are treated, separately.

The first cyclic process is as follows: At each cycle, the axial stress $\bar{\tau}$ goes up from zero to a given value $\bar{\tau}_0$ and then goes down from $\bar{\tau}_0$ to zero. An example of such a cyclic process is given as follows:

$$\begin{cases} \bar{\tau} = \bar{\tau}_0 \left(\frac{t}{t_0} - 2s + 2 \right) , & (2s - 2)t_0 \leq t \leq (2s - 1)t_0 , \\ \bar{\tau} = -\bar{\tau}_0 \left(\frac{t}{t_0} - 2s \right) , & (2s - 1)t_0 \leq t \leq 2st_0 , \end{cases} \tag{23.19}$$

where the t_0 is a given duration of time and $s = 1, 2, \ldots$ and the s is the cycle number.

At each cycle of the above process, only tensile loads are involved. The second cyclic process is concerned with both tensile and compressive loads: At each cycle, the axial stress $\bar{\tau}$ changes from zero to a given tensile stress, $\bar{\tau}_0$, and then from $\bar{\tau}_0$ to zero and, after that, changes from zero to a compressive stress, $-\bar{\tau}_0$, and then from $-\bar{\tau}_0$ to zero. An example of such a cyclic process is given as follows:

$$\begin{cases} \bar{\tau} = \bar{\tau}_0 \left(\frac{t}{t_0} - 2s + 2 \right) , & (2s - 2)t_0 \leq t \leq (2s - 1)t_0 , \\ \bar{\tau} = -\bar{\tau}_0 \left(\frac{t}{t_0} - 2s \right) , & (2s - 1)t_0 \leq t \leq (2s + 1)t_0 , \\ \bar{\tau} = \bar{\tau}_0 \left(\frac{t}{t_0} - 2s - 2 \right) , & (2s + 1)t_0 \leq t \leq (2s + 2)t_0 , \end{cases} \tag{23.20}$$

where s is again the cycle number.

Let h_s and κ_s be the axial Hencky strain and the plastic work at the start of the sth cycle. Then, with these initial values the response for the sth cycle may be calculated by numerically integrating the system of the two rate equations (23.15) and (23.16). Note that $h_1 = 0$ and $\kappa_1 = 0$ for the first cycle. The initial values h_s and κ_s at the sth cycle are just the values of h and κ at the end of the last cycle, i.e., the $(s - 1)$th cycle.

It may be expected that either of the two cyclic process at issue should be terminated at a certain number of cycles, denoted N, and fatigue failure should follow immediately after the Nth cycle, since the stress limit q goes rapidly down to vanish as the plastic work κ accumulates to exceed a critical value, as indicated before. The cycle number N at failure may be very large for a relatively small $0 < \bar{\tau}_0 < 1$, while it may be fairly large for a $\bar{\tau}_0$ close to 1. This and other salient features of both high and low cycle fatigue behavior may be predicted from the simple model proposed, as will be shown in the numerical results presented in the next section.

23.4 Numerical Results

In this section, responses of the proposed model will be numerically studied for the two cyclic processes as specified by Eq. (23.19) and (23.20), separately. Toward this end, the material parameters, including E, m, q_0, β and κ_c etc., are first be determined by simulating a uniaxial stress-strain curve with monotonically increasing strain up to failure and then numerical results for fatigue failure may be obtained by integrating the system of the two equations (23.15) and (23.16) for the two types of cyclic loadings at issue.

The uniaxial curve in the foregoing should be obtained from experimental test. Here, the curve given in Fig. 23.1 in Giroux et al. (2010) for the strain rate 2.5×10^{-5} is taken into account. The material parameters are found by fitting this curve and given below:

$$E = 120\,\text{GPa},\ \ \nu = 0.3,\ \ m = 1,\ \ q_0 = 395\,\text{MPa},\ \ \kappa_c = 90\,\text{MPa},\ \ \beta = 3.8/\text{MPa}.$$

The simulation with the above parameters is shown in Fig. 23.2.

Model predictions are calculated for $\bar{\tau}_0 = 150/395,\ 250/395$ for the two cyclic loading processes given by Eqs. (23.19) and (23.20), separately. For these cases, the stress amplitudes $\bar{\tau}_0$ considered and the corresponding cycle numbers N at failure are as follows:

$$\bar{\tau}_0 = 0.633, \quad N = 2698;$$
$$\bar{\tau}_0 = 0.380, \quad N = 29957;$$

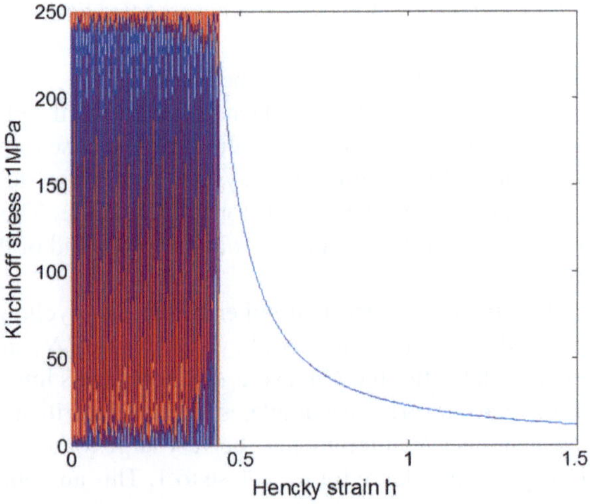

Fig. 23.1 Fatigue failure under the cyclic process Eq. (23.19) with $\bar{\tau}_0 = 0.633$

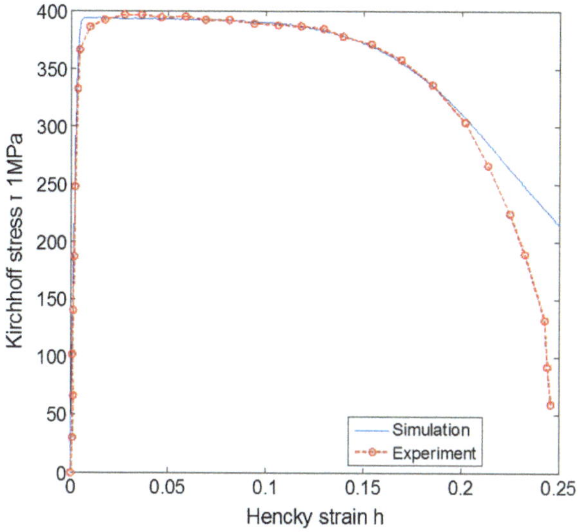

Fig. 23.2 Simulation of uniaxial stress-strain *curve* up to failure

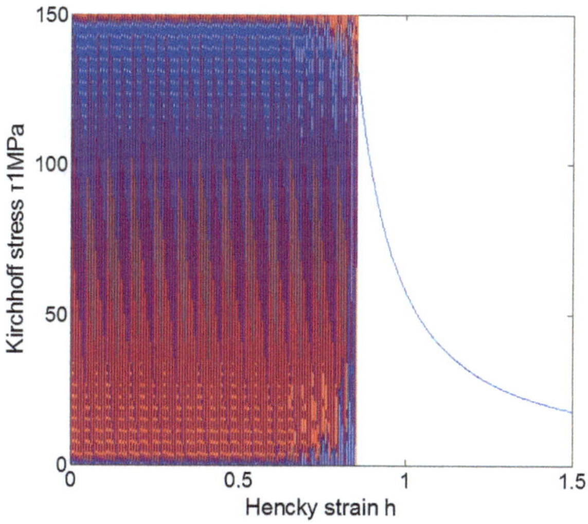

Fig. 23.3 Fatigue failure under the cyclic process Eq. (23.19) with $\bar{\tau}_0 = 0.380$

for the cyclic process Eq. (23.19), and

$$\bar{\tau}_0 = 0.633, \quad N = 1347;$$
$$\bar{\tau}_0 = 0.380, \quad N = 14942;$$

for the loading process Eq. (23.20). Detailed results are shown in Figs. 23.1 and 23.3 for the cyclic process Eq. (23.19) and in Figs. 23.4 and 23.5 for the cyclic process Eq. (23.20).

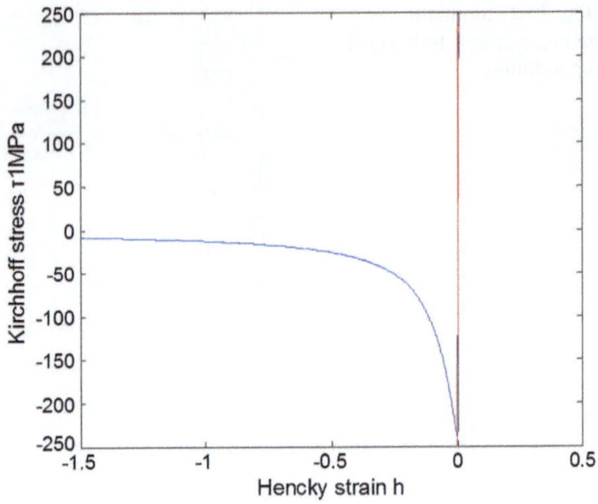

Fig. 23.4 Fatigue failure under the cyclic process Eq. (23.20) with $\bar{\tau}_0 = 0.633$

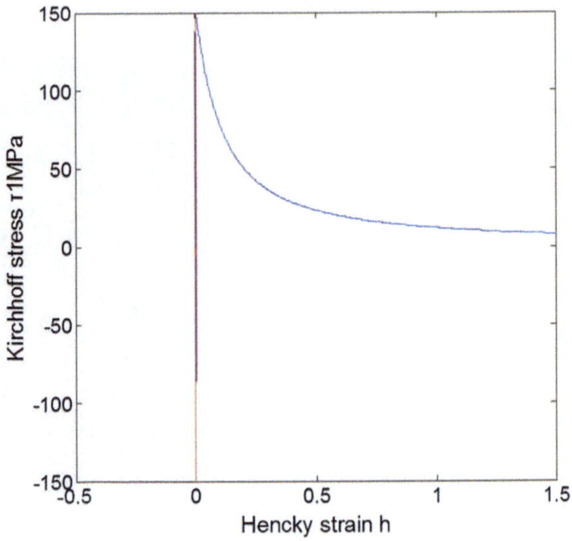

Fig. 23.5 Fatigue failure under the cyclic process Eq. (23.20) with $\bar{\tau}_0 = 0.380$

23.5 Concluding Remarks

In the preceding sections, it has been demonstrated that certain salient features of both high and low cycle fatigue failure behaviors may be predicted from new elastoplastic equations and, accordingly, may be incorporated into inherent features of elastoplastic models. It is noted that the parameters characterizing the plastic behavior, in particular, the parameters m and q_0 and κ_c, play significant roles in simulating the fatigue failure behavior. It may be of interest to study their effects in a detailed sense. Moreover, the elastoplastic equations proposed here are of simple forms, in the sense that neither isotropic nor anisotropic hardening is treated. Both should be incorporated in further developments. Results will be reported elsewhere.

Acknowledgments This work was carried out under the support of the start-up fund (No.: S.15-B002-09-032) for the 211-plan of the Education Committee of China and the fund (No.: 11372172) from the Natural Science Foundation of China. This support is gratefully acknowledged.

References

Benzerga A, Sturovik D, Keralavarma S (2012) On the path-dependence of the fracture locus in ductile materials–analysis. Int J Plast 37:157–170

Bonora N, Gentile D, Pirondi A, Newaz G (2005) Ductile damage evolution under triaxial state of stress: theory and experiments. Int J Plast 21:981–1007

Bronkhorst C, Cerreta E, Xue Q, Maudin P, Masson T, Gray G III (2006) An experimental and numerical study of the localization behavior of tantalum and stainless steel. Int J Plast 22:1304–1335

Bruhns O, Xiao H, Meyers A (1999) A self-consistent Eulerian rate type elastoplasticity models based upon the logarithmic stress rate. Int J Plast 15:479–520

Bruhns O, Xiao H, Meyers A (2003) Some basic issues in traditional Eulerian formulations of finite elastoplasticity. Int J Plast 19:2007–2026

Bruhns O, Xiao H, Meyers A (2005) A weakened form of Ilyushin's postulate and the structure of self-consistent Eulerian finite elastoplasticity. Int J Plast 21:199–219

Brünig M (2003a) An anisotropic ductile damage model based on irreversible thermodynamics. Int J Plast 19:1679–1713

Brünig M (2003b) Numerical analysis of anisotropic ductile continuum damage. Compt Meth Appl Mech Eng 192:2749–2776

Brünig M, Gerke S (2011) Simulation of damage evolution in ductile metals undergoing dynamic loading conditions. Int J Plast 27:1598–1617

Brünig M, Ricci S (2005) Nonlocal continuum theory of anisotropically damaged metals. Int J Plast 21:1346–1382

Brünig M, Chyra O, Albrecht D, Driemeier L, Alves M (2008) A ductile damage criterion at various stress triaxialities. Int J Plast 24:1731–1755

Brünig M, Gerke S, Hagenbrock V (2013) Micro-mechanical studies on the effect of the stress triaxiality and the Lode parameter on ductile damage. Int J Plast 50:49–65

Chaboche J (2008) A review of some plasticity and viscoplasticity constitutive theories. Int J Plast 24:1642–1693

Chung K, Ma N, Park T, Kim D, Yoo D, Kim C (2011) A modified damage model for advanced high strength steel sheets. Int J Plast 27:1485–1511

Giroux P, Dalle F, Sauzay M, Malaplate J, Fournier B, Gourgues-Lorenzon A (2010) Mechanical and microstructural stability of p92 steel under uniaxial tension at high temperature. Mater Sci Eng A 527:3984–3993

Jansen Y, Logé R, Milesi M, Massoni E (2013) An anisotropic stress based criterion to predict the formability and the fracture mechanism of textured zinc sheets. Trans ASME J Mater Process Technol 213:851–855

Khan A, Liu H (2012a) A new approach for ductile fracture prediction on Al 2024–T351 alloy. Int J Plast 35:1–12

Khan A, Liu H (2012b) Strain rate and temperature dependent fracture criterion for isotropic and anisotropic metals. Int J Plast 37:1–15

Kim J, Sung J, Piao K, Wagoner R, (2011) The shear fracture of dual-phase steel. Int J Plast 27:1658–1676
Malcher L, Pires F, César de Sà J (2012) An assessment of isotropic constitutive models for ductile fracture under high and low stress triaxiality. Int J Plast 30–31:81–115
Ohno N (1990) Recent topics in constitutive modeling for cyclic plasticity and viscoplasticity. Appl Mech Rev 43(11):283–295
Pook L (2007) Metal fatigue. Springer, Berlin
Shojaei A, Voyadjis G, Tan P (2013) Viscoplastic constitutive theory for brittle to ductile damage in polycrystalline materials under dynamic loading. Int J Plast 48:125–151
Shyam A, Allison J, Szczepanski C, Pollock T, Jones J (2007) Small fatigue crack growth in metallic materials: a model and its application to engineering alloys. Acta Mater 55:6606–6616
Stoughton T, Yoon J (2005) Sheet metal formability analysis for anisotropic materials under non-proportional loading. Int J Mech Sci 47:1972–2002
Stoughton T, Yoon J (2011) A new approach for failure criterion for sheet metals. Int J Plast 27:440–459
Stoughton T, Yoon J (2012) Path independent forming limits in strain and stress spaces. Int J Solids Struct 49:3616–3625
Stoughton T, Zhu X (2004) Review of theoretical models of the strain-based FLD and their relevance to the stress-based FLD. Int J Plast 20:1463–1486
Wu P, Graf A, MacEwen S, Lloyd D, Jain M, Neale K (2005) On forming limit stress diagram analysis. Int J Solids Struct 42:2225–2241
Xiao H (2014) Thermo-coupled elastoplasticity model with asymptotic loss of the material strength. Int J Plast 63:211–228
Xiao H, Bruhns O, Meyers A (1997) Logarithmic strain, logarithmic spin and logarithmic rate. Acta Mech 124:89–105
Xiao H, Bruhns O, Meyers A (2000) The choice of objective rates in finite elastoplasticty: General results on the uniqueness of the logarithmic rates. Proc Roy Soc Lond A 456:1865–1882
Xiao H, Bruhns O, Meyers A (2006) Elastoplasticity beyond small deformations. Acta Mech 182:31–111
Xiao H, Bruhns O, Meyers A (2007) Thermodynamic laws and consistent Eulerian formulations of finite elastoplasticity with thermal effects. J Mech Phys Solids 15:338–365
Xiao H, Bruhns O, Meyers A (2014) Free rate-independent elastoplastic equations. Z Angew Math Mech 94:461–476
Yoshida K, Kawabara T, Kurada M (2007) Path-dependence of the forming limit stresses in a sheet metal. Int J Plast 23:361–384

Chapter 24
Maximization of Strengthening Effect of Microscopic Morphology in Duplex Steels

Ikumu Watanabe, Gaku Nakamura, Kohei Yuge, Daigo Setoyama and Noritoshi Iwata

Abstract An inverse analysis method based on nonlinear finite element analysis is developed to find an optimized morphology of periodic microstructure for improving the macroscopic mechanical properties in duplex elastoplastic solids. Here a gradient-based computational optimization method and two types of homogenization methods are employed. In this study, the optimization problem is defined as the maximization of the sum of macroscopic external works for several macroscopic deformation modes, enabling us to obtain a high strength material. The morphologic strengthening effect is discussed through a comparison with experiments and classical theories.

Keywords Homogenization method · Finite element method · Optimization theory · Morphology · Strengthening effect · Metals

24.1 Introduction

Macroscopic material properties appear as the average of material behavior on a finer scale. One of the major subjects in materials research and development (R&D) is to understand the effect of microscopic information on the macroscopic material

I. Watanabe (✉)
National Institute for Materials Science, 1-2-1 Sengen, Tsukuba, Ibaraki 305-0047, Japan
e-mail: WATANABE.Ikumu@nims.go.jp

G. Nakamura · K. Yuge
Seikei University, 3-3-1 Kichijoji-Kitamachi, Musashino, Tokyo 180-8633, Japan
e-mail: yuge@st.seikei.ac.jp

D. Setoyama · N. Iwata
TOYOTA Central R&D Labs., Inc., 41-1 Yokomichi, Nagakute, Aichi 480-1192, Japan
e-mail: daigo@mosk.tytlabs.co.jp

N. Iwata
e-mail: noriiwata@mosk.tytlabs.co.jp

H. Altenbach et al. (eds.), *From Creep Damage Mechanics to Homogenization Methods*, Advanced Structured Materials 64,
DOI 10.1007/978-3-319-19440-0_24

responses. In metallic materials, it is well known that control of microscopic factors as represented by the volume of the second phase and grain size is as important as the alloy design for determining the potential strengthening effects (Dieter and Bacon 1989). Recently, approaches based on tailored microstructures have received much attention in R&D of structural metals in order to achieve a balance between strength and elongation. Kimura et al. (2008) created an innovative steel by fabricating a fibrous microstructure with a caliber-rolling process, Ameyama and Fujiwara (2012) produced a distinctive bimodal microstructure by using powder metallurgy and ball milling, and Koseki et al. (2014) developed a class of multi-layered metallic sheets on the basis of a general rolling process.

Micromechanics is a research field about the relationship between microstructure and macroscopic material response, which has progressed remarkably from original analytical approaches (Voigt 1910; Reuss 1929; Hashin and Shtrikman 1963; Mori and Tanaka 1973; Weng 1990; Chen et al. 1992) to computational approaches using high-performance computers and discretization theory (Guedes and Kikuchi 1990; Ohno et al. 2001; Terada et al. 2003). A classical mixture rule (Voigt 1910; Reuss 1929) provides theoretical upper and lower bounds on properties such as the elastic modulus for two components in terms of the relationship between the volume fractions and their bulk properties, which means the assumption of homogeneous strain or stress states at the heterogeneous body. Hashin–Shtrikman's upper and lower bounds (Hashin and Shtrikman 1963) are narrow in comparison with the counterparts of the classical mixture rules (Voigt 1910; Reuss 1929), which are understood as a conceptual problem characterized by a matrix and an ellipsoidal inclusion on the basis of the Mori–Tanaka theory (Mori and Tanaka 1973; Weng 1990; Chen et al. 1992). The above fundamental knowledge in classical micromechanics indicates the strengthening effect of microscopic morphology, i.e. the continuous distribution of strong components leads to high strength. Nowadays, we can confirm this explanation of the morphologic strengthening effect with the computational approaches that enable us to deal with an arbitrary microscopic morphology and calculate its averaged responses. Such computational approaches are applicable to various problems and materials. However, such approaches require the detailed information about the objective to make the computational models.

A class of steels is unquestionably one of the most important structural materials. Steels typically have very complicated microstructures containing a number of hierarchic heterogeneities, where the interaction between iron and carbon atoms plays a complex role. For example, ferrite–pearlite duplex steels, which are commonly used carbon steels, are literally composed of ferrite and pearlite parts in the microstructure. Ferrite part is a polycrystalline aggregate characterized by crystal orientations for each grain, and each ferrite crystal typically contains carbon as an interstitial solid solution at the atomic scale. On the other hand, pearlite part is composed of numerous pearlite colonies characterized by a lamellar structure of ferrite crystal and cementite which is a metastable ferrous carbide. Watanabe et al. (2012) approached the multi-scale modelling of the ferrite–pearlite duplex microstructure in consideration of the above hierarchic system. Actually, the modeling of steel microstructures has been less well developed despite the importance of this material in industry. It is still challenging to understand the comprehensive mechanism of the strengthening effects in steels. To predict only macroscopic material responses, it is possible to illustrate the macroscopic stress-strain curve of a duplex steel from the stress-strain curves of single-constituent steels (Ohata et al. 2010), where strengthening microstructural characteristics, e.g. grain size, dislocation density, alloy elements and so on, must be aligned almost equal between the two constituents in the objective duplex steel and the single-constituent steels to eliminate these underlying strengthening effects at the finer scale.

In this study, we develop a computational approach to discover an optimized microscopic morphology that would have a strengthening effect in a duplex steel by combining the finite element (FE) analysis method for periodic microstructures and a gradient-based computational optimization method in an attempt to develop a design direction in materials R&D with computer-aided engineering technologies.

24.2 Finite Element Analysis Method for Periodic Microstructures

Use of the finite element analysis method for periodic microstructures is well established on the basis of mathematical homogenization theory (Allaire 1992, 1996). Within the framework of mathematical homogenization theory, the micro/macroscale coupling boundary value problem (BVP) is derived by purely mathematical operations. In the two-scale FE analysis method, finite element method is employed to solve the derived two-scale BVP (Guedes and Kikuchi 1990; Ohno et al. 2001; Terada et al. 2003).

The mathematical framework of the two-scale BVP has a unique feature in that the microscopic BVPs act as constitutive models for macroscopic BVP. Then the microscopic BVPs are simultaneously solved at every macroscopic material points in the computations, which is a source of massive computational efforts in its practical application. By reducing the macroscopic BVP to a point-wise stress-strain relationship, the two-scale BVP turns into a problem to evaluate the material behaviour of a numerical model of a periodic microstructure while controlling the macroscopic stress or strain (Watanabe et al. 2012; Watanabe and Terada 2010). Here the FE model is prepared as a representative volume element (RVE) of the objective material in consideration of the microscopic morphology.

24.2.1 Boundary Value Problem on a Microscale

In this approach, the BVP on a microscale is represented as

$$\int_{\Omega_Y} \boldsymbol{P} : \nabla_Y \boldsymbol{\eta}^{(1)} d\Omega_Y = 0 \quad \forall \boldsymbol{\eta}^{(1)} \in W_{\text{periodic}}, \tag{24.1}$$

where $\boldsymbol{Y}$ is the corrdination system on a microscale, $\boldsymbol{P}$ is the first Piola-Kirchhoff stress, $\boldsymbol{\eta}^{(1)}$ is the variation of the periodic displacement $\boldsymbol{u}^{(1)}$, $d\Omega_Y$ denotes the differential volume of the overall RVE Ω_Y and W_{periodic} is the Sobolev space of the periodic functions. The displacement field ($\boldsymbol{w}$) on a microscale is defined as

$$\boldsymbol{w} = \bar{\boldsymbol{H}}\boldsymbol{Y} + \boldsymbol{u}^{(1)}, \tag{24.2}$$

where $\bar{\boldsymbol{H}}$ is the macroscopic displacement gradient. Then the displacement gradient ($\boldsymbol{H}$) on a microscale is given as

$$\boldsymbol{H} = \nabla_Y \boldsymbol{w} = \bar{\boldsymbol{H}} + \nabla_Y \boldsymbol{u}^{(1)}, \tag{24.3}$$

where ∇_Y is the gradient operator. Also, the macroscopic nominal stress is defined as the volume average of the corresponding microscopic variable in the overall RVE.

$$\bar{\boldsymbol{P}} := \frac{1}{\Omega_Y} \int_{\Omega_Y} \boldsymbol{P} \, d\Omega_Y. \tag{24.4}$$

24.2.2 *Setting of Analysis Condition*

Following the definition of periodicity and Eq. (24.3), the difference of the displacements at points A and B that satisfies periodicity is written as follows:

$$\boldsymbol{w}_A - \boldsymbol{w}_B = \bar{\boldsymbol{H}} \left(\boldsymbol{Y}_A - \boldsymbol{Y}_B \right). \tag{24.5}$$

Equation (24.5) relates the macroscopic displacement gradient $\bar{\boldsymbol{H}}$ to the displacement field $\boldsymbol{w}$ on a microscale. Using Eqs. (24.4) and (24.5), the FE analyses of periodic microstructure are carried out by controlling of macroscopic displacement gradient $\bar{\boldsymbol{H}}$ or macroscopic stress $\bar{\boldsymbol{P}}$.

24.3 Strengthening Effect of Microscopic Morphology in Duplex Steels

In this section, the morphologic effect of a ferrite–pearlite duplex microstructure on the macroscopic mechanical behavior is investigated by the comparison of the material responses between morphologically different finite element models. This section is based on our previous work published in Japanese (Setoyama et al. 2012).

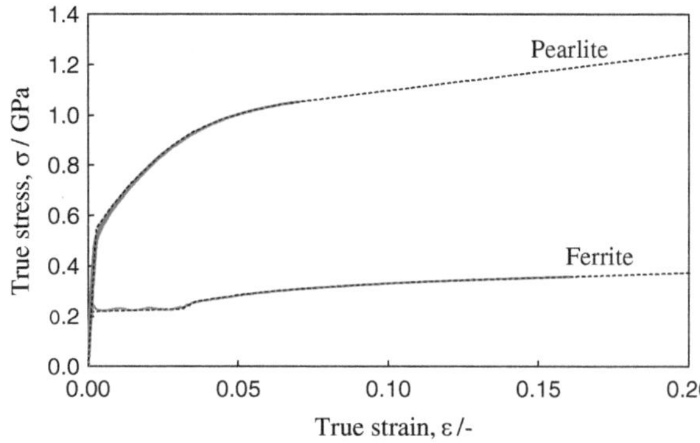

Fig. 24.1 Stress-strain curves of ferrite single-phase steel and full-pearlite steel. The *solid line* of the experimental results is extended to the finite strain range described by the *dashed line* for the computations

24.3.1 Simulations and Experiments

Ferrite-pearlite duplex steels contain two constituents characterized by different mechanical behaviors on a microscale, as mentioned in Sect. 24.1. In this study, an isotropic elastoplastic constitutive model is employed for each constituent to reduce computational efforts and simplify the modeling. The stress-strain relationships of the ferrite and pearlite constituents are shown in Fig. 24.1, where the experimental results of tensile tests are extended to the finite strain range for the computations. These extended stress-strain curves are described with a piecewise linear hardening law in FE simulations. The difference of strengths between these is wide enough to consider this difference as the major heterogeneity. However, the heterogeneity in each constituent becomes dominant in the cases of their increasing volume fractions.

In this study, three kinds of FE models are prepared for the ferrite–pearlite duplex microstructure, as illustrated in Fig. 24.2. A series of model A and model B possess the so-called core-shell structure composed of connecting and distributed components, i.e., the matrix and its inclusions. The ferrite part, which is the lower-strength constituent, is distributed continuously in model A and also discretely in model B. In model C, each part is randomly distributed. The purpose of this case study is to characterize the dependency of the microscopic morphology and volume fraction on the macroscopic material response.

We also conducted experimental tests for the comparison study. Specimens of ferrite–pearlite duplex steels were fabricated in different volume fraction of pearlite constituent by controlling the carbon content. Herein we assumed that the perlite constituent comprised almost all the carbon. Also, in order to remove the major strengthening factors except the volume fraction, the grain size of the ferrite grains and the lamellar spacing of the pearlite colonies were carefully adjusted to be approximately 10.0 and 0.15 μm, respectively, with the heat treatments.

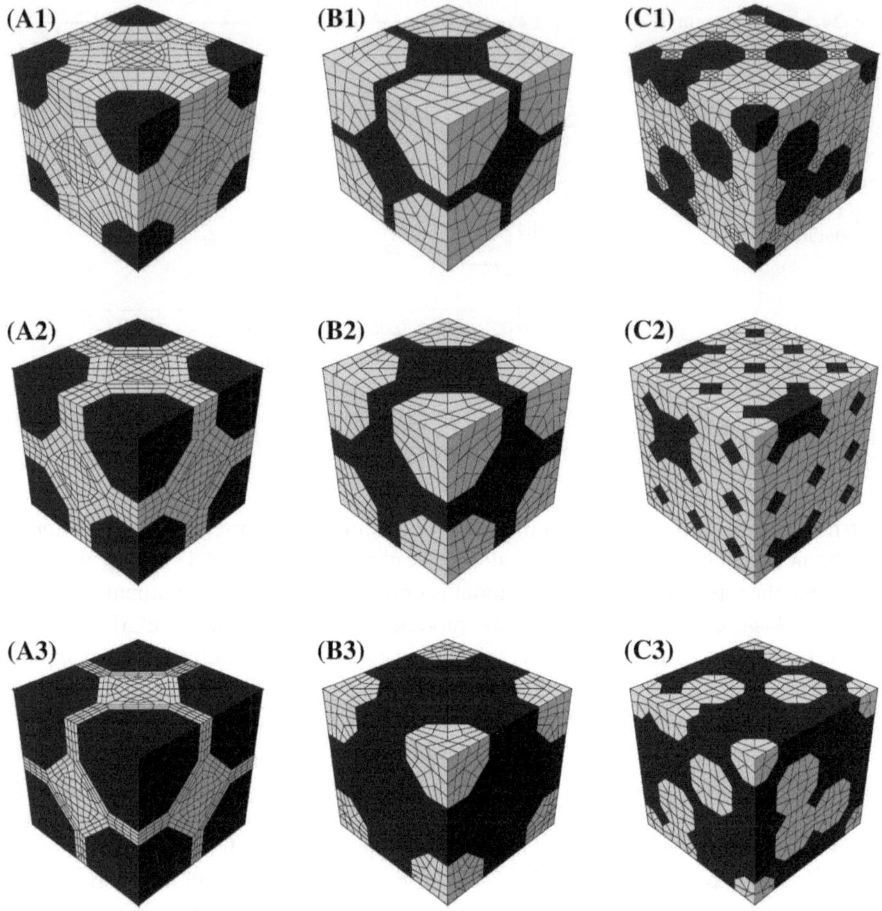

Fig. 24.2 Finite element models of ferrite–pearlite duplex microstructure with different microscopic morphology and volume fraction of pearlite part. *Bright* and *dark* areas represent respectively ferrite and pearlite parts. The volume fractions of pearlite part are defined as follows: *Model A* **A1** 21.6 %, **A2** 51.2 %, **A3** 72.9 %. *Model B* **B1** 27.1 %, **B2** 48.8 %, **B3** 78.4 %. *Model C* **C1** 22.2 %, **C2** 55.6 %, **C3** 77.8 %

24.3.2 Results and Discussions

The macroscopic stress-strain curves of the volume fractions of approximately 22 and 52 % pearlite part for macroscopic uniaxial tensile stress are shown in Fig. 24.3, where the different yield strengths and work-hardening behaviors are obtained as the macroscopic material responses. The differences of the macroscopic material responses are enlarged in the range around 50 % because the heterogeneity of the ferrite–pearlite duplex microstructure becomes dominant in this range.

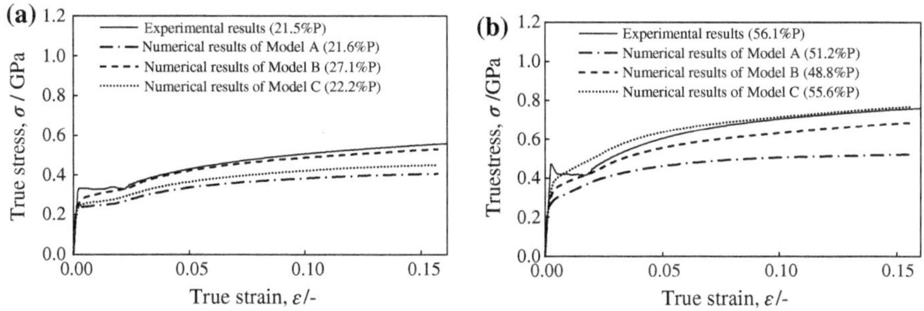

Fig. 24.3 Stress-strain curves of ferrite–pearlite duplex steels for uniaxial tensile stress. The *solid* and *dashed lines* indicate numerical and experimental results, respectively. **a** $V_p \approx 22\,\%$. **b** $V_p \approx 52\,\%$

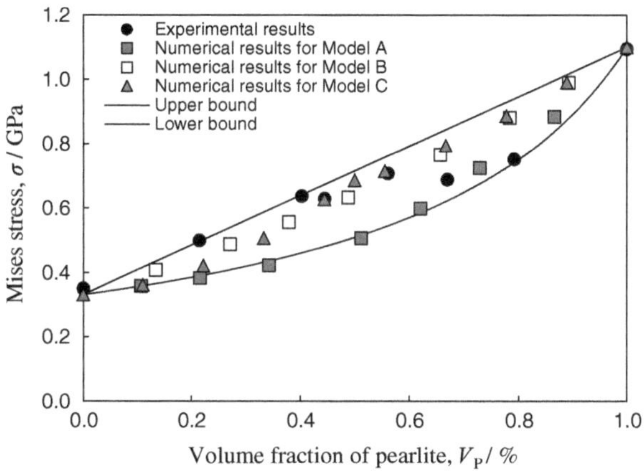

Fig. 24.4 Relationship between the Mises stress at axial strain 0.10 and the volume fraction of the pearlite part. The *upper* and *lower* bounds are drawn on the basis of Mises stresses of ferrite and pearlite single-constituent steels

The relationships between the macroscopic Mises stress at 10 % axial strain and the volume fraction of pearlite part V_p are illustrated in Fig. 24.4, in which the upper bound σ_{upper} and the lower bound σ_{lower} are drawn on the basis of the same concept as the classical mixture rule (Voigt 1910; Reuss 1929).

$$\sigma_{upper} = \left(1 - V_p\right)\sigma_f + V_p\sigma_p, \tag{24.6}$$

$$\sigma_{lower} = \left(\frac{1 - V_p}{\sigma_f} + \frac{V_p}{\sigma_p}\right)^{-1}, \tag{24.7}$$

where σ_f and σ_p are the Mises stresses of ferrite and pearlite single-constituent steels, respectively. Figure 24.4 shows that the upper and lower bounds are effective for all of the data in this case study. Following the abovementioned knowledge of classical micromechanics, the results of model A and model B form the narrower upper and lower bounds corresponding to Hashin–Shtrikman's upper and lower bounds (Hashin and Shtrikman 1963). The results of model C are basically located between these bounds and shifts from the lower bound to the upper bound with increasing volume fraction of the pearlite part, where the results are partly beyond these bounds because of the effect of anisotropy of the microscopic morphology. This result means that the topology of the pearlite part, i.e. the strong part, changes gradually with increasing volume fraction.

On the other hand, the tendency of the corresponding experiments shifts inversely from the upper bound to the lower bound with increasing volume fraction of the pearlite constituent. In our understanding, this result is caused by the incompleteness of both our simulation models and the preparation of the specimens in the experiments. For example, the constitutive model employed in these calculations cannot follow the stress dropping at the yield point shown in Fig. 24.3. Also, the yield-point phenomenon is typically caused by the effect of the interstitial carbon remaining in the ferrite crystals. Therefore, the assumption that the carbon is used only for the pearlite constituent is not strictly satisfied. This result represents the difficulty of the modeling and the validation of carbon steels. Nonetheless, it is concluded that the FE analysis method for periodic microstructures is useful for associating the microscopic morphology with the corresponding macroscopic material response and for characterizing the strengthening effect of the microscopic morphology.

24.4 Maximization of Morphologic Strengthening Effect

The finite element analysis method for periodic microstructures is a computational approach for evaluating the macroscopic material response from a representative volume element of a microstructure. As shown in Sect. 24.3 the microscopic morphology has the potential to improve the material properties. In this section, a framework of an inverse computational approach is developed to find the optimized morphology of the microstructure for the objective macroscopic material response by combining the finite element analysis method and a gradient-based computational optimization method.

24.4.1 Computational Approach

In the research field of FE analysis, the topology optimization method (Suzuki and Kikuchi 1991; Matsui and Terada 2004) is a well-known approach using a mathematical optimization method that optimizes the material layout in a design space for given boundary conditions. The topology optimization of a microstructure has

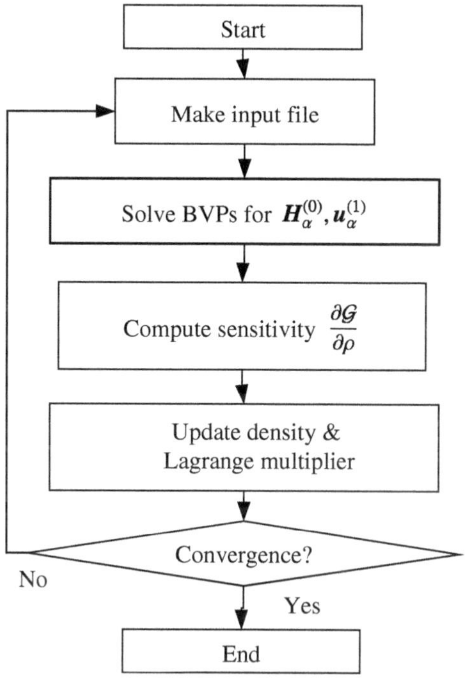

Fig. 24.5 Flowchart of gradient-based optimization analysis

been reported in application studies for small-strain elasticity and thermal conductivity (Sigmund 1994; Giusti et al. 2010). In this study, the topology optimization approach is extended to our objective, which is a filled periodic microstructure composed of two nonlinear deformable solids.

For the optimization calculations, a gradient-based optimization algorithm is utilized, which is an iterative procedure that includes nonlinear FE analysis in solving BVPs and computing the sensitivity of the objective function, as shown in Fig. 24.5. It is noted that this computational procedure is available even for nonlinear problems; however the computational efforts increase significantly. As seen in Sect. 24.3, an isotropic elastoplastic constitutive model is employed to describe the material response of each component, and the material constants are given from experiments of the corresponding single-constituent materials.

The optimization problem is generally described as

$$\text{Maximize}\,[\mathcal{G}(\rho)]\,, \tag{24.8}$$

where $\mathcal{G}$ is an objective function and ρ is a design variable. The definitions of the objective function $\mathcal{G}$ and the design variable ρ are definitely key elements in such an optimization problem.

24.4.1.1 Design Variable

A density representing the existence of a material is typically employed as a design variable of an optimization problem in topology optimization and the integration of the density in the design space is controlled with a constraint equation as

$$\int_{\Omega_Y} \rho d\Omega_Y - \Omega_0 = 0, \tag{24.9}$$

where Ω_0 is the constraint volume fraction of the material. Our objective is a filled duplex microstructure; therefore, one of difficulties is how to deal with the mixture state of the materials during the optimization process. Here we append a constraint equation for the density ρ as

$$\rho + \rho_{\mathrm{B}} = 1, \tag{24.10}$$

where the physical meaning of ρ is the volume fraction of a component and ρ_{B} is the volume fraction of another component. A simple mixture rule (Voigt 1910) is applied to handle the mixture state during the optimization process:

$$\boldsymbol{P} = \rho^m \boldsymbol{P}_{\mathrm{A}} + \left(1 - \rho^m\right) \boldsymbol{P}_{\mathrm{B}}, \tag{24.11}$$

where $\boldsymbol{P}_{\mathrm{A}}$ and $\boldsymbol{P}_{\mathrm{B}}$ are the first Piola-Kirchhoff stresses of the two components and m is a stabilization parameter in optimization analysis. The stabilization parameter m contributes to segregate the two components. Following previous studies in topology optimization (Matsui and Terada 2004), the design variable is interpolated in a finite element with shape functions of the finite element.

24.4.1.2 Objective Function

Our objective is to draw out the strengthening effect of the microscopic morphology. Here we define the objective function as the sum of the macroscopic external works $\mathcal{G}_\alpha$ for several macroscopic strain modes $\boldsymbol{H}_\alpha^{(0)}$:

$$\mathcal{G} := \sum_{\alpha=1}^{N_{\mathrm{mode}}} \mathcal{G}_\alpha(\rho; \boldsymbol{H}_\alpha^{(0)}, \boldsymbol{u}_\alpha^{(1)}) = \sum_{\alpha=1}^{N_{\mathrm{mode}}} \int_0^T \bar{\boldsymbol{P}}_\alpha : \dot{\boldsymbol{H}}_\alpha^{(0)} dt, \tag{24.12}$$

where N_{mode} is the number of macroscopic strain modes. That is, the objective function is calculated on the basis of the results of finite element analysis for some normalized macroscopic strain modes $\boldsymbol{H}_\alpha^{(0)}$. In this case, the maximization of the objective function means to find a high-strength microscopic morphology under the constraint of volume fraction (24.9).

Based on Eqs. (24.8)–(24.12), the optimization problem is summarized as follows:

$$\text{Maximize}\left[\sum_{\alpha=1}^{N_{\text{mode}}} \mathcal{G}_\alpha(\rho;\ \boldsymbol{H}_\alpha^{(0)}, \boldsymbol{u}_\alpha^{(1)})\right] \tag{24.13}$$

$$\mathcal{G}_\alpha := \int_0^T \bar{\boldsymbol{P}}_\alpha : \dot{\boldsymbol{H}}_\alpha^{(0)} dt \qquad \forall \left(\boldsymbol{H}_\alpha^{(0)},\ \boldsymbol{u}_\alpha^{(1)}\right) \in \{\text{solution of BVP}\}$$

$$\text{subject to} \quad \int_{\Omega_Y} \rho d\Omega_Y - \Omega_0 = 0,\ \rho_{\text{B}} = 1 - \rho, \quad \rho \in [0, 1]$$

24.4.2 Demonstration and Discussion

The proposed approach is demonstrated for a duplex microstructure. Here, Swift-type hardening law is employed to describe the material response of each component:

$$k = a\,(c + \alpha)^n\,, \tag{24.14}$$

where k is the yield stress, α is the accumulated plastic strain, and a, c, and n are material constants. Material constants are given from the literature (Rudiono and Tomota 1997) and are based on the experimental data of ferrite and martensite single-constituent steels. We here consider the design variable of this optimization problem as the volume fraction of the martensite constituent and set up the constraint volume fraction in this calculations as 50 %, i.e. $\Omega_0 = 0.5$. The design space at the undeformed state is defined as the $17\times17\times17$ hexahedral finite elements shown in Fig. 24.6. The optimization calculation starts from a mixture state containing an imperfection at the centre of the design space. In the following calculations, each result was obtained after approximately 40 iterations of the optimization analysis shown in Fig. 24.5

Figure 24.7 shows the optimization results for a macroscopic uniaxial tensile deformation mode along the X-axis shown in Fig. 24.6. In Fig. 24.7a, only the morphology of constituent A, i.e. the stronger of the two components, is illustrated in the undeformed design space. Figure 24.7b shows the macroscopic material responses of the resulting microstructure for various deformation modes. Both the microscopic morphology and the corresponding macroscopic material responses are anisotropic, corresponding to the imposed macroscopic deformation mode. This result clearly shows that the resulting microstructure definitely depends on the imposed macroscopic strain modes.

Next, let us impose macroscopic uniaxial tensile deformation modes for three orthogonal directions. The results of this optimization analysis are shown in Fig. 24.8. Figure 24.8 also shows the consistent microstructure and material responses with the imposed macroscopic deformation modes, that is, it is symmetric for the three orthogonal directions but anisotropy for the shear modes still remains.

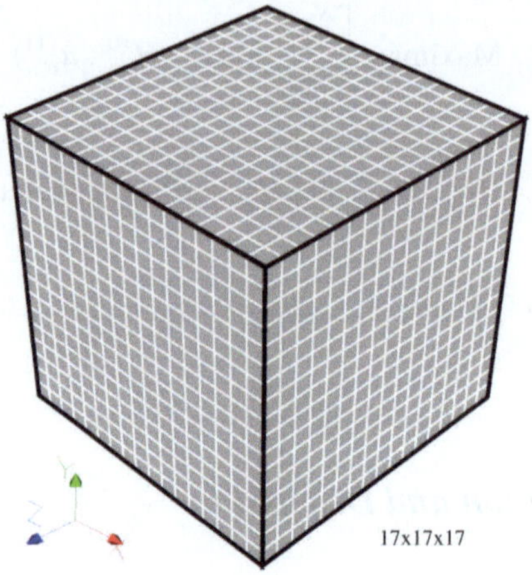

Fig. 24.6 Design space of periodic microstructure composed of hexahedral finite elements

In this context, by imposing six independent macroscopic deformation modes containing three uniaxial tensile and three shear deformation modes, the optimized microstructure that provides the isotropic macroscopic material responses is obtained as shown in Fig. 24.9. The resulting microstructure represents trivially the isotropic and continuous distribution of the stronger component, which is consistent with the knowledge of micromechanics.

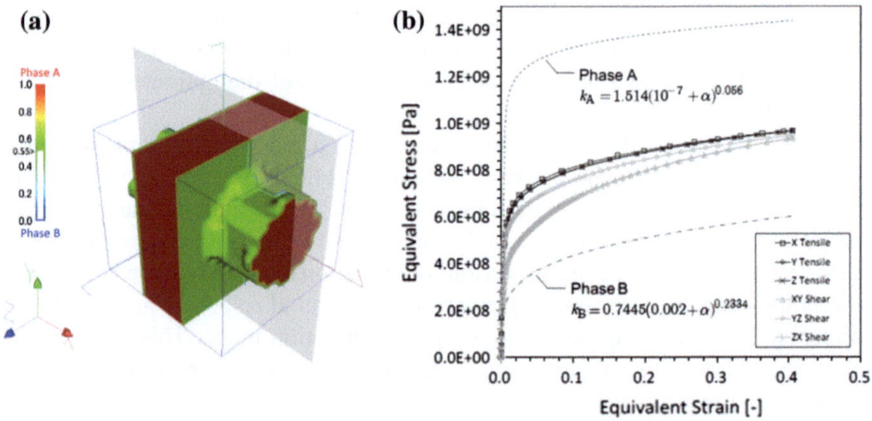

Fig. 24.7 Results of optimization analysis for macroscopic uniaxial tensile deformation mode ($N_{\text{mode}} = 1$). **a** The morphology of the stronger component in the undeformed design space. **b** The macroscopic material responses of the resulting microstructure for various deformation modes

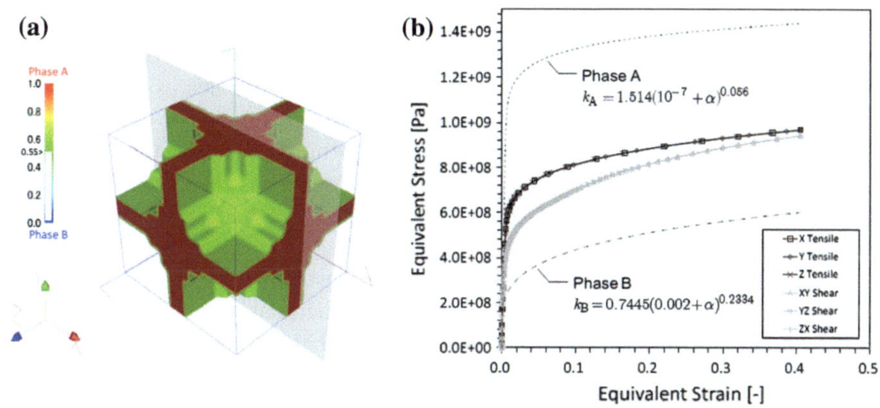

Fig. 24.8 Results of optimization analysis for three orthogonal macroscopic uniaxial tensile deformation modes ($N_{\text{mode}} = 3$). **a** The morphology of the stronger component in the undeformed design space. **b** The macroscopic material responses of the resulting microstructure for various deformation modes

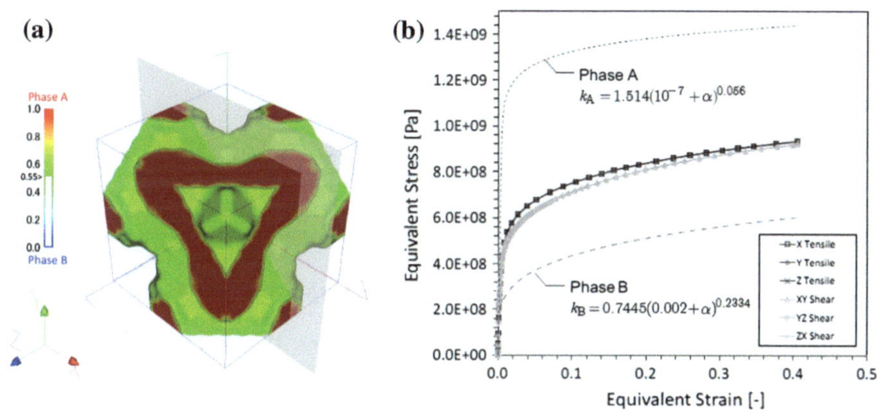

Fig. 24.9 Results of optimization analysis for six independent macroscopic deformation modes ($N_{\text{mode}} = 6$). **a** The morphology of the stronger component in the undeformed design space. **b** The macroscopic material responses of the resulting microstructure for various deformation modes

24.5 Conclusion

A computational optimization approach was proposed to draw out the strengthening effect of the microscopic morphology in a deformable inelastic duplex microstructure as an extension of the topology optimization method. As a demonstration of the proposed approach, an optimization analysis was carried out to find the microscopic morphology of a high-strength material, where the objective function was defined on the basis of the macroscopic material responses obtained as results of FE analysis.

Such a computational optimization approach has enormous potential in engineering design fields because the framework is founded on nonlinear FE analysis. For example, employing a ductile fracture model supposedly enables us to discuss the ductility and toughness in relation to the microscopic features. The computational framework can be extended further to consider the forming process of the microstructure by coupling it with phase-field calculations (Chen 2002; Moelans et al. 2008) and calculation of phase diagrams (CALPHAD) based on a thermodynamic database (Kaufman 2008; Lukas et al. 2007).

Physics-based constitutive modeling and the reasonable way to determine the material constants are inevitable to characterize the material behavior on continuum scale in relation to the nanoscopic feature. However these are still challenging in practical structural metals in both theoretical and experimental approaches as mentioned in Sect. 24.1. We believe that mutual collaborations between theoretical and experimental researches lead to a breakthrough in this research field.

Acknowledgments This research was supported by the Japan Science and Technology Agency under collaborative research based on industrial demand "Heterogeneous Structure Control: Towards Innovative Development of Metallic Structural Materials" and Grant-in-Aid for Young Scientists (No. 25102711, 25820359).

References

Allaire G (1992) Homogenization and two-scale convergence. SIAM J Math Anal 23:1482–1518

Allaire G (1996) Mathematical approaches and methods. In: Hornung U (ed) Homogenization and porous media. Springer, New York, pp 225–250

Ameyama K, Fujiwara H (2012) Creation of harmonic structure materials with outstanding mechanical properties. Mater Sci Forum 706–709:9–16

Chen LQ (2002) Phase-field models for microstructure evolutions. Ann Rev Mater Res 32:113–140

Chen T, Dvorak G, Benveniste Y (1992) Mori-tanaka estimates of the overall elastic moduli of certain composite materials. Trans ASME J Appl Mech 59:539–546

Dieter G, Bacon D (1989) Mechanical metallurgy, 3rd edn. In: Materials science & engineering. McGraw Hill, London

Giusti S, Novotny A, de Souza Neto E (2010) Sensitivity of macroscopic response of elastic microstructures to the insertion of inclusions. Proc R Soc A 466:1703–1723

Guedes J, Kikuchi N (1990) Preprocessing and postprocessing for materials based on the homogenization method with adaptive finite element methods. Comput Meth Appl Mech Eng 83:143–198

Hashin Z, Shtrikman S (1963) A variational approach to the elastic behavior of multiphase minerals. J Mech Phys Solids 11:127–140

Kaufman L (2008) Computational thermodynamics and materials design. Calphad 25:141–161

Kimura Y, Inoue T, Yin F, Tsuzaki K (2008) Inverse temperature dependence of toughness in an ultrafine grain-structure steel. Science 320:1057–1060

Koseki T, Inoue J, Nambu S (2014) Development of multilayer steels for improved combinations of high strength and high ductility. Mater Trans 55:227–237

Lukas H, Fries S, Sundman B (2007) Computational thermodynamics—the calphad method. Cambridge University Press, Cambridge

Matsui K, Terada K (2004) Continuous approximation of material distribution for topology optimization. Inter J Numer Methods Eng 84:733–756

Moelans N, Blanpain B, Wollants P (2008) An introduction to phase-field modeling for microstructure evolution. Calphad 32:268–294
Mori T, Tanaka K (1973) Average stress in matrix and average elastic energy of materials with misfitting inclusions. Acta Metall 21:571–574
Ohata M, Suzuki M, Ui A, Minami F (2010) 3d-simulation of ductile failure in two-phase structural steel with heterogeneous microstructure. Engrg Fracture Mech 77:277–284
Ohno N, Matsuda T, Wu X (2001) A homogenization theory for elastic-viscoplastic composites with point symmetry of internal distributions. Inter J Solids Struct 38:2867–2878
Reuss A (1929) Berechnung der fließgrenze von mischkristallen auf grund der plastizitätsbedingung für einkristalle. Z Angew Math Mech 9:49–58
Rudiono, Tomota Y (1997) Application of the secant method to prediction of flow curves in multi microstructure steels. Acta Mater 45:1923–1929
Setoyama D, Watanabe I, Iwata N (2012) Comparative study of microscopic morphology and mechanical behavior between experiments and numerical analyses for ferrite-pearlite dual component steel (in Japanese). Tetsu-to-Hagane 98:290–295
Sigmund O (1994) Materials with prescribed constitutive parameters: an inverse homogenization problem. Int J Solids Struct 31:2313–2329
Suzuki K, Kikuchi N (1991) A homogenization method for shape and topology optimization. Comput Methods Appl Mech Eng 93:291–318
Terada K, Saiki I, Matsui K, Yamakawa Y (2003) Two-scale kinematics and linearization for simultaneous two-scale analysis of periodic heterogeneous solids at finite strain. Comput Meth Appl Mech Eng 192:3531–3563
Voigt W (1910) Lehrbuch der Kristallphysik. Teubner, Leipzig
Watanabe I, Terada K (2010) A method of predicting macroscopic yield strength of polycrystalline metals subjected to plastic forming by micro-macro de-coupling scheme. Int J Mech Sci 52: 343–355
Watanabe I, Setoyama D, Nagasako N, Iwata N, Nakanishi K (2012) Multiscale prediction of mechanical behavior of ferrite-pearlite steel with numerical material testing. Int J Numer Methods Eng 89:829–845
Weng G (1990) Average stress in matrix and average elastic energy of materials with misfitting inclusions. Acta Metall 21:571–574

Chapter 25
Molecular Dynamics Simulations on Local Lattice Instability at Mode I Crack Tip in BCC Iron

Kisaragi Yashiro, Yuta Tsugawa and Hiroshi Katayama

Abstract Molecular dynamics simulations are performed on the [001](010), [001](110) and [11$\bar{2}$](111) through cracks in bcc Fe under mode I loading, in order to discuss about the local lattice stability at the crack tip within the framework of FS potential function. The crack width is set to $2c = 0.1L_x, 0.2L_x$ and $0.5L_x$, respectively, against the periodic cell length of $L_x = 20 - 30\,\mathrm{nm}$. The [001](010) crack shows ductile behavior of blunting by dislocation emission, resulting in the remarkable nonlinearity in the stress–strain curve. Both the [001](110) and [11$\bar{2}$](111) cracks propagates in a brittle manner, showing the abrupt stress drop by the rupture. Then the local stability is discussed by the positiveness of the determinant of 6×6 matrix of elastic stiffness coefficients, B_{ij}^{α}; that is, Wang's B-criteria is applied to each atom. Negative atoms emerges at far smaller strain than the peak or rupture one, corresponding to the onset of local "plastic deformation", i.e. dislocation emission or rearrangement at the crack surface. For the detail of the unstable mode, we have evaluated the eigenvalue of B_{ij}^{α} of each atom. Since the negative 1st eigenvalue leads the negative det B_{ij}^{α}, it is natural the tendencies of the former and the later are almost same. However, a few atoms at the [11$\bar{2}$](111) crack tip turn to negative in their 2nd eigenvalue just at or just after the stress–strain peak.

Keywords Local lattice instability · Crack tip · Bcc iron

K. Yashiro (✉)
Gifu University, 1-1, Yanagito, Gifu 501-1192, Japan
e-mail: kisaragi@gifu-u.ac.jp

Y. Tsugawa · H. Katayama
Graduate School of Engineering, Kobe University, 1-1, Rokkodai, Nada, Kobe 657-8501, Japan
e-mail: tsugawa@mail.mm4.scitec.kobe-u.ac.jp

H. Katayama
e-mail: katayama.hiroshi@mail.mm4.scitec.kobe-u.ac.jp

H. Altenbach et al. (eds.), *From Creep Damage Mechanics to Homogenization Methods*, Advanced Structured Materials 64,
DOI 10.1007/978-3-319-19440-0_25

25.1 Introduction

Inspired by Wang's stability criteria (B-criteria) for crystal under finite strain and temperature (Wang et al. 1993), we have discussed the onset of local deformation or rearrangement of local lattices based on the atomic elastic stiffness coefficients (AES) at each atom point (Yashiro and Tomita 2001; Yashiro et al. 2006; Nishimura et al. 2010; Yashiro et al. 2012; Yashiro and Yamane 2014). In order to pursue the unique atomistic criterion beyond the conventional theory for continuum body, e.g. stress intensity factor, and also to discuss the fundamental characteristics between system instability and local ones, we have changed our target from homogeneous system of bulk perfect lattice, to nanowire with surfaces and bamboo structure of grain boundaries, etc. by adding the system inhomogeneity. In the case of brittle Si, the AES reveals that the system deforms toward mechanical homogeneity, then reaches the homogenization limit and the system instability or sudden stress drop is triggered by the emergence of negative AES atoms (Yashiro et al. 2012). That is, even though there is a large difference first in the AES between inhomogeneous defects and the other bulk part, the later lost the stiffness under external loading and finally the difference vanishes at the "homogenization limit". Contrary to this smart story between the system instability and AES in Si, AES in the fcc and bcc metallic system could not capture the system instability yet. Certainly many negative AES atoms can be observed in the transition from elastic limit to the onset of local deformation, e.g. dislocation burst, in many cases it is rather reasonable to consider the system already starts its unstable change accompanying the emergence of the negative AES atoms. Especially in the case of bcc Fe, local deformation does *not* occur even all the AES becomes negative in perfect bulk and some surface plate model under the strain controlled tension, resulting in the smooth stress–strain peak (Yashiro and Yamane 2014). Thus the bcc Fe would start unstable elongation without local deformation under the stress controlled loading. This reflects the bcc Fe tends to reach the instability for massive and collective atomic motion, such as martensitic transformation rather than the local deformation.

The present study is the first report that we applied our AES analysis to the mode I crack tip in the bcc Fe. Molecular dynamics simulations of the [001](010), [001](110) and $[11\bar{2}](111)$ through cracks are performed with different crack length of $2c = 0.1L_x, 0.2L_x$ and $0.5L_x$ against the periodic cell length of $L_x = 20-30$ nm. In the framework of FS potential function, the 1st and 2nd order derivatives can be formulated for the atomic stress and elastic coefficients so that we can evaluate AES only with the atomic configuration at each time step. After observing the crack blunting/propagation, we discuss its mechanism first by the positiveness of the determinant of B^{α}_{ij}, which is the 6×6 matrix representation of the 4th order tensor of the atomic elastic stiffness, B^{α}_{ijkl}, as same as our previous studies. As more detailed study on the unstable mode, the eigenvalue of B^{α}_{ij} is investigated in the 6 dimension space of ε_i.

25.2 AES Evaluation in FS Potential Function

FS potential function (Finnis and Sinclair 1984, 1986) is adopted here for bcc Fe interaction. As the EAM type function, the system energy E_{tot} is represented as follows

$$E_{\text{tot}} = \sum_{\alpha} E^{\alpha} = \sum_{\alpha} \left\{ \frac{1}{2} \sum_{\beta \neq \alpha} V(r^{\alpha\beta}) - Af(\rho^{\alpha}) \right\} \quad (25.1)$$

here superscripts α and β indicate atoms α and β, $r^{\alpha\beta}$ is atomic distance, $V(r)$ is pair potential, $f(\rho)$ is embedding function and ρ^{α} is atomic density at atom α. The function forms and parameter A are written in the original paper (Finnis and Sinclair 1984, 1986). The system energy is the summation of the energy contribution of each atom, E^{α}, as shown in the bracket. In the thermodynamics of crystal lattice, the stress and elastic coefficients are defined as the 1st and 2nd order derivatives of the energy per volume, against strain perturbation (Wallace 1972). Thus we define the atomic stress and elastic coefficients as follows

$$\sigma_{ij}^{\alpha} = \frac{1}{\Omega} \frac{\partial E^{\alpha}}{\partial \varepsilon_{ij}} = \frac{1}{\Omega} \sum_{\beta \neq \alpha} \left\{ \frac{1}{2} V'(r^{\alpha\beta}) - Af'(\rho^{\alpha})\phi'(r^{\alpha\beta}) \right\} \frac{r_i^{\alpha\beta} r_j^{\alpha\beta}}{r^{\alpha\beta}} \quad (25.2)$$

$$\begin{aligned} C_{ijkl}^{\alpha} = &\left[\frac{1}{2} \sum_{\beta \neq \alpha} \left\{ V''(r^{\alpha\beta}) - \frac{V'(r^{\alpha\beta})}{r^{\alpha\beta}} r \right\} \frac{r_i^{\alpha\beta} r_j^{\alpha\beta} r_k^{\alpha\beta} r_l^{\alpha\beta}}{(r^{\alpha\beta})^2} \right. \\ &- Af'(\rho) \sum_{\beta \neq \alpha} \left\{ \phi''(r^{\alpha\beta}) - \frac{\phi'(r^{\alpha\beta})}{r^{\alpha\beta}} \right\} \frac{r_i^{\alpha\beta} r_j^{\alpha\beta} r_k^{\alpha\beta} r_l^{\alpha\beta}}{(r^{\alpha\beta})^2} \\ &\left. - Af''(\rho) \left\{ \sum_{\beta \neq \alpha} \phi'(r^{\alpha\beta}) \frac{r_i^{\alpha\beta} r_j^{\alpha\beta}}{r^{\alpha\beta}} \right\} \left\{ \sum_{\beta \neq \alpha} \phi'(r^{\alpha\beta}) \frac{r_k^{\alpha\beta} r_l^{\alpha\beta}}{r^{\alpha\beta}} \right\} \right] / \Omega \end{aligned} \quad (25.3)$$

Here subscripts $i \sim l$ indicate free indices in the Cartesian coordinate (x, y, z), $r_i^{\alpha\beta}$ is the i component of position vector $\boldsymbol{r}^{\alpha\beta}$, the prime $'$ indicates the derivative and Ω atomic volume. In the no-load equilibrium and linear elasticity region, the stresses in two different states **x** and **X** are related with the strain ε_{ij} of **x**→**X** by the elastic coefficients; however, it is not valid in the nonlinear elasticity region so that the elastic stiffness coefficients, B_{ijkl}, is derived for general relationship of $B_{ijkl} \equiv \Delta\sigma_{ij}/\Delta\varepsilon_{kl}$ with the stress σ_{ij} and elastic coefficients C_{ijkl} (Wallace 1972). Referring to the original form of the B_{ijkl} for whole crystal and adding the Voigt symmetry (Wallace 1972), we defined the atomic elastic stiffness as follows

$$B_{ijkl}^{\alpha} = C_{ijkl}^{\alpha} + (\sigma_{il}^{\alpha}\delta_{jk} + \sigma_{jl}^{\alpha}\delta_{ik} + \sigma_{ik}^{\alpha}\delta_{jl} + \sigma_{jk}^{\alpha}\delta_{il} - \sigma_{ij}^{\alpha}\delta_{kl} - \sigma_{kl}^{\alpha}\delta_{ij})/2 \quad (25.4)$$

where δ_{ij} is the Kronecker delta. Using the Voigt notation for the Cartesian indices set of $(xx, yy, zz, yz, zx, xy)\Rightarrow(1, 2, 3, 4, 5, 6)$, the 4th order tensor B^{α}_{ijkl} can be represented as 6×6 matrix B^{α}_{ij} in the 6 dimensional phase space of ε_i. As same as our previous study, we check first the positiveness of det B^{α}_{ij} as the local stability at each atom point. Furthermore, we evaluate the eigenvalue of B^{α}_{ij} or the solution of the following equation

$$\boldsymbol{B}^{\alpha}\boldsymbol{\varepsilon} = \beta^{\alpha}\boldsymbol{\varepsilon} \tag{25.5}$$

For each atom at each time step, 6 eigenvalues of $\beta^{\alpha(k)}$ and corresponding eigen vectors of $\boldsymbol{\varepsilon}^{(k)} = (\varepsilon_1, \varepsilon_2, \varepsilon_3, \varepsilon_4, \varepsilon_5, \varepsilon_6)^{(k)}$ are obtained. Here (k) indicates the kth eigenvalue and eigenvector.

25.3 Simulation Procedure

As shown in Fig. 25.1, three crack lengths of $2c = 0.1L_x, 0.2L_x$ and $0.5L_x$ are considered with thin slab periodic cell, changing the crystal orientations for the [001](010), [001](110) and $[11\bar{2}](111)$ cracks. The cell dimensions of L_x, L_y and L_z are

$$100a \times 100a \times 6a, \quad 70\sqrt{2}a \times 70\sqrt{2}a \times 6a, \quad 70\sqrt{2}a \times 50\sqrt{3}a \times 3\sqrt{6}a$$

for each cracks, respectively, with about 120,000 atoms. Here $a = 0.287\,\text{nm}$ is the equilibrium lattice length under the temperature $T = 0\,\text{K}$. Crack displacement h and tip curvature R are set to $h = 2R = 0.6\,\text{nm}$, satisfying the h is greater than the cutoff radius of FS potential (0.357 nm) and the R is less than the Griffith crack condition of $R < \sqrt{8a/\pi}$. However, the crack tip shape is not round but square due to the discrete atomic configuration as shown in the lower figures in Fig. 25.1.

As same as our previous studies, we first aim to eliminate the effect of thermal vibration in the fundamental discussion, so that all simulations are performed at very low temperature of $T = 0.1\,\text{K}$. The initial configuration is relaxed during 10 ps controlling the cell size to cancel the normal stresses of $\sigma_1, \sigma_2, \sigma_3$ $(\sigma_{xx}, \sigma_{yy}, \sigma_{zz})$ under periodic boundary condition in all the directions. Then small strain increment of $\Delta\varepsilon_{yy} = 1.0 \times 10^{-7}$ is applied at every MD steps of 1 fs. The normal stress in the x direction is controlled to zero during the tension, while that in the z direction is not controlled, i.e. plain strain condition. Although we have checked the difference of plain strain and plain stress conditions in the thickness direction, there is little difference so that we omitted the plain stress condition in the present paper.

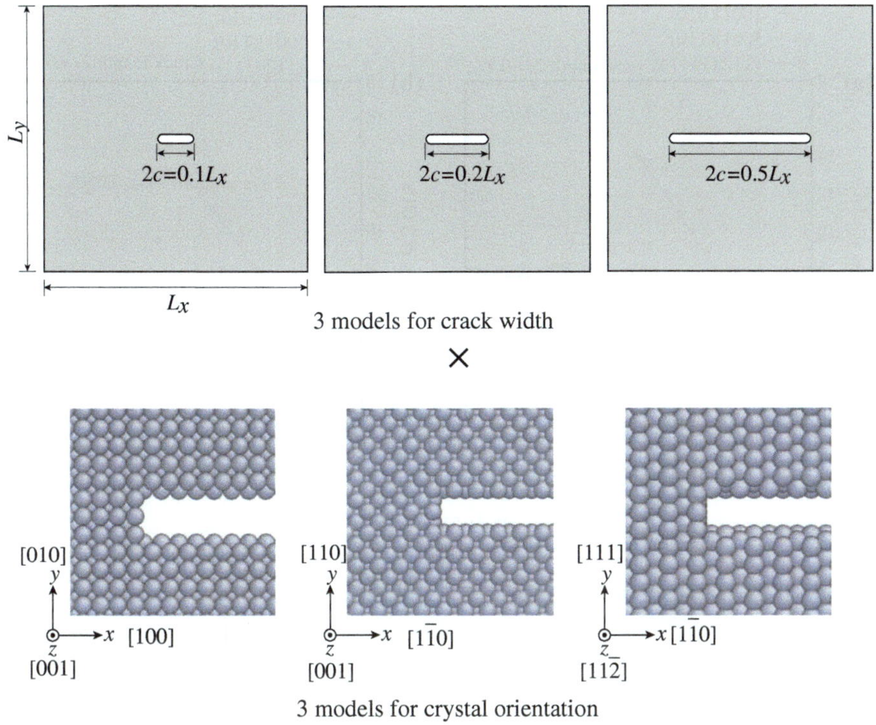

Fig. 25.1 Dimensions and crystal orientations for 2D through crack in periodic slab cell

25.4 Results and Discussion

25.4.1 Stress–Strain Curves and Deformation Behavior

Figure 25.2 shows the stress–strain curves. The [001](110) and [11$\bar{2}$](111) cracks show monotonic stress increase and then sudden stress drop to zero or the rupture. Their stress–strain curves are very close each other until the peak point, despite of the different tensile axis of [110] and [111]. Both cracks has same crystal orientation of [1$\bar{1}$0] in the lateral or the crack direction, so that the same response could owe to it. It also should be noted that the [11$\bar{2}$](111) cracks of $2c = 0.1L_x$ and $0.5L_x$ don't trace the shortest way to the rupture after the stress peak, i.e. there is some different deformation in these systems before the immediate rupture. The [001](010) crack does not show drastic stress drop but show small stress peak or inflection point, then the stress–strain curve lose the smoothness and has small oscillation. For all the crack orientation, the wider crack results in the smaller strain for the elastic limit.

Figures 25.3, 25.4 and 25.5 illustrate the snapshots of atoms after the stress–strain peak/flexion point for the crack length of $2c = 0.2L_x, 0.1L_x$ and $0.5L_x$, respectively. The atoms are colored by the von Mises local shear strain equipped in Atomeye (Li

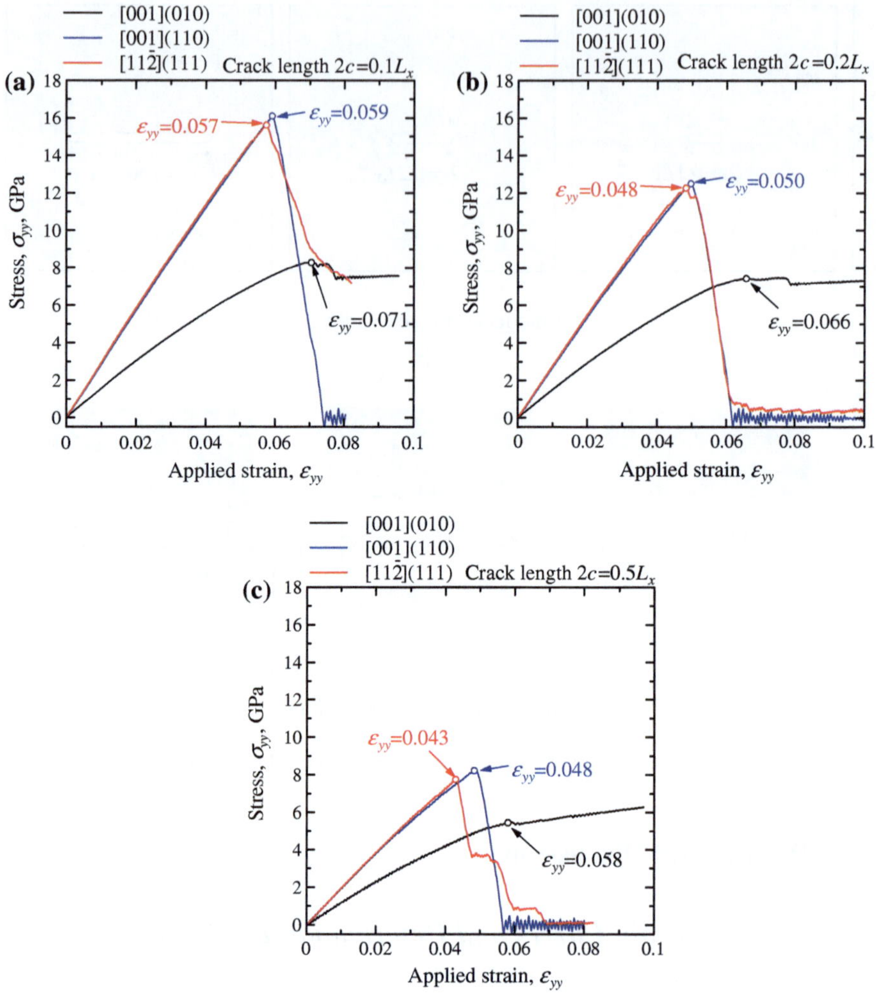

Fig. 25.2 Stress–strain curves. The *open circles* indicates the stress peak or flexion point. The curves are drawn for each crack lengths of **a** $2c = 0.1L_x$, **b** $2c = 0.2L_x$ and **c** $2c = 0.5L_x$

2003). There is no remarkable difference in the deformation morphology in the [001](010) cracks, so that we omit them in Figs. 25.4 and 25.5. In the case of the [001](010) crack, dislocations are emitted from crack tip *before* the stress peak and then they propagate and extend the deformation twins across the periodic cell. The crack tip is blunted by this dislocation emission and deformation twin formation; the [001](010) crack never propagate so that there is no drastic stress drop. On the [001](110) crack, high von Mises strain is observed at the shoulder of the opened crack. The high strain regions are inclined about 6–8° against tensile direction; however, these directions may not be selected by some special crystal orientation but

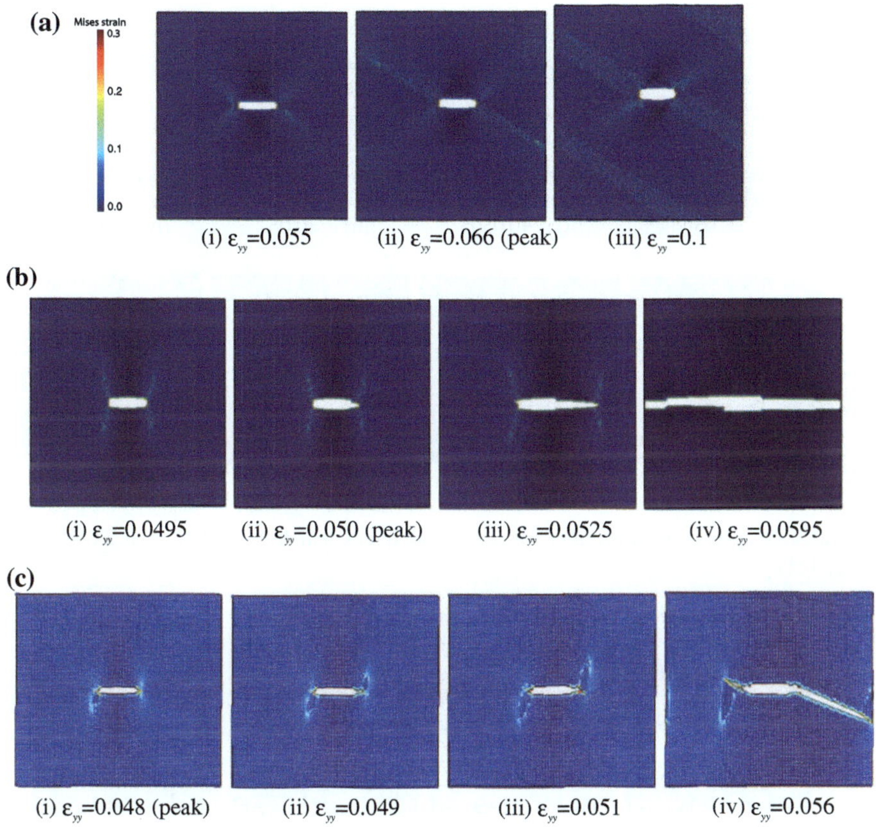

Fig. 25.3 Snapshots of atoms after the stress peak (crack length $2c = 0.2L_x$), **a** [001](010) crack, **b** [001](110) through crack, **c** [11$\bar{2}$](111) through crack

normal to the round shoulder at the corner of the opened crack tip. At the peak point, cleavage cracking occurs between the (110) planes and propagates to final rupture in a very short period. Finally in the case of the [11$\bar{2}$](111) crack, high von Mises strain can be seen at crack tips (see Fig. 25.3c(i)) and propagates like dislocation loops (see Fig. 25.3c(ii),(iii)). At the same time, cleavage cracking starts first in the x-direction then the inclined direction about $-23°$ against x-axis. There is no corresponding plane parallel to both the [11$\bar{2}$] and this direction; however, the (110) is close to the plane so that we can deduce that this cleavage is the multiples of different (110) planes. In Fig. 25.4b, the $2c = 0.1L_x$ crack has much plastic deformation and shows staggering propagation while the $2c = 0.5L_x$ has no loop-like deformation but shows straight propagation in Fig. 25.5b. In both cases the propagation direction is same as the crack of $2c = 0.2L_x$, i.e., the multiples of the (110) planes.

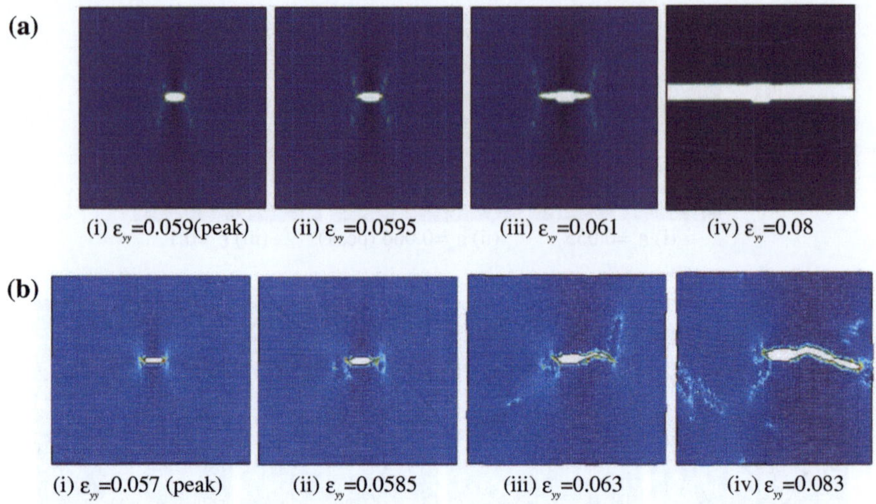

Fig. 25.4 Snapshots of atoms after the stress peak (crack length $2c = 0.1L_x$), **a** [001](010), **b** [11$\bar{2}$](111)

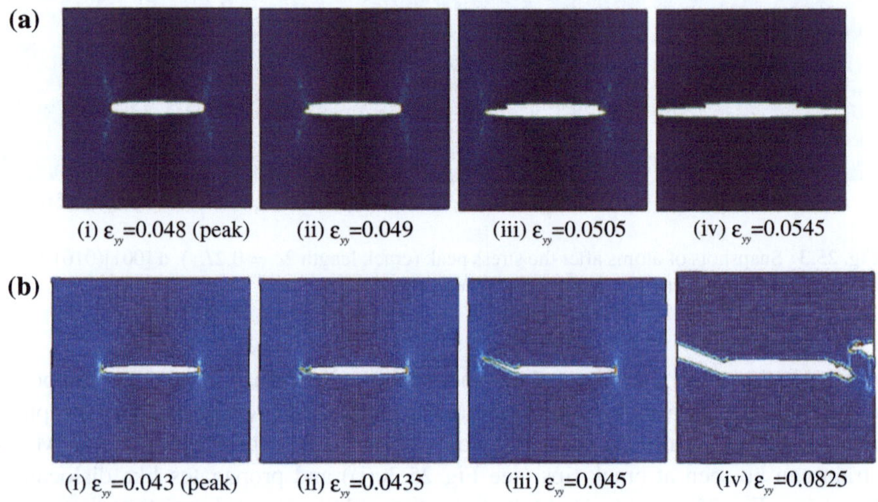

Fig. 25.5 Snapshots of atoms after the stress peak (crack length $2c = 0.5L_x$), **a** [001](010), **b** [11$\bar{2}$](111)

25.4.2 Change in Negative AES Atoms

All the change in the negative AES atoms are summarized in Fig. 25.6. The ordinate is the ratio of negative AES atoms in whole the simulation cell. The left column figures show the overall change throughout the simulation for each crack length.

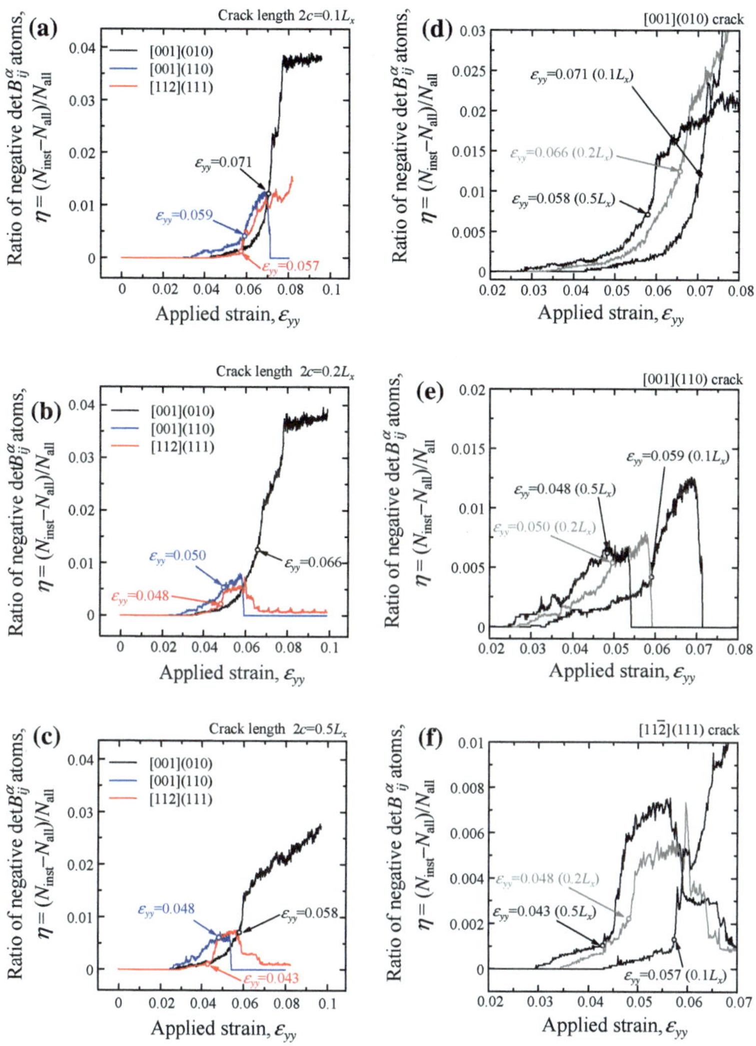

Fig. 25.6 Change in the ratio of negative det B_{ij}^{α} atoms. Left column figures **a**–**c** are drawn for each crack length, right ones **d**–**f** for each crack orientation with magnified scales, **a** crack length $2c = 0.1L_x$, **b** crack length $2c = 0.2L_x$, **c** crack length $2c = 0.2L_x$, **d** [001](010) crack, **e** [001](110) crack, **f** [11$\bar{2}$](111) crack

The right ones magnify the changes for each crystal orientation. In all cases, there is no negative AES atoms at the no-load equilibrium and the initial loading up to the strain of $\varepsilon_{yy} = 0.02 - 0.04$. Open circles and arrows indicate the stress peak or flexion point in Fig. 25.2. The snapshots of atoms at those points are drawn again in Fig. 25.7, indicating the negative AES atoms by red circles. Here, the crack part is magnified for the [11$\bar{2}$](111) cracks while whole the simulation cell is drawn for the others. The [001](010) cracks show smooth exponential increase in the negative AES atoms until the open circle point as shown in Fig. 25.6d. From the top figures in Fig. 25.7, we can understand that the dislocation emission already occurs before the stress peak, and the peak or system instability is brought by the connection of the "plastic deformation" under periodic boundary.

In the case of the [001](110) crack, the morphology of the negative AES distribution seems to be proportional to each other as seen in Figs. 25.7d–f. Contrary to the previous [001](010) cracks, the negative AES domains are not connected under the periodicity. The ratio of the negative AES atoms at the peak point, η_{cr}, is also in the order of the crack length as seen in Fig. 25.6e. Thus we can conclude that the system instability or crack propagation for this system could be predicted by some threshold, proportional to the crack length and the local instability domain.

Finally the $[11\bar{2}](111)$ cracks have much smaller η_{cr}, 0.001 for $0.1L_x$ and $0.5L_x$, 0.002 for $0.2L_x$, respectively, as evaluated in Fig. 25.6f. For the case of $0.1L_x$ and $0.5L_x$ cracks, the open circles are located at near the cross point of two different lines, suggesting that the deformation mechanism changes at the point. On the other hand, the change in the $0.2L_x$ crack seems to be composed of three lines. Careful observation of Fig. 25.2b reveals that there is a very small changes in the stress–strain curve at the first connection point of the three lines around $\varepsilon_{yy} = 0.044$. The ratio η at this point is around 0.001. Thus we can deduce that the $[11\bar{2}](111)$ crack reaches propagation limit around the constant rate of $\eta = 0.001$ despite of the crack length. The $0.1L_x$ and $0.5L_x$ cracks start the brittle cleavage at this point while the $0.2L_x$ have some small local change near the crack tip. The morphology of negative AES distribution of Fig. 25.7g–i supports this deduction; negative AES atoms can be observed only at the crack tips in Fig. 25.7g, i. On the other hand, we can also see negative AES atoms widely in the lower left of the crack tip of Fig. 25.7h.

25.4.3 Change in Eigenvalue of AES

Figure 25.8 illustrates the change in the minimum eigenvalue of B_{ij}^{α} in the $2c = 0.2L_x$ cracks. Here, we evaluate 6 eigenvalues of all the atoms, $6N$, and pick up 6 at each strain or time step; that is, the change is *not* trace nor set of the eigenvalue of some target atom. Rather we can discuss the emergence of the negative eigenvalue in the whole simulation cell by this diagram. In all the crystal orientation, the 1st eigenvalue, $\beta^{(1)}$, becomes negative at far smaller strain than the peak point. The emergence of $\beta^{(1)} < 0$ coincides with that of the negative det B_{ij}^{α} atoms, i.e. takeoff point from the horizontal line of $\eta = 0$ in Fig. 25.6b. The other eigenvalues never become negative in the [001](010) and the [001](110) cracks, however, the 2nd eigenvalue $\beta^{(2)}$ becomes negative in the $[11\bar{2}](111)$ crack. It is also same for the other crack length of $2c = 0.1L_x$ and $0.5L_x$. The strain at the emergence of $\eta^{(2)} < 0$ is just on or just after the peak strain. Figure 25.9 shows the magnified snapshots of atoms near the crack tip, indicating the atoms of $\beta^{(2)} < 0$ by red circle.

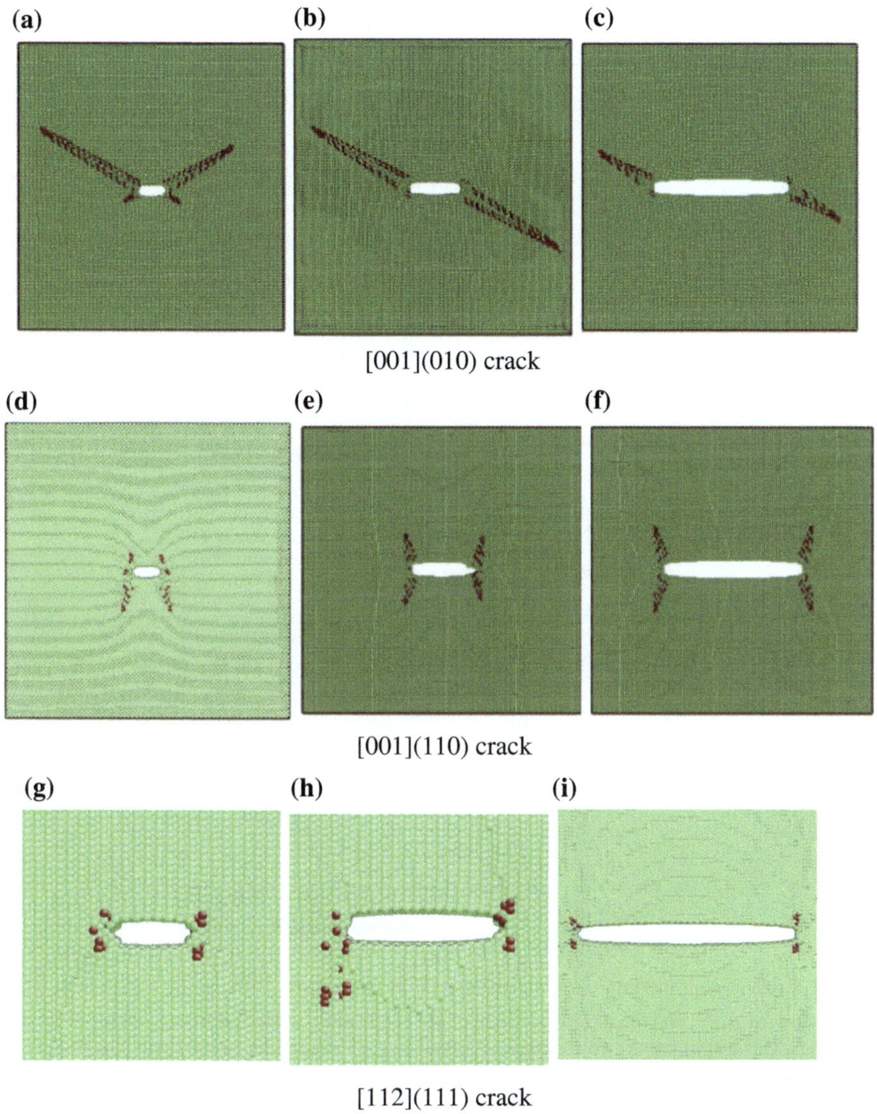

Fig. 25.7 Snapshots of atoms at the peak points. *Red circles* indicate the negative AES atoms, **a** $2c = 0.1L_x$, $\varepsilon_{yy} = 0.071$, **b** $2c = 0.2L_x$, $\varepsilon_{yy} = 0.066$, **c** $2c = 0.5L_x$, $\varepsilon_{yy} = 0.058$, **d** $2c = 0.1L_x$, $\varepsilon_{yy} = 0.0059$, **e** $2c = 0.2L_x$, $\varepsilon_{yy} = 0.050$, **f** $2c = 0.5L_x$, $\varepsilon_{yy} = 0.048$, **g** $2c = 0.1L_x$, $\varepsilon_{yy} = 0.059$, **h** $2c = 0.2L_x$, $\varepsilon_{yy} = 0.050$, **i** $2c = 0.5L_x$, $\varepsilon_{yy} = 0.048$

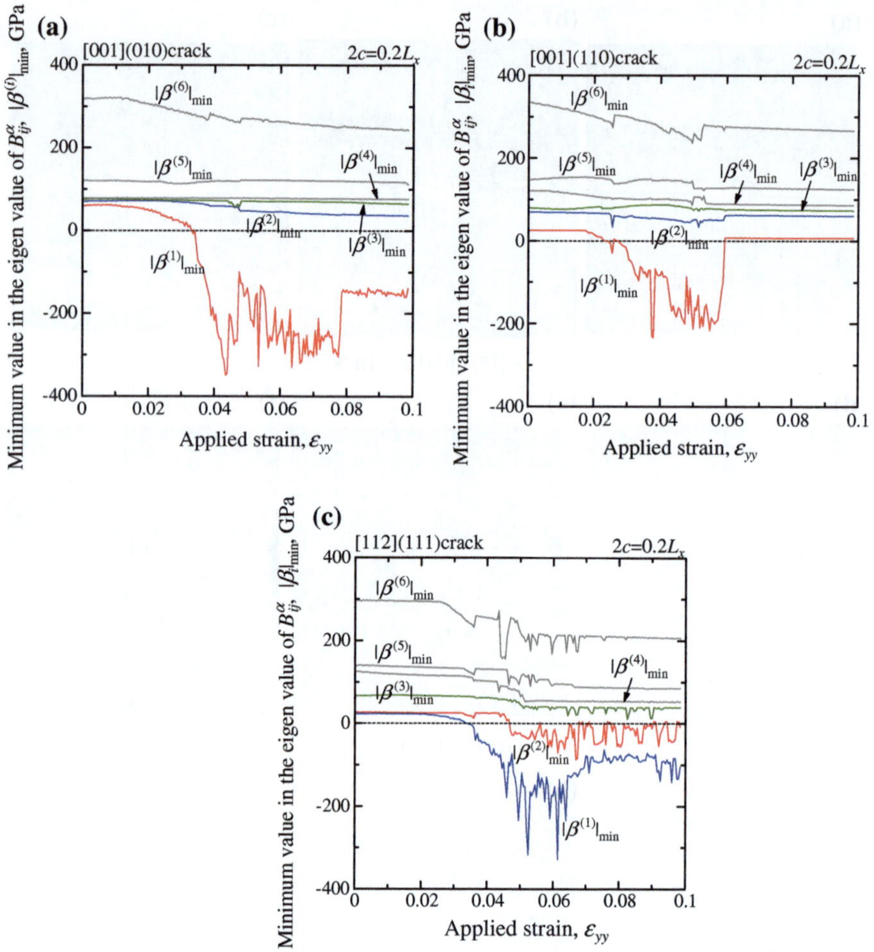

Fig. 25.8 Change in the minimum eigenvalues of atomic elastic stiffness B_{ij}^{α}. Each atom has 6 eigenvalues of $\beta^{\alpha(k)}$, and the minimum value of kth eigenvalue is selected from all atoms N at each time step. The graphs are drawn for three different crystal orientations, **a** [001](010), **b** [001](110) and **c** $[11\bar{2}](111)$ cracks of $2c = 0.2L_x$

The size of atom sphere is changed by each crack length for visibility. The $0.2L_x$ crack has disorder around the shoulder, and the negative $\beta^{(2)}$ can be found at the boundary of the order/disorder. The $0.1L_x$ and $0.5L_x$ cracks show brittle propagation without remarkable disorder by "plastic deformation", and the negative $\beta^{(2)}$ is located just at the frontedge of the propagating crack tip. In all cases of the crack length and left/right crack tip, only three atoms have negative $\beta^{(2)}$ and arrayed in the z-direction. This number coincides with that of the rectangular parallelepiped unit lattices in the z-direction. Figure 25.10 shows the parallelepiped while the x, y, z is changed in the same family of the Miller index for the visibility. There is 6 atoms in the

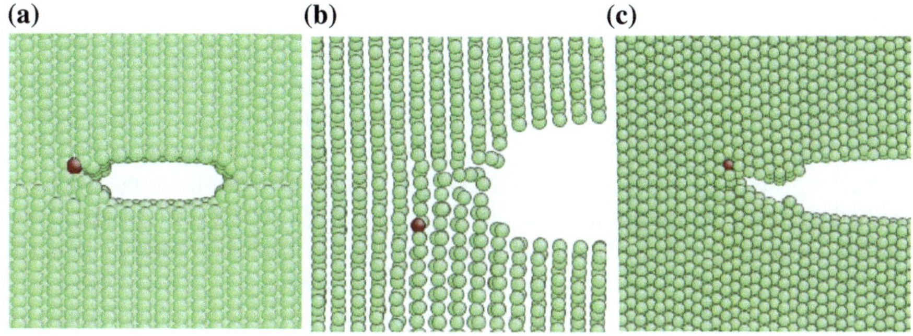

Fig. 25.9 Snapshots of atoms near the $[11\bar{2}](111)$ crack tip. The red circles indicate the atoms with negative 2nd eigenvalue of B_{ij}^{α}, **a** $2c = 0.1L_x$, $\varepsilon_{yy} = 0.057$, **b** $2c = 0.2L_x$, $\varepsilon_{yy} = 0.049$, **c** $2c = 0.5L_x$, $\varepsilon_{yy} = 0.0435$

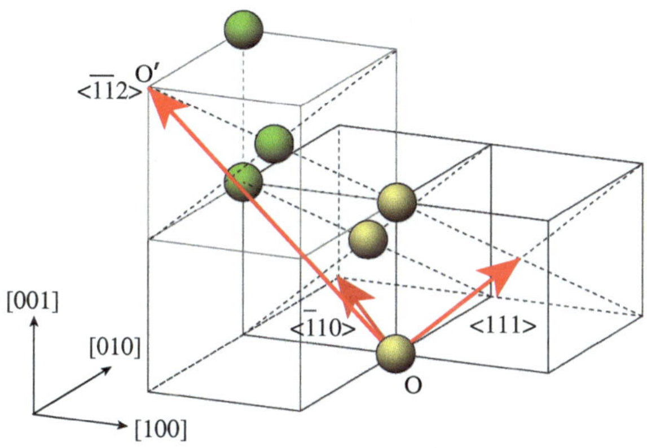

Fig. 25.10 Bcc unit lattice in the [111], $[\bar{1}10]$, $[\bar{1}\bar{1}2]$ crystal orientation

Bravais lattice formed by the vector of $(\sqrt{3}a/2, \sqrt{3}a/2, \sqrt{3}a/2)$, $(-\sqrt{2}a, \sqrt{2}a, 0)$, $(-a, -a, -2a)$. The position relationship of the negative $\beta^{(2)}$ is O-O' lattice points in the figure. On the eigenvector of the negative $\beta^{(2)}$, the fifth component of ε_5 has the greatest value of about 0.5, however, the other components have also large value so that we can not identify the unstable mode by the eigenvector.

25.5 Conclusion

We have first investigated the local lattice instability at mode I crack tip in bcc-Fe under tension; that is, we evaluated the positive definiteness of the atomic elastic stiffness, B_{ij}^{α}, and its eigenvalue $\beta^{\alpha(k)}$ for each atom α in the molecular dynamics simulations of the [001](010), [001](110) and $[11\bar{2}](111)$ cracks with three different crack length. In all cases, negative det B_{ij}^{α} and negative 1st eigenvalue $\beta^{\alpha(1)}$ emerge at far smaller strain than the stress–strain peak; i.e. we can not predict the onset of *system instability* with these quantity. In the [001](010) cracks, which never show brittle propagation but ductile blunting, the increase in these negative atoms corresponds to the onset of local deformation, i.e., dislocation emission and twin propagation from the crack tip; and the stress–strain peak is brought when these deformations connected each other under the periodic boundary condition. The [001](110) cracks shows immediate rupture by cleavage propagation in the (110) planes. Similar butterfly distribution of det $B_{ij}^{\alpha} < 0$ atoms can be found in the snapshot at the stress–strain peak, despite of the different crack length. Thus we could predict the propagation limit of the [001](110) crack by some ratio of det $B_{ij}^{\alpha} < 0$ atoms against the crack length. Finally, only the $[11\bar{2}](111)$ cracks have negative 2nd eigenvalue just at or just after the stress–strain peak. That is, the *system instability* can be predicted by the emergence of the negative $\beta^{\alpha(2)}$ in this system. The $[11\bar{2}](111)$ cracks show brittle propagation in the 23° inclined plane against the crack plane, by the cleavage cracking on the multiple (110) planes. The negative $\beta^{\alpha(2)}$ atoms are located at the frontage of the propagating crack tip, as a single line connected in the $[11\bar{2}]$ direction, in the 2D-like periodic slab cell in the present study.

References

Finnis MW, Sinclair JE (1984) A simple empirical n-body potential for transition metals. Philos Mag A 50(1):45–55

Finnis MW, Sinclair JE (1986) Erratum. Philos Mag A 53(1):161

Li J (2003) Atomeye: an efficient atomistic configuration viewer. Modell Simula Mater Sci Eng 11:173–177

Nishimura M, Yashiro K, Arai M (2010) Local lattice instability analysis on amorphous metals: stability switching between stable and unstable. J Solid Mech Mater Eng 4(11):1550–1562

Wallace DC (1972) Thermodynamics of crystals. Wiley, New York

Wang J, Yip S, Phillpot SR, Wolf D (1993) Crystal instabilities at finite strain. Phys Rev Lett 71(25):4182–4185

Yashiro K, Tomita Y (2001) Local lattice instability at a dislocation nucleation and motion. Journal de Physique IV 11(PR5):Pr5-3-Pr5-10

Yashiro K, Nishimura M, Tomita Y (2006) Deformation analysis of amorphous metals based on atomic elastic stiffness coefficients. Modell Simul Mater Sci Eng 14:597–605

Yashiro K, Nishimura M, Tomita Y (2012) Molecular dynamics study on atomic elastic stiffness in Si under tension: Homogenization by external loading and its limit. Modell Simul Mater Sci Eng 20(4):045,002

Yashiro K, Yamane T (2014) Global versus local instabilities of pure bcc iron. In: Proceedings of ICCM2014, Sientech Publisher

Chapter 26
Modeling of Large-Strain Cyclic Plasticity Including Description of Anisotropy Evolution for Sheet Metals

Fusahito Yoshida, Takeshi Uemori and Hiroshi Hamasaki

Abstract The present paper describes a framework for the constitutive modeling of a large-strain cyclic plasticity to describe the evolution of anisotropy and the Bauschinger effect of sheet metals that is based on the Yoshida-Uemori kinematic hardening model. In the model, the shapes of the yield and the bounding surfaces are assumed to change simultaneously with increasing plastic strain. An anisotropic yield function that varies continuously with the plastic strain is defined by a non-linear interpolation function of the effective plastic strain using a limited number of yield functions determined at a few discrete points of plastic strain. With this modeling framework, any type of yield function can be used and the convexity of the yield surface is always guaranteed. A set of kinematic parameters can be identified experimentally independent of the anisotropic parameters.

Keywords Constitutive model · Yoshida-Uemori model · Cyclic plasticity · Anisotropy evolution · Sheet metal

26.1 Introduction

The use of constitutive models that properly describe the elastic-plastic deformation behavior is essential for accurate numerical simulation of sheet metal forming. The anisotropy of sheets is of great concern to the forming industry because it strongly

F. Yoshida (✉) · H. Hamasaki
Department of Mechanical Science and Engineering, Hiroshima University,
1-4-1, Kagamiyama, Higashi-hiroshima 739-8527, Japan
e-mail: fyoshida@hiroshima-u.ac.jp

H. Hamasaki
e-mail: hamahiro@hiroshima-u.ac.jp

T. Uemori
Graduate School of Natural Science and Technology, Okayama University,
1-1-1, Tsushima-naka, Kita-ku, Okayama 700-0082, Japan
e-mail: uemori@okayama-u.ac.jp

H. Altenbach et al. (eds.), *From Creep Damage Mechanics to Homogenization Methods*, Advanced Structured Materials 64,
DOI 10.1007/978-3-319-19440-0_26

influences the formability of sheets. Thus, many types of anisotropic yield functions have been proposed in the past, e.g., Hill (1948, 1979, 1990); Gotoh (1977); Barlat and Lian (1989); Barlat et al. (1991, 2003, 2005); Cazacu and Barlat (2001, 2003, 2004); Karafillis and Boyce (1993); Bron and Besson (2004); Banabic et al. (2005); Hu (2005, 2007); Leacock (2006); Vegter and Boogaard (2006); Comsa and Banabic (2008); Steglich et al. (2011); Desmorat and Marull (2011), etc.

Another important issue in material modeling is describing cyclic plasticity behavior. Descriptions of the Bauschinger effect and workhardening have been intensively investigated within the framework of a combined isotropic-kinematic hardening model for the past few decades, e.g., Armstrong and Frederick (1966); Mróz (1967); Krieg (1975); Dafalias and Popov (1976); Ohno (1982); Chaboche and Rousselier (1983); Ohno and Wang (1993); MacDowell (1995); Geng and Wagoner (2002); Yoshida (2000); Yoshida et al. (2002, 2013, 2015); Yoshida and Uemori (2002, 2003); Haddadi et al. (2006); Taleb (2013); for more details, refer to reviews by Chaboche (2008); Ohno (2015). Before 2000, most cyclic plasticity models were constructed within the theory of infinitesimal deformation without considering material anisotropy because they were applied mainly to structural analyses for predicting low-cycle fatigue life and ratcheting. In the early 2000s, some researchers pointed out that the Bauschinger effect of materials greatly affects springback behavior, especially for high-strength steel (HSS) sheets, and several cyclic plasticity models were proposed for springback simulation, e.g., Yoshida and Uemori (2002, 2003); Geng and Wagoner (2002). The present authors previously proposed a model of large-strain cyclic plasticity, so-called 'Yoshida-Uemori (Y-U) model' (Yoshida et al. 2002; Yoshida and Uemori 2002, 2003) to describe the following cyclic plasticity behaviors (see Fig. 26.1) together with the anisotropy of materials:

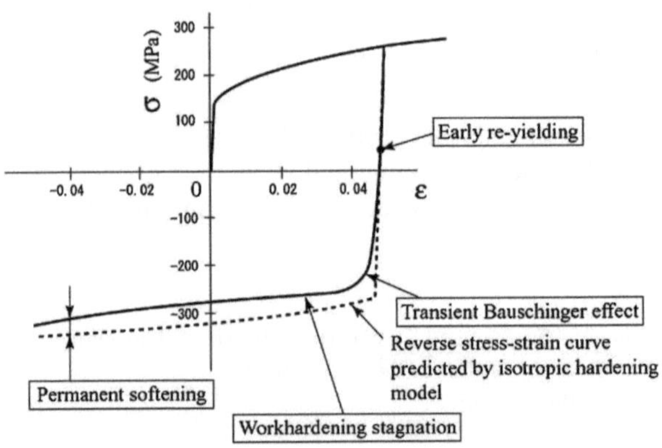

Fig. 26.1 Elastic-plastic behavior in a reverse deformation: early re-yielding, transient Bauschinger effect, workhardening stagnation and permanent softening (Yoshida and Uemori 2003)

- Two stages of the Bauschinger effect: (i) *the transient Bauschinger deformation* characterized by early re-yielding and smooth elastic-plastic transition with a rapid change in the workhardening rate; and (ii) *the permanent softening* observed in a region after the transient period;
- *Workhardening stagnation*, which appears at a certain range of reverse deformation;
- The strain-range and mean-strain dependency of cyclic hardening, i.e., the larger cyclic strain range induces the larger saturated stress amplitudes.

A recent topic of plasticity for sheet metals has been the modeling of anisotropic hardening. Conventional plasticity models assume that the shape of the yield surface does not change during a plastic deformation; consequently, the r-values and flow stress directionality calculated with these models remain constant throughout the deformation. However, some metallic sheets exhibit significant changes in r-value anisotropy and flow stress directionality (e.g., Hu 2007; Stoughton and Yoon 2009; An et al. 2013; Safaei et al. 2014) and the shape of the yield surface (e.g., Tozawa 1978; Kuwabara et al. 1998; Yanaga et al. 2014; Yoon et al. 2014) as plastic strain increases. Although there are some models of the anisotropic hardening (e.g., Hu 2007; Plunkett et al. 2008; Stoughton and Yoon 2009; An et al. 2013; Safaei et al. 2014; Yanaga et al. 2014; Yoon et al. 2014), most of them exclude a description of the Bauschinger effect. Distortion yield function modeling is another type of formulation used to represent the Bauschinger effect and stress-strain responses under non-proportional cyclic loading (e.g., Shiratori et al. 1979; Voyiadjis and Foroozesh 1990; Kurtyka and Życzkowski 1996; Francois 2001; Feigenbaum and Dafalias 2007; Barlat et al. 2011, 2013, 2014). However, to the best of the present authors' knowledge, only Barlat et al.'s homogeneous anisotropic hardening (HAH) model (Barlat et al. 2011, 2013, 2014) reproduces the Bauschinger effect well together with the anisotropy evolution of sheet metals.

The present paper proposes a model of large-strain cyclic plasticity that describes the evolution of anisotropy and the Bauschinger effect of sheet metals based on the Y-U kinematic hardening model. This modeling framework has great advantages over other models. It allows any type of yield function to be used, and the convexity of the yield surface is always guaranteed. A set of kinematic parameters can be identified from experimentally independent of anisotropic parameters.

26.2 Framework of Combined Anisotropic-Kinematic Hardening Model

With the assumption of small elastic and large plastic deformation, the rate of deformation $\boldsymbol{D}$ is decomposed into its elastic and plastic parts, $\boldsymbol{D}^{\mathrm{e}}$ and $\boldsymbol{D}^{\mathrm{p}}$, respectively, as follows:

$$\boldsymbol{D} = \boldsymbol{D}^{\mathrm{e}} + \boldsymbol{D}^{\mathrm{p}} \tag{26.1}$$

The decomposition of the continuum spin $\boldsymbol{W}$ is given as follows:

$$\boldsymbol{W} = \boldsymbol{\Omega} + \boldsymbol{W}^{\mathrm{p}}, \tag{26.2}$$

where $\boldsymbol{W}^{\mathrm{p}}$ denotes the plastic spin and $\boldsymbol{\Omega}$ is the spin of substructures. The constitutive equation of elasticity is expressed as follows:

$$\overset{\circ}{\boldsymbol{\sigma}} = \dot{\boldsymbol{\sigma}} - \boldsymbol{\Omega}\boldsymbol{\sigma} + \boldsymbol{\sigma}\boldsymbol{\Omega} = \boldsymbol{C} : \boldsymbol{D}^{\mathrm{e}}, \tag{26.3}$$

where $\boldsymbol{\sigma}$ and $\overset{\circ}{\boldsymbol{\sigma}}$ are the Cauchy stress and its objective rate, respectively, $\boldsymbol{\Omega}$ is the spin tensor; and $\boldsymbol{C}$ is the elasticity modulus tensor. The initial yield criterion is expressed by the following equation:

$$f = \phi_0(\boldsymbol{\sigma}) - Y = \overline{\sigma}(\boldsymbol{\sigma}) - Y = 0, \tag{26.4}$$

where Y is the initial yield stress and $\overline{\sigma}$ is the effective stress. To describe the Bauschinger effect, as well as the evolutionary change of anisotropy, the subsequent yielding is expressed by the following equation:

$$f = \phi(\boldsymbol{\sigma} - \boldsymbol{\alpha}, \overline{\varepsilon}) - Y = \overline{\sigma}(\tilde{\boldsymbol{\sigma}}, \overline{\varepsilon}) - Y = 0, \qquad \tilde{\boldsymbol{\sigma}} = \boldsymbol{\sigma} - \boldsymbol{\alpha}, \tag{26.5}$$

where $\boldsymbol{\alpha}$ denotes the backstress. Based on the following definitions of the effective plastic strain and its rate

$$\overline{\sigma}\dot{\overline{\varepsilon}} = \tilde{\boldsymbol{\sigma}} : \boldsymbol{D}^{\mathrm{p}}, \qquad \overline{\varepsilon} = \int \dot{\overline{\varepsilon}} \mathrm{d}t, \tag{26.6}$$

the associated flow rule is written as follows:

$$\boldsymbol{D}^{\mathrm{p}} = \frac{\partial f}{\partial \tilde{\boldsymbol{\sigma}}}\dot{\lambda} = \frac{\partial \phi}{\partial \tilde{\boldsymbol{\sigma}}}\dot{\lambda} \tag{26.7}$$

where $\dot{\lambda} = \dot{\overline{\varepsilon}}$.

Most kinematic hardening models assume the following form of the evolution equation of the back stress:

$$\overset{\circ}{\boldsymbol{\alpha}} = \left\{ \frac{A}{Y}(\boldsymbol{\sigma} - \boldsymbol{\alpha}) - \boldsymbol{x} \right\} \dot{\overline{\varepsilon}} = \left(\frac{A}{Y}\tilde{\boldsymbol{\sigma}} - \boldsymbol{x} \right) \dot{\overline{\varepsilon}}, \tag{26.8}$$

Here $\overset{\circ}{(\ldots)}$ denotes the objective rate. For example, in the linear kinematic hardening model:

$$\overset{\circ}{\boldsymbol{\alpha}} = \frac{H'_{LK}}{Y}(\boldsymbol{\sigma} - \boldsymbol{\alpha})\dot{\overline{\varepsilon}} = \frac{H'_{LK}}{Y}\tilde{\boldsymbol{\sigma}}\dot{\overline{\varepsilon}}, \qquad A = H'_{LK}, \boldsymbol{x} = \boldsymbol{0} \tag{26.9}$$

In the Armstrong-Frederick model (Armstrong and Frederick 1966):

$$\mathring{\boldsymbol{\alpha}} = \left\{\frac{\gamma_1}{Y}(\boldsymbol{\sigma} - \boldsymbol{\alpha}) - \gamma_2\boldsymbol{\alpha}\right\}\dot{\bar{\varepsilon}} = \left(\frac{\gamma_1}{Y}\tilde{\boldsymbol{\sigma}} - \gamma_2\boldsymbol{\alpha}\right)\dot{\bar{\varepsilon}}, \qquad A = \gamma_1, \boldsymbol{x} = \gamma_2\boldsymbol{\alpha} \tag{26.10}$$

The Y-U kinematic hardening law has the same form (for details, see the following section). The constitutive equation is given by the following form:

$$\mathring{\boldsymbol{\sigma}} = \boldsymbol{C}^{\text{ep}} : \boldsymbol{D}, \tag{26.11}$$

$$\boldsymbol{C}^{\text{ep}} = \begin{cases} \boldsymbol{C} & \text{if } \dot{\lambda} = 0 \\ \boldsymbol{C} - \dfrac{\boldsymbol{C} : \dfrac{\partial \phi}{\partial \tilde{\boldsymbol{\sigma}}} \otimes \boldsymbol{C} : \dfrac{\partial \phi}{\partial \tilde{\boldsymbol{\sigma}}}}{\dfrac{\partial \phi}{\partial \tilde{\boldsymbol{\sigma}}} : \boldsymbol{C} : \dfrac{\partial \phi}{\partial \tilde{\boldsymbol{\sigma}}} + H' - \dfrac{\partial \phi}{\partial \bar{\varepsilon}}} & \text{if } \dot{\lambda} > 0 \end{cases} \tag{26.12}$$

where

$$H' = A - \frac{\partial \phi}{\partial \tilde{\boldsymbol{\sigma}}} : \boldsymbol{x} \tag{26.13}$$

26.3 Cyclic Plasticity Model to Describe the Bauschinger Effect and Workhardening Stagnation: Yoshida-Uemori Model

The Y-U model was constructed within the framework of two-surface modeling (Krieg 1975), wherein the yield surface moves kinematically within a bounding surface, as schematically illustrated in Fig. 26.2. To describe anisotropic hardening (i.e., expansion of the surface with shape change) and also kinematic hardening, the bounding surface F is expressed by the equation:

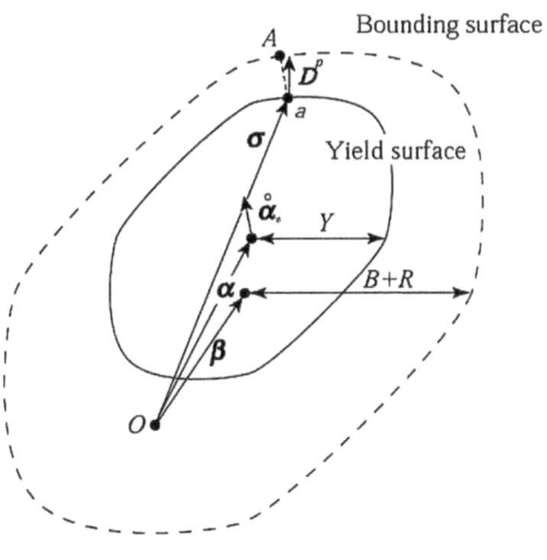

Fig. 26.2 Schematic illustration of the Yoshida-Uemori two-surface model

$$F = \phi(\boldsymbol{\sigma} - \boldsymbol{\beta}, \bar{\varepsilon}) - (B + R) = 0, \tag{26.14}$$

where $\boldsymbol{\beta}$ denotes the center of the bounding surface, and B and R are the initial size of the surface and its workhardening component, respectively. To include the description of anisotropic hardening in the model, it is assumed that the shapes of both the yield and bounding surfaces vary simultaneously.

The kinematic hardening of the yield surface describes the transient Bauschinger deformation, which is characterized by early re-yielding and subsequent rapid change in workhardening rate. The relative kinematic motion of the yield surface with respect to the bounding surface is expressed by the following equation:

$$\boldsymbol{\alpha}_* = \boldsymbol{\sigma} - \boldsymbol{\beta} \tag{26.15}$$

The evolution of $\boldsymbol{\alpha}_*$ is given by the following equation:

$$\mathring{\boldsymbol{\alpha}}_* = C\left\{\left(\frac{a}{Y}\right)(\boldsymbol{\sigma} - \boldsymbol{\alpha}) - \sqrt{\frac{a}{\bar{\boldsymbol{\alpha}}_*}}\right\}\dot{\bar{\varepsilon}} = \left\{\left(\frac{Ca}{Y}\right)\tilde{\boldsymbol{\sigma}} - C\sqrt{\frac{a}{\bar{\boldsymbol{\alpha}}_*}}\right\}\dot{\bar{\varepsilon}}, \tag{26.16}$$

$$\bar{\alpha}_* = \phi(\boldsymbol{\alpha}_*), \qquad a = B + R - Y \tag{26.17}$$

An Armstrong-Frederick-type evolution equation is used to express the kinematic hardening of $\boldsymbol{\beta}$

$$\mathring{\boldsymbol{\beta}}_* = k\left\{\left(\frac{b}{Y}\right)(\boldsymbol{\sigma} - \boldsymbol{\alpha}) - \boldsymbol{\beta}\right\}\dot{\bar{\varepsilon}} = \left(\frac{kb}{Y}\tilde{\boldsymbol{\sigma}} - k\boldsymbol{\beta}\right)\dot{\bar{\varepsilon}} \tag{26.18}$$

Thus, in Eq. (26.8),

$$A = Ca + kb, \qquad \boldsymbol{x} = C\sqrt{\frac{a}{\bar{\alpha}_*}}\boldsymbol{\alpha} - \left(C\sqrt{\frac{a}{\bar{\alpha}_*}} - k\right)\boldsymbol{\beta} \tag{26.19}$$

and in Eq. (26.13).

With respect to the expansion of the bounding surface, i.e., the evolution of R, in the first version of the Y-U model (Yoshida and Uemori 2002, 2003), the following equation based on the Voce hardening law (Voce 1948) was proposed:

$$R = R_{\mathrm{Voce}} = R_{\mathrm{sat}}\{1 - \exp(-k\bar{\varepsilon})\}, \tag{26.20}$$

written as

$$\dot{R} = \dot{R}_{\mathrm{Voce}} = k(R_{\mathrm{sat}} - R_{\mathrm{Voce}})\dot{\bar{\varepsilon}} \tag{26.21}$$

However, it is not necessary to use the Voce-type formulation. For example, based on the Swift law (Swift 1952):

$$R = R_{\mathrm{Swift}} = K\{(\varepsilon_0 + \bar{\varepsilon})^n - \varepsilon_0^n\}, \tag{26.22}$$

the following evolution equation can be obtained

$$\dot{R} = \dot{R}_{\text{Swift}} = nK^{1/n}(R_{\text{Swift}} + K\varepsilon_0^n)^{(n-1)/n}\dot{\bar{\varepsilon}} \tag{26.23}$$

Furthermore, a combination of the above two hardening laws, Eqs. (26.22) and (26.23), is also possible and can be expressed as follows:

$$\dot{R} = \omega\dot{R}_{\text{Swift}} + (1-\omega)\dot{R}_{\text{Voce}}, \qquad 0 \leq \omega \leq 1 \tag{26.24}$$

where ω is a weighting coefficient. This model has high flexibility in describing various levels of workhardening at large strain levels.

One of the features of the Yoshida-Uemori model is that it is able to describe the *workhardening stagnation* that appears in a reverse stress-strain curve for a certain range of reverse deformation (see Hasegawa and Yakou 1975; Christodoulou et al. 1986). This phenomenon is closely related to the strain-range and mean-strain dependency of cyclic hardening. Specifically, the larger the cyclic strain range is, the larger the saturated stress amplitudes are. This dependency is expressed by the stagnation of the expansion of the bounding surface for a certain range of reverse deformation. The states of hardening ($\dot{R} > 0$) and non-hardening ($\dot{R} = 0$) of the bounding surface are determined for a so-called non-IH (isotropic hardening) surface, g_σ, defined in the stress space as follows and schematically illustrated in Fig. 26.3a, b:

$$g_\sigma = \phi(\boldsymbol{\sigma} - \boldsymbol{q}, r) - r = 0, \tag{26.25}$$

where $\boldsymbol{q}$ and r denote the center and size of the non-IH surface, respectively. It is assumed that the center of the bounding surface $\boldsymbol{q}$ exists either on or inside of the surface g_σ. The expansion of the bounding surface takes place only when the center point of the bounding surface, $\boldsymbol{q}$, lies on the surface g_σ (see Fig. 26.3b), i.e., when

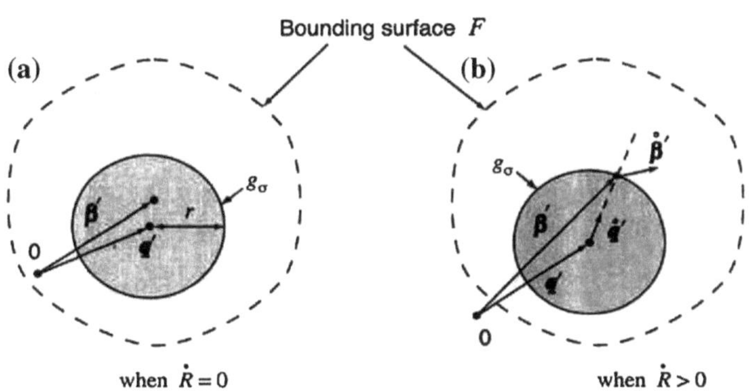

Fig. 26.3 Schematic illustration of the non-IH surface defined in the stress space, when expansion of the bounding surface **a** stops, and **b** takes place (Yoshida and Uemori 2003)

$$\dot{R} > 0 : g_\sigma(\boldsymbol{\beta} - \boldsymbol{q}, r) = \phi(\boldsymbol{\beta} - \boldsymbol{q}, r) - r = 0 \tag{26.26}$$

and

$$\dot{R} = 0 : \Gamma = \frac{\partial g_\sigma(\boldsymbol{\beta} - \boldsymbol{q}, r)}{\partial \boldsymbol{\beta}} : \overset{\circ}{\boldsymbol{\beta}} > 0 \tag{26.27}$$

otherwise. In an analysis of some experimental data, the plastic strain region of workhardening stagnation was found to increase with the accumulated plastic strain. To describe this phenomenon, it was assumed that the surface g_σ moves kinematically as it expands. The governing equations of the kinematic motion and expansion of the surface are given by Eqs. (26.28) and (26.29), respectively.

$$\overset{\circ}{\boldsymbol{q}} = \frac{(1-h)\Gamma}{r}(\boldsymbol{\beta} - \boldsymbol{q}), \tag{26.28}$$

$$\dot{r} = h\Gamma \tag{26.29}$$

Here, h is a parameter that controls the strength of the workhardening stagnation characteristic. A larger value of h corresponds to a larger strain region within which workhardening stagnation occurs, and as a result, a larger value of h leads to weaker cyclic hardening of a material. We may assume that the shape of the surface g_σ, is fixed $\phi = \phi_0$, or even $\phi =$ von Mises type, throughout the deformation, because the shape of g_σ has not been measured experimentally yet, and its effect on the stress-strain calculation would be rather minor.

Models of workhardening stagnation were recently reviewed by Ohno (2015). It should be noted that Ohno's model of non-isotropic-hardening, where the non-hardening region is expressed in the plastic strain space, is identical to the infinitesimal-strain Yoshida-Uemori model when assuming a linear kinematic hardening of the bounding surface.

In the proposed model, the size of the yield surface is held constant. However, if we carefully observe the stress-strain response during unloading after plastic deformation, we find that the stress-strain curve is no longer linear but rather is slightly curved due to very early re-yielding and the Bauschinger effect. To describe this phenomenon, in the model, the following equation for *plastic-strain-dependent Young's modulus* is introduced (Yoshida et al. 2002):

$$E = E_0 - (E_0 - E_\alpha)\{1 - \exp(-\xi\bar{\varepsilon})\}, \tag{26.30}$$

where E_0 and E_α are Young's modulus for virgin and infinitely large pre-strained materials, respectively, and ξ is a material constant.

Figure 26.4 shows stress-strain responses of 780 MPa high strength steel sheets under cyclic straining and uniaxial tension, calculated by the Y-U model, together with the corresponding experimental results.

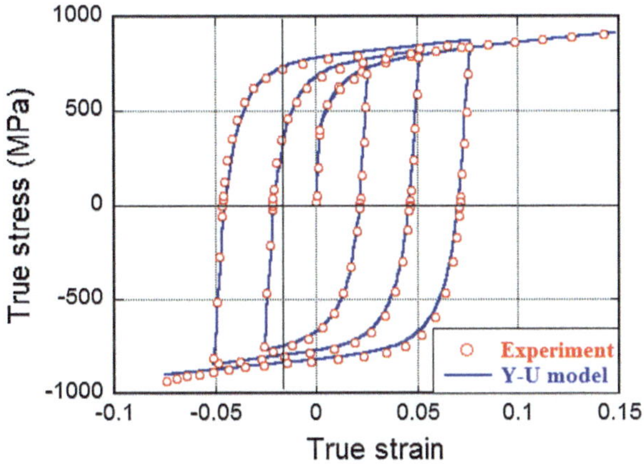

Fig. 26.4 Stress-strain curves for 780 MPa HSS sheet calculated with the Y-U model and the corresponding experimental data

26.4 Description of Evolution of Anisotropy

The evolution of anisotropy is expressed by the anisotropic hardening of the yield surface, as follows (refer to Yoshida et al. 2015):

$$\phi(\tilde{\boldsymbol{\sigma}}, \bar{\varepsilon}) = \mu(\bar{\varepsilon})\phi_A(\tilde{\boldsymbol{\sigma}}) + (1 - \mu(\bar{\varepsilon}))\phi_B(\tilde{\boldsymbol{\sigma}}) \quad \text{for} \quad \bar{\varepsilon}_A \leq \bar{\varepsilon} \leq \bar{\varepsilon}_B \tag{26.31}$$

Here, $\phi_A(\tilde{\boldsymbol{\sigma}})$ and $\phi_B(\tilde{\boldsymbol{\sigma}})$ are two different yield functions defined at the effective plastic strains $\bar{\varepsilon}_A$ and $\bar{\varepsilon}_B$, respectively, i.e., $\phi_A(\tilde{\boldsymbol{\sigma}}) = \phi(\tilde{\boldsymbol{\sigma}}, \bar{\varepsilon}_A)$ and $\phi_B(\tilde{\boldsymbol{\sigma}}) = \phi(\tilde{\boldsymbol{\sigma}}, \bar{\varepsilon}_B)$, and $\mu(\bar{\varepsilon})$ an interpolation function of the effective plastic strain, where

$$1 = \mu(\bar{\varepsilon}_A) \geq \mu(\bar{\varepsilon}) \geq \mu(\bar{\varepsilon}_B) = 0 \tag{26.32}$$

Note that the types of these two yield functions, $\phi_A(\tilde{\boldsymbol{\sigma}})$ and $\phi_B(\tilde{\boldsymbol{\sigma}})$, do not need to be the same. An advantage of this modeling framework is that, if the two yield functions $\phi_A(\tilde{\boldsymbol{\sigma}})$ and $\phi_B(\tilde{\boldsymbol{\sigma}})$ are convex, $\phi(\tilde{\boldsymbol{\sigma}}, \bar{\varepsilon})$ always satisfies the convexity. The derivatives are expressed as follows:

$$\frac{\partial \phi}{\partial \tilde{\boldsymbol{\sigma}}} = \mu(\bar{\varepsilon})\frac{\partial \phi_A(\tilde{\boldsymbol{\sigma}})}{\partial \tilde{\boldsymbol{\sigma}}} + (1 - \mu(\bar{\varepsilon}))\frac{\partial \phi_B(\tilde{\boldsymbol{\sigma}})}{\partial \tilde{\boldsymbol{\sigma}}}, \qquad \frac{\partial \phi}{\partial \bar{\varepsilon}} = (\phi_A(\tilde{\boldsymbol{\sigma}}) - \phi_B(\tilde{\boldsymbol{\sigma}}))\frac{\partial \mu(\bar{\varepsilon})}{\partial \bar{\varepsilon}} \tag{26.33}$$

Several linear and nonlinear functions can be used for the interpolation function $\mu(\bar{\varepsilon})$. Assuming that $\bar{\varepsilon}_A = 0$ (at initial yielding) and $\bar{\varepsilon}_A = \infty$ (at infinitely large strain) and that $\phi_A(\tilde{\boldsymbol{\sigma}}) = \phi_0(\tilde{\boldsymbol{\sigma}}) = \phi_0(\boldsymbol{\sigma})$ and $\phi_B(\tilde{\boldsymbol{\sigma}}) = \phi_\infty(\tilde{\boldsymbol{\sigma}})$, Eq. (26.31) reduces to the following

$$\phi(\tilde{\boldsymbol{\sigma}}, \bar{\varepsilon}) = \mu(\bar{\varepsilon})\phi_0(\tilde{\boldsymbol{\sigma}}) + (1 - \mu(\bar{\varepsilon}))\phi_\infty(\tilde{\boldsymbol{\sigma}}) \quad 0 \leq \bar{\varepsilon} \leq 0 \tag{26.34}$$

Some examples of forms of interpolation functions are as follows:

$$\mu(\overline{\varepsilon}) = \exp(-\lambda\overline{\varepsilon}), \tag{26.35}$$

$$\mu(\overline{\varepsilon}) = a\exp(-\lambda_1\overline{\varepsilon}) + (1-a)\exp(-\lambda_2\overline{\varepsilon}), \tag{26.36}$$

where λ, a, λ_1 and λ_2 are material constants.

If we have M sets of experimental data ($\sigma_0, \sigma_{45}, \sigma_{90}, \sigma_b, r_0, r_{45}, r_{90}$, etc.) for material parameter identification corresponding to M discrete plastic strain points, $\overline{\varepsilon}_1 (= 0), \overline{\varepsilon}_2, \ldots, \overline{\varepsilon}_i, \overline{\varepsilon}_{i+1}, \ldots, \overline{\varepsilon}_M$, we can determine M sets of yield functions $\phi_1(\widetilde{\boldsymbol{\sigma}}), \phi_2(\widetilde{\boldsymbol{\sigma}}), \ldots, \phi_i(\widetilde{\boldsymbol{\sigma}}), \phi_{i+1}(\widetilde{\boldsymbol{\sigma}}), \ldots, \phi_M(\widetilde{\boldsymbol{\sigma}})$. Using an interpolation function $\mu(\overline{\varepsilon})$, the yield function $\phi(\widetilde{\boldsymbol{\sigma}}, \overline{\varepsilon})$ can be defined by the following equation:

$$\phi(\widetilde{\boldsymbol{\sigma}}, \overline{\varepsilon}) = \mu(\overline{\varepsilon})\phi_i(\widetilde{\boldsymbol{\sigma}}) + (1-\mu(\overline{\varepsilon}))\phi_{i+1}(\widetilde{\boldsymbol{\sigma}}) \quad \text{for} \quad \overline{\varepsilon}_i \le \overline{\varepsilon} \le \overline{\varepsilon}_{i+1} \tag{26.37}$$

The following nonlinear equation is proposed for use as the interpolation function:

$$\mu(\overline{\varepsilon}) = 1 - \left(\frac{\overline{\varepsilon} - \overline{\varepsilon}_i}{\overline{\varepsilon}_{i+1} - \overline{\varepsilon}_1}\right)^{p_i} \qquad \overline{\varepsilon}_i \le \overline{\varepsilon} \le \overline{\varepsilon}_{i+1}, \tag{26.38}$$

where $p_i (i = 1, 2, \ldots, M-1)$ are material constants.

Among the various types of anisotropic yield functions available, stress polynomial-type models (e.g., Hill 1948; Gotoh 1977; Soare et al. 2008; Yoshida et al. 2013) are suitable for use in modeling anisotropy evolution. A polynomial-type yield criterion is given by the following equation:

$$f = \phi^{(m)}(\boldsymbol{\sigma}) - Y^m = \overline{\sigma}^m - Y^m = 0, \tag{26.39}$$

where $\phi^{(m)}(\boldsymbol{\sigma})$ denotes the mth order stress polynomial-type yield function. For example, when $m = 6$ (Yoshida et al. 2013) under plane stress condition,

$$\begin{aligned}\phi^{(6)} &= C_1\sigma_x^6 - 3C_2\sigma_x^5\sigma_y + 6C_3\sigma_x^4\sigma_y^2 - 7C_4\sigma_x^3\sigma_y^3 + 6C_3\sigma_x^2\sigma_y^4 - 3C_6\sigma_x\sigma_y^5 \\ &+ C_7\sigma_y^6 + 9(C_8\sigma_x^4 - 2C_9\sigma_x^3\sigma_y + 3C_{10}\sigma_x^2\sigma_y^2 - 2C_{11}\sigma_x\sigma_y^3 + 2C_{12}\sigma_y^4)\tau_{xy}^2 \\ &+ 27(C_{13}\sigma_x^2 - C_{14}\sigma_x\sigma_y + C_{15}\sigma_y^2)\tau_{xy}^4 + 27C_{16}\tau_{xy}^6\end{aligned} \tag{26.40}$$

In the same manner as Eq. (26.31), when the following equation is assumed

$$\phi^{(m)}(\widetilde{\boldsymbol{\sigma}}, \overline{\varepsilon}) = \mu(\overline{\varepsilon})\phi_A^{(m)}(\widetilde{\boldsymbol{\sigma}}) + (1-\mu(\overline{\varepsilon}))\phi_B^{(m)}(\widetilde{\boldsymbol{\sigma}}), \tag{26.41}$$

it reduces to an interpolation for material parameters C_k, $k = 1, 2, \ldots, N$, as follows

$$C_k = \mu(\overline{\varepsilon})C_{k(A)} + (1-\mu(\overline{\varepsilon}))C_{k(B)} \tag{26.42}$$

Here, $C_{k(A)}$ and $C_{k(B)}$ are material parameters determined at the effective plastic strains, $\overline{\varepsilon}_A$ and $\overline{\varepsilon}_B$, respectively. Assuming that $\overline{\varepsilon}_A = 0$ (at initial yielding) and $\overline{\varepsilon}_B = \infty$ (at infinitely large strain) and that $C_{k(A)} = C_{k(0)}$ and $C_{k(B)} = C_{k(\infty)}$,

Eq. (26.42) reduces to the following:

$$C_k = \mu(\bar{\varepsilon})C_{k(0)} + (1 - \mu(\bar{\varepsilon}))C_{k(\infty)}, \qquad 0 \leq \bar{\varepsilon} \leq \infty \tag{26.43}$$

In discretization form:

$$C_k = \mu(\bar{\varepsilon})C_{k(i)} + (1 - \mu(\bar{\varepsilon}))C_{k(i+1)}, \qquad i = 1, 2, \ldots, M - 1 \tag{26.44}$$

To validate the model, calculated stress-strain responses were compared with the corresponding experimental data for AA6022-T43 aluminum sheet (Stoughton and Yoon 2009). As for the yield function, sixth-order polynomial model is employed. One of advantages of this model is that, flow stresses $\sigma_0, \sigma_{45}, \sigma_{90}, \sigma_b$ and r-values r_0, r_{45}, r_{90} are calculated by using the material parameters $C_1 \sim C_{16}$ explicitly. Thus the material parameters are easily identified.

$$\sigma_{90} = \left(\frac{C_1}{C_7}\right)^{\frac{1}{6}} \sigma_0, \quad \sigma_{45} = \left(\frac{C_1}{S + 9T + 27U + 27C_{16}}\right)^{\frac{1}{6}} \sigma_0, \quad \sigma_b = \left(\frac{C_1}{S}\right)^{\frac{1}{6}} \sigma_0, \tag{26.45}$$

$$r_0 = \frac{C_2}{2C_1 - C_2}, \quad r_{45} = \frac{-S - 3T + 9U + C_{16}}{2S + 12T + 18U}, \quad r_{90} = \frac{C_6}{2C_7 + C_6} \tag{26.46}$$

$$\begin{aligned} S &= C_1 - 3C_2 + 6C_3 - 7C_4 + 6C_5 - 3C_6 + C_7, \\ T &= C_8 - 2C_9 + 3C_{10} - 2C_{11} + C_{12}, \\ U &= C_{13} - C_{14} + 3C_{15} \end{aligned}$$

On AA6022-T43 aluminum sheet, r-value planar anisotropy remains fixed throughout the plastic deformation. In this calculation the kinematic hardening was excluded since in monotonic loading the stress-strain calculation is not affected by the kinematic hardening. The results of flow stresses, $\sigma_0, \sigma_{45}, \sigma_{90}, \sigma_b$ calculated using Eqs. (26.44) and (26.38), with $M = 3$, are compared to the experimental data, as shown in Fig. 26.5. Here, three discrete plastic strain points, $\bar{\varepsilon} = 0, 0.1$ and 0.5 are selected to define $\phi_1(\boldsymbol{\sigma})$, $\phi_2(\boldsymbol{\sigma})$ and $\phi_3(\boldsymbol{\sigma})$. The calculated results agree well overall with the experimental results for the stresses.

The model was also validated by comparing the calculated results of stress-strain responses with experimental data on r-value and stress-directionality changes in an aluminum sheet (Hu 2007) and a stainless steel sheet (Stoughton and Yoon 2009), as well as the variation of the yield surface of an aluminum sheet (Yanaga et al. 2014). Furthermore, anisotropic cyclic behavior was examined by performing experiments of uniaxial tension and cyclic straining in three sheet directions on a 780 MPa advanced high-strength steel sheet. For details of the results, refer to Yoshida et al. (2015).

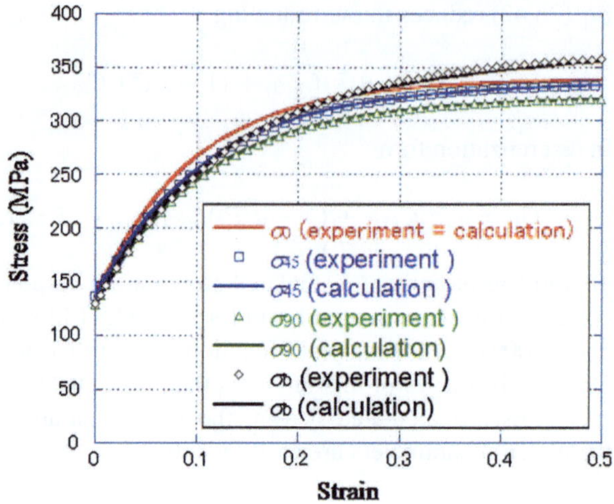

Fig. 26.5 Flow stresses of AA6022-T43 aluminum sheet predicted using the anisotropy evolution model

26.5 Concluding Remarks

The present paper describes a framework for the constitutive modeling of large-strain cyclic plasticity that describes the evolution of anisotropy and the Bauschinger effect of sheet metals based on the Y-U kinematic hardening model. The Y-U model predicts the springback much more accurately than the classical isotropic hardening model (e.g., refer to Yoshida and Uemori 2002; Eggertsen and Mattiasson 2009, 2010; Ghaei et al. 2010; Wagoner et al. 2013; Huh et al. 2011). It has gained popularity in the sheet metal forming industry because it has already been implemented into several FE commercial codes (e.g., PAM-STAMP, LS-DYNA, StamPack) and is widely used for springback simulation. The highlights of this modeling are summarized as follows.

- The Y-U model is highly capable of describing various cyclic plasticity characteristics such as the Bauschinger effect, the workhardening stagnation, strain-range-dependent cyclic workhardening, and the degradation of unloading stress-strain slope with increasing plastic strain. Furthermore, any type of anisotropic yield function can be used.
- It requires a limited number of material parameters (seven or eight plasticity parameters and three elasticity parameters including Young's modulus). The scheme for material parameter identification and testing have been clearly presented, see Yoshida and Uemori (2002).
- The evolution of the anisotropy can be described by incorporating the proposed anisotropic hardening model in the Y-U model. In this modeling framework a set of kinematic hardening parameters can be identified experimentally independent of anisotropic hardening parameters, and their values remain fixed throughout the plastic deformation
- An anisotropic yield function that varies continuously with the plastic strain is defined by a nonlinear interpolation function of the effective plastic strain using a

limited number of yield functions determined at a few discrete points of the plastic strain. In this modeling framework, it is possible to use any type of yield function, and the convexity of the yield surface is always guaranteed.
- This approach, which requires only one interpolation equation, offers a great advantage over other approaches in that it involves fewer material parameters.

Acknowledgments The authors are sincerely grateful to Professor Nobutada Ohno of Nagoya University for fruitful discussions on cyclic plasticity modeling for so long years.

References

An YG, Vegter H, Melzer S, Triguero PR (2013) Evolution of the plastic anisotropy with straining and its implication on formability for sheet metals. J Mater Process Technol 213:1419–1425

Armstrong PJ, Frederick CO (1966) A mathematical representation of the multiaxial bauschinger effect. Technical Report GEGB report RD/B/N731, Berkley Nuclear Laboratories

Banabic D, Aretz H, Comsa DS, Paraianu L (2005) An improved analytical description of orthotropy in metallic sheets. Int J Plast 21:493–512

Barlat F, Lian J (1989) Plastic behavior and stretchability of sheet metals. Part 1. a yield function for orthotropic sheets under plane-stress conditions. Int J Plast 5:51–66

Barlat F, Lege DJ, Brem JC (1991) A 6-component yield function for anisotropic materials. Int J Plast 7(7):693–712

Barlat F, Brem JC, Yoon JW, Chung K, Dick RE, Lege DJ, Pourgoghrat F, Choi SH, Chu E (2003) Plane stress yield function for aluminum alloy sheets—Part 1: theory. Int J Plast 19:1297–1319

Barlat F, Aretz H, Yoon JW, Karabin ME, Brem JC, Dick RE (2005) Linear transformation-based anisotropic yield functions. Int J Plast 21:1009–1039

Barlat F, Gracio JJ, Lee MJ, Rauch EF, Vincze G (2011) An alternative to kinematic hardening in classical plasticity. Int J Plast 27:1309–1327

Barlat F, Ha J, Gracio JJ, Lee MJ, Rauch EF, Vincze G (2013) Extension of homogeneous anisotropic hardening model to cross-loading with latent effects. Int J Plast 46:130–142

Barlat F, Vincze G, Gracio JJ, Lee MJ, Rauch EF, Tome CN (2014) Enhancements of homogeneous anisotropic hardening model and application to mild and dual-phase steels. Int J Plast 58:201–218

Bron F, Besson J (2004) A yield function for anisotropic materials, application to aluminum alloys. Int J Plast 20:937–963

Cazacu O, Barlat F (2001) Generalization of Drucker's yield criterion to orthotropy. Math Mech Solids 6:613–630

Cazacu O, Barlat F (2003) Application of the theory of representation to describe yielding of anisotropic aluminum alloys. Int J Eng Sci 41:1367–1385

Cazacu O, Barlat F (2004) A criterion for description of anisotropy and yield differential effects in pressure-insensitive metals. Int J Plast 20:2027–2045

Chaboche JL (2008) A review of some plasticity and viscoplasticity constitutive theories. Int J Plast 24:1642–1693

Chaboche JL, Rousselier G (1983) On the plastic and viscoplastic constitutive equations, Part I and II. Trans ASME J Press Vessel Technol 105:153–164

Christodoulou N, Woo OT, MacEwen SR (1986) EEffect of stress reversals on the work hardening behaviour of polycrystalline copper. Acta Metall 34:1553–1562

Comsa DS, Banabic D (2008) Plane-stress yield criterion for highly-anisotropic sheet metals. In: Hora P (ed) Proceedings of the 7th International conference and workshop on numerical simulation of 3D sheet metal forming processes (NUMISHEET 2008), pp 43–48

Dafalias YF, Popov EP (1976) Plastic internal variables formalism of cyclic plasticity. Trans ASME J Appl Mech 43:645–651
Desmorat R, Marull R (2011) Non quadratic Kelvin modes based plasticity criteria for anisotropic materials. Int J Plast 27:328–351
Eggertsen PA, Mattiasson K (2009) On the modelling of the bending-unbending behaviour for accurate springback predictions. Int J Mech Sci 51:547–563
Eggertsen PA, Mattiasson K (2010) On constitutive modeling of springback analysis. Int J Mech Sci 52:804–818
Feigenbaum HP, Dafalias YF (2007) Directional distortional hardening in metal plasticity with thermodynamics. Int J Solids Struct 44:7526–7542
Francois M (2001) A plasticity model with yield surface distortion for non-proportional loading. Int J Plast 17:703–717
Geng LM, Wagoner RH (2002) Role of plastic anisotropy and its evolution on spring-back. Int J Mech Sci 44:123–148
Ghaei A, Green DE, Taherizadeh A (2010) Semi-implicit numerical integration of Yoshida-Uemori two-surface plasticity model. Int J Mech Sci 52:531–540
Gotoh M (1977) Theory of plastic anisotropy based on a yield function of 4th-order (plane stress state)-1. Int J Mech Sci 19:505–512
Haddadi H, Bouvier S, Banu M, Maier C, Teodosiu C (2006) Towards an accurate description of the anisotropic behavior of sheet metals under large plastic deformation: modelling, numerical analysis and identification. Int J Plast 22:2226–2271
Hasegawa T, Yakou T (1975) Deformation behaviour and dislocation structures upon stress reversal in polycrystalline aluminium. Mater Sci Eng 20:267–276
Hill R (1948) A theory of the yielding and plastic flow of anisotropic metals. Proc R Soc Lond A 193:281–297
Hill R (1979) Theoretical plasticity of textured aggregates. Math Proc Cambridge Philos Soc 85:179–191
Hill R (1990) Constitutive modeling of orthotropic plasticity in sheet metals. J Mech Phys Solids 38:405–417
Hu WL (2005) An orthotropic yield criterion in a 3-D general stress state. Int J Plast 21:1771–1796
Hu WL (2007) Constitutive modeling of orthotropic sheet metals by presenting hardening-induced anisotropy. Int J Plast 23:620–639
Huh H, Chung K, Han SS, Chung WJ (eds) (2011) NUMISHEET 2011, Part C: Benchmark problems and results, BM4—Pre-strain effect on spring-back of 2-D draw bending. The Korean Society for Technology of Plasticity, KAIST Press, Daejeon
Karafillis AP, Boyce MC (1993) A general anisotropic yield criterion using bounds and a transformation weighting tensor. J Mech Phys Solids 41:1859–1886
Krieg RD (1975) A practical two surface plasticity theory. Trans ASME J Appl Mech 42:641–646
Kurtyka T, Życzkowski M (1996) Evolution equations for distorsional plastic hardening. Int J Plast 12(2):191–213
Kuwabara T, Ikeda S, Kuroda K (1998) Measurement and analysis of differential work hardening in cold-rolled steel sheet under biaxial tension. J Mater Process Tech 80:517–523
Leacock AG (2006) A mathematical description of orthotropy in sheet metals. J Mech Phys Solids 54:425–444
MacDowell DL (1995) Stress state dependence of cyclic ratchetting behavior of two rail steels. Int J Plast 11:397–421
Mróz Z (1967) On the description of anisotropic workhardening. J Mech Phys Solids 15:163–175
Ohno N (1982) A constitutive model of cyclic plasticity with a non-hasrdening strain range. Trans ASME J Appl Mech 49:721–727
Ohno N (2015) Material models of cyclic plasticity with extended isotropic hardening: a review. Bull JSME: Mech Eng Rev 2(1):14-00425
Ohno N, Wang JD (1993) Nonlinear kinematic hardening rule with critical state of dynamic recovery. Part I: Formulation and basic features for ratchetting behavior. Int J Plast 9:375–390

Plunkett B, Cazacu O, Barlat F (2008) Orthotropic yield criteria for description of the anisotropy in tension and compression of sheet metals. Int J Plast 24:847–866
Safaei M, Lee MG, Zang S, Waele WD (2014) An evolutionary anisotropic model for sheet metals based on non-associated flow rule approach. Comput Mater Sci 81:15–29
Shiratori E, Ikegami K, Yoshida F (1979) Analysis of stress-strain relations by use of anisotropic hardening plastic potential. J Mech Phys Solids 27:213–229
Soare S, Yoon JW, Cazacu O (2008) On the use of homogeneous polynomials to develop anisotropic yield functions with applications to sheet forming. Int J Plast 24:915–944
Steglich D, Brocks W, Bohlen J, Barlat F (2011) Modelling direction-dependent hardening in magnesium sheet forming simulations. Int J Mater Forum 4:243–253
Stoughton TB, Yoon JW (2009) Anisotropic hardening and non-associated flow in proportional loading of sheet metals. Int J Plast 25:1777–1817
Swift HW (1952) Plastic instability under plane stress. J Mech Phys Solids 1:1–18
Taleb L (2013) About cyclic strain accumulation of the inelastic strain observed in metals subjected to cyclic stress control. Int J Plast 43:1–19
Tozawa Y (1978) Plastic deformation behavior under conditions of combined stress. In: Wang NM, Koistinen DP (eds) Mechanics of sheet metal forming. Plenum Press, New York, pp 81–110
Vegter H, van den Boogaard AH (2006) A plane stress yield function for anisotropic sheet material by interpolation of biaxial stress states. Int J Plast 22:557–580
Voce E (1948) The relationship between stress and strain for homogeneous deformation. J Inst Metals 74:537–562
Voyiadjis GZ, Foroozesh M (1990) Anisotropic distortional yield model. Trans ASME J Appl Mech 57:537–547
Wagoner RH, Lim H, Lee MG (2013) Advanced issue in springback. Int J Plast 45:3–20
Yanaga D, Takizawa H, Kuwabara T (2014) Formulation of differential work hardening of 6000 series aluminum alloy sheet and application to finite element analysis. Trans JSTP 55:55–61 (in Japanese)
Yoon JW, Lou Y, Yoon J, Glazoff MV (2014) Asymmetric yield function based on the stress invariants for pressure sensitive metals. Int J Plast 56:184–202
Yoshida F (2000) A constitutive model of cyclic plasticity. Int J Plast 16:359–380
Yoshida F, Uemori T (2002) A model of large-strain cyclic plasticity describing the bauschinger effect and workhardening stagnation. Int J Plast 18:661–686
Yoshida F, Uemori T (2003) A model of large-strain cyclic plasticity and its application to springback simulation. Int J Mech Sci 45:1687–1702
Yoshida F, Uemori T, Fujiwara K (2002) Elastic-plastic behavior of steel sheets under in-plane cyclic tension-compression at large strain. Int J Plast 18:633–659
Yoshida F, Hamasaki H, Uemori T (2008) A user-friendly 3D yield function to describe anisotropy of steel sheets. Int J Plast 45:119–139
Yoshida F, Hamasaki H, Uemori T (2013) A user-friendly 3D yield function to describe anisotropy of steel sheets. Int J Plast 45:119–139
Yoshida F, Hamasaki H, Uemori T (2015) Modeling of anisotropic hardening of sheet metals including description of the bauschinger effect. Int J Plast. doi:10.1016/j.ijplas.2015.02.004

Chapter 27
A New Kinematic Hardening Rule Describing Different Plastic Moduli in Monotonic and Cyclic Deformations

Yilin Zhu, Guozheng Kang and Qianhua Kan

Abstract To describe the different plastic moduli of the metal materials presented in the monotonic and cyclic deformations, a new nonlinear kinematic hardening rule is proposed by modifying the Chaboche's one (Chaboche 1989). In the proposed rule, the back stress is assumed to be decomposed into three components as done by Chaboche (1989), but the linear hardening and dynamic recovery terms of each back stress component are further divided into two parts, respectively, and a part in each of them is only activated when the reverse loading occurs so that the cyclic stress-strain hysteresis loops can be predicted more accurately; moreover, a rachetting coefficient is introduced into one part of dynamic recovery term to describe the ratchetting. The proposed rule can be reduced to the Chaboche's one under the monotonic loading conditions, or by setting some material parameters as zero. Finally, the proposed model is verified by comparing the predicted results with corresponding experimental ones. It is seen that the predicted results are in good agreement with the corresponding experimental ones.

Keywords Cyclic plasticity · Kinematic hardening rule · Stress-strain hysteresis loops · Ratchetting

Y. Zhu · G. Kang (✉)
State Key Laboratory of Traction Power and School of Mechanics and Engineering, Southwest Jiaotong University, Chengdu 610031, People's Republic of China
e-mail: lin5210feng@163.com

G. Kang
e-mail: guozhengkang@home.swjtu.edu.cn

Q. Kan
School of Mechanics and Engineering, Southwest Jiaotong University, Chengdu 610031, People's Republic of China
e-mail: qianhuakan@foxmail.com

H. Altenbach et al. (eds.), *From Creep Damage Mechanics to Homogenization Methods*, Advanced Structured Materials 64,
DOI 10.1007/978-3-319-19440-0_27

27.1 Introduction

In various engineering applications, many structural components are often subjected to a complex cyclic loading. The cyclic deformation features, e.g., the cyclic hardening/softening and rachetting of materials should be realized, since they play very important roles in the safety assessment and fatigue-life estimation of such structural components. Cyclic plastic constitutive models suitable for describing such cyclic deformation features were developed extensively in the last few decades by including a nonlinear kinematic hardening rule, such as those done by Chaboche (1989, 1991), Ohno and Wang (1991, 1993), Ohno and Abdel-Karim (2000), Bari and Hassan (2000, 2001, 2002), Kang and Gao (2002), Kang et al. (2002), Kang (2004), Bai and Chen (2009), Abdel-Karim (2009), Zhu et al. (2014) and so on, by extending the Armstrong and Frederick nonlinear kinematic hardening rule (Armstrong and Frederick 1966), simplified as the A-F one. More detailed reviews for the state-of-arts of kinematic hardening rules can be referred to Ohno and Abdel-Karim (1990); Ohno (1997), Kang (2008) and Chaboche (2008).

From the existing experimental observations to the cyclic deformation of metal materials (Mizuno et al. 2000; Kang and Gao 2002; Kang et al. 2002; Dong et al. 2014), it is found that the plastic modulus presented during cyclic plastic deformation differs from that occurred in the monotonic one, especially at the initial stage of plastic yielding. It is seen from Fig. 27.1 that a lower plastic modulus generally occurs during cyclic loading. However, such difference has not been concerned in the existing models mentioned above. For instance, the predicted plastic modulus by Ohno-Wang's rule I (Ohno and Wang 1993) at the initial stage of plastic yielding during cyclic deformation is larger than the experimental ones since the material parameters are determined from the monotonic tensile stress-strain data; although the Chaboche's model (Chaboche 1989) gives a reasonable prediction to the plastic modulus occurred during cyclic deformation by determining the material

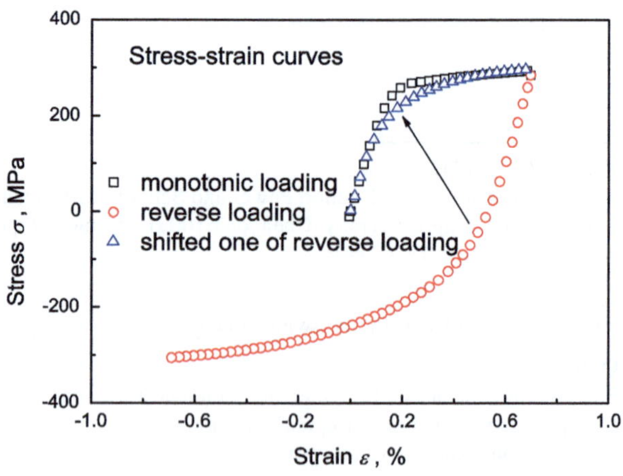

Fig. 27.1 Comparison of plastic moduli obtained during the monotonic and repeated loading tests of 316L stainless steel (Dong et al. 2014)

parameters directly from the cyclic stress-strain data, it underestimates the plastic modulus obtained in monotonic tension. Thus, it is necessary to consider the different plastic moduli occurred at the initial stages of monotonic and cyclic deformations so that the monotonic stress-strain responses and cyclic hysteresis loops can be simultaneously and accurately predicted by the constitutive model.

Therefore, here a new kinematic hardening rule is developed by modifying the Chaboche's rule in order to describe the different plastic moduli occurred in the monotonic and cyclic deformations, especially at the initial stage of plastic yielding. In the proposed rule, the back stress is also assumed to be decomposed into three components, i.e., the short-range, middle-range and long-range component, as done by Chaboche (1989), but the linear hardening and dynamic recovery terms of each back stress component are further divided into two parts, respectively, and a part in each of them is only activated when the reverse loading occurs so that the cyclic stress-strain hysteresis loops can be predicted more accurately; moreover, a rachetting coefficient is introduced into one part of dynamic recovery term to describe the ratcheting of materials. It is noted that the proposed rule can be reduced to the Chaboche's one under the monotonic loading condition, or by setting some specific material parameters as zero. Finally, the proposed model is verified by comparing the predicted results with corresponding experimental ones (Kang and Gao 2002). It is seen that the predicted results are in good agreement with the corresponding experimental ones.

27.2 Constitutive Model

27.2.1 Main Equations

To formulate an elastic-plastic constitutive model in the framework of small deformation, the additive decomposition of strain tensor $\boldsymbol{\varepsilon}$ is usually postulated as:

$$\boldsymbol{\varepsilon} = \boldsymbol{\varepsilon}^{\mathrm{e}} + \boldsymbol{\varepsilon}^{\mathrm{p}}, \tag{27.1}$$

where, $\boldsymbol{\varepsilon}^{\mathrm{e}}$ and $\boldsymbol{\varepsilon}^{\mathrm{p}}$ represent the elastic and plastic strain tensor, respectively. The elastic strain tensor $\boldsymbol{\varepsilon}^{\mathrm{e}}$ is commonly characterized by the Hooke's low, i.e.,

$$\boldsymbol{\varepsilon}^{\mathrm{e}} = \frac{1+\nu}{E}\boldsymbol{\sigma} - \frac{\nu}{E}(\mathrm{tr}\boldsymbol{\sigma})\mathbf{1}, \tag{27.2}$$

where ν and E are Poisson's ratio and Young's modulus, respectively; $\mathbf{1}$ is second-order identity tensor; $\boldsymbol{\sigma}$ is stress tensor. Rate-independent plasticity is assumed by using a von-Mises-typed yielding surface:

$$F_{\mathrm{y}} = \sqrt{\frac{3}{2}(\boldsymbol{\sigma}' - \boldsymbol{\alpha}) : (\boldsymbol{\sigma}' - \boldsymbol{\alpha})} - Q, \tag{27.3}$$

where F_y is yield function; $\boldsymbol{\sigma}'$ is deviatoric stress tensor; $\boldsymbol{\alpha}$ is back stress tensor; Q is isotropic deformation resistance (Q is considered as a constant since the aim of this work is focused on proposing a kinematic hardening rule). For the evolution of plastic strain tensor $\boldsymbol{\varepsilon}^{\mathrm{p}}$, associated plasticity and normality postulate together lead to a flow rule as following:

$$\dot{\boldsymbol{\varepsilon}}^{\mathrm{p}} = \dot{\lambda}\frac{\partial F_{\mathrm{y}}}{\partial \boldsymbol{\sigma}}, \tag{27.4}$$

where $\dot{\lambda}$ is a plastic multiplier which can be determined by Kuhn-Tucker's conditions, i.e.,

$$\dot{\lambda} \geqslant 0, \qquad F_{\mathrm{y}} \leqslant 0, \qquad \dot{\lambda} F_y = 0 \quad (\text{if} \quad F_y = 0). \tag{27.5}$$

27.2.2 Modified Chaboche's Kinematic Hardening Rule

As done by Chaboche (1989), the back stress tensor is divided into three components, i.e., so-called short-range, middle-range and long-range ones,

$$\boldsymbol{\alpha} = \sum_{i=1}^{3} \boldsymbol{\alpha}_i. \tag{27.6}$$

Furthermore, in order to describe the experimental observation that the plastic modulus during the cyclic deformation differs from that obtained in monotonic one (shown in Fig. 27.1), a new evolution rule of $\boldsymbol{\alpha}_i$ (i.e., a new kinematic hardening rule) is proposed here, i.e.,

$$\dot{\boldsymbol{\alpha}}_i = \frac{2}{3}\zeta_i(r_i^{\mathrm{m}} + r_i^{\mathrm{c}})\dot{\boldsymbol{\varepsilon}}^{\mathrm{p}} - \frac{r_i^{\mathrm{c}}}{r_i^{\mathrm{m}}(1-\exp(-\zeta_i q))}\zeta_i \boldsymbol{\alpha}_i \dot{\mathrm{p}} - \left(\frac{\|\boldsymbol{\alpha}_i\|}{r_i^{\mathrm{m}}(1-\exp(-\zeta_i q))}\right)^{m_i} \zeta_i \boldsymbol{\alpha}_i \dot{\mathrm{p}}. \tag{27.7}$$

where ζ_i is a positive dimensionless material parameter; r_i^{m} and r_i^{c} are material parameters with a dimension of stress (r_i^{m} is positive, and a restriction of r_i^{c} is set on $r_i^{\mathrm{c}} \geqslant -r_i^{\mathrm{m}}$ to ensure the coefficient of the first term in the right side of Eq. (27.7) is positive, which will be discussed in details in the next paragraphs); m_i is so-called ratchetting coefficient, which control the evolution of ratchetting; $\dot{\mathrm{p}}$ is the accumulated plastic strain rate defined by $\dot{\mathrm{p}} = \sqrt{\frac{3}{2}}\dot{\lambda}$; q is the radius of memory surface F_q in plastic strain space, i.e.,

$$F_q = \sqrt{\frac{2}{3}\boldsymbol{\varepsilon}^{\mathrm{p}} : \boldsymbol{\varepsilon}^{\mathrm{p}}} - q. \tag{27.8}$$

The evolution rule of q is

$$\dot{q} = \langle \boldsymbol{N} : \boldsymbol{N}_q \rangle \dot{\mathrm{p}}, \tag{27.9}$$

where $\langle\rangle$ is the MaCauley operator; $\boldsymbol{N}$ and $\boldsymbol{N}_q$ are the unit normal vectors to the yielding and memory surfaces in current state (corresponding to $\boldsymbol{\sigma}$ and $\boldsymbol{\varepsilon}^{\mathrm{p}}$), respectively:

$$\boldsymbol{N} = \frac{\boldsymbol{\sigma}' - \boldsymbol{\alpha}}{\|\boldsymbol{\sigma}' - \boldsymbol{\alpha}\|}, \qquad \boldsymbol{N}_q = \frac{\boldsymbol{\varepsilon}^{\mathrm{p}}}{q}. \tag{27.10}$$

It should be noted that, under the monotonic loading condition, the evolution rule shown in Eq. (27.7) can be reduced as

$$\dot{\boldsymbol{\alpha}}_i = \frac{2}{3}\zeta_i r_i^{\mathrm{m}} \dot{\boldsymbol{\varepsilon}}^{\mathrm{p}} - \zeta_i \boldsymbol{\alpha}_i \dot{\mathrm{p}}. \tag{27.11}$$

which is the Chaboche's model (Chaboche 1989). On the other hand, if the parameters r_i^{c} and m_i are set to be zero, Eq. (27.11) can also be obtained from Eq. (27.7). It means that the new proposed kinematic hardening rule (Eq. (27.7)) can be reduced to the Chaboche's one (Eq. (27.11)) under the monotonic loading condition, or in case that the parameters r_i^{c} and m_i are set to be zero.

As mentioned before, the new kinematic hardening rule, i.e., Eq. (27.7), is proposed in this work to capture the different plastic moduli occurred during the monotonic and cyclic deformations of materials, especially at the initial yielding stages of forward and reverse loadings. According to Chaboche (1989), the plastic modulus at the initial yielding stage is mainly controlled by the short-range component of back stress. Therefore, only activating r_1^{c} is enough to fit the experimental results well (i.e., the parameters r_2^{c} and r_3^{c} can be set to be zero for simplicity).To demonstrate the role of the parameter r_1^{c} in describing the different plastic moduli occurred in the monotonic and cyclic deformations, the predicted cyclic stress-strain curves are obtained by the proposed model with various r_1^{c} and keeping other parameters being the same as those listed in Table 27.3 in the next section, and are shown in Fig. 27.2 (a symmetrical axial strain-controlled cyclic loading with a strain amplitude of 0.8 % and in 1st cycle). It is seen that lower plastic modulus is obtained with smaller value of r_1^{c}. It implies that the experimental stress-strain hysteresis loops can be fitted well by adjusting the parameter r_1^{c}.

On the other hand, a new parameter named as ratchetting coefficient is also introduced into the new proposed model, which makes the proposed model be able to describe the ratchetting of materials by setting a suitable ratchetting parameter m_3 since it is also concluded by Gaudin and Feaugas (2004) that the racthetting of metal materials is mainly controlled by the long-range back stress, i.e., the third component of the Chaboche's model. It implies that the parameters m_1 and m_2 can be set to be zero, for simplicity. Thus, the effect of varied m_3 on the predicted ratchetting is discussed here, and the predicted results are shown in Fig. 27.3 by adopting an

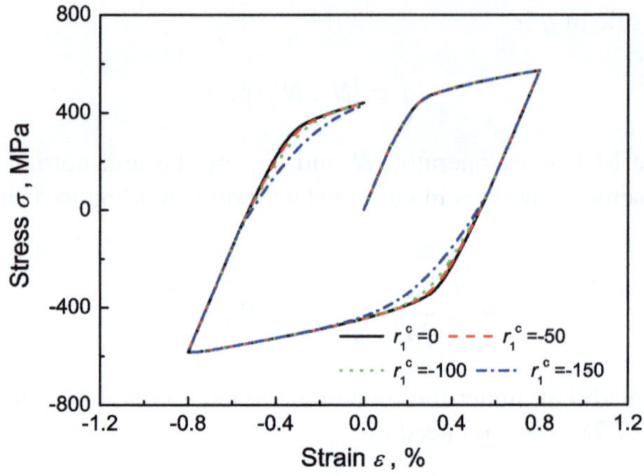

Fig. 27.2 Effect of varied r_1^c on the plastic modulus in the cyclic deformation

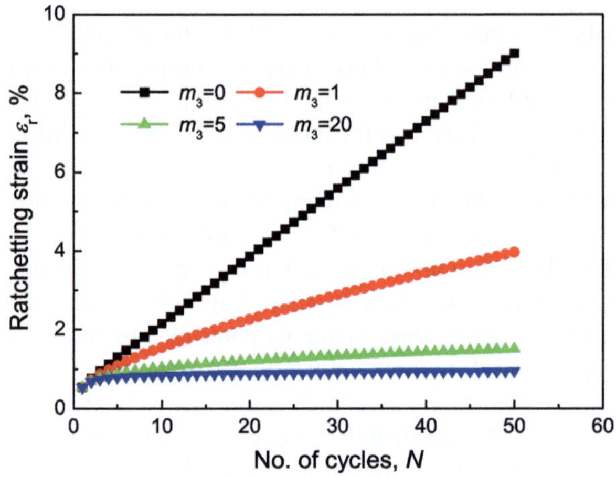

Fig. 27.3 Effect of varied m_3 on the predicted ratchetting

asymmetrical stress-controlled cyclic loading case with an applied mean stress of 100 MPa and stress amplitude of 500 MPa. To illustrate the evolution of ratchetting more clearly, a ratchetting strain ε_r per cycle is defined as:

$$\varepsilon_r = \frac{1}{2}(\varepsilon_{max} + \varepsilon_{min}), \tag{27.12}$$

where, ε_{max} and ε_{min} are the maximum and minimum strains per cycle, respectively. It is seen from Fig. 27.3 that smaller ratchetting strain is obtained with larger m_3, and the predicted ratchetting strain rate is no longer a constant, but decreases with the increasing number of cycles, if the m_3 is non-zero.

27.3 Simulation and Discussion

In this section, the capability of the proposed model to predict the monotonic and cyclic deformations of materials is verified by comparing the predictions with the corresponding experimental results obtained by Kang and Gao (2002) for a cyclically stable material, annealed U71Mn rail steel. The monotonic tensile test was performed at a constant strain rate of $2 \times 10^{-3}\ \mathrm{s}^{-1}$ and the applied strain is up to 6 %. For strain-controlled cyclic loading tests, two loading cases (listed in Table 27.1) were conducted to investigate the cyclic softening/hardening features of the material:

(i) single-step symmetrical strain-controlled cyclic loading with a constant strain amplitude (denoted as A);
(ii) multi-step symmetrical strain-controlled cyclic loading with varied strain amplitude (denoted as B).

For stress-controlled cyclic tests, three loading cases shown in Table 27.2 were performed to examine the ratchetting of the material:

(i) single-step asymmetrical stress-controlled cyclic loading (denoted as C);
(ii) multi-step asymmetrical stress-controlled cyclic loading with a constant stress amplitude and varied mean stress (denoted as D);
(iii) multi-step asymmetrical cyclic stressing with a constant mean stress and varied stress amplitude (denoted as E).

The strain rate in strain-controlled cyclic tests was $2 \times 10^{-3}\ \mathrm{s}^{-1}$, while the stress rate in stress-controlled cyclic tests was 51 MPa/s. More details about the experimental procedure can be referred to Kang and Gao (2002).

Table 27.1 Load cases in strain-controlled cyclic tests

Load cases	Step 1	Step 2	Step 3	Step 4	Step 5
A	±0.8 %				
	(2c)				
B	±0.4 %	±0.6 %	±0.8 %	±0.6 %	±0.4 %
	(20c)	(20c)	(20c)	(20c)	(20c)

Table 27.2 Load cases in stress-controlled cyclic tests

Load cases	Step 1	Step 2	Step 3	Step 4
C	358 ± 447			
	(MPa) (80c)			
D	205 ± 411	256 ± 411	308 ± 411	256 ± 411
	(MPa) (20c)	(MPa) (20c)	(MPa) (20c)	(MPa) (20c)
E	359 ± 321	359 ± 385	359 ± 449	
	(MPa) (20c)	(MPa) (20c)	(MPa) (20c)	

27.3.1 Determination of Material Parameter

Under the monotonic tension, the tensor equation, Eq. (27.11), can be rewritten as a scalar equation, i.e.,

$$\dot{\alpha_i} = \frac{2}{3}\zeta_i r_i^{\mathrm{m}} \dot{\varepsilon}^{\mathrm{p}} - \zeta_i \alpha_i \dot{\mathrm{p}}. \tag{27.13}$$

Then, by integrating Eq. (27.13), and combined with Eq. (27.3), it yields

$$\sigma = \sum_{i=1}^{3} r_i^{\mathrm{m}}(1 - \exp(-\zeta_i \varepsilon^{\mathrm{p}})) + Q, \tag{27.14}$$

where σ is the axial stress.

Apparently, if the isotropic deformation resistance Q is determined, the parameters r_i^m and ζ_i can be readily determined by the least square method from the experimental tensile curve of $\sigma \sim \varepsilon$. Further, the material parameters r_i^{c} and ratchetting coefficient m_i can be determined from one strain-controlled cyclic stress-strain curve and one evolution curve of ratchetting by trial-and-error method, respectively.

The values of material parameters obtained by using the procedure mentioned above are listed in Table 27.3. It should be noted that the parameters r_i^{m} and ζ_i are determined from the experimental stress-strain curve of monotonic tension; r_1^{c} and m_3 are obtained from the experimental data of loading cases A and C, respectively. The simulated and predicted results obtained by the proposed model are also compared with those by the Chaboche's rule and Ohno-Wang's rule I. The material parameters of the Chaboche's rule are the same as those used in the proposed model, but the parameters used in the Ohno-Wang's rule I are listed in Table 27.4. It should be noted

Table 27.3 Material parameters for U71Mn rail steel in the proposed model

$E = 215\,\text{GPa}$,	$\nu = 0.33$	$Q = 250\,\text{MPa}$
$\zeta_1 = 5000$	$\zeta_2 = 500$	$\zeta_3 = 45$
$r_1^{\mathrm{m}} = 192$	$r_2^{\mathrm{m}} = 50$	$r_3^{\mathrm{m}} = 494\,(\text{MPa})$
$r_1^{\mathrm{c}} = -172$	$r_2^{\mathrm{c}} = 0$	$r_3^{\mathrm{c}} = 0\,(\text{MPa})$
$m_1 = 0$	$m_2 = 0$	$m_3 = 22$

Table 27.4 Material parameters for U71Mn rail steel in the Ohno-Wang's rule I

$N = 8$	$E = 215$ GPa	$\nu = 0.33$	$Q = 250$ MPa;				
$\zeta_1 = 9828.6$	$\zeta_2 = 2502.2$	$\zeta_3 = 525.0$	$\zeta_4 = 200$	$\zeta_5 = 76.5$	$\zeta_6 = 34.4$	$\zeta_7 = 24.8$	$\zeta_8 = 17.9$;
$r_1 = 86.0$	$r_2 = 88.5$	$r_3 = 37.1$	$r_4 = 28.9$	$r_5 = 79.4$	$r_6 = 114.8$	$r_7 = 82.0$	$r_8 = 150.5$ (MPa)

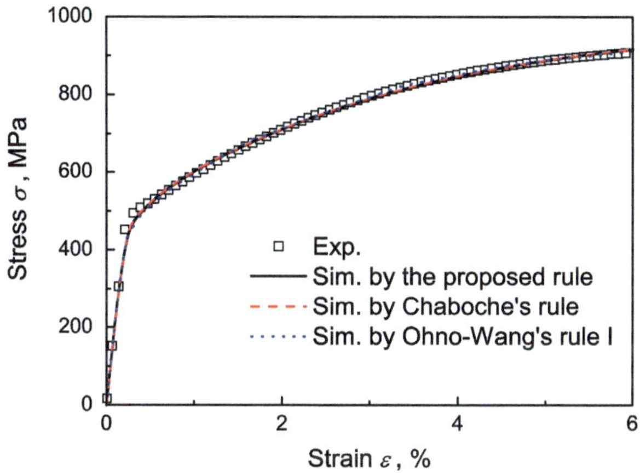

Fig. 27.4 Experimental and simulated tensile stress-strain curves of U71Mn rail steel

that the predicted ratchetting cannot be provided by the Ohno-Wang's rule I, since it gives a closed stress-strain hysteresis loop in the framework of elasto-plasticity.

27.3.2 Simulation and Discussion

Figure 27.4 shows the experimental and simulated stress-strain curves of U71Mn rail steel. It is seen that the monotonic tensile stress-strain response of U71Mn rail steel is reasonably described by the proposed rule, and also by the Chaboche's and Ohno-Wang's ones.

Figure 27.5 gives the experimental and simulated stress-strain hysteresis loops of U71Mn rail steel in the uniaxial strain-controlled cyclic tests. Apparently, the proposed model provides a reasonable prediction to the different plastic moduli occurred in the monotonic and cyclic deformations by setting a suitable r_1^c; while the Chaboche's and Ohno-Wang's rules can not give a reasonable description.

Figure 27.6 illustrates the experimental and simulated stress-strain curves of U71Mn rail steel in uniaxial stress-controlled cyclic test (i.e., loading case C). It is indicated that the ratchetting is reasonably simulated by the proposed model; while a much larger predicted ratchetting is obtained by the Chaboche's rule. It is also seen from Figs. 27.7 and 27.8 that the proposed model can describe the dependence of ratchetting on the applied loading level and history; while the prediction by the Chaboche's rule is far from the experimental results.

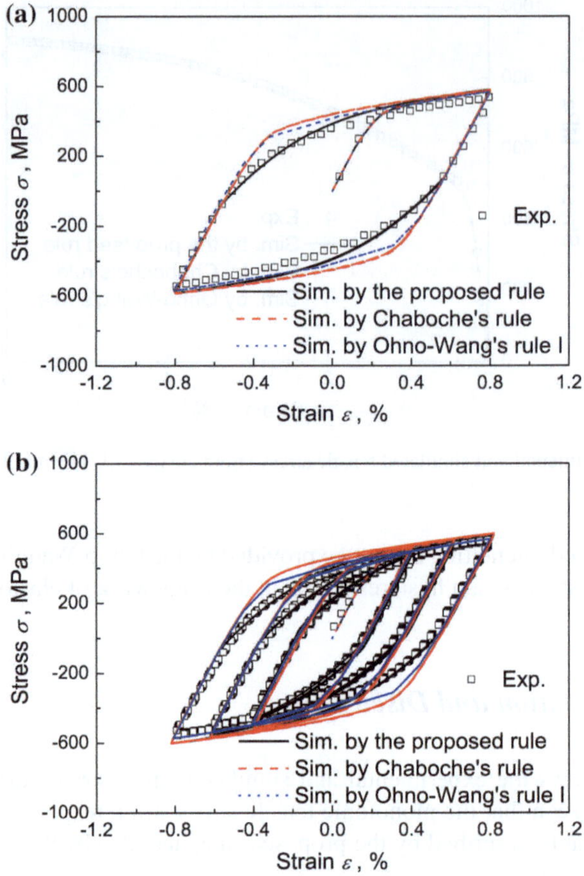

Fig. 27.5 Experimental and simulated stress-strain hysteresis loops of U71Mn rail steel in the uniaxial strain-controlled cyclic tests: **a** loading case A; **b** loading case B

It should be noted that, only the monotonic and cyclic deformations of cyclically stable materials, e.g., annealed U71Mn rail steel are simulated reasonably by the proposed model, but the cyclic softening/hardening features of materials and their effects on the ratchetting can not be described yet, since a constant isotropic deformation resistance is assumed in this work. The proposed model will be improved by considering such features in a future work.

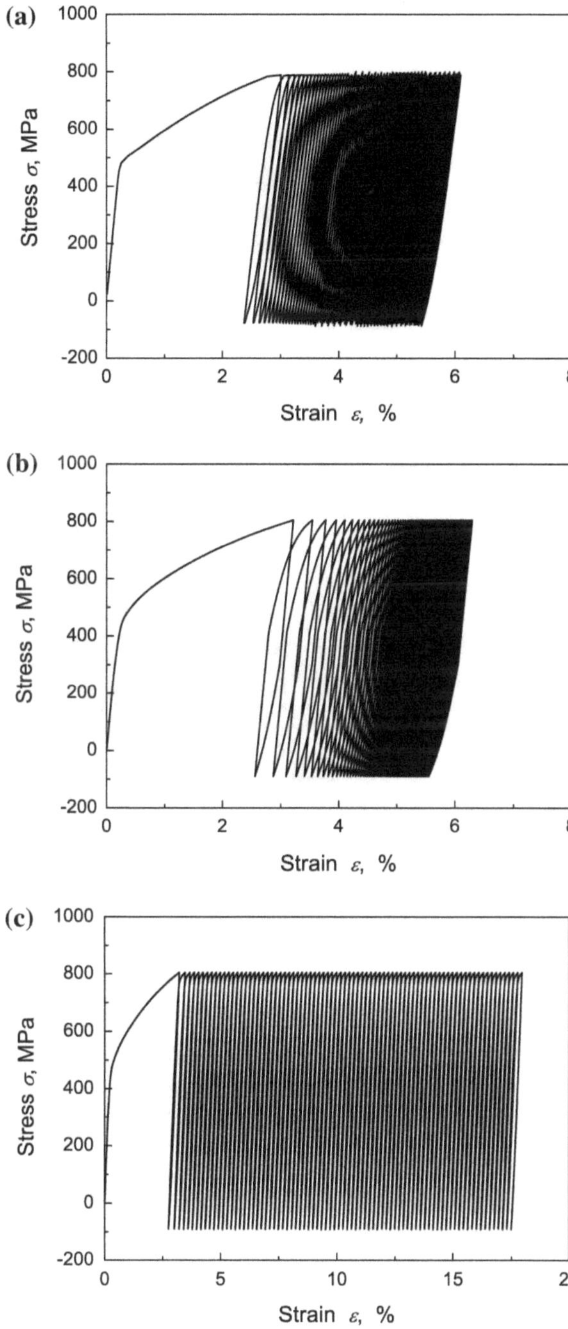

Fig. 27.6 Experimental and simulated stress-strain curves of U71Mn rail steel in single-step stress-controlled cyclic test (loading case C). **a** Experiment. **b** Simulation by the proposed rule. **c** Simulation by the Chaboche's rule

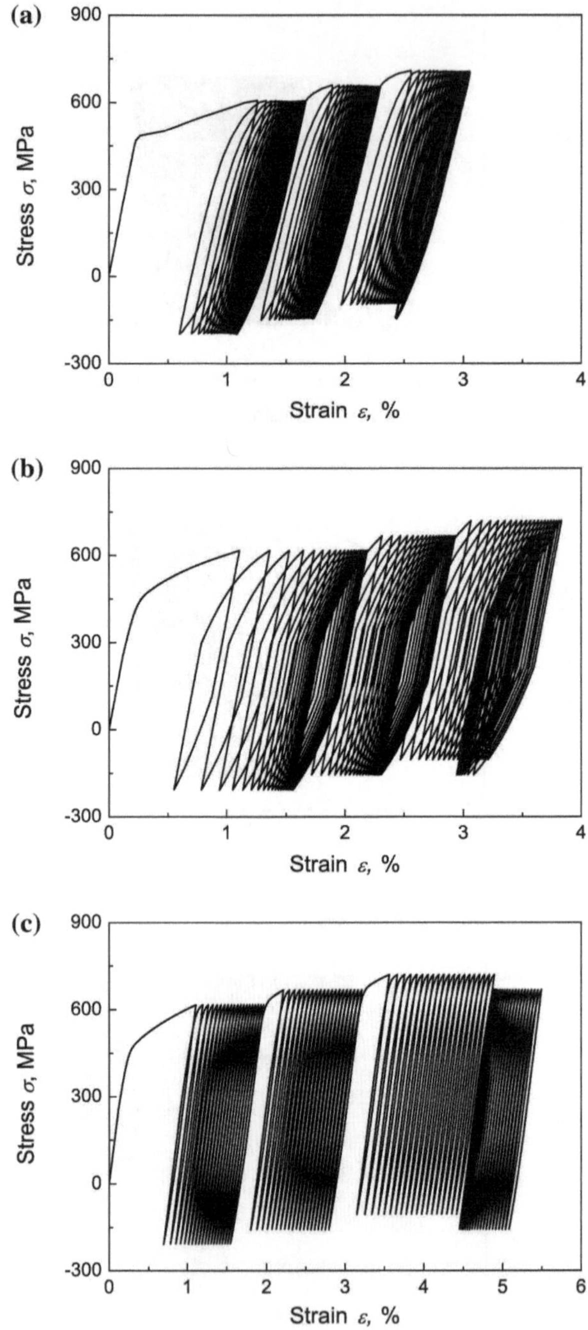

Fig. 27.7 Experimental and simulated stress-strain curves of U71Mn rail steel in multi-step stress-controlled cyclic test (loading case D). **a** Experiment. **b** Simulation by the proposed rule. **c** Simulation by the Chaboche's rule

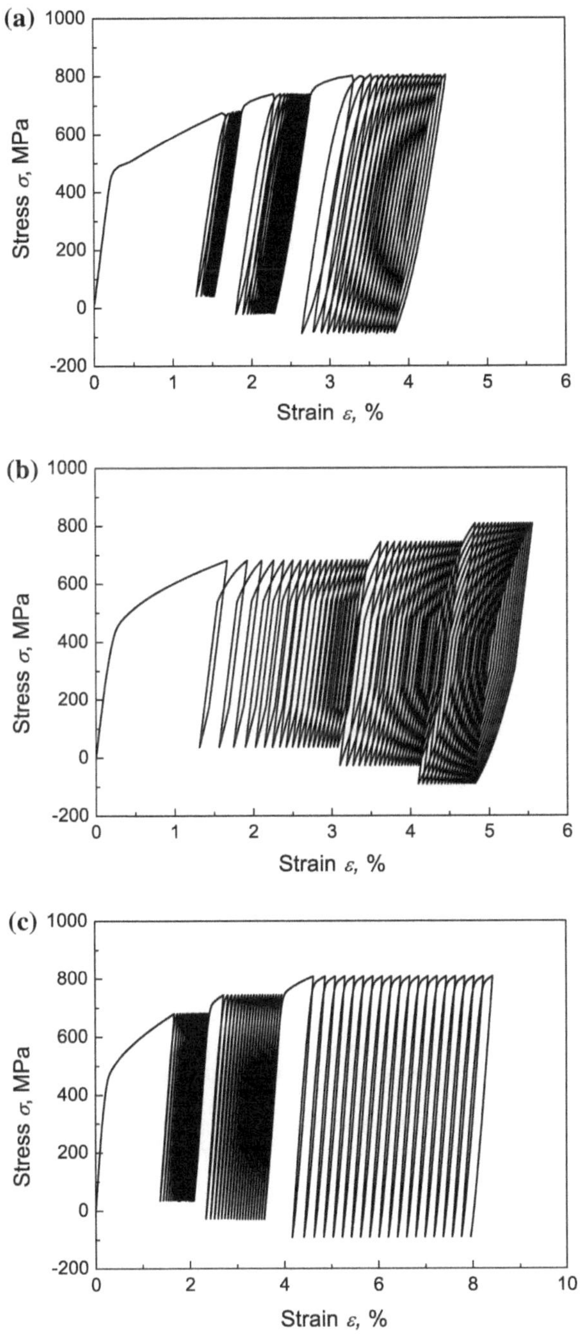

Fig. 27.8 Experimental and simulated stress-strain curves of U71Mn rail steel in multi-step stress-controlled cyclic test (loading case E). **a** Experiment. **b** Simulation by the proposed rule. **c** Simulation by the Chaboche's rule

27.4 Conclusions

By modifying the Chaboche's kinematic hardening rule (Chaboche 1989), a new kinematic hardening rule is proposed for describing the different plastic moduli of the metal materials presented in the monotonic and cyclic deformations in this paper. It is then verified that the proposed model is capable of modeling the differente plastic moduli occurred in the monotonic and cyclic deformation, and then provides more reasonable simulations to the cyclic stress-strain hysteresis loops and ratchetting strain. The simulated results are in good agreement with the corresponding experimental ones.

Acknowledgments Financial supports by the National Natural Science Foundation of China (11025210), the project for Sichuan Provincial Youth Science and Technology Innovation Team (2013, China), the 2013 Doctoral Innovation Funds of Southwest Jiaotong University and the Fundamental Research Funds for the Central Universities are gratefully acknowledged.

References

Abdel-Karim M (2009) Modified kinematic hardening rules for simulations of ratchetting. Int J Plast 25:1560–1587

Armstrong P, Frederick C (1966) A mathematical representation of the multiaxial Bauschinger effect. Technical Report RD/B/N731, Central Electricity Generating Board, Berkeley, UK

Bai N, Chen X (2009) A new unified constitutive model with short-and-long-range back stress for lead-free solders of Sn-3Ag-0.5Cu and Sn-0.7Cu. Int J Plast 25:2181–2203

Bari S, Hassan T (2000) Anatomy of coupled constitutive models for ratcheting simulation. Int J Plast 16:381–409

Bari S, Hassan T (2001) Kinematic hardening rules in uncoupled modeling for multiaxial ratcheting simulation. Int J Plast 17:885–905

Bari S, Hassan T (2002) An advancement in cyclic plasticity modeling for multiaxial ratcheting simulation. Int J Plast 18(7):873–894

Chaboche J (1989) Constitutive equations for cyclic plasticity and cyclic viscoplasticity. Int J Plast 5(3):247–302

Chaboche J (1991) On some modifications of kinematic hardening to improve the description of ratchetting effects. Int J Plast 7(7):661–678

Chaboche J (2008) A review of some plasticity and viscoplasticity constitutive theories. Int J Plast 24(10):1642–1693

Dong Y, Kang G, Yu C (2014) A dislocation-based cyclic polycrystalline visco-plastic constitutive model for ratchetting of metals with face-centered cubic crystal structure. Comput Mater Sci 91(10):75–82

Gaudin C, Feaugas X (2004) Cyclic creep process in AISI 316L stainless steel in terms of dislocation patterns and internal stresses. Acta Materialia 52:3097–3110

Kang G (2004) A visco-plastic constitutive model for ratcheting of cyclically stable materials and its finite element implementation. Mech Mater 36:299–312

Kang G (2008) Ratchetting: recent progresses in phenomenon observation, constitutive modeling and application. Int J Fatigue 30:1448–1472

Kang G, Gao Q (2002) Uniaxial and non-proportionally multiaxial ratcheting of U71Mn rail steel: experiments and simulations. Mech Mater 34:809–820

Kang G, Gao Q, Yang X (2002) A visco-plastic constitutive model incorporated with cyclic hardening for uniaxial/multiaxial ratchetting of SS304 stainless steel at room temperature. Mech Mater 34:521–531

Mizuno M, Mima Y, Abdel-Karim M, Ohno N (2000) Uniaxial ratchetting of 316FR steel at room temperature—Part I: experiments. J Eng Mater Technol 122:29–34

Ohno N (1997) Recent progress in constitutive modeling for ratchetting. Mater Sci Res Int 3(1):1–9

Ohno N, Abdel-Karim M (1990) Recent topics in constitutive modeling of cyclic plasticity and viscoplasticity. Appl Mech Rev 43:283–295

Ohno and Abdel-Karim(2000)]Ohnokarim2000 Ohno N, Abdel-Karim M (2000) Uniaxial ratcheting of 316 fr steel at room temperature. Part II: constitutive modeling and simulation. J Eng Mater Technol 122:35–41

Ohno N, Wang J (1991) Transformation of a nonlinear kinematic hardening rule to a multisurface form under isothermal and nonisothermal conditions. Int J Plast 7:879–891

Ohno N, Wang J (1993) Kinematic hardening rules with critical state of dynamic recovery, Parts I and II. Int J Plast 9:375–403

Zhu Y, Kang G, Kan Q, Bruhns O (2014) Logarithmic stress rate based constitutive model for cyclic loading in finite plasticity. Int J Plast 54:34–55

MIX
Papier aus verantwortungsvollen Quellen
Paper from responsible sources
FSC® C105338

If you have any concerns about our products,
you can contact us on
ProductSafety@springernature.com

In case Publisher is established outside the EU,
the EU authorized representative is:
Springer Nature Customer Service Center GmbH
Europaplatz 3, 69115 Heidelberg, Germany

Printed by Libri Plureos GmbH
in Hamburg, Germany